AN HISTORICAL GEOGRAPHY OF EUROPE
1800–1914

An historical geography of Europe 1800–1914

NORMAN J. G. POUNDS
UNIVERSITY PROFESSOR OF HISTORY AND GEOGRAPHY EMERITUS,
INDIANA UNIVERSITY

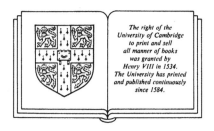

CAMBRIDGE UNIVERSITY PRESS

CAMBRIDGE
NEW YORK   NEW ROCHELLE
MELBOURNE   SYDNEY

CAMBRIDGE UNIVERSITY PRESS
Cambridge, New York, Melbourne, Madrid, Cape Town, Singapore, São Paulo, Delhi

Cambridge University Press
The Edinburgh Building, Cambridge CB2 8RU, UK

Published in the United States of America by Cambridge University Press, New York

www.cambridge.org
Information on this title: www.cambridge.org/9780521358910

© Cambridge University Press 1985

This publication is in copyright. Subject to statutory exception
and to the provisions of relevant collective licensing agreements,
no reproduction of any part may take place without the written
permission of Cambridge University Press.

First published 1985
First paperback edition 1988
Re-issued in this digitally printed version 2009

A catalogue record for this publication is available from the British Library

Library of Congress Catalogue Card Number: 84–23054

ISBN 978-0-521-26574-4 hardback
ISBN 978-0-521-35891-0 paperback

Contents

	List of maps and diagrams	page vi
	List of abbreviations	xiii
	Preface	xix
1	From Waterloo to the First World War	1
2	The resource pattern of Europe	37
3	The population of nineteenth-century Europe	66
4	Urban development in the nineteenth century	119
5	Agriculture in the nineteenth century	187
6	Agricultural regions	248
7	Manufacturing in the nineteenth century	298
8	The growth of industrial regions	354
9	Transport and trade	427
10	Europe in 1914	495
	Notes	533
	Index	586

Maps and diagrams

Page		
17	1.1	The political boundary of 1814–15 in north-western France
19	1.2	The Habsburg Empire in the nineteenth century. The Austrian part is shown by diagonal shading, the Hungarian by horizontal. Lombardy and Venezia were lost in 1861. Bosnia and Hercegovina were occupied in 1878 and were incorporated into the Empire in 1908. The Sanjak was occupied militarily 1878–1908
20	1.3	The language boundary in Belgium and the partition of Limburg and Luxembourg
24	1.4	The formation of the German customs unions, until 1834
24	1.5	The formation of the German customs unions, 1834–88
25	1.6	Italy before unification, 1861
27	1.7	The Balkan peninsula before and after the Congress of Berlin, 1878
30	1.8	Gross National Product per head, about 1830. The units are 1960 U.S. dollars. Based on P. Bairoch, 'Europe's Gross National Product: 1800–1975', *Jl Eur. Ec. Hist.*, 5 (1976), 273–340
31	1.9	Increase in Gross National Product per head, 1830–1910, as percentage of 1830 figures
32	1.10	The increase in the Gross National Product per head in the principal European countries, 1830–1913
33	1.11	The value of agricultural and industrial production in France, 1830–1910
34	1.12	The Gross National Product of Germany, by sector. The value of transport services was too small before 1886 to show on the graph. Based on statistics in W. G. Hoffmann, *Das Wachstum der deutschen Wirtschaft seit der Mitte des 19. Jahrhundert* (Berlin, 1965)
35	1.13	Growth of employment by sector in Germany, 1849–1925. Based on F. B. Tipton, *Regional Variations in the Economic Development of Germany during the Nineteenth Century* (Middletown, Conn., 1976); Gerhard Bry, *Wages in Germany 1871–1945* (Princeton, N.J., 1960); Gustav Stolper, *The German Economy 1870 to the Present* (London, 1967)
35	1.14	Employment in Austria, 1869–1910. Based on G. Otruba, in *Viert. Soz. Wirtgesch.*, 62 (1975), 40–61
35	1.15	Growth of the Hungarian economy, 1850–1912
36	1.16	Growth of population, by employment sector, in Sweden. After H. Osvald, *Swedish Agriculture* (Stockholm, 1952)
39	2.1	The distribution of mineral fuel resources in Europe. Oil resources were not exploited on a significant scale until the last decade of the nineteenth century
40	2.2	Active coalfields in France, 1814, and the movement of coal (figures in metric quintals). Based on *Jl Mines*, 36 (1814), 321–94
40	2.3	Coal production in Europe (Great Britain excluded), 1820–1913. This graph illustrates the increasing importance of Germany. See also fig. 10.5
42	2.4	The coal market in Germany, late nineteenth century. The importance of British coal in the North German market reflects the relative cheapness of maritime transport. After *Das Östliche Deutschland* (prepared by the Göttinger Arbeitskreis) (Würzburg, 1959), 678
42	2.5	Coal production in Europe (including Great Britain) in 1840. Note the importance of Central Belgium and Saint-Étienne
44	2.6	Mining concessions along a north–south transect across the Ruhr coalfield. Note

		the increasing size of concessions towards the north. Based on Hans Spethmann, *Das Ruhrgebiet* (Berlin, 1933)
46	2.7	Production of brown coal and lignite, 1850–1913
48	2.8	The development of hydroelectric power in the French Alps. Based on G. Veyret-Verner, *L'Industrie dans les Alpes Françaises* (Grenoble, 1948)
50	2.9	Iron-ore production in Europe (excluding Great Britain), 1830–1910
55	2.10	Production of zinc, 1820–1913
56	2.11	Production of lead, 1822–1913
57	2.12	Production of copper, 1830–1913
57	2.13	Production of bauxite, 1885–1913
58	2.14	Production of mercury, 1800–1913
67	3.1	Growth of population in Europe, 1800–1910, by country
68	3.2	Birth-rates, by country, 1810–1910
69	3.3	Fertility rates in England, France, Germany, Italy and Sweden, 1780–1930
69	3.4	Net reproduction rates in England, France, Germany and Italy, 1800–1920
70	3.5	Death-rates, by country, 1810–1910
70	3.6	Crude birth- and death-rates in Germany, 1870–1915. After K. Witthauer, 'Die Bevölkerung der Erde', *Pet. Mitt.* Erg., 265
71	3.7	Population pyramids for France
72	3.8	Ratio of coarse grains to wheat in the French diet about 1830. Based on J.-P. Aron and E. Le Roy Ladurie, *Anthropologie du conscrit français* (Paris, 1972)
76	3.9	Diffusion of cholera in Europe. Based on G. Melvyn Howe, *Man, Environment and Disease in Britain* (Harmondsworth, 1976), and P. Bourdelais, J.-Y. Raulot and M. Demonet, 'La marche du choléra en France: 1832–1854', *Ann. E.S.C.*, 33 (1978), 125–42
80	3.10	The contraction of rural population in France with the growth of towns. The difference between the 'rural' and the 'urban' graphs is a rough measure of internal migration; a = maximum estimate, b = minimum estimate. After J.-C. Toutain, *La Population de la France de 1700 à 1959*, *Hist. Quant. Éc. Fr.*, 3 (1963)
80	3.11	Percentage of the population of France considered rural in 1851 and 1911. Based on A. Armengaud, *La Population française au XIXe siècle*, Paris, 1971
82	3.12	Migration within Germany to the industrial provinces of Rhineland and Westphalia, 1880. For key, see fig. 3.13. Based on W. Kollmann, 'Die Bevölkerung Rheinland-Westfalens in der Hochindustrialisierungsperiode', *Viert. Soz. Wirtgesch.*, 58 (1971), 359–88
83	3.13	Migration within Germany to Rhineland and Westphalia, 1907
83	3.14	Migration from Germany, 1820–60, and the price of rye. After Mack Walker, *Germany and the Emigration 1816–1885* (Cambridge, Mass.), 1964
85	3.15	Migration from Hungary, 1907–13, correlated with the price of grain. This shows how dependent migration continued to be on the quality of the harvest. Based on *Ann. Est* (1951)
85	3.16	Migration from Italy, 1876–1913. Note the increasing importance of migration from southern Italy and Sicily
88	3.17	The chief foreign communities in France, about 1850. After J. Grandjonc in *Arch. Sozgesch.*, 15 (1975), 211–300
89	3.18	The Pale of Jewish Settlement within Russia and the migration of the Jewish people into central and south-eastern Europe. After H. G. Wanklyn, 'Geographical Aspects of Jewish Settlement East of Germany', *Geog. Jl* 95 (1940), 175–90
90	3.19	Distribution of Jewish population in Europe, 1825–1900. Based on U. Z. Engelman, 'Sources of Jewish Statistics', in L. Finkelstein, ed., *The Jews: Their History, Culture and Religion* (New York, 1949)
96	3.20	Language boundaries in Switzerland. The extent of the Sonderbund is also shown. Note how the latter conformed with the mountainous areas, but cut across the language boundary. Based on K. B. Mayer, *The Population of Switzerland* (New York, 1952), and H. Ammann and K. Schib, *Historischer Atlas der Schweiz* (Aarau, 1951)
99	3.21	Population of the Habsburg Empire, 1810–1910. Figures are for provinces, not for ethnic communities. Based on P. Horska, in *Ann. Dém. Hist.*, 1967, 173–95, and *Raum und Bevölkerung in der Weltgeschichte* (Würzburg, 1956), 217–32
104	3.22	Population of France; (a) 1821 (after C. H. Pouthas, *La Population française pendant la première moitié du XIXe siècle* (Paris, 1956)) and (b) 1911 (persons per hectare)

viii Maps and diagrams

105	3.23	Illiteracy in France, about 1830. Based on J.-P. Aron and E. Le Roy Ladurie, *Anthropologie du Conscrit Français* (Paris, 1972)
108	3.24	Percentage increase in the population of Germany, 1816–1914
110	3.25	Population of Poland
121	4.1	The urban net in Alsace, about 1850. Based on M. Rochefort, *L'Organisation urbaine de l'Alsace* (Paris, 1960)
122	4.2	The urban net in Białystok province, Poland, 1895–7. Based on L. Kosiński, *Miasta województwa Białostockiego* (Warsaw, 1962)
124	4.3	Model of nineteenth-century development of old urban centre
128	4.4	Growth of aggregate urban population in Europe, excluding Russia, 1600–1950; (a) aggregate, (b) as percentage of total population. Based on P. Bairoch, 'Population urbaine et taille des villes en Europe de 1600 à 1970', *Rv. Hist. Éc. Soc.*, 54 (1976), 304–35
129	4.5	Settlements by size in Germany, 1875–1910. Based on O. Buchner, 'Der Einfluss der wirtschaftskrise auf die Wanderungsbewegung in den deutschen Städte', *Rv. Inst. Int. Stat.*, 4 (1936), 1–26
130	4.6	Population by settlement size in France. Note, in contrast with Germany, the predominance of small settlements
130	4.7	Population by settlement size in Belgium. Based on P. Meuriot, *Des agglomérations urbaines dans l'Europe contemporaine* (Paris, 1897)
131	4.8	The size of the primate city as a percentage of national population, 1800–1910
132	4.9	Aggregate population of cities, 1800–1970, as a percentage of national population. Based on P. Bairoch, 'Population urbaine et taille des villes en Europe de 1600 à 1970', *Rv. Hist. Éc. Soc.*, 54 (1976), 304–35
133	4.10	Rank–size graph of the towns of Białystok province, Poland (see fig. 4.2)
134	4.11	Rank–size graph of the larger towns in Germany in 1801, 1871 and 1910. Based on P. Meuriot, op. cit., and *Pet. Mitt.*, 57 (1911), i, 131
135	4.12	Rank–size graph of cities and towns in France in 1801, 1851 and 1911. Number 9, left blank on the graph, represents territory not part of France at this time. Based mainly on *Ann. Stat. de la France*
136	4.13	The urban pattern in Southern Germany, including the *Reichlsand* of Alsace-Lorraine, in 1910. The area shown is approximately that covered by Christaller's study
138	4.14	The occupational structure of Toulouse in 1809, 1851 and 1872. Based on M.-T. Plegat, 'L'évolution démographique d'une ville française au XIXe siècle: l'exemple de Toulouse', *Ann. Midi*, 64 (1952), 227–48
139	4.15	The occupational structure of Strasbourg in 1866, 1882 and 1895. Based on A.-E. Sayous, 'L'évolution de Strasbourg entre les deux guerres (1871–1914)', *Ann. E.S.C.*, 6 (1934), 1–19, 122–32
141	4.16	Sources of immigrant population in Cologne, 1906–12. Based on R. Haberle and F. Meyer, *Die Grossstädte im Ströme der Binnenwanderung* (Leipzig, 1937)
142	4.17	Migration to Paris; the source of migrants, 1833 and 1891. The figures represent those born in each *département* and living in Paris, expressed as the number per 1,000 of those living in the *département*. After L. Chevalier, *La Formation de la population parisienne au XIXe siècle* (Paris, 1950)
142	4.18	Migration to (a) Marseilles, (b) Lyons. The figures represent the number per 1,000 of population in the home *département*. After Abel Chatelain, 'L'attraction des trois plus grandes agglomérations françaises: Paris–Lyon–Marseille en 1891', *Ann. Dém. Hist.* (1975), 27–41
143	4.19	Population pyramid for Dijon, 1851. After J.-P. Viennot, 'La population de Dijon d'après le recensement de 1851', *Ann. Dém. Hist.* (1969), 241–60
144	4.20	Population pyramid for Arras in 1876. It suggests a low urban birth-rate and a considerable net in-migration. After A. Cornette, 'Arras et sa banlieue', *Rv. Nord Livr. Géog.*, 9 (1960)
148	4.21	The distribution of municipal gasworks in Germany in 1850 and 1860. After I. von Karolyi, 'Verhandlung über den Bau einer Gasanstalt in Göttingen 1836–1861', *Gött. Jb.*, 1969, 109–26
151	4.22	Changes in a single city block in Łódź, 1827–1914. The diagram for 1827 represents agricultural land. After A. Ginsbert, *Łódź, Studium Monograficzne* (Łódź, 1962)
155	4.23	The urban growth of Roubaix and Tourcoing. After F. Delgrange-Vancomerbeke, 'Le développement spatial de Roubaix–Tourcoing', *Rv. Nord Livr. Géog.*, 4 (1955), 9–26
156	4.24	Growth of population in Leipzig. That of the Old City is seen to contract with movement from the centre to the suburbs. (The column indicates the total

Maps and diagrams

		population of the city; the portion in black, that of the *Altstadt* or Old City.) Based on G. Brause, 'Entwicklungsprobleme von Grossstädtzentren unter besonderer Berücksichtigung Leipzigs', *Jb. Reggesch.*, 3 (1968), 184–203
156	4.25	Distribution of (a) domestic servants in Paris, 1872 and (b) working-class population. The figures represent percentages of total population. After L. Chevalier, *La Formation de la population parisienne au XIXe siècle* (Paris, 1950)
158	4.26	The ethnic composition of urban population in the Danubian province (Bulgaria) of the Ottoman Empire, 1866. Based on N. Todorov, 'The Balkan Town in the Second Half of the 19th Century', *Ét. Balk.*, 5 (1969), 31–50
160	4.27	The contrasted occupational structures of the Czech and German populations of Prague, about 1900. Based on J. Havranek, 'Social Classes, Nationality Ratios and Demographic Trends in Prague, 1880–1900', *Hist.*, 13 (1966), 171–208
162	4.28	Urban development in Europe, about 1840
164	4.29	Paris in the first half of the nineteenth century, showing the development of its railway system. The zigzag line represents the *Mur des Fermiers Généreaux*.
166	4.30	Haussmann's rebuilding of Paris, 1850–6. After D. H. Pinkney, *Napoleon III and the Rebuilding of Paris* (Princeton, N.J., 1958)
170	4.31	Expansion of the Berlin metropolitan area. Based on I. Thienel, *Industrialisierung und Städtewachstum*, Hist. Komm. Berlin, 6 (1971), 106–49. Vertical shading, incorporated, 1841; horizontal, 1861.
171	4.32	The growth of population in Greater Berlin
174	4.33	Growth of the chief cities of Switzerland. Note the absence of a primate city
179	4.34	Towns of the Balkan peninsula, about 1910
180	4.35	Rank-order graph of towns in Bulgaria
182	4.36	Cities and towns in Spain and Portugal, about 1855. This may be one of the earliest attempts to use proportionate symbols to represent city size. Reproduced from T. E. Gumprecht, 'Die Städte-Bevölkerung von Spanien', *Pet Mitt.* (1856), 303
184	4.37	Towns of Italy, about 1800 and in 1910
189	5.1	Growth in production of wheat and rye in France, Germany, Austria-Hungary and Italy
189	5.2	Grain production in France. Based on J.-C. Toutain, *Le Produit de l'agriculture française de 1700 à 1958*, Hist. Quant. Éc. Fr.
191	5.3	The average monthly price of wheat in France. After E. Labrousse, *Le Prix de froment en France au temps de la monnaie stable* (1726–1913), Éc. Prat. Htes. Ét., Monn.-Pr.-Con. (1970)
192	5.4	Growth in cereal production in Prussia. Based on H. W. Finck von Finckenstein, *Die Entwicklung der Landwirtschaft in Preussen und Deutschland, 1800–1903* (Würzburg, 1960)
192	5.5	Growth in production of rye and wheat in Sweden, 1866–1926. After H. Oswald, *Swedish Agriculture* (Uppsala, 1952)
199	5.6	Size of agricultural holdings in Germany, about 1880. Based on table in *Jb. Ges. Verw. Volksw.*, 11 (1887), 1011–25
202	5.7	Size of farm holdings in Serbia, Bulgaria and Romania, 1905–8. Based on J. R. Lampe and M. R. Jackson, *Balkan Economic History, 1550–1950* (Bloomington, Ind., 1982), 185
203	5.8	Distribution of estates in Spain. (a) Minute holdings, of less than family size; (b) medium-sized farms, held on long leases; (c) medium-sized farms, held on short leases; (d) *latifundia*, large estates worked mainly by casual labour. After G. Brenan, *Spanish Labyrinth* (Cambridge, 1943), 334
206	5.9	Land use in France, 1835–1905. Based on J.-C. Toutain, op. cit.
207	5.10	Land use and agricultural production in Prussia. Based on H. W. Finck von Finckenstein, op. cit.
208	5.11	Changes in land use in Bohemia, 1848–1900, showing the diminution of fallow. Based on J. Purš, 'Die Entwicklung des Kapitalismus in der Landwirtschaft der böhmischen Länder in der Zeit von 1849 bis 1879', *Jb. Wirtgesch.* (1963), part 3, 31–96
209	5.12	Land use in Denmark, 1866–1901. Note the considerable increase in the area under fodder-crops and grazing. Based on R. J. Thompson, 'The Development of Agriculture in Denmark', *Jl Roy. Stat. Soc.*, 49 (1906), 374–419
229	5.13	Distribution of the chief types of plough in Europe during the nineteenth century. Based on A. G. Haudricourt and J.-B. Delamarre, *L'Homme et la charrue à travers le monde* (Paris, 1955); P. Leser, *Entstehung und Verbreitung des Pfluges*, Int. Sam. Eth. Mon., 3 (Münster/Westfalen, 1931)
233	5.14	Percentage of arable land under bare fallow between 1852 and 1882. After *Histoire*

x   Maps and diagrams

économique et sociale de la France, ed. F. Braudel and E. Labrousse, vol. III, part 2 (Paris, 1976), 672

235 5.15 Increase in potato production in France, Poland and Prussia
240 5.16 Changes in the area under vineyards in France as a result of the *Phylloxera* epidemic. The figures indicate the number of hectares lost in each *département*
242 5.17 Yield-ratios (a) and yields in quintals per hectare (b) of wheat in France, about 1840. After M. Morineau, 'Y a-t-il eu une révolution agricole en France au XVIIIe siècle?', *Rv. Hist.*, 239 (1968), 299–326
242 5.18 Yield-ratios (a) and yields in quintals per hectare (b) of rye in France, about 1840. After M. Morineau, op. cit.
245 5.19 Transhumance in Spain; the migration of the chief flocks and the migration routes. After A. Fribourg, 'La transhumance en Espagne', *Ann. Géog.*, 19 (1910), 231–44
256 6.1   Physical divisions of the North Italian Plain. After M. I. Newbigin, *Southern Europe* (London, 1949), 157
257 6.2   Map of rivers and irrigation canals in Lombardy. Reproduced from R. Baird Smith, *Italian Irrigation, being a Report on the Agricultural Canals of Piedmont and Lombardy* (Edinburgh, 1855)
260 6.3   Physical divisions of Flanders. After E. de Laveleye, *Essai sur l'économie rurale de la Belgique* (Paris, 1875)
262 6.4   Distribution of rye, wheat and cattle in Flanders. After R. Blanchard, *La Flandre* (Lille, 1906). Data relate to about 1900
265 6.5   Land enclosure in Denmark. After H. Thorpe, 'The Influence of Inclosure on the Form and Pattern of Rural Settlement in Denmark', *Trans. I.B.G.*, 17 (1951), 11–29. The maps relate respectively to 1769, 1805 and 1893.
269 6.6   Irrigated huertas in the Valencia region of Spain. After R. Courtot, 'Irrigation et propriété citadine dans l'Acequia Réal del Jucar au milieu du XIXe siècle', *Ét. Rur.*, 45 (1972), 29–47
274 6.7   The drawing made about 1875 by Sir Arthur Evans of a *zadruga* in Bosnia; published in A. J. Evans, *Through Bosnia and Herzegovina on Foot* (London, 1877), 57
276 6.8   Distribution of *chiflik* estates in the Balkans. In the inset, a marks the position of cooking hearths, b of the threshing floor. After J. Cvijič, *La Péninsule balkanique*, 223–4
278 6.9   Length of the growing season (in days) in Sweden. After H. Oswald, *Swedish Agriculture* (Uppsala, 1952), 18
280 6.10 Farmer-ironworkers in Västmanland. The zones are (A) peasant miners, (B) peasant carters, (C) food production. After David Gaunt, in *Chance and Change: Social and Economic Studies in Historical Demography in the Baltic Area* (Odense, 1978), 69–83
281 6.11 The pattern of villages and *saetars* in the Siljan region of Sweden. After Hans Aldskogius, 'Changing Land Use and Settlement Development in the Siljan Region', *Geog. Ann.*, 41 (1959), 250–61
283 6.12 Size of holdings in the Beauce, 1821–44 and 1913. Note the increase in the area in small holdings and the decline of that in large. Based on Gregor Dallas, *The Imperfect Peasant Economy: The Loire Country, 1800–1914* (Cambridge, 1982)
286 6.13 Diminution of arable and increase in the area of grazing in the Pays de Caux, 1823–1901. After M. Lévy-Leboyer, *Le Revenu agricole et la rente foncière en Basse-Normandie* (Paris, 1972), 72
290 6.14 Distribution of the cultivation of wheat and rye in the Limousin. Figures indicate the percentage of the total cropland. After A. Corbin, *Archaisme et modernité en Limousin au XIXe siècle* (Paris, 1975)
293 6.15 Land use in Lower Alsace, 1837 and 1938. After E. Juillard, *La Vie rurale dans la plaine de Basse-Alsace* (Paris, 1953)
296 6.16 Agricultural regions in Europe in the mid-nineteenth century
310 7.1   The use of the steam engine in Bohemia and Moravia, 1841–1902. This graph shows the diminishing *relative* importance of its employment in mining and the manufacture of textiles, and the increasing diversification of its uses. Based on J. Purš, 'Použití parních strojů v průmyslu v českých zemích v období do nástupu imperialismu', *Česk. Čas. Hist.*, 3 (1955), 254–90, 427–84
315 7.2   The distribution of weavers in Germany, about 1885. The principal cloth-making areas – the lower Rhineland, Baden and South Germany, Saxony and Silesia –

Maps and diagrams

		are readily apparent. After a map in P. Kollmann, 'Die gewerbliche Entfaltung im Deutschen Reiche', *Jb. Ges. Verw. Volksw.*, 12 (1888), 437–528
335	7.3	The iron-producing region of *dépt* Marne
344	7.4	The flow of materials in the European iron and steel industry, early twentieth century. The scale is only approximate
344	7.5	The increase in the production of iron and steel in Europe during the later nineteenth century (1900 = 100). After I. Svennilson, *Growth and Stagnation in the European Economy*, U.N., Ec. Comm. Eur. (1954)
347	7.6	The increase in production of textiles and chemicals in Germany, 1850–1913 (1913 = 100). Note the contrast between the linear expansion of the textile industry and the exponential growth in the manufacture of chemicals
353	7.7	Model to represent the transition from proto-industry to modern factory industry
358	8.1	The industrial region of north-west Europe
365	8.2	The Ruhr industrial region, 1850–70. About half the smelting capacity was built for the purpose of using coal-measures ore. The rest aimed to use ore imported into the area
370	8.3	Growth of population in the north-west European region. Based on statistics in E. A. Wrigley, *Industrial Growth and Population Change* (Cambridge, 1961)
371	8.4	Urban development in relation to the Ruhr coal resources, about 1800. Based on K. Olbricht, 'Die Städte des Rheinisch-Westfalischen Industriebezirks', *Pet. Mitt.*, 57 (1910)
375	8.5	Iron-working in the Ruhr in relation to coal and iron-ore working, 1850–70
383	8.6	Urban development in the Ruhr industrial region, 1910. Source as for fig. 8.4
385	8.7	The Ruhr, Belgian and Saar coalfields and the iron-ore resources of Lorraine and Luxembourg
386	8.8	The conceded orefield in Lorraine, about 1900
389	8.9	Ironworks on the Lorraine–Luxembourg orefield and their affiliation with iron and steel companies elsewhere in Western Europe, about 1910
390	8.10	The movement of ore from the Lorraine–Luxembourg orefield and the import of coal, about 1910
393	8.11	The industrial region of Saxony and the Ore Mountains
399	8.12	The Upper Silesian–Moravian industrial region
400	8.13	The spread of coal mining in Upper Silesia–northern Moravia. The distribution of mines in 1860
401	8.14	Iron-working in Upper Silesia, 1857
402	8.15	Zinc mining and smelting in Upper Silesia, about 1860
404	8.16	Graph showing the increase in coal production in Upper Silesia–Dąbrowa–northern Moravia, 1800–1920
406	8.17	Coal mining in the Upper Silesian–Moravian coalfield in 1900
407	8.18	Iron-working in Upper Silesia, 1912
409	8.19	The city of Łódź and its region
411	8.20	The growth of population in Łódź, probably the fastest growing city in nineteenth-century Europe. After A. Ginsbert, *'Łódź, Studium Monograficzne* (Łódź, 1962)
413	8.21	The Loire coalfield and the industrial region of Saint-Étienne. Based mainly on L. Babu, 'L'industrie métallurgique dans la région de Saint-Étienne', *Ann. Mines*, 9e série, 15 (1899), 357–460; M. Perrin, *Saint-Étienne et sa région économique* (Tours, 1937)
414	8.22	The silk-weaving region of Lyons. Based on G. Garrier, *Paysans du Beaujolais et du Lyonnais 1800–1970* (Grenoble, 1973); M. Lévy-Leboyer, *Les Banques européennes et l'industrialisation internationale* (Paris, 1964)
421	8.23	The textile region of Catalonia. After P. Vilar, 'La vie industrielle dans la région de Barcelone', *Ann. Géog.*, 38 (1929), 339–65
423	8.24	The North Italian industrial region
428	9.1	A simple model of the development of transport during the nineteenth century
428	9.2	Internal transport in France. Based on J.-C. Toutain, *Les Transports en France de 1830 à 1965* (Paris, 1967)
432	9.3	Internal navigation in France, late nineteenth century. The classification into 'main' and 'secondary' waterways is that adopted in the Freycinet Plan. Based mainly on *Report on Canal Traffic in France*, For. Off. Misc. Ser., 342, 1895
437	9.4	Rhine navigation about 1910. Based mainly on *Der Rhein: Ausbau, Verkehr, Verwaltung* (Duisburg, 1951)
440	9.5	Freight handled in the port of Duisburg-Ruhrort. After *Der Rhein*

Maps and diagrams

440	9.6	Coal shipped on the River Ruhr. The abrupt decline about 1870 marks the closure of the more southerly mines of the Ruhr coalfield
445	9.7	Navigable waterways in the North German Plain. After *Royal Commission on Canals and Waterways*, vol. VI, *Foreign Inquiry*, Cmd 4841 (1909), 491–735
448	9.8	The Danube delta and its navigable channels. After Sir Edward Hertslet, *Map of Europe by Treaty* and Erzherzog Heinrich Ferdinand, *Die Wasserstrasse Mitteleuropas* (Vienna, 1917)
451	9.9	Railway development in Europe, 1840
451	9.10	Railway development in Europe, 1850
452	9.11	Railway development in Europe, 1880
452	9.12	Growth in the use of railways for freight, 1851–1911; 'a' indicates a change in the method of calculating the volume of freight
455	9.13	Friedrich List's plan for a unified German railway system. After F. Lenz, *Friedrich List* (Berlin, 1936), 429
462	9.14	Freight handled in Norwegian ports, about 1860. Based on *Reports by Her Majesty's Secretaries of Embassy and Legation...of the Countries in which they Reside*, No. 7, *Parl. Pap.*, 1864, vol. LXI, p. 126
463	9.15	Ports named in chapter 9
472	9.16	Port of Hamburg, about 1910, showing dock basins excavated in the alluvial deposits
474	9.17	Ports of the Rhine delta
484	9.18	Port of Genoa, about 1910, showing a port constructed on a rocky coast
500	10.1	Types of population density: a simple model
501	10.2	Diagram showing the relationship of population density to the economy
501	10.3	Crude population density, 1910
504	10.4	An urban map of Europe, 1910. Only cities of more than 50,000 are shown
508	10.5	Coal production in Europe, 1912. The overwhelming importance of Great Britain and Germany is apparent
509	10.6	Iron and steel production, 1912
515	10.7	Dairy cattle about 1913. Each dot represents 5,000 head. Source: V. C. Finch and O. E. Baker, *Geography of the World's Agriculture*, U.S. Department of Agriculture (Washington, D.C., 1917)
516	10.8	Sheep, about 1913. Each dot represents 10,000
517	10.9	Pigs, about 1913. Each dot represents 5,000
519	10.10	Wheat, about 1913. Each dot represents 5,000 acres (2,024 ha)
519	10.11	Wheat, about 1913. Each dot represents 100,000 bushels
520	10.12	Rye, about 1913. Each dot represents 5,000 acres (2,024 ha)
521	10.13	Oats, about 1913. Each dot represents 5,000 acres (2,024 ha)
521	10.14	Barley, about 1913. Each dot represents 5,000 acres (2,024 ha)
522	10.15	Maize, about 1913. Each dot represents 5,000 acres (2,024 ha)
523	10.16	Potatoes, about 1913. Each dot represents 2,000 acres (810 ha)
524	10.17	Sugar beet, about 1913. Each dot represents 1,000 acres (405 ha)
528	10.18	Gross National Product per head, 1913 (in U.S. dollars)
529	10.19	Gross National Product per head by sector for (a) industrialised countries, (b) non-industrialised countries. Data for 1913

Abbreviations

A.A.G. Bij.	Afdeling Agrarische Geschiedenis Bijdragen (Wageningen, Neth.)
Ac. Prem. Congr. Balk.	Actes du Premier Congrès International des Études Balkaniques (Athens)
Ac. Roy. Belg. Mém.	Académie Royale de Belgique, Mémoires (Brussels)
Ac. Sci. Bes.	Académie des Sciences de Besançon (Besançon)
Acta Hist. Dac.	Acta Historica Academiae Dacoromanae (Bucharest)
Acta Hist. Neer.	Acta Historiae Neerlandica (Leiden)
Acta Pol. Hist.	Acta Poloniae Historica (Warsaw)
Actes Coll. Int. Dém. Hist.	Actes du Colloque International de Démographie Historique (Liège)
Agr. Hist.	Agricultural History (Chicago)
Agric. St., F.A.O.	Agricultural Studies, Food and Agriculture Organization (Rome)
Am. Hist. Rv.	American Historical Review (Bloomington, Ind.)
Am. Jl Soc.	American Journal of Sociology (Chicago)
Anc. Pays État	Anciens Pays et Assemblées d'État (Luxembourg)
Ann. A.A.G.	Annals of the Association of American Geographers (Washington, D.C.)
Ann. Am. Acad. Pol. Soc. Sci.	Annals of the American Academy of Political and Social Science (Philadelphia, PA)
Ann. Bourg.	Annales de Bourgogne (Dijon)
Ann. Bret.	Annales de Bretagne (Rennes)
Ann. Dém. Hist.	Annales de Démographie Historique (Paris)
Ann. E.S.C.	Annales: Économies–Sociétés–Civilisations (Paris)
Ann. Est.	Annales de l'Est (Nancy)
Ann. Est. Mém.	Annales de l'Est, Mémoires (Nancy)
Ann. Fac. Nice	Annales de la Faculté des Lettres et Sciences Humaines de Nice (Paris)
Ann. Géog.	Annales de Géographie (Paris)
Ann. Midi	Annales du Midi (Toulouse)
Ann. Mines	Annales des Mines (Paris)
Ann. Norm.	Annales de Normandie (Caen)
Ann. Norm. Cah.	Cahiers des Annales de Normandie (Caen)
Ann. Sil.	Annales Silesiae (Wrocław)
Ann. Stat. de la France	Annuaire Statistique de la France (Paris)
Antem.	Antemurale (Rome)
Arch. Frank. Gesch. Kunst	Archiv für Frankfurts Geschichte und Kunst (Frankfurt)
Arch. Öst. Gesch.	Archiv für Österreichische Geschichte (Vienna)
Arch. Sozgesch.	Archiv für Sozialgeschichte (Bonn)
Aust. Hist. Jb.	Austrian History Yearbook (Houston, Texas)
Balt. Scand. C.	Baltic and Scandinavian Countries (Toruń)
Beitr. Gesch. Öst. Eis.	Beiträge zur Geschichte des Österreichischen Eisenwesens (Vienna)
Berg. Hütt. Rd.	Berg- und Hüttenmännische Rundschau (Kattowitz)
Berg. Hütt. Ztg	Berg- und Hüttenmännischen Zeitung (Leipzig)
Bibl. Fac. Liège	Bibliothèque de la Faculté de Philosophie et Lettres de l'Université de Liège (Paris)
Bibl. Fac. Lyon	Bibliothèque de la Faculté des Lettres de Lyon (Paris)
Bibl. Gen. Éc. Prat. Htes. Ét.	Bibliothèque Générale de l'École Pratique des Hautes Études (Paris)

Abbreviations

Bibl. Hist. Rom.	Bibliotheca Historica Romaniae (Bucharest)
Bibl. Hist. Rv.	Bibliothèque de l'Histoire de la Révolution de 1848 (Paris)
Boh.	*Bohemia* (Prague)
Bonn. Geog. Abh.	*Bonner Geographische Abhandlungen* (Bonn)
Bul. Com. Cent. Stat.	*Bulletin de la Commission Centrale de Statistique* (Brussels)
Bul. Inst. Arch. Liège	*Bulletin de l'Institut Archéologique de Liège* (Liège)
Bul. Inst. Rech. Éc. Soc.	*Bulletin de l'Institut de Recherches Économiques et Sociales* (Louvain)
Bul. Soc. Ant. Norm.	*Bulletin de la Société des Antiquaires Normands* (Rouen)
Bul. Soc. Belge Ét. Géog.	*Bulletin de la Société Belge d'Études Géographiques* (Louvain)
Bul. Soc. Hist. Paris	*Bulletin de la Société de l'Histoire de Paris et de l'Île-de-France* (Paris)
Bulg. Hist. Rv.	*Bulgarian Historical Review*
Cah. Brux.	*Cahiers Bruxellois* (Brussels)
Cah. Hist.	*Cahiers d'Histoire* (Grenoble)
Cah. Hist. Mond.	*Cahiers d'Histoire Mondiale* (Neuchatel)
Cah. Int. Hist. Éc. Soc.	*Cahiers Internationaux d'Histoire Économique et Sociale* (Geneva)
Cah. I.S.E.A.	*Cahiers de l'Institut de Science Économique Appliquée* (Paris)
Cent. Ét. Éc., Ét. et Mém.	Centre d'Études Économiques, Études et Mémoires (Paris)
Camb. Econ. Hist. Eur.	*Cambridge Economic History of Europe* (Cambridge)
Cent. Hist. Éc. Soc. Rég. Lyon.	Centre d'Histoire Économique et Sociale de la Région Lyonnaise (Lyons)
Cent. Nat. Rech. Sci.	Centre National de la Recherche Scientifique (Paris)
Cent. Rech. Hist. Dém. Soc.	Centre des Recherches Historiques: Démographies et Sociétés
Česk. Čas. Hist.	*Českolovensky Časopis Historicky* (Prague)
Ciba Rv.	*Ciba Review* (Basel)
Col. Univ. St. Hist. Ec.	Columbia University Studies in History and Economics (New York)
Coll. Int. Cent. Nat. Rech. Sci.	Colloques Internationaux du Centre National de la Recherche Scientifique (Paris)
Colln Cah. Hist.	Collection des Cahiers d'Histoire (Paris)
Comm. Hist. Éc. Soc. Rév. Franç.	Commission d'Histoire Économique et Sociale de la Révolution Française (Paris)
Comp. St. Soc. Hist.	*Comparative Studies in Society and History* (Cambridge)
Conf. Int. Hist. Éc.	Conference Internationale d'Histoire Économique
Congr. Agric. Int. Paris	Congrès d'Agriculture Internationale (Paris)
Congr. Int. Ét. SE Eur.	Congrès International des Études du Sud-Est Européen (Athens)
Congr. Int. Géog.	Congrès International de Géographie (various places)
Congr. Int. Sc. Géog.	Congrès International des Sciences Géographiques (various places)
Cuad. Hist. Ec. Cat.	*Cuadernos de Historia Economica de Cataluña* (Barcelona)
Der Don.	*Der Donauraum: Zeitschrift fur Donauraum-Forschung* (Vienna)
Deutsch. Akad. Wiss.	*Deutsche Akademie der Wissenschaften zu Berlin*
Ec. Comm. Eur.	Economic Commission for Europe (United Nations)
Ec. Geog.	*Economic Geography* (Worcester, Mass.)
Ec. Hist.	*Economic History* (New York)
Ec. Hist. Rv.	*Economic History Review* (London)
Ec. Jl	*Economic Journal* (London)
Éc. Prat. Htes. Ét.	École Pratique des Hautes Études (Paris)
Civ. et Soc.	Civilisations et Sociétés
Dém. et Soc.	Démographie et Sociétés
Ét. et Mém.	Études et Mémoires
Ind. et Art.	Industries et Artisans
Monn.-Pr.-Con.	Monnaie-Prix-Conjoncture
Soc. et Id.	Société et Idéologies
Ec. Devel. Cult. Change	*Economic Development and Cultural Change* (Chicago)
Ec. Soc. Hist. Jb.	*Economisch- en Sociaal-historisch Jaarboek* (The Hague)
Econ.	*Economica* (London)
Econ. Hist.	*Economy and History*
Est. Rv. Hist. Mod.	*Estudis: Revista de Historia Moderna* (Barcelona)
Ét. Balk.	*Études Balkaniques*
Ét. Hist. Mod. Cont.	*Études d'Histoire Moderne et Contemporaine* (Paris)
Ét. Rur.	*Études Rurales* (Paris)
Expl. Ec. Hist.	*Explorations in Economic History* (New York)

Abbreviations

Explns Entrepr. Hist.	*Explorations in Entrepreneurial History* (Cambridge, Mass.)
Fac. Univ. Strasb.	Publications de la Faculté d l'Université de Strasbourg (Paris)
F.A.O.	Food and Agriculture Organisation (Rome)
Féd. Arch. Hist. Belg.	*Fédération Archéologique et Historique de Belgique* (Antwerp)
For. Off. Misc. Ser.	Foreign Office Miscellaneous Series (London)
Forsch. Soz. Wirtgesch.	*Forschungen zur Sozial- und Wirtschaftsgeschichte* (Stuttgart)
Francia	*Francia: Forschungen zur westeuropaischen Geschichte* (Munich)
Fr. Hist. St.	*French Historical Studies*
Geog.	*Geography* (Manchester)
Geog. Ann.	*Geografiska Annaler* (Stockholm)
Geog. Jl	*Geographical Journal* (London)
Geog. Rv.	*Geographical Review* (New York)
Giess. Abh. Agr. Forsch.	*Giessener Abhandlungen zur Agrar- und Wirtschaftsforschung des Europäischen Ostens* (Wiesbaden)
Glück.	*Glückauf* (Essen)
Gött. Jb.	*Göttinger Jahrbuch* (Göttingen)
Greifsw. Jb.	*Greifswald-Stralsunder Jahrbuch* (Weimar)
Grotius Soc. Pubns	*Grotius Society Publications* (London)
Harv. Ec. Ser.	Harvard Economic Series (Cambridge, Mass.)
Heid. Geog. Arb.	*Heidelberger Geographische Arbeiten* (Heidelberg)
Hess. Jb. Lgesch.	*Hessisches Jahrbuch fur Landesgeschichte* (Marburg)
Hist. (P)	*Historia* (Prague)
Hist. Carp.	*História Carpatica* (Bratislava)
Hist. Éc. Soc. France	*Histoire Économique et Sociale de la France* (Paris)
Hist. Komm. Berlin	Historische Kommission zu Berlin (Berlin)
Hist. Quant. Éc. Fr.	*Histoire Quantitative de l'Économie Française* (Paris)
Hist. Stud.	*Historické Štúdie* (Bratislava)
Inst. Ét. Slaves	Institut des Études Slaves
Inst. Nat. Ét. Dém.	Institut National des Études Démographiques (Paris)
Ét. Doc.	Études et Documents
Trav. Doc.	Travaux et Documents
Inst. Rech. Éc. Soc.	Institut des Recherches Économiques et Sociales (Louvain)
Int. Cong. Ec. Hist.	International Congress of Economic History
Int. Geog. Congr.	International Geographical Congress
Int. Geol. Congr.	International Geological Congress
Int. Lab. Rv.	*International Labor Review* (Geneva)
Int. Rv. Agric.	*International Review of Agriculture* (Rome)
Int. Rv. Soc. Hist.	*International Review of Social History* (Amsterdam)
Int. Sam. Eth. Mon.	Internale Sammlung Ethnologischer Monographien (Münster/Westfalen)
Jb. Fränk. Lforsch.	*Jahrbuch für Fränkische Landesforschung* (Neustadt)
Jb. Gesch. Mitt. Ostd.	*Jahrbuch für die Geschichte Mittel- und Ostdeutschlands* (Berlin and Tübingen)
Jb. Ges. Verw. Volksw.	*Jahrbuch für Gesetzgebung, Verwaltung und Volkswirtschaft im Deutschen Reich* (Schmollers Jahrbuch) (Leipzig)
Jb. Nat. Stat.	*Jahrbücher für Nationalökonomie und Statistik* (Jena)
Jb. Reggesch.	*Jahrbuch für Regionalgeschichte* (Leipzig)
Jb. Westd. Lgesch.	*Jahrbuch für Westdeutsche Landesgeschichte* (Koblenz)
Jb. Wirtgesch.	*Jahrbuch für Wirtschaftsgeschichte* (East Berlin)
Jl Cent. Eur. Aff.	*Journal of Central European Affairs* (Boulder, Colo.)
Jl Ec. Bus. Hist.	*Journal of Economic and Business History*
Jl Ec. Hist.	*Journal of Economic History* (New York)
Jl Eur. Ec. Hist.	*Journal of European Economic History* (Rome)
Jl Fam. Hist.	*Journal of Family History* (Minneapolis, Minn.)
Jl Hist. Med. All. Sci.	*Journal of the History of Medicine and Allied Sciences* (Minneapolis, Minn.)
Jl I.S. Inst.	*Journal of the Iron and Steel Institute* (London)
Jl Intdis. Hist.	*Journal of Interdisciplinary History* (Cambridge, Mass.)
Jl Mines	*Journal des Mines* (Paris)
Jl Mod. Hist.	*Journal of Modern History* (Chicago)
Jl Roy. Ag. Soc.	*Journal of the Royal Agricultural Society* (London)
Jl Roy. Stat. Soc.	*Journal of the Royal Statistical Society* (London)
Jl Soc. Arts	*Journal of the Society of Arts* (London)

Abbreviations

Köln. Forsch.	Kölner Forschungen zur Wirtschafts- und Sozialgeographie (Wiesbaden)
Kw. Hist.	Kwartalnik Historyczny (Warsaw)
Kw. Hist. Kult. Mat.	Kwartalnik Historii Kultury Materialnej (Warsaw)
Kyklos	Kyklos (Bern)
Mém. Agr.	Mémoires d'Agriculture (Paris)
Mém. Soc. Géol. Bret.	Mémoires de la Société Géologique de Bretagne (Rennes)
Mém. Soc. Émul. Liège	Mémoires de la Société d'Émulation de Liège
Metron	Metron (Rome)
Mitt. Ver. Gesch. Osnbr.	Mitteilungen des Vereins für Geschichte und Landeskunde von Osnabrück (Osnabruck)
Mitte	Die Mitte: Jahrbuch für Geschichte, Kunst und Kulturgeschichte des mitteldeutschen Raumes (Hamburg)
Mont. Rdsch.	Montanistische Rundschau (Berlin)
Münst. Btr. Gesch.	Münsterische Beiträge zur Geschichtsforschung (Paderborn)
Nass. Ann.	Nassauische Annalen (Wiesbaden)
New Camb. Mod. Hist.	New Cambridge Modern History (Cambridge)
Niedersachs. Jb.	Niedersachsisches Jahrbuch für Landesgeschichte (Hildesheim)
Nor.	Norois (Bourges)
Nouv. Ét. Hist.	Nouvelles Études Historiques (Budapest)
Od. St. Hist. Soc. Sci.	Odense Studies in History and Social Science (Odense)
Ost. Inst. Qu. u. St.	Osteuropa Institut, Quellen und Studien (Leipzig)
Pac. Hist. Rv.	Pacific Historical Review (Glendale, Calif.)
Pamphl.	The Pamphleteer (London)
P. & P.	Past and Present (London)
Pap. Mich. Acad. Sci.	Papers of the Michigan Academy of Science (Ann. Arbor, Mich.)
Paris I. de F.	Paris et l'Île de France (Paris)
Parl. Pap.	Parliamentary Papers (London)
Pet. Mitt.	Petermanns Mitteilungen (Jena)
Pet. Mitt. Erg.	Petermanns Mitteilungen Erganzungsheft (Jena)
Pop.	Population (Paris)
Pop. St.	Population Studies (London)
Prac. Inst. Bud. Miesz.	Prace Instytutu Budownictwa Mieszkaniowego (Warsaw)
Prac. Kom. Nauk. Hist.	Prace Komisji Nauk Historycznych (Kraków)
Proc. Roy. Soc. Med.	Proceedings of the Royal Society of Medicine (London)
Prz. Geog.	Przegląd Geograficzny (Warsaw)
Prz. Hist.	Przegląd Historyczny (Warsaw)
Prz. Nauk. Hist. Społ.	Przegląd Nauk Historycznych i Społecznych (Łódź)
Prz. Zachod.	Przegląd Zachodny (Poznań)
Pub. Fac. Let. Paris	Publications de la Faculté des Lettres de l'Université de Paris (Paris)
Pub. Fac. Strasb.	Publications de la Faculté de l'Université de Strasbourg (Paris)
Quart. Rv.	Quarterly Review (London)
Rec. Trav. Hist. Phil.	Recueil de Travaux d'Histoire et de Philologie (Louvain)
Rhein. Vbl.	Rheinische Vierteljahrsblätter (Bonn)
Rocz. Biał.	Rocznik Białostocki (Białystok)
Roy. Inst. Int. Aff.	Royal Institute of International Affairs (London)
Rum. St.	Rumanian Studies (Leiden)
Rv. Deux Mondes	Revue des Deux Mondes (Paris)
Rv. Éc. Int.	Revue Économique Internationale (Brussels)
Rv. Écon.	Revue Économique (Paris)
Rv. Est	Revue de l'Est (Paris)
Rv. Ét. SE Eur.	Revue des Études Sud-est Européennes (Bucharest)
Rv. Géog. Alp.	Revue de Géographie Alpine (Grenoble)
Rv. Géog. Lyon	Revue de Géographie de Lyon (Lyons)
Rv. Hist. Éc. Soc.	Revue d'Histoire Économique et Sociale (Paris)
Rv. Hist. Mod.	Revue d'Histoire Moderne (Amsterdam)
Rv. Hist. Mod. Cont.	Revue d'Histoire Moderne et Contemporaine (Paris)
Rv. Hist. Sid.	Revue d'Histoire de la Sidérurgie (Jarville)
Rv. Hist.	Revue Historique (Paris)
Rv. Inst. Int. Stat.	Revue de l'Institut International de Statistique (Edinburgh)
Rv. Inst. Soc.	Revue de l'Institut de Sociologie (Brussels)
Rv. Mines	Revue des Mines (Paris)
Rv. Mines, Mém.	Revue des Mines, Mémoires (Paris)

Abbreviations

Rv. Nord	Revue du Nord (Lille)
Rv. Nord Livr. Géog.	Revue du Nord, Livraison Géographique (Lille)
Rv. Roum. Hist.	Revue Roumaine d'Histoire (Bucharest)
Rv. Univ. Mines	Revue Universelle des Mines (Liège)
Sc. Ec. Hist. Rv.	Scandinavian Economic History Review (Stockholm)
Sc. Geog. Mag.	Scottish Geographical Magazine (Edinburgh)
Sc. Jl Hist.	Scandinavian Journal of History (Stockholm)
Schr. Inst. Allg. Gesch.	Schriften des Instituts für Allgemeine Geschichte (Berlin)
Schr. Dresd.	Schriftenreihe des Staatsarchivs Dresden (Weimar)
Schr. Ver. Sozpk	Schriften des Vereins für Sozialpolitik (Berlin)
Schr. Rein-Westf. Wtgesch.	Schriften zur Rheinisch-Westfälischen Wirtschaftsgeschichte (Cologne)
Schw. Zt Gesch.	Schweizer Zeitschrift für Geschichte (Zurich)
Slav. Rv.	Slavic Review (Seattle)
Soc. Hist.	Social History (London)
Soz. Welt.	Soziale Welt (Göttingen)
St. Düss. Wtgesch.	Studien zur Düsseldorfer Wirtschaftsgeschichte (Düsseldorf)
St. u. Eisen	Stahl und Eisen (Düsseldorf)
St. Hist. Acd. Sci. Hung.	Studia Historica, Academia Scientiae Hungarica (Budapest)
St. Hist. Oec.	Studia Historiae Oeconomicae (Poznań)
St. Hist. Slov.	Studia Historica Slovaca (Bratislava)
Stat. Int. Rétr.	Statistiques Internationales Rétrospectives (Brussels)
Staats. Soz. Forsch.	Staats- und sozialwissenschaftliche Forschungen (Leipzig)
Stud. Fanfani	Studi in Onore di Amintore Fanfani (Milan)
Stud. Gen.	Studium Generale (Berlin)
Stud. Slov.	Studia Slovenica (New York)
Studi Sapori	Studi in Onore di Armando Sapori (Milan)
Stud. Stor.	Studi Storici (Rome)
Südost Forsch.	Südost Forschungen (Munich)
Trans. I.B.G.	Transactions, Institute of British Geographers (Oxford)
Ung. Jb.	Ungarn Jahrbuch (Mainz)
Univ. Lond. Hist. Stud.	University of London Historical Studies (London)
Univ. Copen. Inst. Econ. Hist.	University of Copenhagen, Institute of Economic History (Copenhagen)
Univ. Manchester Econ. Ser.	University of Manchester Economic Series (Manchester)
Univ. Michigan Pubns Hist. and Pol. Sci.	University of Michigan Publications in History and Political Science (Ann Arbor, Mich.)
U.S. Bureau of Mines Bul.	United States Bureau of Mines Bulletin (Washington, D.C.)
Veröff. Hist. Kom. Berlin	Veröffentlichungen der historischen Kommission zu Berlin (Berlin)
Veröff. Schwab. Forsch.	Veröffentlichungen der Schwabsiche Forschungsgemeinschaft (Munich)
Viert. Soz. Wirtgesch.	Vierteljahrschrift für Sozial- und Wirtschaftsgeschichte (Wiesbaden)
Wien. Hist. St.	Wiener Historische Studien (Vienna)
Weltw. Arch.	Weltwirtschaftliches Archiv (Jena)
Yorks. Bull. Ec. Soc. Res.	Yorkshire Bulletin of Economic and Social Research (Leeds)
Ymer	Ymer (Stockholm)
Zt Aggesch.	Zeitschrift fur Agrargeschichte und Agrarsoziologie (Frankfurt on Main)
Zt Berg. Hütt. Sal.	Zeitschrift fur das Berg-, Hütten- und Salinenwesen im Preussischen Staate
Zt Gesch. Oberr.	Zeitschrift für die Geschichte des Oberrheins (Karlsruhe)
Zt Ges. Staatsw.	Zeitschrift für die Gesamte Staatswissenschaft (Tübingen)
Zt Hist. Ver. Niedersachs.	Zeitschrift des Historischen Vereins für Niedersachsen (Hanover)
Zt Hist. Ver. Steiermk	Zeitschrift des Historischen Vereins für Steiermark (Graz)
Zt Oberschles. B.H.V.	Zeitschrift des Oberschlesischen Berg- und Hüttenmannischen Vereins (Kattowitz)
Zt Stadtgesch.	Zeitschrift für Stadtgeschichte, Stadtsoziologie und Denkmalpflege (Stuttgart)

Preface

The publication of this volume completes a project conceived nearly forty years ago. It sprang from a discussion of the nature of historical geography which the author was privileged to have with the late Professors E. G. R. Taylor and S. W. Wooldridge. The original suggestion was for a series of 'horizontal' pictures of Europe at critical or important stages in its history, but this has undergone a number of changes. It has been demonstrated by the late Professor J. O. M. Broek and by Professor H. C. Darby that the resulting studies are likely to be too static, and in their own writings they have interspersed the horizontal pictures with 'vertical' narratives showing the dynamics of change between them. The first volume of this study of the historical geography of Europe consisted essentially of five horizontal studies. The addition of linking chapters, which were in fact sketched, would have made a long book far too long, and they were omitted. In retrospect this may seem to have been a mistake.

The second volume remedied this error, if error it was, by interpolating between studies of Europe in the early sixteenth century and on the eve of the Industrial Revolution, long vertical chapters which traced the changing spatial pattern of the main categories of human activity. The present book follows the example of the latter. Its starting-point is, in fact, the last chapter of the second volume, which attempted to present a picture of Europe after the Napoleonic Wars, but before the intensive development of modern industry, which marked the middle years of the nineteenth century. It, begins, in fact, a little earlier in time, since it seemed desirable to present the immediate antecedents of the revolutionary changes which occurred. The vertical chapters, dealing with population and urban growth, agriculture, manufacturing and commerce, are drawn together in a concluding chapter, which summarises the chief features of that Europe which was overwhelmed by the cataclysm of the First World War.

Historical geography is concerned with the changing spatial pattern of human activity, and the author has tried to keep this objective always in view. The sources are historical; their presentation geographical. There must, however, always be a tendency or temptation to explain development in terms of environment or of resources. One cannot ignore the importance of Europe's resource base in coal and iron-ore in examining the rise of nineteenth-century manufacturing, nor the significance of the great rivers in water-borne transport, but the random and irrational factors in human decision-making have been emphasised throughout. The case of Łódź, for example, is used to demonstrate the growth of manufacturing in an area which, whatever contemporaries may have thought, proved to be totally devoid of natural advantages. Agricultural developments and specialisations also are not necessarily those which an examination of the resources would lead one to expect. Rural change and adaptation were hindered by a complex web of social obligations and relationships. Who would have supposed that a common right to glean after harvest could possibly have inhibited the consolidation of fragmented strips and the introduction of modern harvesting equipment? Jean François Millet's famous painting of *The Gleaners* takes on an added significance in these circumstances.

The century covered by this book was one of rapid change, the detail of which is too great and too complex to present. For this reason sample studies have been included. In the case of agriculture, as varied a group of regions as possible was selected, including two for which the chief and in some ways most penetrating evidence is literary. In the case of manufacturing, the development is traced of all major industrial regions and also of some of only minor importance.

It would have been ideal to illustrate with maps every area and region discussed. This proved to be impossible, and the reader is urged to have an atlas at hand. In the spelling of place-names, the conventional English usage is followed wherever possible. In other cases the spelling that would have been usual in the nineteenth century has been employed, as with Breslau (Wrocław) and Christiania (Oslo).

The author wishes to thank all who have in varying ways assisted him. They are too numerous to name, but he wishes to record his debt in particular to his friends and colleagues Professors W. B. Cohen and George Alter who between them read much of the manuscript. He is also grateful to those who made it possible for him to visit industrial works in all parts of Europe; to those in the Patents Library and elsewhere who procured obscure journals for him, to the library of Cambridge University which made available photographs of Sir Arthur Evans' drawing of a *zadruga*, of R. Baird Smith's map of irrigation works in the Lombardy Plain, and of T. E. Gumprecht's map of Spanish cities. Indiana University provided the maps of agricultural distributions used in the last chapter, and also help and encouragement during the whole period of gestation for all three volumes.

N. J. G. POUNDS

Department of History
Indiana University
Bloomington
Indiana
January 1985

1
From Waterloo to the First World War

The century which elapsed between the battle of Waterloo and the outbreak of the First World War witnessed changes more profound than in any comparable period in the span of human history. It was one of unprecedented economic growth. The population of continental Europe more than doubled. There was a comparable growth in agricultural production; industrial output must have increased at least tenfold, and the gross product of Europe probably increased six times.[1] This expansion of economic activity was accompanied by a radical shift in its location. An older, 'pre-industrial' pattern of manufacturing slowly and reluctantly decayed and was replaced by another, which grew in response to factors which the pre-industrial age had scarcely known. The geography of Europe, when the Napoleonic Wars at last ended, differed fundamentally from that which saw the lights go out in August 1914.

The intervening century was for most of Europe mainly one of peace. There were of course international conflicts, but even the Franco-Prussian War, the most significant of them, lasted only six months, and compared with the more disastrous wars that occurred both before and after, cost relatively little in lives or money. Most other conflicts were related to people's democratic aspirations, like the risings of 1848, or to their demands for political independence, like the Italian *Risorgimento* and the recurring struggles within the Balkans.

In these circumstances only a very small part of Europe's productive effort was diverted to military purposes or squandered in the destruction of war. Before the early years of the twentieth century military expenditure consumed in most countries of western and central Europe only a relatively small part of their national budgets. Most governments were thus able to concentrate as their primary objective on creating the conditions for economic growth.

But economic growth in Europe was a highly localised phenomenon. Much of the continent was almost untouched by progress – industrial, agricultural or commercial – until late in the century. At the same time there were areas where growth was precocious and rapid, and from which the new technology was diffused to other parts of the continent. The contrast between the developing regions of early nineteenth-century Europe and the least developed grew more extreme as the century progressed. The difference between central Belgium, the lower Rhineland and the Swiss Plateau on the one hand, and the mountains of Macedonia and Romania or the forests of eastern Poland and northern

Scandinavia on the other was as great in the mid-nineteenth century as that between the developed and the Third World today.

What, then, were the factors which permitted or encouraged certain restricted areas of the continent to, in Rostow's phrase, 'take off into sustained growth'? It is tempting to relate the distribution of these advanced areas of the European economy to the continent's endowment in solid fuel and metalliferous ores. One is reminded of Keynes' judgement that the power of Germany at the end of this century of growth was founded more 'on coal and iron than on blood and iron'. Indeed it was, but that particular example of economic growth came late in the nineteenth century, not during the early phases of the Industrial Revolution. Some of the early centres of economic growth were well endowed with fossil fuel. But that was coincidental. The earliest developments on the coalfields of the Ruhr and Upper Silesia, the best endowed in the continent, were based on water-power and charcoal. Manufacturing was attracted to Saxony and to Łódź in Poland, in part because of the potential which these areas offered for the development of water-power. The earliest developments in northern Italy and Catalonia were based on water-power. Elsewhere, in Bohemia, Switzerland and the Austrian Alps, water continued to power the mills until, in many cases, the giant water-wheels were replaced by electric motors which derived their energy from newly built hydroelectric plants.

Although fossil fuel had been first used for smelting iron in 1709 in England, its adoption in continental Europe was very slow indeed. It was first used in the Ruhr 140 years later, and charcoal smelting remained important in the Rhineland, central France and Austria well into the second half of the century. Only in the Saint-Étienne region in *dépt* Loire and in Upper Silesia was coal of some significance from the start of modern industrial growth.

However important the fossil fuels became later in the history of industrial growth their significance was clearly small in the early stages. One must look for other factors. Those which favoured the early growth of manufacturing were more social and economic than they were physical. Foremost amongst them was a social climate which encouraged and rewarded the innovator and the entrepreneur. The supply of investment capital was of some importance, but early industry was not on the whole capital-intensive. More important was the existence of a market accessible to the products of factory or industry. It was the absence of a market in much of pre-revolutionary France or in eastern Europe in the first two-thirds of the nineteenth century which, more than any other factor, inhibited economic growth. Arthur Young had observed that 'a large consumption among the poor [was] of more consequence than among the rich'. This fact was clearly realised at the very end of our period by Thomas Bat'a in eastern Europe when he built a factory for making coarse, cheap, mass-produced footwear. The market was almost unlimited once an effective demand began to emerge amongst the peasantry.

Specialised manufacturing, as distinct from the generalised and domestic production of rough cloth and leather, called for investment capital and a commercial infrastructure. The former seems to have come initially from the profits of agriculture and of trade. As manufacturing progressed, it was able in some measure to generate its own investment surplus. Specialised manufacturing was predicated on trade, not only in the raw materials which it used, but also in the finished products. It was its nature to produce at a fixed point far more goods than could be absorbed locally. A factory, even one of modest size, had to be in a position to distribute its output over a relatively wide area. This called for an infrastructure consisting of routes and means of transport, as well as of warehouses, shops and merchants.

In the late eighteenth century this infrastructure was feebly developed or was even non-existent over large areas of Europe. Regions which lacked usable roads, rivers and canals, which had few specialised merchants and commercial outlets, could not easily initiate an expansion in industrial production. By the same token, there could be no large and effective demand for factory products.

Economic and, in particular, industrial growth called for a relatively dense population with the capacity to absorb the products of manufacture, and also an infrastructure by which goods could be moved and demand satisfied. This clearly places the causal factors in *early* industrial growth in continental Europe on the demand side. *Early* manufacturing, it might be said, could have grown up almost anywhere. It actually appeared where there was an effective demand for its products. This ceased to be the case later in the century when a greater dependence developed on highly localised resources, such as coal and minerals; when the railway facilitated the distribution of goods, and intense competition between producers gave an added importance to marginal costs in production and distribution.

These conditions were satisfied most fully in north-western Europe, in the southern Low Countries, the Lower Rhineland and parts of northern France. This region was the most populous and highly urbanised. Its commercial importance in medieval and early modern times had left as its legacy to the nineteenth century a network of roads and navigable waterways, a system of financial and commercial houses and an economy based on money and exchange. Such an infrastructure was totally lacking from most of southern Europe and eastern Europe, and in central Europe was developed in only a few favoured areas such as Saxony and around the great cities of southern Germany and Switzerland.

Innovation and diffusion

The process of economic growth during the nineteenth century was, from a geographical perspective, one of innovation and diffusion. Innovations were made in manufacturing, agriculture, transport and management. This in itself implied a receptivity to change on the part of the entrepreneur, but the willingness to change varied greatly within Europe. Innovations spread from their earliest centres to others which were receptive to them. The diffusion of innovation was at first an almost random process. New techniques never remained secret for long. They were carried by workmen who had used them, and their migrations were dictated primarily by the prospect of employment and profit. One of the most unlikely instances of such diffusion was the spread of the English technique of making coke to Upper Silesia (see p. 343). More probable was the diffusion of spinning machinery from England to northern

France, central Belgium and Switzerland, in all of which there was a fertile soil in which the new technology could grow. From Belgium it spread to the Aachen district of western Germany and thence to the Rhineland; from Switzerland it spread into Alsace and southern Germany, and from the latter to Saxony. Entrepreneurs from Saxony carried the new technology to Łódź, and from both Saxony and southern Germany into Bohemia. A similar process of innovation and diffusion can be traced in the metal industries, notably in the spread of the puddling process and of improved methods of blast-furnace design and operation. The speed of diffusion increased during the century with improvements in communication. One has only to contrast the slowness with which the puddling process spread across Europe with the alacrity of the adoption of the basic method of steel-making (see p. 343).

Diffusion of innovation was influenced by a multitude of factors over and above a natural, human dislike of change. Amongst them were the ways by which new ideas could be disseminated, the attitude of governments to industry and agriculture, the cycles of boom and depression as they influenced the profitability of economic activity, and, late in the century, the impact of overseas sources of supply. To these factors should be added the influence, admittedly of only minor importance, of boundary changes.

Communications and the spread of ideas

The rapid expansion of agricultural and industrial production, which marked the middle and later years of the nineteenth century, was the consequence, in part, of superior organisation and management, but, above all, of the application of science to agriculture and manufacturing. Scientific speculation had tended previously to be divorced from the practical world of farmer and craftsman. The latter had made many and significant technological advances, but these were based entirely on the craft experience.[2] The spinning wheel and loom had been improved and made to operate faster and more evenly. Blast furnaces were built larger and operated at higher temperature; refining of iron was transferred from a small hearth to a large reverberatory furnace. The steam-engine, swifter, more powerful, more reliable, replaced the power drawn from the wind, running water and the strength of man and beast. In all these respects the scale and speed of production were increased. But there was no radical change in the technology of the processes themselves. In all of them there was an appalling waste of energy. Little was known of the strength of materials, and in many ways wasteful and unnecessary concessions were made to the demands of safety and tradition. The craftsman could only operate within the sphere of his own knowledge and experience. It took a theoretical scientist to broaden this field and to make possible the fundamental advances of the later nineteenth century.

By the middle years of the century a scientific attitude was beginning to penetrate agriculture and most branches of manufacturing, and the gulf beween scientist and craftsman was narrowing appreciably. This was in large measure due to the setting up of scientific institutes for teaching and research, at first in Germany, but later also in France, Switzerland, the Netherlands and Great Britain. In Prussia this emphasis on science emerged from the trauma of military defeat and French occupation. Many of the older and more conservative German universities disappeared during the Napoleonic wars. In 1809, however, the University of Berlin – the Royal Frederick Wilhelm University – was founded. It incorporated the older Berlin Academy of Sciences, and its curriculum was shaped by one of the greatest scientists of the day, Wilhelm von Humboldt. From the first the University of Berlin became a centre for scientific teaching and research, firmly based upon observation and laboratory experiment. Other scientific institutes followed, and laboratory work became an essential feature of scientific training. Research laboratories were established at Munich, Giessen, Bonn, Heidelberg and elsewhere, and it was at Giessen that Justus von Liebig worked on the chemistry of agriculture and organic growth. Much of the work at these centres was in chemistry, which was in the early nineteenth century the science which offered the greatest practical advantages, but mining and metal-working also benefited, with the creation of *Bergakadamien* and schools of metallurgy. Then research in the generation and use of electricity came to the fore with the invention of the first voltaic pile at the beginning of the century. Henceforward progress was made by the scientist rather than the craftsman, in the laboratory rather than the workshop.[3] Nevertheless, progress was slow. Electricity was not used significantly for lighting until the mid-century (see p. 48), and before it could be widely used in the home it required the production of lighting bulbs of low intensity. The filament bulb appeared in 1881, and with it a growing demand for a public supply of electric power. The following decade saw a sharp increase in the number of generating stations and in the use of electricity for lighting. This was followed by fundamental improvements in the electric motor and in its use for machines, both mobile, as in tramways, and fixed.

Early progress in electrical engineering had been mainly in Great Britain and France, but in the 1880s the lead passed to Germany, where the most determined efforts were made to solve the problems, both theoretical and practical, in the production and use of electric power. Developments at the close of the nineteenth century demonstrated, if this was still necessary, that significant progress in the practical and applied fields was dependent on heavy investment in theoretical research. No country benefited from this more than Germany, where research institutes were most developed.

These high-level research institutions came to be served by lower-level polytechnics and technical schools. Between them they trained industrial management in scientific methods and equipped the skilled workman to use the methods and tools developed. In 1884 the Royal Commission on Technical Education[4] reported on the status of the subject in continental Europe. It wrote:

Technical High Schools now exist in nearly every Continental State, and are the recognised channel for the instruction of those who are intended to become the technical directors of industrial establishments...the success which has attended the foundation of extensive manufacturing establishments, engineering shops, and other works, on the continent, could not have been achieved to its full extent in the face of many retarding influences, had it not been for the system of high technical instruction in these schools...

Before the Industrial Revolution the diffusion of industrial technology had been erratic and unpredictable. Innovators were, in the absence of any form of legal protection, highly secretive. Governments, notably that of Great Britain, prohibited the migration of skilled workmen, familiar with the industrial processes which they sought to protect, and industrial espionage became a common and even lucrative occupation. The diffusion of technological innovations was further slowed by the difficulty of describing them. Gabriel Jars was sent by the French government to enquire into, amongst other matters, the method developed in Great Britain of making metallurgical coke and using it to smelt iron. The results of his enquiries were published as *Voyages métallurgiques*.[5] In them he described the processes as he witnessed them, and countless metallurgists must have tried to follow his recipe as if from a cookbook. Almost without exception they failed. The problem was that neither Jars nor those who used his text had any accepted technological vocabulary; they could not analyse and describe the fuels used, nor could they measure the furnace temperature or estimate the quality of the ore and the effect of impurities in it. Theirs was a wholly empirical procedure, and when they were successful they had no means of understanding why.

Both these obstacles to the diffusion of technical innovations were gradually removed in the course of the nineteenth century. In the first place, the need for secrecy disappeared when the innovator gained legal protection for his invention through the patents law. In Great Britain this had been developing during the eighteenth century, but its operation remained imperfect and costly until the Patents Office was set up in London in 1883. In Germany and France, the only other countries apart from the United States in which it was of great significance, a patents law also evolved during the century, gradually removing the apparent need for secrecy and subterfuge in technological development. In the year in which the Patents Office was opened in Great Britain an international patents convention was signed by most of the developed countries. It became the practice for the successful innovator in one country to grant a licence to patentees elsewhere in the world to use his invention on payment of an agreed royalty. The speed with which, for example, the Gilchrist-Thomas process for making basic steel from high-phosphorus iron was adopted under licence by the leading steel-manufacturers in continental Europe is in marked contrast to the slow diffusion of coke-smelting or even of the puddling process.

If the improved patents law made science more open, it was the emergence of a technological press which encouraged the rapid diffusion of the new technology. Scientific publications assumed two forms. In the first place there were books, sometimes of very considerable length, which surveyed, often with a wealth of crude steel engravings, the state of technology and the economic prospects of particular industries. In a sense these works derived from that uneven collection of technical monographs published by Duhamel de Monceau in the later eighteenth century under the name of *Descriptions des arts et des métiers*.[6] Technological descriptions were also embodied in the contemporary *Encyclopédie* of Diderot and d'Alembert and in its British imitator, the successive editions of *Encyclopaedia Britannica*.

Both Swendenborg[7] and Réaumur[8] published treatises on the iron industry, but most of the early 'craft' studies were concerned with agriculture. The reason probably was that the landowner was more likely to be able to read than the artisan. Indeed the multiplication of such handbooks in the middle years of the nineteenth century had to await the increasing literacy of the artisan classes. The best of them passed through a series of editions, in which their authors attempted – sometimes successfully – to keep abreast of an advancing technology.

It fell, however, to the periodic literature to record the changes and advances from day to day in particular branches of industry. Here, too, some of the earliest of such serials were devoted to agriculture.[9] The first important journal to concern itself exclusively with industry and industrial technology was the *Annales des Mines*. It first appeared in 1792 as the *Journal des Mines*, later changing to the more familiar title under which it has continued to appear until the present. The *Annales* was an official publication of the French government. A not dissimilar journal, the *Berg- und Hüttenmännische Zeitung*, began in Germany in 1842. It was, however, published privately, and lasted until 1904. In 1853 the Prussian government began the *Zeitschrift für das Berg-, Hütten- und Salinenwesen im Preussischen Staat*. Three years later the *Revue Universelle des Mines* began publication in Paris and in 1853 the *Oesterreichische Zeitschrift für Berg- und Hüttenwesen* followed.

These are merely the earliest and most important journals in the fields of mining, smelting and metallurgy. During the second half of the century the number of such journals increased and they became more specialised. They extended to the textile, leatherworking and ceramic manufactures and to a vast range of craft industries. There is unfortunately no published catalogue of industrial and craft journals, many of which were published for only a very short period and are extremely difficult to locate.

The spread of scientific technology on which most of the industrial and agricultural advances of the later nineteenth century were based was conditioned by developments in other fields than laboratory research. The freedom of information on which it was based has already been discussed. Related to it was the freedom to communicate, first by means of the postal system and then by telegraph and telephone. These were supplemented and in part replaced by radio communication, but the effective use of the latter did not come before the First World War and is thus beyond the scope of this book.

The first modern postal system was initiated in Great Britain in 1840, and was quickly imitated in most European countries. Problems in the international despatch of mail were first examined at a conference in Paris in 1863, and in 1874 the International Postal Union was formed. By the second half of the century it was possible to communicate by mail as quickly as at present within almost the whole of Europe except the remoter parts of the south-east and east.

During the middle years of the century an even more speedy means of communication came into use, the telegraph. The idea of communicating by means of electric impulses sent along a wire had originated in the eighteenth century. By the 1840s coded messages were being sent on an experimental basis.[10] Soon afterwards networks of overhead wires were constructed in parts of Europe for the transmittal of messages coded in Morse. Cables began to be placed beneath the streets of cities, and in 1850 the first submarine cable – between Calais and Dover – was laid. It broke, but in the next year was replaced by another, which continued to give good service. The first transatlantic cable was laid in 1858.

The telegraph called for operators able to encode and decode the messages transmitted. The telephone, which made its appearance a quarter of a century later – Graham Bell's American patent was taken out in 1876 – used electrical impulses generated by the human voice. People could *speak* by telephone, thus eliminating the need for coding the message. Private telephone systems multiplied and were linked into national and subsequently international systems with the construction of publicly owned telephone exchanges. The first opened in London in 1879 and at about the same time the telephone was adopted in rural areas in Germany, because here the volume of 'traffic' was too small to justify the employment of a *telegraph* operator. Elsewhere, however, the use of the telephone was a response to the intensity of traffic. A public telephone system was initiated in France in 1881. In Switzerland, a privately owned system in Zurich (1880) was followed by the building of public exchanges in most of the larger cities. During the following decade the telephone was in use in the Scandinavian countries, followed by the Low Countries, Austria and Italy and eventually by the countries of eastern and south-eastern Europe. From 1890 to the outbreak of the First World War, the number of telephones in use increased very greatly, and the network of telephone wires was extended into the most remote and least populous regions.

The adoption of the telegraph and then the telephone did nothing directly to hasten the diffusion of innovation. Their importance lay in the fact that they allowed decisions – in particular commercial decisions – to be communicated rapidly. They increased the speed of flow of information and they hastened the completion of commercial transactions. The rapid growth of manufacturing industry and of industrial production in the closing years of the nineteenth and at the beginning of the twentieth centuries was not unrelated to the increasing speed with which information of all kinds could be transmitted not only across a city but from one end of Europe to another. It is perhaps in this intensification of linkages, both internal and with other countries, and in the greater speed of reaction that the geography of Europe in 1914 differed most significantly from the Europe which carried on its slow deliberations in 1815.

At the beginning of the nineteenth century there were parts of Europe which could, and did, exist in almost complete isolation from the rest of the continent.[11] Their economy was self-sufficing, and they were not dependent either on supplies from elsewhere or on the markets which other parts of the continent could offer for their products. Unless engulfed by war, they were, for good or ill, unaware of the problems which faced the rest of Europe. Such areas had formerly been far more extensive than they were in 1815. Only some twenty-five years earlier there was even a community in the hills of eastern France which claimed in answer to the government's enquiry to be uncertain whether it lay in France or the German Empire and thought itself independent of both. In the course of the nineteenth century, with improvements in transport and communications, such patches of isolation and self-sufficiency were to disappear from the European map. The reindeer-herder of Lappland and the plum-grower in the forests of the Serbian Šumadija were being drawn into a European economy, as their fortunes were gradually influenced by the demand for newsprint and by breakfast habits in the cities of western Europe.

The nineteenth century saw the creation of a complex pattern of interlinked demand and supply. The disparate economies of Europe were drawn together, so that any disturbance anywhere sent reverberations through the whole system. Local self-sufficiency gave way gradually to monopoly production; domestic craft industry to the factory with its mass production methods. The thin veneer of manufacturing, integrated at the local level with agriculture and spread lightly over much of the continent, became concentrated into knots of highly specialised and intensive production. The gain of a few industrialised regions was offset by the impoverishment of extensive rural areas. A pattern of regional boom and depression was created, with human migration as a kind of adjustment between them. Historians have tended to emphasise the areas of growth. The historical geographer, to whom all geographical space is significant, must inevitably look also at these forgotten areas of Europe.

Free trade and protection

The economic growth of the latter half of the nineteenth century and the resulting changes in the geography of production were shaped by the play of market forces. But these were in varying degrees formed and influenced by government policy. This assumed two forms. On the one hand a government could offer direct encouragement, usually financial, to any branch of manufacturing or agriculture that it wished to help. On the other, it could by manipulating quotas and import duties, shield any or all economic activities from competition from other and cheaper products.

There was nothing new in this. In pre-industrial Europe governments had regarded the protection of agriculture and craft industries as a major object of policy. This objective was, it is true, linked with the question of the balance of trade and the prevention of the outflow of bullion, but it nevertheless tended to create a series of closed economic systems. This mercantilist view of the economy was vigorously attacked in 1776 by Adam Smith in his *Inquiry into the Nature and Causes of the Wealth of Nations*. His argument that protection tended to prevent economic activities from being established in those places and areas best suited to them, was irrefutable. Without protection goods would be produced more cheaply and more abundantly, because production would adapt itself to the availability of the factors of production. In a well-known passage Adam Smith argued that an excellent wine could be produced in Scotland. It is, indeed, doubtful whether it could, but Smith was a patriotic Scot and must be allowed this point. That it was not produced, he said, was due to the fact that the factors of production were combined so abundantly in France that the resulting cheapness of French wine put it beyond the competition of any Scottish vintage. And so, in keeping with the analysis of comparative advantage, Scotland turned to the distillation of another liquor, in which its factor costs were more favourable.

The gospel of Adam Smith, modified and extended by his prophets, David Ricardo, John Stuart Mill and others, was widely accepted, but implemented only very slowly. A start was made in the United Kingdom in the 1820s. Some duties were reduced and the jungle of tariff legislation simplified, but the real battle for free trade was not fought for another twenty years. In 1846, at the end of a long campaign for their repeal, the duties on imported breadgrains were drastically reduced. Successive budgets further reduced tariffs, and in 1860 the Cobden–Chevalier treaty with France led to a reciprocal reduction of duties. By 1874 the United Kingdom had abandoned almost all protective restraints on trade.

The 1860 commercial treaty with France had contained one important innovation, the 'most favoured nation' clause, which obligated the two participants to extend to any other country with which they might conclude a similar agreement the most favourable treatment that they accorded one another. The effect of this was the gradual extension of the principles of free trade, for the nineteenth-century trading system was built upon a network of commercial treaties.

The progressive reduction of tariffs had begun even earlier in Germany. In 1818 the Prussian government drastically reduced import duties. German liberals had, by and large, been converted to the free-trade ideals of Adam Smith, but it was not entirely for this reason that the government acted as it did. Its reasons were in part nationalistic, the reduction and elimination of the barriers between the states which made up Germany at this time. It was, however, supported by the landed interests of eastern Germany, looking for markets for their growing wheat surpluses.[12] In 1834 the formation of the Zollverein marked the beginning of a large free-trade area in central europe. The Zollverein itself pursued a policy of low tariffs as it sought to bring the remaining states into the German Confederation.[13]

In the 1860s the United Kingdom, France and Germany all pursued policies of reducing tariffs and were followed by most of the other advanced European countries, including Belgium, the Netherlands, Spain and Italy. In none, however, was the practice of free trade pursued as thoroughly and as consistently as in the United Kingdom. After 1870, however, conditions began to change. The wars of the 1860s, culminating in the Franco-Prussian War of 1870–1, disturbed a delicate European balance and threatened the conditions of co-operation and mutual trust that had developed. At the same time the opening up of new lands outside Europe led to the import of increasing quantities of bread-grains and other foodstuffs, thus undercutting the prices received by the European farmer. The large-scale capital investment associated with railway construction was of diminishing importance. The market for manufactured goods began to contract, and for the first time there was significant unused industrial capacity.

One country after another abandoned its lightly held principles of free trade. In France attacks on free trade and the most-favoured-nation clause began in the early 1870s, and were followed in the next decade by sharp increases in import duties, especially on agricultural goods. In Germany the grain exports from the eastern provinces were threatened by the growing supplies of cheap wheat from the New World and elsewhere, and landowners demanded protection for German agriculture. At the same time the general recession began to affect the metallurgical and other branches of manufacturing. These too asked for protection from foreign competition. The tariff which Bismarck introduced in

1879 served to unite the diverse interests of eastern landowner and western industrialist, and, indeed, served to hold them together in an uneasy partnership for the whole duration of the Second Reich.

Both Russia and Austria–Hungary maintained high tariffs. The former included 'Congress' Poland within its customs boundary, thus tempting western entrepreneurs, prevented from selling direct to Russia and Russian-held Poland, to establish factories just within the Russian customs barrier. The textile industries of Łódź originated in this way. After the Polish rising of 1863–4 the Tsarist government, in its desire to punish the Poles, erected a tariff barrier between Russian Poland and Russia itself. The Łódź industrialists, thus foiled in their attempt to dominate the Russian textiles market, again jumped the tariff wall and established mills, though on a very much smaller scale, at Białystok (see p. 410).

The Habsburg Empire, like the Russian, maintained high tariffs. Little effort was made to absorb Austria into the German customs union, largely because German nationalists feared the problems that this multinational empire would create for themselves. The great extent and variety of resources within the Habsburg Empire should have stimulated trade between its several regions. That it did not at first do so was due mainly to internal barriers to trade. Until 1848 customs were levied between the Austrian and the Hungarian divisions of the Empire (see p. 26). Then, following the Hungarian rising of 1848, the customs barrier was abolished. The purpose of the Austrians was to punish the Hungarians by exposing their inefficient agriculture and youthful industry to the full weight of Austrian competition. In this, however, the failed, and Austria was itself opened up to the growing agricultural production of Hungary.[14] The customs union of Austria and Hungary survived the *Ausgleich* of 1867, and until the outbreak of the First World War the Hungarian Plain remained the granary for industrialised Bohemia and Austria, as the German eastern provinces were for the industrial Rhineland.

Only the United Kingdom obstinately maintained its policy of free trade in the face of a Europe which had without exception returned to a policy of protection. The United Kingdom was, however, enabled to do this only by sacrificing its agriculture. Its factor advantages in manufacturing were still sufficient to give it some protection against foreign industrial production. But cheap grain imports changed completely the traditional pattern of 'high' farming. Cheap food for the industrial worker was reckoned to be adequate compensation for the long depression of British agriculture.

The commercial cycles

Closely linked with the trend towards and then away from free trade were the cyclical changes in European manufacturing and commerce. These had long been apparent, and until late in the nineteenth century they had occurred with such regularity that some sought an external and physical cause for them.[15] They were essentially due to the unstable relationship of supply and demand, and they became more intense as these two parameters fluctuated more widely. The end of the Napoleonic Wars was followed by a period of depression when an economy geared in a certain respect for war had to adapt itself to the demands of peace. Conditions were made worse by the bad harvests and high food prices of 1816–17 and by the accompanying reduction of demand for consumer goods. A revival and succeeding decline in the 1820s arose from over-great expectations and excessive investment in manufacturing. Another cycle of boom and depression followed in the 1830s, and yet another in 1847–8, intensified by bad harvests and over-investment in the railway industry. Political developments were closely related to economic, and the risings of 1848 owed much to popular suffering and discontent. Fluctuations followed in the 1850s and 1860s with such regularity that there were those who even linked them with the newly discovered eleven-year cycle of sunspot activity. Then in 1873 came the largest and most severe depression experienced hitherto in Europe, followed in the early 1880s and again in 1893 by more limited recessions. Amongst the reasons for these complex phenomena were the import of bread-grains and other foodstuffs from the 'new' countries, the end of the railway boom and resulting decline in demand for steel, and a rapidly changing industrial technology with consequent structural changes in factor demand and markets. These were further complicated by developments in the fields of banking and credit facilities. After the crisis of 1893–6 there were fluctuations in the level of business activity, but nothing that could be called a depression before preparations for the First World War led to an expansion in business and in industrial production.

In the present context the significance of the cyclical variations in business and manufacturing is that they influenced the profitability and hence the viability of various productive activities. Each depression was followed by the extinction of a group of activities; that of the later 1840s led to the decline, in particular, of the Flemish linen industry (see p. 318). More significant was the decline of the traditional iron industry, especially in Champagne and Burgundy, as a consequence of the crisis of the 1870s.

A century of imperialism

Some European governments and peoples, notably those of France and Great Britain, diverted part of their energies into overseas expansion. Together with Belgium, the Netherlands, Germany, Spain and Portugal, they either acquired overseas territory or extended that which they already held. The French moved into Algeria in 1830, and within half a century had annexed almost a third of the African continent as well as a significant part of South-east Asia.[16] The Dutch extended their control over the East Indies and used their authority ruthlessly to acquire the raw materials of their commerce. Belgium converted the holdings of a trading company into an empire over much of Central Africa. Portugal and Spain, the oldest among the European imperial powers, lost their vast empire in the New World, but elsewhere were able to hold their island and coastal bases, and in some instances even to extend their authority. Italy and Germany did not achieve political unity until respectively 1861 and 1871, and thus came late into the colonial race. They none the less acquired tracts of the African continent which had hitherto gone unclaimed, as well as islands and island groups in Asia and the Pacific.

Conflict between European states was, as it were, translated into conflict for colonial dependencies in Africa, Asia and the oceans of the world. The motive for European imperialism in the nineteenth, as in previous centuries, was primarily economic. 'Whereas various real and powerful motives of pride, prestige and pugnacity, together with the more altruistic professions of a civilising mission, figured as causes of imperial expansion', wrote J. A. Hobson, a witness to many of these events, 'the dominant motive was the demand for markets'.[17] All the industrialising countries of nineteenth-century Europe desired an unrestricted access to the raw materials of the rest of the world, and were prepared to fight to acquire them. No less important were the markets which an overseas empire offered. The expansion of manufacturing industries in Great Britain would have been inconceivable – at least on the scale on which it actually occurred – without large and insatiable markets overseas for the cotton cloth and hardware, pottery and chemicals that poured from British factories.[18] The weakness of France's industrial 'take-off' can be explained, in part at least, by the fact that there was not until late in the century a captive overseas market for its products.

Dutch colonial policy was, until early in the present century, the most blatantly commercial of any. It was geared to the production of commodities of tropical origin for sale through Dutch intermediaries to European consumers. It did not so much support Dutch manufacturing industry, which was in any event, very slow in starting, as employ one of the world's larger merchant marines.[19] The Belgians stumbled into empire almost by accident. The Congo, now Zaïre, came to them in 1905 as part of an unwanted legacy from their king, Leopold II, the foremost royal entrepreneur in nineteenth-century Europe. The purpose of the Congo Association which he headed had been the exploitation of the resources of the vast Congo Basin. It fell short of the Dutch achievement in the East Indies, not in its crass commercialism, but only in the lower level of efficiency with which this objective was pursued.

Germany came late into the imperial race. 'I want no colonies,' wrote Bismarck in 1871; 'they are only good for providing offices.'[20] Within a few years he had changed his mind, and for a few years in the 1880s and 1890s German energies were concentrated on forging an empire from such scraps and slivers of territory as were still available for the coloniser. Germany's motives, like those of her predecessors in the imperial field, were in the main commercial, and if Germany's economic benefits failed to measure up to the efforts she had expended, this was chiefly because the lands acquired had little of immediate value to offer her. Italy became an imperial power at about the same time as Germany, but her policy was less forceful, and the lands acquired, though extensive, had little of value to a modern state in the early stages of industrial expansion.

The role of the dependent empire in the economic development of western and central Europe has been hotly debated in recent years, along with that of the whole 'periphery'. This is not the place to expand on this theme, but it is relevant to ask certain questions. How significant, in the first place, was the trade of European countries with their dependent colonies during the century? To what extent did colonial markets encourage manufacturing by absorbing the marginal product of Europe's industries? And, lastly, to what extent were the terms of trade between Europe and its dependencies biased in favour of European countries?

When the century began only a relatively small proportion of the trade of European countries was, with the exception of that of Great Britain, with their colonial dependencies. Overall only 15.7 per cent of European exports in 1830 went to what would now be regarded as Third World countries, and over half of this was from Great Britain. By 1910, the fraction of Europe's trade, inclusive of that of Britain, which went to the colonies had risen to about one-fifth. By far the largest element in this commerce was Britain's trade with India. At this time only France and Portugal amongst continental countries carried on a significant part of their trade with their own colonial empires.

Relative to the gross national product of European countries, the colonial trade was very small. Nevertheless, it did contribute some important commodities to Europe, including beverages and foodstuffs of tropical origin, hardwoods, textile materials and metalliferous ores. The European exports with which these goods were requited *were* marginally significant, and in a few instances were more important than that. The Catalan textile industry, for example, suffered severely, when, in 1898, Spain lost the few remains of her colonial empire in Central America and Asia. Italy's lack of a secure overseas market was a source of continual disquiet.

It is difficult to say to what extent European countries benefited from favourable terms of trade with their colonial empire, but there seems to be little doubt that they did. The manipulation of the terms of trade – buying cheap and selling dear – seems to have been the most common and the most subtle form of exploitation of the colonial world.[21] As long as colonial powers were able to exercise any kind of monopoly over colonial trade they were able to benefit themselves in this way.

The concept of the 'open door' to dependent territories, the right, that is, of all countries to trade there, became a matter of public law in the Congo Act of 1885. Its implementation was gradual and imperfect, but in the later nineteenth century not even the Dutch were able to make good their claim to have an exclusive control over the resources and trade of their colonial dependencies. It was difficult for a European imperial power to monopolise the trade and resources of its colonies, but its right to invest in them was rarely disputed.[22] The opening up of colonial territories occurred very broadly at a time when the construction of European railway systems was nearing completion. The production of railway equipment had been an essential preoccupation of the European iron, steel and engineering industries (see p. 339). Now these activities were extended at a useful level into the twentieth century. The same held good for the building of docks and ships and the opening of mines. Investment in the infrastructure of colonial territories gave a new lease of life to many a capital-goods industry in the developed European countries.

Important though the role of overseas dependencies was in the economic growth of much of Europe during the nineteenth century, its significance was a great deal less than that of other parts of the periphery which had escaped from or had never been under European domination and control.[23] These included the former empires of Spain and Portugal in the New World, the United States and Canada, and the Tsarist empire in Europe and Asia.[24] These territories taken together accounted for a far larger proportion of Europe's trade than did colonies in the strict sense. They were far more important in the supply of foodstuffs, especially bread-grains, and industrial raw materials, foremost amongst them raw cotton and softwood lumber. These imports were requited by the export from Europe not only of consumer goods, but also of those capital goods which were essential for economic development. The developing but non-colonial world of the nineteenth century thus served as a market and stimulus for European industrial output on a scale far greater than the colonial empires themselves.[25]

The European periphery

It has been customary to include within the periphery most of southern, south-eastern and eastern Europe. It is difficult to define the periphery in geographical terms because, unlike colonial dependencies, it was not delimited by political boundaries and was, for statistical purposes, included along with areas which did not belong to it. Romania and Bulgaria belonged without question to the periphery, as did the Russian-occupied Kingdom of Poland. But not all of Prussia, Hungary, Italy and Spain can be so defined.[26]

The European periphery resembled the extra-European periphery in certain important respects. Urban development had made little progress in both. Manufacturing industry had not developed, and there was even little proto-industrial development (see p. 303). Exports were almost exclusively primary goods: grain, wool, lumber and metalliferous minerals. These were exported almost exclusively to the 'core area' of north-western Europe, and were paid for by the import of factory products. The peripheral market was of some considerable importance to the manufacturer in the more developed regions of Europe, and the western capitalists invested on a not inconsiderable scale in developing an infrastructure in peripheral Europe.

But there were important differences between the periphery as it developed in Europe and in the rest of the world. The economy in the former was managed, not by the capitalist and entrepreneur from the 'core area', but by the 'feudal' elements of society who had not ceased to control the agricultural and mineral resources of the region. If there was exploitation, it was by the local 'feudal' classes, not by the western bourgeoisie. In several countries of the European periphery the great landowners, who controlled most of the resources, were able to influence the government in their own favour. The classic instance is the role played by the Junkers of eastern Germany in the government of the *Reich*, maintaining import duties on imported foodstuffs which competed with the products of their own estates. The same is broadly true of the Hungarian magnates who succeeded in keeping the Austrian markets open to their products, and also of the owners of the great estates of southern Spain and southern Italy.

Indeed the great estate and the consequent urgency of land reform was an almost universal feature of the European periphery. Only in Serbia and Bulgaria, where Turkish owners had been expropriated, and in 'Congress' Poland, where serfdom had been abruptly terminated by the Russians for their own political ends, was there a truly peasant society. Peripheral Europe was poor and backward for a variety of reasons, but amongst the most important was the agrarian system, which depressed the purchasing power of the agricultural population and thus restricted the development of manufacturing industry. Manufacturing *was* established within the periphery, but comparatively few enterprises were successful, and most were restricted to the production of the humblest of consumer goods, such as might sell in a poor, peasant society. Such was the textile industry of Łódź. Terms of trade between the developed, industrial countries and the primarily agricultural countries of eastern Europe seem in general to have favoured the former. This was certainly the case in trade between the Austrian and the Hungarian sectors of the Habsburg Empire.[27]

In general, however, the market demand within the periphery was satisfied by entrepreneurs from outside the area, especially from central Germany, Bohemia and Moravia. In the decades immediately preceding the First World War, Italy adopted a policy of investment in the railroads of the Balkans and of marketing Italian manufactured goods within the Ottoman Empire.[28] It failed largely because it was overtaken by the Balkan Wars and then by the First World War. After that war was over the east European periphery came to be divided into independent states. Land reform of varying degrees of effectiveness was introduced, and a modern programme of industrial development was implemented, using Western capital. Parts of the periphery were thus absorbed, if not into the core area of western Europe, at least into its partly industrialised penumbra.

The political map of Europe

The statesmen who gathered at Vienna in 1814 and reconvened the next year after the final defeat of Napoleon set themselves to restore the conditions that had been so rudely overthrown in the Revolutionary and Napoleonic wars. In this they failed, as, indeed, they had to; far too much had happened during the intervening years to both the social and political systems for it to be possible to put the clock back. It was, however, easier to restore the political boundaries and re-establish the states system of pre-revolutionary Europe than the social system, and the map that emerged from the Second Treaty of Paris (1815) bore a superficial resemblance to that of 25 years earlier.

Napoleon had obliterated almost every boundary in Europe, but the political units that he created, like the Kingdom of Westphalia or the Grand Duchy of Berg, had no historical depth. They attracted little loyalty and could not outlive their creator. It was to the credit of those who shaped the treaties that they did little to penalise France for the disasters to which the French Revolution had contributed. On balance, France lost nothing territorially by the First Treaty of Paris, which did little more than straighten out the feudal debris that cluttered the country's north-eastern boundary (fig. 1.1). The second treaty was marginally less favourable and gave to the Low Countries and Germany certain territories that had previously been allowed to remain in France. It so happened that these

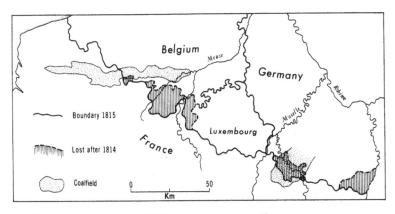

1.1 The political boundary of 1814–15 in north-western France

included part of the Saar coalfield and also of the northern coalfield near Mons, though there is no reason to suppose that the Allied statesmen knew of this, or would have been greatly infuenced by it if they did. The first treaty of Paris had also allowed France to retain part of Savoy, occupied in 1792. In 1815 this was restored to the Kingdom of Sardinia-Piedmont.

In western Europe the most significant change in the political map was the incorporation of the southern or Austrian Netherlands into a Kingdom of the United Netherlands. The purpose was to create a strong state, able to resist any possible French expansion in this direction; the result was to join under one crown the most developed country in western Europe with one of the least, and to bring together two peoples who, despite a shared history and language, now had little in common.

In Scandinavia, the decline of Sweden's influence in the Baltic was recognised in the transfer of Finland from Swedish sovereignty to that of the Tsar, who, as Archduke of Finland, ruled it in a personal union with Russia until 1917. At the same time Norway, which had previously been linked uneasily with Denmark, elected to be joined with Sweden, a union which endured until 1905.

Most of central Europe had been occupied by the German Empire, the titular headship of which had for some 500 years been vested in the Austrian Archdukes. In 1806 the Empire ceased to exist, and the event passed almost without notice. The Empire as such was not revived. The title of Emperor was retained by the Habsburgs, but applied only to their lands in Austria, Hungary and their dependencies. Instead the German Confederation, or *Deutsches Bund*, was established. In most directions its boundaries were those of the defunct German Empire; only on the west had they been tidied up by the elimination of numerous feudal enclaves. The *Bund* was an untidy and politically impotent body. Not only did its constitution render it almost powerless; its most important members, Prussia and Austria, derived their influence in part from territories which lay outside its boundaries.

In the west the Duchy of Luxembourg, a personal possession of the House of Orange of the United Netherlands, was included in the *Bund,* and in the north the royal house of Denmark similarly held the German duchies of Holstein and

Lauenburg. To add to the prevailing sense of disunity, the Kingdom of Hanover continued until 1837 to be a personal possession of the King of England. If the Vienna Congress had sought deliberately to prevent the achievement of German unity it could not have been more successful. On the other hand, however, it did simplify the political geography of Germany. Before the French conquest there had been from 300 to 400 political units, all of them claiming if not effectively exercising political independence. Most were swept away by Napoleon, and the larger amongst them – Bavaria, Württemberg, Baden in south Germany; the Kingdom of Saxony in central, and Hanover and Mecklenburg-Schwerin in north Germany – survived, their territory enlarged in some instances by the incorporation of enclaves within their boundaries and dwarf states bordering them. Altogether, the number of states within the *Deutsches Bund* was reduced to 39, including those which also embraced extensive territories beyond its borders. But the chief beneficiary of territorial changes within Germany was Prussia. Its nucleus, the March of Brandenburg, lay between the Elbe and Oder. To this it added Pomerania and Neumark beyond the Oder, East Prussia and the province of Silesia. A new orientation to Prussian policy was given in the early seventeenth century by the inheritance of small territories in the lower Rhineland. One of them, the County of Mark, included the more easterly part of the Ruhr coalfield as well as part of the iron-working Sauerland. It was primarily the possession of these Rhineland territories that was to make Prussia the foremost industrial state in continental Europe. In the course of the seventeenth century Prussia acquired former episcopal lands in north Germany and, while still under Napoleonic control, added part of the lands of the Bishops of Münster and Paderborn. Even before the Congress of Vienna Prussia had thus come to embrace, in addition to its vast but thinly peopled lands in the east, many important but scattered territories in central and west Germany. In 1815 these were rounded out by the inclusion of the lands of a number of suppressed states and also of important lands to the west of the Rhine, including Jülich and the church lands of Cologne and Trier. More important, these were already the scene of considerable development in the textile and metal industries, thus complementing the developments that had taken place in the County of Mark.

The Habsburg Empire was the most extensive state in Europe, with the exception of Tsarist Russia and the Ottoman Empire, both of which extended far into Asia. In 1815 it consisted of two distinct and contrasted parts. The smaller but much the more important was made up of the historic Austrian lands. these consisted (fig. 1.2) of Austria in the narrow sense, including the provinces of Tirol, Krain or Carniola, and Dalmatia, and the Czech lands of Bohemia and Moravia together with the small but economically important territories of Austrian Silesia and the Duchies of Teschen, Auschwitz and Zator. It was in these territories that much of the Moravian coal and iron industry was located. The city and territory of Kraków, which had gone to Austria in the Third Partition of Poland, was in 1815 constituted a small city-state under the general supervision of the powers. In 1846 it was annexed to Austria.

All these territories, with the exception of Kraków, were included in the *Deutsches Bund*. The rest of the Habsburg lands, consisting by and large of lands conquered from the Turks or annexed since the end of the seventeenth century,

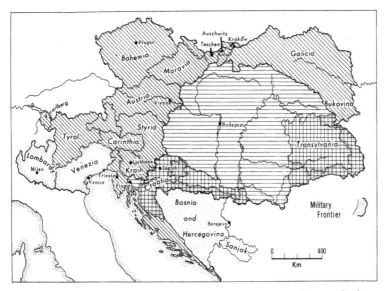

1.2 The Habsburg Empire in the nineteenth century. The Austrian part is shown by diagonal shading, the Hungarian by horizontal. Lombardy and Venezia were lost in 1861. Bosnia and Hercegovina were occupied in 1878 and were incorporated into the Empire in 1908. The Sanjak was occupied militarily 1878–1908

lay outside the Confederation. The largest part was the titular Kingdom of Hungary. There had been no independent King of Hungary since 1526, and the Habsburg emperors, who had reconquered Hungary from their base in Austria, had tacitly assumed the title. To the east was Transylvania, previously a dependency of the Hungarian crown, and, to the north, Galicia and Lodomeria represented what remained to Austria of the spoils of the Partitions of Poland. Lastly, to the south of Hungary lay Croatia and the Military Frontier, a narrow strip of land from the Adriatic Sea to Transylvania, organised to protect the Empire from attack by the Ottoman Turks.

Prussia and Austria had both been deprived by the Treaty of Tilsit (1807) of substantial areas of Poland which they had occupied in the course of the Partitions. These were formed into the Grand Duchy of Warsaw by Napoleon, who thus sought to appease the Poles. In 1815, the Grand Duchy, deprived of the Posen (Poznań) region, which was handed back to Prussia, was placed under the personal rule of the Tsar, but in 1847 was fully absorbed into the Russian state. The Prussian provinces of west and east Prussia, together with Posen, most of which derived from Poland, lay outside the German Confederation.

There was little change in the boundaries of the Swiss Confederation, other than some smoothing out of its ragged boundary with France. Nor was there any significant political change in Italy. The Venetian Republic had been suppressed by Napoleon, and its territory, together with that of the Duchy of Milan, was annexed by Austria. The Kingdom of Sardinia, which included Savoy and Nice, was enlarged by the addition of that other famous sea-state, the Republic of Genoa. The rest of the Italian peninsula remained substantially as it had been in the eighteenth century, with the States of the Church extending

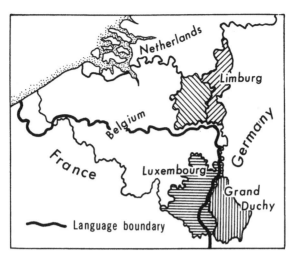

1.3 The language boundary in Belgium and the partition of Limburg and Luxembourg

from sea to sea and separating the Kingdom of the Two Sicilies in the south from Tuscany, Lucca, Modena and Parma-Piacenza.

To the south-east lay the Ottoman Empire. The boundary which ran from the Adriatic coast to the Carpathian Mountains was in the main that established by the Treaty of Beograd as far back as 1739. The only significant changes had been in the north-east, where Bukovina had been ceded to Austria in 1775 and where Russia had absorbed Little Tartary on the Black Sea coast in 1791 and Bessarabia in 1812. This allowed the Russians to advance to the Danube delta, where they held the northern shore of the Kilia Channel, a fact of considerable importance when the Danube was opened up to navigation by western powers.

Within the Balkan peninsula the Ottoman government exercised varying degrees of control. Relatively few Turks had settled there, but garrisons were maintained at strategic points within the region. In Romania and in mountainous areas south of the Danube, such as Montenegro, the local peoples enjoyed a *de facto* independence. Early in the century the Serbs living in the Šumadija region to the south of the Danube had rebelled against Ottoman rule, and by 1817 were exercising a sort of self-government within the *pashalik* of Beograd.

A century of nationalism

The Vienna Congress made no concessions to the spirit of nationalism, which was beginning to be an active force in European politics. Indeed, it violated what were to become its most sacred principles. The French-speaking Walloons of the southern Netherlands were attached to the Dutch Kingdom of the Netherlands, as also was German-speaking Luxembourg (fig. 1.3). The Austrians received Italian Lombardy and Venetia, and Germany, in which the flame of nationalism was to burn most brightly, was left divided and with boundaries which coincided at scarcely any point with those of the German nation. The only suggestion of nationalist sympathies – wholly unintentional – was the perpetuation of Napoleon's Grand Duchy of Warsaw, its boundaries modified in favour of Prussia, as the 'Congress' Kingdom of Poland under the rule of the Tsar.

The next 60 years saw an outburst of national feeling that was to change fundamentally the map of Europe and to bring political boundaries more closely into line with national feelings. Complete accord between the two was not, and indeed could not be achieved. Nationalism was itself a fluctuating emotion, and national movements that were scarcely apparent in 1815 were to become powerful forces before the end of the century. Nowhere could it have been possible to draw a simple line on the map to separate neighbouring peoples. If language is regarded as the primary criterion of nationhood, as indeed it came to be, each nation merged gradually with its neighbour through a transition zone of variable width. In theory a boundary could have been drawn that did the least violence to national aspirations; in fact there were many other considerations influencing Europe's statesmen when at last they came to the redrawing of political boundaries.

Foremost amongst them was the desire for so-called 'natural' boundaries. It was a very old idea that a good boundary was one which coincided with a well marked and easily defended physical feature.[29] Rivers were preferred, because they were clearly marked in nature and could usually be described verbally in unambiguous terms. Sometimes mountains were preferred because in the thinking of the military they assisted national defence. Scarcely less important were historic claims to territory. A country, it was argued when such an argument seemed relevant, should embrace those lands with which it had been endowed at some distant period of time. The claim to a 'greater Bulgaria', made in the 1870s, could not be justified on any other ground than the supposed extent of the early medieval Bulgarian state. Such historic claims might also be 'legitimist' in that they were based upon descent and inheritance by a ruling family, irrespective of whether the country making the claim had ever had possession of the territory in question. Prussia's claim on the Duchies of Schleswig, Holstein and Lauenburg was basically legitimist. Lastly, though boundaries represented in some sense a compromise between these considerations, the determining factor in most instances was the relative power of the states which were separated by them. Germany took from Denmark (1864) and France (1871) and lost territory in 1919. A particular boundary can best be regarded as the resultant of opposed political forces, justified or excused by historical and ethnic considerations.

The principal changes that occurred in Europe between 1815 and the outbreak of the First World War concerned the Low Countries, Germany, Italy and the Balkans. Some of them were of such fundamental importance for the development of the resources of the continent that they are discussed here in some detail.

The Low Countries

No country in 1815 reflected contemporary political considerations more clearly than the Kingdom of the United Netherlands. The United Provinces, or Netherlands in the narrower sense, had long been concerned for the stability and security of its fragmented southern neighbour, the Austrian Netherlands. It had by agreement garrisoned towns in the south – the so-called Barrier Fortresses[30] – against the possibility of French attack. In 1815 Dutch concern was shared by most of the Allies, who removed the southern Low Countries from Austrian rule and joined them with the United Provinces to form the United Netherlands. At the same time the Duchy of Luxembourg, which had been part of the German Empire, was transferred from the Habsburgs to the Dutch House of Orange, which ruled it in a personal union with the United Provinces, while the Duchy itself remained a member of the German Confederation.

The settlement in the Low Countries was an untidy compromise, and did not last 25 years. The northern and southern Low Countries were separated by tradition and by their different economic developments. The King, William I, did much to encourage the expansion of manufacturing in the south, but this did little to recommend him to his Belgian subjects. In 1830 the July Revolution in Paris prompted the Belgians to follow the French example. A year later Belgium became an independent kingdom. Its right of navigation by way of the Scheldt to the sea was guaranteed (see p. 477), but many other issues remained unresolved until 1839. Prominent amongst them was the question of Limburg and Luxembourg, which belonged to the House of Orange. In the end both were partitioned. Eastern Limburg remained under the direct control of the Orange dynasty until in 1864 it was fully incorporated into the Kingdom of the Netherlands.[31] The Grandy Duchy of Luxembourg was similarly partitioned, along broadly linguistic lines. The larger western part, most of it Walloon- or French-speaking, became a province of Belgium. The eastern, wholly German in speech, retained the title of Grand Duchy and continued to be part of the German Confederation, though ruled by the Orange family. The important iron-ore deposits were almost wholly in the Grand Duchy. This must be one of the earliest attempts to align a political boundary along a linguistic divide. This situation was terminated only in 1890 when the succession of Wilhelmina, Queen of the Netherlands, was precluded by Luxembourg law. The Grand Duchy passed to a male heir of the last Dutch king, and thus became an independent state.

A curious feature of the Belgian revolt is that the Flemings joined with the Walloons to oppose Dutch rule. This harmony was, however, short-lived. The Walloons dominated the economy and politics of Belgium. At once the Flemish movement took shape, first as a literary and romantic cult, then as a political movement, demanding for Flemings a bigger role in the country's administration and for their language similar rights to those enjoyed by French.[32] The Flemish question was to embitter the internal politics of Belgium for the next century, with significant consequences for the development of agriculture and manufacturing.

The unification of Germany

At the Paris and Vienna Conferences of 1814–15, the principle of legitimism triumphed over that of nationalism, but it was impossible to restore the pre-revolutionary situation in Germany. Most of the tiny city-republics and

Zwergstaate had been swept away irretreivably in the Napoleonic reforms. But the statesmen rescued and revived all that they could. The number of states was reduced. Six of them either had non-German rulers, or extended far beyond the bounds of the German *Bund*. All were grouped into an ineffective Confederation under the presidency of Austria, which thus perpetuated its role as head of the German Empire. Each of the states was in effect sovereign. It had its own tariffs and commercial policy, its own judicial system and military forces, so that within each of them there were considerable vested interests in its perpetuation.

Such an arrangement was, needless to say, the despair both of German nationalists, eager to achieve a united Germany, and of the growing industrial and commercial interests which looked for a nation-wide market. For the former the leaders of the Confederation had only hostility, but for the latter there was much sympathy. Indeed, many states openly advocated some form of economic union. For no state, however, was freedom of movement and trade within Germany more necessary than for Prussia. It stretched from the borders of Russia to those of France in no less than six discrete areas, among which the agricultural lands in the east were complemented by the developing industrial regions of the Rhineland. A solution to this problem, without making any concession to German nationalism, was the formation of a customs union. During the 1820s there were movements in this direction in many quarters.[33] In south Germany a customs union was formed between Bavaria and Württemberg. Prussia negotiated tariff agreements with the independent enclaves, such as Anhalt, which it embraced and also with Hesse-Darmstadt. Furthermore, Prussia arranged to build a road through Meiningen, in Thuringia, to link it with south Germany. Lastly a Middle German Commercial Union was formed by Saxony, Brunswick, Hanover, Oldenburg and the small Thuringian States. Its purpose, however, was more to thwart Prussia than to gain any great commercial advantage for its members. It proved to be the least stable of the unions, and some of its members were weaned away by the greater advantages of membership of the Prussian union.

By 1828 the situation in Germany was that shown in fig. 1.4. Only a few peripheral states remained outside one or other of the customs unions. By 1834 the Prussian Union had been joined by Bavaria, Württemberg, Saxony, the Thuringian States, Hesse-Kassel and most of the small states of central Germany to form the Zollverein. The Middle German Union, now called the Tax Union, was reduced to Hanover, Brunswick and Oldenburg (fig. 1.5). There were further accessions to the Zollverein: the fragmented state of Brunswick; Nassau, Baden and Luxembourg; then Hanover and Oldenburg, and in 1864 the northern group of states: Holstein and Lauenburg, as a result of the Austro-Prussian War; Mecklenburg-Schwerin and Mecklenburg-Strelitz, and lastly the Free City of Lübeck. This left only the two Hanseatic ports of Bremen and Hamburg, which joined the Zollverein in 1888, and the Dutch province of Limburg which never joined at all.

Throughout, however, there was one significant omission. Austria never became a member of the Zollverein, and every attempt to bring her in proved abortive. There were good reasons for this. The Habsburg Empire was a high-tariff region, whereas Prussia and the Zollverein tended at this time towards

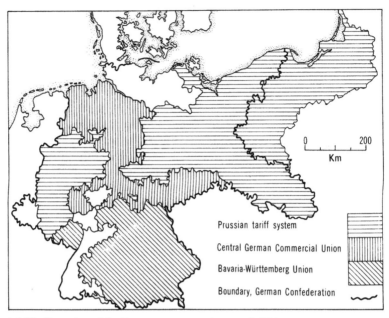

1.4 The formation of the German customs unions, until 1834

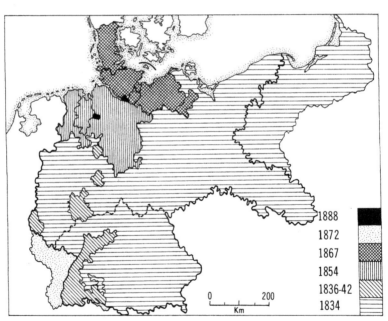

1.5 The formation of the German customs unions, 1834–88

low tariffs. Austria sought a customs union within the Austro-Hungarian lands, and in 1850 the tariff barrier between the Empire of Austria and the Kingdom of Hungary was abolished. German feeling might have tolerated the Czech lands within the Zollverein; they certainly would not have accepted the Hungarian

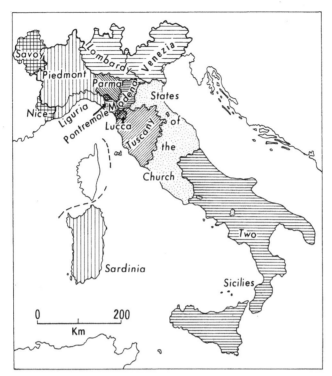

1.6 Italy before unification, 1861

lands as well. Prussia, on the other hand, had established an economic hegemony within the Zollverein to match the political supremacy enjoyed by Austria. She was in no mood to share economic power with the Habsburgs, and set her face firmly against the admission of any part of the Austrian Empire to the German Zollverein.

Political unity came in 1871 and with it the extension of German boundaries to include the *Reichsland*, consisting of Alsace and part of Lorraine. The ostensible reason for the latter was the alleged national sympathies of the local population. The course of the boundary was almost certainly dictated by military considerations, but the effect of the change was to give to Germany all the prospected and at least half the total reserves of the *minette* iron-ore field, the best endowed in Europe.

Italy

The situation in Italy resembled in many ways that in Germany. Hopes for the creation of a national state, raised by the Napoleonic conquest, were dashed at Vienna. The principle of legitimism prevailed, modified here and there to suit the aims of the allied statesmen. Most of the former rulers returned (fig. 1.6). Austria took the lands of the suppressed Venetian Republic, together with most of the Duchy of Milan, thus acquiring a base from which the peninsula could be dominated and controlled and all nationalist tendencies resisted. To the south lay the duchies of Parma-Piacenza and Modena and the Grand Duchy of Tuscany, all of them politically and economically weak, and to the west the Kingdom of Sardinia which included Piedmont, Savoy and the former Republic of Genoa, which had been, like Venice, suppressed by the French. A strong Piedmont-Savoy would, it was hoped, serve as a buffer against France. The Holy See regained all its territories, extending from the River Po to Terracina, with exclaves even farther south. The whole south of the Italian peninsula formed part of the Kingdom of the Two Sicilies, which, little disturbed by the Napoleonic reorganisation of Europe, was allowed to survive in all its squalid splendour. It is hard to say to what extent this chaotic political situation postponed the economic development of modern Italy. It is a fact that almost no progress was made until after unification in 1861.

The Habsburg Empire

Though defeated both militarily and diplomatically, the Empire of the Habsburgs was able to survive little altered territorially until the First World War. Internally, however, it underwent significant changes. Until 1866 the Austrian part of the Empire remained largely within the German Confederation. The rest, made up of Hungary and its dependencies, together with Galicia and Lodomeria, lay outside the Confederation, just as they had been excluded from its predecessor, the Holy Roman Empire.

Both Austria and Hungary contained national minorities. Within the Austrian sector were the Czechs, the Poles of Teschen, Kraków and western Galicia and the Ruthenes of eastern Galicia, and, to the south, the Italians of southern Tyrol, Lombardy-Venezia and the so-called Küstenland, the Slovenes of Carinthia and Krain and the Croats of Dalmatia (p. 101). The Hungarian kingdom included Slovaks,[34] Romanians, Croats and the miscellaneous peoples who had settled the Banat and neighbouring parts of the Plain. In its resistance to the national claims of the Hungarians, the Austrians found natural though far from reliable allies in the subject peoples of Hungary, notably the Croats. The major nationalist dispute within the Habsburg lands, that between the Austrians and their Hungarian subjects, was settled at least for the remainder of the pre-war period by the *Ausgleich*, or Compromise, of 1867, which established the Hungarian Kingdom as an equal partner with Austria within the Habsburg Empire. Other nationalist movements survived and intensified, and in 1918 tore the Empire apart.

To the Austrians the great enemy was still the Turk, though the Ottoman Empire had long ceased to be any kind of threat to their security. The danger came rather from its weakness, which gave scope for intrigue by the powers, eager to profit from its disintegration. None the less, the borderland of the Habsburg Empire was organised as the Military Frontier, a narrow strip of territory extending from the Adriatic Sea to northern Transylvania (see fig. 1.2). The frontier survived until the 1870s, by which time it had long outlived its usefulness.

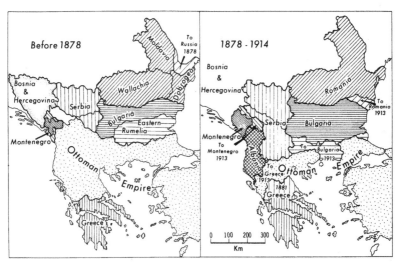

1.7 The Balkan peninsula before and after the Congress of Berlin, 1878

The Ottoman Empire and its successors

The Habsburg boundary was established along the line of the Sava, the Danube and the Transylvanian Alps early in the seventeenth century, and here it remained with little significant change until late in the nineteenth (fig. 1.7). To the east and south the Ottoman Empire was weak and vulnerable to attack both from the great powers and internally from its own subject peoples. The Turks were always a small minority in south-eastern Europe. They held much of the fertile land which they organised in estates (*chifliks*) and cultivated with servile labour; they established garrisons in many of the larger cities, and they provided local governors and administrators.[35] The latter were inefficient and corrupt, and the Sultan's government in Constantinople had neither the will nor the ability to control them. The Christian population, most of it Slav, provided labour on the Turkish-held lands, as well as paying most of the taxes (see p. 275).

It would be wrong, however, to suppose a perpetual struggle between Christian and Moslem. Indeed religion played only a minor role in the clash between the Orthodox peasant and the Muslim Turk. There were Muslim peasants whose plight was only a degree less severe than that of the Christian *raya*. The Turks, for their part, were obliged to employ Christians – many of them Greeks – in positions of authority because they needed the skill and experience which only the latter could provide. Greek, Macedonian and Dalmatian merchants were to be found all over the Balkan peninsula. They encouraged the development of crafts by providing the means by which the products of local industries could be transported and sold.

In 1815 the Ottoman Empire in the Balkans covered a vast area, some 160,000 square kilometres, about a twentieth of the area of non-Russian Europe. It reached from the borders of Bukovina and Bessarabia in the north to the southern capes of Greece. It was a varied and mountainous region through which the rivers had opened up a number of valley routes. Apart from the Danube, which flowed from west to east *across* the line of the Transylvanian Alps–Balkan Mountains, the most important led from the Hungarian Plain towards the Mediterranean, from Beograd to Thessaloníka and, by way of Sofia and Plovdiv, to Constantinople. There were other, less important routes. West of the trough formed by the Morava and Vardar rivers, lay the less convenient but historically important routeway which followed the valley of the Ibar (see p. 458). East of the Vardar, the river Struma opened up a routeway from the Aegean Sea to Sofia.

West of the Ibar lay the most rugged and intractable area of mountains in the Balkan Peninsula. Towards the north-west it narrowed and became lower, and could be crossed with a minimum of difficulty between the Hungarian Plain and the head of the Adriatic Sea. Austria had long had firm control of this area, and was to use it to gain access to the port of Trieste and the Hungarian port of Fiume (Rijeka). To the south-east the Dinaric Mountains became a formidable barrier between the Adriatic coast and the plains of the interior, and the only break in their continuity, before central Albania was reached, was the Naretva valley. These mountains served as a cultural and political divide. On the seaward side the Italian cultural influence was paramount in the coastal cities, and from their base in Slovenia the Austrians extended a long finger of political control as far as the narrow, rock-bound harbour of Kotor (Cattaro).

Within the areas that were nominally Ottoman the degree of the Sultan's authority varied greatly. In the most rugged region of the Dinaric Mountains, Crna Gora or Montenegro, it had been ineffective since the sixteenth century. Throughout the nineteenth century this state was ruled by the princes of the Njegoš family from their village-capital of Cetinje. Their boundaries fluctuated according to the degree of Turkish pressure, but Montenegro held firm until it was merged with the kingdom of the Serbs, Croats and Slovenes (Yugoslavia) after the First World War.

In the neighbouring territory of Albania Ottoman authority was openly flouted. Earl Granville wrote in 1880:

The state of the country in north-east Albania is little short of anarchy. The Turkish officials are powerless to execute justice; murder, violence and forced exactions are prevalent, and the peaceable population is at the mercy of the armed committees, who, under the name of the Albanian League, have been allowed to assume absolute authority.[36]

Elsewhere defiance of the Turk had reached the stage of open rebellion.[37] Serbia achieved effective independence in 1830, though this was not recognised by the powers until 1856. The Romanian provinces of Walachia and Moldavia were in effect governed by their native boyars headed by a phanariot[38] Greek official of the Sultan.

The forces which shaped the Balkan peninsula during the century from the Vienna Conference to the First World War were fourfold. In the first place, the weakening authority of the Turks was matched by a growing sense of national identity among the subject peoples of the region. One by one, with or without active help from the western powers, they were to throw off Turkish rule and assert their independence. Defeating the Turks was the easier part of their task; the amicable settlement of territorial claims amongst themselves was far more difficult and, but for the overriding Soviet presence, could lead to conflict even today.

The second factor in Balkan affairs was the ambitions of the Habsburgs. For almost three centuries their policy had been one of advance on a broad front against the Turk. This was unlikely to change, and Austria–Hungary tended to regard the Balkan peninsula, especially the western Balkans, as its designated sphere of influence. Late in the eighteenth century, a third force had become apparent in the affairs of the Balkans, the Empire of the Russian Tsars. The Russians had under Peter the Great completed their occupation of the northern shore of the Black Sea. In 1812, they occupied Bessarabia, between the Dniestr and the Pruth, and in 1829 they added part of the Danube delta and were at once in a position to control the growing shipping on the lower river. They exerted a powerful influence in the Romanian principalities and the Tsar even claimed in the Treaty of Kuchuk Kainarji (1774) to be the spokesman for and protector of the Orthodox subjects of the Sultan.

The fourth and final factor in the political pattern of the Balkans was Turkey's uncertain allies. The west European powers, particularly France and Great Britain, had no desire to see the Balkan peninsula dominated by either Austria–Hungary or Russia, because this might threaten their own sea routes through the Mediterranean and their trade with the Danube. They preferred to prop up the shaky Ottoman Empire, even though public opinion repeatedly condemned the arbitrary and tyrannical rule of the Turks. The Crimean War was thus fought by France and Great Britain, with the participation of other west European powers, to hold back the forces of the Tsar from the Balkans and the Turkish Straits. On the other hand, the fierce denunciation of the 'unspeakable Turk' in 1876 showed where British sympathies really lay, even though it was clearly to Britain's advantage to oppose Russian aims in Bulgaria.

The history of the Balkans is one of conflict between these irreconcilable forces. Austria–Hungary triumphed with the occupation and then annexation of Bosnia-Hercegovina; Russia with her advance into the Danube delta; the forces of nationalism with the creation of the Serbian, Romanian and Bulgarian states, and the western powers when they checked Russia's advance to the Mediterranean. But a century of almost continuous strife inhibited economic growth and stored up a legacy of hatred which broke out in the Balkan Wars of 1912–14 and again in the First World War.

Gross National Product

A century of economic growth, however uneven it may have been spatially, came to an end in 1914. Innovations made first in Britain and adopted in the Southern Low Countries or in Switzerland, had been diffused to the ends of the continent. The overall increase in productivity was by a factor of about five. In north-western Europe it was far higher; in south-eastern and much of southern a great deal lower.

The nineteenth century saw the transition from a kind of statistical prehistory to a period when the collecting and publication of reliable statistics became a major task of government. When the century began statistics of population and production were at best rough approximations, at worst uninspired guesses. By 1914 they still left much to be desired, but those made available by public authorities were by and large acceptable. We are here concerned with the growth

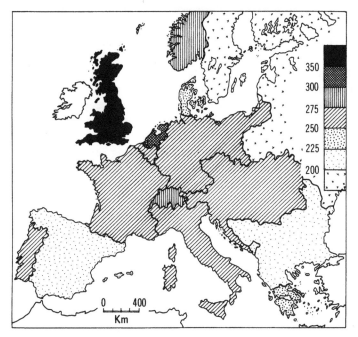

1.8 Gross National Product per head, about 1830. The units are 1960 U.S. dollars

in economic production in the continent as a whole, and with regional variations of the growth pattern. It must be stated at the outset that adequate statistical data do not exist for the whole period; that information is more satisfactory for western and central Europe than for eastern and south-eastern, and that it tended everywhere to improve in both scope and quality as the century progressed. For the earlier part of the period in particular many extrapolations and imperfectly supported estimates have had to be made. The result is a picture of the growth of the European economy that may well be accurate in its broad outline, but is more than a little suspect in detail.

A further reservation is that the discussion is conducted and the maps have been drawn on the basis of states. No other course is possible. The statistical data used have mostly been compiled by governments for their own national territories. Boundaries are seen to assume a sharpness which in reality they did not possess, and each country takes on a homogeneity which was entirely lacking (fig. 1.8). There was not a single European country in which significant regional differences in wealth and economic growth did not exist during the nineteenth century, and in some these regional disparities tended to intensify rather than disappear.[39] For some parts of Europe, the Balkan peninsula for example, the evidence for the regional pattern of economic growth consists of little more than the unsubstantiated and subjective judgments of travellers. For others, like France and Sweden, there is a wealth of local statistical data.

The choice of units of measurement is important in any comparative study of economic growth. National units of currency, volume and weight must be converted to common measures, not an easy task when purely local units of

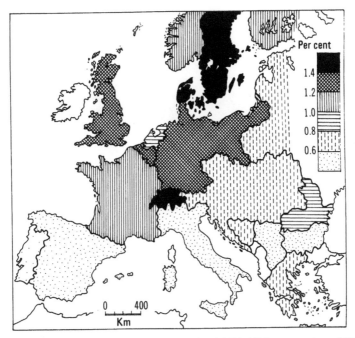

1.9 Increase in Gross National Product per head, 1830–1910, as percentage of 1830 figures

measurement continued in use, as they did in parts of Germany and the Habsburg Empire, well into the second half of the century. The use of contemporary values also means that an element of inflation is incorporated into the figures. Bairoch, in one of the most complete national income series ever attempted, used the United States dollar at its 1960 value as his unit of monetary measurement, and his figures are used here.[40]

On this basis the gross national product of Europe is held to have increased five times – from 58.2 to 256.9 billion dollars – between 1830 and 1913 (fig. 1.9). During this period however, the population of Europe increased more than twofold. The per capita increase in gross national product thus becomes a more significant figure. It is presumed to have increased from about $200 per head about 1800 to $240 in 1830 and to $534 in 1913. Growth was uneven through the century and also between the countries of Europe. Very broadly the *rate* of growth accelerated until the recessions of the 1870s and 1880s, and increased again in the 1890s. But it is doubtful whether much of eastern and south-eastern Europe, imperfectly linked with the markets of western Europe, shared in the depression of the 1870–80s, just as they took little part in the growth that preceded and followed it.

It is generally assumed that economic growth was rapid during the century and very rapid towards its close. There were indeed sectors in which growth was very fast indeed, notably the steel industry, but overall it was consistent rather than rapid.[41] Its most significant feature was that it was not interrupted by war on any important scale. The annual rate of growth for Europe as a whole was less than 1 per cent annually between 1830 and 1910. Since growth in the United

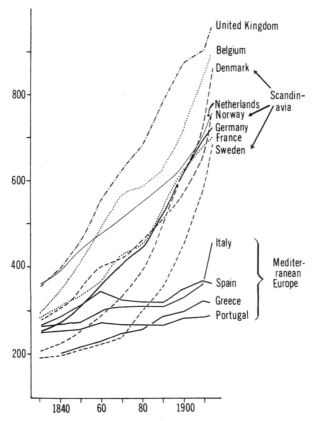

1.10 The increase in the Gross National Product per head in the principal European countries, 1830–1913

Kingdom, particularly rapid in the earlier part of the period, amounted to 1.21 per cent, that of the rest of Europe must have been less than 0.9 per cent annually (fig. 1.10).

Growth was most rapid in the Scandinavian countries and Germany; least so in southern and south-eastern Europe. Russian growth was well below the European average before 1860, and close to the average thereafter, and it is presumed that growth in Poland was close to or perhaps somewhat above the Russian level.

Fig. 1.8 shows the gross national product of Europe per head of population in 1830. Not unexpectedly Great Britain is seen to have been the wealthiest country by far, and was followed by the Netherlands and Switzerland. Next came Norway, Austria-Hungary, Belgium, France, Italy and Germany. In the bottom half of the table were the rest of Scandinavia and of southern Europe and as far as one can say from the totally inadequate data, eastern Europe and the Balkans.

The most marked feature of the map is the relatively small range shown in these national averages, with Germany and France closely matched not only by the Habsburg Empire but also by Portugal and Italy. It is noteworthy also that,

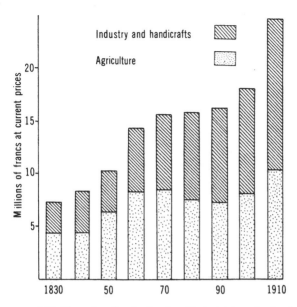

1.11 The value of agricultural and industrial production in France, 1830–1910

with the exception only of Norway, the whole of Scandinavia fell within the lower half of the table. A comparable map prepared for 1913 (fig. 10.18) shows a distribution of wealth closer to that of the present. The United Kingdom remains first, but her lead has been drastically cut. Then come, fairly closely bunched, Belgium, Switzerland, Germany and most of Scandinavia. Low in the table are southern, south-eastern and eastern Europe. These two maps taken together summarise the spatial changes that are the subject of this book.

A greater insight into the nature of these changes is obtained by examining the contribution of the several sectors to the economy. It does not need to be emphasised that agriculture made the greatest contribution to the gross national product of the whole of Europe at the beginning of the period, nor that the share of manufacturing increased through the century and became by the time of the First World War the largest sector in much of the continent. At the same time, however, there was a steady increase in the tertiary sector comprising trade, transport and services. Indeed, it can be claimed that the more advanced a country became the more significantly did its tertiary sector develop.

If immense difficulties surround the compilation of national income statistics, those relating to its several sectors almost defy analysis for much of Europe. Employment figures, which are as a general rule more easily obtained, may be, and often are, used as a surrogate. The problem is that the reward for agricultural labour was, by and large, less than that for industrial, and that by using the *numbers* employed in agriculture a somewhat exaggerated idea is obtained of the *value* of its contribution to the economy. Whichever is used there nevertheless remains the problem of allocating certain marginal activities, such as building, transport and mining, to the several sectors. For only France (fig. 1.11) and Germany (fig. 1.12) is it possible to trace the changing balance between

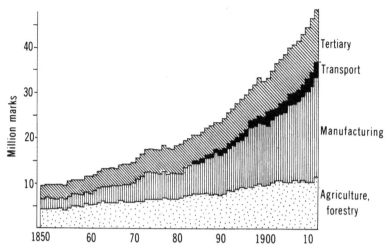

1.12 The Gross National Product of Germany, by sector. The value of transport services was too small before 1886 to show on the graph

sectors of the economy, and for Germany the data become adequate only after the formation of the German Empire in 1871.[42]

Fig. 1.11 shows the growth of the gross national product of France.[43] At the beginning of the century the value of agricultural products exceeded that of industrial by a wide margin. It was not until after 1870 that, helped by the recession in agricultural production, the value of manufactured goods, including building construction, first exceeded that of farm products and forestry. By 1910 it was more than 40 per cent greater.

The study of German economic growth has been greatly facilitated by the exhaustive work of Walther Hoffmann.[44] Fig. 1.12, which has been based on his statistics, shows by major sectors the growth of the economy. The sharp increase in 1871–3 reflects the incorporation of Alsace–Lorraine into the Reich. The flattening of the graph during the following years results from the depression of the later 1870s. The agricultural sector grew somewhat more rapidly than in France, but the expansion of the industrial sector, especially after the mid-1980s, is conspicuous. The very important role played by transport made it desirable to distinguish it within the tertiary sector from 1886, before which it was too small to be shown.

The German labour force expanded much more slowly than the gross national product (fig. 1.13). After declining in the closing decades of the nineteenth century, the agricultural labour force increased again, as it did in France, around the turn of the century. Despite the rapid expansion of manufacturing production in the late nineteenth century, it was not until the 1890s that the industrial labour force in Germany began to exceed that in agriculture.

In Austria also there was an increase in the labour force of about 25 per cent between 1869 and 1910, with a small decline in the agricultural sector and a large increase in the industrial (fig. 1.14).[45] Hungary witnessed a far more rapid economic growth during these years. In the mid-nineteenth century it was a

From Waterloo to the First World War

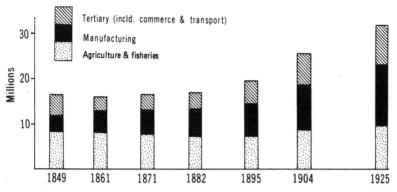

1.13 Growth of employment by sector in Germany, 1849–1925

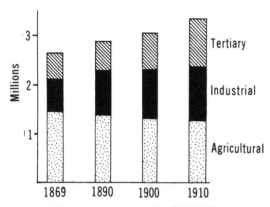

1.14 Employment in Austria, 1869–1910

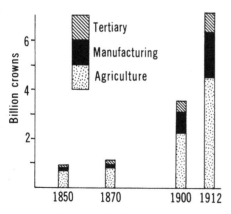

1.15 Growth of the Hungarian economy, 1850–1912

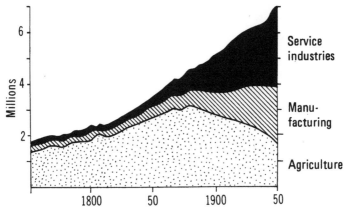

1.16 Growth of population, by employment sector, in Sweden

backward, agricultural country. The Austrian market was then opened up to Hungarian farm produce, and, at the same time, Hungary received considerable investments from Austria and Germany (fig. 1.15). The result was a rapid growth of the economy, though the agricultural sector never ceased to be dominant and much of the manufacturing was related to agriculture. Throughout eastern and south-eastern Europe the economy remained predominantly agricultural, and economic growth was very slow ineeed.

Sweden, lastly, shows (fig. 1.16) a pattern of growth typical of that of the rapidly developing countries of north-western Europe. Employment in agriculture declined sharply after about 1880, and that in manufacturing and tertiary occupations rose in no less marked a fashion.

2
The resource pattern of Europe

Fundamental to the economic development of Europe during the nineteenth century were the resources of the continent in fuels, metals and other minerals. Without them a machine-based civilisation could not have been developed, and the revolution in industrial production and transport would have been inconceivable without abundant and easily worked reserves of coal and ferrous metals. Increased agricultural production was, in part at least, dependent on metal tools and the fertilisers which were amongst the by-products of the mineral industries. It was not only the German Reich which was 'built...on coal and iron',[1] but the whole of nineteenth-century industrial civilisation.

Europe's reserves of solid fuel were large in total volume and varied in their physical properties. The continent, excluding Russia, but including the British Isles, was subsequently found to contain no less than 13 per cent of the world's resources. This fact was unknown to the pioneers of European industry, and would certainly not have influenced their actions. What mattered to them far more than the volume of resources was the fact that many lay in areas already important for their manufacturing activity, that they were easily accessible, and could be transported cheaply by water to the chief consuming countries.

Scarcely less important was the fact that Europe contained very large reserves of iron ore, amounting to about one-fifth of the world's known resources. Apart from two very large deposits – in eastern France and northern Sweden – there were countless small ore bodies scattered throughout the continent. Some had already been worked out by the beginning of the century; others were of too low a grade to attract the miner and smelter, but the rest supported in the early years of the century countless furnaces which supplied a growing quantity of metal.

To the resources of iron-ore should be added the not inconsiderable reserves of non-ferrous ores, particularly those of zinc, lead and copper, as well as of potash and other salts, which in the later years of the nineteenth century began to support a vast chemical industry. Nowhere else, except perhaps in the north-eastern United States, was so rich and so varied a mineral resource concentrated in so small an area and under conditions as favourable for exploitation by proto-industrial technology.

The scale of investment in a mine should bear some relationship to the size of the ore body or fuel reserve that was to be worked.[2] In the early nineteenth century the true extent of the resources could be only vaguely sensed. There were many instances of over-investment in reserves which proved to be too small to justify it. The many very small coalfields in France were cases in point. An eighteenth-century report on the diminutive coalfield of La Machine, in the Bourbonnais, claimed, with what in retrospect must seem the height of unwarranted optimism, that it could satisfy the needs of all France.[3] If, on the other hand, the true extent of the Ruhr and Upper Silesian coalfields, or of the iron reserves of Lorraine could have been known in advance their exploitation is unlikely to have followed a very different course. In fact, these resources were revealed only gradually as the century progressed and the need for them developed.

Fuel resources

It is generally accepted that an abundant supply of fossil fuel was an essential condition of the growth of modern industry in Europe.[4] It is none the less paradoxical that in the early stages of industrialisation little significant use was made of it. Water-power continued far into the nineteenth century to be the most important source of industrial energy, and was superseded chiefly because the amount which could be generated from each production unit was finite. It was only gradually displaced by steampower, generated by burning solid fuel, of which there was no limit (fig. 2.1). During the nineteenth century the chief industrial uses of coal were twofold. In the first place it came gradually to supply almost all the fuel used in the blast furnace, as well as that consumed in the voracious puddling furnaces which were used for refining iron. The other dominant use for coal was in the steam-engine, without which the larger factories could not have been activated or the mines kept clear of water. To these must be added from the mid-nineteenth century onwards a growing consumption in steam locomotives on the railways.

All these uses for coal were at first remarkably inefficient, and vast quantities were consumed in relation to the energy generated or the metal produced. Gradually, however, economies were effected, and by the end of the century both the iron and steel industry and also the steam-engine had become much more efficient in their use of fuel. This had important consequences in the location of those branches of industry which were heavily dependent on coal. Nevertheless, coal remained a bulky commodity and was difficult and costly to transport, and the fact that it was entirely consumed in the course of being used tended to orient manufacturing industries towards the coalfields.[5]

Coal production in continental Europe increased from less than three million tonnes a year at the end of the Napoleonic Wars[6] to about 280 million tonnes on the eve of the First World War. In the earlier decades of the century coal was obtained in very small quantities from an immense number of small coal basins, some of them with very restricted resources. In France there were no less than 50 active coalfields, most of which yielded only a very small quantity of inferior coal and remained in production only because they were protected from competition by distance and the poor means of transport (fig. 2.2).[7] In Germany

The resource pattern of Europe

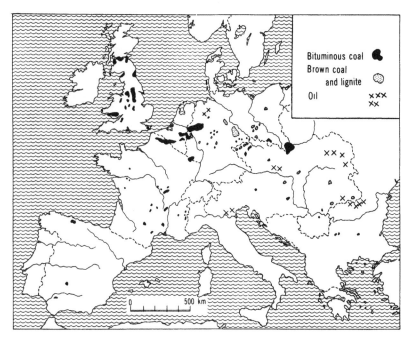

2.1 The distribution of mineral fuel resources in Europe. Oil resources were not exploited on a significant scale until the last decade of the nineteenth century

and central Europe also there were numerous small coalfields, important in the nineteenth century only for their local regions.[8]

The history of coal-mining in Europe is that of the eclipse of these small and uneconomic workings and of the gradual domination of the market by a small number of richly endowed coalfields. Of these the most important were the coal basins of northern France, central Belgium, the southern Netherlands and the Aachen region of Germany, which Wrigley has collectively termed the *Austrasian* basin. Next came the Ruhr field, of the Lower Rhineland and Westphalia, and the Upper Silesia coal basin. It is not easy to compute the contributions of each of these coal basins to the total European production throughout the nineteenth century, but it seems likely (fig. 2.3) that these three immense fields provided not more than a quarter of the total in 1815 and almost 75 per cent in 1913. This increased proportion of a sharply rising total was due in part to the relative ease of transport and the high quality of the coal available but also in part to the wide extent of the deposits, which made it practicable to build an infrastructure and exploit them on a large scale.

The first coalfield to achieve major importance in western Europe was that which lay along the Meuse valley near Liège. Coal seams outcropped on the valley sides and the river itself provided a cheap and convenient means of transport. Eighteenth-century writers regarded this as the richest and best developed of European coalfields.[9] The steam-engine – Newcomen's atmospheric engine – had been introduced for pumping early in the century, and when modern

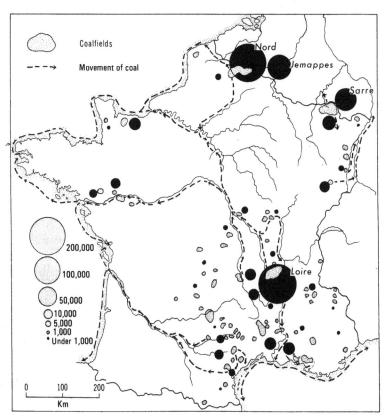

2.2 Active coalfields in France, 1814, and the movement of coal (figures in metric quintals)

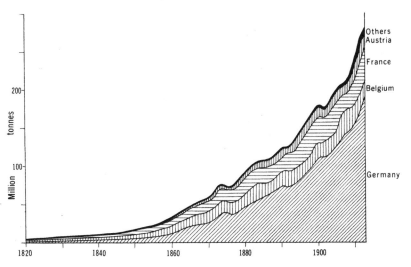

2.3 Coal production in Europe (Great Britain excluded), 1820–1913. This graph illustrates the increasing importance of Germany. See also fig. 10.5

industry was developed near Liège there was already a large output from a fully explored field. The seams were, however, thin, contorted and severely faulted. The better amongst them were quickly worked out and others gradually abandoned in the course of the century. The Liège field thus supplied a steadily diminishing fraction of the output from the northern field (see p. 361).

This coalfield stretched westward from near Namur into northern France, but beyond Charleroi the coal seams lay beneath a thick cover of younger rocks. That part of the basin which lay in the Belgian province of Hainault – the Borinage, as it was called – began to be opened up in the eighteenth century, but reached its greatest importance in the later nineteenth, after canals had been built across the plateau to move the coal to market. Coal was discovered close to Valenciennes in northern France in 1734, but the thick cover of secondary rocks made it difficult to trace the outline of the basin, and it was not until about 1850 that its true extent became known.

The northern coalfield quickly became the most productive in France. From an output of about 430,000 tonnes in 1830, about a quarter of the French production, output rose to about 2,500,000 tonnes (30 per cent) in 1861, to 44 per cent in 1880, and 54 per cent in 1900.[10] The only other Franch coalfield that ever rivalled that of Nord and Pas de Calais was the Saint-Étienne field which extended from the River Loire in the west almost to the Rhône in the east. Its seams were thin and folded and their quality poor. Nevertheless, the convenience of river-borne transport gave them an immense importance. By river and canal (see p. 416) coal from the Loire field supplied Paris before the northern field had been developed, and was displaced only when coal began to arrive from the richer and more conveniently placed coalfield. Not until the 1860s did production from the northern field begin to exceed that from the Loire.

The history of coal-mining in Germany followed a similar course to that in France. The many small coalfields, notably in Saxony and Lower Saxony, were gradually extinguished by the competition of the two giants of the German coal industry: the Ruhr and Upper Silesia.[11] In both, reserves were immense, and ranged from 'lean' or anthracitic, through coking coal, to gas and long-flame coals. In both, also, the exposed coalfield was relatively small and the more useful qualities of coal had to be pursued beneath a cover of later rocks. They differed, however, in the natural provision of means of transport. The River Rhine, the best and most used navigable waterway in Europe, flowed across the western margin of the Ruhr coalfield and from the Middle Ages had served to distribute the coal. The Upper Silesian field, situated on the watershed between the Oder and Vistula rivers, had no such advantage. To the east lay Russian-held Poland, and to the south the Habsburg Empire, in neither of which was there any significant demand for Silesian coal (fig. 2.4). Within Germany the nearest important market was in Berlin, 400 km away, and, despite the construction of the Klodnitz (Kłodnicki) canal to the Oder, little coal moved in this direction. Indeed, the chief market for Upper Silesian coal lay in the smelting industries which grew up on the coalfield itself.

Until the early years of the nineteenth century the coalfields of Lower Saxony, of Saxony itself and of Lower Silesia were regarded at least as highly as that of the Ruhr (fig. 2.5). The Ibbenbüren field, near Osnabrück, supplied coal to

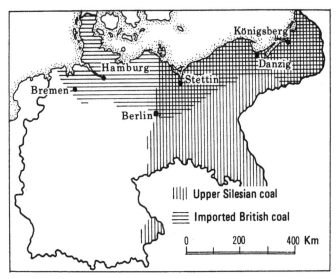

2.4 The coal market in Germany, late nineteenth century. The importance of British coal in the North German market reflects the relative cheapness of maritime transport

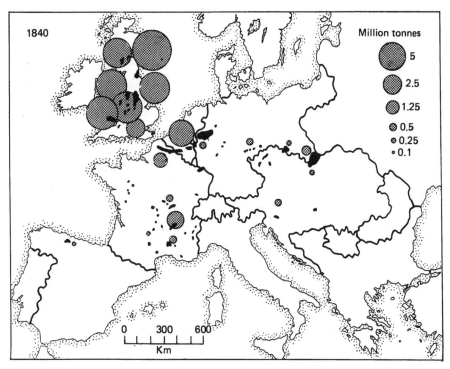

2.5 Coal production in Europe (including Great Britain) in 1840. Note the importance of Central Belgium and Saint-Étienne

much of north-western Germany.[12] There were also small coal basins near Magdeburg and Halberstadt,[13] and, farther to the east, near Zwickau and Dresden,[14] and in the Giant Mountains which separated Silesia from Bohemia. Here were the not unimportant coalfields of Waldenburg (Wałbrzych) and Glatz (Kłodzko[15]).

The Habsburg Empire contained, not only the southward extension of the Upper Silesian coalfield[16] (see p. 403), but also smaller coalfields near Plzeň and Kladno in Bohemia, at Brno in Moravia and also in the Giant Mountains. Very little fossil fuel was to be found within the Alpine system. In the Balkan peninsula were a number of small deposits of bituminous coal, notably in eastern Serbia and near Pernik in Bulgaria, together with numerous occurrences of lower quality sub-bituminous coal.[17] Some of these deposits were worked by the peasants for local use, but none was developed on any significant scale before the First World War.

Fossil fuels were notably absent from southern Europe. Both the Greek and the Italian peninsulas held nothing more than small reserves of sub-bituminous coal, and the Spanish peninsula, the best endowed of the three had only a small coalfield near the coast of the Asturias and a few even smaller basins within the Meseta.[18]

By the time of the First World War, the extent of Europe's coal resources was roughly known. The extension of the Ruhr field beneath the north German Plain had been proved, as had that of the Saar basin under the younger deposits of Lorraine. The limits of the Austrasian field had been explored, and the extension of the Upper Silesian into the Habsburg Empire and Russian-held Poland had been mapped. Only the Campine field in northern Belgium had not been fully explored or exploited. It was part of a coal basin which lay to the north and north-east of that of Liège. It had long been worked in the east, near Aachen, and it was suspected that richer resources lay beneath the Meuse valley and the sandy heaths of the Campine. In 1895 the Dutch opened up a mine in Limburg,[19] and their example stimulated the Belgians to look for coal in Brabant, especially as their resources in the Liège region were nearing exhaustion.[20] In 1901 coal was found. By 1906 the limits of the Campine coalfield had been established and concessions granted, but the first mine did not come into production until 1917.[21]

Fig. 10.5 shows the distribution of coal production on the eve of the First World War. Compared with fig. 2.4, it shows how completely European coal production had come to be dominated by the few very large coalfields. Most of the small basins, which had been so active a century before, had gone out of production, their reserves too small to justify the heavy capital investment required by the modern industry.

Nor were the large coalfields entirely free from such problems, though these arose rather from the small scale of early mining than from a poverty of resources. The coal-mining industry was more heavily burdened than most with the legacy of its own past (fig. 2.6). The earliest mining concessions, made in the eighteenth century or even earlier, had been very small. As the nineteenth century wore on new concessions became larger, and the most recent in northern France and the Ruhr extended over 100 sq.km or even more. Such large concessions permitted a more efficient layout of the underground workings than

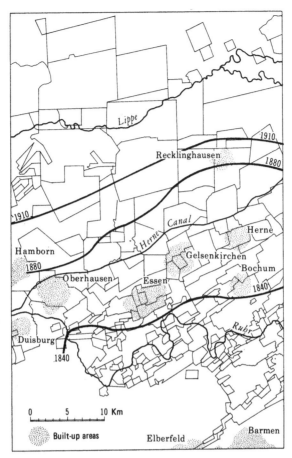

2.6 Mining concessions along a north–south transect across the Ruhr coalfield. Note the increasing size of concessions towards the north. The thick lines show the advancing limit of mining at the dates indicated

had been possible on the early concessions, as well as a more effective use of pumping and winding gear. It was the utter confusion created by a multitude of small claims, rather than the exhaustion of the seams themselves, that led to the abandonment of mining in the south of the Ruhr coalfield and in parts of central Belgium.

The development of European coalfields was strongly influenced by the availability of the means of transport by river or canal. In this respect only the Ruhr coalfield and that of central Belgium were really well placed, though the Saint-Étienne field succeeded, despite certain navigational problems (see p. 416), in using both the Rhône and the Loire for the transport of coal. From the seventeenth century rivers had been supplemented by canals, many of which were constructed for the primary purpose of transporting coal, and, despite a flurry of railway building in the middle years of the century, the carriage of coal remained a primary task of Europe's canals and navigable rivers.[22]

Throughout the century European coal production was dominated by the vast output from Great Britain. In 1850 this made up about two-thirds of the total, and on the eve of the First World War, after the rapid expansion of production from the Ruhr, it still formed 43 per cent. The coal reserves of Great Britain were a great deal smaller than those of either the Ruhr or Upper Silesia, but some of the richest of them lay close to the coast, conveniently placed for the export trade. Given the relative cheapness of water-borne transport, British coal from Northumberland or South Wales could be sold in the ports of north-western Europe more cheaply than coal brought overland from the principal continental coalfields. Neither the improvement of inland navigation nor the building of the railways entirely drove British coal from the ports of northern and north-western Europe. The German ports, particularly Bremen and Hamburg, continued to rely on British coal, and small amounts were distributed by barge through their hinterlands.[23] Berlin, for example, found the cost of domestic coal from the Ruhr or Upper Silesia and that from Northumberland and Durham evenly balanced (fig. 2.4). Scandinavia obtained most of its coal from Britain. Although little was used in the Low Countries, which were supplied more conveniently from the Ruhr and the Belgian fields, there was a large market for British coal in the ports and coastal regions of France. British coal formed a large and steadily increasing proportion of the coal consumed in France. From less than half a million tonnes, or about 9 per cent of local demand in 1841–5, it rose to three million tonnes or 12 per cent in 1876–80, and to more than 11 million tonnes, or 18 per cent, in 1911–13.[24] Normandy, Brittany and the west of France relied heavily on British coal. It was imported in considerable quantities through the ports of Portugal, Spain and Italy, and towards the end of the century stocks were being held at many points in the Mediterranean for bunkering ships.

The movement of coal *within* Europe was determined in large measure by the availability of inland waterways. Inevitably the coalfields of the Loire, the Ruhr and Upper Silesia assumed the greatest importance, followed, as canals were constructed in northern France, by that of the Borinage, Nord and Pas de Calais. The pattern of movement fluctuated from year to year, but its salient features remained unchanged up to the First World War.

France provides an instructive lesson in the problems of distributing coal. In 1865 Jevons contrasted the ease of access of British coalfields with the difficulties experienced by the French, 'situated in mountainous districts, difficult of access, where lines of communication have penetrated' slowly and at great cost. This, he added, was why at present 'the price of coal at market exceeds, in a very high proportion, the wholesale price at the pit mouth'.[25] It was, he added, 'the absence of easy lines of carriage and communication, which enable English coal to be sold on the French coast at a profit'. And not only English: a great deal of Belgian coal was distributed by canal over the whole north-east of France.

Brown coal

Bituminous coal was not the only solid fuel to be found in Europe. Deposits of sub-bituminous coal – lignite or brown coal – were widespread. This fuel had a low calorific value and a high water content, but occurred in level and often thick beds, close to the surface of the ground, the ease of working compensating in some measure for its small intrinsic value. The largest deposits were in the

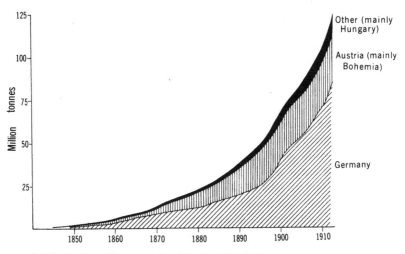

2.7 Production of brown coal and lignite, 1850–1913

lower Rhineland, in Saxony and in Bohemia, but small deposits were to be found in many other parts of Europe. They were numerous in the Balkans and Italy, where bituminous coal was very scarce, and were to be found also in southern France, Austria, Hungary and Spain. Although attempts were made to use brown coal in the eighteenth century, it attracted little attention until late in the nineteenth.[26] The principal reason was that it had so little value as a fuel that it could not bear the cost of distant transport from the place where it was extracted. Not until 1877 was a method devised of compressing it into briquettes, expelling most of the water, and greatly improving its quality as a fuel.[27] Transport of brown coal thus became more practicable. At about the same time it began to be used in chemical works, primarily for the extraction of tar, and as a boiler fuel. It really came into its own about 1900, when it began to be burned in thermal-electric stations, erected close to the workings. Complex machines were developed for extracting and briquetting the fuel and for feeding it into the furnaces, thus ensuring that only the largest deposits of brown coal would be developed.[28]

Although there were hundreds of small lignite and brown-coal deposits in Europe, most of the rapidly expanding output came from only three. The first to be developed was that which underlay the plain of Saxony between Halle and Leipzig.[29] This was quickly followed by the huge Bohemian field which occupied a trough along the foot of the Ore Mountains.[30] The third formed a narrow belt lying parallel with the Rhine to the west of Cologne.[31] The rapid expansion of lignite and brown-coal production during the last two decades before the First World War is shown in fig. 2.7. Two-thirds of the world production was in Germany, and almost three-quarters of this was in Saxony.[32] There were small deposits elsewhere: in Nassau,[33] Brandenburg,[34] Bavaria and other places.[35] The Habsburg lands were well endowed but outside Bohemia reserves were small.[36] It was mined near Esztergom in Hungary.[37] In the Balkans were numerous small deposits which began to be exploited late in the century, but output remained slight and the workings primitive.[38] Lignite, lastly, was said to occur 'on nearly all the great plains of Greece' but little attempt was made to use it.[39]

Petroleum and natural gas

Europe was not supposed, in the nineteenth century, to have any considerable reserves of natural oil, and most were thought to occur in Romania and Galicia where oil had been produced from wells since the Middle Ages and, in the eighteenth century, was burned in lamps in the peasants' cottages.[40] Oil was also obtained at this time in very small quantities from wells in Alsace, from oil-shales in Burgundy, and from bituminous lignite in Saxony.[41] The discovery of oil at Titusville, in Pennsylvania, by E. L. Drake in 1859 led to research into methods of refining and using crude oil.[42] It began to be used to fire boilers, but its most important use came when it was found that the lighter distillates of crude oil could be used in an internal combustion engine. It was the automobile that did most to encourage both prospecting and the development of oil technology.

The Romanian oil industry began to grow in the 1870s. Oil was found to occur in a zone some 400 km in length and about 20 km wide, on the southern edge of the Transylvanian Alps.[43] Five major centres of production emerged by the end of the century: the Olt valley in western Walachia, the Dimboviţa and Ialomiţa valleys to the north-west of Bucharest, the Prahova valley near Ploieşti, the Buzău valley and, in Moldavia, the Trorus and Tazlău valleys. Of these the Prahova valley quickly became and remained the most important, with more than one-third of the total wells.

Growth in the Romanian industry was slow. There was a lack of technical skill; the landowners, preoccupied with the export of grain, did little to facilitate oil-working on their lands; transport was rudimentary, and capital scarce. 'On the whole', wrote the British consul, 'the extraction of petroleum...is defective from all points of view'.[44] Oil was available 'in enormous quantities [but] primitive methods of extraction, want of initiative, of capital, and of specialists, combined with the indifference of the natives, conduce to render the production far smaller than could otherwise be made'. Conditions improved in the 1890s. With the collapse of grain prices the landowners became more interested in petroleum, and this in turn obliged the government to encourage foreign investment. Production rose sharply from 80,000 tonnes in 1894–5 to more than half a million ten years later and 1,352,000 in 1910.

The largest producer, apart from the United States, was Russia, where the Caucasus field, near Baku on the Caspian Sea, appeared to hold untold resources. At the turn of the century oil production was beginning in Austrian Galicia, where the oil fields, like those of Romania, bordered the arc of the Carpathian Mountains. But elsewhere in Europe it remained insignificant. Natural gas was, of course, obtained along with the oil, but it was not until the end of the century that any attempt was made to conserve and use it. The first country to record the production of natural gas was Italy in 1899.

At no time did the oil production of Europe, Russia excepted, cover actual needs, and there was, at least from the mid-nineteenth century, an import of crude oil, at first from the Baku field and later from the United States.

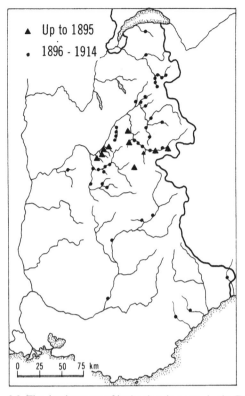

2.8 The development of hydroelectric power in the French Alps

Hydroelectric power

During the middle years of the nineteenth century the electric generator was developed and improved to the point at which it became possible to build power-stations for the public supply of electricity. The first power-stations, built about 1880, used steam-engines to operate the dynamos. But already a turbine which could convert the pressure of steam or flowing water into rotary motion had been developed. As early as 1870 it was proposed to harness the turbine to the dynamo, an operation greatly facilitated by the newly developed technique of creating a powerful 'head' of water, delivered by an iron pipe from a great height. The first true hydroelectric plant was established on the Valserine, an Alpine tributary of the Rhône, in 1882.[45] Improvements quickly followed in both turbine and generator, and were paralleled by the increase in the number of hydroelectric stations (fig. 2.8).

Physical conditions were most suitable in the Alps of Dauphiné, where rainfall was heavier and more prolonged than farther to the south.[46] By 1914 hydroelectric stations had begun to succeed one another along the Drac and the valleys of Maurienne and Tarantaise, to the east of Grenoble.[47] The energy generated was used to light Grenoble and other places as far as Lyons; it was used in sawmills and industrial works and was, above all, being applied to metallurgical and chemical processes (see p. 418).

The Pyrenees were developed later than the Alps. They were climatically less suitable and the local demand for electric power was smaller than in the Rhône valley. The first works in the central Pyrenees was begun in 1885, and three years later the energy produced was being used to illuminate the city of Toulouse. The first in the drier eastern Pyrenees did not come until about 1900.[48]

Switzerland was the first European country to be extensively electrified. The development of hydroelectric power was encouraged by the high price of coal, and here the predominance of the coal-fired steam-engine was indeed short-lived. From the closing years of the nineteenth century the use of hydroelectric power spread rapidly, and by 1911 only 20 per cent of the energy used had been generated by burning coal.[49]

The Scandinavian countries were in a similar position to Switzerland, with an immense potential for hydroelectric power and an almost complete lack of solid fuel. Development came, however, somewhat later. Sources of power were mostly distant from the centres of population and industry, and their use had to await the invention of methods of transmitting power over great distances. Steam-generated energy thus remained important, and the import of coal into Scandinavia continued to increase until the First World War.

The growing use of hydroelectric power brought about a revolution in industrial location only a degree less important than that of steam-power a century earlier. Electric power, whether hydro- or thermal-, led to a renewed dispersal of manufacturing and to a revival of small units of production, for which the electric motor was far more suited than the steam-engine. Hydroelectric power, furthermore, attracted industry to mountainous areas which in the past had known little manufacturing activity. The chief beneficiaries were Switzerland, the French Alps, Sweden and Norway. The use of hydroelectric power also encouraged a range of industries which made heavy demands on energy, especially the electro-chemical and electro-metallurgical. These were to become especially important in the western Alps and on the Norwegian coast where power resources were most abundant.

The ferrous metals

No two branches of industry were more intimately linked than coal-mining and the smelting and processing of iron. Neither could have achieved its late-nineteenth-century importance without the other.

Iron ores are amongst the most widespread of economic minerals. No major region of Europe was entirely without deposits of some kind, though these varied greatly in size and in the quality and grade of the ore (fig. 2.9). Some – those of the Siegen district of Germany and of central Sweden, for example – were naturally suited, owing to their admixture of trace elements and lack of impurities, to producing metal of a high quality. Others, including some of the largest in Europe, contained so much impurity that refining the metal was difficult.

A low-grade ore, that is one in which the percentage of metal is low – less, let us say, than 25 per cent – was costly both to transport and to smelt, since the waste had also to be melted down. This did not greatly matter when the ore

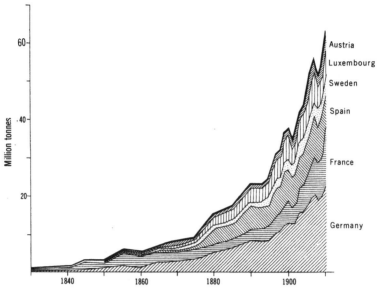

2.9 Iron ore production in Europe (excluding Great Britain), 1830–1910

was mined close to the forests which provided an abundance of charcoal, and was smelted only to satisfy a local market. Indeed, there were hundreds, even thousands, of small charcoal-burning furnaces scattered across Europe from Spain to Poland (p. 340), most of them dependent on a local supply of ore of indifferent quality and on charcoal from the neighbouring woodlands.

Among the hundreds of ore bodies that were exploited during the century were a few of superior quality, distinguished by their higher iron content, by the presence of a trace of manganese and by an absence of phosphorus and sulphur. Such were the ores of the Marne, Siegen, Styria, Dauphiné and of the *Bergslagen* of central Sweden. These ores, or the iron produced from them, acquired a high reputation at an early date. In most instances, however, individual ore bodies were quite small and total reserves in each of these areas were far from large. In consequence many such deposits were worked out in the course of the nineteenth century, while others, too small to justify the most modern mining and ore processing equipment, were simply abandoned.

Some of these ores acquired great importance during the middle years of the century. The Bessemer process, introduced in 1856, allowed steel to be made more quickly and in far greater quantities than previously, but it was soon found to be discriminating in the ores which it could use. The presence of phosphorus in the ore led to the production of a steel which was brittle or 'cold-short'. Until this problem had been overcome by the introduction of the basic or Thomas process in 1879, there was a frantic search for low-phosphorus ores. Most of the superior haematites contained very little phosphorus, a fact which – however little the earlier ironmasters understood it – gave them their importance. The most abundant of these ores were to be found in the eastern Pyrenees,[50] in northern Spain, on the island of Elba, in the Sieg-Lahn-Dill district of western Germany, in Styria, Carinthia and central Sweden. These deposits had long been known, and since the Middle Ages had been used to produce the best quality steel. Some had become exhausted, and most others held only small reserves. In the 1860–70s attention was focused on the few that remained. Foremost amongst them were the ores which lay in the hills around Bilbao in the Basque country of northern Spain. They had the advantage of being near the coast and could be exported by sea at little cost. Production from the Bilbao mines increased sharply in the late 1860s.[51] Most of the ore was exported, the bulk of it going to Great Britain.[52] By the end of the nineteenth century the Basque ores and most others with a low phosphorus content were nearing exhaustion.[53] Other ore bodies of high quality continued to be exploited, but most were far too small to satisfy the growing demands of the Bessemer iron industry. Amongst them were the scattered deposits of Brittany and Normandy, individually too small to support an industry and some too remote to export their ore.[54] Good-quality ores were worked near Allevard, in the Alps of Savoy,[55] and from the residual clays which covered the plains of Berry, Burgundy and Champagne.[56] During the eighteenth century the latter had supported a flourishing iron industry, but during the nineteenth the better deposits were worked out and the smaller ceased to be profitable (see p. 335).

Amongst the most important of the low-phosphorus ores were those found in the Rhineland. They varied in quality, but the best of them were of a very high quality indeed, low in sulphur and phosphorus and containing that small amount of manganese which added greatly to the quality of the metal. Many of the deposits were small and unsuited to mechanised mining, but the larger amongst them, notably the Stahlberg at Müsen, continued to support a smelting industry throughout the century.[57]

Ore deposits, smaller in volume and generally less desirable in quality, were to be found scattered through the hills which lay between the Siegerland and Saxony. In some areas, Thuringia for example, they occurred in greater abundance, and continued to be mined and smelted up to, and even after the First World War. A small iron-mining industry also survived in the Ore Mountains of Saxony and in the old rocks of the Bohemian massif.

There were small but historically important ore reserves in the Italian Alps which at one time were used in the not unimportant smelting industry of Piedmont and Lombardy.[58] The Austrian Alps were more richly endowed and the largest of the deposits, the Erzberg of Styria, continued to be worked in open quarries throughout the century, supplying important ironworks near Leoben.[59] Similar ores were to be found in the mountains of Upper Hungary, where also they supplied a small iron industry.[60]

South-eastern Europe was not well endowed with ferrous ores. A number of deposits, unremarkable for either their extent or their quality, were to be found in Slovenia and Bosnia in the west of the Balkan peninsula,[61] and near Sofia, in the east.[62] Most important were those in the Transylvanian Alps. These were mined during the nineteenth century and were smelted in primitive charcoal furnaces, some of which remained in use until after the First World War.

Sweden was from the later Middle Ages until the nineteenth century an important source of quality bar-iron, used in western Europe for making tools and weapons. This iron came mainly from the *Bergslagen* district of central

Sweden, where many of the deposits were high-grade and suited for steel production.[63] It was on these that the Swedish export of *osemund* iron and later the Swedish quality steel industry were founded.

Important amongst the smaller ore deposits was the coal-measures or blackband ironstone. It occurred in the coal beds; sometimes it actually incorporated coaly materials, thus carrying, as it were, its own fuel for smelting. It was inevitably brought to the surface along with the coal. Most coalfields contained some blackband ore, but it was truly important only in the Ruhr and Upper Silesia. In the former it occurred most abundantly in the lower part of the coal series, which outcropped near the southern margin of the field. So important was it in the third quarter of the century that no less than eight blast furnace works were built to use it between 1850 and 1870 (fig. 8.2).[64]

With the exhaustion or abandonment of most of these small deposits of good-quality ore, production came to be concentrated more and more on large deposits of poorer ore. The chief problem facing the iron industry in the latter half of the nineteenth century was to develop a technology that could overcome their obvious shortcomings. Most of them were low-grade, highly phosphoric, bedded ores, and were found in the Secondary rocks, mainly Jurassic, from France to Poland. Their low grade – few had more than 30 per cent iron – made it uneconomic to transport them far. Fuel costs were high because of their large amount of slag. But they yielded a very fluid iron, highly suitable for castings, and the metal obtained from them could be puddled to an acceptable quality of bar-iron. But it was steel, not wrought iron, that was in demand in the second half of the century, and in both the Bessemer and the open hearth processes the phosphorus remained in the metal, making it cold-short.

Until this disadvantage had been overcome by the basic process, these ores had no great advantage and comparatively little use was made of them. In the meantime the few important reserves of non-phosphoric ore were exploited to the full. Nevertheless, the supply of ore suitable for making iron for steel-making was becoming critical by the 1870s, and this was reflected in the alacrity with which ironmasters took out patents for the Thomas process. The low-grade phosphoric ores now came into their own, with profound consequences for the industry as a whole.

Most extensive of these ores were the *minette* deposits of Lorraine, Luxembourg and southern Belgium. They occurred in beds of varying thickness which came to the surface in the Jurassic escarpment of eastern France, but dipped westward until they passed beyond the limits of profitable mining.[65] Reserves in southern Belgium were small, and were effectively worked out early in the present century. Those in Luxembourg were larger and have since supported a considerable smelting industry. The Lorraine ores, however, were immense and contained almost a half of the total reserves in continental Europe[66] (see p. 337). Ores of broadly similar composition and occurrence were found in the North German Plain, near Osnabrück and at Peine-Salzgitter, though little use was made of them before the First World War. Lastly, similar ores were to be found in the Jurassic formation near Częstochowa in Poland. The total reserves in the north German and Polish deposits amount to no more than a third of those to be found in Lorraine.

The only ore body which could rival that of Lorraine in the volume of its reserves was to be found in northern Sweden. It was, however, of a quite different origin. It occurred as a number of intrusive masses in the Palaeozoic rocks of Swedish Norrland, where it could be quarried rather than mined. The ore was a high-grade – about 63 per cent – magnetite, low in sulphur, but with a variable but generally high phosphorus content. Its disadvantage was that it lay within the Arctic Circle, where living conditions were hard and transport difficult. Although these ores had been known at least since the seventeenth century, attempts to exploit them continued to be defeated by the harshness of the environment and the lack of transport. In 1891 a railway was completed, linking the orefield both with the Baltic port of Luleå and with Narvik on the Atlantic coast of Norway.[67] The ore was thus able to reach the industrial centres of north-western Europe, but, as the northern Baltic was ice-bound for a large part of the year, most of the ore eventually moved through Narvik.

Throughout the period up to the First World War Europe possessed on balance reserves of iron ore which were fully adequate for its developing iron and steel industry. Problems in their exploitation and use were threefold. In the first place, a very great number of the deposits were small. This meant that, as mining became increasingly highly capitalised it ceased to be profitable to work them, and many of the mines were abandoned before they had become exhausted. There was, secondly, a relative scarcity of ores with a low phosphorus content, and as these ores tended to occur in small deposits they were exhausted at a comparatively early date. This fact alone was responsible for the import of ores of non-European, mainly African, origin after the introduction of the Bessemer process. Lastly, few ores were located close to the coalfields which supplied fuel for smelting, and overland transport of ore and also of fuel became a very significant factor in costs. Indeed, there was a tendency in the years before the First World War to smelt ores imported by sea at coastal sites owing to the relative cheapness of maritime compared with overland transport.

Non-ferrous metals

Reserves of the non-ferrous metals were noteworthy more for their variety than for their richness and abundance. They were, however, worked intensively until, with their small reserves running out, they yielded place to larger resources in Africa and the Americas. Their importance cannot, however, be overestimated. Not only did they play an essential role in providing materials for industry, but the mines themselves were a stimulus to industrial development. They created a demand for factory products, and the steam-engine was itself in part a response to the needs of metalliferous mining.

With few exceptions the non-ferrous ores occurred either in the ancient rocks which made up most of Scandinavia or in the rocks of somewhat younger geological age which formed a belt of hills from Spain to Poland and Transylvania. Most of the metalliferous minerals were found in lodes and masses which filled cracks and interstices in the country rock. On rare occasions they had been dissolved or eroded and then redeposited in very much younger beds. In order to be economically significant an ore had to be large and rich enough to be worth the high cost of mining. Most ores become unstable after long exposure to the weather and break down into their soluble constituents, which are in turn carried away to the sea. There were, however, two exceptions, gold and cassiterite, the ore of tin. These were inert and, once eroded from the rocks, tended to lie in the sediments which accumulated in stream beds. Here they could be obtained by alluvial or placer-mining. Most of the small quantity of gold obtained in Europe was from such deposits, and alluvial tin continued to be of some small importance in England throughout the century. Metalliferous mining was a far more risky undertaking than iron- or coal-mining. There were few large ore bodies and the behaviour of lodes was unpredictable. They might thin out and disappear, and the discovery of their continuation was often a matter as much of luck as of judgement. The technical problems were often greater than in iron-mining; the mines were frequently deeper, the rocks harder, and difficulties of drainage more acute.

A lode was made up of minerals of economic importance, intimately mixed with others, which had to be discarded. The ratio of ore to waste was its grade, and for any deposit there was a grade below which, under existing technological conditions, it was not economic to mine.[68] The limits of a workable ore became lower with improvements in ore-crushing and dressing. Sometimes more than one economic mineral was present in a lode. Occasionally one was found higher in the system than another, as copper tended to give place to tin with increasing depth. Sometimes their respective crystals were intermixed; sometimes they occurred in separate aggregates. In Upper Hungary silver was thus associated with lead; in Italy, lead with antimony; in Bohemia, silver with lead, and in Upper Silesia, lead with zinc. The existence of two or more minerals which were thus in joint production did much to increase the profitability of mining, as market conditions favoured now one, now the other.

Although a very wide range of metalliferous ores was mined in nineteenth-century Europe, only some half dozen were of any great commercial importance. Of these tin was produced in significant amounts only in Great Britain, and gold was found only in very small quantities either in association with silver or in alluvial deposits.[69] Silver, which had long been the most valuable non-ferrous metal produced in Europe, continued to be mined on a small scale in Saxony, Bohemia and the Harz Mountains, but such ancient centres of silver-mining as Jachymov and Kutná Hora were in rapid decline,[70] and the nineteenth-century revival in Slovakia was short-lived.[71]

The base metals were of far greater importance, and in zinc, lead and copper Europe remained on balance very nearly self-sufficing until the end of the century (fig. 2.10). Resources were largest in zinc. There were many small deposits and two of world importance: the Moresnet-Eupen ores on the borders of Belgium and Germany (see p. 380) and those of Upper Silesia.[72] Together they produced in 1870 80 per cent of the European output of zinc ore and three-quarters of the world production. This proportion tended to decline, but in 1900 Europe was still yielding 60 per cent of the world's output of zinc ore. Much came from the deposits of Upper Silesia, the richest of which at this time lay within Germany. The reserves at Vieille Montagne in Belgium were running out in the middle years of the century, but Belgium had in the mean time developed a large

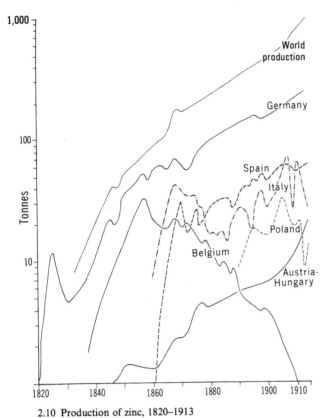

2.10 Production of zinc, 1820–1913

smelter capacity which continued to be fed with imported ores.[73] During the years before the First World War Belgium was producing almost half the world's metallic zinc.[74] Elsewhere, in Spain and Italy as well as Upper Silesia, production continued to grow during the years before the First World War.

The uses for both lead and zinc were increasing during the century. The process of galvanising – the coating of mild steel with a protective film of zinc – was discovered in 1836, and from the mid-century the production of galvanised sheet and wire were increasingly important.[75] Both zinc and lead were used in electric batteries and in the manufacture of paint. Lead continued to be used as a roofing material, in water pipes and in printers' type.

Lead-mining and smelting have a very much longer history than zinc. Lead was commonly found in association with other economic minerals, including ores of silver, zinc and antimony. The most abundant resources lay in Spain, which on the eve of the First World War was producing two-thirds of the European output,[76] Upper Silesia, Slovakia[77] and the Balkans.[78] Until about 1850 some 90 per cent of the world's production was from Europe. By 1900, despite the growth of the North American industry, it was still nearly 60 per cent (fig. 2.11).

Copper was in increasing demand during the nineteenth century. In addition to its traditional uses in the manufacture of brass and bronze, it was used in

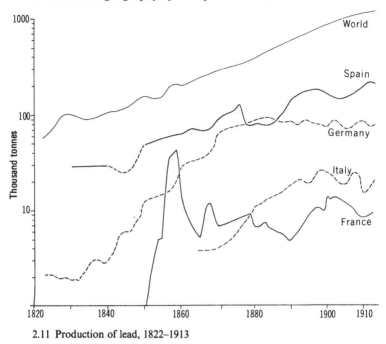

2.11 Production of lead, 1822–1913

the generation and distribution of electricity. At the same time the ability of copper to resist corrosion led to its use in water-supply systems and in the pipework of steam-engines and other machines. Continental Europe was, however, able to supply only a small part of its needs (fig. 2.12). Although Sweden continued to produce copper, the great Kopparberg deposits were nearing exhaustion, and the Spanish mines of Rio Tinto and Tharsis did not come into full production until the second half of the century. Apart from south-western England, which remained until the 1860s the world's chief source of copper, only the German mines at Mansfeld in the eastern Harz were a significant and continuous source of copper ore.[79]

Other non-ferrous metals which achieved importance before the First World War were aluminium, manganese and mercury. There was also a very small and generally intermittent production of chrome in Greece, of nickel in Norway, and of tungsten in Germany. Aluminium was first isolated early in the century, but was not produced commercially until the 1870s. It was derived from a complex silicate of aluminium, known as bauxite, from Les Baux in Provence, where it was first worked on a large scale.[80] Indeed Les Baux was the only significant source of the ore in Europe and by far the most important in the world until the eve of the First World War (fig. 2.13). In 1886 the electrolytic method of reducing refined bauxite or alumina to metallic aluminium was discovered and at once production expanded. The process, however, made very heavy demands on electric power, and until 1914 most of the output was from Switzerland and the French Alps, where hydroelectric power was available. Bauxite was far from being a rare mineral. It usually occurred close to the earth's surface, and, unlike most metalliferous ores, was a product of the decomposition of other rocks. In

The resource pattern of Europe

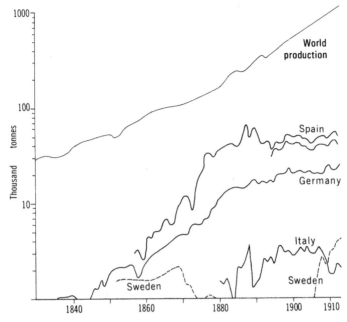

2.12 Production of copper, 1830–1913

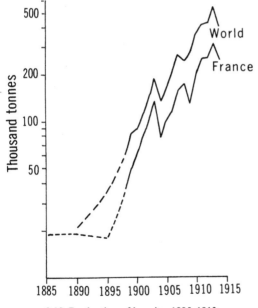

2.13 Production of bauxite, 1885–1913

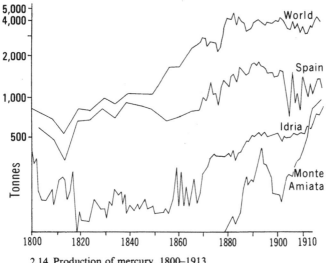

2.14 Production of mercury, 1800–1913

addition to the abundant reserves in the south of France, there were also large deposits in the Hungarian plain, opened up on the eve of the First World War.

Manganese had been used in the refining of steel long before it was consciously exploited. It had a 'cleansing' effect on iron, by combining with and removing sulphur. In larger amounts, it produced a hard alloy-steel, used for weapons and tools. The iron ores of central Sweden, the Siegerland and Austria owed their high quality largely to the presence of manganese. Later in the nineteenth century it became the practice to add manganese, in the form of either ferro-manganese or *Spiegeleisen*, to the metal in the steel-making process. Manganese ore was also found independently of iron at a number of places, and was mined from early in the century. The chief sources were France and Spain, and also Austria and Bosnia.

Until the second half of the nineteenth century there were effectively only two sources of cinnabar, the ore of mercury: the mines of Almadén in the southern part of the Spanish Meseta, and of Idrija (Idria), to the west of Ljublijana (Laibach). To these was added in the middle of the century Monte Amiata in central Italy, which increased steadily in relative importance after about 1880 (fig. 2.14).

Non-metallic minerals

Non-metallic minerals have played a role quite different from that of metallic. They include rock-salt and potash, sulphur and silica, barytes, gypsum and a host of chemical substances of less importance. Most are relatively abundant but highly localised. Several are associated only with the Triassic beds and represent a precipitate formed in the shallow seas which existed during this geological period. This series, and hence the minerals which it embraced, occurred most extensively in eastern France, central Germany and southern

Poland, and it was here that most of the industries based on them grew up in the later nineteenth century.

Apart from silica and clay, used in the glass, ceramic and brick industries, these minerals were of no importance during the earlier phases of industrialisation. It was only with the development of chemical industries in the latter half of the century that they came into their own as the raw material of fertilisers and the basic chemicals.

Rock-salt

The first of such minerals to gain importance was rock-salt or sodium chloride. It had been obtained by evaporation along the coasts of southern Europe as far north as the 'Bay' of Bourgneuf in western France. Inland brine springs were tapped at Lüneburg and in Saxony,[81] Lorraine and the Jura, and rock-salt was quarried in mines at Wieliczka and Bochnia in Austrian Galicia.

Potash salts were closely associated with rock-salt, since both were formed as precipitates in the shallow Triassic sea. Both occurred near Magdeburg and along the northern margin of the Harz mountains, especially near Stassfurt. The potash salts had long been known. Shafts were sunk into the beds in 1856, and from 1860 the salts were exploited on a significant scale.[82] Their chief use was at first in the preparation of agricultural fertilisers. With the expansion of the central German chemicals industry they became the basis for the manufacture of alkalis. Throughout most of the period up to the First World War Germany held a near monopoly of the European production of potash salts. Similar deposits were discovered about 1900 between Mulhouse and Colmar in Alsace, but their exploitation was not pressed until the territory was returned to France in 1919.[83]

Sulphur

This also became an important industrial chemical in the later nineteenth century (see p. 346). Apart from a small amount derived from the smelting of pyrites, chiefly in Spain and Norway, it was obtained exclusively from Italy. It occurred in shallow deposits which derived ultimately from volcanic activity. The largest resources were in Sicily, where they lay in a broad belt across the centre of the island, but there were also important deposits in the volcanic areas of the mainland. In 1830 the Sicilian production was of the order of 40,000 tonnes. Fifty years later it had increased tenfold. In 1886 there were said to be 564 pits and an employment within them of about 32,000.[84] By 1910 there were about 400 mines, employing only 21,000, but output had reached 2.8 million tonnes.

Resources for agriculture

The agricultural potential of the European continent was severely restricted. Much of the land lay in high latitudes with a severe climate and short growing season, while in southern Europe the summer drought made much of the land suitable more for tree- than for field-crops. Between the two, high altitude on the one hand and glacial sands and gravels on the other greatly restricted the extent of good agricultural land.

Yet in pre-industrial Europe these environments embraced closed societies, totally dependent for their food supply upon what could be grown locally. Almost every community produced bread-grains and legumes which together constituted their staple diet. Animal proteins were eaten sparingly. The yield-ratios of all crops, with the exception of such newly introduced species as maize and the potato, were very low. There was no surplus production of food, and crop failure, from whatever cause led directly to famine. Over much of Europe a situation that could be called Malthusian prevailed, with the restricted resources used inadequately or improperly. How then could such an agricultural system support not only an increasing population but also an industrial and tertiary superstructure?

The agricultural revolution of the nineteenth century is examined in chapter 5. Here it is necessary only to discuss the physical basis of agriculture in Europe. In 1913 only about 6 per cent of the total land area was under regular cultivation. The proportion was almost certainly larger a century earlier, though statistics are not available except for a few restricted areas. Broadly stated, the area of agricultural land expanded during the first half of the nineteenth century and during the second contracted under the impact of cheap bread-grains imported from the 'new' counties.

Apart from human agency – government protection, systems of land tenure, farm size and market influences – the chief constraints on agriculture were the quality of the soil, the climate, which was predictable, and the vagaries of the weather, which no human ingenuity could anticipate.

The only really extensive areas of high-quality soils formed a belt from the Paris Basin in the west, through central Belgium to the lower Rhineland. From here it stretched eastwards in a narrow zone, through Saxony, Silesia and southern Poland into Russia. These were the *terres de culture facile* of Vidal de la Blache. They were composed mainly of beds of loess or *limon*, laid down in varying thicknesses during the glacial period. To some extent they were sorted by running water and redeposited as brick-earths, such as are widely found in southern England. More restricted areas of such soil were to be found farther south, in the plain of the upper Rhine, in parts of southern Germany, Bohemia and the Danube valley. Elsewhere alluvium, the accumulated deposits of river floods, constituted narrow bands of good soil along the valleys of many rivers. The natural fertility of such soils was, however, often modified by their poor drainage and their tendency to become waterlogged. Furthermore they were often interrupted by terrace-gravels, which also owed their origin to deposition by the rivers when their level stood higher than at present. The natural productivity of the Lombardy Plain, one of the most extensive of such alluvial lowlands, was reduced in this way (see p. 255). The intensively cultivated land of the Netherlands is basically alluvial soils modified by the growth of peat. Similar soils border the lower courses of most of the rivers of northern Europe. In southern Europe they are often limited to small patches between the mountains and the sea, a factor of great importance in the history of the region.

North of the loess belt the quality of the soil was reduced by deposits laid down by the ice and by the flood waters which flowed from it. They varied from heavy clay – the 'boulder clay' – to light and infertile sands and gravels. The latter gave rise to the waste of Brandenburg, with its 'starveling pine-plantations and its sandy fields'; to the hungry soils of Prussia and the heathlands of Denmark, West Germany and the Netherlands. The former produced the lake-studded landscapes of Mecklenburg, Great Poland and Masuria. In neither instance were the soils at best of more than mediocre, quality, and over large areas they were almost incultivable.

Farther north, where the land had been free of ice for only a relatively short period, soils were even less developed and the climate more severe. This was the northern frontier of agriculture, where a small climatic fluctuation could produce a wide latitudinal shift in the limit of cultivation, and it was here that the greatest advances were made during the climatic amelioration which occurred in the latter half of the nineteenth century.

To the south of the loess belt the quality of the soil was greatly influenced by the nature of the underlying rock from which it was derived. Areas of highland, built of Palaeozoic rock, with a thin and generally acid soil, alternated with areas of lowland, floored with clay, and with ridges of limestone with a dry alkaline soil.

South again lay the Alpine belt. Much of the area rose above the limit of agriculture. The valleys offered scope for farming though this in turn varied with orientation and local climatic conditions. Between the valleys and the bare rock of the higher slopes lay an intermediate zone, a climatically marginal area which provided only seasonal grazing for transhumant animals. Overall the Alpine belt was a food-deficit region though one with a considerable potential for dairy farming.

Lastly, the Mediterranean region suffered from a double disadvantage. Its hot summers coincided with drought which varied in duration and intensity from one part of the region to another. The land was mountainous; much of the rock was limestone which intensified the consequences of the summer drought and commonly supplied only a thin soil. The coarse, drought-resistant vegetation was ill-suited to supporting animals or to providing humus for the soil, and without animals there could be no manure. It was a vicious circle from which Mediterranean people extricated themselves only partially and locally, on the pockets of alluvial soil which bordered the sea, where rivers brought irrigation water from outside the region.

Despite its obvious inadequacies in both soil and climate, European agriculture, supplemented by only a small net import of foodstuffs, was able to support a population of about 130 million early in the century. A century later this population had risen to 294 millions and the production of cereals alone had increased threefold (see p. 188). This was achieved by higher yields and yield-ratios rather than by any increase in the cultivated area. The improvement in yields was more marked in western than in central Europe, and was least in eastern Europe and the Balkans. It was, indeed, a remarkable achievement even if it still left Europe with a sizeable import of cereals and animal products.

This increase in agricultural production was achieved in three ways: by changes in the system of agriculture, by modifying the physical environment, and, where that was not practicable, by adapting to it. The most significant change in agricultural methods was the gradual elimination of fallow from much of Europe. This in turn allowed rotations to be changed and adapted more closely to physical conditions. Fodder-crops tended to replace fallow and these in turn led to larger herds and a more abundant supply of manure. At the same time, especially in Germany, the use of fertiliser from the growing chemicals industry allowed marginal land to be cultivated and permitted increased yields on that of better quality.

The heavy use of fertiliser must be regarded as a means of modifying, if only marginally, the quality of the physical environment. So also were underground drainage in northern Europe and irrigation projects in Spain, Provence and Italy. Lastly, there was a progressive adaptation of systems of agriculture to physical conditions. Bread-grains ceased gradually to be grown in quantity on many heavy clay soils, such as those of Normandy, and were replaced by permanent grass and fodder.

The area of agricultural land was increased to a small extent by the reclamation of coastal and riverine marshes. The primary purpose of such works was in most instances not the increase of cropland, but flood control, coastal protection and the eradication of diseases associated with marshland. In France land reclamation was encouraged and controlled by a series of legislative acts from 1807 onwards. During the middle years of the century important drainage works were carried out along the shores of the Gironde and near the mouths of the Loire and Rhône.[85] On the lower Seine about 10,000 hectares are said to have been reclaimed, and some 30,000 were drained along the Lower Rhône where 'fevers almost completely disappeared'. Attempts were made to drain the fever-ridden Pontine marshes to the south-east of Rome, but the task was a formidable one, and little progress was made before 1914.

The greatest scope for land reclamation lay in the Low Countries. Here there was an urgent need both to control the discharge of the Rhine, Meuse and Scheldt and to protect the low-lying land from storm-floods from the North Sea. Coastal marshes in *dépt* Nord, drained in the 1820s, were inundated again by a surge of the sea in 1829. Farther north, progress was made in the islands of the Rhine delta, and most of the remaining *meers* of Holland and Zeeland, including the Haarlem Meer, were reclaimed. The introduction of the steam-engine for pumping greatly assisted the work, but in 1882, out of about 80,000 hectares over 53 per cent had been drained by windmill drainage, and only 36 per cent wholly by steam-driven pumps.[86]

It was proposed as early as 1849 to drain the Zuider Zee by building a vast enclosing dyke and pumping out the water.[87] It was not until half a century later, however, that the first steps were taken to implement the plan. In 1901 work was authorised by the Dutch Parliament, its purpose being to improve the quality of the land, to control the discharge of water and to reduce the danger of storm-floods.[88] Soon afterwards work began on the reclamation of the Wieringen polder and the building of the enclosing dam. The whole project was expected to be completed within 18 years. In fact, little progress had been made by 1914 and the undertaking is still unfinished in the 1980s.

From the Netherlands to Jutland the North Sea is bordered by coastal marshes at risk from flooding by rivers and from storm-surges. Small areas were reclaimed along the lower Weser[89] and Elbe and much larger areas made safe by the strengthening of sea defences. North of the Elbe mouth, where slow progress had been made in reclaiming the Ditmärschen since the Middle Ages, small areas were added during the nineteenth century.[90]

Climatic change in the nineteenth century

Climate and its fluctuations constitute the last element to be discussed in the environment of nineteenth-century Europe. The century began with two years of unparalleled severity and ended with a climate as benign as any in modern times. Between these extremes were short spells of severe weather as well as periods with mild winters and warm summers. The overall trend was from severe to less severe conditions.

A period of relatively harsh weather, which had begun in the sixteenth century, came to a close in the middle years of the nineteenth. The term commonly given to it, the 'Little Ice Age', greatly exaggerates its severity, but it was a period when average temperatures over much of Europe were a degree or two lower than those prevailing during the first part of the present century. Summer temperatures fluctuated greatly, but showed no clear, long-term trend. Severe winters characterised much of the Napoleonic period and culminated in the severe conditions of 1816–17, when summers were cool and wet and winters cold.[91] Conditions were worst in central Europe, but everywhere there were crop failures and acute distress. Although they occurred within the period of the so-called Little Ice Age, the immediate cause was almost certainly the eruption of a volcano, Mount Tomboro in the East Indies.[92] The dust which was spread through the atmosphere cut down the heat received from the sun and led to a drastic shift in the pattern of winds.[93] A relatively mild period followed in the 1820s, and was in turn succeeded by periods of acute severity in the 1830s and by severe winters again in the late 1840s. Acute conditions recurred in the 1850s and again in the late 1870s and 1880s. So severe was this prolonged cold spell of the mid-nineteenth century that there was a marked readvance of the Alpine glaciers which had been in retreat for many years.[94]

The extreme variability of winter temperatures is illustrated by the fact that the canal between Haarlem and Leyden in the Netherlands was frozen over for 89 days in 1830 and 58 days in 1838, but for only two in 1835 and none in 1834.[95] The amelioration of the winter climate appears to have been more marked in higher than in lower latitudes. In northern Sweden winter temperatures are said to have risen on average by 1 °C,[96] a fact of great significance in the northward extension of farming. In Vienna, by contrast, little change was recorded and in Rome there was even a slight deterioration.[97] It would appear that the climatic improvement of the second half of the nineteenth century was brought about by a tendency for the wind belts to take a more northerly course, especially in winter.

The year-to-year variations in temperature were greater than any long-term changes, and it was the former which had the greatest impact on the economy. 'Most short-term climatic oscillations', wrote Butzer, 'are of a regional nature, and temperature or precipitation anomalies in one area are compensated by excess or deficit in another.'[98] thus it can be expected that harvest failure in one area will be compensated by abundance elsewhere. This, broadly speaking, was the situation in the nineteenth century.

There appears to be general agreement on the course of climatic change during the last century and a half. There is no consensus at all on its economic impact. De Vries, for example, has emphasised that there is no linear correlation between winter temperatures and the quality of the ensuing harvest, and Le Roy Ladurie regarded any attempt at a synthesis between climatic history on the one hand and economic or agricultural on the other as premature.[99] Yet the influence of weather on harvest yields is undisputed. The problem is that there are so many climatic and meteorological variables that it is difficult to isolate them to examine their impact. A severe winter, such as those of 1816–17 or of 1846–7 might be disastrous for human beings, owing to the resulting increase in infections and pulmonary diseases; it would not necessarily be so for crops, especially if they were protected by a thermal blanket of snow. The first great German emigration of the nineteenth century followed the severe weather and bad harvests of 1816–17, and reached its peak after the disastrous crop failures of 1846–7.[100] A mild and very wet winter, such as that of 1878, might be far more destructive by washing out crops already sown and preventing spring ploughing.[101] A hot, dry summer, as in 1846, could be as disastrous as a cool, wet one, like that of 1816. Climatic historians tend to emphasise the qualities of summer and, above all, of winter.[102] Of scarcely less importance for agriculture, however, is a wet autumn which interrupts the harvest and prevents ploughing, or a cool and late spring which delays the germination of crops and nips the fruit buds.

'Most historians', wrote Post, 'have not only neglected striking meteorological abnormalities, but have even resisted or downgraded their significance in the face of obvious evidence to the contrary.'[103] How right he is. Meteorological conditions do not in themselves cause an economic crisis except perhaps on a very local scale, but reinforcing, as they did in 1816 and again in 1846, conditions which are already adverse, they can have profound historical consequences.

Conclusion

Except in fossil fuels, the natural resources of Europe were not exceptional, and much of the fuel endowment was restricted to two immense coal basins. Other resources were characterised more by their range and variety than by their volume. There were few metalliferous ores that did not occur in Europe, but deposits of most of them were either quickly exhausted or were too small for profitable exploitation. The resources of Europe were calculated more to encourage innovation than to sustain growth. Despite the richness of the coal resources grave fears were expressed in the later nineteenth century even of their imminent exhaustion, and a number of small coalfields were indeed effectively worked out.

The opening up of European coal resources was in part a reaction to the growing scarcity and high price of the preferred fuels, timber and charcoal. The development of a technology for using coal was the essential feature of the Industrial Revolution. It reshaped the industrial structure of the continent. There was in fact never any danger of a shortage of fossil fuel, despite the fears of W. S. Jevons and others.[104] Not so, however, with ferrous and non-ferrous metals. Ores of the types that were used at the beginning of the century were largely exhausted before its end, and technological advances consisted in discovering substitutes.[105] By and large high-phosphorus iron ores replaced those with a low phosphorus content, blende gradually replaced the more tractable calamine in the zinc smelter, and overseas resources were used to supplement and then replace the diminishing European production of most non-ferrous ores.

Only coal remained in the aggregate both abundant and cheap. Alternative sources of energy were developed more because they were convenient than because they were necessary. Hydroelectricity had of geological necessity a very different spatial distribution from that of coal. To this extent it supplemented rather than replaced it, permitting a whole new industrial system to develop in areas which under the earlier method of energy supply would have been incapable of industrialisation. Petroleum was, as far as Europe was concerned, a scarce and expensive energy source. If it had not been for the invention and development of the internal combustion engine, its importance might have remained small. Despite these supplementary sources of energy, the Industrial Revolution in Europe is inconceivable without the continent's rich reserves of coal. Other mineral resources were but incidental.

Europe's agricultural endowment was no more exceptional than its mineral resources. Areas of first-class soil were very restricted, and much of the continent had only a limited value for growing crops. A consequence was that over much of the area agriculture was incapable of yielding a surplus which could support urban development and the infrastructure of economic growth. It can only be said that Europe had enough highly localised areas of good soil for the development process to begin. Once started, however, the process of growth could become self-generating, as a developing technology was able to produce good crops from an indifferent environment. How this was achieved is examined in chapter 5.

The climate of Europe in the nineteenth century did not differ greatly from that of the late twentieth. There were times when temperature and precipitation departed significantly from the average, but the long-term trend, at least in the second half of the century, was towards warmer winters and higher average temperatures. Over most of Europe this change was too small to have any measurable influence on agriculture and wealth. Only on the northern frontier of human settlement was its influence felt. Here a slight increase in the length of the growing season made it possible to advance a significant distance into the wastes of Swedish Norrland and Finland.

3
The population of nineteenth-century Europe

The century between Waterloo and the outbreak of the First World War witnessed a more rapid growth in population than this continent had ever known before. Indeed, it was steeper than in either Africa or Asia at this time and was exceeded only by those lands which were receiving a vigorous inflow of immigrants. The precise rate of increase is difficult to establish, because early censuses are unreliable and in some parts of Europe no attempt was made to enumerate the population until late in the century.[1] In compiling the totals used here the figures given by B. R. Mitchell have mainly been used,[2] and extrapolations made where no totals are cited. In 1800 continental Europe, including Poland, which did not exist as a separate state, and Finland, but excluding Russia, had a population of about 123 millions. In 1910 one can say, with a somewhat greater pretension to accuracy, that the population within the same area had risen to 294 millions, an increase of 138 per cent.

The rate of population growth during the intervening years was neither regular or uniform. As might have been expected, it was relatively slow before 1820 – less than 7 per cent in each decade. Then, from 1820 to 1840 it was higher than during any other comparable period in the century. Thereafter the rate of growth fell and was at its lowest in the 1850s. The rate of growth again began to increase in the 1860s and continued until 1910. Variation in the rate of growth between countries was even more marked. Hungary, for which the statistics are amongst the least reliable, appears to have shown an increase between the beginning of the nineteenth century and the eve of the First World War of almost 320 per cent. Finland, with somewhat more reliable statistics, showed an increase of over 260 per cent. But if these two countries are ignored, we find that most European countries showed an increase which ranged from about 125 to somewhat over 200 per cent. Only the 'Latin' countries – France, Portugal, Spain and Italy – departed significantly from this range. Their increase was less than 100 per cent, except in Italy where the population just doubled. The countries of eastern and south-eastern Europe have been omitted from this tabulation because of the unsatisfactory nature of the few available statistics. It would appear, however, that their rates of increase were closer to that of Hungary than to the levels shown in the 'Latin' countries (fig. 3.1).

This unevenness both spatially and through time calls for an explanation. Yet the reasons are far from clear-cut. In Italy, Spain and Portugal, where growth

The population of nineteenth-century Europe 67

Table 3.1. *Estimated population in continental Europe, excluding Russia*

	Total (millions)	Percentage increase during previous decade
1800	123	—
1820	140	14 (over 20 years)
1830	156	11
1840	170	9.5
1850	184	7.8
1860	194	5.6
1870	210	7.5
1880	225	8.1
1890	244	8.5
1900	267	9.3
1910	294	10.0

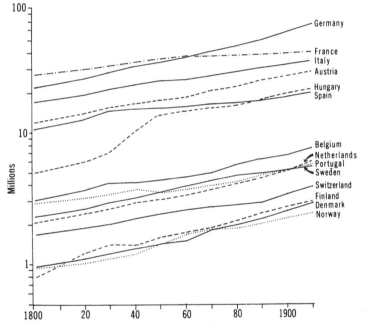

3.1 Growth of population in Europe, 1800–1910, by country

was at first very slow, industrialisation came late and had not made any great progress by the end of the period. On the other hand, France, which had the lowest rate of growth of any, was neither backward nor unindustrialised. Hungary, where there never really was an industrial revolution, showed one of the most rapid rates of growth to be found in nineteenth-century Europe.

Factors in population growth

The rate of population growth was, of course, dictated primarily by the relationship of birth-rate and death-rate, though influenced also by the level of

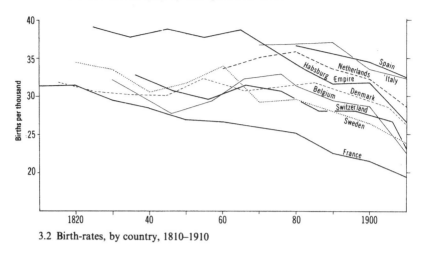

3.2 Birth-rates, by country, 1810–1910

in- and out-migration. There are good grounds for 'believing that falling mortality accounted for the observed rate of population growth in England',[3] and the same can be said of all other European countries. The improvement in the expectation of life was continuous, and was in fact sufficient in some countries to offset a marked decline in fertility (fig. 3.2). This was, in turn, due very largely to improvements in the level of public health, to a drastic reduction in the incidence of epidemic disease, and to improvements in diet and food supply. The level of population was further influenced by two other groups of factors which became increasingly significant as the century progressed: migration both within Europe and between Europe and the rest of the world, and the deliberate restriction of births. Neither was new, but their importance was probably not great before the later nineteenth century.

Birth- and death-rates

In most parts of Europe the birth-rates declined irregularly through the century. For the first half this decline was slight except in France and also in Sweden, where it recovered in the middle years of the century. The third quarter of the century was marked by a general stability in birth-rates with locally, as in Denmark and the Habsburg lands, a small drop. The last quarter, however, showed a general decline. In most countries the birth-rate fell to less than 30 per thousand, and even in Spain, Italy and probably also Portugal, the decline became manifest. After 1900 this trend accelerated. In all countries illustrated in fig. 3.2, except Spain and Italy, it fell to below 30 per thousand, and in several below 25. The birth-rate in France began to fall about 1820 and declined continuously, thus anticipating by about half a century the trends that were later to be demonstrated elsewhere in western and central Europe.

More significant than the crude birth-rate are measures of 'total fertility', defined as the average family size, and of the net reproduction rate. The latter is a measure of the number of female children born on average to each woman. The former was already declining in France when the century began,[4] but

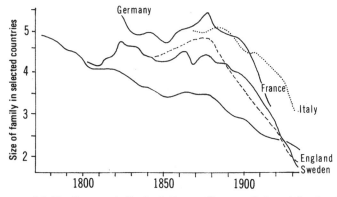

3.3 Fertility rates in England, France, Germany, Italy and Sweden, 1780–1930

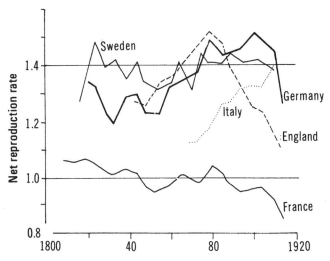

3.4 Net reproduction rates in England, France, Germany and Italy, 1800–1920

elsewhere it fell significantly only after 1880 (fig. 3.3). The net reproduction rate fluctuated greatly, but a downward trend did not become conspicuous, except in France and Great Britain, much before the end of the century (fig. 3.4).[5]

The pattern of death-rates through the century shows a wider spread between the highest rates and the lowest, and also a greater degree of consistency. The decline began early – by 1830 or soon afterwards – and continued until the last quarter of the century, when it became more precipitate. The short-term increase in mortality in Switzerland in the 1860s was due to severe epidemics of cholera and typhoid,[6] and the same may have been the case in Belgium. A comparison of birth- and death-rate graphs in figs. 3.2, 3.5 and 3.6 shows a growing divergence through much of the century, with death-rates dropping earlier and more rapidly than birth-rates. Only in the last two decades before the First World War, when birth-rates fell more sharply, did the two graphs again approach one another.

The demographic history of Europe during the century can thus be divided

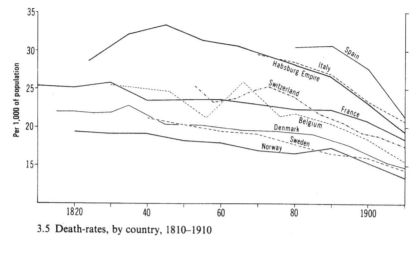

3.5 Death-rates, by country, 1810–1910

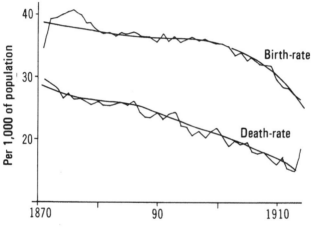

3.6 Crude birth- and death-rates in Germany, 1870–1915

into three phases, with the date of the transition from one to the next varying considerably between countries. Until the mid-century the demographic picture was very broadly similar to that of the eighteenth century. Death-rates remained high, though modified to a small extent by progress in medicine and public health. Major epidemics continued into the 1850s, but the last great *crise de subsistance* was, it can be argued, that of 1846–7. The birth-rate had already fallen significantly in France, and there was a temporary drop in some other countries in the 1840s, which can also in all probability be attributed to the epidemics and food shortages of this period.[7]

The period from about 1850 until about 1880 marked a transition from a pre-industrial to a modern pattern. The downward trend of the death-rate, despite the high mortality which followed the last major epidemics of the century, became accentuated, but birth-rates overall showed a slight downward trend. The resulting rate of population increase was marginally less than during the first half of the century.

The population of nineteenth-century Europe

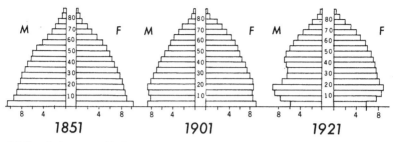

3.7 Population pyramids for France

The years from about 1880 until the outbreak of the First World War were a period of more rapid change. In all countries the death-rate tumbled from about 1880 or shortly before. Major epidemics had disappeared; living conditions, especially in the larger cities, had greatly improved, and there were no longer recurring food crises though it cannot be said that diet was everywhere adequate. At the same time, in the 1880s and 1890s, the birth-rate dropped sharply. The size of the completed family, which had stood at more than four in all the developed countries except France, fell on average by at least one child. Yet the decline in the birth-rate was not sufficient, even in France, to compensate for the fall in the death-rate. Overall population growth was greater in the aggregate than ever before, and in percentage terms was equalled only by that in the 'pre-industrial' years before 1840.

These fluctuations in birth- and death-rates had as their inevitable consequence important changes in the age-structure of the population. The decline in the death-rate meant that people were living longer and that the older age groups became proportionately larger. At the same time the decline in the birth-rate reduced the size of the younger cohorts. The population on average was becoming older. This is reflected in the age-pyramid of national populations (fig. 3.7). The population pyramid for France in 1851 already begins to show in its 'beehive' shape the contracted age groups that had been born during the middle years of the century. By the end of the century these groups had become relatively even smaller. The same pattern was apparent in every other European country, though the chronology of the change would have come a decade or two later in most of them.

Diet and food supply

The fall in the birth-rate could, by and large, have been caused only by the human decision to reduce the size of the family. The decline in the death-rate, on the other hand, resulted from external factors, from a reduction in the incidence of disease and the improvement in diet. Progress in these respects was slow and halting, which explains why the death-rate fell so slowly. In general, it remained significantly higher in the towns, where living conditions were commonly worse, than in the countryside. It is, however, impossible to separate statistically the consequences, for population growth, of epidemic disease, inadequate diet and poor housing and physical conditions.

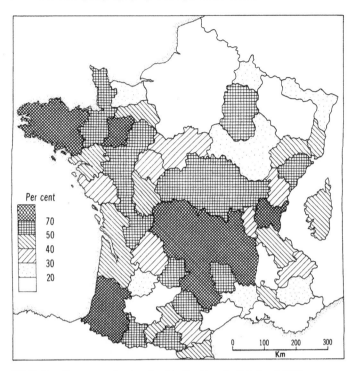

3.8 Ratio of coarse grains to wheat in the French diet about 1830

An increase of more than 100 per cent in total population was made possible by a more than proportionate increase in the available food supply. This derived in part, especially during the later years of the century, from imports from outside the European area, but in the main it was provided by the increased productivity of agriculture within Europe.

During the first half of the nineteenth century the rural population continued to live at much the same standard as it had done in the eighteenth. Its food supply was in large measure generated locally, and remained subject to the same vagaries of weather and harvest. Subsistence crises had occurred at irregular but none the less frequent intervals. The last crisis of continental proportions was that of 1816–17,[8] but there were less serious and more localised crises in 1830[9] and 1846–7,[10] and even in the 1850s poor harvests caused widespread distress and increased mortality. Not only were crop failures less frequent and less catastrophic in the second half of the century, but the improved means of transport and the increased government efficiency permitted food supplies to be moved about the continent with greater speed and efficacy.

Nevertheless, diets remained severely limited in variety and in food value. The bread-grains, varying from one region to another, continued to be the basis of the diet, supplemented in many areas by the potato. It was not only in Ireland that the latter became a mainstay of the population. It was of immense importance in Scandinavia in supporting an increasing population, and Drake claimed that it was vital to the population growth in Norway.[11] In Sweden it was used for distilling, thus sparing cereals for human and animal consumption.[12] The food crisis in Flanders in 1846–7, and the resulting high level of mortality were a consequence of, more than anything else, the potato blight.[13] A diet of bread-grains, whether baked into bread or eaten as a porridge or soup, and potatoes was supplemented inadequately by green vegetables, milk, cheese, meat and, in coastal areas, fish (fig. 3.8). Northern France was reckoned to be an area of heavy meat consumption, but even here, in the middle years of the century, consumption in the Paris Basin and Normandy amounted on average only to about 30 kg a year, or a little more than 500 grammes a week, and throughout central France it was less than half of this.[14] Elsewhere the consumption of protein was appreciably less than in France. In eastern Europe, it is claimed, the increase in food production failed to keep pace with that of population before about 1860, and the level of welfare declined.[15] The same has been said of Portugal.[16] It is probable that all parts of Europe where agriculture failed to make vigorous progress suffered in the same way. In the Nivernais, by no means one of the poorer regions of France, the diet did not begin to improve until after about 1850, when the consumption of proteins increased and wine came to be drunk more widely by the peasants.[17]

The small towns were heavily dependent on their rural environments for food supply, and shared the fortunes of the latter. Larger towns, however, faced more serious difficulties (see p. 157). Although the middle-class citizens lived well, the bulk of the urban population probably fed less well than the rural. Whereas a large proportion of the rural population produced its own food and to this extent was not vulnerable to price fluctuations, the urban was wholly dependent on the market. Small fluctuations in supply were reflected in large price changes, and in times of scarcity food was priced beyond the reach of the urban proletariat. This was well shown in the crisis in the Low Countries of 1846–7. A study of food consumption at Ghent in the mid-nineteenth century showed a diet dominated by the bread-grains.[18] Out of an average of 2,435 calories, 1,479 were supplied by cereals and potatoes, and 428 by drink, almost certainly beer brewed from grain. Nearly 80 per cent of the calorie intake thus derived from cereals, and only about 20 per cent from animal proteins. Vegetables were of negligible importance, reflecting in all probability the difficulty experienced in supplying a large city with perishable foodstuffs. There had been, it was claimed, a decline in the level of urban consumption, and this was not reversed until after 1875.

For the great majority of the population during the nineteenth century the available food supply was sometimes inadequate and almost always ill-balanced. For many it was short of protein, and for most deficient in certain vitamins, notably A, D and probably also B_1, B_2 and C. Much, however, would have depended upon the intake of green vegetables and dairy produce, which were likely to have been deficient at least during winter and in cities during much of the year. These dietary inadequacies fell a long way short of famine conditions. The latter occurred locally in 1816–17 and again in 1846–7 as a result of harvest failure, and in eastern Europe and the Balkans as a consequence of insurrection and war. The problem was rather that deficiencies in diet weakened the resistance of the poorest segment of the population and exposed it to epidemic and other diseases.

The effects of inadequate or unsatisfactory diet were reinforced by the poor physical conditions under which very many lived. There were few growing cities which did not have vast complexes of insanitary and ill-lighted alleys and courts where the poor lived in the utmost squalor (see p. 150). Those least able to resist infection were most exposed to it. This is shown in the higher death-rate and the lower expectation of life of urban dwellers compared with the rural population. As a general rule, the rural population had a more assured food supply than the urban and were less prone to epidemic disease because, being spread more thinly on the ground, they were less at risk of infection.

Public health, like the quality of the diet, improved only slowly through the century. When the century began, all the well-known epidemic diseases, including even plague, were present, and increasing mobility on the one hand and the crowding of people into large cities on the other meant that disease could be spread more readily. Add to this the inadequate means of sewage disposal and of water and food supply, and we can fairly expect a deterioration rather than an improvement in public health. This was, indeed, the situation in the rapidly growing industrial cities, and it was not until late in the century that developing medical science on the one hand and civil engineering on the other revolutionised the situation and contributed to a significant fall in the death-rate.

Medical diagnosis was an uncertain skill in the nineteenth century, and in many instances the causes given for death cannot necessarily be accepted. There was much confusion between the respiratory complaints, and the distinction was not always accurately drawn between scarlatina, diphtheria and other fevers. Nevertheless, the broad picture is clear. There was no mistaking epidemic diseases such as smallpox, cholera and typhus, which were as frequent in their occurrence as ever they had been, and tuberculosis, pulmonary complaints and dysentery were all too familiar. Each claimed its toll every year until late in the century, when medical science and public hygiene at last began to triumph over some of them.

One disease, however, had disappeared from most of Europe, the bubonic plague. Its last major outbreak in the west had been in Provence in 1720, but it remained endemic in the Ottoman Empire, and there was a severe outbreak in the Balkans in 1812–16. It was diffused by the rat and carried to human beings by the flea, though the aetiology of the disease was not known until the present century. Nevertheless, the Austrian government, like the Venetian before it, had learned from long experience that the movement of the plague could be controlled, provided a rigorous sanitary cordon was thrown around an infected area. The quarantine arrangements devised by the Austrians left much to be desired, and rats could have passed through them wherever they wished.[19] Nevertheless, they worked. The plague never crossed the Military Frontier, and, in 1871, the quarantine control was ended.

Another disease which declined in importance during the century was malaria. Its vector was the anopheles mosquito, which bred in the still, shallow water of lakes and marshes. Malaria was common throughout Europe, except Scandinavia, and was early in the century a major cause of death. A report, probably prepared for Stanislas Potocki, placed it first amongst illnesses in Poland in the 1820s.[20] The marshes of the Roman Campagna – the Pontine

Marshes – had become infected with malaria, and ill-conceived attempts to reclaim them early in the century served only to extend the lakes and intensify the danger from the disease.[21] The 'Roman fever' wrought heavy casualties on the French army in Rome between 1860 and 1870 and was eradicated close to the city only when the lakes had been drained and their soil colonised by the expanding suburbs. In Flanders, where countless lakes remained, the disease was endemic, breaking out at intervals in epidemic proportions. In the second half of the century the danger of *paludisme* was reduced by draining the *meers*, and there was no major outbreak after 1857–8.[22] In France also, malaria was widespread wherever marshes and lakes were numerous: the Loire valley, Sologne, Dombes and along the Mediterranean coast. The close association of malaria with marshes was recognised, though the agency of the mosquito was not known until the present century. The disease was attributed to a 'miasma' which rose from the wet land. There were numerous epidemics, localised because the vector could not live and breed away from lakes, but they gradually disappeared as the lakes and marshes were drained.[23] In southern Italy the incidence of malaria actually increased because deforestation and the resulting soil erosion led to the formation of coastal and valley marshes.

Smallpox, which had first become significant in the sixteenth century, continued to afflict the population of Europe throughout the century. Its ravages had, however, been diminished first by the use of inoculation and then by vaccination, which had been introduced at the end of the eighteenth century. There can be no question of the efficacy of Jenner's method of vaccination, and several governments in the west made it obligatory, but smallpox continued its ravages in eastern and southern Europe. Before the nineteenth century it had been one of the more significant causes of death, and its elimination or at least reduction was a major factor in the increase of population during the nineteenth century.[24] The conquest of smallpox was aided by the fact that, alone amongst the great epidemic diseases of the nineteenth and earlier centuries, it required no vector. Its virus was not carried in the bloodstream of a rat or the abdomen of a mosquito or a louse; nor could it survive in sewage or polluted water. It was a person-to-person disease, communicated by the breath, and isolation of a patient was an adequate means of preventing its spread.

This was not so with other diseases which became more menacing during the early years of the nineteenth century, as people crowded into cities and conditions became increasingly congested and insanitary. Foremost amongst them and certainly the least tractable were typhus and cholera. Typhus was essentially a disease of the Industrial Revolution. Its vector was the body louse, which carried the bacterium from person to person. A prerequisite was dirty bodies, unwashed clothing, crowding and congestion, which permitted the louse to breed and to travel from person to person. Liability to the disease was increased by undernourishment and deprivation. It had at times flourished in gaols and in the armed forces – hence its alternative names of 'gaol' and 'trench' fever. In nineteenth-century Europe every subsistence crisis, such as those of 1816–17 and of 1846–7, was accompanied by an epidemic of typhus. It swept through Flanders in 1846–8,[25] and is said, along with smallpox, to have been a major cause of the increased mortality in 1816–17. 'Wherever typhus appeared

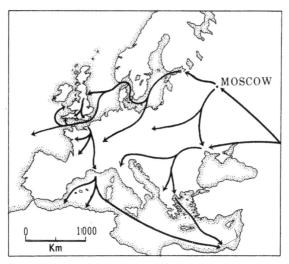

3.9 Diffusion of cholera in Europe, 1831–3

in the post-war years, the contagion had been preceded by hunger and distress', wrote Post,[26] adding that 'no large district in the Italian peninsula escaped either famine or typhus'.

Cholera was for most of Europe a new disease. It had long been known in southern Asia, but the slowness and infrequency of contacts had kept it at bay from Europe. But in 1831–2 it arrived by way of the caravan routes of Central Asia and struck a virgin population with the virulence of the Black Death, which had travelled the same route nearly five centuries earlier (fig. 3.9).[27] Russian troop movements occasioned by the Polish Rising of 1830 contributed to its spread. It was carried by ship to the ports of north-western Europe, whence it was diffused overland to all parts of the continent and to the Middle East.[28] In France it spread along the rivers, and was prevalent amongst the washerwomen who used them. It was the first great epidemic of the post-Napoleon era, and, furthermore, its symptoms were unfamiliar and the resulting mortality immense. In Paris the death-rate was doubled, and one in 20 of the population died in the poorer quarters of the city, where it showed a high correlation with the degree of crowding.[29] At Lille, Roubaix and Tourcoing there was one case of cholera for every 42 of the population, and here too the heaviest incidence was concentrated in areas of extreme congestion and deprivation. It was especially severe amongst the lace-makers, many of whom lived under the most squalid conditions. Well-to-do quarters of the town enjoyed an almost complete immunity.[30] In rural areas its incidence was sporadic, and many districts escaped completely.[31] Port-towns were particularly vulnerable, and it became apparent that conditions on ships did much to encourage the multiplication of the bacterium. For this reason attempts were made to use a quarantine system.[32] Cholera entered by the ports of the Balkan peninsula, where it was sometimes mistaken for plague, but its spread was checked by the sparse population and lack of movement in much of the region.[33]

Although the association of cholera with overcrowded and insanitary areas of the cities quickly became apparent, its mode of transference went undetected for more than 20 years. In the meantime there were recurrent localised outbreaks in many parts of Europe and more serious epidemics in 1832, the 1840s, 1867, 1873 and 1883. Not until 1854 did John Snow establish a correlation between an outbreak of that year and those who used the Broad Street pump in Soho, London.[34] The pump water had in fact been contaminated by sewage. It was sealed and this localised epidemic disappeared without any recurrence. It was not, however, until 1883 that the bacterium was isolated and the cause of the disease demonstrated beyond doubt. The bacterium, in fact, multiplied in the human intestines and was passed with excreta into the sewage. Any failure – and they were countless – to keep the water supply separate from sewage disposal provided the means for its spread. The remedy was simple: on the one hand, to create a supply of pure water piped from areas free of contamination, and, on the other, to replace the countless septic tanks and soak-aways with sewers built of masonry or of iron pipes. A supply of pure water, wrote Thuillier, was a privilege reserved for the few, and its absence was until the end of the century an important factor in the recurring epidemics.[35] The high cost of such projects prevented the immediate implementation of remedial measures, but during the half-century before the First World War progress was made in all large cities, and in many small ones in creating the two essentials of urban living: a safe system of sewage disposal and a piped water supply (see p. 144). Concurrently the ravages of cholera were reduced, and its significance was small by 1914.

The psychological impact of cholera was immense, comparable with that of the bubonic plague in the later Middle Ages. Fear of its recurrence probably did more than any other factor to encourage the public health movement in the later years of the century.[36] Improvements made in living conditions, whether or not they were motivated by fear of cholera, went far towards reducing the impact of other diseases. Typhoid fever, scarlatina, dysentery and other abdominal ailments were also spread mainly by infected water and food supplies, and their incidence became less severe as the century wore on. Urban conditions – crowding, factory smoke, lack of ventilation, damp and squalor – also helped to spread tuberculosis, especially pulmonary, throughout much of the century. It was *une maladie 'sociale' par excellence*,[37] nurtured in crowded, damp and ill-ventilated rooms, and passed from person to person on the breath or in clothing. It was common amongst lace-makers and domestic spinners and weavers, and took a very heavy toll throughout the century.

The incidence of disease, especially of epidemic disease, diminished slowly during the century owing to a succession of small incremental changes in the environment. Indeed, it had begun to do so even before the aetiology of the diseases themselves had been studied and analysed. It is impossible to correlate disease mortality with wealth and social standing with any precision; it is clear, however, that the relationship was close. Severe epidemics were mostly restricted to working-class areas, and were always far more serious in towns than in the countryside.[38] In Paris mortality rates were highest in the poorest quarters of the city, and there is confirmatory evidence from other French cities.[39]

Industrial towns fared far worse than others. The death-rate at Le Creusot,

Table 3.2. *Death-rate per thousand*

	Le Creusot	All France
1851–55	33.0	24.1
1856–60	31.5	23.8
1861–65	31.7	22.9

far from being one of the worst of industrial towns, was greatly in excess of the average for France (table 3.2).[40]

The death-rate amongst children, especially during their first year of life, was high and declined only very slowly towards the end of the century. In France it was consistently above 15 per cent of live births until about 1900,[41] and in much of Germany was even higher.[42]

It is apparent that without high mortality from epidemic disease the population would have risen far more rapidly in the nineteenth century than in fact it did. The maximum rate of increase was less than 1 per cent a year, and during the middle years was not much above 0.5 per cent. If the major infections and contagious diseases could have been eliminated this might have raised the rate of population increase in a level close to that of some Third World countries today. But, if that had been the case, it is likely that social pressures to reduce the rate of growth would have operated earlier.

Restrictions on births

The birth-rate began to drop significantly in the 1860s, though there is evidence that the level of conceptions had fallen during earlier periods of high grain prices.[43] This may have been due directly to undernourishment, but is equally likely to have been the consequence of the decision to reduce the number of births. But the decline in the later years of the nineteenth century can only have been due to the human decision to have smaller families. The lower birth-rate was not an automatic reaction to a diminished death-rate; it occurred only after the latter had become clearly apparent and it was socially acceptable to have fewer children. This trend towards smaller families first became significant in France soon after the end of the Napoleonic Wars, though it had been apparent during the later years of the Ancien Régime.[44] Frédéric Le Play attributed the decline to the testamentary provisions of the Napoleonic Code, which prescribed the division of an inheritance between direct heirs. The peasant, it was said, anxious to maintain the integrity of his holding, deliberately restricted the number of those who would inherit it. There can be little doubt that amongst the peasantry this was a powerful motive for keeping the birth-rate low, even though they learned to evade the testamentary rules.[45] But it was almost certainly only part of the explanation, and it could not have been an important consideration with the non-agricultural population. A significant role must also have been played by the aspirations and ambitions of the middle class for material comforts, education and leisure, and what was good for the middle classes would in the course of a time be seen to be good for the lower. In this

Migration

The only factor influencing the size of the population, apart from the ratio of births to deaths, was migration. There had always been migration within Europe, and from the beginning of modern times external migration became increasingly important. Internal migration merely transferred people from one place or one region to another. It made little difference to aggregate population, though it commonly led to a postponement of marriage and to smaller completed families. External migration, on the other hand, provided that it was permanent, meant an absolute loss of population. It might have been compensated by movement into Europe from without, but this, before 1914, was of little importance.

Much of the internal migration was over relatively short distances, from rural areas to the growing industrial town. Sometimes it occurred in two or more stages, first to a small town and then to a large. It was probably as much a movement of females as of males. Large number of young girls from the villages sought 'positions' with urban families, many of them returning to their villages to marry.

The pattern of migration was complex. It included the movement of land workers to the 'promised land' of the big industrial city, and also a return flow of those who failed to adapt to urban conditions; it included the millions of temporary or seasonal workers who gathered the harvest, built the canals and railways and put up buildings, in every growing city.[47] It is extremely difficult to quantify the extent of internal migration, in part because permanent movement cannot easily be separated from temporary or seasonal. It might be thought that statistics of overseas migration would be more adequate if only because immigration authorities kept records. But even here the picture is complicated not only by large numbers who returned to their country of origin but also by the *seasonal* migration to the New World which developed with the coming of steamships on the North Atlantic route. Of the two forms of migration, internal and overseas, the former was in statistical terms by far the more important. The numbers moving to areas of concentration within Europe were far greater than those who went in search of the wide, open spaces of the New World.

It is posible, within Europe, to distinguish only the more important directions of migration, and, since these were by and large within the limits of individual countries, it is very difficult indeed to measure them. France illustrates the widespread trend of migration from rural areas to the cities and industrial regions. It had begun long before the nineteenth century, but increased in volume after 1815. Nevertheless, rural population and specifically population living by agriculture continued to increase until the middle of the century.[48] Thereafter the flight from the land became more rapid (fig. 3.10). The pattern of migration varied from region to region. At first it was mainly from the mountainous areas and those with poor soil and unrewarding agriculture, such as the Pyrenees, the Central Massif and the French Alps. Even within regions generally judged to

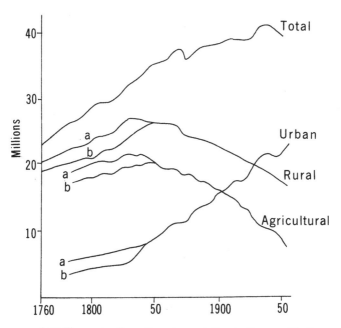

3.10 The contraction of rural population in France with the growth of towns. The difference between the 'rural' and the 'urban' graphs is a rough measure of internal migration; a = maximum estimate, b = minimum estimate

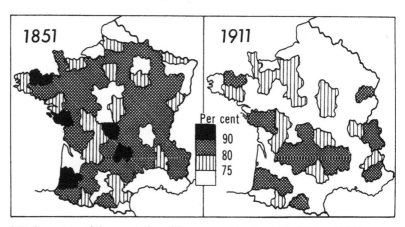

3.11 Percentage of the population of France considered rural in 1851 and 1911

be fertile and productive there were less good areas from which migration became significant during the early years of the century.[49] In Savoy population growth was virtually restricted to the shores of Lake Geneva and the lowlands near the Rhône. Almost the whole of the mountainous area was depopulated.[50] Much of its population went to swell that of the city of Lyons, and of the industrial area of *dépt* Loire,[51] where it met the flow of migrants from the Central Massif (fig. 4.18). In the south-west also, the peak of migration from the Pyrenean foothills and the barren Causses was reached in the middle years of the century.[52] The heaviest migration appears, however, to have been from the Central Massif which supplied much of the immigrant population of Paris.

A change came over French migration after the Franco-Prussian War. It ceased to be only from the areas of poor soil and backward agriculture. It became significant also from the fertile areas of the Paris Basin and the north. The basic reason was that agriculture was becoming more mechanised and less labour-intensive, and that with every improvement workers were forced off the land and into the cities. It is a curious fact, however, that rural out-migration appears to have been least significant in those areas, such as Lower Normandy, where craft industries provided an important alternative employment.[53]

Permanent migration was preceded and accompanied by an intensive seasonal movement. In part this consisted of agricultural workers attracted by the grain harvest in the north or the wine harvest in Languedoc; in part also by the construction industry in the towns and on the railways. A large part of this yearly movement was from the Central Massif.[54] *Dépt* Creuse alone supplied between 30,000 and 45,000 seasonal workers, and overall there were said to have been 878,600 in 1852, of whom nearly a half went to the Paris Basin or northern France.[55]

Despite the superabundance of labour in some rural areas, France was by the end of the century short of labour in many parts. There were certain occupations, including coal-mining, to which the French peasant did not readily turn. This left the field free for immigrants. As early as 1844 complaint was made that there were many foreigners working in the coal mines of *dépt* Loire.[56] By 1900 there were more than 6,000, most of them Italian. As early as 1851 there were nearly 400,000 foreign workers in France.[57] From this date the foreign population increased steadily, from less than 1 per cent of the total at the mid-century to 2.8 per cent in 1911. Most were Belgians and Italians, with Spaniards in the south-west, Germans in the east, and a large population of mixed origin in Paris.[58] It is difficult to show how many of the foreign population had made France their permanent home. Certainly most of the Italians in Nice and more widely in Provence had no intention of returning to Italy.[59]

Overseas migration was of small importance in France, since overall there was no population pressure. Available statistics are inadequate, but the total before 1914 was a great deal less than half a million, many of whom went to the French dependencies in North Africa, especially after the *Phylloxera* epidemic. The period of most intensive emigration appears to have been the last two decades of the nineteenth century.[60] Much of this emigration was from the Basque region, the Central Massif and the Pyrenees, areas from which out-migration was in any case most vigorous.[61]

The second major example of large-scale internal migration was provided by Germany, where internal migration was on an even larger scale than in France (fig. 3.12). The birth-rate was higher and rural conditions, especially in the east and south, were worse. In the former the end of serfdom had left a mass of peasants with little land and no recourse but to work on the estates of the nobility. In the latter, a system of partible inheritance had created a plethora of micro-holdings, each too small to support a farming family.[62] Migration began early, before in fact the development of modern industry had created a

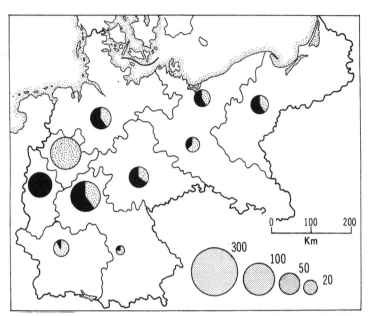

3.12 Migration within Germany to the industrial provinces of Rhineland and Westphalia, 1880 (in thousands). For key see fig. 3.13

demand for labour. In consequence overseas migration played a far greater role in Germany than it did in France (see p. 87). Almost 800,000 are said to have migrated overseas between 1815 and 1849,[63] and of these more than 90 per cent came from the provinces west of the Elbe. The first impulse came from the bad harvests and consequent distress of 1816–17. Indeed, the rate of overseas migration correlates closely with the price of grain. The eastern provinces, which were to become the great reservoir of German labour, were still able at this time to absorb their increasing population. A wave of migration followed the failure of the risings of 1848. It is commonly assumed that it was made up largely of disillusioned or disappointed liberals. Many, however, came from areas which had not been deeply involved in the political movements, and their motivation was far more likely to have been poverty, harvest failure in 1846–7, and the inadequacy of their holdings.[64] Harvests continued to be poor in south-west Germany, and the agricultural crisis did not end until 1854. The volume of emigration seems to have varied inversely with the size of holdings.[65] In the second half of the century the expansion of industry and of tertiary activities in southern Germany led to a gradual falling-off of migration (fig. 3.14).[66]

The next wave of overseas migration was from central rather than western Germany and occupied the 1860–70s. By this date, however, the industrial towns were beginning to grow and the proportion of migrants making for Bremen and the ships which sailed to America was proportionately less. After 1880 the third and final wave of migration before the First World War gradually built up. It was overwhelmingly from the eastern provinces, and was directed mainly to greater Berlin and the rapidly growing industrial cities of the Lower Rhineland. The reasons for the heavy migration were, in part, the high birth-rate and lack

The population of nineteenth-century Europe 83

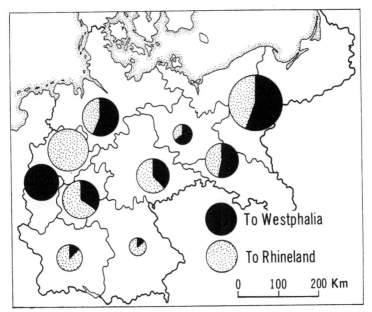

3.13 Migration within Germany to Rhineland and Westphalia, 1907. Scale as in Fig. 3.12

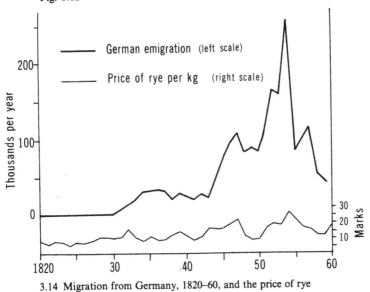

3.14 Migration from Germany, 1820–60, and the price of rye

of economic opportunity in the east, and, more important, the collapse of grain prices in the face of growing imports from the New World (see p. 191). Migration to Westphalia and the Lower Rhineland provides an index of this movement (table 3.3).[67]

In figs 3.12 and 3.13, which shows the origin of immigrants into Lower Rhineland-Westphalia in respectively 1880 and 1907, the growing predominance of the German East is apparent. Migration was, in fact, so great that it left a

Table 3.3. *Immigrants from East Germany to the Rhineland and Westphalia as a percentage of total immigrants*

	Lower Rhineland		Westphalia	
	Number	Per cent	Number	Per cent
1880	32,147	12.4	28,192	15.1
1890	67,673	17.5	89,693	29.2
1900	153,411	24.1	252,573	42.2
1907	211,599	27.3	301,080	44.9

serious labour shortage on the estates of Prussia and in the province of Posen. This was in turn filled by Poles who came across the boundary from Russian-held Poland.[68] To them even work on the great estates of Prussia was likely to bring rising standards. Many of the migrants from Posen, and even from west and east Prussia were Polish-speaking, and formed colonies in Saxony, Westphalia and many other parts of Germany. The migrants were predominantly male. They tended to marry German women and in time were assimilated to the German population, only their Polish names showing their origin. By 1910 there were said to have been more than 250,000 Poles in the Ruhr area, but this may underestimate their numbers, since many who also spoke German had merged with the local population.[69]

Internal migration was on a smaller scale in the rest of Europe, in part because there were fewer centres of attraction; but partly also because local resources were more adequate. There was a continuing out-migration from the mountains which ringed Bohemia and from the Carpathians of Slovakia and Romania.[70] Most migrants made their way to the large cities of Prague, Vienna and Budapest, or to developing industrial regions like northern Moravia.[71] Some even migrated to Bosnia, where a rudimentary industrial development offered them some opportunities.[72] Most were poor and illiterate, and went to swell the urban proletariat. In Italy peasants migrated from rural areas to the towns, some of which, such as Naples, Palermo and Bari, achieved an immense size without at the same time developing commensurate urban function. They also moved from the south of Italy to the north. Opportunities for migrants within Italy were fewer than in Germany and France, and the urge to go overseas was greater. The scale of migration increased considerably after about 1860, when the peninsula became politically united and the population increased more sharply. To begin with, migration was heaviest from northern Italy (fig. 3.16), but at the beginning of the present century the number of migrants from southern Italy and the islands began to exceed that from the north. The reasons were not far to seek.[73] Industrial employment was increasing in northern Italy, while in the south, physical conditions were actually deteriorating. Soil erosion had become severe and resulting marshes in the valleys and along the coast contributed to the spread of malaria. Farm holdings were small and fragmented; and the whole social structure of agriculture was backward and feudal. Heavy migration came first from the overcrowded province of Campania and the

The population of nineteenth-century Europe

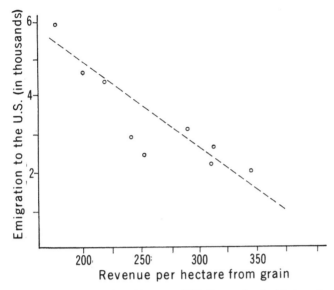

3.15 Migration from Hungary, 1907–13, correlated with the price of grain. This shows how dependent migration continued to be on the quality of the harvest

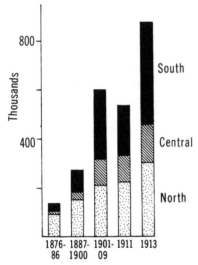

3.16 Migration from Italy, 1876–1913. Note the increasing importance of migration from southern Italy and Sicily

mountains of Basilicata and Calabria, and later from Sicily. On the eve of the First World War Sicily had become by far the most important source of migrants and, it is said, 5 per cent of the population of some districts migrated.[74]

Not all of this migration was permanent. Many migrated as far as South America, only to return after a few years. In many instances the purpose was to earn enough money to buy and equip a farm back home. A problem for the poor peasant of the south, however, was to accumulate enough money for the boat passage from Naples or Palermo to the New World. For this reason many migrated into neighbouring European countries, especially France, Switzerland and the Habsburg Empire. In fact this, rather than overseas, was their primary destination until late in the century. In 1881 there were about 360,000 Italians living within Europe, but outside Italy, of whom two-thirds were in France. By 1891 this had risen almost to 450,000. Many, perhaps most, considered themselves to be temporary migrants, and had every hope of returning to their homelands.[75]

Migration both within and from the Iberian peninsula was never on the scale of that in Italy, though agricultural and demographic conditions were not dissimilar. Internal movement was small before the second half of the century, not because conditions of extreme distress did not exist, but rather because there was nowhere to go. After about 1850 the growth of the Catalan textile industry attracted labour from neighbouring provinces, and a decade or two later, the Basque iron industry had a similar effect. The only other pole of attraction was Madrid, where, however, growth was appreciably less rapid than in many other European capitals.[76] Overseas migration increased after about 1866, and reached its maximum in the decade before the First World War. The crowded lands with small, fragmented holdings in northern and north-western Spain were the chief source of migrants, a large part of whom went to Latin America. At the same time, however, the Spanish colony in France grew steadily, reaching a million in 1880 and 1,133,000 in 1911.[77]

There was an intense movement of peoples within the Balkan peninsula. There was nothing new in this. Revolt and war had always been accompanied by migration and flight. Every unsuccessful rising against the Turks was followed by savage repression and a stream of refugees which moved northward to escape Ottoman rule. In addition to such periodic waves of migration there was a steady outflow from the plains of Macedonia and Thrace towards Serbia and the Danube plains of Christian *rayas*, ejected from their lands by Ottoman officials or no longer able to tolerate their abuses. At the same time there was an outflow, probably larger in total numbers, from the barren Karst plateau of Bosnia, Montenegro and Raška. This was, at best, a marginal environment. The depredations of Turkish soldiers or drought and the bad harvest which followed were enough to drive peasants from the land. In 1890, for example, famine in Montenegro was followed by the emigration of 10,000 of its inhabitants as well as of many others from Hercegovina.[78] Remains of the fields and hamlets abandoned during the century are visible all over the karstlands of the Dinaric region. In Greece also there was a steady migration through at least the latter half of the nineteenth century from the rural areas. This went in part to Athens and other developing towns, but also in part overseas.[79]

The Scandinavian countries, wrote Hovde,[80] 'in proportion to their populations...have contributed more heavily to the stream of migration than almost any other region in Europe'. It was strongly influenced by the fact that Scandinavia had a marginal environment, and that weather fluctuations which in western or central Europe would have caused merely short-term scarcities, could here become disastrous. Migration was directly related to recurring periods of hardship and suffering, and in retrospect may not appear to have been

Table 3.4. *Overseas migration from Europe*

	Total migration in thousands	As a percentage of total population in 1900
Austria-Hungary	4,878	10.7
Belgium	172	2.6
Denmark	349	13.9
Finland	342	12.7
France	497	1.3
Germany	4,533	8.0
Italy	9,474	29.2
Netherlands	207	3.9
Norway	804	36.5
Portugal	1,633	32.7
Spain	4,314	23.2
Sweden	1,145	22.5
Switzerland	307	9.3

wholly necessary. 'It does not seem', wrote Jensen,[81] 'that the great emigration from the Scandinavian countries was an economic necessity.' Most went to the United States. In Finland there was a relatively small movement to the towns, notably Helsinki and Turku, and a relatively large migration overseas.[82] Eastern Finland lay close to St Petersburg, and came within that city's sphere of influence. Farmers sold their produce in St Petersburg, and itinerant Finns worked there. Indeed, much of the income of the Finnish province of Karelia derived from hauling goods into Russia. There thus came to be a large Finnish colony in the Russian capital, most deriving from the harsh region of Karelia and many hoping eventually to return to their native land.[83]

It is difficult to form any precise estimate of the numbers who migrated from Europe before the First World War and impossible to assess the scale of internal migration. It is certain, however, that the latter always exceeded the former.[84] It is claimed that about 55 millions migrated from Europe, including the British Isles and Russia, before 1924, or about 29 millions if the latter are excluded (table 3.4). Such figures exclude movement from Russian-held Poland and all migration before 1846, for which there are no valid statistics. Furthermore, they take no account of returning migrants. Net migration is likely to have been appreciably – perhaps up to 25 per cent – less in some countries. Lastly, the country from which the migrant actually travelled may not have been his country of origin. This was especially the case with the very important Jewish migration, much of which came ultimately from Russia (see p. 89).[85]

What matters, however, is not the aggregate number of migrants from each country but the relationship of this number to total population. Ideally this should be worked out for each year or period of years. This is an impossible task, so the last column merely shows estimated total migration as a percentage of population in 1900. This changes greatly the picture of migration. The country which, in proportion to its size, contributed most to the flow of migrants was Norway, followed by Portugal and Italy. The countries which contributed least were those which were first to develop modern industry: France, the Low Countries and Germany.

3.17 The chief foreign communities in France, about 1850

The vigorous migration movement within most of the countries of Europe was from countryside to town, from agriculture to industry. The scale of this movement has already been discussed; it was an essential condition of urban and industrial development. Migration *between* European countries was, however, rather limited in scale. It was restricted by problems of language and culture, as well as by conditions imposed on immigrants by some countries. The chief intra-European movements have already been mentioned. But 'seasonal migration and frontier movements within Europe were far more important than permanent emigration',[86] and in the years before the First World War from one to two millions moved annually across national boundaries. They included a few skilled craftsmen who pursued traditional crafts like masonry, but most were unskilled. They came mainly from the peripheral areas – Spain, Italy, the Habsburg lands and Poland – and found employment in France, the Low Countries, Switzerland and Germany (fig. 3.17).

Two other groups of migrants call for discussion: the migratory or transhumant pastoralists of the Alpine region and of the Balkans, and the Jews. The former migrated seasonally, but always had permanent homes to which they returned, sometimes in winter, sometimes in summer. They did not cease at any time to be pastoralists, and their migratory movements were merely a means of using marginal land. Their numbers were few; their impact on the demographic structure of Europe negligible, and a discussion of their way of life is deferred to chapter 5.

The population of nineteenth-century Europe

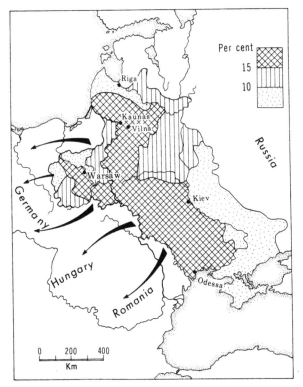

3.18 The Pale of Jewish Settlement within Russia and the migration of the Jewish people into central and south-eastern Europe

Jewish migration

The movement of the Jewish people during the century is quite a different matter. Both in scale and in social and economic significance it was very important indeed and brought about fundamental changes in the geographical pattern of Jewish settlements.[87] At the beginning of the century, European Jewry was divided into two distinct communities which had taken shape early in the Middle Ages. The Sephardic Jews had settled in the western Mediterranean. They were identified in particular with Spain, from which they were driven at the end of the sixteenth century. Some moved to Great Britain and the Low Countries; others eastwards to the Ottoman Empire. Dispersed in this way, they tended to lose their coherence, and many were assimilated to the people amongst whom they lived. During the earlier Middle Ages a second body of Jews had moved from the Byzantine Empire to settle in a broad belt from the Black Sea to central Europe. The same intolerance which had driven them from Spain now forced them back towards the east of Europe, where at the beginning of the nineteenth century they were settled in White Russia, the Ukraine and 'Congress' Poland (fig. 3.18). Here their numbers began to increase more rapidly than those of other Jewish communities and far more rapidly than the non-Jewish peoples among whom they lived. At the end of the eighteenth century about 44 per cent of the

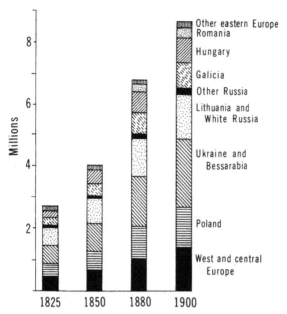

3.19 Distribution of Jewish population in Europe, 1825–1900

Jewish people lived east of Germany; by 1880, it was three-quarters of a much larger number.[88]

In 1791 the government of Catherine II established the Pale of Settlement, the restricted area of western Russia, within which the Jews were required to live (fig. 3.18). The number of Jews in Prussian and Habsburg territories at this time was negligible. Their settlement within Russia was predominantly rural; and they lived scattered through the countryside, a few families in each village.[89] They were forbidden to settle in the larger cities, but came to make up a large proportion of the population of the smaller. They could not hold land, and were thus excluded from agriculture. Jews were thus left with petty trade and crafts as their only pursuits. They dealt in locally produced materials: hides, leather and wool; they handled much of the grain export; they served as tax- and rent-collectors; they distilled liquor and managed the taverns – a seigneurial prerogative in most of the Pale; and they produced cloth for sale mainly to the Russian peasantry.[90]

There was little scope for expanding these activities in so poor and backward a society. As the numbers of the Jewish population increased, they were obliged to migrate in the only direction open to them: westwards. They moved on a broad front which stretched from Prussia, through Posen to Galicia and Bessarabia. They settled in large numbers in Posen province, from which they filtered westwards to Berlin and the cities of central Germany, to Prague and Vienna. From Galicia and Bessarabia they spread into Hungary and Moldavia. The Danube served as a barrier to their movement. Social conditions in the Ottoman Empire were not encouraging and the brash nationalism of the new Balkan states was hostile. The map (fig. 3.18) shows with as much precision as is possible the spread of the Jewish population during the century.

As they migrated so the economic basis of their life gradually changed.[91] They remained mainly rural in Galicia and Moldavia where they were the 'petty shopkeepers, pedlars and hawkers'.[92] They filled a vacuum in this heavily underdeveloped region. As the trade in grain, cotton and iron goods, timber, furs and skins and, above all, petroleum, developed, there they were waiting to profit from it. Some Moldavian and Walachian towns – Iaşi and Botoşani, for example – became very largely Jewish. In central Europe, however, the Jews became almost entirely urban-dwelling. They came at the right time to profit from advances being made in manufacturing industry. In particular they dominated the factory manufacture of cottons in Saxony and then in Poland. They became fully emancipated in the Habsburg lands in 1867 and in Germany in 1871. This opened the liberal professions to them, and they became increasingly prominent in medicine, law and the universities. At the same time they pursued their traditional commercial and financial activities, though in an increasingly up-market direction.

Within the Pale of Settlement and its borderlands the Jews had as a general rule spoken one of the Slav languages, but as they progressed towards the west they took over the German language, modifying it to form Yiddish. From Germany Yiddish was carried back to Poland and other parts of the Pale, where it became a means for identifying and perpetuating Jewish culture.

Not all Jews who migrated from the Pale succeeded in adapting themselves to western society, nor did all who wished to settle in the west succeed in doing so. At least a fifth migrated in fact from Europe to the New World. There was no effective Jewish migration to America before 1870, and immigrants, mainly from Russian Poland and Austria-Hungary, numbered little more than 20,000 at this date.[93] Up to 1899 American Jews numbered fewer than half a million. Then followed an upsurge in Jewish migration. Between 1900 and 1914 no less than 1,500,000 Jews left Europe, including Russia, for America. The reasons were complex. Jewish birth-rate was high, and their numbers were increasing, especially in eastern Europe and Russia, far faster than the economic opportunities open to them. A series of bad harvests not only reduced their business, but, through the resulting suffering, led the Gentile population to turn against them. The pogroms of the early years of the present century were a powerful factor in this second diaspora. Lastly, in Poland and Russia, from which most of the migrants came, the restrictions of the Pale survived until the Russian Revolution, long after Jews had gained political and economic independence in most of Europe. Unquestionably a revolution of rising expectations was stirring in the breasts of Polish and Russian Jews, and these, it seemed, could be realised best in the New World. There were, however, the first stirrings of Zionism as an alternative way of escape for east European Jewry (fig. 3.19).

The Jewish migration differed from that of other European peoples in that there was no significant return flow. Furthermore, whole families migrated, and, once established in America, often remitted money for their relatives and friends to make the journey. Lastly, the Jews who migrated to the New World were not peasants, like most others from central and eastern Europe. They were craftsmen and traders, and this influenced powerfully their settlement and the economic role which they pursued. The New York garment district is a lasting monument to the Jewish migration of the late nineteenth and early twentieth century.

The ethnic and linguistic pattern

In the strict sense the ethnic pattern of Europe had been established long before recorded history began. The waves of invaders – Celts, Germans, Asiatic Barbarians – which flowed across the continent in historical times were but ripples on its ethnic surface. Their numbers were too small to bring about any fundamental change in the physical make-up of European peoples. The latter remained, as they had been for thousands of years, divided by inconspicuous differences into three broad, west-to-east belts or zones. Europeans had grown so accustomed to them that literary references to *racial* distinctions scarcely existed before the late nineteenth century. If this had continued to be the case, there would be no point in including in a chapter on European population in the nineteenth century a section on its racial basis.

In 1853–5 the Comte de Gobineau published an *Essay on The Inequality of Human Races*. In it he developed, if he did not altogether invent, the myth of the superior race, the Aryans. His was not altogether a nationalist work; it was rather an aristocratic and reactionary piece in which he argued from false premises towards untenable conclusions. It would have been of little consequence if it had not been manipulated by his successors for more sinister ends. He never identified any *national* grouping as racially superior. His Aryans were a patrician group, whose blood might have been found in many nations. They were more a superior class than a superior race. Gobineau's successors, armed with an immense body of cranial measurements, identified the superior race as long-headed (*dolichocephalic*) and fair in skin and hair pigmentation, and identified it with the 'Nordics'. Still superior and inferior races were not identified with particular national groupings; rather, elements of superiority and inferiority were, it was said, to be found in all European peoples. It was not until late in the century, when national rivalries had reached a dangerous level, that Gobineau's racism came home to roost. The 'race of heroes' was identified with specific national groupings. In this the French led, but were, in arrogance and flagrant disregard for the evidence, quickly outshone by the Germans. Several German writers had laid claim to that superiority which Gobineau had postulated, but all their views were subsumed in Houston Stewart Chamberlain's *The Foundations of the Nineteenth Century*.[94] This, one of 'the most confused, pretentious, and overwritten books that racism has produced',[95] had a great vogue in Germany and was widely read elsewhere. The superior or 'Aryan' race was redefined in terms that would include all Germans but exclude almost everyone else. This new racism was of some importance in France in the later years of the century, and in Germany had the effect of encouraging hostility to the Jews, now entering the country in growing numbers, and intensifying German arrogance and xenophobia. It was on this foundation that Hitler and Rosenberg erected their monstrous theories of race and super-race.

The reality of race in Europe was far more complex than Gobineau and his successors admitted, or could, indeed, have understood. During the closing decades of the last century a great deal of progress was made in comparative measurements of the human frame. Most concentrated on the skull, though other parameters of race, including pigmentation and stature, were collected.

But blood-groups, in some ways the most critical indicator of racial affiliation, were not studied until after the First World War. On the basis of these limited criteria a belt of relatively long-headed and lightly coloured people was seen to spread across northern Europe.[96] These have been termed Nordic. To the south, from western France to the Balkans, was a belt of darker, broad-headed (*brachycephalic*) peoples, usually known as Alpines, and lastly, around the Mediterranean Basin, was a people of generally slighter build and darker pigmentation, with again relatively long heads. These were the Mediterraneans.[97] It is not relevant to discuss how this pattern emerged. More important is the fact that these groups merged where they bordered one another, producing a host of distinctive combinations of racial characteristics. 'Intermixture between Nordic and Alpine stocks', wrote Fleure, 'has spread the dominant broad-headedness of the Alpine over most of what is now Germany, but it is often combined with characters derived from the Nordic side.'[98] A similar hybridisation, but between Alpine and Mediterranean, is to be found in the southern peninsulas of Europe. Elsewhere, islands of yet earlier peoples, deriving from the first human inhabitants of the continent, are to be met with, and conversely, there are groups of recent immigrants to Europe who have scarcely begun to merge with the people they found there. Characteristic of the latter are the mongoloid Lapps of northern Scandinavia. Other 'ethnic' islands can be said to have resulted from migration within historical times: the Siebenburger Germans in Transylvania; Nordic types in the upper Rhineland and Switzerland; Mediterranean characteristics along north-western seaways. But the outlines of such groups had long been blurred, and as time passed the mosaic of European peoples became ever less sharply faceted.

Even as late as 1914 most Europeans were supremely unconscious of racial differences between themselves. What they recognised were linguistic and cultural variations, and these were of great and steadily growing importance in their perceptions of their fellow men. Attempts were made, as has been seen, to underpin these cultural contrasts by postulating racial differences, and in Germany this error assumed tragic proportions. But, in fact, it is impossible to demonstrate any degree of correlation – excepting only the small Lappish community in northern Scandinavia[99] – between racial and cultural features. Cultural and linguistic characteristics show sharp distinctions and discontinuities, 'whereas the physical characters show no corresponding hiatuses, but gradual transitions and continuity'.[100]

By the beginning of the nineteenth century most of the major languages in Europe were each represented by a particular political unit which can be called a nation-state. While language is only one of many aspects of culture, it is that which facilitated communication within the group and thus became the most important criterion of social grouping. In this respect religious adherence was its only significant rival. The idea that the state should correspond spatially with the area within which a particular language was spoken gained in acceptance during the century. It was only through language that many aspects of culture – history, folklore, tradition and song, for example – could be expressed. So 'cultural nationalism became the foundation for political nationalism'.[101] The fulfilment of the programme of nationalism called for the unification of cultural groups, like the Germans and Italians, which had been politically divided; it demanded changes in boundaries and the creation of states where none had existed before. It called for the overthrow of the old political order and its replacement by a new, liberal regime which reflected what were presumed to be the aspirations of peoples in their cultural groupings. Unfortunately for the liberals the aspirations of most people did not conform with this ideal. A coalescence of self-interest, respect for tradition and outright reaction meant that the liberal, national revolutions were less than successful. Progress towards the nation-state was slow and halting. The First World War seemed in retrospect to have been fought in order to achieve it, and the Wilsonian doctrine of self-determination called for the readjustment of boundaries 'along clearly recognisable lines of nationality'.

The slow development of the nation-state during the nineteenth century involved two contrasted and in some ways conflicting trends. In the first place it involved the extension of boundaries to embrace all members of a particular national group. A policy of cultural unification had deep emotional appeal, but its satisfaction was dependent on the possession of a sufficient political and military weight. Germany was able to annex Holstein and Schlesvig, which was only partly German, and also the *Reichsland* of Alsace-Lorraine to which the claim on cultural grounds was far from perfect. Italy, on the other hand, could cherish only *irredentist* claims to the Trento and Alto Adige. In no instance, either before or after the First World War, can a boundary claim be said to have been established on strictly cultural and linquistic grounds.

At the same time, within the territory of each of the major national groups, there lay small cultural enclaves. Some differed in no fundamental respect from the culture which surrounded them, like the Kaszubs in Prussian-held Poland or the Frisians in north-western Germany. They might, on the other hand, be culturally distinct, with a language of their own, like the Bretons in France and the Sorbs in eastern Germany. The overall trend was towards the absorption of these cultural minorities by their surrounding cultural groups. But cases arose of resistance to this trend and of the use of political means to maintain the integrity of the threatened group. One cannot, however, generalise. There is no good reason why Kaszubs, Sorbs and Frisians should have been lost with scarcely a murmur, while Basques, Bretons and Catalans have been able to develop and sustain a vigorous movement for the independence of their national minorities.

Despite the fact that some incipient 'nations' foundered, the nineteenth century was a period of growing nationalism, when the strength of national feelings within each state tended to increase rather than diminish. Yet nationalism was a double-headed monster. It also embraced the liberal demand that little nations should be given the opportunity to develop their cultural institutions within the limits of an independent state. Devolution of political power within the Habsburg monarchy in the interests of Czechs, Slovaks, Slovenes, Croats and others would have been supported by all liberals. But the French who cast their eyes towards the 'blue line of the Vosges' and the Germans who achieved the unification of the Second Reich in 1871 were conservatives from the far right of the political spectrum.

The boundaries between neighbouring languages and cultures were rarely, if ever, simple lines of demarcation. The two merged into one another so that there was a zone of transition. Where a political boundary should run if it was to separate them had often to be a matter of judgement and compromise. In no boundary settlement before 1914 were linguistic and cultural considerations of primary importance in delimiting a boundary, though in several instances they were ostensibly the reason for the change. No such change came about without conflict, and, as a general rule, the victorious party to the dispute incorporated the whole of its own national group, and with it a sizeable minority of the oppposing group. This, whatever the legal technicalities, is what Germany did in South Schleswig in 1864 and in the 'Reichsland' of Alsace-Lorraine in 1871. The same happened in the course of the many changes in the Balkans before 1914 and throughout eastern and south-eastern Europe after the First World War.

We can thus postulate a simple model for the development of the nation-state during the nineteenth and twentieth centuries. The dominant national group tended or attempted to assimilate cultural and linguistic minorities within its territory and at the same time to extend its boundaries to include closely related peoples in neighbouring countries. Neither process was likely to go unresisted. There was, however, a small number of states within each of which the linguistic-cultural balance was so finely drawn that there could be no question of the one dominating or assimilating the other. Two of them – the Habsburg and Ottoman empires – fragmented into their constituent national components, but two others – Belgium and Switzerland – survive, their internal problems far from resolved.

The Low Countries

Belgium was divided by a language boundary which ran from west to east, separating the Flemish-speaking provinces of east and west Flanders, Antwerp and Belgian Limburg from the Walloon- or French-speaking provinces of Hainault, Namur, Liège and Luxembourg. Brabant was linguistically divided (fig. 1.3). In 1846 the Flemings made up 95 per cent or more of the population in the northern provinces, and the Walloons an even higher proportion in the southern, except in Luxembourg, where there was a significant German minority.[102] The association of the two peoples in a single state was a matter of historical accident. Each had closer ties culturally with its respective neighbour to north and south – the Flemings with the Dutch, the Walloons with the French – and there were proposals early in the century that Belgium might be partitioned between the Netherlands and France. Not only did such a suggestion meet with fierce opposition from the powers which feared any further strengthening of France, but it aroused little enthusiasm within Belgium. The Catholic Flemings distrusted what they perceived to be the Calvinist Dutch and the Walloons were no less suspicious of the government of Paris.

Thus the two national groups were obliged to live together. Assimilation of one by the other was impossible, and their mutual antipathy was strengthened by social differences. French was the language of public affairs and of the upper

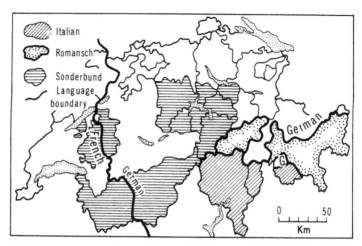

3.20 Language boundaries in Switzerland. The extent of the Sonderbund is also shown. Note how the latter conformed with the mountainous areas, but cut across the language boundary

classes. To rise in the social scale was to become first and foremost a French speaker. In Brussels, for example, Flemish was the *langue* of the lower town, French of the *beaux quartiers*. The domestic servants spoke Flemish and their mistresses, French.[103] This correlation between language and class was, it is claimed, repeated wherever there was a mixed population.[104] Less apparent was a tendency for the Walloons to conform with a French pattern of social behaviour. In particular, their birth-rate fell below that of the Flemings. This may have been nothing more than the common tendency for the middle classes to restrict births before the lower classes began to do so. In Belgium, however, it led to a shift in numbers in favour of the Flemings and thus, in turn, to a southward migration of the latter. Isolated groups of Flemings in the industrial towns of southern Belgium tended to become assimilated, but along the divide between the two cultures there was a slow southward shift of the sharp boundary between them.[105]

Switzerland

Switzerland provided a more complex example of a state divided between two or more linguistic-cultural groups. Here there were no less than four: German, French, Italian and Romansch, in proportions which remained fairly constant between 1850 and 1914. The French-speaking Swiss, in conformity with the behavioural pattern in France itself, had a somewhat lower birth-rate than the rest, and their numbers rose less sharply (fig. 3.20). The three exclusively French-speaking cantons – Geneva, Vaud and Neuchâtel – had consistently lower birth-rates than German- and Italian-speaking cantons, lower even than most urban birth-rates (table 3.5). Romansch speakers declined relatively, as many of them found it desirable or expedient to become German- or Italian-

Table 3.5.[106] *Live births per thousand married women*

		1899–1902	1909–12
Uri	German	399	326
Obwalden		336	312
Nidwalden		338	303
Schwyz		304	257
Lucerne		305	262
Bern		299	246
Ticino	Italian	293	250
Vaud	French	242	188
Neuchâtel		240	177
Geneva		150	109

Table 3.6.[107] *Percentage composition of the population of Switzerland*

	German	French	Italian	Romansch	Other
1850	70.2	22.6	5.4	1.8	—
1910	69	21.1	8.1	1.1	0.6

speaking. This was, in effect, a case of assimilation of a small cultural minority by a stronger culture. Only the Italian sector grew appreciably, and this was due to a significant immigration from northern Italy (table 3.6).

All four languages were recognised by the Swiss constitution, but most cantons had one official language, used in the schools and courts. Migration within Switzerland tended therefore to be quickly followed by the assimilation of the migrant. A small migration from German- to French-speaking Switzerland has thus had the effect of supporting the French proportion of the total population. Most language boundaries are abrupt and several follow cantonal boundaries. The language map is underwritten and stabilised by the constitution of the Confederation. Cultural disputes tended to follow religious rather than linguistic lines. It is a feature of Swiss cultural geography that there is little correlation between linguistic and religious affiliation. Indeed, the reformed faiths were adopted in urban areas, such as Geneva, Bern and Zurich, but were rejected in rural, such as Valais, Luzern and the Forest Cantons. One canton, Appenzell, was in the sixteenth century partitioned along religious lines into Protestant Ausser-Rhoden and Catholic Inner-Rhoden. In the 1840s the two faiths served to polarise attitudes to the proposed liberal reform of the Swiss constitution. This was fiercely opposed by most of the Catholic cantons, which formed the short-lived Sonderbund of 1847. Its forces were routed in a brief war and the reforms adopted. But most of the cantons which made up the Sonderbund remained ultra-conservative in a republic which became increasingly liberal.

There was a tendency for the rigorous division of the cantons along religious lines to break down. In 1837 eight cantons were exclusively either Catholic or

Protestant, and only four were at all evenly balanced. By 1910 only eight still had a large majority for one faith.[108] Internal migration had blurred the confessional lines, without any significant change in the linguistic.

Central Europe

Both the Habsburg and the Ottoman Empires were made up of numerous linguistic and cultural groups, held together in mutual fear and distrust by the military strength of the imperial power. The Habsburg Empire originated in the wholly German-speaking Duchy of Austria, but marriage, inheritance and conquest had turned it into a vast, polyglot empire. In 1815 it covered about 666,550 sq km and embraced a population of almost 20 millions. The earlier censuses contained no record of linguistic affiliation, and not until 1880 does this become generally available. This is unfortunate, because the balance between the ethnic groups which made up the Habsburg Empire was continually shifting. Not only did birth- and death-rates vary between them, but there was also a good deal of internal migration and also assimilation of some groups to the Austrian or Hungarian majorities. In 1867 the empire was divided along traditional lines into an Austrian Empire and a Hungarian Kingdom. Henceforward the Hungarians pursued a vigorous policy of Magyarisation and within 30 years, according to their own census figures, the Magyar population of Hungary increased by more than 50 per cent.

In the middle years of the century the German-speaking population was in a minority even in the Austrian segment of the Habsburg Empire. There were linguistic minorities even in the southern border areas of Austria proper (fig. 3.21). In Trento, or South Tyrol, Germans were to be found only in the north, and in the rest the Italian population was increasing steadily with immigration from northern Italy. Within a measurable period of time they 'would have accomplished peacefully and without disturbance that transformation of all "South Tyrol" into an Italian area'.[109] To the east were the Slovenes of Carinthia and Styria, numerous but in slow process of assimilation to the German-speaking majority. Kranj (Krain or Carniola) and the Küstenland were heavily Slovene, with a scattering of Italians along the Adriatic coast. Austrian Croatia lay to the south, extending a long finger along the Dalmatian coast. Its population was mainly Croat, with Italian-speakers in the coastal towns. The latter were steadily diminishing in number, and must have embraced many 'disguised Croats' who shook off the Italian influence and returned to their own culture when the opportunity offered.[110]

North of Austria proper lay the lands of the Bohemian crown with, in Bohemia and Moravia, a population which was two-thirds Czech and the rest German-speaking. Czech Silesia was evenly divided between Germans on the one hand and Czechs and Poles on the other. To the east lay the vast province of Galicia or Lodomeria, with, tacked on to its eastern extremity, the Duchy of Bukovina. The west was mainly Polish; the east, Russian or Ruthene, with a large and ever growing Jewish population and a sizeable minority of Poles, Romanians and Magyars. Of all the minority peoples under Austrian rule, only the Czechs were able to present any effective opposition to their Austrian

The population of nineteenth-century Europe

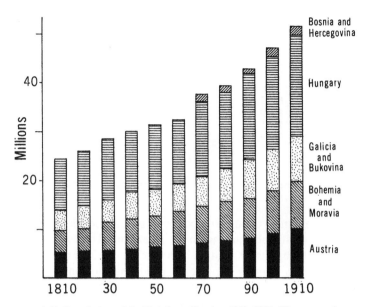

3.21 Population of the Habsburg Empire, 1810–1910. Figures are for provinces, not for ethnic communities

masters. The others were too dispersed, too jealous of one another, or too immature politically for any opposition to have been effective. The Austrians were thus able to pursue their policy of 'divide and rule' with little interruption.

Within Hungary the Magyars were also at the time of the *Ausgleich* in a minority. They occupied the plain of the middle Danube, giving way to other peoples with a remarkable consistency along the foothills of the encircling mountains. Even within the plain they shared the land with numerous colonies of Germans, most of them settled here in the wake of the retreating Turks during the eighteenth century. In the south of the plain, in the Baranya, Bačka and Banat, the population was more mixed than in any other comparable area of Europe. The region had been depopulated in the course of the wars of reconquest and had been resettled by peoples brought in from all quarters. There were Serb, Croat and Romanian villages alongside colonies of 'Mariatheresa' Germans, with no linguistic group dominant.[111] The encircling mountains and hills were settled mainly by Slovaks, Ruthenes, Romanians and the southern Slavs, Slovenes, Croats and Serbs. It is impossible on a small map to represent the complexity of the linguistic pattern, but Count Teleki's large map[112] is a fair representation of the situation in 1910. The interdependence of mountain and plain has often been stressed. A series of small market towns, predominantly Magyar and German, lay near the margin of the plain and served as places of exchange.

To the east of the plain the hilly region of Transylvania had from an early date been settled by a Magyar people, known as Szeklers, and by the Transylvanian Germans. These lived in fairly compact groups, but around them the population was Romanian. To the south of the plain the Hungarian kingdom

Table 3.7. *Ethnic composition of the Kingdom of Hungary*

	1880		1910	
	Total	Percentage	Total	Percentage
Magyars	6,403,687	46.65	9,944,627	54.5
Germans	1,869,877	13.62	1,903,357	10.4
Slovaks	1,845,442	13.52	1,946,357	10.7
Romanians	2,463,035	17.5	2,948,186	16.1
Serbs and Croats	631,995	4.6	656,324	3.6
Ruthenes	353,226	2.57	464,270	2.5
Others	211,366	1.54	401,412	2.2

embraced the predominantly Croat provinces of Croatia and Slavonia, which were to the Magyars what Bohemia was to the Austrians. The composition of the population of the Hungarian Kingdom is shown in table 3.7.[113] An absolute rise of 55 per cent in the Magyar population in the space of only 30 years is far more than can be accounted for by natural increase. Significant numbers of the subject peoples had become sufficiently Magyarised to be included in the total.

The Balkan peninsula

The Danube valley, inhabited mainly by Germans, Magyars and Romanians, served as a barrier between the northern, or western Slavs and the southern. South of the river the land was overrun by the Turks in the fourteenth century and remained under Ottoman rule until this was overthrown by national revolts in the nineteenth. The Ottoman Turks seized the land, which was then worked on their behalf by the Christian *rayas*, but settled in considerable numbers only in the larger towns (see p. 276). The Turks were Moslem, and nowhere were they assimilated by the Christian majority amongst whom they settled. On the other hand, some Christian groups, notably the Bosniaks, the Pomaks of Bulgaria and many Albanians were converted to Islam. Apart from the Turkish minority, the population of the Balkans was basically Slav, with Greeks in the southern peninsula and islands and Albanians in the mountainous region to the west. There were also other minority groups, including Tatars, relics of early invasions from the Steppes, in Dobrudja, semi-nomadic Vlachs, Germans and Magyars.

The Vlachs, Kutso-vlachs or Aromani were a relict people, descendants of the half-Latinised Thracians of the classical period. Cvijić in 1918 put their total number at only 150,000–160,000, grouped into 154 communities. They were transhumant pastoralists, maintaining their permanent homes in the mountains but descending to the plains with their animals for the winter. Most were to be found in the Pindhós Mountains of Macedonia and northern Greece, with winter quarters in the Thessalian and Macedonian plains.[114] They mostly spoke their own Aromanian language, a dialect of Romanian together with the Macedonian dialect of Slav. Their relationship to the Romanian people is far from clear; were the latter merely Vlachs who had wandered north of the Danube, or were the Vlachs Romanians who had moved south with the decline of Byzantine power, or were both just descendants of romanised provincials? Their number had long been declining; Cvijić put them at 400,000–500,000 in the eighteenth century. They had been recognised as a separate *millet*, or ethno-religious group, by the Turks, and a feeble national movement took root amongst them in the later nineteenth century.[115] The Romanian government tried to take them under its protective wing, but warfare and new boundaries within the Balkans closed their migration routes, and once they had ceased to be transhumant they were quickly assimilated to the Slav population.

The Slavs themselves were divisible into a western and an eastern group. To the former belonged the Slovenes, Catholic and relatively westernised; the Croats, mainly Catholic with a sense of nationhood which had been fostered by conflict with the Magyars, and the Serbs, Orthodox and treasuring memories of their lost independence. The Montenegrins were close to the Serbs, but regarded themselves as a separate people. The eastern group consisted only of the Bulgars who were settled between the Danube and the Aegean Sea. This is, however, to simplify the ethnic and social system. Life for many of the southern Slavs centred in 'small tribal cells',[116] in which language, religion, local traditions and historical experience of both Turks and their own Slav neighbours wove a constantly shifting pattern of loyalty and hostility. There was, in Vucinich's words, 'an enormous and intricate network of social subsystems...At any one time the [Ottoman] empire amounted to a vast congeries of discrete cultural and societal elements, which shifted and related much like the particles in a kaleidoscope.' Nowhere was 'the complex mosaic of Ottoman society' more devastatingly apparent than in Albania and Macedonia. Indeed *macédoine* has become a culinary term for a confusing mixture of ingredients. There were Catholic, Orthodox and Moslem Albanians, 'Albanianised' Slavs and Greeks and, amongst the Albanians themselves, deeply rooted tribal animosities and, at the village level, a society split by blood-feuds.[117] The Macedonian language was distinct from both Serb and Bulgarian, though grammatically closer to the latter. The people, heavily intruded by Ottoman Turks as well as by Greeks and Albanians, could not be said to belong indubitably to either the Serb or the Bulgarian nations which contended for their loyalty. The national movement, known as the Internal Macedonian Revolutionary Organisation (I.M.R.O.), which developed in the 1890s, was linked with Bulgaria, and its claims for Macedonian autonomy were a thin disguise for annexation by Bulgaria[118]. The nineteenth and indeed the twentieth centuries have been marked by a welter of claims, based on every conceivable criterion, to this unhappy territory.[119]

A 'nation' so indeterminate as the Macedonians could not have precise territorial boundaries. In the nineteenth century they merged in one direction into the Serbs; in another, into the Bulgarians. The Bulgarians may have been more numerous and homogeneous than Serbs or Croats, but their settlement area was deeply intruded by Turks, who were far more numerous here than anywhere else in the Balkans. In Thrace and over the Deli Orman plateau in the north-east there were large Turkish communities. In the Rhodope Mountains was a sizeable community of Pomaks, or Bulgarians who had been converted to Islam. Small Bulgarian communities were to be found scattered through eastern Thrace and north-eastwards into Dobrudja and even, beyond the Danube delta, in southern Bessarabia.

Drawing political boundaries (see p. 27) through this ethnic confusion, with fragmented peoples and uncertain loyalties, would have been a matter of the gravest difficulty even if there had been goodwill on all sides. This there never was, and the boundaries drawn in the Balkans, from the independence of modern Serbia to the creation of the Albanian state in 1913, reflected the balance of political power rather than the ethnic realities of the region.

The lack of clear ethnic boundaries has made it impossible to estimate with precision the size of the linguistic communities which inhabited the Balkans. The early censuses are quite untrustworthy, and it was always to someone's advantage to minimise or exaggerate the numbers of any particular people. The Ottoman government levied no direct tax on Moslems, but assessed a hearth-tax, later changed to a head-tax on all adult males who were not of the faith. The records for the assessment of 1815 gives a total of about 935,000 for the Ottoman lands south of the Sava and Danube, suggesting a total population of about three millions.[120] To these totals must be added the Islamic population – small in the western parts of the Balkan peninsula, but very considerable in Bulgaria, Macedonia and Thrace. The available statistics do not allow us to trace the relative changes which must have occurred between the peoples of the Balkans. It is clear, however, that the Vlachs declined sharply and that there was a loss amongst the Turkish population as one area after another gained independence and the Turks, whether officials, military or landholding *rentiers*, left. There is, however, no evidence for a migration of Islamicised Slavs, such as the Bosniaks and Pomaks, and in Macedonia, Thrace and north-eastern Bulgaria large Moslem comunities survived into the post-war world to confuse the boundaries and politics of the region.[121] At the same time the steady migration within the Balkan peninsula (see above, p. 86) had the effect of blurring what hard lines there were between languages and culture. Serbs migrated from the Karst of Bosnia, Montenegro and Raška into the richer land of the Croats, bringing with them their southern dialects as well as their Orthodox religion. Macedonians moved northwards into Serbia, and Bulgarians from the Rhodope and Srednja mountains into the plains of both Serbia and Thrace.[122] In the long run the effect was to weaken the cultural contrasts within the Balkans, but the immediate result was to sharpen rivalries and intensify hostility between the several ethnic groups.

The pattern of population during the nineteenth century

The increase in Europe's population by about 140 per cent in the course of a century was accompanied by significant changes in its distribution. At the beginning of the century the pattern of population was still basically medieval. It was dense in northern France, the southern Low Countries and the Rhineland, and also in the North Italian Plain and Tuscany. There were islands of dense population in Saxony, the Naples district and in northern Portugal. But elsewhere it rarely exceeded 70 to the sq km and the great European periphery, from Spain, through the Balkans and eastern Europe to Scandinavia had generally less than 25. By 1914 a profound change had taken place (fig. 10.3). The dense population which a century earlier had characterised only the southern and maritime Low Countries and the Rhineland, now spread eastwards to Brandenburg, Saxony, Bohemia and Silesia, and beyond it a belt of relatively dense population extended eastwards to Warsaw, Vienna and Budapest. The high density of northern Italy had spread south to most of the peninsula as well as to Sicily. Areas of dense population had developed in southern Sweden and in northern Spain. In every direction the nuclear area of dense population was spreading out towards Europe's demographic frontier.

France

The burden of more than 20 years of warfare weighed heavily on France. The total casualties have been put at over a million,[123] but it is unlikely that mortality in the actual fighting was more than 860,000 during the period.[124] Nevertheless the years of peace which followed began with an unbalanced sex structure and continued with a steadily declining birth-rate.[125] In consequence the population of France grew more slowly than that of any other European country. Many rural areas were drained of their population, and France admitted immigrant workers on a larger scale than any other country.

France has a series of censuses unequalled in continental Europe, even though earlier totals can be considered no more than approximations.[126] Their imperfections, however, cannot disguise the trends, both aggregate and regional, in the French population. At the end of the Napoleonic Wars the only parts of France which could be considered densely (over 100 per sq km) peopled were the northern *départements* from Normandy to Nord, northern Alsace, and the Lyons and Paris regions. The vast heart of France, from Champagne south-westwards to the Central Massif and the Pyrenees, was sparsely (less than 50 per sq km) populated. Yet France was commonly thought of as overpopulated during the first half of the nineteenth century, with fragmented farm holdings and a growing number of landless labourers.[127] Nevertheless, the rural population continued to have a higher birth-rate than the urban, and until the 1840s increased both absolutely and as a proportion of total population. From about 1846 it began to decline,[128] at first slowly; then, after about 1872 very rapidly.[129] Rural decline was relieved in some areas, such as Lower Normandy, by the practice of domestic crafts, where cotton-weaving and lace-making long remained important. Those who migrated to the towns from areas where such craft industries were important were less than half those from the purely agricultural areas.[130] It was these crafts alone that allowed the land to support its relatively dense population.[131] In the Auvergne, it is claimed, depopulation was the consequence almost wholly of the 'decline, even the disappearance of industrial activities which were complementary to the agricultural and pastoral'.[132] The economic unity of the rural community was broken, and with this '*ruralisation*' of the countryside, the village became almost exclusively agricultural.[133]

No part of France was immune from these migratory movements. The flow of migrants first became marked from the marginal lands of the Central Massif,

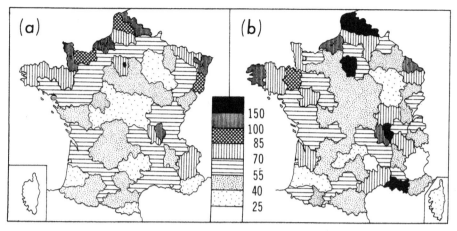

3.22 Population of France; (a) 1821 and (b) 1911 (persons per hectare)

the Pyrenees and the Alps, which had long been a source of seasonal and harvest labour to the towns and richer agricultural regions.[134] Paradoxically, migration was greater from the more fertile and productive areas of France than from the marginal, because they lent themselves more readily to labour-saving devices.[135] Even in areas, such as *dépt* Nord, where population was increasing throughout the century, growth was in effect limited to the towns.[136] Only in those areas of southern France where intensive viticulture and fruit-growing were expanded late in the century, can one point to a reversal of the general downward trend in rural population before 1914.

Migration was mainly to the cities, and in particular to Paris itself, which grew from about 550,000 in 1801 to 2,800,000 in 1901. Next in importance came the industrial region of *dépts*. Nord and Pas de Calais, then Lyons and the industrial region of *dépt* Loire. French provincial towns, in general, grew little during the century, and many were scarcely larger in 1914 than they had been a century earlier. The graph in chapter 4 (fig. 4.11) represents superimposed the populations of the 20 largest towns in France in 1801, 1851 and in 1911. It shows that a small number of large cities increased very greatly in size during the century, but that in most growth was very small (see p. 128). This failure of urban population to increase, as it had done in both Great Britain and Germany, is a reflection of the slow rate of industrial growth.

That the population fell to a lower level than was economically or socially desirable is generally admitted. By the later decades of the century, migration, coupled with agricultural expansion in some areas, had relieved what degree there was of overpopulation, and locally an acute labour shortage had developed There was a certain reluctance on the part of the French peasant to enter the coal and iron mines, and these in particular drew quite heavily on immigrant labour. The number of foreigners resident in France increased from about 380,000 in 1850 to a million in 1887,[137] and to about 1,160,000 by 1911, or almost 3 per cent of the total population of the country.[138] In addition there was a daily movement of Belgian workers from their homes in Flanders and

The population of nineteenth-century Europe 105

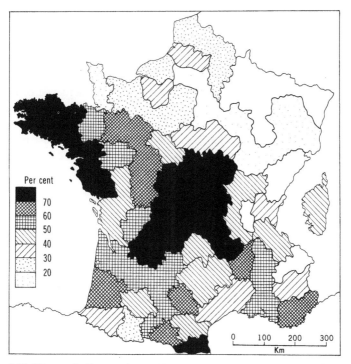

3.23 Illiteracy in France, about 1830

Hainault to the factories and farms of northern France. In 1913 this was said to amount to 50,000–60,000.

By 1911 the distribution of population in France was that shown in fig. 3.22b. In its broad outlines this resembles the map for 1821 (fig. 3.22a), but the contrast between the densely populated north and the empty heart of France had become greater. The population of France was becoming increasingly peripheral as agriculture claimed a diminishing share of the workforce.

In any discussion of the population of France a division quickly becomes apparent between the north-eastern third of the country and the remainder, between approximately the area where population was increasing during the nineteenth century and that where it was stable or diminishing. This was also very roughly the division between the regions of secondary rock and *limon* soil (see p. 60) and those of palaeozoic rocks or high, folded mountains with their thin and acid soils. The latter region, certain islands of better soil and more prosperous agriculture excepted, was characterised by its greater poverty. It was the area from which most of the migrants came. This contrast between the two faces of France is powerfully reinforced by two maps, prepared as early as 1836, of diet (fig. 3.8) and literacy (fig. 3.23).[139] The first shows by *départements* the ratio of inferior cereals to wheat in the diet of the population. It was highest in the Central Massif and Brittany, but nevertheless high throughout the south-western two-thirds of the country. The only exceptions were the plains of Aquitaine and of Provence. It is noteworthy also that diet was poor in the *dépts.* of Marne and Aube, in Champagne. The pattern of illiteracy, itself a function of poverty, is similar, with the highest levels in the Central Massif and Brittany.

The evidence compiled from medical reports on conscripts to the French army shows a similar contrast between the relatively well developed young men from the north-east and the stunted population with a high incidence of disease in the rest of the country.[140] Nor were these conditions quick to improve during the century. The lack of immigrants into these remote rural communities meant that there was a high level of inbreeding. A study made in the Maurienne (Savoy) showed a large number of consanguineous marriages, with a consequently high level of infant mortality as well as of hereditary diseases such as goitre and cretinism.[141]

The Low Countries and Switzerland

Belgium and at least the western provinces of the Netherlands were already densely populated when the century began, and until the time of the First World War population continued to increase. The Netherlands, in fact, had one of the sharpest rates of increase to be found in Europe. In the Netherlands the trend of the birth-rate was upwards during the first two-thirds of the century, and not until after 1890 was there any significant downturn. Death-rates had begun to show a downward trend by the mid-century and fell sharply after 1890.[142] The rate of growth of the population was slower in Belgium, but nevertheless amounted to 71 per cent between 1846, the year of the first census, and 1910. In neither country was the rate of growth uniform. Northern, or Flemish Belgium approximated to the demographic pattern of the Netherlands, whereas the Walloon south showed a low and declining birth-rate similar to that in France.[143] In some parts of the Walloon provinces of Hainault, Namur and Luxembourg a peak population was reached early in the second half of the nineteenth century, after which there was a slight decline. The picture is, however, complicated by migration from the Flemish-speaking north to the industrial belt of central Belgium and by the assimilation of Flemings by the Walloons. But for this the Flemings would by 1914 have greatly outnumbered the Walloons. The reason for the differential birth-rates lies probably in the fact that the Walloon population accepted more readily the values and mores of the middle class than the Flemings. The size of the household was consistently higher in Flanders and Brabant than in Walloon Belgium, though nowhere was the extended family at all common.[144]

The distribution of population was characterised by two regions of very high density, framed by areas of sparser settlement. The older of the two was spread across the low plateau of Hainault and Brabant and included Brussels, Ghent, the port of Antwerp and most of the textile centres of the Belgian plain. The second developed in the course of the century. It consisted of a narrow belt of dense and heavily industrial population, which followed the valleys of the Sambre and Meuse, expanding in the west to embrace Mons and the Borinage. Its basis was coal-mining and the industries based on coal.

Density was significantly lower in the maritime region of Flanders and over the infertile sands of the Kempenland (Campine) which bordered the Netherlands.[145] A belt of lower density also lay to the south of the urbanised and industrialised region of Flanders and Brabant. But the lowest densities were to be found to the south of the Sambre and Meuse. The plateau of the Ardennes had a cooler and wetter climate than central and northern Belgium. Its soils were poor and agriculture less intensive, and the scattered metal industry was in decline from the earliest years of the century.

The pattern of population growth in the Netherlands was similar to that in Flemish Belgium. The birth-rate remained high until late in the nineteenth century and the rate of increase of the population was one of the highest in western Europe.[146] In the absence of any considerable programme of industrial development, much of the increase had to be absorbed in agriculture. This was extended on two fronts, by the reclamation of polderland in the west and of the sandy heaths in the south and east. In Overyssel, for example, there was an increase of nearly 30 per cent in the agricultural population between 1849 and 1899 and a more than commensurate increase in food production.[147] The Veluwe, the region of sandy soil lying to the west of the IJssel, appears to have had in percentage terms an even greater increase in population.[148] This growth was maintained, even after the import of New World cereals had begun to bring down the price of grain. The use of fertilisers allowed a more intensive agriculture to develop, and at the same time increasing numbers were absorbed into the textile and other consumer goods industries of the eastern and southern Netherlands. The effect of these changes was to raise the population density of the eastern provinces closer to that of the urbanised west.

In Switzerland the population grew somewhat more slowly, especially in the latter half of the century, than in the rest of central Europe. The death-rate began to drop soon after 1870, and the birth-rate showed a significant decline only a few years later. But the speed at which these changes occurred was strongly influenced by cultural considerations. It was, as was to be expected, slower amongst the urban than the rural population, but also amongst the Protestant than the Catholic, and the French- than the German-speaking.[149] The only French-speaking cantons where this was the case were Fribourg (only partly French-speaking) and Vaud, where most of the population was strongly Catholic. In Belgium, the lower birth-rate amongst the French-speaking population could be explained by their more strongly middle-class attitudes. Such an explanation does not appear to be applicable to Switzerland. It seems rather that *La Suisse romande* shared the same cultural values in this respect as France itself. The Italian-speaking canton of Ticino, which was both catholic and rural, might similarly have been expected to share the values of Italy. It did not do so, and its birth-rate was amongst the lowest. These differentials might have been expected to result in a change in the balance of nationalities within Switzerland. Although there was a small percentage decline in the French-speaking Swiss, any more significant change was prevented by internal migration and assimilation (see p. 96).

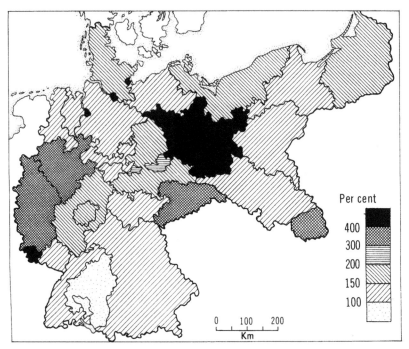

3.24 Percentage increase in the population of Germany, 1816–1914

Germany

The population of Germany increased more rapidly during the nineteenth century than that of any other major European country. Although statistics for the first half of the century are defective it is evident that there was for Germany as a whole a very nearly threefold increase. The birth-rate (fig. 3.6) was very high in the middle years, and did not decline appreciably until after 1900. Even so, it remained very high in eastern Germany and also in the industrialised region of the lower Rhineland and Ruhr, where also there was a large immigrant population from eastern Germany, in the Saarland, Palatinate and south-eastern Bavaria. It is evident that birth-rates were highest in two contrasted regions of Germany: the developing industrial regions of the west, where the age-structure of the population favoured it, and in the strongly Catholic, peasant lands of Bavaria and the German east. It was from these lands, especially the latter, that the flow of migrants, both internal and overseas, was most vigorous.[150] It is noteworthy that the birth-rate was appreciably lower in the predominantly Lutheran province of East Prussia.

Fig. 3.24 shows the rate of population growth. Apart from the Hanseatic cities of Bremen, Hamburg and Lübeck, the highest rates of growth – more than 300 per cent – were in the industrialised Rhineland and Saar, Berlin and Brandenburg, the Kingdom of Saxony and Upper Silesia. All were areas within which modern industry was developing throughout the century, and all benefited from a vigorous in-migration. The periods when this growth took place varied. In upper Silesia it was strongest early in the century. In the Hanseatic cities it

Table 3.8. *Birth rates and migration*

	Excess of births over deaths		Migration ±	
	1856–80	1881–1910	1856–80	1881–1910
East Prussia	890,430	759,888	−98,928	−629,649
West Prussia	444,765	789,816	−133,199	−492,240
Posen	549,725	1,028,059	−238,964	−631,625
Pomerania	518,625	645,862	−267,725	−468,975
Silesia	992,705	1,710,185	−167,276	−492,148
Brandenburg	616,465	984,042	−142,619	+841,749
Berlin	176,320	469,017	+484,722	+479,910
Net balance in East Germany			−563,989	−1,392,978

Table 3.9. *Growth of population in eastern Germany (percentages)*

	1855–80	1880–1910
East Prussia	25.4	6.7
West Prussia	28.5	21.2
Posen	22.3	23.3
Pomerania	19.5	11.5
Silesia	25.9	30.4
Brandenburg	26.4	80.5
Berlin	143.3	84.5

was most rapid after the creation of the German Empire in 1871, when overseas trade was developing strongly. In Westphalia and the Rhineland the fastest growth was in the two decades before the First World War.

The lowest rates of growth were to be found in south Germany, and in some of the small states of Thuringia and central Germany. The political basis on which the statistics were compiled disguises the fact that in the latter areas and also in some other parts of Germany there were districts which suffered a net loss of population. The high plateaux which border the Rhine and hilly areas of south Germany were in fact less populous in 1914 than they had been a century earlier.

Only about half of the population enumerated in the census of 1907 lived in the places where they had been born.[151] A third of those born elsewhere had come a considerable distance, many of them from the German east. Indeed it is impossible to overestimate the importance of this migration both in relieving population pressure in the eastern provinces and in providing labour in Berlin, the Ruhr and the Rhineland (table 3.8).[152] Despite the immense net migration from eastern Germany, the population grew by the percentages shown in table 3.9.

At the beginning of the nineteenth century the population map showed only three areas of dense population: the highly urbanised areas of the lower Rhineland, the middle Rhineland, and the Kingdom of Saxony. Elsewhere the population was relatively sparse and across the north, from Oldenburg in the

3.25 Population of Poland

west to East Prussia, its density was particularly low. Industrial and demographic growth during the century took place mainly in areas which were already populous. Density increased throughout the Rhineland, and the Rhine-Ruhr and Mannheim-Frankfurt regions grew to be amongst the most densely peopled in Europe. The Saxon region, restricted at first to the hilly areas where industry first developed, intensified and extended northwards to include the brown-coal region near Halle, and a broad belt of fairly dense settlement developed to link the Rhine-Ruhr with Saxony. Berlin grew into a vast island of dense population, the largest agglomeration in continental Europe, set improbably amid the heaths of Brandenburg. Population density intensified in Silesia, along the foothills of the Ore Mountains, and, above all, in the coalfield region of Upper Silesia. In south Germany, lastly, growth was restricted to a few large urbanised areas, despite the prevailing high birth-rate. Stuttgart became the focus of a region of light and mechanical industries at the end of the period, but Munich, Nuremberg and Augsburg remained large islands of population in a thinly settled countryside.

Poland

The history of population in the Kingdom of Poland paralleled very closely that in the eastern provinces of Germany. In both it was heavily rural and predominantly Catholic; the birth-rate was very high and, despite periods of very high mortality, the population increased rapidly. In the Polish Kingdom, both birth- and death-rates fluctuated greatly until about 1855 (fig. 3.25). Then the death-rate fell steadily. The birth-rate, however, remained high – consistently above 40 per thousand – until after 1900.[153] The population had undoubtedly suffered in the course of the warfare which accompanied the last Partitions and of the Napoleonic campaigns which followed, as well as in the famine of 1816–17 and the Rising of 1830. Thereafter its growth was steady and continuous, and between 1830 and 1913 its overall increase was of the order of 250 per cent.[154] The proportion engaged in agriculture, over 80 per cent at the beginning of the century, fell only to about 60 per cent by 1916. Food production increased at

Table 3.10. *Population of Poland by religious affiliation (percentages)*

	1827	1848	1870	1897	1913
Catholic	86.0	82.8	76.3	74.8	75.4
Orthodox	—	0.2	4.7	6.6	3.7
Protestant	4.5	5.5	5.4	4.5	5.4
Jewish	9.3	11.4	13.5	14.0	14.9
Others	0.2	0.1	0.1	0.1	0.6

least as fast as the population, especially after serfdom had been ended in 1863 (see p. 200), but the growth of the landless labouring class led to a strong migration of peasants into eastern Germany.[155] The distribution of a population so heavily dependent on agriculture reflected by and large the quality of the soil, densest on the good loess soils of the south and of Mazowsze, to the west of Warsaw; least in the region of moraine and outwash in the north and north-east.[156] A feature of the population of the Polish Kingdom was the sharp increase in the Jewish population, and the commensurate decrease in percentage terms of the Catholic (table 3.10).[157]

Scandinavia

At the beginning of the nineteenth century the Scandinavian countries were the most sparsely settled and the least developed in Europe. The northern half of Sweden, the mountains of Norway and the vast Lake Plateau of Finland were virtually uninhabited. The small population was mostly settled on the Danish peninsula and islands, in the coastal regions of southern Sweden and Finland, and on the narrow shelves of land between the Norwegian mountains and the sea. It was engaged almost exclusively in agriculture except in Denmark, where urban craft industries and commerce were more vigorously developed.[158] Throughout Norway, Sweden and Finland the practice of agriculture was as backward as anywhere in Europe. In Western Norway, plough and harrow were scarcely known, and cultivation was mainly by spade and hoe.[159] Samuel Laing, the English traveller in Scandinavia in the 1830s, described the Norwegian peasant as farming, 'not to raise produce for sale so much as to grow everything they eat, drink, and wear in their families. They build their own houses, make their own chairs, tables, ploughs, carts, harness,... A body of small proprietors, each with his thirty or forty acres, scarcely exists elsewhere in Europe...'[160] Although there was a greater degree of prosperity in the mining and iron-working areas, Sweden as a whole, as well as Finland, lived at the same impoverished and self-sufficing level.[161] A population living at the edge of subsistence and in an environment as marginal as that of Scandinavia was vulnerable to every fluctuation of weather. Laing found 'unusual distress' in Dalarna, where crops had been poor for several years, whereas in Norway the harvest had failed 'in consequence of early frost in autumn'.[162] Only in Denmark were conditions more propitious, and here, in a gentler climate and on better soil, a denser population was supported in greater comfort.[163]

Here, it might have been supposed, was scope for the operation of the Malthusian checks on population growth in all their intensity. And yet the population grew steadily through the century, increasing in Norway, Sweden and Denmark together by about 170 per cent, as well as improving its living standard. This increased level of consumption was achieved without any significant industrial and commercial development, at least until the later years of the century. The death-rate declined steadily in both Norway and Sweden from the beginning of the century (fig. 3.5). The reasons are not readily apparent. Drake has persuasively argued that the incidence of smallpox was greatly reduced. Inoculation was introduced in the eighteenth century, and vaccination became compulsory in Norway in 1810 and in Sweden six years later.[164] It seems unlikely, however, that smallpox had ever been a major cause of death in the Scandinavian countries. It is more likely that the introduction of the potato was here, as in Poland, a major factor in the increase in the food supply (see p. 235). It might be added that neither Norway nor Sweden, except in 1809, was directly involved in the Napoleonic Wars, and Denmark and Finland only marginally so, and these countries were thus spared the epidemics spread by marching armies, as well as the destruction which they wrought. Furthermore, death-rates were lowest in rural areas, and most of Scandinavia was very rural.[165]

Birth-rates remained high in Sweden until the 1860s and did not begin to drop significantly until 1880, while in Norway it was not until after 1900 that a decline in the birth-rate became apparent (fig. 3.2). The time-lag between the beginning of the decline in the death-rate and that in the birth-rate seems to have been longer in Scandinavia than elsewhere in western and central Europe. The system of partible inheritance may have been a contributory factor in maintaining a relatively high birth-rate, as was also the northward advance of the frontier of settlement. Nevertheless, the class of landless peasants, with no significant urban or industrial employment to which to turn, increased, especially in Sweden, and in the middle years of the century there appears to have been a degree of overpopulation.[166] This contributed to migration both internal and overseas. Internal migration has been defined as either progressive, towards more complex and higher-order environments, or regressive, towards less complex or more backward societies.[167] Both occurred within Scandinavia, the former consisting of movement to the towns, mines, pulp-mills and factories; the latter towards the frontier of northern Scandinavia.

There was in addition a large and at times an excessive overseas migration which tended to 'increase when times are bad and decrease when times are good'.[168] Migration was particularly intense in 1869–70, from 1880 to 1895 and again in 1903. In all, about 1.1 millions migrated from Sweden between 1840 and 1914, and about 750,000 left Norway in the same period.[169]

The Habsburg Empire and south-eastern Europe

Any study of the demographic history of the Habsburg lands and the Balkans is inevitably bedevilled by inadequate and unreliable statistics, by migration and by changes of boundary. Data are best for Austria. Those for Hungary are, in

Table 3.11. *Estimated population of the Balkan peninsula (in millions)*

	c. 1845	1890	1910
Bosnia and Hercegovina	1.1	1.34 (1885)	1.9
Serbia	1.1	2.2	2.9
Bulgarian Province	3.0	—	1.7
Bulgaria	—	3.3	4.3
Albania	—	—	—
Rest of the Ottoman Empire in Europe	5.2	—	—
Montenegro	0.13	—	0.24
Greece	0.861 (1841)	2.187 (1889)	2.632 (1907)

the opinion of B. R. Mitchell, subject to an error of 10 or more per cent,[170] and figures for the Balkans are little better than intelligent guesses. The Austrian lands, including Bohemia and Moravia, are thought to have had a population of about ten millions at the beginning of the century, together with about four millions in Galicia and Bukovina. By 1869 the total had risen to about 20 millions, of whom more than 7,500,000 lived in Bohemia and Moravia, while the almost wholly rural population of Galicia and Bukovina had increased to about six millions.[171] The rate of increase was greatest – about 66 per cent – in the Czech lands, which became during these years the chief industrial region of the Habsburg monarchy. But in Galicia and Bukovina, which remained almost exclusively agricultural, the increase was of the order of 50 per cent, and farm holdings became notoriously small and the region as a whole one of the most overpopulated and poverty-stricken in Europe.

Birth-rates remained high until late in the century,[172] though death-rates fell sharply from about 35 per thousand in the 1870s to less than 25 early in the next century.[173] The impact of a falling birth-rate was first felt in the industrialised lands of Bohemia and Moravia,[174] where peasants were drawn to the factories and mines with inevitable postponement of marriage. It remained very high in the wholly rural provinces of Galicia and Bukovina. The population of the Austrian lands as a whole increased to about 23.9 millions in 1890 and 28.5 by 1910, despite the loss of some two millions by emigration.

The population of the Hungarian Kingdom grew rapidly, from about 11 millions in 1820 to nearly 21 millions in 1910, despite a large out-migration.[175] Hungary was much more agricultural than Austria, and the decline in the death-rate late in the century was not paralleled by any commensurate fall in the birth-rate.[176] This, however, varied between the national groupings, and was highest amongst the Slavs and lowest in the small German population. There was a heavy migration from Hungary towards the end of the century. This was greatest amongst the Slovaks who lost, between 1899 and 1913, the equivalent of 15 per cent of their 1910 population.[177]

Estimates of population in the Balkan peninsula remained quite unreliable until after the First World War. No census was ever held in the lands under Ottoman rule, and for those which had gained their independence the available figures can only be regarded as approximations. Austria–Hungary conducted

Table 3.12. *Population of the Iberian peninsula*

	Spain	Portugal
1800	c. 10,500,000	2,932,000
1830	c. 12,000,000	
1850	c. 14,500,000	
1857	15,455,000	3,665,000 (1860)
1880	16,830,000	4,260,000
1910	18,618,000	5,423,000

a census in Bosnia and Hercegovina in 1885, and totals for Greece are acceptable. Table 3.11 presents the most reasonable estimates made. The totals represent the population within the national boundaries at the dates indicated. The frequent territorial changes combine with migration to make comparisons between different periods both difficult and unrewarding.

The Iberian peninsula

The Iberian peninsula began the century with a population of about 14 millions, not greatly in excess of the probable total during the period of Spanish greatness in the sixteenth century.[178] Growth was slow during the next quarter of a century. Warfare, it is said, caused the loss of 1.5 million lives between the French invasion in 1808 and 1823, and to this must be added the very heavy mortality from epidemic disease, especially the cholera epidemic of 1833–4.[179] Early nineteenth-century travellers in Spain, notably George Borrow, who hawked his bibles around the country between 1835 and 1840, presented a picture of utter desolation in many parts. To Widdrington the whole Meseta was *despoblado*.[180] To this period of depression succeeded, despite the interruption of the Carlist Wars, one of agricultural reform, industrial growth (see p. 210) and population increase.[181] The rate of population growth began to slacken during the last third of the century, owing, in all probability, to the heavy wave of migration to the New World as well as to epidemic disease and localised famines, which continued to disturb Spain as late as the 1880s (table 3.12).[182]

When the century began, the distribution of the population was heavily peripheral, concentrated in the most favoured coastal regions: Catalonia, Andalusia, northern Portugal and the Biscay coast. The contrast between these and the empty heartland intensified as the century progressed. The Meseta lost 11 per cent of its population during the first 60 years of the century, and a further 4 per cent before it ended. From about 42 per cent of the Spanish total in 1800, it had fallen to 36 per cent in 1910.[183] Migration was largely to the developing coastal regions, where population growth was rapid, and to the capital city of Madrid. Both birth- and death-rates were amongst the highest in Europe before the First World War, but there were strong regional variations. They began to decline first in the more liberal and economically advanced areas bordering the Mediterranean. The contrast between Catalonia and the more backward areas

Table 3.13. *Birth- and death-rates, – Italy*

	Birth-rate per 1,000		Death-rate per 1,000	
	c. 1875	c. 1895	c. 1875	c. 1895
North Italy	36.2	33.2	28.7	23.7
Central Italy	34.4	34.8	20.8	24.0
South Italy	40.8	37.2	30.8	29.0

of the Meseta was as strong as any to be found. About 1900 Barcelona province had a crude birth-rate of 27.8, but Cuenca province in New Castile, of 41.6.[184] This illustrates that polarisation in Spanish thought and behavioural patterns which had contributed to the Carlist Wars and was to continue unabated into the Civil War of the 1930s.[185]

Italy

Italy was, after France and Germany, the most populous country in Europe. Although the first national census was not held until 1861, the year in which unity was achieved, a number of the Italian states had taken censuses – of varying degrees of reliability – long before this. At the beginning of the century the population of the peninsula within the limits of 1914, was of the order of 18 millions,[186] and by 1840 about 23.3 millions.[187] The first census in 1861 gave a total of 25 millions. Thereafter the increase was more rapid: 28.5 millions in 1881, 32.6 in 1901 and 34.8 in 1911.

Human life faced graver hazards in Italy than perhaps in any other European country. Population was denser and the normal sanitary precautions even less observed than elsewhere. Both birth- and death-rates were amongst the very highest in Europe, but, as in Spain, there were sharp regional variations. Both remained very much higher in southern than in northern Italy, and at the beginning of the present century the Sicilian birth-rate was possibly the highest in Europe (table 3.13).

The result was an alarmingly rapid increase of population in southern Italy and Sicily, the region of Italy least able to support it. The population of Sicily increased from about 1,750,000 at the beginning of the century to 2,391,800 in 1861 and to 3,695,000 in 1911 – one of the fastest rates of growth to be met with in late nineteenth-century Europe. The rate of increase was scarcely less in Calabria and Apulia. The environment suffered severely from over-cultivation and over-grazing. Erosion choked the rivers, producing conditions ideally suited to the spread of the malarial mosquito. Malaria, it was claimed, abounded in coastal regions of Calabria and Basilicata, 'particularly where the gathered waters rest upon a substratum of clay'.[188]

Everywhere the traveller from northern Europe found symptoms of poverty and deprivation. Nor did these conditions originate with the rapid rise in population late in the century. Samuel Laing found before the mid-century that in 'this earthly paradise' of the plain of Capua 'the people are not merely in rags and wretchedness; it is difficult even to conceive humanity in so low a condition'.[189]

Population growth was relieved only by an intense out-migration. This was, in the first instance, to the towns. Naples grew to be one of the largest European cities without any commensurate increase in manufacturing activity. Overpopulation was transferred from the countryside to the towns. Smaller numbers migrated to the towns of northern Italy, where alone there was an industrial development which could absorb them, and to neighbouring countries in western Europe.[190] But the biggest movement was overseas, and southern Italy became the chief source of migrants to the New World. Migration was greatest from the poor hill country of northern Sicily, Calabria and the southern Apennines; least from Apulia and Campania, where agricultural resources were greatest (fig. 3.16).[191]

Although northern Italy in general supplied few migrants, there was a significant movement out of the mountainous areas. This was at first only seasonal. Craftsmen like builders and carpenters went to the towns, but came back 'for the winter to their native homes. There is hardly an instance', wrote Gallenga in 1850, 'of a Piedmontese mountaineer settling permanently abroad.'[192] This, however, was soon to change. In the last third of the century permanent migration became general, not only from the Piedmontese Alps, but also from the Dolomites and Alps of Venezia. This phenomenon closely paralleled the exodus from the French and Swiss Alps (p. 79) and movement from the Dinaric mountains (p. 86).

The spatial pattern of population which emerged by the early years of the present century was one of islands of very dense and mainly agricultural population, separated by sparsely settled mountains. The densest was to be found on the coastal plains of southern Italy and Sicily, especially Campania, the rich lowland which enclosed the city of Naples. A narrow strip of dense rural population followed the Adriatic coast northwards until it merged with that of the Lombardy Plain. Within the peninsula were pockets of dense population wherever alluvial soils had been formed in the mountain basins. Amongst these was the Arno Valley of Tuscany, but not the Roman Campagna. Here poor soil had combined with corrupt government, marshes and malaria to restrict population growth since the Middle Ages.

The Plain of Lombardy was by far the largest and most populous of such lowlands. But its density was not consistently high. The quality of the soil varied greatly: too ill-drained in some parts, too sterile in others. The alluvial land close to the River Po, potentially the most fertile of all, was ravaged at intervals by disastrous floods. Gallenga even claimed that population was 'in an inverted ratio to the fertility', and found the 'broad deep plain...almost a blank, only dotted here and there at wide distances with closely-built towns and villages'.[193] This was the problem: the peasants 'crowd together in their dingy villages, and have often a day's journey to travel before they reach the fields'. The settlement pattern did indeed change in the course of the century as agriculture became more intensive and the peasants made their homes amid the fields.

Conclusion

The rate of population growth in most of Europe during the nineteenth century was the most rapid in the whole course of human history up to this date. The basic pattern of human behaviour had changed little. Age at marriage and size of completed families during the first half – even two-thirds – of the century were much as they had been in the pre-industrial age. Only late in the century, and then very slowly, did birth-rates begin to fall significantly. The sharp increase in population was due almost wholly to the fall in the death-rate. This was most marked, as it was also most significant, amongst infants and children. In most of Europe the pattern of births slowly adjusted itself to that of deaths. With fewer deaths, especially juvenile deaths, the number of births necessary to replace them became smaller. This adjustment came first in France and those parts of western Europe which were strongly influenced by French culture; slowest in eastern and southern Europe.

The ethnic pattern of Europe changed little during the century and ethnic boundaries shifted only to a small extent. The low birth-rate within the French cultural area allowed Flemings and German-speakers to encroach to a very small degree. There was an extensive Magyarisation of the population of the Hungarian Plain and in the Balkans there was some degree of migration and assimilation. There was, however, especially late in the century, an extensive migratory movement within Europe. Three streams of movement can be distinguished, of which two were within the same culture area. The first was the relatively short-distance migration from rural areas to the towns. This was especially marked in France, Italy and Spain, but was also important in the Low Countries and Germany. The second was the longer-distance migration from southern Italy and Sicily to the North Italian Plain; from the Danube valley to Vienna and Budapest, and, above all, from the German east to Berlin and the industrial Rhineland. The third was of a quite different order. It was the slow spread of the Jewish people from the Pale of Settlement to east-central and central Europe. To these movements must be added the overseas migration, predominantly to North America, which drew away no less than 30 millions of the population of Europe. There is, however, no means of knowing how many returned home.

The spatial pattern of population also changed during the century. The densest population at the end, as at the beginning, was in north-western Europe, where it formed a belt from Pas de Calais eastwards approximately to Dortmund. From here a belt of lower, but still relatively high density reached through Saxony and into Bohemia and Silesia. A branch from this populous region extended up the Rhine valley to Switzerland. The only other area of dense population early in the century was the Plain of Lombardy and Venezia. In the course of the century population increased in all these areas, and other regions of dense population emerged, including the Saint-Étienne region, Upper Silesia-Moravia, the capital cities and the great ports. Growth was mainly in and near the cities, and some rural areas, including much of France, parts of Bohemia and eastern Germany and mountainous areas generally showed either a stable or a declining population. Some rural areas, on the other hand, showed a considerable increase, greater in some cases than that for Europe as a whole. These included the lightly urbanised areas of Denmark, the eastern Netherlands, Hanover, Poland and Galicia. There was also a sharp increase in the Vienna Basin and the Hungarian Plain, and, south of the Alps, in Italy and Sicily. In all these areas there was a spread of settlement from the good soils to the less good, encouraged by a growing population pressure, and made possible by a slowly improving agricultural technology.

4
Urban development in the nineteenth century

The nineteenth century was a period of rapid and continuous urban growth, marked not only by the foundation of new towns but also by the expansion of most which existed already. The towns which had grown up during earlier periods of urban growth, in classical times and the Middle Ages, had been intimately bound up with the life of their surrounding areas, for which they served as market and administrative centres.[1] The urban revolution of the nineteenth century changed all this over much of Europe. A new kind of city came into existence. It was larger in population, more specialised in function, and it lacked that intimate relationship with surrounding rural areas which had been so important a feature of the pre-industrial town. For its food supply it was increasingly dependent on distant fields in Russia and the Americas. The pre-industrial city had been integrated with the landscape amid which it grew. It was built of local materials, in local styles and its citizens were fed from the surrounding fields. The new industrial town, whether or not it grew out of a small country town, was superimposed upon the landscape, built in styles that were eclectic and international and of concrete and brick of no particular origin.

Some of the new towns of the industrial period arose on the foundations of the old. In the heart of nineteenth-century Essen or Lille, Marseilles or Saint-Étienne lay an ancient core, made up of the old, tightly built city which had existed here before the Industrial Revolution. Architectural fragments from that different though not very distant age survived, but were succumbing to the prevailing urge to destroy and to rebuild. Others, like Roubaix and Tourcoing, Oberhausen and Gelsenkirchen, Łódź and Kattowitz, belonged wholly to the nineteenth century. They were at most villages when that century began, and their growth during the century which followed was uncoordinated and uncontrolled. They were laced, as if with reinforcement, by railways and canals, and scattered through them, almost at random, were the factories which were the mainspring of their growth. Around the factories there sprang up serried rows of houses and naked tenement blocks.

Such towns offered few amenities. They grew fast in response to the increasing demand for the products of their factories. Health conditions were invariably bad, but the liability to epidemic disease, such as the cholera epidemic of 1832, eventually compelled the authorities to make some provision for a supply of fresh, running water and for the disposal of sewage, if only to pass it on to the next town down the river.

The growing size of towns made it necessary to establish some system of public transport within them, to supply gas and electricity for industrial use and for domestic and public lighting, and to facilitate the building of wholesale and retail markets for the provision of food. The rapidly growing and industrialising town of the nineteenth century had never relied upon or owed much to the surrounding countryside. Fresh vegetables and milk continued to be supplied from its farms, but this had to be supplemented increasingly from more distant sources. From early in the nineteenth century a positive antithesis began to emerge between town and country,[2] and the industrial town became, in Juillard's phrase, *la ville insulaire*.[3]

We can thus distinguish three broad and far from mutually exclusive categories of city and town in nineteenth-century Europe. There were, first, the pre-industrial towns which continued little changed in size and dominant function through the century. In the second place there was the category of pre-industrial towns, which, thanks to favourable location or the possession of some resource or other, had developed new functions and grown to a far greater size. And there were, lastly, the new towns of the nineteenth century which had grown from nothing to urban size if not also to urban status with the advent of modern manufacturing and transport.

The pre-industrial town

In much of Europe the majority of towns belonged to the first category. They were characterised by the intimacy with which most citizens knew most others. They were the 'home towns' of Mack Walker, who claimed that no less than 4,000 German towns conformed with this ideal at the beginning of the century.[4] Apart from its small size and 'face-to-face' character, the pre-industrial town was distinguished by its integration with its immediate neighbourhood and its 'distance' from both regional capital and industrial centre.[5] Balzac, who has left us the supreme literary description of society in the pre-industrial town, pictured it as lying off the mainstream of nineteenth-century development, without active links with the capital or even with its regional centre.[6] This picture may have been overdrawn, at least for the middle years of the century, for the small town was, with the building of railways and the improvement of roads, gradually losing its sense of isolation, its economic independence, and its local self-sufficiency.[7]

Over much of eastern Europe, however, the pre-industrial small town not only outlived the century, but continues to exist today, even though the line which separates it from the *ville industrialisante* has become increasingly blurred. The small town was part of a hierarchy of central-places. Below it were the farms and villages which used it as a market and service centre. Above it were towns which combined local and administrative functions with a more up-market type of business, and above these were regional capitals. All *could* have been classed as pre-industrial at the beginning of the century, but before its end the regional capitals and in all probability many of those at an intermediate level in the urban

Urban development

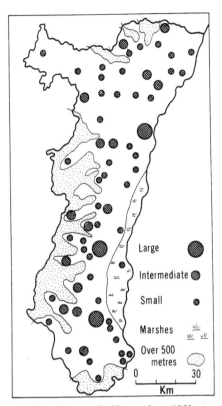

4.1 The urban net in Alsace, about 1850

hierarchy were developing what were at the same time broader and more specialised economic functions. They were becoming industrialised and were beginning to serve the needs of distant regions as well as of their own tributary areas. In the course of the century these broader functions were spreading downwards through the urban hierarchy, so that by the time it ended only the smallest and most remote of urban centres failed to perform some service or manufacture some article for a distant, unseen and unknown market. The 'face-to-face' character of the 'home town' was breaking down, though the number that had by 1914 been assimilated to the category of *villes industrialisantes* was still relatively small.

At the same time, with improving transport and better means of communication, people were drawn not to the nearest but to the best appointed market centre, where they not only had a wider choice of goods but could combine marketing with other tasks normally performed at the central-place. Some towns thus failed to expand their functions through the century and may even have lost in importance. Throughout Europe very small towns were tending to revert to the function and status of villages, first in western and central, more recently in eastern and south-eastern Europe, where the problem of the economic viability of small towns remains acute.

The study of urban development during the nineteenth century is bedevilled

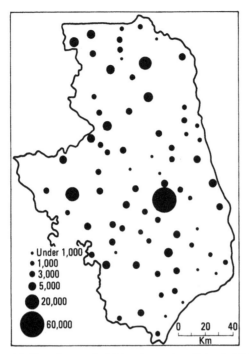

4.2 The urban net in Białystok province, Poland, 1895–7

by the prevailing confusion between legal and functional concepts of the town. Most towns were held to be such by charter, grant or prescriptive right and had developed urban functions so that there could be no question of their status. There were rare cases, notably in Switzerland, of legal towns which had lost or were in process of losing every vestige of urban function. More importantly, some places which had never been legally incorporated nevertheless acquired urban functions, notably manufacturing. The industrial village became quite common in, for example, Flanders and in regions such as the Ruhr and Upper Silesia. As a general rule, administrative law managed to catch up with reality, but it remained the case that places like Gelsenkirchen, Łódź and Kattowitz had developed far towards industrial towns long before they gained a corresponding legal status.

Fig. 4.1 shows the distribution of towns and places with some urban functions in Alsace during the middle years of the nineteenth century.[8] If one excludes the hilly area of the Vosges to the west and the marshy area of the *reids* along the banks of the Rhine, one is left with a not uneven spread of towns. As late as about 1850 in this populous and progressive region, the traditional relationship of small town to its regional centres still survived, the region itself being divided 'en petites unités à l'intérieur desquelles la ville locale assure la plupart des relations nécessaires et produit même, dans ses ateliers, la plupart des objets réclamés sur le marché'.[9] Above the level of the small town, clearly recognisable, were those next in the urban hierarchy: Haguenau, Sélestat, Colmar and Mulhouse, and, rising above them all, like the spire of its own cathedral, the regional capital of Strasbourg. The political and social changes which followed the war of 1870 eroded the urban structure and made the whole province more directly dependent upon the few large towns.

The Polish province of Białystok underwent no such change.[10] Here there were some 75 places with urban status or urban pretensions, and for most of them a series of population figures is available for the period from the early nineteenth century to 1910.[11] Fig. 4.2 shows the distribution and size of the towns of the province in 1895–7. The hierarchy is readily seen.[12] Above the mass of small towns, each with a population of from one to three thousands rose the few larger towns, but towering above them all was the regional capital, Białystok. Next were the regional centres like Suwałki, Lomża and Ełk. Some were centres of the Russian provincial administration; others of the new railway network. But the mass of the small towns of the *województwo* did little more than double their population.

Industrialising towns

Most of the urban growth during the nineteenth century took place in those cities and towns which had already established themselves as regional centres of some importance. They were already served by a network of routes and had built up the infrastructure of commerce. Some lay on navigable waterways and could thus be supplied with raw materials for their further industrial growth. It is impossible to list such towns; only to cite examples which typify them: in France, Lyons, Rouen, Nancy, Metz, Toulouse, Orléans; in the Low Countries, Ghent, Liège and Utrecht; in Germany, Cologne, Breslau (Wrocław), Hanover, Essen, Stuttgart and Frankfurt. Within the Habsburg Empire, Brno, Graz, Prague and Budapest. These were the places which were linked together by the earliest railway nets and there is no question but that rail links served to hasten their growth and to attract industry, business and service functions.

These cities had much in common in their patterns of development during the century. Almost without exception they had been walled and could be entered only by a limited number of gates. The demolition of the walls had begun in the seventeenth and eighteenth centuries when they were thought to have no more military value, but their destruction in most instances was left until the nineteenth. They had almost wholly disappeared by 1900, when such fragments as survived were more likely to be preserved as historical monuments. The circle of walls had an important influence on urban growth even after they had been demolished. In few instances, amongst them Paris and Vienna, did a city expand beyond the line of its walls before the nineteenth century. Everywhere, walls acted as a constraint on urban growth. Streets were in general narrower and building more congested within the walls than outside them. Inside the walls the few open spaces were generally used as markets. There were public buildings, gild-houses and churches, more of the latter in fact than were commonly needed in the nineteenth century, when people were beginning to leave the old city for its expanding suburbs. The line of the walls was almost always marked by the curving line of streets, which perpetuated the open space which previously lay outside and sometimes also inside the walls. Not all nineteenth-century cities had

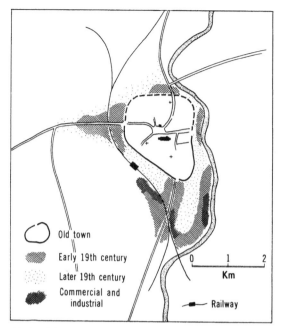

4.3 Model of nineteenth-century development of old urban centre

the vision and the wealth to replace the walls with a planned open space on the scale of the *Ring* in Vienna, but a great many laid out tree-lined boulevards like the *Planty* in Kraków and those of countless French provincial towns.

This was in very many instances the stage of urban growth when the railway was built, often in the 1840s or 1850s. Both track and station were excluded from the old city by the prevailing congestion and high price of land (fig. 4.3). The railway was most often built tangentially, as it were, to the city, in the open space a short distance beyond the line of the walls, and that is where, in a spacious forecourt, bordered by neo-baroque hotels, the railway station was built. In large cities, such as Paris, Berlin and Vienna and, of course, London, several termini were established in a ring beyond the line of the vanished walls, as if besieging the old city.

The area outside the walls was in the main that of nineteenth-century growth. Here development was unplanned and generally rapid. It spread along the roads like fingers thrusting into the open countryside. The roads themselves were wider than city streets; they had evolved from tracks along which peasant carts and animals had once been driven to the city. Even today one is aware of a sudden widening of the road as one passes beyond the line of the walls and of the boulevards which succeeded them. The intervening space was gradually filled in with factories, houses and tenement blocks. A limit was set to the outward spread of the town by the need to commute to its central areas for work or marketing. The introduction of horse-tramways in the middle years of the century, followed by the bicycle and other means of transport such as electric trams and buses, facilitated the development of better-quality housing for the

Urban development

middle classes on the outermost fringe of the growing city by the early years of the present century.

At the same time a change was coming over the old city. The wealthy merchants and urban patricians had once lived near the centre. Their opulent houses survive at Brussels, Frankfurt, Antwerp and Augsburg, and, rebuilt and restored, at Gdańsk, Posen and Warsaw. In the latter half of the century, however, they began to desert their old homes and to build new on the expanding periphery of the city. Their former residences, rebuilt or adapted, came to be used for business or commerce.[13] At Leipzig, which was in this respect typical of west European cities, the population of the *Altstadt* fell from 45 per cent of the total urban population in 1834 to only 2 per cent in 1910, and in absolute terms declined by over 40 per cent.[14]

New towns of the nineteenth century

The third category of towns in nineteenth-century Europe consisted of those which had known no previous urban existence. Their sites had been, at most, villages at the beginning of the century. Their rural roads and tracks developed into town streets. A field became the site of a factory merely because road, canal or railway lay near at hand. Or a mine might be sunk only because the uncertain geological surveys suggested that coal might be found in depth. The railway was put through in the early stages of growth or even before development had begun. In consequence the station was central to the town. Indeed in some, Gelsenkirchen and Kattowitz, for example, the station yard served as a central point, a kind of market-place for the town. There were no constraints, like town walls, on growth. There were no monuments to catch the eye and give an historical depth to the settlement. The town had no patrician class; it was geared wholly to industrial production, and its population was, with scarcely an exception, proletarian. Buildings were hastily erected, functional and unpretentious. The town itself spread out along the lines of transport; factories, mills and mines were linked by spurs from the railways or branches from the canals. There was to begin with not even a market to which peasants could bring their produce. Arrangements developed for the supply of foodstuffs from both local and distant sources, but the town nevertheless fitted uneasily into its landscape. That sense of isolation from the rural environment, which was growing in the older towns, was present from the start in the new.

The new towns began their urban lives under another disadvantage. Whereas the older towns had been incorporated for centuries, with a system of local government and a mechanism, however rudimentary, for environmental control, the new towns had no such advantage. They fitted uneasily and unevenly into the framework of parish, *commune*, *Gemeinde*, which had been designed and created by rural people for essentially rural purposes. Eventually the new towns acquired urban status, with defined boundaries, elected local government, a gild-hall and even a coat-of-arms, but for many decades they were merely overgrown villages with public services which scarcely rose above the level of the parish pump.

The new towns of the nineteenth century were to be found in most parts of

Europe. They were especially numerous and unusually squalid in those industrial regions which had developed on the basis of coal: the Ruhr, Upper Silesia and parts of northern France and central Belgium. They were typified by the mushroom growth of Gelsenkirchen, Herne and Wanne-Eickel; Zabrze, Königshütte (Chorzów) and Kattowitz, Ostrava and Karviná. But they also grew up far from the coalfields, amid the tranquillity of the untouched countryside, at Decazeville, Le Creusot and Łódź, and around the mills of Catalonia and of the Lombard plain.

But new towns were founded in the nineteenth century not only to serve the needs of expanding industry. They were also established for recreational and health purposes. The phenomenon of the resort town belongs essentially to the second half of the century, though there are many instances of incipient resorts before 1850. Its growth came with the appearance of a mass market for holidays and recreation amongst the middle classes. For the rich and distinguished there had always been places where they could disport themselves with whatever degree of privacy they needed. The new resort town was sometimes built on the foundations of an older, pre-industrial urban settlement; more often it was brash and new. Resort towns were, very broadly, of two kinds: the inland spas and health resorts, and the coastal towns. The former had been frequented long before the nineteenth century. As a general rule they had developed around a spring whose curative powers had come to be highly regarded. There were dozens of such health resorts. Spa, in the Ardennes to the south of Liège, was their archetype. Baden-Baden, in the Black Forest, and Karlsbad (Karlovy Vary), in Bohemia, were similar in having given rise to a small pre-industrial town before the nineteenth century. Marienbad (Mariánské Lázně), a neighbour of Karlsbad, developed later and was not incorporated until 1868. The elegance of the 'spa' towns placed them in a different world from that of the new industrial towns. Their neo-baroque architecture, their casinos and salons made them the haunts of the rich and the favoured meeting-places of the pampered politicians of nineteenth-century Europe.

The seaside resorts were different. Their development was a response to the growing sense of well-being amongst the middle classes, for whom a summer holiday became normal during the second half of the century. The cult of the sea coast had begun earlier, probably in England. In 1834 Lord Brougham stayed at Cannes on the French Riviera and subsequently popularised it by his eulogies. Biarritz owed its fame to the Empress Eugénie, who stayed there and praised its charm. The intrinsic merits of a coastal resort were clearly less important than the aristocratic accolades which it received. No less important, however, was the railway, without which the number of visitors would have been drastically reduced and many seaside towns could never have developed. A pleasant climate, attractive surroundings, ease of access by rail and a strong recommendation from a supposedly reliable quarter were the essentials for growth, and no part of the European littoral had them more fully than the French Riviera and its continuation, the Italian.

The towns that grew up along the Mediterranean coast were built facing the sea. They turned their backs on their hinterlands; no new industrial town was more effectively cut off from its rural surroundings. Here a new form of

Urban development 127

economic activity developed – *l'exploitation des étrangers*.[15] In 1861 the urban population of the *département*, which contained Cannes, Antibes, Nice and Menton, was only about 93,600; by 1911 it had risen to 262,500. During this period the amenities of the area were greatly developed. The Corniche roads were built,[16] and hotels spread along the waterfront behind the broad esplanade. After about 1850 winter visitors became more numerous, and, with the prospect of a year-round business, the number of hotels rose sharply. By 1858 no less than 8 per cent of the permanent population of Nice was made up of domestic servants.[17]

No other part of the European littoral could equal the development which took place on the French Riviera after about 1850. But resorts on a humbler scale were established on the coast of the English Channel, at Deauville about 1860, Paris-Plage in 1880 and Le Touquet in 1902, all of them more accessible than the Mediterranean coast from the populous areas of north-western Europe.[18]

Urban population

Estimates of the size of urban population in nineteenth-century Europe and of its rate of growth are subject to two kinds of error: uncertainty as to what constituted a town and, secondly, changing and generally expanding urban boundaries.[19] In France an urban area was in 1840 defined as any *commune* with a *chef-lieu* of more than 2,000.[20] A similar limit was adopted in Germany, but the threshold was set at 5,000 in Belgium and Spain; at 6,000 in Italy; at 10,000 in Switzerland and Greece and at 20,000 in the Netherlands. Elsewhere the status of town was less definite and the contradiction between 'legal' town and functional town was not fully resolved. In Hungary, for example, 'there were in 1890 thirteen legal "cities" having less than 3,000...while 38 other places that had more than 10,000...had not attained the dignity of "cities"'.[21] Everywhere in the peripheral regions of Europe the phenomenon was to be found of vast and highly nucleated settlements, urban in size but rural in function.[22] How these were represented in the national statistics is not always clear. A second source of error lies in the fact that boundaries of growing towns were extended at intervals during the century, thus including in later estimates of their population suburban areas which were omitted from the earlier.

About 1800 the urban population of Europe was, in percentage terms, little higher than it had been three centuries earlier. According to Bairoch's estimates, about 14.5 per cent was urban (fig. 4.4). During the first half of the century it increased approximately twofold, but as a fraction of the total population it rose only to 22.3 per cent. Bairoch included only settlements in which the population was thought to have been at least 5,000, thus excluding many, but not all the 'agrarian' towns, as well as many small, traditional towns. After the mid-century the growth, both relative and absolute, in the urban population was rapid: 32.4 per cent of a greatly increased total population in 1880; 40.9 per cent in 1900; and 43.5 in 1910, when the urban population of Europe is estimated to have been about 135,600,000.

Fig. 4.4 shows the growth of urban population as a percentage of total

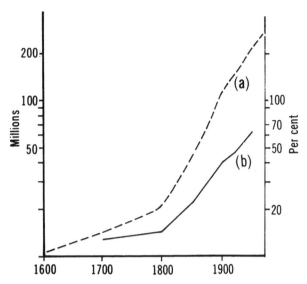

4.4 Growth of aggregate urban population in Europe, excluding Russia, 1600–1950; (a) aggregate, (b) as percentage of total population

population for all Europe, except Russia. Italy and Spain appear to have been relatively highly urbanised, largely because a considerable proportion of their agricultural population lived in large, agglomerated settlements which were urban in size if not wholly so in function. The rapid urban growth in the Low Countries is also to be noted, as well as the 'late-developers', countries in which significant growth did not begin until late in the nineteenth century. Amongst them were Sweden, Norway, Finland, Portugal and the countries of the Balkan peninsula.

Growth was far from evenly balanced between large towns and small. Broadly speaking, it was most marked in the towns which were already large; least in the small. Large towns grew larger, and the small stagnated. The basic reason why the latter are in general picturesque and attractive today lies in the fact that there was never any great need to pull down their ancient buildings and replace them with others more functional but less aesthetic. In both France and Germany, countries in which the pattern of urban growth was sharply contrasted, small towns grew proportionately very little, while the large expanded rapidly. In Germany in 1845 only about 15 per cent of the population lived in cities and towns of over 20,000. By 1910 this had increased to 37 per cent (fig. 4.5, table 4.1).[23] In France the growth of towns, in particular of large towns, was very much slower than in Germany. In 1910, towns of over 100,000 contained only 14 per cent of the population, in contrast with 21 per cent in Germany, and altogether only 34 per cent in towns of more than 20,000, as against 40 (fig. 4.6).

A feature of the growth of large cities was the emergence of what Mark Jefferson has called the 'primate city'.[24] In every country and region a single city not only achieved a position of primacy but also reached a population very much larger – in some instances several times larger – than the second-ranking

Urban development

Table 4.1. *German population by settlement size (in thousands, and as indices)*

	2,000–5,000	5,000–20,000	20,000–100,000	Over 100,000
1875	4,785	4,607	3,049	2,666
1900	6,099	6,695	6,552	8,712
1910	6,471	7,854	8,001	13,236
1875	100	100	100	100
1900	127.5	145.3	214.9	326.8
1910	135.2	170.5	262.4	496.5

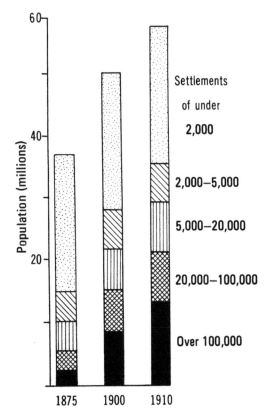

4.5 Settlements by size in Germany, 1875–1910

city. The population of Brussels, for example, rose from about 66,000 at the beginning of the century to 720,000 in 1910, from less than 2 per cent of the total population to 10 per cent (fig. 4.7). In Germany the phenomenon was yet more marked. At the beginning of the century Berlin contained less than 1 per cent of the estimated German popultion; by 1910 more than 3 per cent of the German population lived in Berlin and its suburbs. Other instances of the rapid growth of the primate city were Madrid, Copenhagen, Rome, Stockholm, Paris, Vienna, Amsterdam and Warsaw.

A discussion of the size of cities cannot be separated from that of the functions

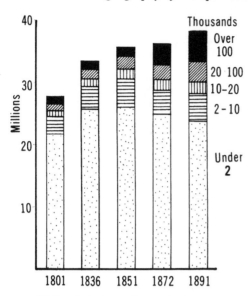

4.6 Population by settlement size in France. Note, in contrast with Germany, the predominance of small settlements

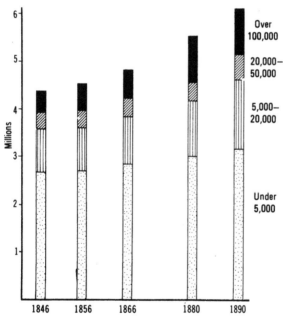

4.7 Population by settlement size in Belgium

which they performed. Large cities served many purposes. They were centres of administration in the widest sense. They sometimes contained military garrisons. They carried on basic industries, that is those which produced goods for the wider community, as well as non-basic industries which catered for the needs of the city itself. They had also a great many commercial and other service

Urban development

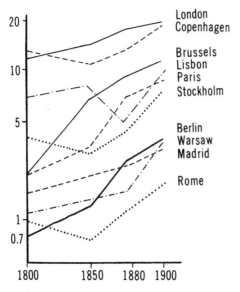

4.8 The size of the primate city as a percentage of national population, 1800–1910

functions. None of these functions could have been developed to the degree reached by the end of the nineteenth century without an elaborate system of transport. Indeed, all large cities grew *pari passu* with the building of the railway net which served them. The railway, followed by the spread of postal and telephone services, allowed urban functions to become more concentrated than ever before. The role of political capital had clearly the effect of aggregating political functions, but these also brought with them a number of others, cultural, commercial, legal and educational, which it was supposed, could be carried on most effectively near the seat of government.

The level of urban development was closely related to that of economic development, as measured by gross national product. The correlation between the two variables was high in those countries for which satisfactory data are available, about 0.8 at the beginning of the century, rising almost to 0.9 in 1900.[25]

Fig. 4.8 shows the growth of the population of capital or primate cities as a percentage of total population. All increased in size relative to the population of their respective countries over the period of a century, but several – notably Rome, Stockholm and Copenhagen – declined relatively during the first half of the century, even though they all showed an absolute increase. Rome did not become the capital of the whole of Italy until 1870, and both Denmark and Sweden showed little industrialisation and only a very slow rate of economic growth until late in the nineteenth century. Lisbon declined both relatively and absolutely, and was the only primate city to do so. On the other hand, the steepest rate of growth during the first half of the century – that experienced in Brussels – accompanied the earliest and most vigorous industrial expansion in continental Europe at that time (fig. 4.9).

Other capital cities which grew rapidly during the nineteenth century were Vienna and Budapest. Indeed, they were the fastest growing of all the great cities

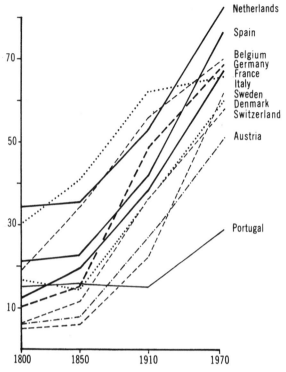

4.9 Aggregate population of cities, 1800–1970, as a percentage of national population

of Europe. Vienna increased eightfold within the period of a century and Budapest no less than sixteen times.

In only three areas was there no clearly discernible primate city: Switzerland, the Netherlands and south-eastern Europe. Although Switzerland had a federal capital in Bern, the country's constitution vested relatively little power in the central government, and left many functions to the cantonal. As there were thus as many capitals as there were cantons, there was little scope for the development of a central giant city. Geneva was the largest city until the 1870s, when Zurich greatly extended its municipal boundaries and drew ahead.[26] By the early years of the twentieth century it was without question the primate city of Switzerland in virtue of its financial, commercial and industrial functions. In the Netherlands, Amsterdam had previously been without question the primate city, but in the later years of the century was closely rivalled by The Hague, seat of the government, and Rotterdam, the most rapidly developing port.

There were few towns in the Balkan peninsula, and no large towns other than Constantinople. The functions which supported large cities were only feebly developed, and the largest owed their size chiefly to their military garrisons.[27] The achievement of national unity by the Balkan states was followed, however, by the development of functions associated with a capital city. Beograd, Bucharest, Sofia and Athens all grew in size, but, with the exception of Athens, less rapidly than the population as a whole until late in the century. Their

Urban development 133

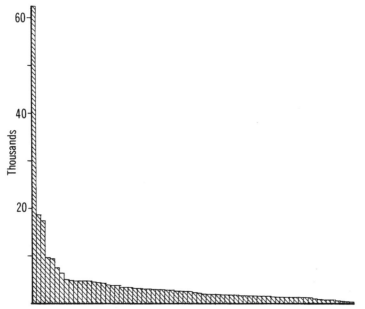

4.10 Rank–size graph of the towns of Białystok province, Poland (see fig. 4.2)

industrial development was slight before the First World War, and it was not until after 1918 that their primacy was clearly established.

The urban hierarchy

In pre-industrial Europe a great many small towns served as centres for their respective market areas. Simple manufacturing and service functions were duplicated in all of them, and the need, rarely experienced by most of the population, for higher-order functions had to be satisfied in regional capitals. Above them all stood the political capital, in almost all instances the primate city. The regional capitals were not always clearly distinguishable by size alone from the more numerous market centres. Placed in rank-order, they all tended to form a smooth curve, only the primate city rising high above the rest. In the province of Białystok, the primate town was Białystok itself (fig. 4.10). One can distinguish six towns which might qualify as regional capitals, and the remaining towns – no less than 63 of them – were only market towns of local importance.

But the *województwo* of Białystok was, even at the end of the nineteenth century, still pre-industrial in its social and economic structure (fig. 4.9). The effect of industrialisation and of the development of transport and other services was to complicate but not to destroy this hierarchical urban structure. Many of the towns which grew up in western and central Europe were too specialised to fit neatly into the three-tiered structure outlined above. The growing concentration of urban functions on the one hand, and the increasing specialisation of many urban centres on the other blurred this simple scheme. Additional levels intruded into the urban hierarchy between primate city and market town,

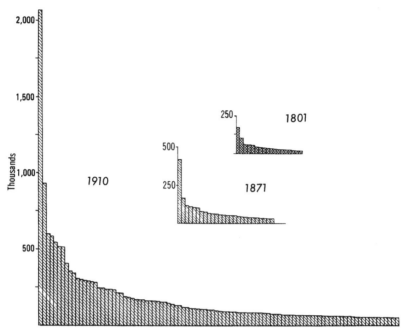

4.11 Rank–size graph of the larger towns in Germany in 1801, 1871 and 1910

so that one found for any country or major region a smooth transition in urban size from the largest to the smallest.

When ordered into rank-size the towns of any country or major region tended to form a concave graph similar to that shown for the towns of Białystok province. The second largest town *tended* to be about half the size of the first; the third to be one-third of its size; the fourth, a quarter, and so on.[28] But this is only a model, a theoretical way of describing an ideal world. The real world might differ in detail from the ideal, but there are no areas of Europe where the rank-size model is totally inapplicable. The Białystok graph, however, is unusual for a region which was still in most respects pre-industrial. The population of the chief town, Białystok itself, was more than three times the size of the second largest, Łomża. This strongly primate character of the largest settlement is a characteristic of industrialised societies, not of agrarian. In this instance it is due to the development, more for political than economic reasons, of a significant textile industry in the vicinity of the town (see p. 410).

The rank-size graph for Germany in 1910 (fig. 4.11) is more typical of that of a developed, industrial country. The tendency in such a country for the largest towns to grow fastest and for the smallest to stagnate has already been noted. This is well shown in the comparison of the rank-sizes of German towns in 1801 and 1871 with that for 1910 (fig. 4.11). The graph for France, intermediate between that for Germany and the graphs for pre-industrial regions, is flatter, with a weaker contrast between the very large and the small centres (fig. 4.12). The rank-size rule is no less applicable to most major regions of these countries,

Urban development

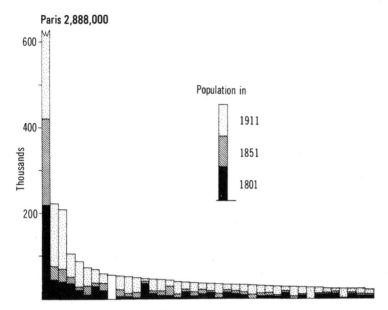

4.12 Rank–size graph of cities and towns in France in 1801, 1851 and 1911. Number 9, left blank on the graph, represents Nice, not part of France before 1860

which can be regarded as samples taken from the urban population of the country as a whole.²⁹

The pattern of towns

Pre-industrial towns had formed a not uneven pattern, such that in any well populated and cultivated area, no inhabited place lay more than the distance of a market journey from one of them. Scattered amongst these lowest-order central-places was a smaller number of larger places, each combining a broader spectrum of functions, and serving a much wider area. The study of the location and interrelationship of central-places was placed upon a scientific basis by Walter Christaller in 1933 (fig. 4.13).³⁰ He postulated an isotropic surface, without physical barriers, uniform in quality and with an evenly distributed population. Under such circumstances, he argued, the lowest-order places would form an even pattern, and, since the whole surface must be divided between their market areas, these must form a pattern of hexagons. Of course, the earth's surface is nowhere isotropic. There are physical barriers to movement as well as man-made obstacles like boundaries, tolls and tariffs. The urban centre itself needs resources which can be satisfied only at a limited number of points. The result is that the theoretical pattern postulated by Christaller and others who have since elaborated his model yet farther has been distorted in the real world. What matters is the extent of this distortion and the reasons why the real world differs from that assumed in the model.³¹

Christaller himself based his study on the settlements of southern Germany.

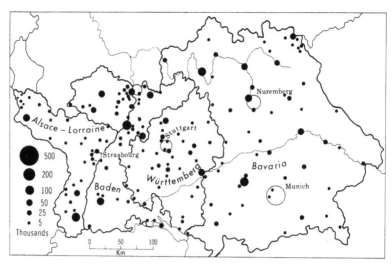

4.13 The urban pattern in southern Germany, including the *Reichsland* of Alsace-Lorraine, in 1910. The area shown is approximately that covered by Christaller's study

The region had developed a number of regional capitals (*Landstädte*), such as Munich and Stuttgart, and, below them in the urban hierarchy, he identified no less than six levels of settlement, each with a progressively smaller service area. Christaller's study was based upon the southern Germany of the Weimar Republic. Here reality came as close as it did to the theoretical model because the region, despite its hilly nature, had no highly localised mineral or fuel resources, no large industrial developments, few new manufacturing centres, and, on the other hand, little empty or negative area. Distortions of the theoretical pattern would have been even less in the nineteenth century than when Christaller wrote. Christaller was able to enumerate and measure the functions served by each centre in his own age, and thus established a hierarchy of settlements. To do the same for the earlier or even later nineteenth century calls for long and careful research in trade directories and other sources. It has not been undertaken. Instead, population about 1900 has alone been used to establish the urban hierarchy portrayed in fig. 4.13. As a *pattern*, it differs only in minor detail from that established by Christaller half a century later.

One can generalise from this study of southern Germany for much of Europe during the nineteenth century. A not dissimilar pattern, as has been seen, emerges for areas as far apart and as different as Alsace[32] and Białystok province. It can be found throughout the plains of northern France and northern Germany, in southern Sweden and northern Italy, over much of the Iberian peninsula and the Habsburg lands. The simplicity of this hierarchical pattern of central-places was, however, interrupted in two ways. In some areas it could never develop, owing to the physical contours of the land. Mountain ranges, as in the Alps and parts of the Balkan peninsula, enforced a linear development along the valleys which separated them. In some areas, such as Norway and Dalmatia, movement *along* the coast almost replaced inland movement. The resultant pattern was again linear, with larger urban centres, like beads on a string, at more or less regular intervals.[33]

The second constraint on the development of a more or less regular pattern of central-places was industrialisation, especially the concentrated form of industrial growth which took place on and near the coalfields during the nineteenth century. The *basic* function of such settlements came to be the mining of fuel and the manufacture of goods. They served a distant market, and there was never any question of that mutuality which characterised centres in a non-industrial region. The nearest approach to a higher-order central-place in such industrial regions was the city in which were concentrated many of the financial and commercial services essential for the area as a whole. Düsseldorf thus became a kind of business 'capital' for the Ruhr; Milan for the industrial towns of northern Italy; Lille and Liège respectively for those of northern France and central Belgium, and Kattowitz for Upper Silesia. At the beginning of the century there was a net of small towns serving the needs of a predominantly rural area. By the second half of the century there was rapid urban growth in the active area of the coalfield as well as in the textile region to the west of the Rhine, and by the early years of the present century a vast conurbation had emerged. A few of the older, pre-industrial towns, such as Essen, Dortmund and Düsseldorf, retained some higher-level functions in marketing and business administration, but a spatially recognisable hierarchy of towns and functions no longer existed.

Urban functions

Towns have always carried on the three types of function between which it is usual to divide all economic activity. Townspeople performed services; they manufactured goods, and throughout the nineteenth century they continued in many instances to cultivate the surrounding fields. The first two of these functions can be regarded as *basic* in so far as they were performed for and requited by a public which lived far beyond the urban limits. The last was non-basic, since the products of urban agriculture were consumed within the town. The balance between these functions not only differed between towns but tended to change through the century (fig. 4.14). Agriculture was of declining importance. It disappeared from the largest towns. In the mid-nineteenth century it employed less than half of 1 per cent of the population of Warsaw,[34] but at the same time Dijon, described as a city 'encore très liée à la terre',[35] had 14 per cent of its active population in agriculture. At Strasbourg the agricultural population was 6 per cent in 1866, but had fallen to less than 4 per cent by 1895.[36] The agricultural component of urban population was very much greater in the *Agrarstädte* of the Hungarian Plain and southern Europe, and there were many settlements of 5,000 and more which were predominantly agricultural in their occupational structure.[37] The service functions of towns tended to increase, but manufacturing played a very variable role in most, and was, broadly speaking, dominant only in some of the 'new' towns of the nineteenth century.

Simple marketing functions, which had been of primary importance in the small, pre-industrial market towns, tended to become of less importance, as

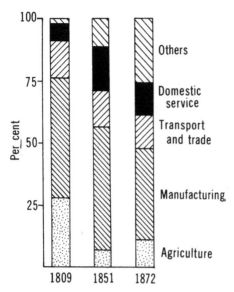

4.14 The occupational structure of Toulouse in 1809, 1851 and 1872

producer and consumer ceased gradually to come face to face. The larger towns outgrew such methods of supply and were obliged to draw on distant sources for foodstuffs and other goods. The same happened in marketing the basic products of a town; they were sold to dealers or agents who transported and sold them elsewhere. The 'face-to-face' market was gradually replaced by commercial houses which bought and sold by catalogue or by sample.[38]

A feature of the service occupations of the nineteenth-century town was the very high proportion of those employed in domestic and personal service. They were mainly women. They had come from the rural districts, while still in their teens, to the towns, where they distorted the age and sex structure (see p. 143) and by their great numbers ensured that their labour was cheap. Numbers varied with the functions of the town. They were most numerous where tertiary activities and a middle-class population predominated. In Warsaw, for example, they are said to have made up 28 per cent of the working population in 1869 and 22 per cent in 1882.[39] These estimates may have been excessive, but at Dijon domestic servants made up 18 per cent.[40] Their numbers varied inversely with the size of the industrial population. At Strasbourg (fig. 4.15), where more than 40 per cent were engaged in manufacturing, domestic service never employed over 4 per cent of the population.

One tends to suppose that the growing town was a centre of manufacturing. The earliest phases of the Industrial Revolution were, however, dependent on water-power and were accomplished in rural areas. The growing use of the steam-engine, however, either brought manufacturing back to the towns or led to the formation of new towns around the factories. The heavy industries tended to concentrate either near the coalfields or at points convenient for the transport of fuel. They disappeared from some towns, and were replaced by light or consumer-goods industries. The proportion of the population employed in

Urban development

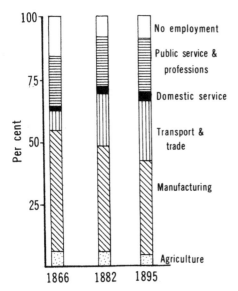

4.15 The occupational structure of Strasbourg in 1866, 1882 and 1895.

manufacturing was always higher in the 'new' towns than in the old. It rose to over a half and not infrequently to more than two-thirds of the active population. In the mining and iron-working town of Königshütte in Upper Silesia, those engaged in manufacturing increased between 1871 and 1907 to more than three-quarters of the actively employed.[41]

In eastern and, particularly, in south-eastern Europe towns retained throughout the century many of the traits of the pre-industrial city. Their occupational structure showed a predominance of crafts. An Ottoman survey of 1866 showed that in some of them cloth-weaving and leather-working on a domestic basis employed up to a third of the active population.[42] At the same time the continued importance of fairs had the effect of reducing the role of towns in wholesale and retail trade. Systems of transport were undeveloped, and 'the peasant [was] still the main means of transportation'.[43] In Pazardzhik near Plovdiv, scene of an important trading fair, more than a third of the active population was engaged in porterage and other forms of transport. The tertiary sector in such towns was large, but economically highly inefficient.

The social geography of the nineteenth-century city

The urban population of continental Europe increased from some 22 millions in 1800 – 14.5 per cent of the total population – to 135.6 millions, or 43.5 per cent, in 1910. Such growth could not have been achieved without a powerful movement to the towns from the rural areas. Indeed, there is good evidence that urban birth-rates were commonly, and especially in the larger cities, considerably less than the average. Most of the increase in population derived from immigration. The motives for migration (see p. 79) were mixed. As agriculture became more capital-intensive, labour, squeezed from the farms, moved to the towns. In many areas the towns were themselves an irresistible attraction; indeed they were the 'promised land'[44] to a peasantry living on the edge of starvation. Such migration was not new. It had been the mainspring of medieval urban growth and it had continued on a small scale during the following centuries. Without it many towns would not have survived. Some migration was seasonal, like that of the masons and construction workers who spent their summers in Paris and other French cities and returned to their villages in the Limousin and Auvergne for the winter.[45] In the course of time such workers found their poverty-stricken homelands less attractive and returned less frequently, until by the end of the century they became permanent city-dwellers. Only a small part of the enormous urban immigration originated in this way. Most resulted from a decision by a peasant, often enough a younger son with little hope of ever succeeding to a tenement, to seek work in the nearest town. Migrants were, it is often said, predominantly male, and most were between the ages of 15 and 30, young enough to establish themselves in their new homes, but not early enough in their lives to found large families.[46] The distorted sex structure of some towns, particularly those with a large middle-class component, shows that there was also a sizeable female migration. The women tended to enter domestic service, and, as has been seen (p. 143), formed in some instances 10 per cent or more of the employed population.

Migration was in general over only a short distance. Most migrants moved to the nearest large city. In Germany, in 1907, 54 per cent of the aggregate population of large towns was immigrant but the majority had moved from, at farthest, the neighbouring province,[47] and a large eastern town, such as Breslau or Königsberg, drew its population almost entirely from its local area. Cities in western Germany were less dependent on their local regions, but even here the proportion which came from distant parts of Germany was small. Fig. 4.16 shows the origin of migrants who were in 1912 settled in Cologne.[48] About a quarter had come from the surrounding *Regierungsbezirk* and over a third from the Rhine province. Numbers fell off sharply with increasing distance, though there was always a tendency for the number of urban migrants to be related to the size of the population of their home regions.

Immigration from distant sources, especially the German East, was more important in areas of rapid industrial growth, such as the Ruhr and greater Berlin. In 1893, 25 per cent of the labour in the industrial towns of the Ruhr was from the East, and by 1907 this had risen to nearly 34 per cent.[49] These immigrants provided much of the mine labour. There were mines in Gelsenkirchen and Recklinghausen in which two-thirds of the workers were from the east, and in the towns of the central Ruhr a significant part of the population was Polish-speaking.[50] Greater Berlin, where about 40 per cent of the population in 1907 were classed as *Fernwanderer*, also drew heavily on the German East. More than half the total number of migrants from Prussia, Posen and the rest of north-east Germany settled in greater Berlin. The growth of both the Ruhr and greater Berlin was so rapid that their local provinces and regions could in no way satisfy their demand for labour. Elsewhere in Germany, however, distant migrants formed a relatively small component in the total urban population.

Foreign migration to German cities was small, and much of it was temporary.

Urban development

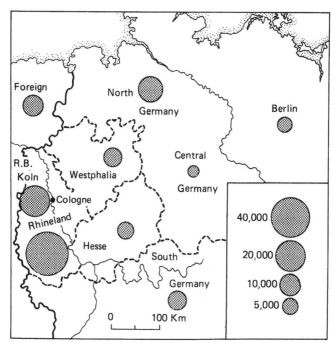

4.16 Sources of immigrant population in Cologne, 1906–12

In 1907 the largest foreign colonies were to be found in Duisburg-Ruhrort and other Rhineland ports, where they were most likely to have been Dutch, engaged in the river traffic; in the northern port cities, and in important commercial towns like Dresden, Chemnitz (Karl-Marx-Stadt) and Plauen, and, of course, in Berlin.

Although there had been an important overseas migration after 1848, internal migration remained small and was carried on over short distances until the 1880s. Long-distance migration then became important. The sharp decline at this time in the volume of overseas migration was a contributory factor, as also was the reduced importance of agriculture in the eastern provinces.[51] The scale of migration rose sharply in the 1890s. In the industrial towns of Rhineland-Westphalia alone, the number of immigrants from the eastern provinces increased nearly fivefold between 1885 and 1900, from about 83,650 to 381,110.[52] This intensive internal movement continued until the First World War. The occupational census of 1907 showed that only a half of the German population had continued to live in the community in which it had been born.

Internal migration in France, the only other country for which there is good evidence, was on a smaller scale than in Germany, and most of it was to the few large conurbations: greater Paris, the North, Lyons and Marseilles.[53] The rate of population growth in France was very much slower than in Germany, and the drift from the land was on a smaller scale.[54] Each of the major centres of urban growth had its own sphere from which it drew most of its immigrants. Migrants to Marseilles came mainly from Provence and the French Alps (fig. 4.17).[55] The Lyons region drew its migrants from the Rhône and Saône valleys

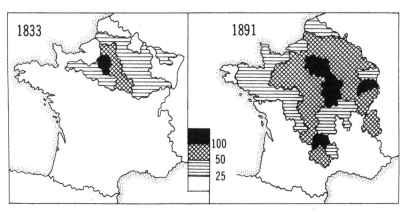

4.17 Migration to Paris; the source of migrants, 1833 and 1891. The figures represent those born in each *département* and living in Paris, expressed as the number per 1,000 of those living in the *département*

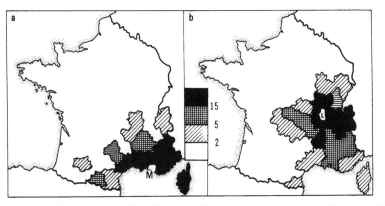

4.18 Migration to (a) Marseilles, (b) Lyons. The figures represent the number per 1,000 of population in the home *département*

and from the mountainous areas to the west and east. Its area of attraction overlapped those of both Marseilles and Paris (fig. 4.18). Paris, such was its attraction, drew from all parts of France except the south-west and Languedoc-Provence. This migration had begun long before the nineteenth century. Most came at first from *dépt* Seine and neighbouring *départements*, particularly those to the east. By 1891 the influence of Paris radiated much more widely. The whole of northern France from Brittany to Burgundy and as far south as the Central Massif contributed to the population of the capital.

A feature of the maps showing migration within France is that few moved from Languedoc and the area between the Loire and the Pyrenees to other parts of France. Migration appears to have been largely within the region, to the main local centres, such as Bordeaux, Toulouse and Montpellier. The reason, it has been suggested by Chatelain,[56] was a psychological one, the mental 'distance' of Paris and the other large urban centres. If this was indeed the case, it merely perpetuated a deep and ancient rift in French society between south and north.[57]

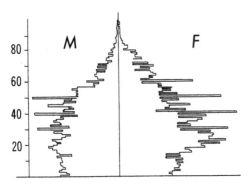

4.19 Population pyramid for Dijon, 1851

Age and sex structure

Both the age and sex structure of urban population was usually distorted in several ways, and was rarely typical of national populations as a whole. The immigration of young adults meant that the youngest age-groups were disproportionately small. Conditions of public health in the cities and liability to epidemic disease increased the death-rate and reduced the expectation of life below that prevailing in rural areas. Lastly, the sex ratio was unbalanced, though this imbalance varied greatly according to the dominant functions of the town. Adna Weber, commenting on the high ratio of women to men, considered it 'an established fact that the cities produce of themselves a larger preponderance of women than do the rural districts'.[58] There is no evidence that this was indeed the case. The reason is rather to be found in the fact that women could find little employment in rural areas, especially after the decline of domestic spinning, and migrated in large numbers to the towns, where they were employed mainly as domestic servants. The town of Dijon in 1851 showed a large excess of women between the ages of about 18 and 35 (fig. 4.19) and also a very high level of female employment. Whereas 71 per cent of males were employed, the percentage amongst females was 58. 'In absolute numbers', wrote Viennot, 'working women were almost as numerous as the men: 9,137 as against 9,390...[and] female immigration almost wholly balanced those who married and ceased to work.'[59] But Dijon, with little industry and a large *rentier* and middle class, was a town likely to attract female labour from the surrounding countryside, much of which would probably in its later years return to the villages from which it had come. Toulouse showed a similar demographic pattern with a sex ratio overbalanced on the female side and small younger age-groups.[60]

Arras was more typical of industrial towns. In 1876 its excess of female inhabitants was small, and it is evident from its population pyramid that the immigration of young males was scarcely less than that of females (fig. 4.20).[61] In truly industrial towns, such as those of the Ruhr, where there was little scope for female employment, the sex ratio was tipped in the opposite direction. The demand was for miners, not maidservants, and a relatively high proportion of the immigrants from the German East was male, and lived in vast dormitory blocks erected by the mining companies.

Table 4.2. *Urban birth- and death-rates in Germany, 1902–8*

	Birth-rate	Death-rate
Berlin	24.5	16.12
8 towns of over 250,000	29.89	17.97
24 other great towns	32.26	18.12
53 towns of over 50,000	32.3	17.64
Towns of over 15,000	31.52	18.31

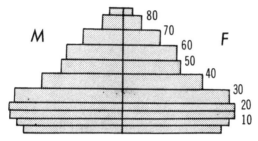

4.20 Population pyramid for Arras in 1876. It suggests a low urban birth-rate and a considerable net in-migration

Urban birth-rates have always been low, commonly two or three births per thousand of population below the national averages, and the larger the city the lower, as a general rule, were the birth-rates (table 4.2).[62] This arose in part from the bad sanitary conditions which prevailed in most towns, but more especially from the large number of immigrant adults, many of whom would be unlikely to marry before establishing themselves in some kind of position. On the other hand, illegitimacy rates were high, especially in rapidly growing industrial towns such as Lille.[63] Natural growth accounted for only a small part of the rapid urban expansion during the nineteenth century. In Paris, for example, immigration made up 71 per cent of the population growth between 1872 and 1901.[64]

If the urban birth-rate was, in general, low, the death-rate was abnormally high. To some extent this was inevitable owing to the large-scale immigration of adults with a shorter expectation of life, but it was due also to the physical conditions which prevailed in all except the smallest towns. Friedrich Engels has left an unforgettable picture of conditions during the first half of the century in the rapidly growing industrial towns of Britain. Conditions differed in detail in those of continental Europe, but were on balance no better. Furthermore, governmental attempts to improve the conditions of life came later than in Britain and were in some areas conspicuously less successful.

Urban health

Urban congestion in all except the smallest towns was extreme. Migration to the growing centres of manufacturing was faster than a slowly growing stock of housing could accommodate. At the beginning of the century instances were cited in Paris of as many as 32 persons living in a house built originally for a single family.[65] The situation became worse as the century progressed, and was relieved towards its end only by the building of vast tenement blocks and the dispersion of housing over a much wider area (see p. 150). Conditions of such acute overcrowding provided ideal conditions for the spread of infections and contagious diseases. No large town was free from outbreaks of cholera, typhoid, typhus and smallpox as well as from the many 'fevers' which sprang from infected food and contaminated water. The history of towns in the nineteenth century, wrote Lavedan,[66] is the history of disease. The provision of a supply of pure water and the disposal of sewage were in this respect the most important tasks facing urban authorities, and few had the knowledge or the resources to cope with them.

In many towns, especially those which grew most rapidly, there was no adequate system of sewage disposal until late in the century. Domestic waste was thrown into the street or deposited in heaps from which it was periodically carted away. Toilets discharged into cesspits or into crudely constructed drains, which in turn led down to the nearest river. In Paris a rudimentary sewer took surface drainage and part, at least, of the domestic sewage. The rest was passed into *'fosses'*, euphemistically translated as septic tanks. In Paris, as late as 1880 there were no less than 70,000 *fosses* which served a much greater number of houses.[67] The aroma of one of them, the Montfaucon cesspool, could, it was said, be detected if the wind was in the right direction over most of the eastern quarters of the city.[68] Part of the sewage was collected and carted to the market gardens which ringed the city. The rest was either allowed to soak into the soil or was discharged into an already polluted river. Disease was endemic, erupting into the all too frequent 'cholera-years' and 'typhoid years', while dysentery, typhus and other fevers took a steady toll. The system of drains was partially rebuilt in the course of Haussmann's rebuilding of the city, and major sewers were constructed to discharge into the Seine *below* the city.[69] But it was not until 1894 that French law required the construction of a comprehensive sewer system in towns. It was opposed, not only by the property-owners who had to pay for it, but also by the *vidangeurs* whose unsavoury task it was to empty the cesspits and transport their contents to the fields. Sewer systems had already been constructed by this date in Brussels and in most of the capital and other large cities. Progress was less rapid in provincial towns. In Toulouse, for example, there was a marked reluctance to comply with the law, and public sanitation as late as 1900 was described as 'archaic'[70] with open sewers along many of the streets. Progress in the provision of a system of sewers was faster in Germany than in France, though it was not until the beginning of the present century that it began to be governed by public law. It was less difficult to develop and maintain a system of public sewers in the Scandinavian countries, where urban growth came late and large cities were few, but in southern Europe little progress was made before the First world War. Many cities, notably those in peninsular Italy and southern Spain, remained notorious for their lack of hygiene.

The question of sewage disposal canot be wholly separated from that of water supply. Indeed, the failure to separate the one from the other in the soil beneath the towns was source of the cholera epidemics which devastated so many of them.

No European city in the early nineteenth century had a water supply as assured as that of many cities of the Roman Empire. They drew it from wells within their limits, from springs beyond them, or they merely dipped it from the rivers. As late as the 1890s Hamburg was taking its water unfiltered from the Elbe, already polluted with the effluent of Altona and of Hamburg itself. The result was the severe cholera epidemic of 1892, which led in turn to a drastic overhaul of the water supply of the major German towns. But Germany's record was better than that of many other European countries. A piped water supply derived from sources as free as possible from contamination had already been established in a number of German cities,[71] at Berlin and Düsseldorf in 1856, at Halle in 1867, at Remcheid in 1884 and at Munich in 1880–93.[72] The technique of sinking deep wells, which tapped layers too deep for contamination, was improved. From 1825 butt-welded tubes were produced for this purpose, and in 1841 a well near Paris was taken down 1,800 feet.[73] Nevertheless, progress was very uneven, even after the connection between epidemic disease and contaminated water supply had been established beyond question. In Paris itself the Seine remained the chief source until 1860,[74] when Haussmann proposed to draw water from springs in Champagne; but his plans proved to be too ambitious, and a pipeline was built instead from springs in *dépt* Aisne.[75] At Rennes, in Brittany, typhoid was the cause of more than 5 per cent of all deaths between 1870 and 1882. In 1883 a piped water supply was laid to parts of the city. At once, cases of typhoid became less numerous here and ultimately ceased.[76]

At Brussels a *machine hydraulique* was used to pump water from springs within the city into a tower-like building, from which it flowed to fountains within the city. This system endured until 1869, when it was replaced with a supply piped from sources beyond the city limits.[77]

In general it can be said that in western and central Europe only the larger towns, and within them only the well-to-do quarters, had an improved water supply by 1850. In the second half of the century an adequate provision became general in the larger towns and was frequently to be found in those of medium size. But progress was none the less highly uneven. In France, a public law of 1902 required each *commune* to ensure a supply of safe drinking water: Toulouse did not even begin to comply until three years later, and the supply as late as 1921 was said to be totally inadequate.[78]

Progress was more rapid where the economic benefits of an abundant supply of pure water were obvious. In the Ruhr, for example, the industrial need was seen to be even greater than the domestic. In 1863 the town of Essen established a waterworks in the Ruhr valley near Steele, 8 km away, and was followed in this by several private companies needing water for cooling and quenching purposes.[79] By 1913 no less than eleven dams had been built on the river Ruhr and its tributaries and a network of 1,582 km of pipeline distributed the water to the towns and factories of the industrial region. At the same time the *Ruhrwasserverband* was created to conserve water and co-ordinate the many projects.[80] Nowhere else in continental Europe was such care taken to provide a piped supply of pure water – but nowhere else was the industrial need as great. In Upper Silesia, the only comparable region, an abundant supply could always be obtained from the water-bearing Triassic beds which bordered and in part underlay the region.

In most of eastern and south-eastern Europe urban water supplies continued to be obtained from local wells and springs, or to be taken from the river. Even in Vienna, where a project to bring water from the Semmering, 80 km away, was begun in 1841, mountain water did not reach the city until 1910.[81]

Urban energy supplies

Scarcely less important than an assured water supply was a means of lighting streets and homes by night. Both safety and convenience required some street illumination, and the need became greater as towns grew larger. The first illuminant to be used was gas obtained by the distillation of coal. Although experiments in the use of gas were made in France and Belgium, the first successful gas installations were probably in England in the 1790s, and the first city to be illuminated by it was London in 1813, followed by Bristol ten years later. Gas lighting was slower in gaining acceptance in most of continental Europe. In Paris the first gas street lighting was not installed until the 1820s and then largely because the unlit streets encouraged crime. In Germany the first city to be lit by gas was Berlin in 1816.[82] During the first half of the century the use of gas spread to most of the larger German cities, to Düsseldorf, Leipzig, Mannheim, Munich and to the port cities of the North Sea and Baltic coasts, where English coal could be obtained relatively cheaply.[83]

By 1850 35 German cities were lit by gas, but their distribution was strongly influenced by the availability of coal. More than half were in the Rhineland, where Ruhr coal could be distributed by barge, and several of the others were well placed to receive sea-borne coal from Britain. Within ten years the situation had completely changed. The primary railway net had been completed and coal could be delivered to dozens of towns inaccessible to river barges. By 1860, the number of towns lit by gas had risen to 250, distributed throughout the country though least numerous in the north-east (fig. 4.21).

The cost of building a gasworks and laying down the necessary pipelines for public and domestic lighting was high, and very few towns in eastern and southern Europe would make such an investment before the twentieth century. Even in France, the municipal gasworks was not found widely before the closing years of the nineteenth century. As a general rule only the main streets were served by gas pipelines and only those homes which lay close to them could be reached. Inevitably it was the well-to-do quarters which received gas first for both public and private lighting. The areas of poor and working-class housing had until late in the century to rely on oil lamps and candles, and, as for street lighting, there was none, with consequent danger to life and property.

The revolution effected by electric power did not come to the cities until the last two decades of the nineteenth century. The generating station was more complex and more costly to build than a gasworks, though problems of distribution were less serious. Although experiments in generating electricity had gone on throughout much of the century, it was not until after 1880 that large generators for public supply began to be built. They were driven by reciprocating

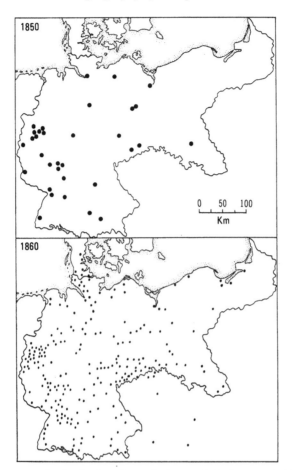

4.21 The distribution of municipal gasworks in Germany in 1850 and 1860

steam-engines and were thus subject to the same problems of fuel supply as the gasworks.[84] In Germany the earliest generators for public supply were built in the 1880s and 1890s,[85] and, as with gas installations, they came first to the cities of the Rhineland and West Germany. Berlin's first power station was built in 1884.[86] Paris began to use electric lighting in 1875, but it was adopted only very slowly in the provinces except where there was an industrial demand for power. From the 1880s hydroelectricity became available in the French Alps and the Pyrenees, but problems of long-distance transmission limited its usefulness. Nevertheless Grenoble, Lyons and Toulouse enjoyed the benefits of electric lighting before the end of the century and Marseilles and Saint-Étienne by the time of the First World War.[87]

Urban transport systems

Transport within the pre-industrial town presented few difficulties. The peasant's wagon and pack animal, supplemented locally by the river barge, brought goods to market and distributed most of the products of the urban crafts. Passenger traffic was by horse-drawn coach over roads of indifferent quality. Most of the towns were themselves too small to require an internal transit system. Only in the largest did the rich move about in their carriages, which, as Mercier said of Paris,[88] made the streets dangerous. But as towns grew larger during the nineteenth century and spread more widely over the surrounding countryside, the need increased for a cheap means of travel about the muddy streets and between the suburbs and the town centre.

This need was at first met by horse-buses and then by horse-trams. The first horse-drawn 'omnibus' operated in Nantes in 1826, and two years later a service 'for all' was established in Paris.[89] It was soon found, however, that, if the vehicle was made to run on iron rails, frictional drag was greatly reduced and the effectiveness of the horses increased. Paris had a rudimentary tramway by 1853–5.[90] But it was extended only slowly because of the government's objection to the use of trams in the congested city centre. By 1893, however, the net embraced 370 km of rails, and was of inestimable importance to the life of the city. A public transit system developed later in Berlin. A horse-bus service began in 1868, and was supplemented a few years later by one of the few systems of steam-trams. Many of the larger German cities established systems of horse-drawn trams in the 1870s: Munich and Königsberg in 1876; Mannheim in 1878, and others during the years immediately following.[91]

While lines were still being laid for horse-trams in the smaller towns, the larger were converting their existing systems to electric power. In this Berlin was in the forefront, owing in larger measure to the presence within the city of Werner von Siemens' electrical engineering works.[92] Elsewhere the conversion of horse-trams to the familiar electric trams was slower. Munich made the change in 1895 and other large towns followed. Most towns of medium or large size in western and central Europe had electric tramways by 1914, but in the east and south their presence was restricted to the largest cities in which alone the necessary generating capacity existed.

It was difficult to draw a clear distinction between the tramway system and the railway, beyond the fact that the former used the public streets and the latter acquired and held land privately. Both used mechanical traction and often the same gauge of track, and sometimes the two systems were interlinked. Tramways were sometimes extended far out into the surrounding countryside, and in Belgium an interurban system of faster and heavier trams, intermediate between a tramway and a railway, was developed between several of the larger towns. Another hybrid form of transport, developed at the end of the century, was the Paris Métro. Its prototype had been London's Metropolitan Railway. It was built in part on the surface, in part underground. Although steam traction was used at first, this was quickly replaced by electricity. Its purpose was to provide rapid transit between the outer suburbs and the city centre, though, in fact it scarcely extended beyond the limits of the city.[93] Lastly, the motor bus began to make its appearance early in the twentieth century, and Berlin had a rudimentary service in 1902.

Urban housing

The rapid growth of towns during the nineteenth century necessitated a corresponding increase in urban housing. Growth of the latter was slow during the first half of the century. Population density, congestion and squalor increased, and were a not unimportant factor in the political disturbances of 1848. In Paris, for example, the number of persons living in the average house is reported to have increased from 21.9 in 1800 to 35.2 in 1851.[94] During the second half of the century there was a more active building programme, but the stock of houses at best only kept up with the influx of people. During the first half of the century the increased population settled as a general rule in the closely built city centres. Buildings were commonly of several storeys and many were built largely if not wholly of wood and plaster. They lacked domestic facilities and amenities; they decayed rapidly and were particularly susceptible to fire. Literary descriptions of this urban squalor are numerous. There was no limit to the congestion and misery described by Balzac and Hugo. Indeed, *Les Misérables* can be looked on as a social geography of the poorer quarters of Paris during the July Monarchy.[95] As late as 1908 a report by the Board of Trade said of the central parts of the older towns that 'conditions as to light and air are those of three centuries ago'.[96] The same could once have been said of every growing industrial town, and was still applicable to many on the eve of the First World War.

During the middle years of the nineteenth century factories and their related housing were springing up not only in the new industrial towns, but also around the tightly built centres of the old. The type of building depended upon the dominant industry and on its mode of organisation. Most common were factories, each employing hundreds of workers, who lived as close as practicable to the scene of their employment (fig. 4.22). Rows of terraced houses were built between surviving pre-industrial cottages. Many were 'back-to-back', especially in Belgium, where they came in for much unfavourable comment from the Commissioners of the Board of Trade.[97] More common in the latter part of the century were the tenement blocks or *Mietskaserne*. They became the dominant form of workers' housing in central Europe and were important in the larger towns of western Europe. They were first built in the middle years of the century, but became most numerous after the formation of the German Empire.[98] 'The large house with a considerable number of tenements', wrote the Board of Trade Commissioners, 'is become more and more predominant... In some of the larger towns these erections often resemble large barracks built round small paved courtyards... Nothing could be more depressing than the sight of the huge structures which surround many of the courtyards in working-class districts... merely a builder's device for exploiting costly sites...'[99]

Such tenement blocks came to dominate the larger towns. Munich was essentially a city of large, barrack-like homes; Stuttgart, Stettin and Barmen were described in similar terms,[100] while at Breslau most of the dwellings were tenements,[101] and at Königshütte 'practically the whole working-class population... may... be said to live in tenements of one or two rooms'.[102] Berlin, which grew within a few years from a city of moderate size to be the largest

Urban development

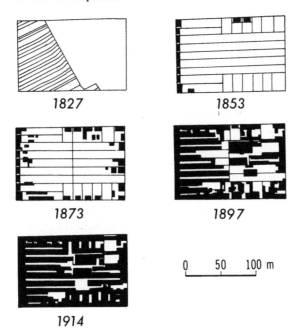

4.22 Changes in a single city block in Łódź, 1827–1914. The diagram for 1827 represents agricultural land

on the continent of Europe, became one of the 'grössten Mietskasernenstädte der Welt', in which a vast population lived in conditions as squalid and overcrowded as any to be found in Europe at this time.[103]

In cities like Berlin most of the tenements were erected by speculative builders, but many industrial firms also built housing in order to attract workers. In this respect Krupp of Essen had a good reputation for the quality of the housing provided.[104] The Bochumer Verein provided about 1,500 housing units at Bochum, and many of the Ruhr coal mines built tenements of a crudely functional kind for their workers, many of them Poles, who had no alternative but to accept what was offered.

Tenements were less prevalent in the Low Countries and France, even though, it was said,[105] 'the tenement system prevails universally' in Paris. The French tenement was less forbidding than the German, and, according to the report of 1909, they had on average only about 13 apartments in each. Outside Paris, the tenement block was rare, largely because both towns and manufacturing units remained small. An exception was Lille,[106] where tenements predominated in the older industrial quarters, and cottages in the newer. Conditions in the former, described in lurid terms by Hugo, were still 'appalling' in 1909.[107]

The worst industrial housing conditions were probably to be found in Upper Silesia. Alongside the surviving peasant cottages were blocks of four or five storeys, even more crudely functional than those of the Ruhr and Berlin. A family dwelling consisted of at most two rooms. Nearby were *Schlafhäuser*, which were no more than their name suggests – barrack-like buildings of the most primitive order, in which unmarried workers, just recruited from the villages, merely slept. In 1890 these provided accommodation for some 3,000. By 1913 there were 220 *Schlafhäuser*, equipped with no fewer than 28,586 beds.[108]

In those industrial towns in which craft industries remained important, neither the tenement nor the back-to-back row became significant. In Elberfeld and Plauen, where hand-weaving and hand-embroidery remained important until the end of the nineteenth century, workers lived mainly in small houses incorporating a workshop. The same was true of the makers of cutlery and small iron goods in Remscheid and Solingen.[109] In the industrial towns of the Low Countries, France, Spain and Italy, industrial housing remained primitive, but in general avoided both the tenement and the rows of back-to-backs. Łódź must stand as typical of the over-rapid growth of a factory town, yet it never developed tenements on the scale of many German towns. The early immigrants merely 'squatted' and, as the opportunity arose, built timber cottages of typically Polish design for themselves. As late as 1860, more than 86 per cent of the housing was still of wooden construction.[110] Even at the end of the century most of the domestic building was still in wood, the only significant masonry buildings being the few dormitories built for temporary occupation, like those to be found in Upper Silesia.

Little control was exercised by the urban authorities over the quality of building before the end of the nineteenth century. Attempts were made in some of the capital cities to limit the height of tenements and other buildings, and occasionally to restrict certain types of business – commonly those of an insanitary or offensive nature – to specified quarters.[111] Governments found it difficult to regulate building standards, when materials and traditional designs varied so greatly, and local authorities were usually unwilling, since the city fathers, as a general rule, wanted nothing so much as complete freedom from regulation. There were, however, exceptions. In the rebuilding of Hamburg after the fire of 1842 it was prescribed that there should be a water-closet on each floor of tenements, and at Basel in 1895 a local ordinance required that all living-rooms built after that date should have a window.[112] But governments in western and central Europe did little more than formulate highly general rules regarding water supply and the disposal of sewage, rules which many of their towns showed considerable reluctance to implement.

The local authorities inherited the obligation to maintain the streets in good order and to cleanse them of the rubbish which daily accumulated. This duty was discharged with varying degrees of care. Most streets and roads tended to be concave in section, so that water made its way to the centre and flowed thence towards the river or sewer. In the 1840s the Paris authorities began to build roads of convex section, with a gutter for drainage purposes on each side. This could only be done successfully if pavements or sidewalks were added to prevent water from flowing into the houses. It served, however, to keep the road surface dry and the sidewalk was a convenience for pedestrians. The road surfaces were generally cobbled or paved with slabs or sets, but as towns grew, road-building rarely kept up with urban expansion, and all too often roads were of earth stiffened with stony rubble, in turn muddy or dusty according to the weather.[113]

The urban plan

Every city and town in nineteenth-century Europe had its own individual plan, and grew in its own peculiar way, guided by the contours of the land and the infinite number of decisions made by its citizens. Yet it is possible to discern beneath the variety of modern urbanism a few regularly recurring patterns. Certain aspects of this urban plan have already been discussed. The pre-industrial town, especially if it had formerly been walled, grew in a very different way from the town which developed *ab initio* during the nineteenth century. In the case of the former, the area within the walls, even if they had been demolished, became increasingly congested until, as if with an explosive force, it burst its constraints and spread outwards over the enclosing fields.

Within the old town there was usually a central *place* or square, often irregular in plan, which focused the town's activities. In a large town there were usually subsidiary market-places, sometimes with specialised names, though their corresponding functions had often disappeared by the nineteenth century. Town- or gild-hall usually fronted upon the chief square, and nearby was commonly to be found the principal church of the town. Any town of even moderate size was divided into a number of parishes each with a church of greater or lesser distinction, and there would have been other churches of non-parochial status, the chapels of vanished gilds or of the orders of friars who had, in most instances, also left the urban scene. The street pattern varied greatly. In some instances it had clearly once been rectilinear, though distorted by rebuilding and encroachments in less orderly times. This plan might have derived from the street pattern of a Roman town, as at Cologne, Bordeaux and Modena, or from a medieval planned layout as at Breslau and the Hanseatic cities of the Baltic coast; or it might be quite irregular, as at Poitiers, Lille and Quedlinburg.[114]

Many towns had not even filled out the space enclosed by their late medieval walls by the early nineteenth century. Even Cologne was little more than half built-up in 1815. But the larger and more rapidly growing cities were becoming congested, with tall, narrow-fronted houses facing on to poorly lit and ill-paved streets. Building outside the line of the walls was usually more spacious because land values were lower. At first there was a zone, developed in the early or middle years of the century, with factories, railways and workers' housing. As the town continued to spread, the quality of building improved with greater wealth available for construction. With the development of means of public transport people moved from the congested quarters of the Old City to the expanding periphery. Then the inner city, lying just outside the limits of the old town, began to decay, and by the early twentieth century there was a slow exodus also from this zone towards the suburbs. The nineteenth- and early twentieth-century town thus came it is said, to consist of a series of concentric rings or belts, the older and more highly industrialised near the centre, the newer towards the periphery.

This model of urban development is associated with E. W. Burgess, who elaborated it in the context of the city of Chicago.[115] But the outward creep of the town and the associated changes in urban function and quality of living can be demonstrated in countless European cities. Growth was also influenced by factors other than simple distance from the city centre. The alignment of railway and canal, liability to flooding and the contours of the land influenced development in countless ways. The result was a number of distinctive belts lying radially across the concentric zones of the city. The pretext for such transverse lines of development was most often a road, railway, or navigable waterway. These attracted factory development, and the latter in turn led to working-class housing. The better housing development, repelled by industrialised areas, tended to be located along axes intermediate between the industrial and working-class developments. This model, which assumes that the development of the various urban functions took place radially, is associated with Homer Hoyt, who first propounded it as a description of urban growth in 1939.[116] The models proposed respectively by Burgess and Hoyt are not mutually exclusive. In most of the nineteenth-century towns there were elements of both. In many, it is as if a radial pattern of industrial belts was laid athwart Burgess's concentric zones, which reflected social development.

The presence of a pre-industrial urban core imposed a kind of order on urban development. Even the suburbs tended to be graded, both functionally and socially, by reference to the city centre. But the 'new' towns never had such a point of reference. Their growth from nothing was unplanned and haphazard, and at no time does it appear that environmental factors were ever taken into consideration. Around it workers' housing grew up, and with it a few shops for the supply of lower-order consumer goods. Sometimes, notably in Upper Silesia, the shops were operated by the factory owners on a kind of truck system. The result was a series of urban nuclei, not an organised urban community. In the course of time these nuclei grew into one another, producing an urban sprawl entirely without focus or character. Łódź, Sosnowiec[117] and Königshütte may have been extreme cases, but there were many other examples of this form of urban development. The Board of Trade report of 1908 commented that factories were scattered through the whole of Chemnitz,[118] and in Roubaix and Tourcoing, where entrepreneurs first erected a factory and then ringed it with 'rows and courts', the situation was similar.[119]

Only in the broadest sense can this pattern of urban development be related to the models of Burgess and Hoyt. Fig. 4.23, showing the growth of Roubaix and Tourcoing, looks superficially like an example of concentric development, as postulated by Burgess, but this disguises, especially within zone 2, a multi-nuclear development around a large number of factory sites. Precision was given to this multi-nuclear model by C. D. Harris and E. L. Ullman. It is clearly not universally applicable, though localised examples of it may be found in the suburban areas of many of the larger towns.[120]

Little provision was made in the course of urban growth for parks and open spaces. Such as there were usually owed their survival to the accident that no one had chosen to build over them. In a few instances the destruction of the former city walls had left an open space which was used for recreational purposes. Berlin in 1920 had 7 per cent of its urban area in parks, and this was much more than most growing cities could boast.[121] The fact that almost a quarter of Paris was in parks and open spaces was due in large measure to the planned rebuilding of much of the city by Haussmann, but most lay near the outskirts and their value as 'lungs' was thus restricted.

In the pre-industrial city particular crafts had often been grouped together,

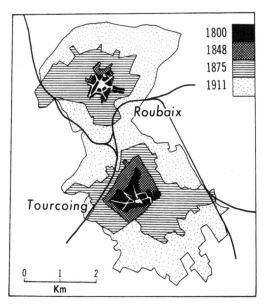

4.23 The urban growth of Roubaix and Tourcoing

but place of work nevertheless was not usually separated from place of residence. A workshop commonly occupied the ground floor or the back quarters of a dwelling house. There was in consequence relatively little flow within the city. 'Parisians did not move about the city', wrote Pickney; 'they lived and worked within a few blocks.'[122] This situation, however, tended gradually to break down, and the city centre became more and more a place of business rather than of residence. Rich and poor, merchant and craftsmen, had lived close together in the old city centre. There was no spatial barrier separating classes. The outward spread of the town, however, changed this (fig. 4.24). The well-to-to were the first to withdraw from the city centres, where their town houses degenerated into warehouses and tenements, and took up residence in the suburbs.[123] Class and wealth began to be expressed spatially and late in the nineteenth century one begins to find suburbs graded by social class. One can measure this by the density of residential buildings, by their tax assessments and by the number of domestic servants employed.

This process of zonation, present in all towns during the century, was most conspicuous in the largest. In Berlin the south-western suburbs, Schöneberg, Wilmersdorf, Steglitz and Dahlem, became the most affluent, and the eastern and south-eastern the most industrialised (see p. 170). In Paris the development of 'quarters' was particularly marked (fig. 4.25).[124] The east and north-east, especially *arrondissements* 11, 19 and 20, came to be pre-eminently working-class quarters, and the western *arrondissements* lying to the north of the Seine, in particular numbers 1, 8, 9 and 16, were favoured by the well-to-do.[125] There was a tendency for the western quarters of great cities to contain better-quality housing, than the eastern, which lay most of the time downwind, characterised by industrial development and working-class housing.

Not only was there a sharp contrast in wealth between different quarters of

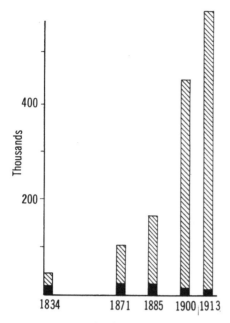

4.24 Growth of population in Leipzig. That of the Old City is seen to contract with movement from the centre to the suburbs. (The column indicates the total population of the city; the portion in black, that of the *Altstadt* or Old City)

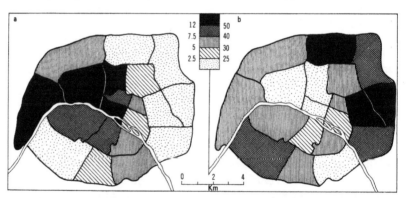

4.25 Distribution of (a) domestic servants in Paris, 1872 and (b) working-class population. The figures represent percentages of total population

any large city, there was also a contrast between cities. The small town continued to be preponderantly middle-class, with an appreciable number of *rentiers* and professional people, and a relatively large proportion of shopkeepers and craftsmen. The larger and more heavily industrialised towns showed a much wider spectrum. The rich were very rich indeed, but there were large numbers of very poor, the true proletariat. Statistics published in 1899[126] show the size of income groups in the twenty largest German cities, and emphasise the wide range between affluent cities and industrial cities.

Urban food supply

The supply of food to the small pre-industrial town had, under normal harvest conditions, never presented any problem. The small town and its rural sphere had developed in mutual dependence. This changed with the growth of the large industrial town, and urban food supply adjusted itself slowly and sometimes painfully to the new circumstances. Paris had been the only really large city before the nineteenth century, and here a mechanism of supply had been developed over the centuries, from grain and animal markets widely scattered through the Paris Basin.[127] This traditional mode of supply was ceasing to be adequate as the city continued to grow during the nineteenth century. The original Paris market, *les Halles*, was supplemented by district markets, which continued to increase in number.[128] This process continued until Haussmann's rebuilding. All through the century the traditional sources of supply within the Paris basin contributed grain and meat; intensively cultivated lands close to the city supplied vegetables, and after the railways had been built, milk and dairy produce were brought from Normandy and Brittany. But these sources had increasingly to be supplemented from overseas, especially with wheat, imported through Rouen and brought up the river by boat.[129]

Elsewhere the pre-industrial system of supply was not seriously taxed until the middle years of the century. In the lower Rhineland, the most highly urbanised part of Europe, the daily supply of milk and produce from the adjoining rural areas was maintained into the second half of the century. As land values rose in the vicinity of the growing cities, agriculture became more intensive, with a sharp increase in the production of vegetables. By the 1860s, however, the supply of bread-grains from local sources had become inadequate,[130] and imports through the Rhine mouth ports increased. Duisburg and Dortmund developed late in the century as the chief flour-milling centres for the region. At the same time the market for meat was extended out into the Münsterland, and that for vegetables into the Netherlands. Food-processing industries developed in the Rhineland cities, notably oil-pressing and sugar-refining. By the early twentieth century dependence on non-local sources for at least the less perishable foodstuffs was overwhelming. This trend was strengthened by the growing use in the late nineteenth century of refrigeration in the transport of some foods, especially meat. At the same time the pasteurisation of milk permitted it to be kept longer and transported farther.

The mechanism for wholesaling and retailing varied greatly with local circumstances. In the smaller towns the direct sale by peasant producer to urban consumer persisted through the medium of the market, though bread-grains seem always to have been handled by dealers, probably because of the need for storage before sale. Even in the largest cities the peasant market continued to contribute to the supply of perishable foodstuffs though its role was of diminishing importance, as large central markets, like the rebuilt *les Halles* in Paris, began to supply the retail shops with produce from distant sources.

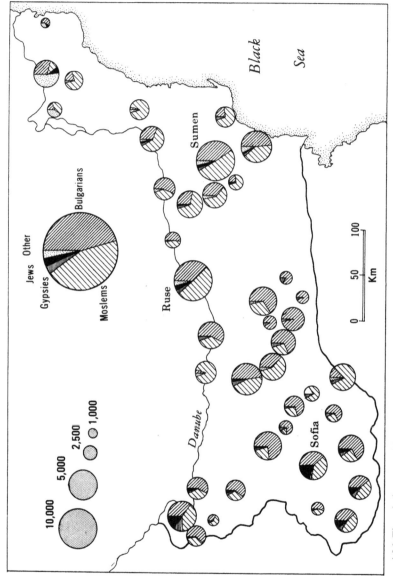

4.26 The ethnic composition of urban population in the Danubian province (Bulgaria) of the Ottoman Empire, 1866

The ethnic structure of the city

The population of the pre-industrial town was at one with that of the rural area which enclosed it. It had been derived from the latter; it spoke the same language and, with modifications, shared the same culture. This changed in many towns during the nineteenth century, and some towns became in effect islands of a foreign culture. The reason was large-scale immigration from distant areas, which was a feature of modern urban growth, and marks yet another step in the breakdown of the age-old symbiotic relationship of town and country. This divorce had, in fact, begun long before the nineteenth century. In the sixteenth century Switzerland was split by the protestant Reformation. The towns became, by and large, Protestant – Calvinist like Geneva, or Zwinglian like Zurich – while rural areas generally remained Catholic: a situation which led to the *Sonderbund* affair in the 1840s (see p. 97).

The Ottoman invaders of the Balkans, acquired much of the agricultural land but they settled mainly in the towns. This, together with the widespread conversions to Islam, especially in Bosnia, and the establishment of large Ottoman garrisons, gave most of them a Moslem majority. Some were, in Todorov's words, 'almost completely transformed ethnically' after the Turkish invasions.[131] Fig. 4.26 shows the ethnic divisions of the urban population in the so-called Danubian province according to the census of 1866.[132] If it could have been extended to include such towns as Üsküb and Sarajevo, they would have been found to be even more strongly Moslem in character. This was, and indeed still is, reflected in the urban landscape, where the minaret is a far more conspicuous feature than the domed cupola of the Orthodox church. The gulf here between town and country became, both culturally and economically, far greater than in Switzerland, and was an important factor in the struggles for independence of the Balkan peoples. Revolution here was rural, not urban. The establishment of free Balkan countries was followed by a withdrawal of a significant part of the Moslem population, especially of the ethnic Turks and military garrisons.

Many cities in central and eastern Europe also had a population which differed ethnically from that of the surrounding region. This was due in large measure to the immigration of a politically or economically dominant class. In the Czech city of Prague this class was mainly German, and had been present since the Middle Ages (fig. 4.27). It became in terms of percentages gradually less important during the century as Czechs migrated to the city from rural Bohemia. Nevertheless, Germans still made up 7.5 per cent of the urban population in 1900, but were especially important in public service and in managerial positions (see p. 99).[133] The situation was not dissimilar in the industrial towns of Upper Silesia, and would have been the same in Łódź if the small immigrant German-Jewish population had not been swamped by Poles from the surrounding countryside. In Warsaw, the Russians, with a large garrison stationed at the Aleksandryiska Citadel, played a not dissimilar role to that of the Germans in Prague, as also did Swedes in Helsinki and Turku on the coast of Finland. In these instances it was the city's élite, the professional and managerial classes,

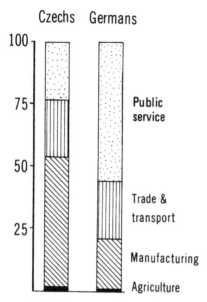

4.27 The contrasted occupational structures of the Czech and German populations of Prague, about 1900

who were the immigrant group. They had come as part of the dispersal of western technical culture during the Middle Ages and early modern times.

During the nineteenth century such migration was largely proletarian, the movement of peasants, squeezed from the land or attracted by the opportunities which the cities were thought to offer. Before 1914 the most significant of such migratory movements were those from eastern Germany and Poland to the lower Rhineland, from Belgium into northern France and from Italy into Provence. The first was by far the largest and socially the most important. A Polish-speaking class was intruded at the base of the social pyramid in the mining towns of the central Ruhr, especially Gelsenkirchen, Recklinghausen and Bochum. They were Polish-speaking in a German environment, and Catholic amid a mainly Lutheran population.[134] They were sufficiently numerous and closely knit to be able to retain their cultural identity, but the fact that they were at the bottom rather than near the top of the social heap and were struggling to establish themselves in an alien environment tended to reduce the political and social tensions that might otherwise have occurred. The migration of Belgians from both Flemish- and French-speaking Flanders into the industrial towns of northern France was similar, but on a much smaller scale; some were French-speaking and most were Catholic, and were more readily assimilated by the local population.[135]

Jews formed a variable but always significant part of the urban population in central and eastern Europe. They had come from several sources, prominent amongst them Spain and the Russian Pale (see p. 90).[136] The Spanish or Sephardic Jews, after their expulsion from Spain, merged with existing communities of Hellenised Jews to form small urban groups in the more important

Balkan towns, such as Thessaloníka and Athens. The more numerous Russian or Ashkanazim community spread westwards into central Europe. The largest towns were barred to them, so the smaller tended to become heavily Jewish. From being mainly rural the Jewish population became heavily urban. Of necessity they turned to trade and the crafts, and chiefly to those which used local raw materials and could command a market amongst the peasant masses of eastern Europe. Many moved on to Berlin and the towns of central and western Germany. Jews dominated the small towns of eastern Poland and Galicia, and made up most of the population of Iaşi and Botoşani in the Romanian province of Moldavia.[137] In Galicia they 'practically monopolise[d] trade, forming 91.2 per cent of the dealers in merchandise in Eastern... and 81 per cent in West', where most were 'merely petty shopkeepers, pedlars and hawkers'.[138] They formed a smaller proportion of the population of the larger cities, but nevertheless their numbers tended to increase through the century as more migrated from the villages of the Russian Pale into the towns. In Warsaw, for example, they are said – there was no census to confirm these estimates – to have made up 31.7 per cent of the population in 1865,[139] and 38.1 per cent in 1914.[140] The proportion was smaller but none the less very significant in Berlin, Prague, Vienna and Budapest, and from here the urban Jewish population declined westwards and southwards with increasing distance from the Russian Pale of Jewish settlement. They were numerous in the towns of Hungary and through the Romanian provinces, but their numbers fell away sharply south of the Danube. Their immigration was clearly restricted by the political insecurity which characterised these regions. Only Sofia and the river towns of Vidin and Ruse had significant Jewish communities in 1866, but even in Sofia, where they were most numerous, they made up less than 20 per cent of the population.[141]

The urban pattern of nineteenth-century Europe

Fig. 4.28 shows the pattern of cities in Europe before their period of rapid growth. Fewer than 20 had a population in excess of 100,000, and there were only four with more than a quarter of a million. No city had reached a million, though Paris was moving steadily towards that total. By 1914 this pattern had been completely transformed. There were no less than 20 towns of over half a million, and four 'million' cities.[142] Such simple statistics fail to show the reality of nineteenth-century urban development, for a new phenomenon, the conurbation or metropolitan area, had come into being. Rapidly growing towns spread into one another, or were separated only by tenuous belts of open country, without losing their separate institutional identities. Greater Paris, Brussels or Berlin far exceeded in area and population the cities which lay at their centres, and in northern France, the Ruhr and Upper Silesia vast urbanised regions were developing, with problems of sanitation and public health, of water supply and amenities that have already been described.

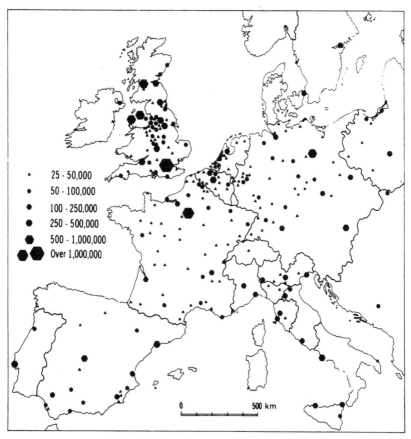

4.28 Urban development in Europe, about 1840

France

In some ways France was an exception to these generalisations. The slow rate of population growth, the high urban mortality rates and the lack of large-scale industrialisation resulted in an urban expansion that was slower than in many other European countries. Growth, furthermore, was conspicuous in only a small number of cities. The number of towns of 20,000 to 50,000 approximately doubled during the century and their combined population increased by some 50 per cent, while the large cities of over 200,000 grew from one to five and their total population increased nearly threefold (table 4.3).[143] Arras was typical of the smaller French towns. Its population grew slowly during the first half of the century; by 1846 it had reached 27,000, and remained approximately at this level until it was decimated during the First World War. Not until 1892 were its ancient fortifications removed to allow the city to spread outwards.[144] The fundamental reason for the slow growth of urban population was the very low birth-rate. This is confirmed by the evidence of another small city, Poitiers, which grew from 29,300 in 1851 only to 41,200 in 1911 (table 4.4). Its rate of natural increase was quite insufficient to maintain its population, and growth was

Table 4.3. *French towns, by population, 1851, 1907*

	1851			1907		
	Number	Population	% of urban population	Number	Population	% of urban population
Paris agglomeration	1	2,375	26	1	4,612	28.9
Over 200,000	—	—	—	4	2,410	15.1
100,000–200,000	4	1,361	14.9	10	2,074	13.0
50,000–100,000	8	1,252	13.7	22	2,521	15.8
20,000–50,000	40	2,777	30.4	87	4,308	27.0

Table 4.4. *Population change in Poitiers*

	Natural increase/decrease	Migration	Actual change	Immigrants per year
1801–41	−1,850	+6,003	+4,153	150
1841–1911	−4,150	+23,016	+18,866	328

possible only with a more than commensurate rate of immigration.[145] Evidence for Toulouse, by no means one of the smaller French cities, shows a similar trend with very high urban death-rates, low birth-rates and large immigration.[146]

Nowhere in western Europe had towns retained their links with the countryside more conspicuously than in France. They remained market towns.[147] Their 'workers were chiefly artisans from traditional craft occupations, working independently or in small shops. In the small and middling towns, where artisans produced mostly for local consumption, the worker was never very far from the countryside, to which he might return for employment during periods of industrial slump'.[148] During the middle years of the century the urban craftsmen were being threatened from two directions. The continuing immigration from the countryside was crowding their ranks. At the same time the worker in the domestic workshop was beginning to feel the competition of factory products, poorer in quality, but cheaper than his own. The result was an ill-directed urban unrest, focused in part in the risings of 1848.[149] Throughout northern France, where the impact of change was most strongly felt, attempts were made to burn factories and break machines, and the wrath of the French craftsman was turned against the immigrant Belgian worker.[150]

Urban growth was by and large significant in only four areas of France: the north and north-east, the Lyons-Saint-Étienne region, Bas Languedoc and Provence, and, lastly, the Paris region. In the first two, growth was due largely to the expansion of the textile, coal-mining and iron and steel industries. Although this centred in ancient towns, such as Lille, Valenciennes and Nancy, which served as administrative and commercial centres, the actual population growth took place mainly in new towns, such as Roubaix, Tourcoing and

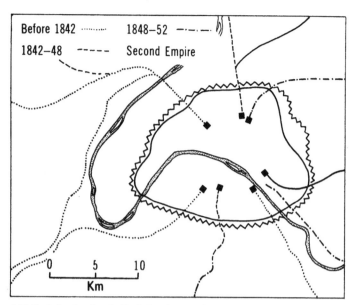

4.29 Paris in the first half of the nineteenth century, showing the development of its railway system. The zigzag line represents the *Mur des Fermiers Généraux*.

Armentières in the north, and Briey, Hayange and Longwy in the east. The Lyons-Saint-Étienne region grew rapidly, especially during the first half of the century. But the physical advantages which in the earlier decades of the century had weighed so heavily in favour of Saint-Étienne gradually lost their momentum, and growth during the last decades before the First World War was slight. The advantages of Lyons, however, were of a more enduring nature, and its growth was uninterrupted. It remained at the end of the century, as it had been at the beginning, the third largest French city after Paris and Marseilles.

A third area of vigorous urban growth lay in Provence and Bas Languedoc. At its centre lay the port city of Marseilles, second only to Paris in size. As a port it stood to benefit from France's links with North Africa and after 1869 from the commerce which passed through the Suez Canal. 'Port' industries were attracted to the city and to its hinterland, from Béziers in the west to Toulon in the east, while between Toulon and the Italian boundary the rise of the tourist industry led to the most vigorous urban growth in nineteenth-century France (see p. 126).

But the most rapid urban growth in Europe, at least until the meteoric expansion of Berlin after 1871, was in Paris itself. It had been one of Europe's largest cities since the Middle Ages. It lay at the hub of a road system, and, when an integrated railway network came to be built it was made to radiate from Paris. It was the capital of the richest country in Europe; its government was strongly centralist, and the city itself had become a centre of extravagant expenditure and luxurious living. Every young peasant in the provinces saw in Paris 'the opportunity of higher wages and easier work',[151] and it became the objective of a stream of migrants throughout the century.

Table 4.5. *Population of Paris* (*thousands*)

	1801	1831	1861	1901
Old city limits	547	786	—	—
Limits of 1860	—	—	1,694	2,714
Dépt Seine	85	149	258	956
Total	632	935	1,952	3,670

By the early nineteenth century the city had spread far beyond its medieval walls, and in 1784–91 a wall, known as the *Mur des Fermiers-Généraux* was cast around the built-up area of the city for the purpose of collecting the *octroi*. In 1841–5 another wall – this time for military purposes – was built around the city, enclosing many areas which at that time were still rural. These were, however, quickly built over, and in 1860 the area within these later fortifications, including the Butte de Montmartre and Mont Géneviève, which had been outside its earlier limits, were brought within the city.[152] In the later decades of the nineteenth century and the early years of the twentieth, the city continued to spread. Villages grew into suburbs and were joined both with one another and with the city itself. The contrast between the extent of the city in 1839 and that on the eve of the First World War is a measure of its spatial growth within the span of 75 years.

The rate of population increase was even more momentous. From about 600,000 in 1801 it rose, including settlements near but not contiguous with the city, to about 935,000 in 1841. Thereafter growth was especially rapid, the city approximately doubling every thirty years (table 4.5). This was achieved only by a large and regular immigration, which in the second half of the century was of the order of 10,000 a year, and at times reached 30,000 and more.[153] At first immigration was largely from *dépt* Seine-et-Oise and the *départements* lying to the east of Paris. The city's catchment area then spread until it embraced the whole of northern France, the Central Massif and even the Alps.

So rapid a growth of population led to conditions which were, even by the standards of the nineteenth century, crowded and insanitary in the extreme. Houses in the poorer quarters were divided into tenements. At the beginning of the century there were said to have been on average about 22 people to every house. By 1851 this had increased to more than 35.[154] As the city spread, more spacious accommodation was built beyond the walls, but as late as 1909 a Board of Trade report noted that Paris was a 'city of large houses...the tenement system prevails universally', and that there were 906,372 tenements in 68,721 houses.[155] The squalor and poverty of early nineteenth-century Paris was beyond belief. Most of the immigrants were penniless: 'almost all the rugged progeny of Cantal...compete for places as tinkers, coalmen, water-carriers, knife-grinders, junkmen. Four thousand masons and sawyers are birds of passage who have their nests in the mountains of Creuse...Women [came] as servants, washerwomen, and cleaning women...from the departments immediately surrounding the capital.'[156] Paris had not enough unskilled work for so vast a horde, and it was they who inhabited the filthy tenements and insanitary courts.[157] The

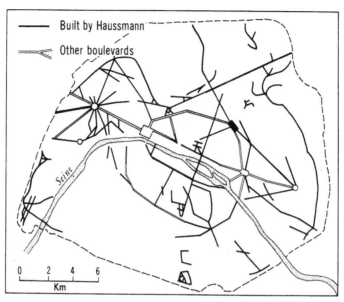

4.30 Haussmann's rebuilding of Paris, 1850–6

description which Mercier gave of Paris at the end of the eighteenth century was still applicable – 'un gouffre où se fond l'espèce humaine'.[158] The gulf between rich and poor, between the elegant quarters of Saint-Jacques and the poorest parts of *la Cité* and of the right bank above the island was daily widening. Nor did these conditions improve significantly by the middle of the century.[159] When the great rebuilding began under Napoleon III, Paris was still 'an overgrown medieval city', with totally inadequate systems of sewage, waste disposal and water supply.[160] The river itself was uncontrolled, and the poorer quarters along the waterfront had in addition to their other problems the periodic flooding of the Seine. It is not surprising that 18,602 died of cholera in the epidemic of 1832, and that the city's death-rate rose to more than 50 per thousand.

In 1852 Louis Napoleon became Emperor of the French. In the following year he presented his plan for the rebuilding of Paris to Georges Eugène Haussmann, who at the same time was appointed *Préfet* of *dépt* Seine (fig. 4.30). Napoleon had long been interested in architecture, and his plans owed much to his appreciation of the planning and the building of parts of West London.[161] He was undoubtedly moved by the congestion, squalor and resulting ill-health of his capital, but he was influenced no less by the fact that its maze of narrow, ill-lit streets gave every advantage to insurgents such as those who five years earlier had threatened the foundations of civil government. Napoleon's plan called for a number of wide boulevards, converging on squares and other open spaces, and, modified by Haussmann, it was put into effect before the fall of the Empire in 1870–1. New roads cut a swathe through the slum quarters of the city as well as through the élite western *arrondissements*.[162] The Louvre was extended, the Opéra built, and Paris was transformed into a city of broad vistas and elegant buildings. The slum quarters were shut away behind the new façades or broken up into segments which, exposed to public gaze, were gradually eroded until their final disappearance.

This spectacular rebuilding of Paris, accompanied and followed by less spectacular slum-clearance, 'fits readily into the pattern of urban reform and public health movements' of the middle years of the century.[163] It was linked with the modernisation of the systems of water supply and sewage disposal, with the provision of improved markets, and the construction of the means of public transport. The cost was immense, and aroused such fierce hostility to Haussmann himself that he was dismissed shortly before the end of the Second Empire. His achievement was however vindicated, not only by the elegance of the city which he had reshaped, but also by the virtual disappearance of cholera and the other epidemics which had previously ravaged it.

Few other French cities received, or indeed needed, so radical a treatment as Paris. Lille and Lyons both underwent a process of 'Haussmannisation' in the 1860s. In parts their density of population had in 1858 been as high as 411 to the hectare. Even after the rebuilding of the 1860s, many such pockets of squalor and congestion survived, and there was a severe cholera epidemic as late as 1866.[164]

In addition to these four areas of rapid urban growth there was also a more than average rate of expansion in the port cities. Le Havre became the chief port serving the Paris region,[165] and Dunkirk that of the North. Bordeaux, without the help of a rich and industrialised hinterland, nevertheless grew to be by far the largest city in south-western France.[166]

The Low Countries

Belgium and the Netherlands had been since the later Middle Ages the most highly urbanised part of Europe. Towns were numerous and closely spaced. Their economic basis was commerce and the traditional craft industries, of which cloth and linen manufacture were the most important. Although Belgium was the first country in continental Europe to develop modern industry, this had remarkably little impact on the older towns of the region. Coal-mining remained a largely rural pursuit and was served mainly by large industrial villages. Even the iron and steel industry, the fastest growing branch of the Belgian economy, cannot be said to have contributed much to urban growth. Seraing in 1910 could still 'hardly be described as a town', and the coal-mining centre of La Louvière was 'essentially a town of recent growth [and could] hardly be said as yet to have taken form...rather a town in outline'.[167] The same was broadly true of the Netherlands, where during much of the nineteenth century most towns remained small, carrying on their traditional functions with little evidence of modern industry.

To this generalisation there were a half dozen exceptions. In Belgium, Brussels, Antwerp, Ghent and Liège grew to be large cities, and in the Netherlands, the port cities of Amsterdam and Rotterdam developed far beyond the level of all other Dutch towns with the exception of The Hague. Brussels became the capital of Belgium in 1830. At this time the area within the late medieval walls was far from built up,[168] and it was not until the middle years of the century that the city began to spread beyond their limit. The river Senne flowed in a marshy valley through the midst of the city, a source of infection to its inhabitants. In 1867–71, following a severe cholera epidemic, the river was covered in and a road constructed above it. A canal, built originally in the sixteenth century, followed the course of the Senne to its junction with the Rupel, a tributary of the Scheldt. It was improved in the 1830s and again in the 1890s, and Brussels thus became a port for the smallest sea-going vessels. The city lay close to the linguistic boundary. At the time of the first census in 1846, 38 per cent of the population was classed as Walloon; 61 per cent as Flemish. This allocation is however, suspect because, unless there was evidence to the contrary, the census-takers classed as Flemings all born in the lower, proletarian part of the city, and as Walloons those from the upper city, within and beyond the walls to the east.[169] In the course of the century the balance tipped in favour of the French-speaking population, though a great many were bilingual.

Antwerp, which in the sixteenth century had been the greatest port in the Low Countries, was in 1609 cut off from direct access to the sea (see p. 476), and though opened again to maritime navigation in 1792, it was not until 1863 that tolls imposed by the Dutch on traffic using the Scheldt were abolished. Only after this date did port and city really begin to develop.[170] Nevertheless, even by the early twentieth century the city had scarcely expanded beyond the line of its seventeenth century barrier of forts, and its population was no more than 275,000.

Both Ghent and Liège were appreciably smaller, though around both lay small towns and industrial villages, administratively separate but economically dependent. It is extremely difficult to allocate a boundary to the agglomerations at the centres of which these cities lay. Indeed the urban development of Belgium was marked by a proliferation of small industrial towns rather than a concentration of urban functions in larger centres. Contributory factors to this were the long survival of domestic industry, especially in textiles, and the Belgium predilection, in sharp contrast with both Germany and France, for low-density, small family homes with garden plots. Only in Antwerp, the Board of Trade report noted, were tenements of any importance.[171]

In the Netherlands, the urban pattern was dominated by the port cities of Amsterdam and Rotterdam, together with The Hague, the seat of government. Their primacy, together with a certain division of functions between them, was apparent in the middle of the century.[172] Since then urban growth has been largely in the provinces of North and South Holland and neighbouring Utrecht. By 1914, a 'ring' of towns, the so-called *Randstad Holland*, had begun to take shape, with a much faster rate of growth than in the towns of the rest of the country.[173] The reasons for this development included, in addition to the denser agricultural population, proximity to the sea, the convenience of water-borne transport, the Netherlands' belated industrial development, and a tendency to fragment the functions of administration, both public and private, between the cities of the region. Over the rest of the country was a pattern of small towns little different in 1914 from that of half a century earlier. There was a conspicuous absence of a truly primate city.

Germany

The rapid expansion of population in Germany was accompanied by an urban growth which was after about 1860 more rapid than in any other country of continental Europe.[174] Germany at the beginning of the century was not highly urbanised. There were no really large cities; apart from Berlin and Hamburg none exceeded a population of 100,000, though Breslau and Königsberg may have approached this total. The ancient centres of German urban civilisation, like Nuremberg, Cologne, Dresden and Frankfurt, each had fewer than 50,000. In all, there were no more than about 60 with a population of more than 10,000,[175] and most of them lay in the Rhineland and Saxony (fig. 10.4). Future industrial towns such as Oberhausen, Salzgitter and Leverkusen were little more than villages.

Growth was slow until the middle years of the century, and became rapid only after about 1870. By 1850 the urban population had grown only to about 15 per cent of the total; only half a dozen cities had more than 100,000 inhabitants, and by 1871 there were still only eight.[176] By the time of the census of 1910 this had changed completely.[177] After 1875 the population resident in settlements of under 2,000 remained constant. The increase was mostly absorbed into cities of medium and large size.[178] Urban growth was most vigorous in five areas: the lower Rhineland and a belt extending eastwards through Westphalia to Hanover; in the middle Rhineland and neighbouring parts of Baden and Württemberg, in Saxony, Silesia and greater Berlin.

Urban growth was due primarily to the vigorous industrial expansion which took place in the latter half of the nineteenth century, but was conditioned by Germany's political structure, which contributed to an over-development of towns. Until 1866 there were no less than 33 separate political units, each with its capital city and administrative structure. The seats of state government were developed as the show-pieces of their respective states, however humble the latter might have been. Karlsruhe, Stuttgart and Potsdam; Dresden, Weimar and Oldenburg; Hanover and Düsseldorf were merely a few of those which tried in their separate ways to become centres of culture as well as of administration, each a veritable Paris in miniature. Then, too, there was a tendency for what can only be termed pre-emptive town-building. The development of a town close to a state boundary led to the creation of another on the opposite side of that boundary. The development of Mannheim in Baden was followed by that of Ludwigshafen across the river in the Bavarian Palatinate; Frankfurt by that of Offenbach beyond the Main in Hesse, and the port of Hamburg by the growth of Altona in Holstein.[179] The statistical study of German urban development is further complicated by changes in municipal boundaries. The rapid growth of towns inevitably meant that many of them were 'under-bounded', consisting for administrative purposes of considerably less than the urbanised area in which they lay. Berlin was the extreme example of this before its boundaries were changed in 1920. The redrawing of municipal boundaries became a recurring event in the decades before the First World War.

Berlin is the supreme example of rapid urban growth during the late nineteenth century. It had grown up on a number of small, marshy islands in

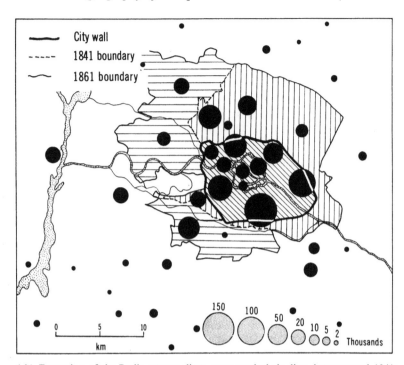

4.31 Expansion of the Berlin metropolitan area; vertical shading, incorporated 1841; horizontal, 1861. Population data for 1875, from I. Thienel, 'Verstädterung, stadtische Infrastruktur und Stadtplanung: Berlin zwischen 1850 und 1914', *Zt Stadtgesch.*, 4 (1977), 55–84

the wide valley of the Spree, near its junction with the Havel (fig. 4.31). The site had the advantage of water-borne transport, but of little else. The early city consisted of a number of quarters, separated by the branches of the river. In the eighteenth century a customs barrier, the *Zollmauer*, was built around the city, and marked its effective limit – until about 1800. In the sixteenth century Berlin became the chief seat of the Hohenzollerns, and grew with the increasing extent and authority of Brandenburg-Prussia. But its most rapid growth came with the building of railways, for which Berlin became a local focus. The population of Berlin was less than 200,000 in 1816 but during the following years it grew steadily (fig. 4.32).[180] The area within the Zollmauer was filled out and suburbs were developed beyond it, in Oranienburg, Rosenthal and Friedrichshain to the north and in Friedrichsvorstadt to the west. In 1841 the municipal boundary was extended to embrace these areas. By 1849 the population of the city within the Zollmauer had risen to about 412,000 and that of its suburbs beyond the wall to some 46,000.[181] Thereafter the suburbs expanded rapidly. The 'inner city' to the south-west, with the suburbs of Schöneberg and Tiergarten, developed as an élite residential area. To the north-west and north, from Moabit, through Wedding and Friedrichshain to Lichtenberg, a vast industrial and working-class quarter developed. Each of these suburbs had its centre and distinctive character. They formed 'une constellation de villes et de villages', which gradually grew into one another until they formed together the biggest

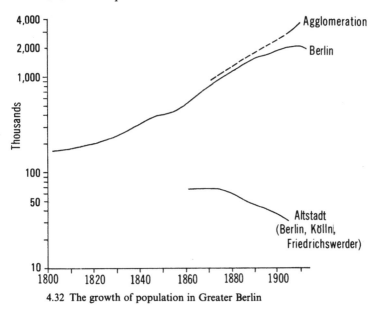

4.32 The growth of population in Greater Berlin

and most populous urban agglomeration in Europe. In 1861 the city boundaries were again extended, this time to include the residential areas of Schöneberg and Tempelhof, and the industrial suburbs of Moabit and Wedding. In the mean time the old city centre had begun to empty itself of its residential population, and developed instead as a commercial and administrative quarter. For a time the *Ringbahn*, a railway built in 1867–77 to encirle the city, set a limit, more psychological than real, to the city's growth. But expansion then became more rapid. The limits of 1861 were quickly exceeded, and population within the official city limits reached a million by 1880, and two million only 25 years later. In 1910 the city itself had 2,070,700 and the agglomeration of *Gross*-Berlin no less than 3,709,500. The old city of Berlin began to decline after the 1860s. A wave of depopulation and decay spread outwards from the *Altstadt*, which in 1905 had less than half the population that had lived there only 50 years earlier. It first reached the western quarters within the Zollmauer, the Friedrichsstadt and Dorotheenstadt; then the northern quarters of Spandau, Oranienburg and Königsstadt, and lastly the south-eastern districts of Stralau and Luisenstadt. In the mean time the more distant suburbs were growing very rapidly indeed. Charlottenburg in the west grew from less than 20,000 in 1871 to more than 300,000 in 1910; Rixdorf to the south-east, from 8,000 to 262,000; and Schöneberg, from 4,500 to 175,000.[182]

This pattern of growth, more like that of London than of Paris in the vast areas over which it spread, was made possible only by a system of public transport that was efficient and fast (see p. 149).[183] In the middle-class suburbs to the west and south-west low-density housing became the most common, but in the populous industrial areas barrack-like tenement blocks, the *Mietskasernen*, became the typical accommodation.[184] They began to be built with the great influx to the city after the war of 1870–1, and continued to house a large proportion of Berliners in growing congestion and discomfort. They were viewed with an increasing concern within Germany[185] long before the Board of Trade turned its condescending attention to them. Little of the old Berlin was allowed to survive within the *Stadtmauer*. It was rebuilt in the pretentious neo-baroque style which became a hallmark of the German Empire.

Without its well developed means of transport and communication Berlin could not have grown as quickly and as successfully as in fact it did. The natural waterways which enclosed and dissected the city were canalised and improved. Earlier links by river and canal both westwards to the Elbe and east to the Oder, were developed so that freight, including fuel, could be brought by barge from the Ruhr and Upper Silesia, as well as from the ports of Hamburg and Stettin. By 1846 a railway net radiated from the city to most parts of Germany, and Berlin became the focus of the railway system of central Europe. Manufacturing industries which had previously been located close to the city centre, moved outwards towards, in particular, the north-west and the east, close to the canals and the principal railways.[186] In the process the nature of manufacturing itself changed, from the luxury manufacturers, the silks and procelain of the Frederickian period, to the weaving, clothing and consumer goods industries of the earlier nineteenth century, and then to the heavy engineering and electrical manufacturing industries which characterised late nineteenth-and early twentieth-century Berlin.[187]

Poland

The Polish Kingdom was noteworthy for the immense number of small towns, each serving as a market centre for a very restricted area and practising the crafts required by a largely rural and agricultural population. Nowhere else in Europe was there so even a pattern of central-places as in the plain of Great Poland, together with a thinner scatter of higher-order towns.[188] Most increased in size during the century, without, however, changing their general economic character. Only in three areas was there urban growth as this was understood in the West: Łódź, which grew from a village into a sprawling industrial town of over half a million; the coalfield area of Dąbrowa in Upper Silesia, and, lastly, Warsaw itself. The growth of Łódź and Dąbrowa are examined later (pp. 398 and 409).

When, in 1596, Warsaw became the capital of Poland, it was only a small town on a bluff above the Vistula. During the eighteenth century the city spread to the west and south. Here were built the palaces from which the Polish magnates sought to dominate the king and overawe the country. Warsaw passed to Prussia in 1793, and to Russia in 1815, and until the First World War was the base for the Russian administration of the Kingdom. To the north of the Old City was built the Aleksandryjska Citadel, symbol of Russian control, effectively blocking any urban development in this direction. There was, in fact, little growth until after the Rising of 1863 and the land reform which followed. By this time a standard-gauge railway had been built south-westwards to Silesia and Moravia, and from the Praga suburb, across the Vistula, a broad-gauge line ran eastwards to Moscow. The city began to spread westwards and south-westwards, and manufacturing industries developed along the railway. The population at the beginning of the century was only about 100,000, and but little more by 1850.

Urban development

But thereafter growth was rapid. Warsaw, with its suburbs reached 684,000 by 1894 and 884,000 by 1914.[189]

Scandinavia

At the beginning of the nineteenth century Scandinavian society was made up largely of small, self-sufficing communities. There was little need for towns. Some had been founded during the Middle Ages, but most had been created in the sixteenth and seventeenth centuries.[190] Very few had been walled and most were built of wood, caught fire readily, and had frequently to be rebuilt. Most remained very small, and only a few port towns had achieved any considerable size by the beginning of the nineteenth century. In Sweden, Stockholm was at this time a city of about 76,000; Göteborg had 13,000 and Norrköping and Karlskrona, the Swedish naval base, had each only about 10,000. Nor had there been any considerable excpansion by the middle years of the century. In 1855, Stockholm, the largest town, had still only 98,000, and Göteborg less than 30,000. The only others which exceeded 10,000 were Norrköping, Malmö and Karlskrona. Uppsala had only 8,000 and Lund was even smaller.[191] Despite their small size, the Swedish towns were notoriously insanitary, and risk of epidemics was great. Without any tradition of urban living, the Swedes were slow to realise that conditions of sanitation and water supply that were proper in a village were quite unsuited to a growing town. Laing found very few craftsmen even in Stockholm, and the business of the small coastal towns was chiefly the export of timber, iron and other metals and ores. He commented on the lack of communication between the capital and the rest of Sweden: 'it receives its supplies from a foreign land [Finland], and consequently wants [i.e. lacks] that communication and interchange with the rest of Sweden which, in other Kingdoms, bind the capital, the government and the country together...The very fire wood for the royal palace...was being landed from Finland out of vessels of that country.'[192] This, it should be added, was due largely to the fact that, as Laing noted, Stockholm lay 'on the verge, or rather on the outside of the country', and transport was cheaper and easier by water than overland. Stockholm grew more rapidly after the mid-century, and had doubled in size by 1880. Swedish industrial development in the nineteenth century was, from the nature of much of the manufacturing carried on, more rural than urban. As late as 1870 no less than 35 per cent of all industrial workers lived in the countryside, and it was not until late in the century that the manufacturing component of the urban population began to increase significantly.[193] In the following years manufacturing industries became increasingly important in and near Stockholm, which by 1910 had grown to be a city of 342,000. Göteborg, as the chief port of Sweden, expanded to 162,500, and the many centres of manufacturing industry in central and southern Sweden became small towns. The urban sector of the total population increased from about 5 per cent in 1800, the lowest in Europe, to almost 23 per cent in 1910.[194]

Norway had, at the beginning of the century, a higher proportion of its population living in towns than Sweden.[195] All the towns were coastal and owed

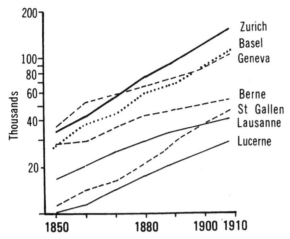

4.33 Growth of the chief cities of Switzerland. Note the absence of a primate city

such importance as they possessed to the fisheries and the lumber trade. They grew slowly during the century. Christiania (Oslo) reached about 50,000 by the mid-century and 228,000 by 1910. Little manufacturing industry was attracted to the towns, and their importance remained, as it had always been, commercial. The rapid growth of the Norwegian merchant fleet, especially after 1865, contributed to their importance.

Finland was even slower in developing than Sweden and Norway, and its urban growth was delayed even longer. By 1910 Helsinki was still only a modest town of about 130,000 and no other town much exceeded 45,000.

To this picture of backwardness and belated development Denmark was a partial exception. Its higher fertility, gentler climate and denser population led to a more vigorous urban growth. The position of the Danish islands astride the entrance to the Baltic Sea gave them a commercial importance denied to most of Scandinavia. Copenhagen benefited from this and developed as a 'merchants' haven' on the eastern coast of the island of Sjaeland, overlooking the Sound, the most easily navigated channel between the Baltic and the North Sea. Until about 1850 Copenhagen was constrained by its seventeenth-century fortifications, and the old town within them became very crowded and congested.[196] By 1850, when it began to spread beyond its fortified perimeter, it had reached 127,000. Its growth thereafter was rapid: 235,000 by 1880 and 462,000 by 1910. Copenhagen came to provide the supreme example of a primate city. It had always been by far the largest in Denmark. In 1910 it contained 16 per cent of the whole population, and was more than eight times bigger than Aarhus, the second largest.

Switzerland

Like South Germany, the Swiss Cantons developed a number of important commercial towns during the Middle Ages. These suffered severely from the changes which took place in trade routes during the sixteenth century, and during

Urban development

the following centuries failed to replace trade with urban industries. The industrial growth which took place in the eighteenth century, particularly in the textile branch, was predominantly domestic and led to no significant urban expansion.[197] At the beginning of the nineteenth century only four cities, Geneva, Bern, Basel and Zurich had more than 10,000 inhabitants and many well known cities had fewer than 5,000.[198] During the first half of the century they lost many of their privileges and with them much of their exclusiveness. This encouraged the growth of population and the expansion of modern industry. Nevertheless, by 1850 there were only eight towns of more than 10,000, and Geneva, the largest of them, had only about 31,250. Altogether, only 11.9 per cent of the population lived in towns of over 10,000 and only 32.6 per cent in settlements of over 2,000. Switzerland was still one of the least urbanised countries in western Europe.[199]

The second half of the century saw a rapid expansion of towns, especially during its last decade, when Zurich and Basel first exceeded 100,000 (fig. 4.33).[200] By 1910 more than a quarter of the population lived in towns of over 10,000, and all together about 60 per cent of the population was urban.

The Habsburg Empire and south-eastern Europe

Town life was only feebly developed in the Danubian and Balkan lands at the beginning of the nineteenth century. Towns were primarily market centres, and manufacturing and administrative functions were of slight importance. It is characteristic of countries in which economic development came relatively late that these functions tended to concentrate in a small number of urban centres. These were most often the foci of the developing transport nets. The result was the concentration of urban growth in a small number of central-places, above all in the political capitals. This development occurred in Scandinavia; it was no less apparent in the Habsburg and Balkan lands.

Vienna, the Habsburg capital, has come to be regarded as the extreme case of the over-expanded capital, the 'big head on little shoulders'. From a population of about 225,000 in 1810, it rose slowly at first to 357,000 in 1840 and about 607,000 at the time of the *Ausgleich*. Thereafter its growth was more rapid: 827,500 in 1890; 1,675,000 in 1900 and 2,086,000 in 1910. Death-rates in Vienna were consistently higher than birth-rates, and growth was sustained entirely by immigration. Even in 1880, towards the end of a period of relatively slow growth, only 35 per cent of the population had been born in the city. Immigrants came mostly from Lower Austria and the provinces of Bohemia and Moravia, but there was also a smaller flow from Germany, Switzerland and Hungary.[201] Jews from Eastern Europe also migrated to the city. Their numbers were negligible at the beginning of the century. By 1850 they made up 3 per cent of the population, but by 1890 they had increased to 10 per cent of a much larger total. The growth of the city has been attributed to its function as capital of an empire of, in the mid-nineteenth century, 31 million people. Government and military service, however, accounted for only a relatively small part of its employment, about 11 per cent in 1869 and less than 10 per cent 40 years later. Vienna developed primarily as an industrial and commercial city (table 4.6).[202]

The city grew up on a terrace above the Danube flood-plain, enclosed on west

Table 4.6. *Employment in Vienna*

	Total employed	Percentage in manufacturing	Transport and trade
1869	377,483	46.7	18.9
1910	1,012,049	48.2	25.1

Source: G. Otruba, 'Wachstumsverschiebungen in der Wirtschaftssektoren Österreichs 1869–1961', *Viert. Soz. Wirtgesch.*, 62 (1975) 40–61.

and south by the hills of the Wienerwald and Wienerberg. At the end of the Middle Ages it was a small walled town which served as principal seat of the Dukes of Austria and administrative centre of their *Ostmark*. From the first its orientation was towards the east. It was a frontier town, twice besieged by the Ottoman Turks, and the base from which was organised the conquest of the Hungarian Plain. By this date, early in the eighteenth century, the city had begun to spread beyond the line of its walls, and by the early nineteenth century its suburbs stretched far beyond the medieval core. In the course of this expansion, however, a broad space was left encircling the walls. It was at least half a kilometre wide, and in total area, far larger than the Old City itself. The military were careful to keep it free from buildings or other obstructions, and when, in 1857, Emperor Franz Josef turned this belt over to civilian uses it was against the advice of his generals. Thus was the ground prepared for the creation of the *Ringstrasse*. The walls were removed and a series of monumental buildings encroached on the open space in the 1870s.[203] The development of the *Ring* must rank beside Haussmann's rebuilding of Paris, with which it was roughly contemporary, as one of the finest examples of urban redevelopment in nineteenth-century Europe.

The growth of Vienna eclipsed that of all other Austrian cities. Most remained small. Only two, Linz and Graz became significant urban centres with extensive, industrialised suburbs.[204]

The urban development of Bohemia and Moravia was more in line with that of western Europe, with the growth of coal-based industrial cities, of which Pilsen (Plzeň), Ostrava and Brünn (Brno) became the largest and most important. In northern Bohemia a number of small mill towns, some of them of medieval origin, developed in the later nineteenth century. Prague was always the largest city in Bohemia, centre of its administration and of its transport net. It grew up as a double city; on the flat, right bank of the Vltáva lay the Staré Město, the Old Town, with its market-place and merchants' quarters. High above the left bank lay the Hradčany, fortress, palace, cathedral and administrative centre of the Bohemian Kingdom, and on the steep slope below it, the Malá Strana, a small and mainly residential quarter where the Charles University was located. At the beginning of the century Prague was a city of some 75,000. By 1850 this had increased to 118,000, but still the city had not filled out the area within the seventeenth-century fortifications. The next 60 years were marked by more rapid growth. The population rose to 162,000 in 1880 and 225,000 in 1910. The built-up area of the city first spread up to the line of the walls and after 1877 beyond them to the east, in the working-class suburbs of Karlin and Zižkov and the residential district of Vinohrady. By 1914 factories and low-quality housing had spread along both banks of the river and had filled out the great bend below the city. In 1910 less than 10 per cent of the population was German-speaking and most of the rest, Czech.[205] It was the Germans, however, who formed much of the middle class and supplied many of the numerous officials, the Czechs supplying most of the workers in manufacturing industry.[206] Growth was maintained largely by immigration, and in 1900, 60 per cent of the city's population had been born elsewhere. Most of the immigrants were Czech and, with a higher birth-rate, the Czech proportion of the population was tending to increase.

Hungary was the less developed and the less urbanised part of the Dual Monarchy. When the nineteenth century began Hungary had still not fully recovered from the wars of the eighteenth century. Urban life had disappeared from much of the Great Alföld and, apart from Budapest, the towns which survived round the western, northern and eastern margins of the plain were all very small. They were of two kinds. Within the mountains were numerous towns which had originated as mining settlements. In their often attractive Renaissance and baroque buildings they showed the influence of the Germans who had founded them, and many still held a substantial German population in the nineteenth century.[207] They included Schemnitz (Banská Štiavnica), Neusohl (Banská Bystrica) and Levoča in Slovakia, and Cluj, Sibiu and Braşov in Transylvania.

The second group of towns lay where the mountains met the plain. They were market rather than mining towns. A reciprocal relationship developed between the treeless, agricultural plain and the infertile, forested mountains, and the products of the two were exchanged at these market centres: Szombathely and Sopron (Odenburg) in the west; Prozsony (Bratislava), Nitra, Košice and Miskolcz in Slovakia; Uzhgorod and Mukačevo in Ruthenia, and Satu Mare, Oradea, Arad and Timisoara (Temesvár) on the east.[208] In the nineteenth century they were predominantly Magyar towns, though there was a Germany-speaking minority and a steadily increasing Jewish community.

In the Great Alföld, the grassy plain which lay to the east of the Danube, towns were of an entirely different character. The settlement pattern which had existed before the Turkish invasion was very largely destroyed in the course of two centuries of fighting. It redeveloped in the eighteenth century as a series of 'giant villages, big enough to be called urban yet with a rustic air: the streets are unpaved, the house are of one storey only, and the occupations of the people are largely agrarian'.[209] Such peasant towns were peculiar to the Great Alföld. Many were of immense size. Debreczen had a population of nearly 30,000, and this increased to 55,000 by the middle of the century without any change in its agricultural basis; Szeged, Szabadka, Kecskemet, Csongrad and Szolnok were only a little smaller.[210] Agricultural work was chiefly in the summer months, when these towns emptied of their younger population. These passed the season in the fields, living in temporary settlements of tents and shacks, known as *tanyák*. In winter they returned to the 'town', taking some of their animals with them, but leaving the rest out on the Alföld in the charge of a few of their number.²¹¹ Around the 'towns' was a ring of small, enclosed parcels of land, known as *kert*, which served the purpose of farmyard and barn. In the course of time the *tanya* became a permanent settlement and the transhumant movement between 'town' and open *puszta* ceased, though the practice long continued amongst the better-off of maintaining a town residence in addition to the farmstead.²¹² The former was often used as a kind of retirement home for elderly members of the family.

Budapest differed from other towns in Hungary. Like Prague, it had a twofold origin. On the hill to the west of the Danube the royal and administrative city of Buda developed, the fortress and cathedral on the summit, the town straggling down the slope to the river. On the flat bank opposite lay Pest, the commercial and industrial city, where fairs were held and animals, driven in from the *puszta*, were marketed.²¹³ To the north of Buda was a third nucleus, Óbuda, which had grown up on the site of the Roman town of *Aquincum*. The site, so similar in many respects to that of Prague, differed in the character of the river. The Vltáva was of no great width. It was easily crossed and the great bridge, the *Karlův most*, was built in the fourteenth century. The two parts of the city were linked economically and administratively from the start. Not so in Budapest. The river was up to 650 m in width. In summer it was swift and difficult to cross. Buda and Pest were as one, it has been said, only in winter, when ice made it possible to cross the river on foot. The first bridge, the Lanchíd or Chain Bridge, was built in 1849, followed in 1872–6 by the Margithíd or Margaret Bridge. The Árpád and Petöfi bridges were constructed before the end of the century. This greater ease of communication between Buda and Pest led in 1872 to their merger to form the single municipality of Budapest.

At the beginning of the century the three settlements had collectively a population of about 50,000, with Buda by far the largest partner.²¹⁴ By the mid-century, Pest had far outgrown its rivals and had about 110,000 and the whole agglomeration, some 150,000. The latter continued to grow, especially after 1867, when it became the political capital of the lands of the Hungarian Crown. It exceeded 300,000 in 1870, 800,000 in 1900 and had reached a million when the First World War began. Budapest grew with the economic growth of Hungary, gathering to itself almost all the country's manufacturing capacity, focusing its transport system and serving as the chief river-port. In no European country, not excluding even Austria, were urban functions more strongly concentrated in one place. The city spread northward and southward, along the river, where docks were built and factory industry developed, and eastwards across the flat, dusty plain. There was comparatively little growth towards the west, where the Budahegy, the eastern extremity of the Bakony, presented a very much more rugged terrain. The city's fortifications, which, until the mid-century, had set a limit to its growth, were swept away, and replaced by a wide and handsome boulevard, the Nagy Korut, which enclosed early nineteenth-century Pest.²¹⁵

The Balkan peninsula was not notable for its urban growth, notwithstanding its precocious development in classical times (fig. 4.34). Its coast was dotted with Greek πόλεις, and Procopius has left a long list of towns founded by the Emperor Justinian in the course of his wars against the invading Germanic and

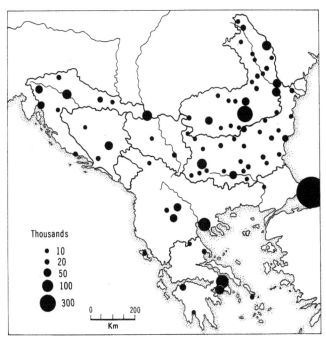

4.34 Towns of the Balkan peninsula, about 1910

Slavic peoples. Most of these have since disappeared or have survived only as villages. In the northern part of the region, in Croatia and Transylvania, towns were founded, most of them by German immigrants, during the middle ages; port cities, such as Zadar and Dubrovnik, were developed under Italian auspices, and around the southern coasts a few cities, Athens, Thessaloníka and, above all, Constantinople survived from classical antiquity. At the beginning of the nineteenth century towns in the Balkans were few and small.[216] The Turks have been blamed for extinguishing urban life. Invasion and war certainly discouraged the trade on which towns had been dependent, but the Turks themselves encouraged town growth for both administrative and military purposes. Sarajevo, for example, was substantially an Ottoman creation, and the modern expansion of Sofia, Bucharest and many other Balkan towns long preceded their independence of the Ottomans. South of the Danube, most of the larger towns were mainly Turkish, or at least Moslem. Novi Pazar was, towards the end of the Ottoman period, about 90 per cent Moslem, though this figure would certainly have included a number of Bosniaks.[217] The larger the town the more likely was it to be dominantly Turkish, and if Moslems were few in Beograd, Niš and other Serb towns, this was only because they had been driven out when Serbia achieved *de facto* independence.

Towns were no larger or more numerous in the Romanian principalities. In Wallachia, Bucharest, by far the largest, had 14,619 families – perhaps 75,000 persons – in 1838.[218] The port-town of Galaţi is reported to have had 12,000, and its neighbour, Brăila, only 4,000–5,000[219] at this time. By 1866 Bucharest had grown to more than 160,000.[220] Thereafter the number of manufacturing

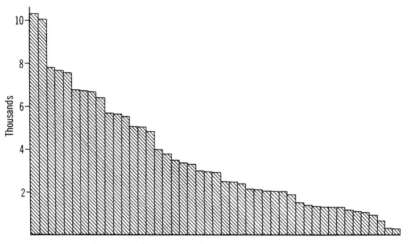

4.35 Rank-order graph of towns in Bulgaria

industries increased; the first railways were built, and Bucharest became the capital of the united principalities in 1859. Population rose rapidly, to 232,000 in 1900 and 300,000 in 1910. Other towns which grew rapidly in the later years of the century with increasing investment of western capital were Ploieşti, centre of the developing oil industry, and the port of Constanţa.

The Bulgarian province, over which the Ottomans retained control until 1876 also experienced a considerable degree of urban and economic development at this time. A number of textile factories was established between the Stara Planina and the Danube, and several small industrial towns grew up (fig. 4.35). Political independence tended to check rather than encourage this incipient urban and industrial growth. In 1866 the largest towns appear to have been the Danubian port towns of Ruse (Ruschuk) and Vidin, the Black Sea port of Varna and the garrison towns of Šumen and Plevna (Pleven). Sofia came sixth. But the record is only of taxable population and probably does not reflect the true relationship of these towns.[221] By 1905, after Bulgaria had annexed Eastern Rumelia, an urban hierarchy had emerged with Sofia at its head and Plovdiv in second place.[222]

Town life was less developed in the western half of the Balkan peninsula, where the terrain was more rugged and transport more difficult. The largest towns lay towards the north or on the route which followed the Morava and Vardar valleys from the Danubian plain to the Mediterranean. There is little certain information on their size early in the century. Beograd is said to have had about 20,000 before the Serb revolt, and, as late as 1884, only about 35,000. By 1910 it had almost reached 90,000, closely comparable with Sofia. Throughout this period Beograd was almost certainly the largest city south of the Danube and Sava, with the exception of Athens and Constantinople.

Town life was no more developed in Greece than in the rest of the Balkans, but the achievement of independence in 1832 was followed by the rapid growth of its capital city, Athens. A nineteenth-century town developed to the north of the classical town centre, and largely outside the line of the Turkish

Urban development

fortifications. By 1853 its population had reached 30,590. It had doubled by 1875, and doubled again by 1900. In 1907 its population of 167,500 was, after that of Constantinople, the largest in the Balkans. Athens lay about 8 km from the sea, and had since classical times been served by the port of Piraeus. Piraeus expanded with the growth of Athens itself. A railway between them was completed in 1869, and, with a growing commerce, Piraeus soon became a city in its own right. Its port facilities were extended (see p. 486) and its population reached 21,600 in 1879 and 73,500 in 1907.

Town life was no more developed in Macedonia, Thrace and the Greek peninsula than in the rest of the Balkans. There was a cluster of small towns in western Macedonia, of which Üsküb (Skoplje), an Ottoman garrison town, was probably the largest. Thessaloníka in the early nineteenth century consisted of the congested Turkish town still contained within its medieval walls and dominated by the citadel. Its population had by 1870 grown to about 50,000. Then the development first of the port and later of the railway up the Vardar valley (see p. 459) led to a more rapid increase. A modern town with wide straight streets was built down to the water's edge and its population grew to 150,000 by 1895 and to 160,000 by 1910.[223] The largest city in the Balkans was, however, Constantinople. Its narrow streets and congested housing spread over the peninsula from the Golden Horn to the Sea of Marmara. At its eastern extremity, looking out across the Bosporus to Skutari, was the Topkapi, palace of the Sultans until in the mid-nineteenth century it was abandoned for the Dolmabahce Palace farther up the Bosporus. On its western edge the wall of Theodosius II still enclosed the city as it had done in the early Middle Ages. Between lived a vast number of people; how many at the beginning of the century no one knew. Weber merely put their number between 300,000 and a million, a figure which was certainly too large.[224] It was one of the least sanitary cities on the European continent; epidemics, including even the plague, were not infrequent, and the use of wood in domestic construction led to frequent and destructive fires. Nevertheless, the city grew during the century. By 1890 it is said to have reached 873,600, and almost certainly it had reached a million when the Ottoman Empire was dragged by its German ally into the First World War.

The Iberian peninsula

In the previous chapter (p. 114) a picture was presented of a population leaving the agricultural plateau of central Spain for the industrialising coastlands. This was reflected in the pattern of urban development, with one exception only. Madrid, in the centre of those depopulating grasslands, grew to be by far the largest city in the peninsula. At the beginning of the century there were about 40 towns with more than 10,000 inhabitants, nearly half of them in Andalusia, which had been since Roman times the most highly urbanised part of the peninsula.[225] Madrid had at this time about 167,000, but other Meseta towns, even the most important of them like Toledo, Valladolid and Burgos, scarcely exceeded 10,000. Apart from the capital, the large cities – Barcelona, Seville, Valencia Cádiz, Málaga, and, of course, the Portuguese city of Lisbon – were all on or near the coast.

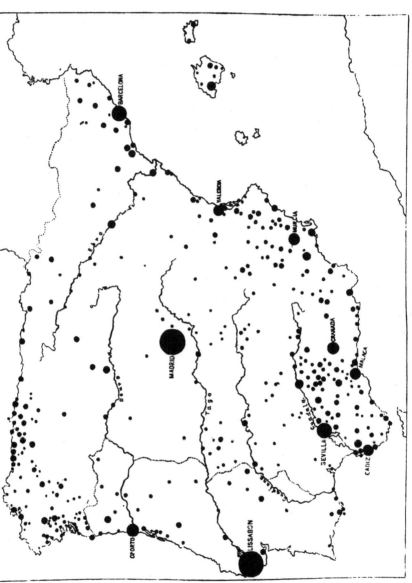

4.36 Cities and towns in Spain and Portugal, about 1855. This map, reproduced from *Pet. Mitt.* (1856), may be one of the earliest attempts to use proportionate symbols to represent city size

Urban development

The situation had changed little when in 1856 Gumprecht produced his map, one of the first to use symbols proportionate to size, of urban development[226] (fig. 4.36). Two urbanised regions stand out, the north-west and the coastal belt from Andalusia to Catalonia, separated by the thinly peopled region in which only two urban centres stand out: Madrid and Lisbon. The source of Gumprecht's figures is not clear, but they are broadly confirmed by those given by Richard Ford.[227] Madrid had about 280,000, and Lisbon 240,000 people. The rest of the century was marked by selective urban growth. The two capital cities grew rapidly, and in 1910 had respectively 572,000 and 436,000. At the same time Barcelona, focus of the most vigorous industrial growth, grew to be a city of 560,000. In all three the medieval and early modern core was transcended by spreading suburbs, built in most cases on a rectilinear plan, with wide boulevards in the style of Haussmann. Urban development in the coastal towns followed a similar, but slower pattern of growth. Their congested cores, Moorish in origin in the Mediterranean cities, were gradually enclosed by less densely built suburbs.[228]

Italy

Modern urbanism had its origins in Italy, and Italy retained at the beginning of the nineteenth century a higher proportion of urban population than almost any other country in continental Europe (fig. 4.37). But by then many of the towns had become empty shells, almost wholly abandoned by commerce and the crafts on which their earlier prosperity had been based, and inhabited chiefly by an agricultural population which made its daily journey to the fields. Charles Dickens, not without a certain Dickensian exaggeration, described a city as eminent as Ferrara as 'more solitary, more depopulated, more deserted... The grass so grows up in the silent streets, that anyone might make hay there',[229] and of Piacenza he wrote: 'a deserted, solitary, grass-grown place, with ruined ramparts; half filled-up trenches, which afford a frowsy pasturage to the lean kine that wander about them...'[230] There were exceptions. Both Genoa and Leghorn (Livorno) were busy ports and centres of manufacturing; Venice and Florence were alive with tourists; the northern cities of Turin and Milan were seats of government, and around them the modern textile industry was beginning to develop, and Rome, set amid 'the brown and bleak and forsaken Campagna, a few flocks of ragged sheep collecting a scanty subsistence', lived by tourists, pilgrims and the tribute of its spiritual empire. In southern Italy, where few tourists went to describe its squalor, conditions were far worse.[231] Dickens wrote of Fondi, the first town which he encountered after crossing into the Kingdom of Naples:

in the name of all that is wretched and beggarly...a filthy channel of mud and refuse meanders down the centre of the miserable street, fed by obscene rivulets that trickle from abject houses. There is not a door, a window, or a shutter; not a roof, a wall, a post, or a pillar, in all Fondi, but is decayed, and crazy, and rotting away.[232]

Most inland towns of Italy were no larger in the 1850s than they had been three centuries earlier, and some were appreciably smaller. Verona, for example,

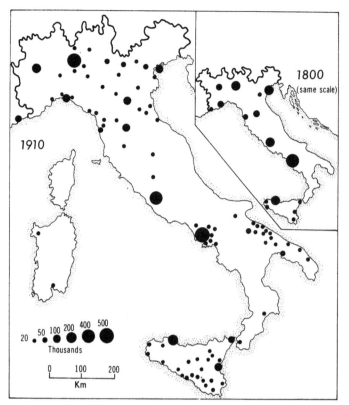

4.37 Towns of Italy, about 1800 and in 1910

had about 56,000 inhabitants in 1593, and only 54,000 in 1855.[233] The largest city at the beginning of the century was unquestionably Naples with 350,000, most of them underemployed and poverty-stricken, living in conditions of the utmost deprivation. Goodwin described 'the meanness of its houses, the wretchedness of its people, and the scarcity or want of all the comforts and necessaries of civilised life'. The only good buildings, he added, 'were castles and gaols, churches and monasteries, mansions and villas, the dwellings of the barons, the clergy, and the local authorities'.[234]

There was little growth in the larger towns and none in the smaller during the first half of the century. By 1836 Florence had reached 97,500, and Rome about 153,700.[235] But Lucca, Pisa, Siena and Pistoia each had considerably under 25,000. In the northern plain, Milan had reached 242,000 and Turin 135,000 by 1850. Thirty years later Naples had reached almost half a million, but Turin, Milan, Rome, Palermo, Genoa and Florence still had fewer than 300,000, and towns of the importance of Bologna, Leghorn and Messina fewer than 100,000.[236]

Urban growth became more marked after the unification of Italy in 1861, but, apart from the continued expansion of such southern cities as Naples and Palermo, it was largely restricted to the northern plain and Tuscany. Urban growth was especially marked near Milan, close to the Alpine foothills, and along the Via Emilia, which extended from Milan south-eastwards to Bologna and Rimini.[237]

There was a marked contrast between urban development in northern and central Italy on the one hand, and that in southern Italy and Sicily on the other. In the former, growth was a response to the expansion of manufacturing and of the tertiary sector. Towns grew largely by immigration from rural areas, and the immigrant population was by and large absorbed into profitable employment. Urban growth in southern Italy and Sicily was largely restricted to a few cities which were already large when the century began. None of them was primarily a manufacturing centre, and such industries as had been established suffered severely from the competition of the more efficient northern industries after 1861. Nor was the tertiary sector significant. Naples, Palermo and the rest of the oversized cities of the Italian South were fed by the immigration of impoverished and often desperate peasants. Employment opportunities were totally inadequate, and the urban population remained underemployed, overcrowded and deprived to a degree that was extreme even for Italy. Not until the mid-twentieth century did the Italian government face up to the problem of *il Mezzogiorno*.

In the later years of the century the larger cities, especially Rome, the political capital after 1870, and the industrial cities of Milan and Turin grew rapidly, but those of intermediate size increased little.[238] The tendency in Italy, as in all countries which developed late in the century, was for the primacy of the few large cities to assert itself more strongly, leaving little room for growth in those lower in the table. The towns of northern and central Italy were becoming increasingly western in both appearance and function. In southern Italy, with only a few exceptions, towns were vast congested villages, rural in function, often perched on hilltops for protection from both human enemies and malaria, without water supply or sewage disposal, a prey to typhoid, diphtheria and smallpox.[239]

Conclusion

The urbanisation of Europe was the most conspicuous change in the geography of the continent between the Napoleonic and the First World War. By 1914 almost half the population of Europe lived in towns, and for north-western Europe and Italy the proportion was very much higher. In 1800 the urban population had amounted only to about 14.5 per cent. Urbanisation closely paralleled and was a direct consequence of the growth in G.N.P., and the rate of urban growth was least in those countries in which economic progress was least marked. Bairoch has demonstrated[240] how high is the correlation between these two parameters. Urban growth was also closely related to the development of the transport network. Although a few towns were able to profit from river- and canal-borne traffic, the coming of the railroad was for most a precondition of rapid expansion.

Cities which grew most rapidly expanded on a broad front, and in really large cities tertiary occupations were at least as important as manufacturing. Narrowly based industrial towns invariably remained in the second rank. In very many instances urban growth was faster than that of the urban infrastructure of water, food and energy supply, of sewage disposal and transport, and even of housing. Physical conditions deteriorated in many cities during the middle years of the century when they were growing most rapidly. Only with the increase in real incomes in the last quarter did living conditions show any significant improvement. Even so, the rise in standards was conspicuous only in western and central Europe. Throughout the century mortality rates remained higher in cities than in rural areas and in small towns, and the series of urban reports commissioned by the Board of Trade in the early years of the present century show in harrowing detail how awful living conditions often were on the eve of the First World War.

5
Agriculture in the nineteenth century

The general course of agricultural development during the nineteenth century bore a certain resemblance to that of industrial growth. Both were highly localised phenomena. Some areas developed an advanced, specialised, capital-intensive mode of agriculture; others remained backward, 'in a state of economic autarchy until the very end of the nineteenth century'.[1] As in manufacturing industry, small nuclei of more advanced technology expanded gradually, drawing other areas into their sphere. The essential feature in this growth was the appearance of a market economy, which provided an outlet for the increased product of a more efficient and specialised agriculture. By and large, the parallel development of manufacturing provided a stimulus for agriculture. It created a localised demand for farm products, and, by providing better tools, equipment and fertilisers, it supplied in part the means by which that demand could be satisfied. The further argument that the profits of agriculture provided the capital whereby manufacturing industry was enabled to develop is, however, less well supported. The resulting development of specialised agriculture and localised demand made necessary a transport net, able to move large quantities of food over great distances. What had previously been a number of local systems of demand and supply was transformed gradually and slowly into a single system which came to embrace first the country, then the continent and lastly the world.[2]

Pre-industrial Europe was made up of a number of small, largely self-sufficing and mainly self-governing communities. That there were 10,000 parishes in England gives some idea of their number. Babeau estimated that there were 44,000 separate communities in the France of the *Ancien Régime*.[3] Their self-sufficiency must not however be exaggerated. The individual community could not produce everything that it needed, and it owed certain obligations beyond its borders to lord and king. But its 'import' and 'export' could never have amounted to more than a small fraction of its communal income. There were, however, exceptions. Even amongst the simplest communities there were specialised groups – miners, metal-workers, charcoal-burners, for example – dependent on some form of exchange. Foremost amongst such specialised consumers was the urban population. It is unlikely to have amounted to more than 15, at most 20 per cent of the European total. All towns, furthermore, including even the largest, were in some degree agricultural, and the smallest amongst them would have been able to satisfy a considerable part of their food needs from their surrounding fields. This suggests that rural communities must in the aggregate have contributed something of the order of a tenth of their total production for the sustenance of the urban minority. Much of this represented only a short-distance movement, since most towns were supported by the rural communities within ten or 20 kilometres of them.

This system of basic local self-sufficiency with only a limited degree of external dependence was slowly eroded during the nineteenth century. On the eve of the First World War it survived only in peripheral areas of Europe. The factors in this breakdown were the growth in population, the increase in urban and industrial demand, improvements in the means of transport and the development of extra-European, including Russian, sources of food supply.

The grain supply

About 1815 the total production of cereals in continental Europe must have been of the order of 40–50 million tonnes a year. By 1913 the average annual production had increased to nearly three times this total. During the same period the population of Europe within the same boundaries had approximately doubled. Its food supply might therefore have been thought to be secure. In fact, cereal production within Europe was in 1913 totally inadequate, and had been so for many years. The countries of continental western Europe had then a net import of 16.4 million tonnes. Eastern Europe, excluding Russia, exported on balance 2.7 million tonnes,[4] so that the continent, as defined in this book, had a deficit of about 13.7 million tonnes, or some 12 per cent of total demand.

There were two significant reasons for this. An increasing share of cereal production was being fed to stock, thus contributing to the growing output of meat and dairy products. At the same time the human diet not only began to include more of cereals, but wheat tended to be substituted for the coarser grains. Countries and regions which could have remained self-sufficing in total cereals found themselves obliged to import when wheat began to be substituted for rye, barley or buckwheat.

There was an increase during the century in the area of cultivated land, though it is difficult to estimate for Europe as a whole the extent to which arable was expanded. In France the area used agriculturally increased by about 4 per cent between 1840 and 1882. In Germany growth in the cultivated area was far greater. In Prussia it increased by 77 per cent in the course of the century, and the area under cereals by 55 per cent,[5] but much of this increase was in the East Elbian lands (fig. 5.4). Elsewhere in Europe there was a marked growth in cropland, where soil and climate made this possible.

The rate of increase in the area of land used agriculturally fell very far short of that of population, and if cropland had not been used more intensively the prospects for Europe's food supply would have been bleak indeed. In France, for example, the area under cereals increased by 3.6 per cent between 1840 and 1882, and then remained fairly stable (fig. 5.1). The production of all cereals increased within the same period by no less than 23 per cent, and over the century the increase was of the order of 65 per cent (fig. 5.2).[6] This remarkable increase

Agriculture

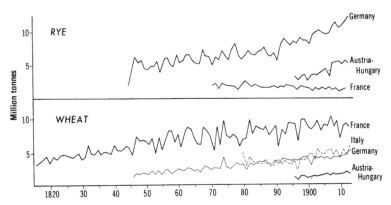

5.1 Growth in production of wheat and rye in France, Germany, Austria-Hungary and Italy

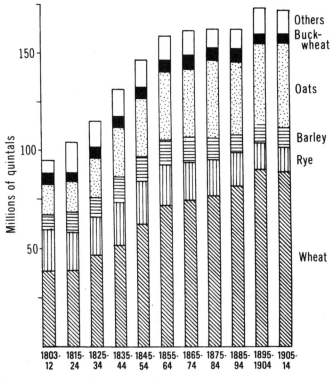

5.2 Grain production in France

in agricultural productivity was achieved by means of improvements in the technology of farming. Land came to be better drained and better fertilised. The use of fallow was reduced to very small proportions by the end of the century. Nitrogenous crops were introduced into the rotations. Superior tools and equipment were employed and better strains of seed became available (see pp. 227–32). The diffusion of these innovations has not hitherto been much studied.

It appears, however, that they were first adopted on estates and on the lands of the richer farmers, but their spread to the lands of the peasantry was at best a slow and gradual process. In many remote areas of Europe little had been done to modernise agriculture on the eve of the First World War, and to this extent there remained the potential for a further increase in agricultural production.

Despite these advances the European production of foodstuffs, of which the bread-grains were by far the most important, became increasingly inadequate. There had been an import of grain into western Europe since at least the sixteenth century, though most had come only from the plains of Poland and Hungary. In the nineteenth century this import increased in volume, especially that from east-central Europe and the Danubian lands. Grain from the Russian Ukraine then assumed an important role, and after the mid-century that from the United States and Canada began to play an ever increasing part.[7] The effect was not only to increase the availability of cereals in Europe, but also to lower their price and to smooth out the fluctuations which had become a feature of the grain market. The large consuming centres – essentially the giant cities and the growing manufacturing regions – received an increasing proportion of their bread-grains from overseas. This was reflected in a price fall in the local markets, and the rural communities which had previously supplied them suffered a loss in income and a decline in purchasing power. There were several ways in which the latter attempted to recoup their losses. One was the perpetuation of domestic crafts, particularly handloom weaving. Another was the conversion of land to uses which seemed to hold out better prospects. These included the cultivation of fodder-crops for stock and the growing of specialties such as vegetables, fruit and root crops.

Although the area used agriculturally grew almost everywhere, that sown with cereals ceased to increase significantly as soon as cheap grain from outside Europe became available. Instead, there was an overall increase in the area under roots, fodder-crops and hay, though the balance between cereals, roots and animal feeds varied greatly from one part of Europe to another.

The extent of this change in the structure of agriculture was to a large extent dependent on the degree of protection which governments were prepared to extend to it. In general this was considerable. 'In no other part of the European economy', wrote Svennilson, 'was the protection of one section of the community at the expense of the rest of the nation carried so far.'[8] A contributor to the Royal Statistical Society's Journal in 1899 described how Germany, 'with the aid of immoderate protection' contrived to produce at home 'the largest part of the food consumed within her borders.'[9] Nevertheless, with a duty of over 20 per cent *ad valorem*, Germany had still to import 11 per cent of her rye and more than a third of her wheat.[10] Even France, 'swathed in protective duties', imported 10 per cent of her wheat supply. In Belgium, which had had high protective duties on bread-grains since 1887, almost three-quarters of the wheat was imported. It is apparent that without a growing substitution of wheat for rye and coarse grains, the dependence on imported grain would have been less heavy.

This trend in agricultural production is illustrated by the graph of cereal prices. Fig. 5.3 shows the trend of average wheat prices in France, which is in this respect

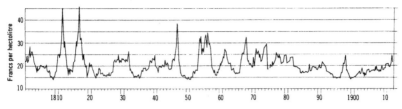
5.3 The average monthly price of wheat in France

the best documented of European countries. Two features stand out: the diminishing amplitude of the fluctuations and the downward trend in prices from the 1870s into the present century.[11] At the same time, the wide regional discrepancies which had characterised the grain market early in the century[12] contracted as a common market was gradually established for the whole of France. The prices for the north, centre and south gradually approximated to one another in the course of the century.[13]

The extreme year-to-year fluctuations in price early in the century were due in the main to the vagaries of the weather and the resulting quality of the harvest. It seems probable that these conditions did in fact improve in the course of the century, and that the cyclical recurrence of severe weather became less frequent and less extreme.

Such in the broadest outline is the 'model' of the behaviour of European agriculture during the nineteenth century. There was, in fact, every possible variation on this theme. In Great Britain import duties on cereals were abolished in 1846. Cheaper bread-grains were secured for the working classes at the expense of lower prices for the farmer. In Germany, France, Belgium and Switzerland high duties were either maintained or restored, and the farming community was protected at the price of a relatively inefficient system of agriculture. Germany, indeed, forms a special case owing to the conflicting interests of its land-owning aristocracy in the east and of the industrial interests in the west. In Denmark a conscious and deliberate change was made in the structure of farming, with the support and encouragement of the government, in order to produce what was, under existing commercial conditions, most profitable.

Increased agricultural production, both arable (fig. 5.4) and pastoral, was not only achieved by an increase in the cultivated area. Yields increased, both the output per unit area and the yield-ratio, or proportionate return on the seed sown. These changes were the result of using fertiliser, better tools and farming practices, and improved seed. In many parts of Europe the traditional practice of fallowing was gradually abandoned, and a catch-crop introduced into the rotation in its place. Various forms of alternate, or long-ley, farming were used in order to achieve an overall increase in production. In a few parts of Europe, notably the Scandinavian countries, the fragmented strips of the open-fields were consolidated into compact holdings with economies in labour and a small increase in the cultivated area. The result of these changes was that total agricultural production increased very much faster than the area of agriculturally used land (fig. 5.5). The aggregate production of cereals in a group of countries

192 An historical geography of Europe 1800–1914

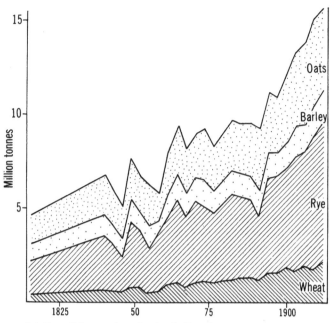
5.4 Growth in cereal production in Prussia

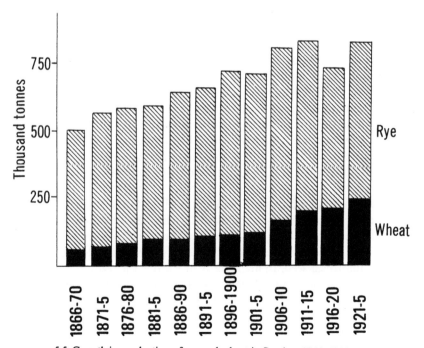
5.5 Growth in production of rye and wheat in Sweden, 1866–1926

Agriculture

shows a very considerable increase – no less than about 120 per cent in the countries represented in Fig. 5.1 between 1845/54 and 1905/14.

The physical conditions of agriculture

The years which immediately followed the end of the Napoleonic Wars were marked by a disaster almost as serious for some parts of Europe as the wars themselves. For two years in succession harvests failed. The cause may have been in some degree the neglect of the land during the previous decades, but primarily it was the weather. The summer of 1816 was cool and wet, and the harvest poor. The following winter was cold, and the summer of 1817 little better than the previous,[14] and the price of wheat soared to levels never before recorded in Europe.

The bad years of 1816–18 were followed by a span of good years, with abundant harvests and low prices. This was interrupted by a spell of poorer weather and smaller harvests from 1827 to 1832. Again several years of good harvests followed, before the fluctuating conditions of 1836–45 culminated in the crisis of 1846–7. This was the most severe to afflict Europe after that of 1816–18. The winter of 1844–5 was a hard one in most of western and north-western Europe, and autumn-sown crops suffered severely. When the potatoes were planted in the following spring they developed a blight and much of the crop was lost. In Flanders, for example, it was reduced to less than a tenth of that usually gathered.[15] The following crop-year was no less severe, though this time the cause was rather the excessive rainfall both in the previous winter and before harvest. The rye rotted in the ground and the potato blight lingered on.[16] Conditions became exceptionally severe where the population depended heavily on wheat, rye and potatoes. Maize and buckwheat withstood the severe conditions better, and people fared less badly where these formed a major ingredient of their diet.[17] In the Limousin and south-western France the situation was no better than in the Low Countries,[18] and the tragedy culminated in the spring of 1847, when the poor harvest of the previous summer was exhausted.

The harvest of 1847, however, was good everywhere. Prices fell sharply, and a succession of five good years restored a degree of prosperity to the countryside and restocked the granaries of Europe. But *la période des grandes crises* was not yet over. Bad harvests again occurred in 1853–6.[19] The situation was not as acute as in 1846–7; prices did not on average rise as high, but the crisis was more prolonged. Crises continued to punctuate the agricultural history of Europe, in 1861, 1867–8, 1874, but they were progressively less severe, amounting to little more than a temporary rise in living costs and a change in consumer demand. Their periodicity at this time prompted the amusing speculations on the link between sun-spot activity and the business cycle. There were further periods of adverse weather in the 1870s, but their influence on prices, though not on the fortunes of the farmer, was much less dramatic. After the 1870s winters were generally milder, the summers drier and the harvests more regular. The period of food crises was over.[20] The smoothing out of the price curve was, however, due no less to the role of imported grain, which served to keep prices down whenever harvests were poor. One tends to think of bad weather as influencing chiefly the grain harvest. It was no less important for root crops and livestock.[21] Cold and damp tended to encourage murrain and other animal diseases.[22]

Corfus has summarised the weather conditions and the quality of the resulting harvests in Romania from 1830 to 1851.[23] There were in this period of 22 harvest years seven good or very good harvests, two medium, seven poor and six bad or disastrous. Interestingly those years which were so disastrous in western Europe, 1846 and 1847, were classed as 'good' in Romania.

The farmer and the grain merchant can anticipate a bad harvest. Those able to do so, hoard grain in the certain expectation that prices will rise, and in doing so ensure that the prices will rise yet more steeply. The graph of grain prices provides the clearest evidence of the seriousness of harvest failure and consequent distress. But weather is a local phenomenon, and the incidence of bad weather is often restricted to a small area and to a narrow range of crops. In such cases the operations of the market were able to distribute supplies, though at an enhanced price, and no major crises resulted. This, however, was not so on the rare occasions when adverse weather covered a very wide area, and afflicted, for example, both winter- and spring-sown crops. The most likely cause of a real *crise de subsistence* was for this reason excessive rainfall rather than drought or very low winter temperatures. This was the case in 1816, as it had been during the previous crisis of 1812. The tendency for root crops, primarily the potato, to be used to supplement the bread-grains was an unconscious insurance against the climatic risks of monoculture. The crisis of 1846–7 was particularly severe in north-western Europe because the period of bad weather, which particularly affected the rye crop, coincided with the apparently unconnected spread of the potato blight.

Why, one may ask, were harvest fluctuations so much less marked in the later years of the century? It seems probable, in the first place, that weather conditions were less adverse than earlier. But there were other factors. The small contraction in arable acreage meant that marginal land went out of cultivation, and this would certainly have included some ill-drained lowlands which would have suffered most during very wet winters. At the same time progress was being made with land drainage by subsurface tiles or other devices, and this in turn would have improved yields. Furthermore, as agriculture became more commercial, so a greater effort was made to adapt the crops to their physical environment, and it became less necessary to grow a particular crop because it constituted the staple food of the local population. Lastly, improved transport and communications within Europe and the growing import of bread-grains, fodder and other crops allowed the market to be satisfied despite any shortfall – local or regional – in the harvest.

The area of land used agriculturally increased in the course of the century by a percentage which varied from less than 5 in France, to more than 70 in Prussia, and may over the continent as a whole have amounted to perhaps 20 per cent. This was achieved in two ways, by the piecemeal extension of cropland and grazing and by major reclamation schemes, which brought extensive areas into some form of agricultural use. The former normally involved the use of marginal land, from which by definition the return for any given input of labour and capital was less. Such areas included the sandy heathlands of the northern plain and the wet, infertile plateaux and hills. Crop-farming advanced into such areas: the sands of the Campine and the Lüneburg Heath, the Ardennes and the Swedish Norrland, for example, under the stimulus of rising population and high prices.[24] The sandy soil of Flanders and the Campine, wrote de Laveleye, 'is the worst...in Europe...A few miles from Antwerp land sells for 20 francs an acre, and those who buy it for the purpose of cultivation get ruined', but that was after the fall in cereal prices had begun.[25] A generation earlier, at a time of extreme population pressure, Colman described the efforts of the Flemish peasant to cultivate these wastes,[26] restoring to the soil, with all the economy of the Chinese, everything that had come from it. It is difficult to document such small-scale reclamation of woodland and waste, but its extent in the aggregate was very large. In France, for example, the area of *landes, pâtis et bruyère* fell from 18 per cent of the total area in 1840 to 12.3 per cent in 1892.[27]

In Germany the trend was similar. The land used agriculturally in East Elbian Prussia rose from 64 per cent of the whole in 1800 to 70 per cent in 1907, and cultivated land from 32 per cent to 55.[28] In addition to the piecemeal extension of cropland was the reclamation of larger areas, such as the Lüneburg Heath, small parts of which were brought under cultivation while the larger were planted with conifers.[29] Such reclamation could not have been accomplished without a generous use of fertiliser, which German factories were then beginning to produce in immense quantities. It is doubtful, furthermore, whether it could have continued without a high tariff protection for the farmer.

In Hungary, which at this time included both Slovakia and Transylvania, the agricultural area increased from 64 per cent of the whole in 1873 to 69 per cent in 1913, while arable land grew from 34.7 per cent to 45.5.[30] In Spain, Italy and the Balkans the area of cropland increased, though there are no statistical data by which to measure it.[31]

In Scandinavia the growth in farmland was more marked than elsewhere in Europe, in part because there was a greater area in which to expand; in part because yields were so low that a large area had to be tilled to secure an adequate return. The total agricultural area in Denmark increased from 5.7 million acres in 1866 to 7.0 in 1901, and, despite the well known conversion from a predominantly arable to a mainly pastoral economy, there was a small increase – from 56.4 to 58.8 per cent – in the cultivated area.[32] In Sweden the increase in cropland was far greater, from less than a million hectares in 1810 to 2.5 millions by 1860.[33] Settlements advanced up the valleys into the highlands of Norrland, and, though many were established for mining or lumbering, there was a very significant addition to the cultivated area.[34] The *landnam* was on a much smaller scale in Norway, where no advance at all was possible on its cold, barren *fjelds*, but in Finland regular burn-beating, as the temporary clearing and burning of the forest was called, was carried on during the nineteenth century over the southern half of the country. This practice became, however, a great deal less extensive in the early twentieth century.[35]

Romania provides an extreme example of the increase in the cultivated area in response to market demand. The province of Wallachia was well placed to take advantage of the need of western Europe for bread-grains.[36] Much of the land was owned by the boyars (nobles), who also enjoyed the rights of *corvée* over their tenants. These rights had not perhaps been used as much as they might have been, but now, with the prospect of large grain sales to the west, they were exacted to the full.[37] This short-lived reversion to feudal practice was similar to the 'second serfdom' in Poland two centuries earlier.[38] There was an immense increase in the cultivated area, difficult to estimate but amounting to several hundred per cent between 1831 and 1851 in those provinces for which comparable data exist.[39] The wheat harvest increased nearly fourfold, and there were also gains in the production of other cereals. This spectacular growth within a period of only 20 years was achieved by ploughing up the steppe. It is doubtful whether it could have continued very much longer; in any case, Russian and then North American grain began to replace Romanian in the Western market.

Much of the increase in the cultivated area in western and central Europe came from the gradual erosion of waste and common land. There were, however, bigger projects, each requiring a large capital investment, which aimed to reclaim wet lands or to irrigate dry. A levée, for example, was built along the Seine near Saint Germain-en-Laye to protect low-lying agricultural land.[40] A more ambitious project was to protect the lands along the upper course of the Rhine from damaging floods. Levées were built during the middle years of the century, and along the River Ill during the later years.[41] A similar work was carried out by Count Szechenyi on the River Tisza in Hungary, and in the 1840s no less than about 272,500 acres (over 110,000 hectares) are said to have been reclaimed for agricultural use.[42] At the same time many of the areas of swamp and lake which dotted the Hungarian Plain were also drained.

The wet lands along the River Po in northern Italy and around the head of the Adriatic Sea were reclaimed and brought under cultivation. No less than 310,000 hectares were said to have been gained in this way between 1860 and 1899.[43] Progress was also made along the Biscay coast of France, in the Ditmärschen, along the lower Elbe and Weser, and in the Vistula delta. All these reclamation works, however, seemed small beside the immense projects undertaken in the Netherlands. Here the shallow lakes had already been reclaimed and only the deeper, for which windmill drainage was inadequate, remained. The steam-engine could, however, cope with a task for which the uncertain windmill had proved inadequate. Work began on the Beemster Polder in 1840 and was completed 12 years later,[44] and smaller polders were reclaimed during the following years.

Irrigated agriculture was not thought of as either practicable or desirable in most of Europe. Water-meadows were 'floated' in England, but the practice was not adopted widely in continental Europe, and over most of the area the rainfall was supposed, however incorrectly, to be fully adequate for the needs of farming. Only in southern Europe was there any great potential for irrigated agriculture, and here capital was inadequate for it to be realised. There had been major irrigation schemes in the North Italian Plain and in coastal districts of Spain since the Middle Ages. These were maintained and even extended in the nineteenth century. A further 7,200 hectares are said to have been brought under irrigation in the 1840s,[45] and soon after unification the Cavour Canal, built to join several Alpine rivers, allowed half a million hectares to be irrigated. By the

Agriculture

end of the century the irrigated area amounted almost to 1.7 million hectares, about 8 per cent of the cultivated area of northern Italy.[46] Attempts to develop irrigation in peninsular Italy met with little success,[47] but more was achieved in southern France, especially in *dépts*. Bouches-du-Rhône and Vaucluse,[48] where melt-waters from the Alps were available in summer. The irrigation works which had survived from the Middle Ages in the Guadalquivir valley of southern Spain and along the Mediterranean coast relied for summer irrigation on rainfall accumulated during the preceding winter.[49] There was little scope, except near the Sierra Nevada and Pyrenees, for the use of melt-water. Much of the progress in irrigated agriculture was necessarily small-scale, the work of separate *communes* or even individuals. Some progress was made in the valleys of the Ebro, Duero and Guadalquivir, especially where the mountains could be used for water storage.[50] The greatest scope lay in Catalonia and the Ebro valley, where the Infante Canal was built early in the century to take water from the Llobregat valley to irrigate the Barcelona district.[51] Early in the present century the Aragon and Catalonia Canal was completed and was expected to irrigate as much as 100,000 hectares.[52]

By 1902, 6 per cent of the agricultural land of Spain was irrigated, and was said to yield no less than 15 per cent by value of the total crops. The importance of irrigation in helping to break down the monocultural systems of Spain was great,[53] but a study of a *huerta* near Valencia showed that during the nineteenth century irrigated holdings were not only very small – over half had less than a hectare of irrigated land – but also severely fragmented.[54] It is noteworthy that scarcely any of the great estates of Spain, which alone had the resources for major developments, had instituted any significant irrigation project (see p. 272).

Agricultural systems in nineteenth-century Europe

The systems by which agriculture was carried on in the nineteenth century were no less significant than the physical conditions, and their range and variety was no less. The term 'agricultural system' is here taken to denote (i) methods of land-tenure, (ii) size of agricultural holding, (iii) field systems, (iv) cropping systems, and, lastly, (v) the supply of capital for agriculture. Any discussion of these interrelated factors must, however, be prefaced by a consideration of land reform which, between the late eighteenth century and the second half of the nineteenth, changed radically the conditions of farming in much of Europe.

Land reform

This is an elusive term for the many separate processes by which feudal methods of land occupation and cultivation were gradually changed into the market-oriented systems of the twentieth century. In no country was land reform a simple act of government; it was a long, slow process of change in the ownership and organisation of the land and in the status and obligations of those who cultivated it. These changes may be summarised as the freeing of the former serfs from hereditary subjection and the abolition or commutation of their obligations in service or in kind; the abolition of seigneurial monopolies, as, for example, in brewing, and other privileges; the transfer to the tenants of some of the allodial right in land they cultivated, coupled with the distribution amongst the peasants of land previously held in common or as part of the demesne, and, lastly, the consolidation of fragmented holdings into compact tenements. Land reform could also include the provision of some modest equipment and farm-stock for the peasant.[55] These several processes rarely proceeded concurrently, and in no country was legislative attention given to all of them. Land reform in France was the achievement of the French Revolution.[56] The peasant gained, if he did not already possess, personal freedom and exemption from feudal dues and obligations, and he also acquired a title to the land which he cultivated. But there was no reallocation of land, no break-up of estates in the interests of the peasantry. The peasant merely acquired title to what he already had. After the Restoration in 1815, the aristocratic landowners, by repurchase and by grants of land which the government had continued to hold, recouped, it is believed, about half of their losses.[57]

Belgium shared the fortunes of France. Here too the peasant was freed from whatever feudal constraints had survived, and gained title to his land. There was no reallocation of land, nor any interference with existing leases and contracts.[58] The Revolution made both countries predominantly lands of small peasants. In the Netherlands, as in Britain, feudalism had been gradually eroded for centuries, and agriculture was already dominated by small to medium proprietors at the time of the Revolution.

Germany, however, presented a more confusing picture, reflecting the political divisions of the country. One can, in fact, distinguish between developments in south Germany, in north-west Germany and in the Prussian-dominated east. In south and south-east Germany serfdom disappeared early; *corvée* and other personal services were restricted, and the region became one of small peasant holdings with or without a rent payment.[59] On the other hand the practice of partible inheritance prevailed in much of the area, and the farm holdings tended to become excessively small as they were subdivided between heirs. The tenurial system thus tended to encourage and perpetuate domestic crafts (see p. 303), which became an essential part of the rural scene. No attempt was made during the period to remedy the problem of micro-holdings.

In north-west Germany, also, peasant servitudes had by and large disappeared by the early nineteenth century, and much of the farmland was held on long leases. But the practice of partible inheritance was not prevalent, and most peasant holdings were adequate in size. Eastern Germany, roughly defined as the area lying to the east of the Elbe, differed in most respects from western, though large estates were not wholly absent from the latter, nor small and medium from the former (fig. 5.6). The greater part of eastern Germany belonged to Brandenburg-Prussia, and those areas which did not, like Mecklenburg, had a similar structure. The manorial system, which had disappeared from most of western Germany, survived little altered in the east, 'giving rise to self-contained economies centred on manors whose lords had concentrated in their own hands the labour [and] judicial and feudal rights which in the west had been divided and attenuated'.[60] In France and Belgium the remaining hereditary servitudes had been abolished by the Revolutionary government; in

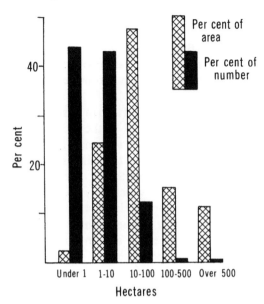

5.6 Size of agricultural holdings in Germany, about 1880

Prussia and its neighbouring states the landowning classes themselves made up the government. They were under great pressure to reform the tenurial system, but were not willing to make any great personal sacrifices, and they made the peasant pay highly for his freedom.

In Prussia land reform consisted of a series of legislative enactments spread over a period of many years. Not all were socially progressive, and some were mutually contradictory.[61] The Stein-Hardenburg edict of 1807 terminated serfdom and at the same time abolished the law of entail (*Fideikommiss*) which had prevented the break-up of the great estates. The intention was to allow land to be sold to form medium-sized holdings for the peasants. The result was the opposite. Subsequent legislation converted the peasant's holding into a freehold, but, since he had held it by services on his lord's demesne, he had to redeem these by a financial payment. Although some money was made available for this purpose, most peasants were obliged to surrender part of their holdings in order to discharge this obligation. Demesnes thus became larger, but peasant holdings smaller. Many conditions, furthermore, were attached to peasant ownership of land, including the possession of a plough-team; they had to be *spannfähig*. Many small peasants failed to satisfy this condition, and gained their personal freedom at the expense of losing their land. A class of landless farmhands was created, living abjectly on the great estates and replacing the labour formerly contributed by the *corvée*. It was from this class that the migrants to the west were largely drawn. It is an open question whether in material terms the east German peasant was greatly benefited by emancipation. Political freedom had been achieved at the expense of greater economic dependence.

In the Czech lands the noble landowners rarely lived on their estates and had little need for demesnes. The right of the lords to labour dues from their peasants was maintained throughout the *Vormärz* period, but in practice was generally commuted for a money payment.[62] The abolition of servile obligations had long been under discussion within the Habsburg Empire when the risings of 1848 gave a new urgency to the problem. In September 1848 peasant obligations were terminated, but with compensation to their lords who had thus been deprived of their services.

Conditions varied in Poland. In the Grand Duchy and also in western Poland controlled by Prussia, personal serfdom was terminated by the decree of 1807. The benefits, however, were illusory, since, without a money income, the peasants had no means of redeeming the *corvée* owed to their lords. In 1846 peasants in the Polish 'Kingdom' gained security of tenure, and a growing prosperity allowed many to commute the services which they owed for their holdings.[63] The solution to their problems came abruptly in 1864. In the previous year the Poles – mainly the *szlachta*, or gentry – rose against the Russian government of their country. They were suppressed, and in retaliation the Russians deprived them of their profitable monopolies such as that of distilling and selling vodka and also of the labour dues owed by their peasants.[64] The latter at once gained what amounted to the freehold of the land they cultivated. The Russians secured peace for themselves by setting the lords against the peasants, who got possession of their lands 'under better conditions than anywhere in Central and Eastern Europe'.[65] The Russian land-reform edict was as carelessly drawn as most others had been. While the peasantry acquired an absolute right to the land it cultivated, no mention was made of their customary rights in domanial forest and common. The lords thus retained control of sources of essential fuel and building timber, and the struggle, always bitter and sometimes violent, on the part of the peasants to terminate these 'servitudes' ran through the rest of the century.[66]

Galicia, or Habsburg-held Poland, shared the fortunes of Austria and Hungary. Since the days of Maria Theresa and Joseph II it had been Habsburg policy to protect and strengthen a class of substantial peasants. In the course of the eighteenth century the *corvée* had largely disappeared, and demesnes had mostly been divided into small farms and let at a cash rent to peasants. Comparatively little land was retained in the hands of the nobles, who lived as absentees on their rents.[67] Although the peasant gained legislative protection and some degree of security of tenure, he did not own his land and his personal status remained in some respects unfree until the reforms of 1848. But reforms did not lead to the creation of a body of substantial peasants. Instead, the practice of partible inheritance encouraged a high birth-rate, and the peasantry decayed into a body of 'smallholders living poorly, oppressed and dominated by capital from the village usurer'.[68] Rural conditions in Galicia were amongst the worst in eastern Europe, and contributed to the bloody rising of 1846.

Large demesnes and servile peasant labour endured longer in Hungary. Unfree status was abolished in 1848, as in the rest of the Habsburg lands, but large numbers of peasants merely became wage-labourers on the estates. The Habsburg policy of *Bauernschutz* was conspicuously unsuccessful in Hungary, in part because the role of the landowning class in government was more powerful than in Austria or the Czech lands.[69]

Agriculture

The Ottoman conquest of the Balkan peninsula had been followed by the confiscation of much of the better land, chiefly in Bulgaria, Macedonia and Thrace, for the creation of *chifliks*. These were large farms worked by subject *kmet* labour for the purpose of supporting the military and the towns.[70] Much of the land was held directly by the Sultan or by Moslem religious foundations. The only private estates belonged either to the Turks or, as in Bosnia, to Slavs who had converted to Islam.[71] The Christian *raya* either lived remote in the mountains from which they emerged to provide seasonal labour on the plains, or they held farms at the price of providing labour on the estates of their masters. It was a feudal system without the well established customs which had softened the asperities of feudalism in the West. The services by which the Turks had held their land of the Sultan ceased gradually to be performed, and the *chiflik* became a *timar*, held almost unconditionally by its Turkish master, who by the nineteenth century enjoyed unrestricted power over his Christian *rayas*.

The destruction of Ottoman power, first in Serbia and Greece; later in Macedonia, Bulgaria and Thrace, was followed by the flight of Turkish landowners. Land reform in much of the Balkan peninsula took the form of that in France and Poland rather than in Austria and Germany. It was a simple expropriation of the rights of the previous landowners without any commensurate compensation, though Serbia did make a small payment to the Ottoman Empire and Bulgaria also agreed after 1878 to pay some compensation. The result was to create a class of landowning peasants. The reform was indeed more complete than in France, because the previous landowning class was no longer present to reclaim or buy back any part of its lands.

Romania, then consisting of the two principalities of Walachia and Moldavia, was not part of the Habsburg lands; nor was it effectively controlled by the Ottoman Empire. As a result it shared in the contrasted land reforms of neither. Here, as in Russia, the landowners or boyars, remained firmly in control both of the government and of their own estates. At a time when, in central and western Europe, the burden on the peasantry was being alleviated, the Romanian boyars were under one pretext or another demanding increased services. The reason, quite simply, was that from about 1833 Romania became a significant exporter of wheat.[72] The boyars conferred on themselves by legislative enactment the right to increase peasant obligations, and also to exchange such services as had little value to them for increased work on the land. In 1848 serfdom ended in Transylvania, and there was a flight of Romanian peasants across the borders into the Habsburg Empire. There was even a movement into the Russian province of Bessarabia. But in the Romanian provinces themselves the boyars continued to veto all attempts at agricultural reform.[73] It was not until 1864, when even the Russian government had yielded, that legislation was passed 'sweeping away all restrictions on the peasants' movements, abolishing all dues and renders, fixing the landlord's compensation and granting land to all peasants'.[74] The contrast in the pattern of landownership between Romania and other Balkan countries is shown in fig. 5.7. But the boyars who had reluctantly enacted the reform had also the task of implementing it. They evaded the limits set to the size of their own estates and the holdings made available to the peasants were based on the number of cattle the latter possessed. Well-to-do peasants thus

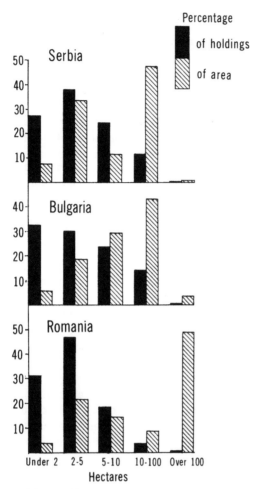

5.7 Size of farm holdings in Serbia, Bulgaria and Romania, 1905–8

benefited and the poorer were left with holdings too small to support a family. The result was to depress much of the peasantry below even the low levels which had prevailed before the reforms. These rural conditions provoked a serious peasant rising in 1907,[75] but went unchanged until after the First World War.

The tenurial system of Spain had been inherited from the days of the *Reconquista*. The southern half of the country was largely composed of great estates (fig. 5.8). The population was increasing and the agrarian problem becoming ever more acute. Proposals for a modest land reform were made during the eighteenth century, but came to nothing.[76] The effect of the French invasion under Napoleon was to increase the demand for land reform and at the same time to postpone its implementation. Early in the nineteenth century entails were abolished, as they had been in Prussia, and personal servitudes were terminated. Disentailment, however, failed to make much land available for the peasants. Although a great deal of land which had belonged to the church and to corporate

Table 5.1. *Size of holdings in Spain*

Size (ha)	Number	Total area	Average size (ha)	Percentage of total Number	Percentage of total Area
Over 250	12,488	7,469	598	0.2	17.6
100–250	16,305	2,340	143	0.1	5.5
10–100	169,472	24,612	27	1.7	58.0
5–10	205,784	1,379	6	2.0	3.3
Less than 5	9,810,331	6,635	0.6	96.0	15.6
Total	10,214,380	42,435			

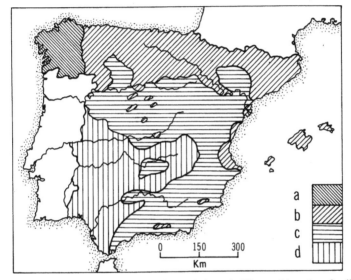

5.8 Distribution of estates in Spain. (a) Minute holdings, of less than family size; (b) medium-sized farms, held on long leases; (c) medium-sized farms, held on short leases; (d) *latifundia*, large estates worked mainly by casual labour

bodies was alienated, most fell into the hands of the middle class, and the reform in the end strengthened the *latifundia* system of Spain. By the early twentieth century about 23 per cent of the total area was in holdings of more than 100 hectares, and some 96 per cent of all holdings occupied only 15.6 per cent of the available land (table 5.1).[77]

In Scandinavia land reform, as understood in central and eastern Europe, had largely been accomplished by the beginning of the nineteenth century. Personal servitude disappeared from Denmark in the eighteenth, and, in so far as it had ever existed, it vanished from the rest of Scandinavia soon afterwards. In Denmark this was followed by the consolidation of the fragmented strips of land which made up each peasant holding,[78] and a system was created of medium-sized peasant farms, most of them held on long-term leases.[79] In Sweden also the open-field system was in decay in the eighteenth century, and with it disappeared any surviving traces of peasant dependence and servile obligations. In Norway there had never been any kind of manorial system, and in Finland, where there were very few estates, much of the agricultural land was held in small peasant farms.[80]

Throughout Scandinavia estates tended to give way to small or medium-sized farms at a relatively early date. A possible reason lay in the fact that, with the exception of Denmark, the Scandinavian countries offered little scope for the kind of commercial grain farming that had developed in eastern Germany and Romania. Indeed, agriculture was so marginal throughout much of the area that independent peasant farming was the only practicable mode of land use. The absence of servitude and *corvée* did not, however, lead to an egalitarian society. In most parts of Scandinavia a class of landless cotters emerged. These were to some extent those who had lost out when village lands were enclosed, but in the main they were younger sons, unable at a time of high birth-rate to acquire land in a system of impartible inheritance. Their numbers increased during the middle years of the century in both Denmark[81] and Sweden.[82] They constituted the foremost agrarian problem of Scandinavia and were the source of the large Scandinavian emigration late in the century.

Land use

Constraints on the mode of land use were set by the physical factors of climate and soil, and by accessibility, transport and demand. But within the limits thus set there was scope for a considerable range of land use. The pattern changed significantly during the century. Not only was there an increase – very difficult to measure but of the order of 10 to 15 per cent overall – in the extent of the land used agriculturally, but there was also a change in the actual use made of farmland. Cropland at first increased as a proportion of the whole, and then, in the later years of the century, contracted in favour of permanent grass. At the same time woodland was reduced in the more densely settled areas, but increased by afforestation in some lands of little agricultural value, such as the heathlands of northern Europe.

It is difficult to present more than a broad picture of the changing pattern of land use because the statistical data for different countries are not strictly comparable and for some they do not even exist. In any regional consideration the whole of Scandinavia appears as an area of forest – very largely coniferous – and waste, with agricultural land only in the coastal lowlands and river valleys. To the south, from France to Poland and including the Danish peninsula, lay a lowland area of predominantly agricultural land. Better soils, a less rigorous climate and a longer growing season made this the chief crop-growing region. Within it, however, lay areas of older and harder rock and of thin, acid soil, like islands of infertility. These included the Ardennes, Vosges, Harz and Ore Mountains. The plain was also interrupted by patches of sand and gravel left by the ice-sheets during their final retreat. Here the poverty of the soil rather than the harshness of the climate repelled the farmer, though many such areas were planted with conifers in the course of the century.

To the south lay the Alpine system, a series of ranges which extended from

Spain to Bulgaria. At their highest altitudes they were a waste of bare rock and snow, of no direct agricultural value. At lower altitudes were forests, coniferous passing downwards into deciduous, interrupted by more open areas which provided summer grazing, an important element in the economy of the mountain-dwellers. In the valleys and on lower ground was agricultural land, immensely varied in quality according to insolation and soil, and used for hardy crops and cultivated grasses. The Alpine system also covered much of the Italian and Balkan peninsulas. Land use changed towards the south where glaciation had been insignificant. The mountain and plateau surfaces had little soil and supported at best a vegetation of grass and scrub. By contrast the lower land had been receiving for thousands of years the waste from the uplands, and had built up a deep and rich though sometimes ill-drained soil. Here agriculture had been practised since the earliest phases of human settlement, restricted only by the uncertainty of rainfall everywhere and its complete absence during summer in the more southerly parts.

Such in highly general terms was the pattern of land use. In reality the differing types of physical environment interpenetrated, forming enclaves of each within the others, like the islands of sandy heath which punctuated the northern plain and the alluvial basins and valleys which lay amid the Alpine system. Every region of good soil and favourable climate was bordered by another where physical conditions were less propitious. With increasing demand and rising prices agriculture tended to spread from areas of optimum conditions into those which could be regarded as marginal, in so far as the total costs of producing and marketing the crop were greater. The cultivation of marginal land became increasingly important during the middle years of the century. De Laveleye has given a picture of marginal cultivation in the sandy areas of the Low Countries: 'Penetrating into the interior of these dunes', he wrote, 'you observe little cottages with a few acres of rye and potatoes around them... With manure they mix seaweed and whatever animal matter the sea throws up, and thus they raise crops of first-rate potatoes and vegetables.'[83] It was a case of marginal cultivation, in which capital in the form of tools and equipment was minimal and the labour input immense. Such an expenditure of labour would be likely to have been made only when it was cheap; in other words, under conditions of acute rural overpopulation. Offer alternative employment in manufacturing, without a parallel increase in population, and such marginal settlements would disappear into the sands of the Flanders plain, as indeed they eventually did.

Throughout western Europe and in those parts of central and eastern Europe which were accessible to commercial influences the margin of cultivation began to recede in the late nineteenth century. The poorest soils and least convenient locations were abandoned for regular cultivation, as the broadening of the market brought down the price of one commodity after another and lessened the reward for the cultivator. Crofts in Scotland, hill farms in Wales, the uppermost terraces on the steep slopes of the Dalmatian coast, the farthest *saetars* in the Swedish forest were gradually abandoned as other opportunities opened up. This retreat at the margin of agricultural land came last to eastern and south-eastern Europe, to southern Italy and parts of Spain, because it was

Table 5.2. *Land use in France (thousands of hectares)*

	1840	1855	1882	1892	1909
Arable	25,682	25,628	25,588	25,385	23,615
Vineyards	1,972	2,102	2,197	1,800	1,687
Gardens, etc.	—	—	1,272	1,322	1,220
Meadow	4,198	5,161	4,115	4,403	4,838
Pasture and grazing	9,901	9,209	7,934	8,037	9,049
Forest	8,805	8,986	9,455	9,522	9,329
Non-agricultural	2,769	1,943	2,296	2,389	3,219

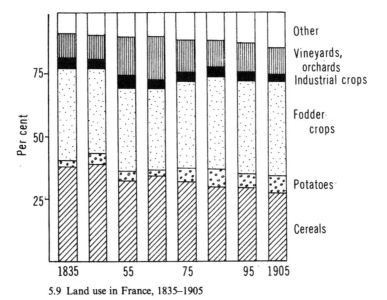

5.9 Land use in France, 1835–1905

not until late in the century that people began to have the choice between back-breaking toil on an unrewarding land and long hours in the heat and dirt of a factory.

In very few countries can this change be documented. In France the area of arable increased until the mid-century,[84] and then very slowly declined (fig. 5.9 and table 5.2). The area of vineyard, after increasing until the 1880s, was struck by *phylloxera* and dropped sharply (fig. 5.16).

In Germany, by contrast, there was an almost continuous expansion of the cultivated area, reflecting in this the continuing growth of the population (fig. 5.10). In Saxony, for example, it increased from 52.5 per cent of the area about 1840 to 54.3 per cent in 1878.[85] In the East Elbian region the increase was spectacular (table 5.3).[86] In Prussia as a whole the area under cereals, roots and fodder crops increased from 18,550 sq km in 1816 to 32,940 in 1912, and cereal land itself grew from 6,310 sq km to 10,250.[87] The eastern provinces were exceptional in the high proportion of their total area which was used agri-

Table 5.3. *Land use in eastern Germany (millions of hectares)*

	Food crops	Fodder crops	Fallow	Total
1800	4.6	—	2.8	7.4
1840	5.9	0.5	3.0	9.4
1861	7.9	0.9	3.2	12.0
1883	9.2	1.1	2.1	12.4
1895	9.5	1.3	1.8	12.6
1907	10.1	1.4	1.1	12.6

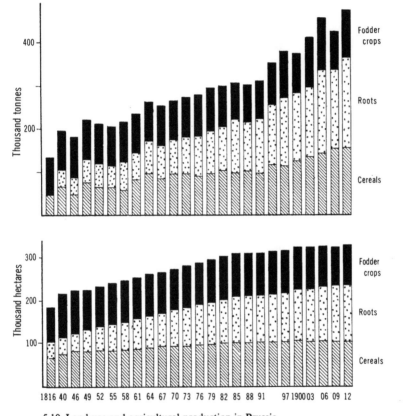

5.10 Land use and agricultural production in Prussia

culturally. It was lower in western and southern Germany, and over Germany as a whole it represented less than 49 per cent at the end of the century.[88]

The situation in Poland was similar to that in the East Elbian lands. The area under crops increased sharply, largely at the expense of the forest (table 5.4).[89]

The extent of agricultural land increased during the century in all parts of the Habsburg Empire, though there are no general statistics before its end (fig. 5.11).[90] It is said, however, to have increased by some 40 per cent in Hungary during the second half of the century alone.[91] The proportion of the land used

Table 5.4. *Land use in Poland (percentages)*

	Agricultural land		Forest
	Under cereals	Total	
1816–20	33.7	39.7	30.1
1839	34.3	40.8	27.9
1848	45.8	56.3	19.5
1859	46.3	61.6	20.1
1894	55.0	74.5	20.6
1909	56.3	75.0	18.0

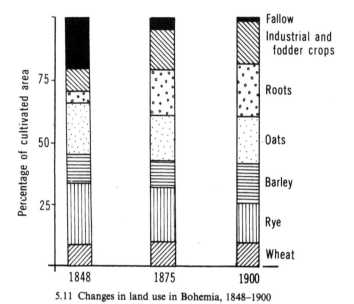

5.11 Changes in land use in Bohemia, 1848–1900

varied greatly. It was little more than 15 per cent in the Alpine region, but rose to 60 per cent and more at the end of the century in the Czech lands, Galicia and the Plain of Hungary. In Hungary as a whole the change in a half-century was as shown in table 5.5.[92]

There are no comparable data for the Balkan countries, though statistics of crop production became available during the closing years of the century.[93] These suggest a not inconsiderable increase in agricultural land at the expense of forest and waste. This appears to have been most marked in Bulgaria and Romania, but was significant also in Serbia. A Foreign Office report of 1897 commented on the great extent of cropland in Serbia, adding, doubtless with some exaggeration, that 'there is no waste land'.[94]

Throughout Scandinavia there was an increase in the area of cropland during the earlier years of the century. This was least in Denmark, where there was little scope for the advance of cultivation (fig. 5.12). Here about two-thirds of the land was already under the plough in 1860,[95] and a somewhat smaller proportion after

Agriculture

Table 5.5. *Land use in Hungary (thousand hectares)*

	1846		1895	
	Area	Percentage	Area	Percentage
Arable	6,959	33.5	10,496	46.9
Vineyards	575	2.8	250	1.1
Meadow	1,861	8.9	2,002	8.9
Grazing	3,199	15.4	2,941	13.1
Forest	6,294	30.3	5,281	23.6
Other	1,922	9.1	1,425	6.4

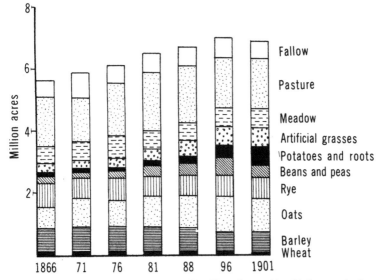

5.12 Land use in Denmark, 1866–1901. Note the considerable increase in the area under fodder crops and grazing

the loss of the Duchies in 1864. Woodland was scanty – less than 5 per cent of the area, but heath and bog covered a fifth of the country. In Sweden and Finland, however, the area increased sharply, in Sweden from less than a million hectares at the beginning of the century to 2.5 millions, or 6 per cent of the total area by 1860. After this, the rate of increase became slower and ceased early in the twentieth century.[96] In Finland also cultivation made progress at the expense of the forest, though largely by the wasteful process of burning the forest and cultivating the soil, enriched by the ashes, for only a few years. This process reached its peak in the mid-nineteenth century, when, however, less than 10 per cent of the land was being used at any one time.

A survey of agricultural land in Italy, made in 1877, showed that overall 37.5 per cent of the land was cultivated and that 12.3 per cent was under forest. The rest must have been made up largely of bare mountain and *maquis*-covered waste. The proportions varied greatly. In the province of Emilia almost 60 per cent of the land was cultivated, and in Lombardy, Umbria and the March, over 40. On the other hand, in the mountainous provinces of Piedmont, Liguria and Sardinia, it was only a quarter or even less. The densely populated province of Sicily had 43.4 per cent of its area under cultivation.[97] The distribution of forest was the inverse of that of cropland; most extensive in Liguria (24.8 per cent) and Sardinia (24.6 per cent), and least in Apulia (7.7 per cent) and Sicily (3.5 per cent). The area of arable had been increasing during the century, but began to contract towards the end with the abandonment of marginal land, especially in the mountainous regions of the south, from which there was a large-scale migration.

In Spain the increase in the cultivated area had been artificially restricted by prohibiting the clearing of the waste and encroachment on lands traditionally used by the transhumant flocks. Between 1813 and 1837 these constraints were gradually removed, and this, together with the rise in the population, led to an explosion of agricultural activity.[98] The area under cereals – that is almost all the arable area – increased from 6.1 million hectares at the beginning of the century to about 9 in 1860. But this expansion was too great. Marginal land was ploughed and cultivated; yields per hectare declined, and land was then abandoned. The cultivated area in 1900 was little more than 15 per cent greater than a century earlier. At the same time, however, there was a continuous increase in the area of vineyards, olive groves and orchards, which became the basis of Spanish agricultural prosperity towards the end of the century.[99] There was some increase in irrigated land, but the crops grown were not in the main cereals, but vegetables, sugar-beet and fruit.

Estates and farms

Any discussion of the relationship of the peasantry to the land is bedevilled by the confusion between ownership and occupation, between estate and tenement. The statistics show how numerous and how large were the great estates in the east and the south of Europe. They commonly fail to indicate, however, that much of their vast area was forest or waste, and that a significant part of the rest was divided into peasant tenements and held by labour service or rent, in lease or *en métayage*. Even in France, where it is commonly assumed that a majority of the peasants actually owned their own farms, there was an important distinction to be drawn between occupation and ownership. The French land-tax of 1858 showed that about 92 per cent of all assessments were for less than 50 francs, suggesting units of not more than about 12.5 hectares.[100] They represented, however, little more than a third of the total land subject to the tax. At the same time units assessed at more than 100 francs, though only 3.7 per cent of the total, accounted for 48 per cent of the area. Thus, it would appear, almost half the land was *owned* in units of above, say, 25 hectares. By contrast, it was found only four years later that about 97 per cent of all actual farms were of less than 40 hectares, and that 87 per cent had under 20. It is clear that estates, some of them very extensive, were divided into medium and small farms which were leased to tenants.

Tenurial systems

The conditions by which a peasant or, indeed, anyone else held the land he cultivated varied greatly. They determined the degree to which he could control his own farming operations and the fixed outgoings which these had to cover. It encouraged, or the reverse, investment in the land, the consolidation of strips and the acquisition of stock and tools. No peasant would be likely to make improvements unless he was sure that the land would be his for at least a term of years. Except for lands cleared or reclaimed during the nineteenth century, all peasant-held land had once borne some form of feudal obligation. It had formerly owed a rent in money or in kind, or some form of service; often it owed both. Services had usually been rendered on the demesne of the lord, who himself had obligations of some kind to a higher level of authority in the feudal hierarchy. The tenurial system over most of Europe in the nineteenth century arose from the breakdown of this feudal relationship.

From it emerged four systems of landholding. In the first place land could be held allodially or, in the common English phrase, in fee simple. This did not mean that it was totally without obligations, though all had been attenuated through time and most had been forgotten. In practice these lands owed only such taxes as the government might impose and, wherever appropriate, tithe to the church. Such freehold tenures might have derived from the demesne, and they generally included much of the forest and waste land. They could also arise from the expropriation by the peasantry of feudally held land, as happened in France during the Revolution, in Poland in 1864 and in Serbia and other parts of the Balkans after the retreat of the Turks.

A feature, normally restricted to estates held in freehold, was the right, even the obligation, to pass them on intact to an heir. This originated in the fear that the division of an estate would so fragment its obligations to prince or king that they would become of little value to him. In the nineteenth century the law of entail prevented the alienation of estate land to peasants. It was abrogated in Prussia and Spain, without, however, having the desired effect of creating more peasant holdings. It survived in Hungary, Italy and, in modified form, in the Balkans.

The converse of entail was the partible inheritance, the right, even the obligation, to divide an estate between all direct heirs. It was in practice restricted to peasant holdings. It was to be met with in many parts of Europe, and its occurrence bore little relationship to physical or other conditions. Even though it was common for one of the co-heirs to a holding to buy out the rights of the others, the practice of partible inheritance led to the formation of very small and severely fragmented holdings, as, for example, in south-western Germany. The rule of primogeniture led, on the other hand, to the formation of a class of landless labourers, as in Sweden, and sometimes to an explosive out-migration. The practice of partible inheritance has been held to contribute to a high birth-rate and a numerous but impoverished peasantry. Habakkuk has pointed to the contrast in size of rural population and degree of fragmentation between Bohemia with its rule of primogeniture and Slovakia with its practice of partible inheritance.[101] It may, however, be questioned whether this was the only factor in the different demographic history of the two provinces in view of the continued contrast in the twentieth century.

The division of land between heirs had been practised before the Revolution in many parts of France. The effect of the *Code Napoléon* was to generalise and legalise this custom.[102] The extreme parcellisation which might otherwise have resulted was limited in part by the low birth-rate, in part by the practice of one son buying out the claims of his co-heirs. Nevertheless, the consequences of partible inheritance have generally been regarded as bad for French agriculture, in the words of Balzac, 'fragmenting the land, breaking up estates and ending by ruining France'. The practice was probably most dangerous in those areas, notably the west and south-west, where there were few alternative industrial or urban occupations.[103]

A second mode of tenure was by payment of rent to a landlord. In a great many instances such tenements derived from the unfree peasant holdings of the feudal period. Labour services had been commuted, the rent payment brought into line with the value of the land, and, of course, the peasants themselves freed from personal servitudes. Alternatively, the tenements could have been part of the demesne which had been let at a money rent. Many changes commonly occurred in the process of transition from feudally controlled land to land held on a modern, commercial basis. Sometimes, as we have seen, a peasant holding was diminished in order to compensate the lord for services no longer rendered. The peasant who had once held at the lord' will, may have acquired a *de facto* security of tenure, or his tenancy may have been given a specific term of years, usually renewable if only at an enhanced rent. An immense variety emerged in the conditions under which rented tenements were held, dependent on local custom and the feudal conditions from which they derived.

Métayage was the third important mode of tenure. It was essentially the form of tenancy known in English as 'share-cropping'. The lord provided the land, tools and equipment and sometimes also farm animals and seed. The tenant, who usually held for a fixed term, provided the labour of himself and his family, and the product was shared in agreed proportions, often half and half, between lord and tenant. *Métayage* was formerly widespread in western Europe. It was roundly condemned by most of the earlier writers on agronomy,[104] and defended by few. A writer in the *Economic Journal* presented a more positive view of the system.[105] It permitted a peasant who lacked capital to work a small farm; it provided an insurance in so far as a bad harvest was partially compensated by a smaller rent payment. On the other hand, it discouraged improvements, since half the increased output went to the lord. The practice of *métayage*, which had been widespread in France and general in Italy, declined during the later years of the nineteenth century and was of little importance in the twentieth. It derived from the wealth of the lord and the relative poverty of the peasant, and must have originated as feudally-held land was broken up and leased to tenants.

The last form of tenancy was land held in common. In origin it had been the forest and the waste which surrounded the rural community, in which the peasantry took timber for building and fuel and grazed such animals, pigs and sheep most often, as could be kept under such circumstances. The common land belonged to the lord, but by custom his rights had become limited by the corresponding rights of the tenants. Clearly there must have been a restriction both on those entitled to use the common and on the extent of the use they made of it. The tendency throughout the nineteenth century was for the common lands, whether grazing or woodland, to be enclosed and divided between those who could prove a right to them. Inevitably not all who used the commons could substantiate such a right, and many were the small peasants and cottagers who lost this important supplement to their income. On the other hand, the importance of common lands in general was tending to diminish. As the quality of livestock improved, so the peasant became less willing to entrust his animals to the poor feed and uncertain fortunes of the commons. He was also unwilling to sacrifice the droppings of his animals, which often formed his only source of manure.[106]

Common in the forests was a different matter. Timber, both for building and firing, was essential to a peasant society. At the same time, the lord, who usually retained a residual right in the forests, had every inducement to maintain his control over them and even to exploit them for his own ends. Clashes between lord and peasant over woodland rights were amongst the bitterest in the long history of rural conflict.

It is difficult to present a picture of the distribution of these tenurial systems. Statistics are, however, available for France, thanks to the *cadastre*, begun by Napoleon, and the property tax (*côtes foncières*) instituted in 1815. Cultivators who held their land in fee simple amounted to at least half and probably 60 per cent of the total.[107] Most of their land derived from that which had passed into peasant ownership at the Revolution. But some of the latter was regained by its previous owners after 1815, and there was also a considerable market in land, both noble- and peasant-owned.[108] Owner-occupied land was to be found in all parts of France. It was greatest in the east and south-east; least in the west and north-west. Most of such holdings were small, but were cultivated by fiercely independent and acquisitive peasants, more intent on adding strip to strip than on developing scientific husbandry. Rented land and land *en métayage* were necessarily important only where estates of some kind had been able to survive. In 1882, when a survey was made of land tenures, rented land (*en fermage*) made up more than half of the total only in eight *départements* in the north and north-west. By and large, these were the Catholic and conservative areas where the results of the Revolution were least permanent.

Métayage was the most highly localised and distinctive form of tenancy. In 1882 it was most important over the area from the river Loire southwards to the Pyrenees, and was significant also along the Mediterranean coast and in Corsica. It was scarcely to be found at all in north-eastern and eastern France, but was especially important in the Landes, Vandée and Bourbonnais. At this time it was in decline.[109] It had at all times been restricted to the small holdings of the poorer peasants, and with rising standards of living the *métayer* tended to become a *fermier*.

The only other country in which métayage (*mezzadria*) was significant was Italy. In 1895 the number of *métayers* exceeded by a wide margin that of independent cultivators.[110] in Tuscany, most of the peasants were métayers;[111] they were numerous throughout central Italy,[112] and in Lombardy they occupied a belt along the northern margin of the plain.[113] In few areas of Italy did lease-holding, rent-paying tenants form more than an insignificant minority. Much of southern Italy continued to be held in great estates, part forest, part waste, and the rest cultivated with hired or tribute labour.[114]

In Germany the line of the River Elbe formed very approximately a divide between western Germany, characterised by small holdings, held either in freehold or at a money rent, and an eastern region dominated by the great estates.[115] Tenurial conditions varied greatly within western Germany, but the region as a whole was characterised by the failure of the manorial system to survive. In consequence peasant dependency took the form of personal services only to a very small extent. Tenements were held by some form of money rent, and tenurial conditions were rarely burdensome. In the Rhineland and much of south Germany, where the practice of partible inheritance prevailed, lands, and thus the obligations attached to them, were divided between heirs. Services became in some instances so attenuated that they ceased to be worth the trouble of collecting.[116] Feudal forms survived more completely in parts of Bavaria and the Alps, and here a system of impartible inheritance served to perpetuate large holdings. In north-west Germany, also, servitudes disappeared relatively early, and farms were generally held on a lease and rent basis. A not dissimilar system spread through Thuringia and Saxony. Everywhere the peasant was relatively free of restrictions and generally paid nothing more than a realistic rent.

The transition from the west German region of small peasant farms was abrupt. East of the Elbe the prevailing mode of land tenure at the beginning of the century was in large estates, worked by peasants, whose holdings were generally burdened with labour dues. It was a strict and all-embracing manorial rule, with hereditary serfdom. Many of the estates were very large though much of their area was of little agricultural value. The liberation of the serfs early in the century was accompanied by the redemption of the services by which they had held their tenements, either by the surrender of part of the peasant holding or by a cash payment made possible by morgaging or selling land. A vigorous debate has turned on the question of which side gained or lost most in the process. It is not disputed that about a million hectares – 6 per cent of the agricultural land – were in one way or another transferred from peasant occupation to that of their lords. Almost half of this came as compensation for lost services and dues, and net sales by the peasants amounted to 100,000–200,000 hectares. The rest exchanged hands when the peasants ran into financial difficulties during the agricultural crisis of the 1820s.[117] Despite assertions by the landowners, it appears that the peasants lost in the transactions. Their holdings were often reduced to an uneconomic size and a class of landless peasants was produced.[118] On the other hand some of the peasants acquired a share in the common lands which were divided and enclosed at the same time.

Eastern Germany thus emerged from the land reforms with its system of great estates strengthened. Almost a century later more than 40 per cent of the *agricultural* land was held in units of more than 100 hectares. At the same time only about 8 per cent of the agricultural land in the rest of Germany was so held (table 5.6).[119] During the previous generation there had been overall a small

Table 5.6. *Land held in estates, 1907 (percentages)*

In units of (hectares)	Prussia and Mecklenburg	Rest of Germany	Whole of Germany
0–5	8.7	21.0	16.2
5–20	21.3	41.0	33.4
20–100	29.5	29.9	29.8
Over 100	40.5	8.1	20.6

decline in the area of farmland held in large units and a corresponding increase in that occupied by farms of up to 20 hectares, created by the leasing of estates.[120]

In Denmark the majority of farms were held in freehold or on long leases.[121] Amongst them were the few very large farms which, it was said in 1860, were invariably better cultivated.[122] They had formerly been worked by servile labour, but feudal services had disappeared by the beginning of the century.[123] In the rest of Scandinavia the abundance of land and the practice of primogeniture ensured that farms were large. There had never been a manorial system, and there were thus few services to commute. Most farms were held in freehold or on long leases. Though free from labour services in the feudal sense, many of the Swedish peasants were until 1865 burdened with the 'posting' system, whereby they were compelled by the state to perform an ill-paid carrier service.[124]

Russian-held Poland was a land of estates, large towards the east, mostly small in the rest of the country. The manorial system was strongly entrenched, as it was in eastern Germany, and the peasants held their small tenements chiefly by service on the *folwarki* of their lords. This was abruptly terminated in 1864, when feudal obligations were abolished. The peasants became in effect owners of their holdings, and the *szlachta* was left with only its demesnes and such rights as it could retain in the commons.[125] By 1886, no less than 41 per cent of the land of Poland had passed into the absolute possession of the peasants.[126] The area held in estates continued to decline, as the gentry were obliged to sell off land to the more prosperous peasants.

Hungary is rightly regarded as the country in which great estates preponderated almost to the exclusion of peasant farms. They had, by and large, been created in the course of the expulsion of the Ottomans and the conquest of the Plain by the forces of the Habsburgs. A modified manorial system was established, in which the peasants held their tenements by a combination of labour services and payments in kind.[127] Serfdom was ended in 1848, and the servile obligations of the peasants redeemed by payments in land or in money. The result was similar to that achieved in eastern Germany and in Spain: large estates became larger, and peasant holdings were often made uneconomically small. This situation changed little. Up to a third of the estates were entailed and indivisible.[128] In the half-century before 1914 the largest of them tended to grow even larger, while small and family holdings increased but less markedly. This was achieved at the expense of 'a small and dwindling *Mittelgrundbesitz*'.

Feudal tenures survived longer in the Romanian principalities than anywhere else in Europe. Labour dues continued to be performed even after 1848. Despite the reform of 1864 peasant holdings became smaller, and much of the land that came onto the market was added to the estates. By 1913 almost a fifth of the land was held in estates of over 100 hectares, while about 55 per cent was in holdings of less than 10 hectares – most of them of less than 5.[129]

In the rest of the Balkans the feudal system of land-holding had largely collapsed.[130] It was a system introduced from Asia by the Ottomans, but was none the less based on the western principle of holding land in return for service. This was the situation when the Ottoman Empire began to crumble beneath the attacks from Habsburgs without the Slav nationalism within. As the Turks withdrew, their lands passed into the hands of the native peasantry. 'The land belongs to those who cultivate it' became the agrarian programme of the Balkan governments. In Serbia *chifliks* were abolished in 1815. Eliot's report of 1897 spoke in eulogistic terms of the 'peasant democracy' of Serbia and 'the entire absence of large properties'.[131] Indeed most of the country was held in small peasant-owned tenements, which grew smaller as the population increased. In 1905 more than 60 per cent of the land was in peasant holdings of less than 10 hectares, and 65 per cent of peasant families had each less than 5.[132] In Bulgaria the estates which had been held by the Turks were appropriated by the peasants, and Bulgaria remained until after the Second World War a land of peasant proprietors, whose holdings were on average even smaller than in Serbia.

In Greece a share-cropping system was developed on some of the *chifliks*. Although many of these were broken up in the course of the nineteenth century, the *chiflik* organisation still survived in the early twentieth century on the newly acquired lands in Macedonia and Thrace.

Between the ancestral lands of the Habsburgs and the Ottoman Empire lay an area which had been fought for by both. It included Kranj, Croatia, Bosnia and Hercegovina. The Turks made good their hold on Bosnia and Hercegovina, and settled the territory in classical *chiflik-timar* fashion. The only significant difference between this and the rest of the Balkans was that a number of the local Christian population, largely derived from the medieval Bogumils, were converted to Islam and became themselves *chiflik* owners.[133] The condition of the Christian *raya* was no better, indeed worse than elsewhere, because after the Habsburg occupation of the area in 1878 the Slav landowners remained, the 'Imperial and Royal' government being unwilling to interfere with their traditional usages.[134] Kranj and Croatia, by contrast, had been under Ottoman rule for only a short time, and throughout the nineteenth century were part respectively of Austria and Hungary, with which they shared the system of large estates and small, dependent peasant holdings.

It is paradoxical that the worst agrarian conditions were to be found in those parts of the Balkans which were most open to western influences. The reason is that in this borderland it was politically and economically desirable for the lords to preserve the old order, whereas in the Balkans proper no one outside the region had any vested interest in preserving the Ottoman system of estates.

The size of farms

Even more important than the conditions by which a peasant held his land was the size of the tenement which he possessed. Leaving aside demesne farms, which were sometimes very large indeed, tenements ranged in size from a fraction of a hectare to 50 or even more. The optimum size of a family holding varied with physical conditions. It was significantly smaller on the good soils of north-western Europe than on the marginal lands of Scandinavia or the hills of central Europe. It varied also with the importance of animal husbandry and with the extent to which agriculture was supplemented by domestic crafts and urban pursuits, with the technical efficiency of the peasant and the cropping system which he practised. Let it suffice to say that most European peasant holdings were too small, and that peasant acquisitiveness was justified in so far as a larger holding could have been managed more efficiently. Not that the peasants normally set much store by efficiency. Most clung obstinately to traditional ways, preferring to increase their production by adding another strip to their holding rather than by more intensive cultivation and better organisation of what they already had.

In France the great majority of peasant holdings were, early in the nineteenth century, very small. In 1826, for example, 'small landowners' with on average only 2.65 hectares, made up almost 90 per cent and occupied near a third of the land.[135] In 1882, 85 per cent of all holdings were of less than 10 hectares, and only 2.5 per cent had more than 40 hectares. The latter, however, accounted for over 40 per cent of the land.[136] The revolutionary settlement had resulted in an immense number of small holdings. In the course of the century, many of the previous estates and farms were reconstituted partly by reacquisition by their former owners, partly by purchase by thrifty peasants. A writer about 1880 commented on the 'continual enlargement' of their properties 'by the purchase of adjoining plots'.[137] This was, however, offset by the contrary tendency for farm holdings to be divided between heirs.

It is generally agreed that the French peasant on average expended an immense labour on a holding that was too small, using poor tools and out-of-date methods. It has been estimated that about 2 hectares of cropland was the absolute minimum necessary to support a peasant family in the mid-nineteenth century, and that 'only in the category of five to ten hectares does one get the beginning of a sense of security'. At this time some 40 per cent of all holdings were of less than 1 hectare, and considerably more than a half had under 5.[138] In consequence a large part of the French peasantry lived at a subsistence level, vulnerable to every *crise de subsistance* and dependent on domestic crafts, like lace-making or embroidery, to obtain the bare necessities of life.

French agriculture, it has been commonly said, was backward, overmanned and undercapitalised, unable either to provide a market for the products of manufacturing or to generate capital for investment. This view is too simplistic.[139] It ignores the fact that not all French farmers lived at this level; that amongst many of them technical innovations were adopted readily, and that the allegedly inefficient peasant nevertheless supplied the urban and industrial population with a considerable volume of food. Development in French agriculture was more likely to have been held back by lack of demand.[140] There was in France a technological dualism. A minority of farms were large enough to provide a good if not abundant living; the peasants made the most of available manure, and they used improved tools and advanced methods. The trouble was that the older and more primitive methods showed little sign of being displaced by the new, because the peasants lacked the knowledge, the resources and the boldness to use them.

If the average farm in France was uneconomically small, in Belgium it was minute, so much so that there was no pretence that most of them could provide a family living. 'The average size of the holdings', wrote Seebohm Rowntree,[141] 'is smaller than in any other country of Europe...[and] there is a steady and continuous movement towards a still further reduction in the average size of a holding.' About 1880 no less than 55 per cent of all farm holdings were of less than 1 hectare, and some 43 per cent had under a half.[142] Only about 8 per cent were of more than 10 hectares. There was some regional variation. Micro-holdings were proportionately more numerous in the north, especially in Flanders and Hainault; least so in the hilly provinces of Luxembourg and Limburg, where small and medium-sized farms were relatively important.

This extreme parcellism was due in part to the prevailing system of partible inheritance; in part to the almost universal practice of supplementing farm income with domestic or factory work. As household industry declined, so labour was drawn to the factories. Most of the rural population lived at no great distance from urban or industrial employment. Some even commuted across the boundary to factories in northern France. Although farm holdings rarely offered a full-time occupation, they were nevertheless intensively cultivated. Fallow had long been abandoned on most of the northern plain; rotation grass was little used except on the damp lands of Flanders. Instead the maximum use was made of animal manure, domestic soil and every form of organic waste. Here farming approximated to gardening, and the closer to the towns one came, the more the rural landscape took on the aspect of a vast allotment.[143]

Germany presented a more strongly regionalised and contrasted picture than either France or the Low Countries. The common generalisation that eastern Germany was characterised by large estates and western and southern by small and medium-sized holdings is broadly true, but needs qualification. Each of the major divisions of Germany contained the whole spectrum from great estates to micro-holdings; only the 'mix' varied. Secondly, many of the large estates in the east were, for purposes of cultivation, divided into smaller units which were commonly held on lease. Prussia, which appears in the statistics as an area of large estates, nevertheless had nearly 30 per cent of its area in farm units of less than 20 hectares, and little more than 40 per cent in estates of over 100.[144] Only in Pomerania and Posen did large estates, here defined as of more than 150 hectares, make up more than a half of the rural area.[145] Within Bavaria, on the other hand, nearly half the farm holdings in Pfalz were of less than 1 hectare, and only 5 per cent had more than 10 hectares, whereas in Oberpfalz these proportions were reversed.[146]

The geographical pattern of large and small estates was far from static. Partible inheritance was practised in parts of western Germany, and in eastern Germany the termination of entails meant that estates too could be divided. The

Table 5.7. *Size of farm holdings in Poland, 1892 (hectares)*

	Below 5	5–20	20–50	50–100	Over 100
Percentage of holdings	38.85	52.77	7.02	0.34	1.02
Percentage of area	5.96	25.43	8.0	1.81	58.8

resulting tendency towards smaller holdings, was, however, offset by the practice, especially in areas where large estates were numerous, for their owners to purchase the small parcels of land that resulted from partition.[147] The high tariffs on imported grain and the determination of the German government to maintain a prosperous agriculture made estate farming a profitable business.

In Poland after 1864 most peasants acquired possession of the lands they cultivated, and the country became one of small to medium-sized holdings, interspersed with small estates which had managed to survive the enfranchisement of 1864 (table 5.7).[148] A considerable fraction of the latter was, however, under forest.

The Habsburg lands presented a more varied picture, ranging from the large estates which dominated the rural scene in Hungary to the micro-holdings of Galicia. The Czech lands resembled eastern Germany, with estates making up a large part of the area. In Bohemia in 1861–72, almost a quarter of the land was held in units of more than 200 joche (115 ha). This area tended to diminish, but remained very large until the First World War.[149] On the other hand, two-thirds of the peasant holdings were of less than 5 hectares, and their number tended to increase during the century as the larger amongst them were divided between heirs.[150]

In Austria proper the situation was broadly similar. Most of the land was held by noble families, and after the eighteenth century was leased to peasants.[151] The lowlands of the Danube valley were mostly in peasant tenements, the majority of them of less than 5 hectares.[152] The north-eastern province of Galicia had by the end of the nineteenth century become a byword for the extreme division of holdings. Here about 80 per cent were of less than 2 hectares, and 98 per cent were under 5; there were, in effect, no rich peasants.[153] Nearly half the land was held in estates, but much of this consisted of mountain and forest.

In Hungary the great estates dominated. In 1867 well over half the land was held in estates of over 200 holds (about 115 ha), whereas nearly 60 per cent of peasant farms were each of under 5 holds (2.9 ha). There was a tendency here, as elsewhere in eastern Europe, for peasant holdings to be absorbed into the demesne and to be cultivated with hired labour.[154] By 1914 the area in large estates had increased appreciably and those of over 1,000 holds (about 580 ha) made up some 40 per cent of the country (table 5.8[155]).[156] The rest was made up of peasant holdings, many of them too small and ill-equipped to support the families which occupied them.[157] In regions bordering the great plain peasant farms, often uneconomically small, prevailed. In Slovakia and Ruthenia the majority were of less than 10 hectares,[158] and in Transylvania their average size was even smaller, the majority being of less than 5 hectares.[159]

According to the land census of 1895 only Transylvania and the south-eastern

Table 5.8. *Size of estates in Hungary*

Size (holds)	1867 Area	1867 Per cent	1885 Area	1885 Per cent	1914 Area	1914 Per cent
Under 5	3,801	14.1 ⎱	9,368	33.2 ⎫	13,852	49.1
5–30	4,847	18.1 ⎰		⎬		
30–200	3,879	14.5	4,262	15.1 ⎭		
200–1,000	3,833	14.3	3,529	12.5	3,193	11.3
1,000–10,000	8,195	30.6	8,511	30.2	5,728	20.3
Over 10,000	2,262	8.4	2,539	9	5,464	19.3

part of the plain, the Tisza-Maros Corner or Banat, had significantly more than 30 per cent of their area in peasant holdings. This reflects the fact that Transylvania never underwent the depopulation and reconquest to which the Alföld had been subjected, and that the Banat was deliberately settled in the eighteenth century with peasants from central Europe.[160]

It is one of the many paradoxes regarding peasant life that in the backward Balkans the lot of the peasant was in material terms better than in central Europe. In Serbia the greater part of the land was held in farms of from 5 to 20 hectares, even though a large number of holdings were smaller than this.[161] In Croatia, Kranj and Dalmatia conditions were similar, with small to medium peasant holdings prevailing. The land was however, hilly, and a holding was often less valuable in real terms than one of comparable size in western or central Europe.[162]

Data are scanty for the rest of the Balkan peninsula, though it appears that small to medium holdings predominated in Bosnia and Hercegovina and probably also in Bulgaria and Greece after the expulsion of the Turks.[163] Romania was of course an exception. Far too little land was made available for the peasants in the course of the land reforms, and restrictions on the size of estates were widely evaded. Indeed, with increasing population the size of peasant holdings actually declined. In 1896 the average size of a peasant holding was 3.4 hectares; in 1907, only 3.[164] On the eve of the First World War rural conditions in Romania were amongst the worst in Europe.[165]

Spain, almost as much as Hungary, was a land of great estates. The system of land tenure derived essentially from the reconquest of the southern half of the peninsula from the Moors, in the course of which large estates were created for the aristocracy, the church and the religious orders. During the middle years of the nineteenth century extensive changes were made in the tenurial system in the name of liberalism. Entails (*mayorazgos*) were broken; crown land and waste were sold and much of the land of the church and municipalities was put on the market. The intention was that it should be bought in small lots by the peasantry. In reality it passed with little exception into the hands of those who could afford to pay for it, the well-to-do middle class and the great landowners.[166] The result of the Spanish land reform was twofold. It led to a small increase in the number of peasant holdings as a result of new clearances, but at the same time it strengthened the system of *latifundia*, or great estates.

Table 5.9. *Size of estates and holdings in Spain, 1930*

	Number	Total area (thousand ha)	Average size	Percentage of Number	Area
Over 250 ha	12,488	7,468.6	598	0.1	33.3
100–250 ha	16,305	2,339.9	143	0.1	10.4
10–100 ha	169,472	24,611.8	27	1.6	20.6
5–10 ha	205,784	1,379.4	6	2.0	6.1
Under 5 ha	9,810,331	6,635.3	0.6	96.0	29.6
Total	10,214,380	22,435.0			

There are no statistics before the tax survey of 1930 of the number and size of estates and farms (table 5.9), but the system of land-ownership could then have differed little from that existing at the end of the previous century.[167]

Yet not all Spain was dominated by great estates. The north was mountainous and here a peasant society had developed. In Galicia their holdings were very small, the result of continuous subdivision, but they were not heavily burdened with rents and services.[168] Across the whole north of Spain, from Asturias to Navarre, medium-sized peasant holdings prevailed, held either in freehold or on long-term leases. Most were adequate in size, and the whole region was relatively prosperous. It was also, by Spanish standards, free from social unrest.[169]

Old Castile, Leon and Aragon, reoccupied from the Moors at an early date, had also by and large escaped the blighting influence of the great estates. Here too both rainfall and soil were adequate for peasant agriculture. Much of the land was in estates of moderate size, which were leased as small peasant holdings.[170] South of the Tagus conditions changed. The land had been reconquered centuries later than northern Spain and had been occupied in very large units. It was made up of the dry plateaux of La Mancha and Estremadura, ill-suited to the traditional peasant agriculture of the humid north. Here the great estate, in part a survival from the *Reconquista*, in part reconstituted from the land sales of the nineteenth century, prevailed almost to the exclusion in some areas of medium-sized and small holdings. The *latifundia* were worked by hired and often seasonal labour which lived in dreary, barrack-like villages at a subsistence level.[171] Not surprisingly this region was the scene of the greatest rural discontent.

The more fertile soils and greater rainfall of the Guadalquivir valley of Andalusia did not, as might have been supposed, contribute to a system of peasant cultivation except in the bordering mountains. A tradition of great estates had, it is said, survived from Roman, through the Visigothic and Moorish occupation, into modern times. They 'provided the revenues of the great aristocratic houses of Castile', even though a majority had been reshaped in the nineteenth century.

The Mediterranean littoral had always been an exception both to the physical conditions and to the economic systems of Spain. The low rainfall and warm climate, complemented by irrigation water from the mountains, encouraged intensive cultivation. Conditions favoured peasant agriculture, whether under

Table 5.10. *Size of holdings in Canton Bern*

Size (hectares)	Number	Percentage	Area (hectares)	Percentage
Under 1	33,320	44.6	11,778.7	4.8
1–2	11,524	15.4	16,544.5	6.8
2–5	15,332	20.5	49,406.8	20.1
5–10	8,824	11.8	61,652.2	25.1
10–20	4,395	5.9	59,480.8	24.3
20–50	1,162	1.6	32,076.3	13.1
Over 50	148	0.2	14,279.7	5.8
Total	74,705	100.0	245,219.0	100.0

Moorish or Christian control, and the good soils were held in small but intensively cultivated holdings.

Italy was also in some degree a land of great estates, which derived from feudal conditions, but in some areas where extensive land reclamation projects had been undertaken, as in Venezia and the Po valley, they resulted from the fact that only the rich could embark on such works.[172] It is broadly true to say that large estates were characteristic of southern Italy, and small peasant holdings, cultivated on a sharecropping basis, of northern and central Italy. In fact, however, the broken nature of the terrain was matched by a fragmented pattern of land-holding.[173] In southern Italy, where large estates occupied at least half the land, there were areas, notably Campania and the coastal tract of Apulia, where peasant holdings prevailed. On the other hand, the Roman Campagna and the Maremma of Tuscany, ill-drained and malarial, were unsuited to peasant farming, and the land, 'wild and sterile', was held in estates, visited only by transhumant flocks.[174]

In the northern plain, large estates predominated only along the damp valley of the Po. On higher and drier land were 'innumerable small farms, each with a farm-house in the centre of it', cultivated on a *mezzadria* basis.[175] The land was mostly owned, as indeed it had always been, by the well-to-do burgesses of the cities. Peasant proprietorship was rare, except in the Alps and northern Apennines. Estates and large farms tended to increase in size and number in the course of the century, not only in central and southern Italy, but also in parts of northern, like the rice-growing parts of the plain, where a heavy capital expenditure was necessary.[176]

The peasant farmer predominated over most of Scandinavia. Except in Denmark, which in this respect resembled north-western Germany, estates were few.[177] A medium-sized farm in Scandinavia was a good deal larger than in the rest of Europe, because the fertility of the soil was generally less and much of it was given over to animal rearing.[178] In Denmark the yeoman's farm was usually of more than 10 hectares, and those which led in the agricultural revolution of the later nineteenth century were a great deal larger than this.[179]

Switzerland was, more perhaps than any other country in the nineteenth century, a land of peasant proprietors and family farms.[180] The data in table 5.10, relating to the Canton of Bern, may perhaps be regarded as typical of the

Fields and field-systems

A field-system is the means whereby agricultural land is divided and organised for the purposes of cultivation and use. It was subject to three kinds of constraint. In the first place, in any particular settlement it reflected the social organisation and tenurial relationships within the community itself; secondly, it was adapted to the crops usually grown and to the tools and methods employed, and lastly, it was limited by the physical conditions of climate, soil and terrain. However well adapted to these conditions the systems may once have been, the tendency in modern times had been towards ossifying them so that they acted as a restraint on progress, preventing modification and change in cropping patterns and methods. Most conspicuous was the tyranny of the common field-system, which imposed a single mode of cropping on all who shared it. Scarcely less significant were the deleterious effects of indiscriminate grazing by stock after harvest and on the fallow. Only slowly was the rigidity of field-systems broken down during the nineteenth century, and in the least progressive areas, such as eastern Poland, they survived into the twentieth.

Europe was broadly divisible into two parts, characterised respectively by open-fields and by a system of enclosed fields, freed from rules of common cultivation. The contrast between the 'champagne' and the 'several' had long been recognised. The boundary between them, however, was never a firm line. There were enclaves of open-field within that of enclosed, and areas of the latter occurred within open-field Europe.[182] To some extent the contrast lay in the soils. The deep, rich soils of, for example, northern France, central Belgium and the loess belt of northern Germany were characterised by open-fields; the thin, poor soils of north-west France and of the upland areas of central Europe, by enclosed fields, but exceptions were too numerous to count. The good soils had been colonised by large village communities, by whose collective efforts the land was cleared and cultivated. The poorer soils, capable of supporting a less dense population, tended to develop scattered settlements around which there could be no common fields. Such an explanation seems to be generally accepted, but developments during the nineteenth century gradually transformed much of open-field Europe into a landscape of enclosed fields, and in part replaced compact nucleated villages with a pattern of scattered farms.

The open-field system was characterised by at least three large fields, divided into an immense number of strips of varying size. Their shape was in some measure a response to the requirements of the heavy plough, but was also a result of the division of farm holdings between heirs. This parcellisation was always most extreme where the rule of partible inheritance prevailed.

In 1828 William Jacob, a sort of latter-day Arthur Young, wrote that 'the greater part of France, a still...greater portion of Germany, and nearly the whole of Prussia, Austria, Poland and Russia, present a wretched uniformity of system...The fields are almost universally unenclosed.' He may have exaggerated its extent, but it was a reason for much of the bad cultivation which he found. It was usually associated with a three-course system of cultivation: two field-crops, followed by a fallow year. Ploughing had ceased to be a communal enterprise in most of Europe, and the system was preserved chiefly by the practice of *vaine pâture*, the communal grazing of the fallow and of the cultivated land after the crops had been gathered.[183] This practice served a dual purpose; it contributed manure to the land and it provided feed in an age when fodder was desperately scarce. As a system it was very difficult to breach. A more rational method was to use the fallow for growing a fodder crop and to dispense with *vaine pâture*.[184] This would in turn have allowed the scattered parcels to be consolidated into compact holdings. This is indeed what happened in some parts of Europe, notably Denmark. In much of open-field Europe fallow was replaced with a catch crop, but peasant opposition prevented the *remembrement* of scattered pieces of land until the twentieth century. Even today large areas remain as open-fields, though the uniform cropping of each has been replaced by a patchwork of different cultures.[185]

Most open-fields were cultivated on a three-course system, in which autumn-sown grain was followed by spring-sown and then by fallow, before the next autumn sowing. There could be – and often were – more than three 'fields' or assemblages of strips and parcels of land; they had only to be grouped into three approximately equal parts for purposes of cultivation. Nor were open fields necessarily cultivated on a three-course system. It is commonly assumed that three-course cultivation developed from a two-course system with alternating crop and fallow. The reason, it is generally supposed, was the possibility which it offered of cultivating two-thirds rather than a half of the agricultural land. It should, however, be emphasised that autumn-sown and spring-sown crops were separate and distinct. The bread-grains only were sown in October, while barley and oats were normally sown in March. In much of southern Europe it was for climatic reasons impracticable to grow spring-sown cereals. Here cultivation was on a two-field system, with alternating bread-grains and fallow. But a two-field system was also to be met with in other climatic areas, the only condition being that there was little need for spring-sown crops. In parts of the middle Rhineland demand was overwhelmingly for bread-grains, which could be produced more abundantly under a two- than under a three-field system.[186]

The open-field system, whether two- or three-, was wasteful of labour, involving as it did repeated journeys between the strips. It also engendered countless disputes regarding the boundaries of and access to particular parcels, but above all it wasted land. It has been estimated that in the province of Białystok, in north-eastern Poland, agriculturally one of the most backward areas of Europe, the area of open-field cultivation surviving until recent years amounted to 204,000 hectares. Of this, it is said, 10,000 or 5 per cent, was made up of 'balks, boundary strips, furrows, access roads'.[187] The same situation must have existed wherever there was a highly fragmented open-field system.

The two features which signalled the demise of the open-field system were, first, the abandonment of fallow and *vaine pâture*, and, secondly, the consolidation of scattered strips and parcels of land into compact holdings. The practice of fallowing began to be abandoned in the eighteenth century in western Europe.

Early in the nineteenth it was little practised in the more densely peopled areas of the Low Countries and northern France. In France fallow, which had once amounted to nearly a third of the arable land, was reduced to a fifth by mid-century and to a seventh in 1882.[188] By the end of the century fallowing continued to be practised only on the poorest soil. Its decline followed a strongly regional pattern, most rapid in the north, in all respects the most progressive part of France; slowest in a belt of territory extending from Vendée in the west to the Auvergne,[189] and here it lingered into the twentieth century.

Its decline in central and eastern Europe was slower. By the end of the century it had almost disappeared from Bohemia, but lasted longer in Hungary and Poland.[190] It was also slow to disappear in southern Europe where the long dry summer made it difficult to find a suitable crop to alternate with bread-grains.

The cultivation of the fallow was quickly followed by the abandonment of the practice of *vaine pâture*. In France it became illegal in 1889. It did not, however, disappear at once, but ceased gradually to be necessary, as fodder crops began to be grown on the fallow.[191] It was abandoned in much of central and eastern Europe towards the end of the century.[192]

The abrogation of grazing rights over the arable removed the last obstacle to the more intelligent organisation and use of the land. Since uniformity was no longer required in each of the open-fields, the peasant could break with tradition and cultivate whatever crops suited his needs without inconveniencing his neighbours. But his lands remained scattered and awkwardly shaped. There was no institutional or social reason why these should not be consolidated; only conservatism and suspicion hindered the process. Progress was first made in Scandinavia. In Denmark a law of 1781 gave to every peasant the right to consolidate his land into a single holding,[193] and by the early nineteenth century few fields of intermixed strips remained. Progress was slower in Sweden and Norway, but even here the process was largely accomplished by the latter half of the nineteenth century.[194] In Sweden it became possible for even a single peasant to demand that the scattered parcels of the community be consolidated.

Little progress was made in Germany until after the final abolition of remaining feudal obligations in 1848. Thereafter most of the German states passed enabling legislation, and slow progress was made during the later years of the century.[195] Change was slowest in areas of extreme fragmentation. Mayhew cites an example from Württemberg, where 1,372 holdings occupied no less than 35,300 separate parcels, an average of more than 25 to a holding. The average size of a strip was only about 0.06 hectares.[196] In areas of such small and intensely fragmented holdings concentration into compact tracts of land was difficult. Most peasants would have some cause to complain that they had suffered in the exchange. There was, however, nothing to prevent the reduction in the number of parcels by exchanges within each field. A case in Hanover showed:[197]

Field 1: 89 parcels reduced to	24
Field 2: 94	33
Field 3: 71	27
Total 264	79

In France and Belgium one finds opposed tendencies, on the one hand towards the subdivision of strips and parcels, partially compensated on the other by the sale and exchange of pieces of land within individual fields. Nevertheless, *morcellement* became more extreme, and the average size of parcels grew smaller. In Belgium it fell from 0.52 hectares in 1834 to 0.51 in 1845 and 0.45 in 1882,[198] less extreme than in parts of Württemberg and Baden, but still uneconomically small. In France the average size of a parcel was less than 0.3 hectares in 17 *départements* and above 0.5 in only 25, and there were many *départements* in which the peasant had on average 15 to 20 such parcels widely scattered through the open fields.[199]

Switzerland presented one of the most extreme cases of fragmentation. In two *communes* in the Val Blenio (Canton Ticino) 88,000 plots were distributed amongst 656 peasants, an average of 134 parcels, each parcel being, on average, only 250 sq metres.[200] Some cantons enacted legislation early in the nineteenth century in order to consolidate severely fragmented holdings, but in most it proved ineffective. Only in 1884 did the federal government assist in the process by giving financial help, and not until 1912 were the procedures finally regularised. Between 1885 and 1910, 107 communities had their fields consolidated, covering an area of 5,195 hectares.[201] In the Balkans the dissolution of the *zadruge* led to an almost unlimited parcellisation. A case was cited of a property of 2.89 hectares divided into no less than 122 separate plots.[202]

Fragmentation was a problem more or less serious in all parts of Europe except Scandinavia, and in much of Europe it was tending to increase with progressive division between heirs. Governments legislated to restrict the practice, but seemed, especially in eastern Europe, powerless to prevent it.

The strips and parcels which made up the open-fields varied greatly in size and shape. Most tended to be long and narrow, with a slight curve, made by the plough as it prepared to turn at the end of the furrow. The use of a heavy plough and a large team favoured a long strip. Shorter and perhaps wider strips were better suited to a light plough, and the shape of the plot was of little significance to the light *araire* in use in much of southern Europe.

Open-field agriculture was associated in most of Europe with compact or nucleated villages. These varied greatly in plan, but all were surrounded by unenclosed fields to which the peasants made their daily journeys. In sharp contrast with the open-field village was the pattern of enclosed fields and paddocks met with in areas of scattered or dispersed settlement. Here the farmstead stood in isolation, or was grouped with at most two or three others in a hamlet. It was surrounded by a small number of fields, generally fairly compact and enclosed by hedges or fences. Occasionally such a field was divided into a small number of parcels, usually in consequence of division between heirs. An important variant of this field pattern was the infield/outfield system, in which the land close to the farmstead was regularly manured and continuously cultivated, whereas the more distant, the outfield, was ploughed and sown only at intervals.

Such a pattern of settlement, with its resulting field system, was commonly found in areas of poor soil, harsh climate and rough terrain. It was to be met with also in areas where animal rearing was as important as arable farming. It occurred, in short, where the density of farm population was low. Scattered settlements and enclosed fields were characteristic of the hilly regions of 'Atlantic' Europe from Spanish Galicia to Norway. They were to be found throughout the hilly region of central Europe and in much of the Alps and Balkans. The heavy plough had never been important in these regions, if only because the community was never large enough to furnish a team. A light plough, pulled by horse or ox in the north, by mule in the south, was used to cultivate the land, and since it could turn easily and reach the corners of a field, a compact shape was as suitable as any. A pattern of scattered settlements, lastly, was a feature of newly settled land, such as Sweden's Norrland, central Finland and the reclaimed polderland of the Netherlands.

In much of Europe there was a tendency for scattered settlement, with its attendant enclosed fields, to replace the nucleated village and open-fields. This process is best demonstrated in Denmark, where, in the course of the century, many of the older villages disappeared, fragmented into dozens of scattered farmsteads. In the Hungarian Plain and in southern Italy, both areas where settlements had been highly nucleated, a dispersal took place, accompanied by the creation of enclosed fields. In northern France, the Low Countries and north-west Germany, wherever in fact there was a degree of consolidation of strips, there was also a tendency for settlement to become more dispersed.

Open-fields and enclosed were merely the opposite extremes of a spectrum of field patterns. Between them were other, more specialised and more localised field types, each responding to particular technical or economic needs. In parts of eastern Europe the village did not develop as a compact settlement, but rather as a row or double row of houses spaced at intervals along a road. Each house had its own land, stretching back from its narrow frontage until it reached whatever boundary or obstacle there may have been. Another field type, sharing in some ways the features of the open-field, was the cultivation terraces which climbed the hillsides in southern France, Italy or Dalmatia. The strip-like terraces were individually owned and cultivated. Often they were used for fruit or the vine. There could be no consolidation here; the only change during the century was the progressive abandonment of the highest and least accessible. Other highly specialised field systems were developed in the rice-growing parts of the Lombardy Plain; in the irrigated *huertas* and *vegas* of Spain's Mediterranean coastlands; the intensively cultivated gardens which ringed every large city, and the vineyards and the orange-and lemon-groves.

Technical progress

The great increases both in crop yields and in total production over a period of a century reflect a corresponding improvement in farming techniques. A threefold increase in production with overall only a small expansion of the cultivated area suggests a breakthrough of major proportions. Yet all that one can point to is a series of small improvements in tools and equipment, in the use of manure and fertiliser, in the cultivation of the fallow and the use of legumes and roots. Little capital investment was made in the land, apart from a few major reclamation schemes. Machines began to be used in the second half

Table 5.11. *Use of ploughs and mechanical equipment in Poland, 1910*

	Manorial farms (%)	Peasant farms (%)
Iron ploughs	22	78
Wooden ploughs	5.3	94.7
Primitive ploughs	14.9	85.1
Steam ploughs	100	—
Steam threshers	93.9	6.1
Hand winnowers	14.5	85.5

of the century, but mainly on the estates whose owners alone had the capital to invest in them. The peasant, who worked most of the arable land in Europe, had neither the money to purchase nor the land to employ fully any piece of mechanical equipment. The co-operative purchase and use of equipment was only beginning at the end of the century. In Poland in 1910 the advanced and the primitive forms of plough were in use side by side (table 5.11).[203] Where there were no demesne farms there was little advanced agriculture. On peasants' land technical advance assumed the form 'not of mechanization but of a switch from lower to higher working capacity hand tools'.[204] Thus the scythe, with its longer blade and greater sweep, gradually replaced the sickle and hook. The sickle was the chief harvesting tool in the Beauce until about 1850, but by 1866, it is claimed, it had been replaced by the scythe in much of northern France.[205] But progress was very local. In the 1870s the sickle continued to be used almost exclusively in Seine Inférieure[206] and it survived much longer in the Gâtinais.[207] Early in the present century an American commission found that in Germany the 'farmer is thoroughly conservative and clings to many ancient habits and customs which are none the better for age. With great difficulty he is brought to introduce new and more profitable practices.'[208]

In Denmark the scythe had by and large replaced the sickle by 1860.[209] In the period 1850–80 it was widely adopted in Germany and Austria, and in Styria its manufacture became an important local industry. By the end of the century the sickle remained important only in remote areas like the mountains of Upper Hungary.[210] The peasant was reluctant everywhere to accept the necessity for technical change. But in the later nineteenth century there was a growing shortage of labour in some areas, and it was probably for this reason that the peasant broke with tradition and used the new implements.[211] With the scythe, it was said, the peasant could do four times as much reaping as with the sickle.[212] On the other hand, most farm tools were made locally, and it must have been at least as difficult to induce the smith to make the new tool as it was to persuade the peasant to use it.

The advantages of using an iron plough were self-evident, and it effected a revolution not unlike that which resulted from the introduction of the heavy plough during the early Middle Ages (fig. 5.13). There were, broadly speaking, two types of plough in use at the beginning of the nineteenth century. Over much of northern Europe a heavy plough was used, consisting of a wooden frame and beam.[213] To the latter was attached a coulter which made a vertical cut through the soil, and to the former a share and mouldboard, which undercut the turf and turned it over. The coulter had to be of metal and the share was commonly

Agriculture

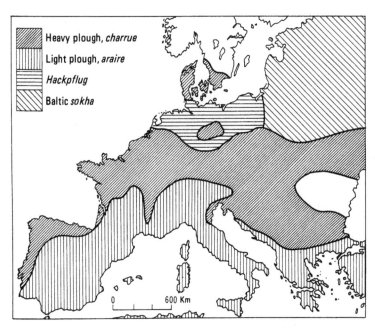

5.13 Distribution of the chief types of plough in Europe during the nineteenth century

tipped with iron. The rest was of wood, heavy to draw and clumsy to handle. In the course of the century the heavy plough was gradually replaced by the lighter iron plough. 'The new plough types', it has been claimed,[214] 'were the most important measure of improvement in agricultural technology.' They cut through the tangled root systems and allowed the soil to be turned without using a large team. They were the 'most decisive factor in early agricultural improvement in Scandinavia'.[215]

The alternative plough was a light wooden implement, easy to carry and turn at the ends of the furrow, capable of being drawn by a single mule, but merely scratching instead of turning the soil. Its construction varied in detail, but consisted broadly of a block of wood, pointed at one end, which was pressed into the soil by means of a handle while being pulled forward by the draught animal. It was used throughout southern Europe, where travellers from the north habitually described it as primitive or even 'primeval'. Beauclerk claimed that the plough which he found in the Val d'Aosta was that described in Virgil's *Georgics*.[216] Some such plough had once been used in much of northern as well as eastern Europe. It survived in remote areas in Spanish Galicia, in eastern Poland and probably elsewhere. Cross-ploughing with such a primitive plough has been described in the Białystok region.[217]

Sowing continued in much of Europe to be by the traditional 'broadcast' method.[218] Though seed drills had long been known in England their use in continental Europe was restricted to a few advanced farms in the north-west. The traditional harrow, used to break up the earth and cover the seed, was also of wood, and it was not until late in the century that it began to be replaced by an iron implement, more effective in operation and easier to handle.[219]

Improved harvesting equipment was no less desirable than better ploughs, since harvesting always placed a severe strain on available labour.[220] Reaping machines had been known since early in the century in parts of western Europe, but were not used in the Gâtinais, admittedly a relatively backward region, until about 1890.[221] Early in the twentieth century they were to be met with in Poland, but always on the estates. Harvesting was succeeded by threshing, normally carried on with a flail and followed by winnowing to separate the grain from the chaff and dust. This was usually carried on in the barns during the winter months, when there was no great demand for labour outdoors. It was hard and exhausting work, and a horse-driven threshing machine was one of the first mechanical appliances to be adopted.[222] In France a stationary steam-engine began to be used for threshing in the 1850s.[223] The use of machines spread in western Europe, but in Poland 66 per cent of the peasants are said to have relied on the flail early in the present century.[224]

The peasant had learned by long experience the value of manure in the improvement of his crops. His problem, however, was to gather it and apply it to his fields. De Laveleye described the careful and protective methods of the Flemish farmer (see p. 260). In Paris 'every species of refuse is husbanded in the most careful manner'[225] and carried to the *maraîchères* which surrounded the city, and the processing of night soil into *poudrette* became a minor local industry. In many areas scope for the use of manure was limited. That from the animals which grazed the common land and the waste was wholly lost, and a benefit of the practice of *vaine pâture* was that it provided at least a little manure. Not until fodder crops were widely grown and stall-fed to the animals did manure become available on a significant scale, and this happened first in the Low Countries, north-west Germany and Denmark. Over most of southern Europe and much of eastern Europe little effective use was made of manure before the present century.

Leguminous crops not only added to the fertility of the soil by allowing increased stock to be carried, but served also to fix atmospheric nitrogen, which they left by way of their root systems in the soil. The spread of legume cultivation and the consequent increase in the nitrogen supply was 'a change of overriding importance'.[226] Where they were grown as a regular part of the rotation, the nitrogen supply was increased by up to two-thirds, and, in the opinion of Chorley, the great increase in crop-yields achieved during the century owed more to this than to any other factor. Unfortunately, available data are insufficient to trace the diffusion of legume cultivation. It does not, however, appear to have spread significantly beyond northern France, the Low Countries and Germany. It had little impact in southern Europe, and none in eastern and south-eastern Europe.

In western Europe use was also made of lime and marl. These were not fertilisers in the strict sense, as they did not normally add nutrients to the soil. They did, however, counteract the acidity of some soils and lightened others. Marl pits were opened up widely, and marl was added to heavy soils.[227] Lime occurred only in restricted areas, and had furthermore to be burned before it could be applied to the fields. It could be used only where transport was sufficiently developed to move it from kiln or quarry to the fields, and over much of rural Europe this was not before the end of the century. Lime became available in the Gâtinais about 1858,[228] and in Anjou arable was increased and wheat replaced rye at about the same time, thanks to the opening up of lime-works at Chalonnes-sur-Loire.[229] As early as 1836 the inhabitants of Chanzeau had demanded a road specifically for the transport of lime.

Little use appears to have been made of such 'sweeteners' in central Europe and none in eastern and south-eastern Europe. On the other hand, chemical fertilisers became available in Germany and Austria from the 1870s. At first they consisted mainly of saltpetre (KNO_3) imported from Chile; then in the 1870s ammonium sulphate and sodium nitrate were made in quantity, mainly as by-products of gas manufacture and coal-tar distillation. Natural salts, particularly sodium nitrate, began to be worked in central Germany, and in the 1880s basic slag with a high phosphorus content became available from the steelworks.[230] The greatest use was made of chemical fertilisers in the Low Countries and Germany. France used little and the countries of southern Europe almost none.

The peasant was everywhere slow to adopt either mechanical or chemical aids to agriculture. Price has commented on 'the extreme reluctance of most to accept the necessity of agricultural or social change'.[231] Everywhere the peasant fought for his traditional usages and privileges, even though to the outsider these may have seemed uneconomic or even retrograde. As late as 1848 the peasants at Conques, *dépt* Aude, demanded the disuse of the scythe and even of the rake because these reduced the profit to the poor peasant from gleaning.[232] Rural unrest was closely linked with attempts to break with such traditions and practices.

Where change came it was most often as a result of the slow emergence of a market economy, and this appeared first on the large estates, which from the very scale of their production, had to be geared to the market. In much of Europe there was a technical dualism; 'ancient and modern techniques coexisted, and the old showed no tendency to be pushed aside by the new'.[233] From France to Poland the contrast is apparent between estates and large farms, relatively well equipped and managed, and peasant holdings, labour-intensive, small in scale and backward in technology, geared only to local self-sufficiency.

A series of local markets spanned most of Europe at the beginning of the century. Most commodities were consumed where they were produced, and few ever travelled more than a few kilometres from the fields where they were grown. Only the surplus grain from the estates and a few specialised products like wine, olive oil and industrial crops ever entered into long-distance trade. The friction of distance not only cut off the local community from distant markets; it also protected it from the competition of other producers, some of them more favourably placed than itself. It was no accident that urban growth first became significant in areas of high natural fertility, where there was a local agricultural surplus to support the town population. Lack of surplus production in the primary sector was a contributory factor to the slow growth of towns and of tertiary activities in large areas of Europe. The railway broke down the isolation of the rural community, but it also brought with it the disappearance of many crops. Vegetable dyestuffs ceased to be grown; the patch of hemp disappeared from the village and with it the craft of the rope-maker, and the community ceased to grow its own supply of flax.[234]

Crops and rotations

Throughout most of Europe at the beginning of the nineteenth century the crops grown were by and large those which had been cultivated for centuries. In northern Europe autumn-sown grains alternated with spring-sown; in southern, wheat succeeded fallow. A few vegetables – beans, peas and the *brassica* – were grown as garden crops, and here and there one found a field of flax or a patch of hemp. A few extraneous crops gradually intruded, despite the resistance of the peasantry – the potato and maize in particular – but they had made little progress by 1815. A few indigenous crops – roots like sugar-beet, turnip, mangold and artificial grasses, legumes and fodder crops – were known but could not easily be fitted into the rigid cropping system that prevailed. This system was geared to the production of cereals; other field-crops were but accidents. Throughout the century and in every part of Europe the bread-grains demanded the best efforts of the peasant. The cereals varied. In northern France wheat was preferred; in Germany and Poland, rye was the dominant crop. In Scandinavia and the Czech lands the bread-grains were more evenly balanced, but in Hungary and Romania wheat was again in the forefront, though not without some rye and barley and a great deal of maize. Throughout southern Europe the chief bread-grain was wheat, though in so varied and broken a terrain one found that every cereal was important somewhere.

The reasons for this pattern varied. To some extent they were climatic, the hardy and quick-maturing barley being grown at higher latitudes and altitudes than other cereals; rye, where winters were hard, and oats, where summers were short, cool and moist. To some degree also the choice of cereal was dependent on the quality of the soil. Wheat preferred a deep loam, whereas rye could outstrip wheat on the poor, acid soils which covered most of the glacial outwash of the northern plain. Barley grew best on a calcareous soil, and oats could yield a crop in soils that would support little else. There were other reasons for preferring one crop rather than another. Wheat produced a more palatable bread and was grown wherever physically possible despite the fact that rye might have cropped more heavily. Oats had to be grown wherever horses were numerous, as they were in much of eastern Europe. Neither barley nor oats were regarded as bread-grains, though both were often used as human food. Barley was the best grain for malting purposes, and was grown wherever beer was drunk. Maize was of increasing importance, chiefly because it cropped much more heavily than the traditional cereals in areas such as the Danube basin and northern Spain and Italy, where summers were hot and moist.

The geographical pattern of cereal cultivation was in process of change. Wheat and maize were increasing in importance and rye was declining. Some grain crops which for one reason or another were regarded as inferior virtually disappeared. Amongst these were spelt, closely akin to wheat, but difficult to separate from its husk, and the primitive *Einkorn*, once widely grown in southern Germany.

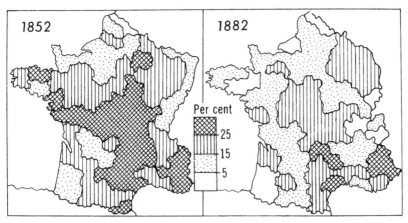

5.14 Percentage of arable land under bare fallow between 1852 and 1882

More widespread were buckwheat or *sarrasin*, which tolerated both poor soil and harsh climate, but yielded only a coarse and bitter grain.[235] Losses in this direction were in some degree offset by an increasing acreage under rice and maize, though rice was restricted for climatic reasons to those plains of southern Europe which could be irrigated, and maize to regions with a hot but far from rainless summer. But the inferior grains tended, most of all, to be replaced by root crops, especially potatoes, and by legumes and other fodder crops.

Fig. 5.2 shows the changing pattern of cereal production in France.[236] Total production increased by about 80 per cent, but wheat by almost 130 per cent, whereas rye declined by 37 per cent. In Germany the cultivation of rye actually increased about fourfold, but wheat showed an appreciably bigger rate of increase. Fig. 5.4 shows the overall changes in the volume of cereal production in the State of Prussia.[237] The increase in wheat-growing at the expense of rye was small in the western provinces, where it was already relatively important at the beginning of the century. In the six eastern provinces, where it was at first of negligible importance, the increase was more than eightfold.

In Poland wheat made little progress, and rye remained the dominant bread-grain. In Denmark there was a relative decline in the cultivation of bread-grains and an increase in that of feed-grains.[238] The expansion of settlement in Sweden was accompanied by a significant increase in the cultivated area. Rye and oats were the chief crops at the beginning of the century. Wheat-growing then expanded, but later in the century the area under oats and barley again increased since the expansion was largely on the harsh soils of Norrland, which could support nothing better.[239] In the Czech lands rye remained the chief cereal throughout the century, though its share in the cultivated land declined steadily as that of wheat and roots increased.[240] As one moved to south and south-east the dominance of rye became less marked. In the Austrian Alps it greatly exceeded wheat, but on the margin of the Hungarian plain and on the Karstlands bordering the Adriatic this preponderance disappeared.[241] In Hungary wheat was the chief crop, despite the growing importance of maize,[242] and on the plains of Romania rye almost disappeared, as wheat became the chief export crop and maize the principal food of the local peasantry, with millet, barley and oats of minor importance. Between 1830 and 1850 the volume of wheat produced approximately quadrupled. There was little long-term change in the production of other cereals, except maize, which had become the most important by the end of the century.[243]

Throughout southern Europe from the Balkans to Portugal the prevailing cereal was wheat, as, indeed, it had been since classical times. Traditionally it alternated with fallow. Most other cereals were grown, but together they amounted to only a fraction of the wheat crop. Maize was the most important, but did not easily tolerate the dry Mediterranean summer, and was chiefly found where summer heat was combined with a moderate rainfall. Amongst such areas were Spanish Galicia and northern Portugal, south-western France, the Lombardy Plain and some of the coastal areas of peninsular Italy where irrigation was possible. It was almost totally absent from the Meseta, and was found only thinly scattered through the Balkans. Here it was grown most often by the Turks, and it was for very good reason that the indigenous population referred to it as 'Turkish Corn'. In Bulgaria at the beginning of the present century wheat made up nearly a half of the cereal crop and maize only a quarter.[244] In some parts of this mountainous country maize was of negligible importance, and wheat made up nearly 80 per cent of the cereal crop.[245]

Statistics are few for the Italian and Spanish peninsulas. 'Wheat', wrote the British consul in the 1880s, 'forms the basis of Tuscan agriculture.'[246] The same could have been said of most of Italy. In 1895 Bolton King estimated that wheat made up two-thirds of the grain crop and maize about 30 per cent.[247] In Spain the position was similar. Little but wheat was grown on the Meseta, with here and there some barley.[248]

Figures cited are aggregates for countries or broad regions; they disguise great local variations, for details of which one is dependent on local studies or the chance survival of local records. Some such exceptions to the general pattern are explicable in environmental terms. For others there is no ready explanation. How to explain the concentration of barley-growing on the Spanish coast near Cartagena, or that of oats in Basilicata and Apulia in southern Italy, or of rye in the Salamanca province of Spain? Why in a community in the Gâtine in central France was no wheat at all grown until the middle of the nineteenth century, when it began to displace rye?[249] Why, on the other hand, did wheat occupy a fifth of the cropland early in the century in South Limburg, in a part of Europe where rye was dominant?[250] Such local practices, once established, could be broken down only very slowly. They would become integrated into the local agricultural system, in which the communal checks and balances would tend to preserve them. A self-sufficing peasantry, living close to the margin of subsistence, is not likely to experiment or innovate, and any change in cropping pattern would have necessitated a capital expenditure – on new seed at least – that might have been beyond the means of the community. Agricultural innovation was left to the rich peasant and the estate farmer who had both a wider vision and a greater willingness to take risks.

Nevertheless, new crops did slowly insinuate themselves into the peasants' routine. They came in two ways. Most often it was on land that would otherwise

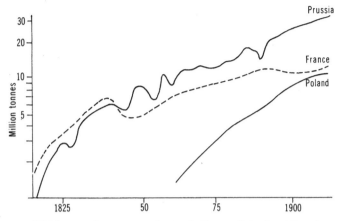

5.15 Increase in potato production in France, Poland and Prussia

have been left fallow. If such a crop should fail, the loss could usually be borne. Secondly they came by way of the garden plots which surrounded the cottages. The potato crept into European agriculture in this way (fig. 5.15). Most important, however, was the replacement of fallow with a catch-crop. The new crops and crop associations which thus spread through much of Europe during the nineteenth century fell into two groups: root crops and fodder crops. The former consisted of potatoes, turnips, mangolds and sugar beet; the latter of clover, cultivated grasses and the so-called 'artificial' grasses.

Of these the potato was in terms of land use and human diet the most important. There was at first considerable resistance to it as a human food.[251] In Scandinavia it was used for distilling before it began to be eaten, but it was nevertheless something to which people could turn in an emergency.[252] At a time of rising population and frequent famine crises it offered great advantages. It cropped heavily and could usually be relied upon in seasons when the bread-grains failed. Its cultivation spread in Great Britain and, in particular, in Ireland during the late eighteenth century. Whether its adoption was a consequence of a rapidly rising population, or whether and to what extent it contributed to that rise may be a matter of dispute, but there can be no doubt that an increasing proportion of the European population came to rely on the potato as the principal component of its diet.[253]

In continental Europe the potato was grown only on a limited scale before the end of the Napoleonic Wars. Then it spread rapidly. From the garden plot it was transplanted to the field. It established itself first in western and central France, where the pattern of small and enclosed fields favoured it; then in Belgium, where the communal system of cultivation had long been abandoned. From here it spread eastwards through Germany and Poland; northwards into Scandinavia and south into the Czech lands and Austria. It never became important in southern Europe, except in a few scattered areas, amongst them Galicia, which was climatically suited, and the Italian Abruzzi and Molise, which were not. As far as northern Europe was concerned it was most grown where sugar-beet was unimportant. For this there were good agricultural reasons;

sugar-beet required a good, loamy soil and deep ploughing; the potato could tolerate a thin, sandy soil and moist conditions. The rich soils of northern France suited sugar-beet; in Brittany and the Central Massif comparative advantages favoured potatoes. In Germany, sugar-beet was grown in the lower Rhineland, Saxony and Silesia;[254] potatoes on the infertile sands of Brandenburg and the east. Everywhere, however, the cultivation of roots hastened the enclosure of the open-fields and encouraged deeper and more careful ploughing.[255]

In France potato-growing continued to increase until about 1890, and then declined in face of competition from the dairy industry.[256] In Belgium the potato established itself early. It was well suited to the sandy soils and 'garden' cultivation of the northern part of the country. It also spread rapidly in the Netherlands, where, in the words of an official report of 1845, it 'should now be regarded as the most general and necessary food for the lower and even part of the middle class'.[257] The potato suffered severely in the Low Countries from the blight in 1845–6. In the Netherlands it never regained its former importance, but on the minute holdings of Belgium cultivation continued to increase until the 1890s.[258]

In Germany the importance of the potato continued to grow until the First World War. It was cultivated almost everywhere, except where sugar-beet was preferred or fodder-crops were grown for dairy cattle. As a general rule it fitted into a rotation with rye and feed-grains; sometimes, on areas of very poor soil, it dominated the cropping system. In the Taunus for example, the cultivated land was made up of (in percentages):[259]

Potatoes	50.7
Rye	18.9
Oats	15.5
Barley	6.9
Other crops	8.0

The potato was introduced into Poland near the beginning of the century, but spread slowly and until the 1860s was confined to the lands of the gentry.[260] After 1864 it was accepted more widely, and the crop increased more than sevenfold between then and 1913.[261] It spread less rapidly in Austria and the Czech lands, and on the loamy soils of Bohemia and Moravia its place was taken by sugar-beet.[262] From here potatoes spread through the Carpathian Mountains and down to the plains of the Little Alföld. A limit was set to potato-growing in Scandinavia by its liability to frost damage. In Denmark it spread during the first half of the century, when potatoes became 'the principal food of the yeoman and the peasant',[263] but thereafter suffered in the general stagnation of arable farming.[264] In Sweden potatoes were for climatic reasons important only in the centre and the south, but here they became the staple food of the crofting population. In parts of Sweden dependence on the potato became almost as great in the middle years of the century as it had been in Ireland a decade or two earlier.[265]

The spread of sugar-beet, 'the greatest agricultural innovation of the early nineteenth century',[266] was always a much more localised phenomenon than that of the potato. Both were used to colonise the fallow, but that was the extent of their similarity. They were grown on different types of soil and served radically different purposes. Potatoes were grown chiefly on peasant holdings as human food, though some were used for distilling. Sugar-beet was exclusively an industrial crop, but produced animal feed as a by-product. Furthermore, sugar-beet-growing had to be linked with processing. It was too heavy and bulky to be transported far and the building of a processing factory served to concentrate the crop in its vicinity.

The beet-sugar industry owed its origin to the Continental System and the cessation of sugar imports from the West Indies. The industry languished for a period after 1815, then revived, first in northern France and Belgium and a few years later in Germany. It was a 'political industry',[267] always dependent on some measure of government protection. It developed first in northern France, and a large part of the French production was from *dépt* Nord.[268] From here it spread to Belgium, but made no great progress in competition with the potato. It was in Germany that sugar-beet achieved its greatest importance, in the lower Rhineland, Mecklenburg and above all in Saxony and Silesia. Both soil and social structure favoured it. A network of small factories was established and in turn defined the beet-growing area.[269] By the mid-century Germany had overtaken France in the volume of beet raised, and from 1850 to 1890 there was a twenty-fivefold increase in production.[270] The industry was established in the Polish Kingdom during the wave of industrialisation which preceded the Rising of 1830. It suffered, however, from its small scale, the general inefficiency of the factories and the economic difficulties of the gentry later in the century.[271]

The only other areas where sugar-beet became important were in Bohemia, Moravia, the Little Alföld of Hungary and the plains near the mouth of the River Po, which were drained and brought into cultivation late in the century. Here, as in western and central Europe, beet-growing was a highly capitalised form of agriculture, carried on largely on the estates.[272]

Germany was quite exceptional in the emphasis that was placed on root crops during the nineteenth century. During the last quarter the area under potatoes or sugar-beet increased from 13.7 to 17.7 per cent of the cultivated land, and by 1914 had 'approximately eight times the proportion of arable that was devoted to such crops in England and three times that in France'.[273] Roots became 'the pivotal crop in sequences'; they required a heavy use of fertiliser and the consolidation of scattered strips, improvements which led in turn to developments in other branches of husbandry. It is noteworthy that in Germany as a whole green fodder played only a minor role in the developing crop rotations. The key probably lies in the contemporary development of the chemical fertiliser industry, which both encouraged root crops and was stimulated by it. The use of fertiliser increased more than 12 times in the last three decades before the First World war (table 5.12).[274] So intensive an application of chemicals made the use of natural manure less necessary. There was thus a smaller inducement to maintain the herds which supplied it and to grow the fodder crops on which they would have subsisted. Cattle and root crops were thus in some degree mutually exclusive.

The cultivation of green fodder was a way of using fallow. The most important feeds were clover and the artificial grasses – sainfoin, alfalfa and

Table 5.12. *Consumption of fertilisers in Germany* (*in kg per hectare*)

	Nitrogen (N)	Superphosphate (P_2O)	Potash (K_2O)
1878–80	0.7	1.6	0.8
1898–1900	2.2	10.3	3.1
1913–14	6.4	18.9	16.7

lucerne – which had the property of fixing atmospheric nitrogen and thus of improving the soil. Secondly there were roots, such as turnip, swede and mangold, and thirdly grass, sown with the last cereal crop and allowed to grow after the latter had been harvested. A crop of grass or lucerne was usually allowed to grow for two or three years, and amounted to an interruption in the sequence of cultivated crops.[275] An increased production of fodder crops, it was argued by the agronomists, allowed more stock to be kept and thus more manure to be made available for the crops. 'The key to a more progressive agriculture was stronger demand for meat and milk.'[276] Unfortunately it was well into the nineteenth century before real incomes had increased to the point at which this demand could become effective and transport had sufficiently developed to satisfy it. Even at the end of the century there were still areas in western Europe where the lack of marketing facilities would have prevented the establishment of such a mixed farming system, even if other factors had favoured it (see p. 427). Nevertheless some parts of Europe turned from cereals to green fodder when conditions of soil and transport favoured it. This was notably the case in Normandy.[277] Even on the micro-holdings of Belgium there was some substitution of feedstuffs for bread-grains. In Germany fodder-crops were widely cultivated in the Rhineland, where French influences were strongest.[278] In the 1840s Banfield described a rotation in use in Cleve in which both clover and potatoes were grown in rotation with barley and bread-grains.[279] In eastern Germany fodder-crops were of negligible importance until, late in the century, cereals ran into competition from imported grain. Fodder then began to be grown more widely, and the area used increased from about 500,000 hectares in 1840 to about 1,400,000 in 1907. Even so, it amounted to only about 10 per cent of the cropped area.[280] In Bavaria in 1890, fodder-crops covered only 7 per cent of the cropland,[281] and in Germany as a whole, green fodder occupied only 3.4 per cent of cropland and rotation grasses a further 7.5 per cent.[282]

As the practice became more general of taking a crop of roots or fodder during the fallow year, so the breakdown of the traditional rotation became complete. Rotations were developed spreading over four, six or even more years. Sometimes root crops alternated with bread-grains; sometimes the land was left under grass or clover for a year or two. Since rotations were now conducted more and more within enclosed fields, they could be adapted to soil, weather and market conditions, and, since either manure or fertiliser could be made available, the same crop could be taken in successive years. This approximation to modern farming methods was achieved first in the Low Countries, northern France and the Lower Rhineland, where commercial agriculture was most developed. It was

Agriculture 239

reached later in central Germany and the Habsburg lands, where tradition was stronger, and not at all in remote areas of the east.

In addition to bread-grains, roots and fodder, there were many highly localised crops, such as tobacco, colza, madder and other dyestuffs, and, of course, citrus fruit, the olive and the vine. Their distribution was in part a matter of climate and soil, in part a result of local developments which lie beyond documentation or explanation. Foremost amongst these crops was flax, widely grown in the coastal regions of north-west Europe, as well as in parts of central and eastern Europe; hemp was grown for local use in much of Europe, and tobacco was intensively produced in Alsace, the Hungarian Plain and a few other areas.[283] Hops, *la vigne des pays froids*, were grown wherever beer replaced wine as the prevailing drink. They were particularly important in Alsace, Bavaria and Bohemia, all of which became noted for their brews. Colza, navette and other forms of rape were also of local importance. They were used early in the nineteenth century as a source of oil for lamps,[284] and were much grown in Belgium and throughout the coastal belt of north Germany, where the marshy soils and damp climate favoured them. Rape was a feature of most estates, and von Thünen grew it at Tellow.[285] From about 1865 its use was undercut by the growing supply of mineral oil, and the acreage under rape gradually declined and was of slight importance by 1914. Vegetable dyes, especially woad and madder (*garance*), had formerly been widespread, but tended to be displaced, first by imported dyestuffs like indigo and brazil-wood, and then by chemical dyes.

Fruit, both that known in England as 'top-fruit' and also the grape and the olive, were grown wherever climatic conditions permitted. There could be little long-distance traffic in so perishable a commodity, and the only significant concentrations of fruit-growing – apart from citrus fruit – were of apples, olives and the vine. The apple was primarily the source of cider, and was important only in areas where the grape vine did not grow well, notably Brittany and Normandy, parts of western Germany and the coastal region of northern Spain. The olive on the other hand, was restricted to areas of southern Europe which were effectively frost-free. It was, in fact, important in the nineteenth century only along the Mediterranean coast of Spain and in southern Italy and Sicily, where it had been grown since classical times. The area under olive groves was tending to diminish as animal fats came to be used increasingly in place of vegetable fats.

A certain mystique surrounded the cultivation of the vine. In terms of climate and soil it is a remarkably tolerant plant, and the qualities which are especially valued in wines are the result of centuries of careful management of the vineyards, rather than of a favourable climate. Some of the best wines were produced near the climatic limit of viticulture, in Champagne and the Rhineland.[286] One must, however, distinguish between the 'noble' vintages, produced from a very small area of long-established vineyards, and the mass production of ordinary wine. The former were well known, even though the volume of production remained small. The interest shown in them in Great Britain was immense, and the quality of the vintage elicited from the British consular officials more frequent and more intensely felt reports than many more

5.16 Changes in the area under vineyards in France as a result of the *Phylloxera* epidemic. The figures indicate the number of hectares lost in each *département*

important matters. The production of inferior wines, on the other hand, grew rapidly. Attention was given to volume rather than quality, and vineyards spread in areas climatically best suited to the vine. Foremost amongst such areas were southern France from the Pyrenees to the Rhône, and, in Italy, the Lombardy Plain, Tuscany, Umbria and the Naples district. In parts of Languedoc the area under vines increased from less than a quarter of the cultivated land at the beginning of the century to 40 per cent after 1840.[287] But the most vigorous expansion of viticulture was probably in Spain. Wine was exported from Andalusia and the Mediterranean coastlands. Production was increasing before the *Phylloxera* epidemic of the 1870s, and after the pest had been overcome the export trade grew very rapidly. Viticulture spread from the favoured coastal regions to the Meseta, and the area of vineyard increased from some 400,000 hectares at the beginning of the century to 1,200,000 in 1860 and 1,450,000 in 1900.[288] During the same period viticulture increased in southern and eastern Austria, on the sands of the Hungarian Plain, and even on the sun-drenched karst of Hercegovina.[289]

This expansion of viticulture was checked by the spread of *Phylloxera* which first appeared in Europe in the 1860s and continued to menace the vineyards until early in the present century. It was an aphid which extracted the sap from the stem of the plant. It first established itself in Languedoc, from which it spread slowly through most of the vineyards of France (fig. 5.16). It reached Spain in the late 1860s and Portugal in 1872.[290] It then spread into Italy, and even reached Hungary, where the vineyards were devastated.[291] Many remedies were attempted, until it was found that grafting European vines on to American stocks reduced their vulnerability. In France, it is said, the area of vineyards was reduced by 30 per cent and the value of the vintage by 44 per cent.[292]

Statistics of wine production are available only for the more important producers. France was without question the largest, though Italian output grew

Table 5.13. *Wine production, 1900*
(*in million hectolitres*)

	Amount of wine produced
France	68.5
Spain	37.8
Italy	36.6
Portugal	5.5
Austria	5.2
Germany	2.0
Hungary	1.8

significantly after the *Phylloxera* outbreak had been halted. German and Austrian production was always small, and the formerly extensive Hungarian vineyards had barely recovered from the *Phylloxera* by the end of the century. Only Spain was a significant rival to France and Italy. Wine production in the major producing countries in 1900 is shown in table 5.13. During the last decades of the century viticulture was developing in Romania, Bulgaria and elsewhere in the Balkans.

Yields and yield-ratios

Despite the decline in crop acreage in the late nineteenth century, total output of crops continued to increase, and in 1910 the combined production of cereals was about three times that in the early nineteenth century. The factors in this increase have already been discussed. It is impossible to assess their relative importance, but their combined effect was to make Europe some 80 to 90 per cent self-sufficient in the basic foodstuffs by the eve of the First World War. There are two ways in which to measure this increase: by the yield-ratio or return at harvest on a given quantity of seed, and by the production of crops per unit area (fig. 5.17). For both the data are at best fragmentary and for some areas they are non-existent. In France, the yield-ratios for wheat varied about 1840 from more than eightfold on the good soils of the north to less than four in much of the Central Massif.[293] The same was true of rye, highest in the north and west, where it often occupied the best soil; lowest in the south and east (fig. 5.18). Yield-ratios seem to have been generally lower in Germany during the earlier years of the century, but, as far as Prussia was concerned, were highest in Saxony, which had the greatest extent of loess soil.[294] But even here, a sevenfold return for wheat and 5.7 for rye were appreciably lower than the best returns in France. Jacob claimed to have met with returns for wheat and rye of more than tenfold near Weimar,[295] but this was quite exceptional at this time.

Yield-ratios improved greatly during the later years of the century, in response to the generous use of fertiliser, but in remote areas and on poor soils it is doubtful whether they much exceeded four except in the best harvest years. In Poland yields were even lower. In the 1830s wheat was said to yield about 4.6 times, but rye generally less than four. Such figures can, however, be deceptive. Wheat, when grown, was sown only on the better soils. Barley cropped so badly –

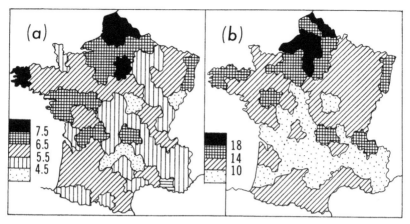

5.17 Yield-ratios (a) and yields in quintals per hectare (b) of wheat in France, about 1840

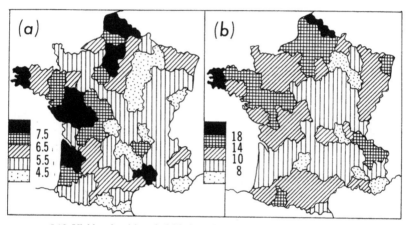

5.18 Yield-ratios (a) and yields in quintals per hectare (b) of rye in France, about 1840

a ratio of only about 2.4 – that one wonders why the peasant continued to grow it.[296] Everywhere the ratios on the estates of the gentry tended to be higher than those on the peasants' land. In Hungary, yield-ratios were as low as any to be found in the Middle Ages.[297] They improved during the century, but six remained high for wheat and five for rye.[298]

Crop-yields per hectare present a similar, though not identical pattern. They take no account of the amount of seed used, and were most significant where land was the scarce factor of production. During the middle years of the century the range between the highest yielding areas and the lowest was immense. In France, where yields can be measured most easily, they ranged from over 20 quintals to the hectare on the good soils of the north and about 16 in Brittany and Alsace, to less than 10 in much of southern France.[299] Rye showed a similar range. Yields increased very slowly.[300] Little fertiliser was used, and in areas of high farming little manure was available. Yields were relatively high in Brittany, never in the forefront of agricultural progress, only because pastoral activities yielded more manure than was available in most of France.

In Germany yields were initially lower than in France, but increased sharply in the last decades of the century. A similar improvement is apparent in most of central Europe late in the nineteenth century. In the Polish Kingdom average wheat yields rose from 7.8 quintals to the hectare in 1867–70 to 9.8 in 1881–5, and for rye from 7 to 8.6.[301] A similar improvement was to be found in Austria and Hungary.[302] Yields were highest on the plains of Bohemia and Moravia; lowest on the limestone of the Dalmatian Karst.

Pastoral activities

Animal farming also underwent significant changes during the century. Some branches contracted; others expanded, and in this they mirrored contemporary changes in arable farming. The enclosure and cultivation of common land, encroachments on the waste, the abandonment of fallowing and the ending of rights of *vaine pâture*, all served to reduce greatly the extent of grazing available to the peasant. This in turn led to a reduction in the number of sheep which had been the chief beneficiaries of the traditional system. On the other hand, the introduction of artificial grasses and fodder-crops into the rotation increased the volume of better-quality feed, which in turn encouraged the raising of dairy and beef cattle.

Sheep were in decline almost everywhere. Only on marginal land, as in the Balkans and southern Europe, where much of the land could offer only rough grazing, did sheep maintain their position. In western and central Europe their numbers fell by almost a half, from about 120 millions to less than 70. Their distribution came to be the inverse of that of cattle, dense in Spain and Portugal, except the moist north-west, in the Italian peninsula and the Balkans. Elsewhere they remained numerous mainly in wheat-growing areas such as the Beauce.

Cattle were widely raised except in Mediterranean and Arctic Europe. They were a multiple-purpose animal, providing both milk when living and meat and hides when dead. In many parts of Europe they were used furthermore as draught animals. Throughout the century they were most numerous where meadow grass was lush and abundant, and this, by and large, was Atlantic Europe, from Galicia, through Brittany and Normandy and the North Sea coastal regions to Denmark. In the course of the century cattle also increased throughout the Rhineland, on the Swiss Plateau and in northern Italy. They were numerous in mountainous areas like the Central Massif of France, but were notably scarce in the main cereal-producing regions, like northern France, eastern Germany and the Hungarian plain.

In the first half of the century there were some 45 million head of cattle in western and central Europe. By 1910 this had increased to about 65 millions. At the same time there was a marked increase in the average weight of cattle. According to von Finckenstein this more than doubled in Prussia's eastern provinces. The increase was less sharp in the Rhineland, where animals were in all probability already better bred.[303]

Dairy farming on anything more than local and self-sufficing scale was

Table 5.14. *Dairy cattle in Brittany*

Dépt	1814	1882	1902
Côtes du Nord	93,000	158,000	217,000
Finistère	97,450	179,000	232,000

dependent on urban markets and rapid transport. In Brittany, for example, the number of dairy cattle increased rapidly after the completion in the 1850s of the railway link with Paris (table 5.14).[304] Normandy also developed intensive dairy farming after the completion of the railway to Paris.[305] In Switzerland the number of cattle, most of them for dairy purposes, increased by 50 per cent between 1886 and 1906. The same development can be traced in most areas which possessed an initial advantage for raising cattle and lay at no great distance from a large city.[306] In some areas where transport was inadequate for the daily movement of liquid milk a local manufacture of butter and cheese developed. Regional cheeses acquired a high reputation in parts of Normandy, the Netherlands and Switzerland.[307] The Parmesan and Gorgonzola cheeses were developed in the grazing areas of the Lombardy Plain, and from the Central Massif of France came less favoured cheeses which were nevertheless sold widely in Mediterranean Europe.[308] Cheese was always of less importance in Germany and eastern Europe because dairying itself was less strongly developed. The most vigorous development of dairy farming in nineteenth-century Europe was, however, in Denmark. Here it became not merely an adjunct to arable husbandry, with fodder-crops replacing the fallow, but rather a conscious attempt to build an agricultural system around the dairy herd and its liquid milk production.

Horses were kept largely for draught purposes. Their use for drawing the plough and the farm wagon was largely a matter of local preference and custom, but for transport by road they had no rival before the coming of the internal combustion engine. This helps to explain their large and increasing numbers in the densely peopled and highly urbanised areas of north-western Europe. Urban transport was horse-drawn before the coming of the electric tram and the motor car. Horses were relatively few in mountainous areas, and rare in southern Europe, where their place was taken by the mule and the donkey.

The largest increase was in the number of pigs, which more than doubled in western and central Europe in the 80 years before the First World War. They were, after sheep, the most numerous of farm animals, and their increase was most rapid in those countries, Denmark, the Netherlands and north-west Germany, in which a dairy industry was developing. The pig was ceasing to be a half-wild forager in the forest, and was becoming a consumer of the waste products of agriculture.

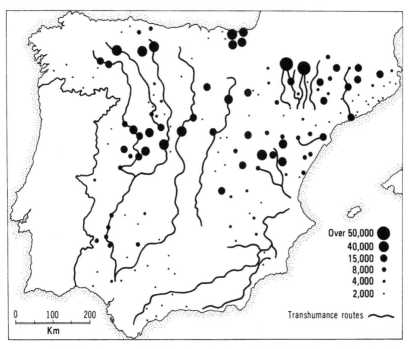

5.19 Transhumance in Spain; the migration of the chief flocks and the migration routes

Transhumance

The seasonal movements of flocks and herds is a special case of the adaptation of pastoral activities to the environment. It is a method of using resources which are unavailable for part of the year, the animals being wintered in one area and taken for the summer months to another. In some cases the journey was the short one from mountain valley to the nearby slopes, or 'alps'.[309] In others the journey was longer, as, for example, between summer grazing in the Pindhos Mountains of Greece and winter pastures in the plains of Thessaly or Macedonia.[310] In Italy sheep migrated between the Apennines and the coastal plains of the peninsula, and their numbers were said to have reached four millions in the first half of the century.[311] The longest of such seasonal migrations was that practised by the flocks of the *Mesta* between the northern and southern Meseta of Spain (fig. 5.19).

Long-distance migration became more and more difficult during the century. In the Balkans new political boundaries separated summer from winter grazing;[312] in Spain, cultivation encroached on the common land and the *cañadas* or tracks used by the migrant flocks; the spread of settlement interrupted the routes between Provence and the Alps of Dauphiné. None the less, the practice continued. In Spain, the size of the transhumant flocks declined from perhaps four millions to only half a million by the mid-century. Then it began to increase again, but the new race of transhumant sheep travelled by rail-car between their summer and their winter homes.[313]

Conclusion

The advances in manufacturing industry during the nineteenth century were both obvious and spectacular. They were accomplished in a series of giant steps, as new technologies were introduced and became accepted in most parts of the continent. The volume of industrial production increased many times (see p. 352), and in much of western and central Europe employment in manufacturing came to exceed that in agriculture. The expansion of agricultural production was less spectacular, though in the aggregate no less impressive. It was not, however, achieved by any sudden innovation, resulting in a spurt in output. It resulted from a multitude of small improvements, each resulting in a marginal increase in production.

In contrast with the relatively rapid spread of innovations in manufacturing, a consequence of the growing volume of technical literature and the publicity given to them by the patents law (see p. 6), was the very slow diffusion of advances in agriculture. The immense period of time needed, even in France, for the suppression of fallow is an illustration of the reluctance to accept improvements despite the fact that their value had been demonstrated beyond dispute. There were many reasons for this. Foremost amongst them was the fact that the practice of agriculture was in varying ways integrated with the whole structure of the rural community. To interfere with the former was to sap the foundations of the latter. To enclose land or to consolidate fragmented strips was to abrogate the rights of gleaning or of grazing stock, and too many people in too much of Europe benefited from such practices for change to be easy. A second reason was that, whereas the industrialist could, and usually did, raise capital from banks, private loans and, late in the period, from the public issue of stock, such opportunities were not open to the peasant. He had little credit and no capital beyond what he could generate in good harvest years. He could not lay drainage tile or apply marl, use a machine for threshing, purchase a new variety of seed or breed stock scientifically unless he could pay for these developments. The co-operative movement, which made headway after about 1890, helped to resolve this impasse by raising money on the collective credit of the rural community. But Raiffeisen and Schulze-Delitzsch banks and similar institutions were most developed and most effective in those areas, like Denmark and the Netherlands, which were already the best developed.[314] Agricultural advances were most significant on some great estates and on lands of the richer peasants. This was in part due to the fact that the richer land-holders were able to read, travel and pick up new ideas; but in part also because only they had the financial resources to make the changes.

Despite these constraints on growth, the cereal supply produced *within* Europe was increased almost threefold, and the total food supply, inclusive of the new crops, potatoes, maize and sugar-beet, grew by a factor of between four and five. The role of new crops in encouraging and supporting an increasing population, notably the potato in Scandinavia, Germany and the Low Countries (quite apart from Ireland), and maize in Romania and the Danubian lands, was of overwhelming importance. In much of Europe, however, the increased production resulted from small changes in farming method: better crop rotations, greater

Agriculture

use of manure and fertiliser, more conveniently shaped fields and hence more effective ploughing. Only to a small extent was it a consequence of cultivating new, and hence marginal, land. Agricultural land was extended by drainage in a few areas, notably the Netherlands, and by the reclamation of forest and heath in northern Germany and Scandinavia. A very slight amelioration of the climate contributed to a significant northward shift in the frontier of settlement.

Towards the end of the period being considered the import of foodstuffs, primarily wheat, and of industrial raw materials like wool, flax and jute (not to mention silk) strongly influenced agriculture in those parts of Europe which were most exposed to it. They contributed to a growing tendency for agriculture to concentrate on animal products in areas which were best suited to their production. Their food value was higher than a predominantly cereal diet, but their higher cost had kept effective demand for them low before the real incomes of the urban population began to increase significantly late in the century.

The local concentration on producing meat and dairy products is merely one of many instances of the widespread tendency in western and central Europe to produce crops best suited to the local area and to rely on a developed transport system to carry them to distant markets. Wine-growing in southern France and Spain; rice in northern Italy, citrus fruit in southern; hops in the Rhineland, southern Germany and Bohemia are other cases in point. The converse of this was the abandonment in much of Europe of the cultivation of certain crops which each community had once required in small quantities. Amongst them were hemp, which almost wholly disappeared before the competition of imported fibres, flax, which remained important in only a few areas, and the many vegetable dyes which had once brightened the lives of pre-industrial people. Woad and madder disappeared from the fields, replaced by the products of the chemists' laboratory.

6
Agricultural regions

The previous chapter has presented a picture of the changing pattern of agriculture in Europe before the First World War. The result of these changes was twofold. On the one hand there was an increase in agricultural production, inadequate to keep pace with the growing demand, but nevertheless remarkable. On the other – and in some respects this was a condition of the growth in output – was an increasing local specialisation. Pre-industrial Europe had presented an almost uniform picture of subsistence agriculture, with the bread-grains forming the dominant crop. Animal husbandry, prominent though it was in certain restricted areas such as the Alps and the Spanish Meseta, was in the aggregate of limited importance, and animal products entered into the human diet only to a small extent. By the beginning of the present century this had undergone a fundamental change. Subsistence agriculture, based on cereals, had become diversified in most parts of Europe, and, with the development of intensive dairy farming, some areas had become no less narrowly based in the direction of animal husbandry.

Fig. 6.16 is an attempt to present a generalised map of agricultural types and regions in the middle years of the last century. Regional boundaries are necessarily very approximate. The basic regional differences are between open-field husbandry with fallow, in which animal-rearing played a comparatively minor role, and farming in enclosed fields with variable rotations and a greater use of farm stock, and, secondly, between peasant and estate farming.

In the following pages a number of studies are presented of farming in different areas of Europe. They have been chosen primarily to illustrate the varied ways in which agriculture was adapted, within the limits set by available technology, to local conditions both physical and social. The choice was, however, constrained by the availability of sources. The Mazowsze village of Lipce in central Poland has been chosen to illustrate primitive, self-sufficing peasant farming and von Thünen's estate at Tellow in Mecklenburg to represent the great estates of eastern Europe, though neither was in all respects typical. The plains of Lombardy and Flanders illustrate a relatively enlightened adaptation to local physical conditions, and Denmark an attempt to take advantage of a developing market for animal products. Central Spain is typical of the unprogressive agriculture practised on some great estates; the *chiflik* in Macedonia is typical of the no less backward farming within the Ottoman

Central Poland

In 1904–9 Władysław Reymont published his novel of peasant life in central Poland – *Chłopi*.[1] Its story is concerned with the loves and hates of the peasants of the village of 'Lipka', but a central theme is provided by their deep attachment to the soil which they cultivated, their peasant acquisitiveness, and their deep suspicion of all change and of everyone from beyond the narrow limits of their own community. The whole is set against the agricultural routine and the changing rural scene in the course of a single year. 'Lipka' is indeed a real place, the village of Lipce, about 80 km to the south-west of Warsaw, where Reymont himself lived on the edge of the village for many years, while working on the railway. The date of the action is not clear. The rising of 1863 remained a vivid though, for political reasons, a barely mentioned memory. On the other hand, the fact that a steel ploughshare was used and that potatoes were very widely grown must advance its writing close to 1890.

This area of central Poland was built almost wholly of glacial and alluvial deposits. The land, cut by numerous small rivers, sloped northward to the Bzura river. Its soils, mostly brown soils of moderate fertility, had been developed on the sandy clays and valley alluvium left when the ice of the mid-Polish glaciation melted away. Across it lay irregular beds of sand, accumulated by the rush of melt-water. Here one autumn morning Reymont looked out over the shallow valley which contained the village of Lipce;

as far as the eye could see lay the drab-hued fields, forming a sort of...basin with a dark-blue rim of forest, a basin across which, like a silken skein glittering in the sunshine, a river coursed, sparkling and winding among the alders and willows on its banks. In the midst of the hamlet, it spread out into a large oblong body of water, and then ran northward through a rift in the hills. At the bottom of the valley, skirting the lake, lay the village with the sunlight playing on the many autumnal hues of its fruit gardens. Thence, even up to the very edge of the forest, ran the long bands of cultivated ground, stretches of grey fields with thread-like pathways between them[2]

The village itself was of the form known in Poland as *owalnica*. Its double row of cottages spread apart towards the middle to enclose an open space where the pond lay and the geese strutted. Six months later:[3]

The mill stood at one end of the village...and at the opposite end the church raised its high white front amongst huge trees, its windows and the golden cross on the steeple shining afar, and the red-tiled roof of the priest's house visible close by...the bluish-grey ring of the forest, the wide expanse of cornland, villages at a distance nestling in their orchards...Nearer...the ground belonging to Lipka – as it were, long strips of canvas or cloth that variegated the sloping uplands. They ran in sinuous bands, one close to the other, separated only by the winding footpaths between them...or by drab fallows...Patches of land sown in autumn, now beginning to turn green, dark-hued potato-fields of last year's crop, bits of newly ploughed soil...Beyond the mill...peaty-coloured meadows...

Lipce was a nucleated, open-field village. Most of the villagers held strips in the fields, where they grew in turn their autumn-sown rye with a little wheat on the better soil, and their spring-sown oats and barley. After this the land was left as fallow for a year, grazed by the cattle and sheep of the villagers and by the inevitable geese. Small patches were planted with potatoes, which were increasingly important in the peasants' diet, and cabbages were grown to be preserved for the winter. Broad beans, clover for the animals, a little maize and 'a few plots of flax in the hollows' completed the inventory of Lipce's cropping.

The cropping pattern at Lipce was typical of that of central Poland, where about three-quarters of the total area was used agriculturally and most of the remaining land was under forest.[4] Of the cultivated land more than 70 per cent was normally under cereals, up to a third of it under rye. Oats were foremost amongst spring-sown crops, and wheat, commonly grown as a cash-crop, occupied a small area on the better soil.[5] Potatoes were increasing rapidly in importance; they had ceased to be merely a garden crop, and were in Reymont's Lipce grown in the open-fields, where the women were busy lifting them in advance of the first autumn frosts.

Since 1864 the peasants had in effect had full possession of their holdings.[6] They owed neither rent nor labour service and could divide and alienate them as they wished. The principle of partible inheritance prevailed, and, with the increase in population, holdings were becoming excessively divided and fragmented. 'I remember the time', said one character in the novel, 'when there were no more than fifteen peasants' farms in Lipka', but with growth in numbers 'the land must be divided again and again. Whether the harvest be rich or poor, the folk must always grow poorer. Ye cannot make the land to stretch. Yet a few more years, and there will be too little for us to live upon.'[7] The actual size of peasant holdings varied from less than a hectare to more than 50, though few were of more than 20.[8] A farm of 'thirty-two acres; also pastures; and a bit of forest; and the outhouses, and the livestock'[9] represented a large holding, and well worth the intrigue which runs through the novel. Nevertheless, the richest peasant cast envious glances on the widow's strip which lay next to his own in the open-field, thinking of what he might do in order to acquire it. There was some clearing of the waste. Cropped land increased in the kingdom as a whole by some 20 per cent in the second half of the century,[10] and in the single year covered by the novel one new settlement was created under the forest edge. Inevitably, however, the class of landless peasants increased. There were in 1886 about 1.5 millions of them.[11] Many derived from the hired workers on the estates; others had inherited only small and fragmented holdings, had incurred debts to the Jewish moneylenders and had been obliged to sell. The novel presents a telling picture of the poverty and distress of the landless *komorniki* and poorer peasants in late winter, when their stocks of food ran low and they had no opportunity to earn by day labour.[12] At this time of the year, too, the community was beset 'with many sicknesses, as is often the case just before spring; for at that time noxious vapours rise up from the thawing ground'. Noxious vapours or not, late winter brought smallpox as well as many other complaints.[13]

The peasants, and in particular the *komorniki*, had been accustomed to rely heavily on the common land. The Tsar's edict of 1864 had been carelessly drawn, and had not made provision for the peasants' rights in these areas, which remained legally in the possession of their lords. The forest was the most important and valuable of the common land. It provided building material for the peasants' huts, which were exclusively of wood, and also their most important fuel. The British consular representative, commenting on the improving conditions of the peasantry, added that 'the principal want of a Polish peasant is fuel. So long as he has enough wood to keep his stove going in winter he is satisfied.'[14]

The gentry or *szlachta*, many of whom were impoverished by the 'reforms' of 1864, turned to the forest as their only realisable asset. Forests were disappearing rapidly.[15] In the kingdom as a whole their extent was reduced by about 35 per cent in the course of the century. In the novel, the lord, a distant and forbidding figure who never entered directly into the drama, planned to sell the forest to timber merchants – Jewish, of course, for the novel has marked anti-semitic overtones. Indeed, the climax of the novel itself is the fight by the peasants to prevent the lord's servants from felling the trees. They scored a temporary victory, but when in late summer the novel comes to an end the question remains unresolved.

Amongst the questions raised was that of the right of the *komorniki* to share in the forests. It was in fact, the old problem of whether those who did not share in the village land could have a part in its other assets. In the novel the landless peasants were allowed to visit the forest twice weekly to gather firewood, and this seems to have been a widespread practice.

The community of Lipce is represented as made up almost wholly of peasants. Only two crafts, those of the miller and the blacksmith, appear in the novel, in addition to the Jewish tavern-keeper and the priest. There was, however, a small group of Jews in the village. They held no land, and were engaged chiefly in moneylending and petty trading. Women in general were engaged in autumn in scutching and preparing flax, and during the winter months they spun it, using both spinning wheel and distaff. The weaver does not figure in the novel, but it appears that much of the cloth was woven by the family which wore it. The assets of one eligible woman included her ability to weave cloth and canvas.

The commercial links of Lipce with the outside world were maintained through the market town of 'Timów', 'a small town whose empty narrow streets were lined with dilapidated houses, like rows of old saleswomen – living gutters full of rubbish, and dirty Jewish children, and pigs'.[16] On market day the street leading to the town square was 'so thronged with carts, one after another and several abreast, that one could drive past only with the greatest care and difficulty'.[17] Much of the business was between peasant and peasant, the sale of a horse or cow, the purchase of seed-corn or the sale of flax. But goods from the outer world made their way to the village through the market or the fair which each autumn attracted merchants from a much greater distance. Here were bought the coloured ribbons, woven in all probability at Łódź, and religious objects which served to brighten the persons and the cottages of the peasants. In most of such business the Jews were the middlemen. They bought up the cash crop and sold the vodka, on which a disproportionate amount of the peasants' cash income was spent. The taverns were, almost without exception, managed by Jews, a survival from the feudal period when distilling was a domainal monopoly and the Jews served as the lord's agents.[18]

The village was made up of huts, timber-framed and weather-boarded, with a thatched roof. The floors were of beaten earth with, in most instances, a stone-built fireplace and sometimes a chimney leading to the outside.[19] Fires were frequent, but attempts to induce the peasants to build in brick or stone were largely unsuccessful. The long side of the hut, divided by a central passageway, fronted on to the village street.[20] Around it, but chiefly to the rear, was a garden where vegetables were grown and fruit trees very nearly obscured the view of the cottage. The peasant house was dark and oppressive, so hot and fly-blown in summer that its occupants commonly slept beneath the trees, and cold and draughty in winter. At Lipce it was the practice in winter to surround the cottage walls with an envelope of straw and branches to serve as insulation.

There was little in the agricultural practice of Lipce to suggest that this was in the closing years of the nineteenth century. A metal ploughshare was used, but the seed was sown broadcast, the peasant 'taking handfuls of grain and scattering them all over the earth with a solemn gesture, as one bestowing a blessing'.[21] There was no evidence of mechanical aids. Even though both reaping and threshing machines were known on the demesnes, the peasant still cut his corn with a sickle and his hay with a scythe, and threshed in the barn with a flail during the dark, wet days of winter.[22] A three-course rotation with fallow was almost universal, but – first sign of modernity – the priest had received some new seed from Warsaw with which he intended to experiment. No fertilisers were used; the supply of manure was quite inadequate, and crop-yields and yield-ratios were but a fraction of those obtained on the demesne.[23]

Mecklenburg-Schwerin

The best known farm in all Germany during the first half of the nineteenth century must surely be that which J. H. von Thünen bought at Tellow in Mecklenburg-Schwerin about 1810. He had been born in East Friesland, had studied under the German agronomist Albert Thaër, and here set out to apply the agricultural methods which he had learned and here he developed his theories regarding land use in *The Isolated State*.[24] Mecklenburg was an unpromising region for such an undertaking. 'Few tracts in any part of Europe', wrote William Jacob in 1828,[25] 'are more miserable in cultivation or more thinly inhabited' than that which lay around Schwerin and Gustrow. Like Mazowsze in central Poland, it was a glaciated landscape, bordered on the south by the ridges which made up the Great Baltic End Moraine. Tellow itself lay on glacial sands and clays, and its prevailing soil was a light loam, of no high fertility, but above the average for Mecklenburg.[26]

Most of the land was held in estates, some of them large and managed by bailiffs, but many of them of medium size and worked by their owners. Peasant farm holdings were the exception. In the neighbouring province of western Pomerania their number and size had been greatly reduced as a consequence of the freeing of the serfs and late in the nineteenth century they made up only

5 per cent of the area. Amongst the medium-sized estates was Tellow, a unit of about 463 hectares, managed by von Thünen himself with evident success. 'I met with no proprietor or farmer', wrote Jacob, 'whose average crop came within a seventh of that of Tellow.'[27] In the late eighteenth century the open fields had been enclosed and the traditional three-course rotation replaced by a kind of short-ley system – *Koppelwirtschaft*. According to Thaër there were four rotations in common use in Mecklenburg, covering respectively six, seven, eight and nine years.[28] All began with a fallow year, followed by either three or four years of cereals. Grass was sown with the last corn crop and was left to provide hay or pasture until the next ploughing. Von Thünen himself favoured a seven-course rotation with three years under cereal and three under grass.[29] The cereals grown were chiefly rye, barley and oats. Wheat was sometimes introduced into the rotation probably in place of rye, and small quantities were exported through Rostock and Wismar.[30] But rye, live animals and animal products like butter and wool were the chief exports of Mecklenburg.[31] Arable farming, von Thünen claimed, was restricted by the shortage of manure, so that only 43 per cent of the area could be planted at any one time with cereals.

Von Thünen was deeply concerned for the efficient management of his estate. It covered so extensive an area – equal to a circle of a little over 1 km radius – that the movement of manure, farm products and timber within it placed a considerable strain on its available means of transport. In fact, as he emphasised, estates in Mecklenburg and western Pomerania were ill-suited to efficient management. The farm buildings were badly sited, bearing 'the traces of their original foundation'. 'The boundaries of farms', he wrote, 'are haphazard and irregular; and on many adjoining properties the fields belonging to the one stretch nearly to the farmstead of its neighbour which, in its turn, owns plots near the buildings of yet another property.'[32] It would make for efficiency if the estate were compact and the farm buildings located near its centre, but 'scarcely a single holding has its farm buildings right at the centre of its land, and almost all of them would gain from an adjustment of their boundaries'.[33] But no such adjustment was possible, and von Thünen was left to theorise about the layout of the fields in his idealised 'state' and to calculate the diseconomies which resulted from excessive and, to him, unnecessary transport and travel.

The zonation of land use, which is the most significant feature of *Der Isolierte Staat*, was never realised at Tellow. We do not have a plan of the estate, but it is implicit that it was as irregular as others in Mecklenburg. Von Thünen is, however, at pains to explain that his model applied, not to farm buildings at the centre of an estate, but to 'a very large town, at the centre of a fertile plain, which is crossed by no navigable river or canal'.[34] Around this focus there would be concentric rings of different forms of land use according to their distance from the centre. The innermost ring was characterised by the intensive production of garden products and milk with relatively little grain cultivation.

The next ring was left under forest since it was necessary to produce a commodity as heavy and difficult to transport as timber as close as possible to the place where it was consumed. Beyond the woodland belt lay intensive crop farming; then successively less intensive arable with fallow, and, lastly, grazing land.

There might be cereal cultivation in this outermost ring, despite the fact that distance made it uneconomic to transport it to the town; '...if the grain is processed, and converted into a product cheaper to transport in relation to its value', part of this last zone might profit from growing grain crops.

It is doubtful whether von Thünen's model could ever have been applied, except in the most general terms, to a very large town and its fertile plain, because it separates – presumably by a very considerable distance – types of farming which in his own Mecklenburg were becoming increasingly integrated. His own favoured rotation system linked arable with pasture, bread-grains with feed-grains. To separate these functions would have been to misuse the soil; Albrecht Thaër would have agreed with that. It might have been logical to locate the woodland as close as possible to the human settlement, but it is very doubtful whether in fact this would have happened unless trees were left as a protective screen.

Nevertheless, there were elements in the pattern of land use in *The Isolated State* which must have derived from von Thünen's practice at Tellow. The intensive, almost horticultural land use close to the farm itself was common practice. There *may* have been a copse nearby, especially if there were a patch of sandy moraine – though such accidents were excluded from von Thünen's isotropic land surface.[35]

Mecklenburg had always been one of the least populous regions of Germany. There were in 1833 fewer than 500 peasant-held farms in the whole province,[36] and much of the population worked as wage labour on the estates. The Tellow estate enjoyed a period of high cereal prices and general prosperity for some three decades following the publication of *Der Isolierte Staat*. During this time further improvements were made on the estates. Fallow was gradually abandoned; the cultivation of clover and other fodder-crops was increased; greater use was made of manure and marl, extensive drainage projects were undertaken on the damper soil, and 'in many cases meadows and pastures were ploughed up and sown with corn and the cattle were fed in stalls'.[37] The number of sheep declined sharply with the abandonment of fallow and the extension of arable farming. From about a million and a quarter in 1867 they declined in numbers to a little over 700,000 by 1892. At the same time the number of cattle increased by more than a fifth.[38]

The prosperity of farming in Mecklenburg was threatened from two directions in the later years of the century. In the first place there was a growing labour shortage. Most of the workforce had no vested interest in the land, and the attractions of the cities lured them away. Attempts were made by the state government to create peasant farms as an inducement to the peasants to stay,[39] and by 1883 their number had increased to more than 5,000.[40] At the same time demand for farm products, in particular cereals, began to weaken, and there was an appreciable shift towards the cultivation of root crops as well as fodder. The number of cattle and pigs continued to increase,[41] but the most far-reaching development was the establishment in 1872 of the first sugar-beet factory in Mecklenburg. Before the end of the century there were twelve, as well as factories for distilling *Brantwein* and making starch from potatoes.[42] Von Thünen would have approved of this processing of agricultural products so close to their place of production.

Table 6.1. *Land use in Mecklenburg, 1878*

	ha	%
Arable	750,243	56.4
Gardens	9,060	0.68
Meadow	103,798	7.8
Pasture and rough grazing	68,417	5.14
Forest and woodland	223,734	16.82
Other	175,122	13.16
Total	1,330,374	100.00

By the end of the nineteenth century Mecklenburg had become a predominantly arable-farming province. In 1878 its land use was as shown in table 6.1.[43]

The North Italian Plain

The plain created by the River Po and its tributaries was one of the most densely settled and intensively cultivated areas of Europe in the early nineteenth century. Despite its Mediterranean setting it was in many respects a non-Mediterranean region, with a considerable summer rainfall and a dairy-farming industry on a scale paralleled only in north-western Europe. The North Italian Plain was a gulf, some 400 km deep and varying in width from 70 km in the west to three times as much in the east. Along the whole of its northern margin the Alps rose abruptly from heights of only 200 or 300 m to their snow-capped summits at above 3,000 m. On the south, the rise to the Apennines was less steep, and their greatest heights were less than half those of the Alps. Rivers in great numbers flowed from both the Alps and Apennines to join the Po, but the more southerly rivers drained an area of summer drought and brought little water to the plain at the season of the year when it was most needed. The northern rivers, by contrast, were fed not only by rain through much of the year, but also by the snow-melt within the Alps, which continued from late spring until autumn. The northern rivers furthermore carried a great deal of silt which they spread along their lowest courses, contributing to the severe floods which were a feature of the Po, and to the advance of the coastline into the Adriatic Sea.[44]

The silt which was currently being added to the Plain was only the most recent of a series of deposits which had been laid down during glacial and post-glacial times. It was preceded by vast spreads of sand and gravel which sloped gently upwards from the floodplain of the Po to the Alpine foothills and again on the south towards the margin of the Apennines (fig. 6.1). These sandy beds were porous and infertile. Towards their outer margins they supported little agriculture, and much of their surface was scrub-covered. Then a remarkable change occurred. At distances of from 10 to 20 km from their outer margin the sands became finer and more fertile and were furthermore covered with a thin deposit of loam, so thin, in fact, that in some parts deep ploughing was liable to bring the sand to the surface. More important than this, however, was the appearance of a line of springs, the *fontanili* line. Water from the Alps and from the rainfall

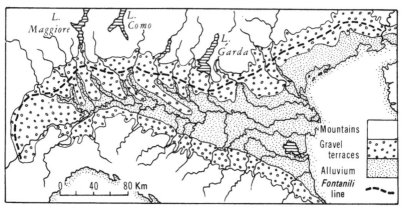

6.1 Physical divisions of the North Italian Plain

soaking into the gravelly soil of the upper terraces was here thrown out in a series of springs by an impervious layer in the deposits. Thousands of small streams flowed southwards towards the Po and its tributaries, creating, as it were, a gigantic water meadow (fig. 6.2). The water, as it issued from the springs, was relatively warm in winter, thus contributing to the remarkable growth of grass at this season.

Agriculture in the North Italian Plain was closely adapted to the physical conditions and was during the nineteenth century one of the most efficient to be found in Europe. The character of farming differed sharply between the three broad zones into which the region was divided. The mountainous belt was characterised by small peasant holdings, like those of Switzerland, held mostly in freehold. The south-facing hills had been terraced, and the narrow strips cultivated with the grapevine, fruit trees and small quantities of cereals and vegetables. 'This mode of cultivation', the British consul wrote, 'could not be kept up were it not for the strong attachment of the cultivator to the soil...[and] the peasant could not submit to such great fatigue without the certainty of enjoying for a long time the fruits of his labours.'[45] Animals were reared – cattle, sheep, goats – and sent higher into the mountains in summer. Cheese and butter, as well as live animals, were sent southwards to the plain in exchange for bread-grains. Close to the lakes the climate was milder in winter; the olive and citrus fruit were grown, and, with growth in the silk industry, the cultivation of the mulberry was spreading during the first half of the nineteenth century. The hills which bordered the mountains were more populous and more intensively cultivated. Here farms were larger and less fragmented, and most were held on a sharecropping basis. Silk production was increasingly important and farming routine was coming to depend on the rearing and feeding of the worms and the reeling of the silk.

The consular report already cited distinguished between the 'high-flat' and the 'low-flat' country, a distinction corresponding with the areas above and below the *fontanili* line. The high-flat region was arid and little suited to arable farming. The rivers were incised into its surface, and underground water could only be reached at great depths beneath the sands and gravels. Although some

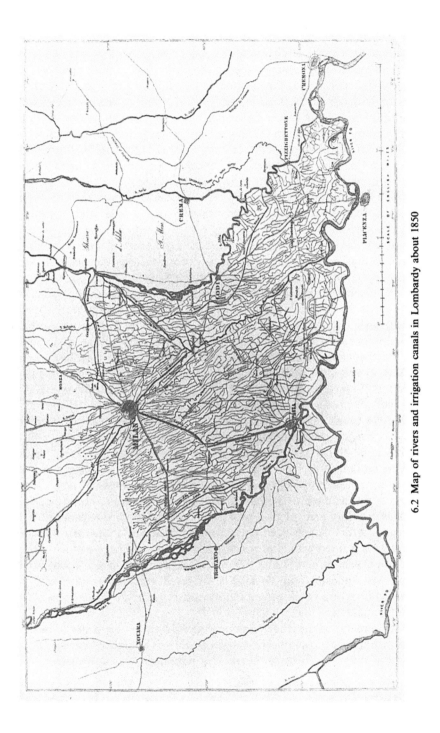

6.2 Map of rivers and irrigation canals in Lombardy about 1850

wheat and rye were grown, crops were in general those, including buckwheat and millet, suited to dry and infertile soil. Largely on account of the poverty of the soil, domestic crafts, chiefly silk, cotton and flax-working, assumed a greater importance than elsewhere in the plain. They contributed to the growth of small manufacturing towns, nuclei of the future Italian cloth industry.

The low-flat country was the most densely populated and intensively used of all divisions of the northern plain, and was also one of the most distinctive agricultural regions in Europe. Its fertile, loamy or silty soil had an abundant supply of water for irrigation. This was brought in part by canals which took off from the rivers in the high-flat country and angled across their interfluves towards the Po. They were especially important for irrigating summer crops, since they drew on the abundant melt-water from the Alps. Additional canals were being built during the early part of the nineteenth century, some of them serving the double purpose of conveying water to the fields and providing a means of transport.[46]

Of greater importance for the characteristic agriculture of the northern plain were the springs, or *fontanili*, 'semicircular excavations in the earth, in which are placed long tubs, from the bottom of which bubble up copious streams of water'.[47] This was carried by an intricate pattern of small canals (fig. 6.2) to the fields towards the Po. The waters varied greatly in agricultural value; 'some are siliceous, some cold, some warm; they are either too fat for the rice grounds, or too lean for the meadows'.[48] Land values were dependent on the supply and quality of the water. In consequence the flow of the water from both springs and from rivers and canals was tightly regulated. Each waterway had its association of users, which controlled 'a vast, closely articulated time-table [by] which each member...was entitled to receive in turn his quota of water'.[49]

The larger canals, which drew their water from the Alpine rivers, were used mainly to irrigate the summer crops, maize, wheat, flax, rotation grass, or *erbatico*, and, above all, rice. A number of rotations had been devised. Characteristic of all of them was a period of three or four years under rice, followed by a year or two of maize, oats and flax and two of *erbatico*. Only on the wettest ground close to the Po was land given over exclusively to rice.[50] The plain was the only area of Europe where rice-growing was significant in the nineteenth century. A survey of 1877 showed that almost a quarter of a million hectares were planted with rice, nearly a half of them in Lombardy and the rest in Piedmont, Venezia and Emilia.[51] The land was held in large units, and peasant farming was almost wholly lacking.[52] Irrigated agriculture, especially rice-growing, called for a far larger capital investment than the Italian peasant was capable of making.

Cattle were raised on the temporary grasslands in the areas of summer irrigation, but were most important where water was available from the *fontanili* throughout the year. This was the *marcito*, where the land was divided

> into so many small parallelograms, sensibly inclined to one side. The water which fills the little canals amongst them overflows [them] slowly: it spreads like a veil over these spaces, and by the inclination of the ground falls again into the opposite canal. From this it is diffused over other parts, so that the whole meadow country is continually flooded: from which there is maintained a rapid and continual vegetation, in the heats of summer and the frosts of winter...[53]

Table 6.2. *Irrigation in Lombardy*

	Summer irrigation (ha)	Year-round irrigation (ha)
Between Ticino and Adda	190,602	3,172
Between Adda and Oglio	118,083	1,012
Between Oglio and Adige	120,820	1,012

Grass was cut no less than five times a year, and was used to feed vast herds of dairy cattle. The *marcite* were most extensive to the south-east of Milan, where the *fontanili* water was most abundant. Here was made the local cheese, known to the Italians as *grana*, and to foreigners as Parmesan. The cattle themselves were generally brought from Switzerland; up to two thousand a year, it was claimed, crossed the St Gotthard. Much of the cheese was made, even in the early nineteenth century, in small co-operative dairies.[54] By the 1890s there were said to have been as many as 313 such dairies in the province of Padua alone. Second only to the Parmesan cheese from the district of Lodi was the soft Gorgonzola made to the east of Milan, and the *stracchino* from the hill country to the north.[55]

The area under *marcito* was increasing during the middle years of the century, but was never more than a fraction of that irrigated in summer from the canals. A limit was set to its expansion, despite the growing demand for Parmesan cheese, by the available water supply from the *fontanili*.[56] In the middle years of the century, the irrigated area within the province of Lombardy was as shown in table 6.2.[57]

If cheese and rice were the most important and noteworthy products of the low-flat country, silk was the most distinguished product of the high-flat and of the mountain foothills. In the early years of the century Italian raw silk was in demand in much of western Europe (see p. 328), and plantations of mulberry trees were spreading through the low hills and sandy terraces which bordered the low plain between Brescia and Lake Maggiore. 'The inhabitants attend principally to the cultivation of silk', wrote the British consul, 'and with the money gained from this production they provide themselves decently with the necessaries of life.'[58] Their homes, he added, were 'large, well-aired and clean' for the convenience of rearing the worms which were 'always more prosperous in good apartments'. It is a curious commentary on life in Italy that the silkworms deserved better accommodation than the peasants themselves. Most of the silk was reeled by the peasants in their homes, but small steam-powered mills were being built, chiefly by the landlords, both for reeling and for throwing or twisting the fibres to make a stronger thread.[59]

Lombardy continued throughout the century to be the chief source of Italian raw silk, and the rearing of the worms to be a domestic or cottage industry. The mulberry plantations, mostly located close to the farms for convenience of transporting the bulky leaves, remained the most valuable land of the peasants, who were content to import part of their bread-grains from the low-flat country, where silk-rearing was of no importance.[60]

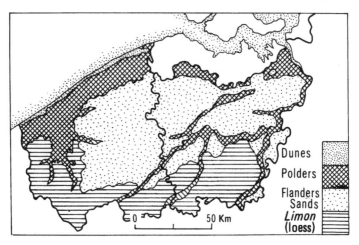

6.3 Physical divisions of Flanders

Flanders

'Flanders', wrote Sir John Sinclair in 1815, 'is the most productive and the best cultivated country on the continent of Europe'.[61] The agricultural system developed here was the envy of people in other parts which were considered less fortunate. It was not always apparent that the high productivity of Flanders was obtained in spite of, not because of a natural endowment which must be reckoned amongst the most niggardly in western Europe. Flanders was made, said the French historian Michelet, 'malgré la nature; c'est une oeuvre du travail humain'.[62] The Flemish soil, in the words of le Laveleye, himself no mean authority on European agriculture, 'is the worst in Europe',[63] made up mainly of sterile sand or heavy, poorly drained clay (fig. 6.3). How then did Flanders develop the densest rural population in Europe?

The two provinces, West and East Flanders, lie in the shallow valleys of the Scheldt and of its tributary, the Lys. In their natural state these rivers were slow-flowing, meandering and liable to severe floods in winter. The greatest altitudes are found in the south, on the border of Hainault, but reach little more than 100 m. The region occupies a downfold in the chalk, strictly analogous to the London Basin. This was infilled with tertiary sands and clays, and in recent geological times the lowest land was several times overflowed by the sea.[64] These circumstances have produced the present pattern of soils in Flanders and have strongly influenced the development of agriculture.

Regional divisions are determined by the variations in Tertiary and Quaternary deposits. Four contrasted regions can be distinguished. Along the coast lay a belt of dunes of recent origin. They yielded at best a dry and infertile soil, and in parts consisted only of blowing sand. Behind them sand, blown across the clay of the Polderland, produced a light soil which was beginning, late in the nineteenth century, to be used for intensive vegetable-growing.

Behind the Dunes lay the Polderland. It formed a belt, some 10 km wide, from beyond the French boundary to the Scheldt, sending long arms of damp clay land up the shallow valleys of the IJzer and Scheldt. The land was flat and low-lying, much of it below the level of high tides. It was gradually reclaimed for agriculture and preserved only by the constant care and watchfulness of the *Wateringues*, or water-boards made up of the local landowners. Although there was no serious marine flooding during the nineteenth century, the rivers regularly overflowed their banks in winter, and much of the land, despite its high fertility, was too damp for cultivation.[65]

Bordering the Polderland to the south lay the Flanders sands, extending in a broad belt from the IJzer in the west to the lower Scheldt in the east. It was the most extensive region, making up nearly half the area of Flanders. Soils were notoriously poor, but the sands were in some areas so thin that underlying clays could be brought to the surface by ploughing. Where the sand was thick, a hard iron-pan had sometimes formed, impeding drainage and making the sands as impracticably damp as parts of the Polderland itself. This, the least fertile region of Flanders, was paradoxically the most densely peopled and intensively cultivated. Blanchard described it as 'one of the foremost agricultural regions of the world'.[66] Lastly, in the southern and south-western parts of Flanders, the sands were partially covered with *limon*, greatly raising the level of fertility.

Agriculture was adjusted to the qualities of the soil, though there were significant changes during the century in the pattern of cultivation in each region. The Polderland was consistently important as a grazing area. Even so, the danger of flooding was too great to leave animals on some of the lower ground through the winter. When de Laveleye wrote in the 1870s, almost two-thirds of the area of the Polderland was under crops.[67] The quality of the soil was high, and Thomas Radcliff claimed in 1819 that no manure was needed.[68] Wheat was the foremost field-crop, followed by barley, legumes and potatoes. Early in the century fallowing was practised but had disappeared by its end. At the same time the cultivated area grew smaller, with the expansion of intensive dairy farming, and, when Blanchard wrote, regular use was made of manure from the dairy herds. Complex rotation systems were developed, and one, cited by de Laveleye,[69] was spread over ten years.

Data on the size of farms and holdings are unfortunately not available on the basis of regions, but West Flanders, which contained by far the greater part of the Polderland, had on average much larger farms than East Flanders, and it is clear that most lay in the Polderland.[70] Even medium-sized farms had more than 30 hectares. On the other hand, a large number of cottagers cultivated small holdings of less than two hectares and supplemented their income with domestic crafts.

By contrast with the Polderland, the sandy region was highly urbanised and densely populated, but its fertility was the creation over the centuries of its inhabitants. All writers on Flanders commented on the minute care which the peasant bestowed on his holding. Nowhere else was weeding as clean or digging more careful. He

scrupulously collects every atom of sewage from the towns; he guards his manure like a treasure, putting a roof over it to prevent the rain and sunshine from spoiling it. He gathers mud from rivers and canals, the excretion of animals along the high roads, and their bones for conversion into phosphate. With cow's urine, gathered in tanks, he waters

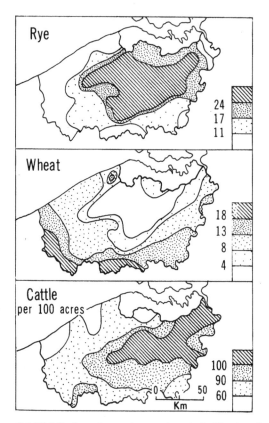

6.4 Distribution of rye, wheat and cattle in Flanders, late nineteenth century

turnips which would not come up without it; and he spends incredible sums in the purchase of guano and artificial manures.[71]

Farm holdings were very small and severely fragmented. The average holding in East Flanders, which consisted largely of sandy soils with an admixture of *limon* in the south, was less than 2.5 hectares. The land was almost wholly arable, and in the 1870s was divided between cereal and industrial crops on the one hand and fodder crops on the other in a ratio of about two to one. There was little grazing land in the sandy region. There are no statistical data for the region as such, but only about 11 per cent of the whole of East Flanders was under grass in 1846, as against about 20 per cent in West Flanders.[72] Nevertheless, the density of cattle was always greater in the sands region than in the Polderland (fig. 6.4). In the latter they were grazed; in the former most were stall-fed.

Complex rotation systems were practised in the sands region, as in the Polderland. Fallow was negligible in East Flanders in 1846, and had disappeared half a century later. Cropping was dominated by rye, which occupied at the same time about a fifth of the agricultural land. Wheat and grass – whether grazed or cut for hay – came next with about 11 per cent each. Oats were grown on 6 per cent, and flax on only about 4.5. Agriculture on the smallholdings of the sands region was in the main subsistence farming, but even so, it was inadequate for local needs, and there was throughout the century a substantial food import into Flanders.

Physical conditions were better in the *Pays de Waes*, the area between the Netherlands boundary and the river Scheldt, and centring in the town of Saint Nicolas. Here the sands had an intermixture of alluvium, making a relatively light and fertile soil. Like the more sterile sands it was divided into minute holdings and intensively cultivated, and its more fertile soil encouraged flax cultivation on a significant scale longer than elsewhere in Flanders.

The region of sandy *limon* which covered southern Flanders and extended into neighbouring Brabant and Hainault was richer and no less intensively cultivated. There was, however, a subtle change in cropping patterns. Wheat came to the fore as the most significant crop in the course of the century, and flax, never as important as farther north, declined and disappeared. In the course of the century wheat began to replace rye in the diet, and the close proximity of Brussels led to a greater dependence on dairy farming and market gardening for the supply of the capital. At the same time, sugar beet spread and became an important crop.[73]

The paradox of Flanders lies in the fact that here the densest rural population in Europe had grown up on a soil which was amongst the less attractive. This was achieved by unremitting toil on the land; 'the fields are never empty; the land never rests'.[74] Despite the climate, double cropping was practised on some lands in East Flanders where, according to Rowntree,[75] nearly half the soil yielded a crop of carrots, turnips or fodder after the cereals had been harvested. The density of rural population remained remarkably high throughout the century. The larger farms were fragmented and the average size of holdings declined.[76] The value of land rose; many peasants could not afford to buy, and leasehold and rented tenancies increased. Nevertheless, agricultural income consistently fell short of the needs of the peasant family, and had to be supplemented by domestic industry.

Population density was consistently higher on the sands and sandy loam than in the Polderland. Its growth has been related both to the practice of partible inheritance and to the development of domestic crafts. Early marriage became institutionalised, and perpetuated the increase long after density had become dangerously high.[77] The population was always less dense in the Polderland. The reason lay in all probability in the fact that the drained and reclaimed land was less suited to the settlement of peasant cultivators. It called for year-round attention to the flocks and herds rather than the seasonal work on the land which left time free for the practice of domestic crafts.

Nowhere else in Europe were domestic crafts pursued with greater vigour. About 1845, on the eve of the crisis of 1846–7, it was said that 328,250 out of a total population of about 1.434 millions were engaged in some way in the linen industry.[78] It was customary for the peasant to grow his flax, ret it himself in a pond or in the River Lys, the 'Golden River', so-called 'on account of the wonderful colour and quality of the linen bleached in its waters'; and spin and weave it 'in the winter, when work on the farm was impossible'.[79] The domestic weavers remained 'at bottom agricultural workers who return to the soil whenever they get the chance'. Each had 'a bit of land...nothing more than an allotment, the produce of which constitutes a welcome, but not very important addition to his income'.[80] The domestic linen industry never recovered from the crisis of the 1840s. The collapse of the linen market (see p. 318) combined with harvest failure to produce acute suffering.[81] In some communes the number of paupers rose to more than a third of the population. For a period the death-rate was more than three times the birth-rate, and the population was further reduced by migration to the expanding industrial centres of northern France or the factories of central Belgium. The decline of domestic industry, especially linen-weaving, continued throughout the century. By 1896 the number of weavers had fallen to 11,000, only a seventh of the number who had plied their looms half a century earlier, and the most important supplement to agriculture was not the linen, but the lace industry.

Domestic workers were most heavily concentrated throughout the century in the sandy region.[82] They were almost wholly absent from the Polderland, and on the sands were most numerous near Courtrai and Roulers and, in the east, near Alost, where holdings were smallest and rural population most dense.

How, one may ask, did the peasantry survive during the second half of the century, with an increasing rural population, holdings of diminishing size and a disappearing domestic industry? Survival was, indeed, difficult, but was made possible by a number of small changes in the rural economy. Agricultural production was improved by increased use of fertilisers and the elimination of fallow. Cash and industrial crops began to replace cereals, and very intensive market gardening developed. The latter, according to Rowntree, was less efficient than the *maraichères* near Paris, and marketing methods were primitive:

a hurrying mêlée of carts of all descriptions, drawn by men, women, dogs, donkeys, ponies or horses, can be seen any morning on the roads leading to the large towns, each cultivator trying to get a good place at the opening of the market. The consignments are often ridiculously small, and a person who might be doing valuable work in the garden spends the best part of the morning in selling a few shillings' worth of produce.[83]

But, what with a stupendous input of labour, the growing of chicory and other produce and even fruit and vegetables under glass, with seasonal labour in France and commuting to urban employment, the rural system of Flanders, one of the most precariously balanced in Europe, survived into the twentieth century.[84]

The Danish Revolution

The agriculture of Denmark at the beginning of the last century was in no way remarkable. William Jacob in the 1820s found nothing to interest or excite him, 'for its production and consumption so nearly approach each other, and both are so insignificant... that no inducement presented itself to view more of it than the small portion between the passage over the Little Belt and the boundary of the province of Sleswick'.[85] Neither Thomas Malthus[86] nor Samuel Laing,[87] half a century later, were more complimentary. Denmark was a land of peasant farmers. Much of the better land had been cultivated in open-field strips, and the rest on some form of the in-field/out-field system (fig. 6.5). There were large estates, and Jacob described one of 4,500 acres, almost a fifth of it in demesne,

Agricultural regions

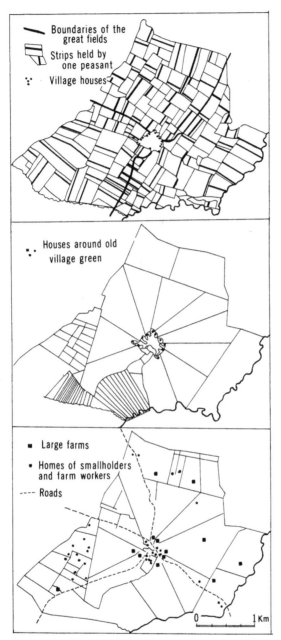

6.5 Land enclosure in Denmark. The maps relate respectively to 1769, 1805 and 1893

with 36 tenants 'paying a fixed rent in corn and labour, with their own draught animals'.[88] But Denmark had two advantages of incalculable importance. There was no population pressure on resources and, secondly, a thorough-going reform at the beginning of the century had consolidated the scattered open-field strips and produced 'a country of small irregular fields, of various shapes and sizes, surrounded by hedges and hedgerow trees... the gently swelling grounds... cultivated to the summits and crowned with woods'.[89] Most were held in freehold, the rest by a rent often paid in kind as late as the mid-century. Consolidation of parcels of land was accompanied by a dispersal of settlement. The nucleated villages were either abandoned or greatly reduced in size and farmsteads built on the individual holdings.[90] Farms were fairly large, by the standards of peasant Europe – from 5 to 60 hectares – and after enclosure were mostly managed on a long-ley system, with a variable rotation over five to seven years followed by a similar period under grass.[91] Agriculture was by and large self-sufficing, though there had long been an export of live cattle to western Europe, mostly driven overland, and of grain to Norway. This was requited by an import of timber of which Denmark was desperately short.

Denmark was an unlikely country to become an agricultural pioneer. The climate was marked by cool and often very wet summers, and long, cold winters, which became increasingly severe towards the east. The growing season, of about six months, was appreciably shorter than in western Europe. The soil of Denmark derived almost exclusively from the deposits left by the ice-sheets. The boundary of the ice stood for a prolonged period from north to south through the Jutland peninsula, and a ridge of morainic material built up, forming 'the barren heathy back of the peninsula'.[92] To the west lay outwash sands and gravels; to the east the undulating boulder clay deposited by the ice as it melted away, broken only by minor ridges of sand. There was never any question that the clay soils of eastern Jutland and the islands were immensely superior to the sands of the west. Cultivators in eastern Denmark, wrote Jacob, were impoverished, but far better off than those in Jutland.[93]

Despite the harshness of the climate and the poverty of much of the soil, the Danish peasant enjoyed a greater degree of prosperity than many in western Europe.[94] The existence of enclosed farms permitted him to make what improvements he chose. In fact he made great use of marl to lighten the heavy soils, and of field tiles to drain the wet. An extensive programme of land reclamation was undertaken along the coast, especially of Jutland. At the mid-century two-thirds of the land area was in arable; by the end of the century this had increased by 20 per cent. Much of the remainder was heath covering the gravelly ridges, and of little agricultural value. At the same time education made great advances, especially through the Folk High Schools, and the Danes became probably the best educated peasant nation in Europe.[95] The Danish yeoman farmer was thus well prepared not only to follow market conditions, but also to adapt his farming system to them.

Long before the agricultural 'revolution' the Danes had been exporting the products of their agriculture, especially live cattle and pigs, poultry and eggs and large quantities of execrable butter, often marketed in Germany as 'Kiel' butter, because 'Danish' had too bad a connotation.[96] It was in order to remedy the ill repute in which Danish agricultural produce was held that the co-operative movement was born. The first co-operative dairy was opened in Jutland in 1882. The institution quickly spread, helped by the widespread adoption of the mechanical cream separator. The improved quality of the farm products enhanced Denmark's competitive position, and the co-operatives became more

Table 6.3. *Number of farm stock in Denmark (in thousands)*

	1881	1898
Horses	348	449
Cattle other than dairy cows	571	678
Dairy cows	899	1,067
Pigs	527	1,168
Sheep	1,549	1,074
Poultry	4,070	8,767

numerous, drawing into their sphere of activity an increasing proportion of the peasant proprietors. When the consular report[97] was prepared in the mid-1880s there were about 200 co-operative dairies, and only three years later there were said to be 500. The quality of the butter which came from them was both high and uniform, and was then supplying more than a quarter of British imports as well as a large market in western and central Europe.

The agricultural crisis of the 1870s was felt severely. It was, however, a crisis induced largely by a surplus of imported grain. In several countries (see p. 10) it led to the raising of tariffs on cereals. In Denmark, which still exported some grain, especially to Norway, it had the effect of hastening the transition to a predominantly dairy-farming economy. 'Between 1883 and 1889', wrote Skrubbeltrang, 'Danish agriculturalists accepted the consequences of the fall in prices and, on a co-operative basis, introduced a dairying system which simultaneously laid the basis of a great expansion in pig production.'[98] Hogs, it was found, could be profitably reared on the waste products of the dairy, and the establishment of bacon co-operatives followed Germany's exclusion of the import of live animals.

The number of cattle and pigs increased, but sheep declined with the progressive elimination of common and fallow grazing (table 6.3). The horse was used for ploughing in Denmark, as it was throughout most of northern Europe, and the increase in its numbers reflects the fact that there was no diminution in the area of arable. Indeed there was an increase in the area under field-crops from 1,026,000 hectares in 1866 to 1,160,000 in 1901.[99] There was, however, a marked change during the period in the crops grown. Wheat declined sharply; rye and barley increased somewhat because they could grow well on the newly reclaimed heaths, but oats and mixed corn, both used primarily as animal feed, increased by more than 40 per cent. The area under potatoes and root crops, which were used as fodder, grew more than fivefold. At the same time the area under clover and sown grasses almost doubled and that used as meadow and pasture also increased (fig. 5.12).

It was not the Danish practice to divide the farm between heirs, and holdings were thus never severely fragmented. A consequence, however, was the creation of a class of landless or near landless cottagers, who worked their own garden holdings and, for the rest, provided labour on the larger farms. The cottager continued throughout the century to present a serious social problem, the converse of the advantages of maintaining an impartible inheritance.[100] There

Table 6.4. *Size of farms in Denmark, 1901*

Size (ha)	Number	Total area (ha)
Less than 1.5	87,157	33,181
1.5 to 5	46,445	143,040
5 to 10	28,892	208,538
10 to 30	52,980	990,302
30 to 60	25,615	1,032,955
60 to 120	6,502	514,939
Over 120	2,392	677,763

was no industrial development to absorb the surplus rural population, and, though there was some domestic weaving, especially of linen, this was only for domestic consumption and never grew into a significant branch of manufacturing, as it did elsewhere. The government did its utmost, especially by the Smallholders' Act of 1899, to provide small farms for the cottagers, and fortunately was helped by the trend towards dairying, a labour-intensive form of agriculture. Although large estates remained numerous, the tendency was for farms to become smaller, and early in the present century almost two-thirds were of less than 10 hectares (table 6.4).[101] The existence of co-operative dairies made the operation of even a small farm economic.

The dairy industry played a dominant role in the economic development of Denmark during the later years of the century. It led to no radical change in the pattern of land use, which remained much the same in 1901 as it had been before the revolution in Danish agriculture. Agriculture was, however, pursued more intensively. The labour input was in absolute terms nearly twice as great in 1901 as a century earlier; only in proportion to the total population had it declined from about 60 per cent to 39.[102] The contribution of agriculture to the gross national product was the largest of any sector until, about 1890, it began to be exceeded by that of the tertiary sector. In 1900 agriculture still contributed 30 per cent, while handicrafts together with the diminutive manufacturing industry, part of which was related to farming, accounted for only 20 per cent.[103]

Farm products continued to dominate Denmark's foreign trade. In the mid-nineteenth century exports had consisted largely of live cattle and grain. After the 1860s butter, cheese and later bacon began to replace these staples. The former export of bread-grains gave place to a large and steadily growing import of wheat, rye and feed-grains.[104] The United Kingdom became during the later years of the century the chief market for Danish farm produce, forcing Germany into second place.[105] By about 1910, agricultural products made up about 80 per cent by value of Denmark's export trade.

Spain: 'Latifundia' and 'Huerta'

Spain, as every traveller asserted, was a land of contrasts. In no European country was the agricultural spectrum from the most intensive cultivation to utter neglect manifested more fully. Spain consists essentially of a high plateau, the Meseta, lying at 750 m and more above the level of the sea, set within a frame

Agricultural regions 269

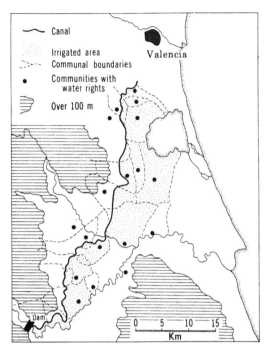

6.6 Irrigated huertas in the Valencia region of Spain

of higher mountains and crossed by mountainous ridges. These plateaux, 'which seem of boundless extent...are not', wrote Borrow, 'tame unbroken flats, like the steppes of Russia. Rough and uneven ground is continually occurring: here a ravine and gully worn by the wintry torrent; yonder an eminence not unfrequently craggy and savage, at whose top appears the lone solitary village.'[106] In sharp contrast with the savage plateaux of Estremadura and La Mancha stood the narrow coastal plains, between the mountains which framed the plateau and the Mediterranean Sea; 'sunny alluvial plains [extending] from Cadiz to Barcelona...except in those parts where the mountains come down abruptly into the sea itself' (fig. 6.6).[107]

The contrast in climate was no less extreme. The Meseta was not a Mediterranean land. Ford warned his Victorian travellers that winters were as severe as those in central and northern Europe, and that the intense heat of summer was interrupted only by violent storms. The coastal plains, on the other hand, alternated between mild winters with a sufficient rainfall for plant growth and hot, dry summers, when only the streams descending from the mountains kept plants alive. Much of the Meseta was covered only with thin skeletal soils, acidic in the west, calcareous towards the east, and everywhere of low fertility. Over parts of the eastern Meseta they had been reduced to the sterile, bright red *tierra rossa* of semi-arid climates. The coastal plains, by contrast, had been built up over recent geological time from the silt brought down from the mountains. Their soil was deep and rich, and, with irrigation and the warmth of the Mediterranean, has always yielded abundantly. Widdrington in 1843 wrote of their 'exuberant fertility'.[108]

Yet the contrast which every traveller was quick to notice between the two regions was not entirely a consequence of differing physical conditions. It arose in part from their contrasted histories and the use and misuse to which they had been subjected. The whole peninsula, except parts of the mountainous north and north-west, had been overrun by the Moors during the eighth century. They were expelled in a relatively short period of time from the northern Meseta and from much of the Mediterranean coastlands. They clung to the southern Meseta much longer, in part because this arid tableland had little to tempt the Christian conquerors from the north. It was not until the thirteenth century that the area lying to the south of the Tagus, known henceforward as New Castile, was occupied.

The problems of the region in the nineteenth century stemmed in part from the circumstances of this conquest. The region was not settled, as was the northern region of Castile and Leon, by the Spanish nobles and their peasant followers. It was allocated in very large tracts to fighting religious orders, which were thought to be better suited to holding it against the Moors than the undisciplined feudal array. It remained in the hands of these non-resident and backward-looking orders until the land reform of the early nineteenth century (see p. 202), when it was taken over by the state and offered for sale. The purchasers were in part the aristocracy, who thus increased the area of their estates; in part the wealthy bourgeoisie. In either case, the new owners were also non-resident and no more progressive than the old.

The southern plateaux are lower than the plains of Old Castile and Leon; their winters are milder, but summers are hotter and drier. Rivers are incised in deep valleys; the water-table lies deep below the surface, and only in the west, in Estremadura, is rainfall really adequate for pasture and occasional cultivation. Population was sparse and there were few towns. Most people lived in large, nucleated settlements. There were villages in La Mancha whose inhabitants were numbered in thousands. They were mostly built of mud-bricks and roofed with tiles. The absence of wood in their construction reflected the treeless character of all except the mountains. The peasants of the southern Meseta lived by cultivating their smallholdings, which they leased from the estates, and providing wage labour on the latter. Their leases were short-term; they had no security of tenure and no protection from arbitrary increases in rent and other obligations. Nowhere were the rights of the peasant safeguarded by a body of law and custom such as was to be found in much of northern Europe. The peasant, constantly in debt, could lose possession of his holding, which might then be subdivided by its lord and leased on yet more onerous terms. Population was increasing and holdings became increasingly divided. The micro-holdings of New Castile were equalled in their extreme division and fragmentation only by those of Flanders or southern Poland. The sharecropping system, widespread in the north, was absent here, and the peasant alone bore the consequences of drought and crop failure.[109] The condition of the peasant of Estremadura and La Mancha had changed little, if at all, since the time of Don Quixote. Similar conditions of abject poverty were probably to be found only in Austrian Galicia. In both there was intense unrest, breaking out periodically in outbursts of extreme and vindictive violence. In the present century Estremadura and La

Mancha were dominated politically by the Socialist Union, in contrast with the more conservative north.[110]

Yet the arid steppes were interrupted by patches of more fertile and better watered soil. Irrigated agriculture was practised along the valley of the upper Tagus, where the Canal de Henares was built in the mid-century to bring water to the fields, and in the Guadiana Valley. Even though investment in agriculture was minimal during the century, there was a tendency for irrigation to be used more fully in the few areas where it was practicable in order to supply produce and fruit to Madrid and other urban markets. Viticulture, which before the nineteenth century had scarcely been known on the southern Meseta, also spread rapidly. Badajos became a wine-growing centre, and vines even spread into La Mancha as, during the *Phylloxera* epidemic, the price of wine rose. At the same time olive cultivation, well suited to the arid steppe, became very much more important.[111] But the basis of agriculture remained, as it had always been, wheat-growing. Wheat was sown wherever the soil was adequate and often enough where it was not. Usually wheat alternated with fallow; sometimes a crop could be taken only every ten or a dozen years. The tools used were primitive: a simple wooden *araire* and a wooden harrow. The wheat was threshed by means of an ox turning on the threshing floor and winnowed in the breeze. Even today, the view from the air in late summer shows the brown villages, set amid the red fields with the threshing floors showing up bright yellow. Fertiliser was not used in the nineteenth century, and animals were too few to yield much manure. The focus of wheat-growing lay on the better soils of Old Castile, but, with a rising population and tariff protection, it increased after about 1834 in marginal regions such as La Mancha. Yields, however, were very low, and after about 1860, when the area under vines and olives began to increase, that planted with wheat decreased. If average yields increased somewhat towards the end of the century that was principally because cultivation had been abandoned on the worst soils.

In Estremadura and La Mancha the peasant had to compete for land with the transhumant sheep. These provinces had provided the principal winter grazing for the sheep of the Mesta. This had largely belonged to the military orders, in whose interest it was to restrict any encroachment by peasant agriculture. In the later eighteenth century the privileges of the Mesta had come in for criticism, and during the following years were gradually eroded. First the common lands used by the sheep were enclosed and subdivided by the local communities. Then, in 1834–6, the Mesta ceased to exist. All that was abolished, however, was the privileged status of the owners of the flocks. The *transhumantes* continued to make their seasonal journeys between the north and the south of the plateau, even if they now encountered greater difficulties along their route.[112] Indeed, their numbers appear to have increased. Great uncertainty surrounds the size of the flocks, but it seems that the numbers making the journey between Estremadura and the summer grazing in the north may have amounted to more than three millions in the 1880s.[113] Nor does there appear to have been any significant decline before the First World War. Cattle remained few, and dairy cows yet fewer. Goats, more suited to the arid environment, were increasing, as were mules, used by the peasant for ploughing and for transport.

There is no simple explanation for the economic backwardness of New Castile. Both soil and climate made this a hard country for the peasant, but this alone cannot account for its condition. A different system of land-holding, a modest investment in better seed, tools and irrigation, and a willingness to experiment with new crops and new methods might have brought about profound changes. Any agricultural system in which the landowner is as much interested in his hunting rights as in the crops his land could support, and in which the peasant is denied the profit from any improvement he might make, is condemned to backwardness and poverty. Nor did the system of multiple rights in a single piece of land contribute to its efficient use. Much of the land in Estremadura was scrubland planted with cork oaks. Here 'the right of winter pasturage may...belong to one person, of cutting cork to another, of collecting the dead branches to a third, of pasturing swine to a fourth and of growing a catch crop every five or six years to a fifth. All these rights may be absolute and hereditary', and the ground landlord quite unable to cut through the legal tangle.[114]

Richard Ford's condemnation of the settlements in La Mancha stands for the whole of New Castile:

few and poverty stricken [with] neither art nor commerce, and...devoid alike of social attractions or interest; one would imagine, looking at the cloaked and listless loungers on the Plazas, that all the work which could be done was done...How great must be that mismanagement when these unemployed hands are not brought in contact with these uncultivated fields...The mud-built villages are the abodes of underfed, ill-clothed labourers; besides the want of water, fuel is so scarce that dry dung is substituted.[115]

Of Extremadura he wrote:

the moral and material obstacles to [prosperity] are painfully exemplified; ignorance, indolence, and insecurity combine with poverty and an absence of small proprietors; here is alike a want of fixed capital in the landlord as of circulating capital in the tenant...The half-employed population vegetates without manufactures or commerce, except in the bacon line, which is brisk, and the sole source of what little wealth there is; all traffic in other matters is merely passive, the smuggler excepted. Each family provides rudely for its limited wants; contented with the barest necessities...they dread all change, well knowing that generally it is for the worse...[116]

And no change of any significance took place before 1914, and little enough before the overthrow of the monarchy and the coming of the Republic in 1931.

Two hundred kilometres to the east and a thousand metres lower lay the *huertas* and *vegas* of the plain of Valencia. At most they occupied a shelf between the mountains and the sea, but it was a shelf containing alluvial soil and watered by countless rivers, most of them seasonal, which descended from the mountains which bordered the Meseta. Amongst the latter was the Jucar, which entered the plain of Valencia some 40 km to the south-west of the city. Its waters, accumulated behind a dam during the rainy winter months, served to irrigate about 13,000 hectares. This was achieved by means of a canal, the *Acequia Réal*, which took off from the Jucar near the mountains and followed the contours northwards towards Valencia. The water was fed from the right or eastern bank of the canal to the fields which lay between it and the Jucar itself (fig. 6.6).[117]

Within this area it supplied no less than 9,000 separate proprietors. The average holding contained about 1.4 hectares of irrigated land together with a variable area of *secano*, or non-irrigated. But the size of holdings varied greatly. In the mid-nineteenth century almost two-thirds were of less than 1 hectare but a small number were of over 20. Much of the land, up to two-thirds in the more northerly *communes*, was owned by the citizens of Valencia, and leased to the peasants who cultivated it. Owner-occupance of the irrigated lands was rare.[118] Holdings were already severely fragmented by about 1850 and became very much more so during the course of the century, largely as a result of division between heirs.

The mode of farming was an 'intensive polyculture', carried on in enclosed and individual plots. Before the nineteenth century field-crops, mainly wheat and vegetables, occupied almost half the cultivated area. Cereals increased greatly in importance during the first half of the nineteenth century; then declined in favour of viticulture and fruit orchards, maize and vegetables, while the olive spread over the less rewarding soils of the *secano*.[119] Farther to the south sugar was grown.[120] Without a steady supply of water the *huertas* would revert to scrub and the land would become almost valueless. For this reason, right of access to the water supply and its division between those who cultivated the soil were carefully managed. The water was owned collectively but was inseparable from the land on which it was used. Problems arose from the year-to-year variation in the supply, and these, together with the maintenance of the hydraulic works, were entrusted to a council which met with appropriate ceremony, and represented the interests of the cultivators.

'Zadruga' and 'Chiflik'

Both *zadruga* and *chiflik* were social institutions created for the purpose of owning and exploiting the land. Both were found during the nineteenth century only in the Balkans, south of the Sava–Danube line, and both decayed and eventually disappeared in the early twentieth century, by which time their social and economic purpose had been accomplished. Both were well known and well documented institutions. Their peculiar and generally non-western character attracted the attention of travellers, who described them at length. They thus appeared to have been more important and extensive than perhaps they were.[121] Together they occupied only a small fraction of the Balkan peninsula, and were embedded in a countryside made up mainly of small peasant-held tenements.

The *zadruga* was the older and at one time much the more widespread institution, and may have derived from the original land settlement. It was, in the words of Mosely, the first to make an intensive study of it, 'an extended family consisting of two or more small or biological families...owning land, livestock and tools in common and sharing the same livelihood'.[122] It was not, as Rebecca West claimed, 'the basis of the Slav social system',[123] but was nevertheless to be found during the nineteenth century over much of the Balkans. Foremost amongst those who described it during the time when it was a major method of social organisation was Arthur Evans, the distinguished archaeologist,[124] who travelled through Bosnia about 1875 and witnessed the rising against the Ottomans. The *zadruge* described by Evans lay in Bosnia (fig.

274 *An historical geography of Europe 1800–1914*

6.7 The drawing made about 1875 by Sir Arthur Evans of a *zadruga* in Bosnia

6.7), but in the late nineteenth century they were found most commonly in western Bulgaria, between the Stara Planina and the Danube and in the mountains around Sofia.[125]

The *zadruga* thus formed a tightly organised, extended family. Its elected headman, or *domakin*, managed it but was always subject to control by the adult members of the community.[126] Normally these were all interrelated, though occasionally an outsider was taken into their company. The *zadruga*, it has been said, was not an institution but a process.[127] It was created by the accretion of nuclear family groups; it expanded within the limits of its agricultural resources, and when pressure on these grew too great, it hived off members who formed either another *zadruga* or merged into the society of peasant smallholders. The size of the *zadruga* varied greatly from region to region and from period to period. In the Šumadija of Serbia, they were generally relatively small. In Bulgaria the great majority had from ten to 15 members in the late nineteenth century with large ones ranging up to 20. It was rare to find a *zadruga* with more than 30 members, and no less rare to find one with fewer than ten. They were largest and most numerous in the Balkan mountains and on the platform of northern Bulgaria; least so in the Marica Valley, where Ottoman influences were greater.[128] Cvijić even claimed that *zadruge* of as many as 50 members were to be found in Albania.[129] They were, in fact, 'like an amoeba, constantly growing, changing, and dividing but never standing still'.[130]

The origins of the institution are far from clear. Mosely appears to regard it as a response to the difficulties and hardships of land-clearance, 'the basic social instrument for the task of pioneering'.[131] It was also a community maintained for the security and defence of its members and its decline accompanied the end of Turkish rule in the Balkans. In the only part of Bulgaria where the *zadruga* still flourished in the 1880s, its survival was attributed to the 'emigration of the men to other districts for work during the summer. Their long periods of absence necessitate[d] stronger family ties.'[132] Other factors influencing the disappearance of the *zadruga* during the later nineteenth century were the decline of patriarchal authority and the rise of a market economy from which the peasant hoped to make profit for himself rather than for the larger community of which he was a member.

The *zadruga* was typically surrounded by its fields, organised, it would appear, chiefly in strips and cultivated with primitive wooden tools. The crops grown, usually on some form of rotation with fallow, were wheat and maize, with oats, barley and rye on higher ground. Custom might dictate a regular sequence of crops, but, since the community was isolated and self-contained, this could be – and certainly was – varied at will.[133]

The *chiflik*, by contrast, developed from the conditions of the Ottoman conquest and land occupation. It derived from the feudally held estate of the early years of Ottoman rule (see p. 216), but by the nineteenth century it had passed into the absolute possession of its Turkish master. To some extent the creation of *chifliks* was encouraged by the Ottoman authorities because they produced a surplus of bread-grains for export or for the supply of Constantinople and other cities. To this extent they can be related to that general model of estate formation and peasant repression in eastern Europe which we call the 'second serfdom'. This accords with the distribution of *chifliks* within the Balkans. According to Cvijić[134] they were to be found almost everywhere at the beginning of the nineteenth century, except in the Dalmatian karst and northern Serbia. But they were concentrated mainly in the Marica valley, in the neighbourhood of Novi Pazar and Skoplje, and across the south from Albania and the Bitola (Monastir) basin in the west, through Macedonia to Thrace, and in the plain

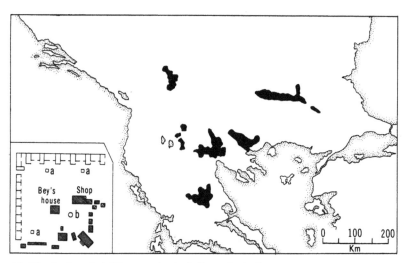

6.8 Distribution of *chiflik* estates in the Balkans. In the inset, 'a' marks the position of cooking hearths, 'b' of the threshing floor

of Thessaly (fig. 6.8). As a general rule they were to be found in areas of good soil, and they never occurred in truly mountainous areas.

The *chifliks* varied greatly in size. Some were large, running to a hundred or more hectares. In Thessaly an average *chiflik* could have been about 300 hectares and the largest ran to 4,000 hectares,[135] but McGowan found that, in Macedonia at least, most were from 50 to 100 hectares.[136] The land had mostly been occupied by Christian peasants before the Turkish conquest. It was then taken over by the Ottomans, and the status of the peasant depressed virtually to that of serfs. The *raya* was technically a free man, and could leave his land and settlement, but in reality he was unlikely to find a living elsewhere under Turkish rule. In fact many were 'forcibly tied to the spot by means of a perpetual and even hereditary debt which their landlord continues to fasten on them. This has practically reduced many of the peasant families to a state of serfdom.'[137] Labour was scarce, and the *chiflik* of little value to its owner without the labour to cultivate it.

The large *chifliks* were run like the great estates in Hungary or Spain. The workers lived in a compound, and worked in the fields in gangs under supervision. Travellers in the Balkans were unanimous in describing the large *chiflik* as essentially a village, roughly square in plan and enclosed by a wall, with the squalid accommodation of the *čifči* ranged along it. In front of it were cooking hearths for the workers, a well, a threshing floor, barns for storage, and dominating it all the tall, Turkish-style house of the bey, from which he could watch over the whole (fig. 6.8). In the small *chifliks*, however, it was not economic for the owner to work the land directly, since the costs of supervision relative to output would have been too high. The practice became general of allowing the *raya* to work the estate on a sharecropping basis. This appears to have been so throughout Macedonia, where *chifliks* were particularly numerous. After harvest a tithe of the crop was first taken for the state, and the rest divided equally between the *raya* and the owner.

The crops grown were chiefly those for which there was a market, preferably an export market. The large *chifliks* tended to be dominated by a single crop, cotton, maize or wheat. Cotton from Macedonia and tobacco from Thessaly were exported by sea to western Europe or overland to central Europe, but maize became the *chiflik* crop *par excellence*.[138]

The *chifliks* were the most important of the many causes of peasant discontent in the Balkan peninsula, and it might almost be said that the frequency of revolts was directly related to the number of Turkish-owned estates. As Ottoman rule was progressively thrown off, first in Serbia and southern Greece; then in Bulgaria, northern Greece, Macedonia and Thrace, the *chiflik* owners, especially if they were Turkish, fled, and the peasants who had worked on them took possession of the land, generally by purchase at a knockdown price.[139] The question of compensation to the former owners inevitably arose, and the treaties of peace included provision for some payment to be made. In the case of Serbia it was relatively small, as also were the *chifliks*. In the case of Bulgaria there was prolonged haggling over the value of the estates to their former owners before agreement was reached.[140] In Bosnia the *chiflik* owners were not Turks but Slavs whose ancestors had been converted to Islam. They did not flee, and the Austro-Hungarian authorities took no steps to remove them. Here the *chiflik* organisation endured until the First World War. A similar situation existed in the Plain of Thessaly. When this area was attached to Greece in 1881 it was dominated by Moslem landlords. Out of 658 villages no less than 466 were said in 1881 to have been *chifliks*. Their owners fled, but many of the estates were purchased by wealthy Greeks, many of them living in Constantinople and other Ottoman cities.[141] The system lasted till after the First World War, even though much of the land was bought by the peasants with, in some cases, help from the government and banks.

Sweden's farming frontier

The development of agriculture in Sweden bore in many ways a close similarity to that in Denmark. Both countries were in the early nineteenth century mainly agrarian, and their land was largely in the hands of peasant proprietors. In neither were great estates either extensive or an inhibiting factor in the development of a progressive peasant society. Lastly, land reform in both brought to an end the system of intermixed parcels in the open-fields and led to a more open and scattered pattern of rural settlement. The course of Swedish agrarian history differed, however, from the Danish in two important respects. It was constrained by a climate of far greater severity, which restricted the range of agricultural production and imposed a need for supplementary activities. Secondly, it operated against a background of almost unlimited space. At the beginning of the century, Norrland, making up almost two-thirds of the country, was virtually uninhabited. Sweden, together with Finland, was the only European country which in the nineteenth century can be said to have had the benefit of a frontier of settlement.

Sweden is made up of four separate and distinct regions. Two are low-lying and moderately fertile, the southern province of Skåne and the central, lake-studded lowlands. Between them lay the forested hills of Småland, and to

278 An historical geography of Europe 1800–1914

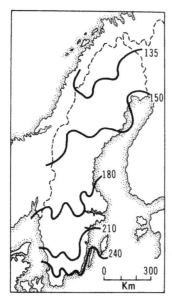

6.9 Length of the growing season (in days) in Sweden

the north of the lake region the vast, uninhabited waste of Norrland. This was Sweden's frontier region. The area of cropland, small in any case, was further restricted by the harshness of the climate. It occurred as a narrow belt of silty soil along the coast of the Gulf of Bothnia, a result of the geologically recent uplift of the land, and as even narrower strips along the lower courses of the rivers. Inland rose the rounded hills, made up of the hardest rock and covered, where covered at all, by a spread of glacial sand and gravel and by the lakes which occupied the depressions gouged by the ice.

The most southerly region of Sweden, the historic province of Skåne, resembled the Danish islands, and was until 1658 a part of the Danish kingdom. The Småland highlands had a more severe climate, and as one progressed northwards the summers became shorter, and the winters longer and more intense. The growing season for crops was reduced to about 200 days in the plains of central Sweden and to a great deal less than half the year over most of Norrland (fig. 6.9). Fluctuations in climate, which in more favoured lands might have led to a poor harvest, could be disastrous in a region as marginal as much of Sweden. This had restricted settlement in most of the country, where advances made in the course of a few mild seasons could be wiped out by one long, hard winter or cool, wet summer. In the later nineteenth century the climate became to a very small degree less severe. In the context of northern Sweden this could have made the difference between a moderate harvest and total crop failure. Settlement and agriculture were thus enabled to advance into parts of Norrland never previously occupied.

Those parts of Sweden, essentially Skåne and the central lowlands, which had been long settled had developed a system of small but nucleated villages and open-fields. In Skåne these were cultivated on a three-course system with fallow, as in Denmark. In the north, however, the climate precluded the growing of a winter-sown crop, and the land alternated between spring-sown grains and fallow. The enclosure movement gained momentum early, and owed much to the English example. At first it consisted of a grouping of strips and parcels of land – the *Storskifte*. A Swedish law of 1757 provided that this regrouping of strips could be carried out even if only one peasant in the community demanded it.[142] This was followed by the more radical *enskifte* and *lagaskifte*, processes in which the village was broken up, and its lands divided into one or more compact blocks for each farm.[143] The enclosure movement was diffused through south and central Sweden from three centres in which foreign influences were strong. These were Skåne, long under Danish influence, the Stockholm-Uppsala area, and the area from Göteborg northwards to Lake Vänern and the Norwegian border. The last areas to come under the influence of the enclosure movement were the hills of Småland and of southern and eastern Norrland.[144] By 1860 the work of enclosing had been terminated in the principal agricultural regions, and open-fields survived only in a few peripheral areas.[145] It was followed by a great improvement in agricultural practice and an increase in crop production. Steel tools replaced wooden ones; a greater use was made of fertiliser; rotations were improved, and legumes were grown increasingly and fed to dairy cattle.[146] Total cereal production increased from about 8 million quintals in 1830 to more than 26 on the eve of the First World War, and the area of arable land increased by some 50 per cent, largely at the expense of meadowland which was ploughed up.[147] The cultivation of potatoes increased fourfold within the same period, and even sugar-beet entered into the rotations on the larger farms.

Sweden was climatically better suited to producing legumes and other forage crops than cereals, and this made it practicable to increase the numbers of livestock, especially dairy cattle and pigs. Sweden, however, never achieved the importance in this respect of Denmark. A factor in Sweden's failure to capitalise on this obvious natural advantage was the smallness of peasant holdings, and the general shortage of agricultural capital.[148] Like Denmark, Sweden suffered through the middle and later years of the century from an abundance of rural labour to the point that it inhibited the adoption of labour-saving equipment.[149] Agriculture continued to employ a large part of the population – 72 per cent in 1870 – until late in the century. The rural proletariat which resulted led to the formation of a class of cottagers even more numerous and less privileged than in Denmark. Without the cultivation of the potato in their cottage gardens, it is difficult to see how they could have survived.[150]

In the absence of any significant degree of industrial expansion before the end of the century there were, apart from overseas migration (see p. 86), only two outlets for the expanding rural population: work in the forests and mines, and migration to the Swedish frontier. In much of Sweden agricultural labour was necessarily seasonal, and in winter part of the labour force had long been employed in the forests and mines. Some became part-time iron-workers or even participated in the management of blast furnaces and forges. In this they were greatly helped by the tendency for iron-ore mining, smelting and refining to be dispersed through the forests and hills of Värmland, Dalarna and

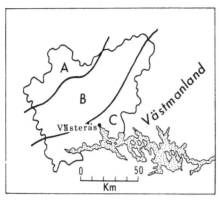

6.10 Farmer-ironworkers in Västmanland. The zones are (A) peasant miners, (B) peasant carters, (C) food production

Västmanland, the provinces bordering the central lowlands. Here, in addition to seasonal work in the mines, the peasants worked in the forests in winter, not only felling and transporting timber, but also making charcoal for the furnaces.[151]

Peasants also served as carters, transporting charcoal and ore to the furnaces, pig-iron to the refineries and bar-iron to the ports and consuming centres. A consequence was that the farmer-miners and farmer-charcoal-burners produced insufficient food for their needs and were in part dependent on larger and more productive farms to the south. A threefold division of functions emerged between peasant-miners, peasant-carters and food-producing estates (fig. 6.10).[152] The charcoal-iron industry continued to increase, and with it the opportunities for supplementary peasant employment, until 1885. Then the exhaustion of the scattered deposits of iron ore, combined with the development of other methods of iron-refining (see p. 341), led to a decline in this symbiotic relationship of farmer and ironworker.[153]

The second outlet for the rural population of central Sweden lay in the frontier region of Norrland. At the beginning of the century there was a narrow ribbon of settlements along the coast of the Gulf of Bothnia as far as the boundary with Finland. Not all were agricultural; some were associated with the development of mining and many with the lumber industry. In the course of the century settlement intensified and spread. It was not 'like a wave [moving] across a homogeneous landscape, but rather a frontier consisting of wedges along the river valleys and outlying spots on the shores of...lakes'.[154] The movement to colonise and develop the northern region of Sweden culminated towards the end of the nineteenth century. The majority of the settlements in Norrland were agricultural, small and largely self-sufficing. Finns were brought in to assist in the process of forest clearance, and they in turn introduced the practice of burn-beating, that is, of cutting and burning the forest and taking a few quick crops while the soil remained enriched with the ashes. Rye was the most important cereal, with oats for the horses, barley and potatoes, without which the peasants could not have lived on this northern frontier.

Agricultural regions

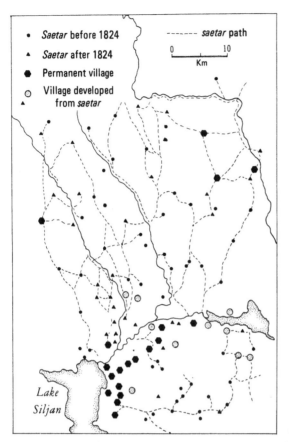

6.11 The pattern of villages and *saetars* in the Siljan region of Sweden

A kind of transhumance was developed in order to make the most of this marginal environment (see p. 245 and fig. 6.11). Settlements were made on the lowest and most sheltered land, but higher ground could often be used for grazing during the summer. Here the *saetar, fäbodar*, or summer settlements were established. The cattle were driven, sometimes over great distances, from the lowlands. The journey was occasionally made in stages, 'to the home seter in May [and] at midsummer... to the distant seter, where the cattle grazed for usually not more than two months'. There were many variations on this migration plan, and in some places a three-stage transhumance evolved to take advantage of small local resources.[155] This *saetar* development reached its peak about 1880, when pressure on the restricted agricultural resources was greatest. Thereafter *saetars* – first of all the most distant – began to be abandoned. By the early twentieth century the system was in decline, though transhumance continued to be practiced until the middle of the century.

After about 1880 the population engaged in agriculture, which had been increasing steadily since at least the early eighteenth century, began to decline. The more distant agricultural settlements in Norrland began to be deserted by their inhabitants. The uncertainty of the climate, the fact that the coarse grains which alone could be grown here were in diminishing demand, opportunities for work in the developing industries of the south, and above all the hardships of life on this climatic frontier were sufficient to explain a retreat which, in the last decade before the war, was beginning to accelerate.[156]

France

The conventional view of French agriculture during the nineteenth century is that of an unprogressive peasantry cultivating diminutive and fragmented holdings with outdated tools and equipment.[157] It has even been questioned whether France ever experienced an 'agricultural revolution' until late in the century. There was, indeed, an agrarian problem in the mid-nineteenth century, which led to acute rural unrest. 'No government since the Revolution', wrote Michelet, 'has preoccupied itself with agricultural interests.'[158] This unflattering view of French agriculture has been severely criticised in recent years, and there can now be no question but that an advanced agriculture was practised on some of the estates and large farms and that even on peasant farms improvements became significant in the second half of the century.[159] This increased willingness to innovate was not unrelated to the growing labour shortage.[160] After about 1860.French agriculture did adjust to the growing demand from the urban and industrial sectors.

A feature of the agricultural geography of France during the nineteenth century was its quite remarkable stability. The balance between arable and grazing, cereals, roots and fodder crops changed little during the century. Rye tended to give place to wheat (fig. 5.2); the area under potatoes increased, and that under vines fluctuated with the depredations of *Phylloxera*, but in percentage terms the change was small.[161]

Nor was there any more significant change in the regional pattern. Whatever parameters are used, the north and north-east of France stand out as an area of more progressive and productive agriculture. The land was largely worked by peasant proprietors or by leasehold tenants; over much of the area *métayage* was unknown. Crop-yields were far greater than in most other areas (figs. 5.17, 5.18), and progress was made in eliminating fallow far earlier than in other parts of France (fig. 5.14).[162] Although the value of agricultural production increased in all parts, the north maintained a clear lead over the rest of the country. To the north-east of a line drawn very roughly from Lower Normandy to Geneva, agricultural income per hectare was more than 25 per cent above the national average. In Brittany, the Central Massif, the south-west, part of the Alps and Corsica, it was more than 25 per cent below. All the evidence confirms that unprogressive agriculture was largely to be found to the south-west of that line, and progressive farming to the north-east.

In the following pages more detailed studies are presented of agriculture in four separate parts of France. The Beauce is chosen as an example of traditional open-field agriculture, slowly adapting itself to conditions of the late nineteenth century, and Normandy of the adoption of a specialised, market-oriented type of farming. The study of the Limousin illustrates a traditional pattern of

Agricultural regions 283

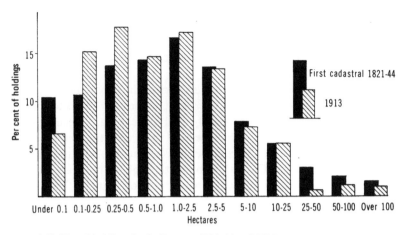

6.12 Size of holdings in the Beauce, 1821–44 and 1913

agriculture where little attempt was made to modernise or innovate, and that of Lower Alsace an agricultural system open and receptive to every innovation.

Beauce

The Marché de Beauce was the chief market in Paris for the sale of grain, showing by its name the source of the commodity which it handled. The Beauce lay some 80 to 100 kilometres to the south-west of Paris. The tertiary limestone here formed an almost level plateau, with a cover of loam to which it owed its fertility.[163] It was a dry region, and much of the rainfall sank into the limestone rock. Streams were few and wells deep. In the *bocage* region to the west, springs broke out at frequent intervals. There was no shortage of water there, and the scattered farmsteads could be supplied without difficulty. In the Beauce, on the other hand, settlements were clustered and 'looked like little islands of stone' amid the variegated sea of the open-fields. They commonly lay in hollows in the land, where water could be reached most easily. Around them 'not a single hill or tree broke up the plain as it faded into the distance, dipping over the horizon which was as firm and rounded as the sea'.[164]

It was near the southern margin of the Beauce, between Châteaudun and Orléans, that Zola located his novel of peasant life, *Terre*. The village of Rognes is imaginary, but the fields with their minutely divided strips, the farm buildings, the animals and, above all, the peasants themselves, were real. Zola spent much time in the Beauce, getting to know both the physical scene and its inhabitants, and the record which he presented, unstatistical and impressionistic though it was, can be taken as a faithful portrayal of the human geography of the Beauce at a time when French agriculture was beginning to adjust to the revolution in transport and to the growth of urban demand.[165]

The Beauce lay mainly in *dépt* Eure-et-Loir, in the fertile and intensively cultivated north of France. A very large proportion of the population, up to 60 per cent, was active in agriculture,[166] and yields per hectare were high, though not the highest in France. Wheat was the dominant crop, though it was grown in association with oats and barley. Fallow had been abandoned in much of Beauce by the time that Zola wrote in the 1880s, and lucerne, rape and colza were grown as catch crops, but sheep were still being grazed on the stubble after harvest.[167]

It is implicit in the pages of Zola that an agricultural revolution was quietly taking place. The seed was still sown broadcast as the sower walked the length of his strips, but the scythe was universally used for reaping. Threshing was still done with a jointed flail, but a rich peasant had recently 'set up a steam threshing-machine, hired from an engineer in Châteaudun, who took it up and down between Bonneval and Cloyes'. Running through the novel are indications of a creeping change. Jean, a *laboureur*, was spreading manure. The same rich farmer addressed him:

'Jean, why haven't you tried phosphates?'... The answer to good crops lay in manure and fertilizers. He tried everything and had just gone through this craze for manure.... He had made a series of experiments, using grasses, leaves, wine-dregs, rape and colza.... 'I've often had good results with phosphates.' Jean replied with a peasant's suspicion: 'people get cheated over them.' To which the rich peasant retorted in a manner which must have crushed the simple *laboureur*, 'Every market should have an expert chemist responsible for analysing these chemical fertilisers, for it's so difficult to get them unadulterated. That's where the future lies...but before the future comes we'll all be ruined. We must have the courage to suffer on behalf of others.'

With that Jean returned to his work and was swallowed up in the stench of manure. One can hardly doubt that this didactic incident derived from Zola's experiences in the Beauce.

A few pages later, when 'the winter ploughing was almost finished...Jean had taken his plough to his big field...He wanted to sow one end of the field with a Scottish variety of wheat, an experiment which had been advised by his former master', the rich peasant, 'who had even put a few bushels of seed at his disposal.' The good Scottish wheat would, of course, have been sown in October, not in March. But one can forgive Zola for a small error such as this. He shows convincingly that innovation came through the rich peasants and small farmers, who had sufficient education and experience to understand the significance of change and the resources to take risks for the sake of the future. Of all the varied characters presented in *Terre* Jean was surely the one most likely to follow the example of his betters.

On the practical side change was in the air, but on the institutional, the dead hand of tradition still rested as heavily as ever on the community of Rognes. Old Fouan had worked his holding during most of his life. He could do so no longer, and, as the law required, it was to be divided equally between his three children. 'I own nineteen *setiers* of land, or nine and a half hectares, as they say nowadays', he said. These were made up of 'the twelve *setiers* of arable land,...the six *setiers* of pasture [and] the meadows beside the river Aigre, where the hay was useless', not to mention a garden plot and a vineyard, for Rognes lay on the outermost edge of the wine-growing region of the Loire. The bargaining between the three parceners and between them and their elderly parents whom they in return had to help support, went on interminably. The partition took place amid the fields which were to be divided and in the presence of the surveyor. The latter had 'contacts with the big farms [which] led him to take progressive views, and he sometimes disagreed with clients who were small property-owners, saying that he was opposed to unlimited division of the land. The only reasonable thing was to come to an agreement and not to slice up a field as though it were a cake... If one of them would be satisfied with the arable land another could have the pasture – in short the three shares could be made equal and ownership would be decided by drawing lots.' The eldest demurred. 'Oh no, I want something of everything, some hay for the cow and the horse and some wheat and wine for myself.' Old Fouan agreed: 'From father to son land had always been handed down and divided up like this; as new acquisitions and marriages followed, the strips were rounded out again.' And so it was agreed, but, in peasant fashion, the dispute continued at a lower level. The surveyor and two of the participants 'wanted to divide the land into three strips running parallel to the Aigre valley, but one insisted that the strips should run at right angles to the valley, for, he alleged, the arable land became thinner towards the slope. In this way everyone would have an equal share of poor ground, whereas the other method would mean that the third strip would be poor throughout.' In the end the division was made, each parcener 'certain that no one could have something that the other two hadn't got', and as the surveyor laid out 'the dividing lines, they kept a sharp eye on him, as though they suspected him of trying to cheat and make one part a centimetre larger than the others'. And so the mingle-mangle of the open-fields was perpetuated and intensified, for within the limits of Zola's account there is no record that strips, once divided, were ever reunited. It was the old story of efficiency sacrificed on the altar of equality, understandable at an earlier stage in western history, but not in the second half of the nineteenth century, when population was growing, expectations rising and market pressures increasing.

The novel ends as the Prussian armies marched into France in 1870. It is doubtful whether the war had much influence on the village community, apart from bringing about a price rise and increasing temporarily the rewards of agriculture. Many small improvements were made possible by co-operatives and credit unions based in Chartres.[168] The use of phosphates, the cultivation of legumes, and the employment of the threshing machine and of other mechanical devices increased only very slowly; the peasants required a lot of convincing that they were worthwhile. Strips continued to be divided equitably if uneconomically, and small holdings grew ever more numerous, mainly at the expense of the larger. The rural economy stagnated through the late nineteenth century.[169] Wheat still dominated the farming system in 1914, and wheat yields remained good by French standards, though not unduly high. The number of cattle increased with the more widespread cultivation of legumes, though the milk and dairy produce were almost wholly for domestic or local consumption, and sheep became fewer, as the last of the fallow was ploughed up and tilled. It was a story of small increments of production, resulting from the slow diffusion of minor innovations, within a broadly traditional and unchanging farming structure.

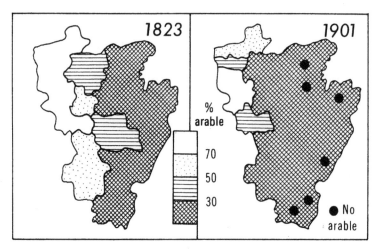

6.13 Diminution of arable and increase in the area of grazing in the Pays de Caux, 1823–1901

Normandy

Northwards the level plains of the Beauce merged into the more broken country of Normandy. The historic province extended from the Pays de Bray in the east westwards to the borders of Brittany. It spanned a wide range of geological structures and rock types. In consequence, its soils were varied, and with them the agricultural potential of the province. The eastern half of Normandy consists of a chalk plateau which reaches the Channel coast in cliffs comparable with those of Dover and Beachy Head. Inland, its treeless, wind-swept surface is covered, deeply in some places, with fertile *limon*. This is the Pays de Caux. Like the Beauce, it has few rivers, and springs are rare. The large nucleated villages were dependent on wells sunk through the chalk. On the eastern border of Normandy the River Béthune had cut through the chalk, exposing the underlying clays, and the Seine has cut deeply incised meanders across the plateau of Caux. South-west of the Seine the chalk continues for another 40 km, until it ends in an irregular scarp. A belt of clays, the Pays d'Auge, intercalated with beds of limestone, extends for another 20 km until the Jurassic limestones come to the surface. These form a region of low broken hills, bordered on the west by another region of clay, the Bessin, which rests against the ancient rocks of the Breton massif.

The striped or banded appearance of the geological map is matched by the pattern of soils. The chalk and limestone, with their variable covering of fine *limon*, provided a light, rich soil. It had long been under cultivation, producing wheat in rotation with coarse grains, roots and fallow.[170] Sheep were reared in large numbers, but cattle were of no greater importance in the early nineteenth century than they were in the Beauce.[171]

The clay land of the Pays de Bray, the Pays d'Auge and the Bessin all showed a different balance between arable and pastoral farming from that on the drier chalk and limestone. Soils were damp and heavy; relatively small areas were

Table 6.5. *Land use in Normandy (in hectares)*

	1830	1930
Plough land	340,000	123,000
Grazing	80,000	370,000
Unused	100,000	25,000

under arable crops, and pastoral activities were important, limited only by the restricted nature of the local markets. But cheese – the only way in which milk could be preserved – was made in quantity. The excellence of Pont l'Évêque, made in the Pays d'Auge, was widely recognised in the seventeenth century.[172] Nevertheless subsistence farming was practised here, as on the drier chalk and limestone plateaux, however unsuited the land may have been for arable. But very little wheat was grown on these heavy soils. The chief bread-grain was rye, with oats and buckwheat. The clay lands had once been heavily wooded, but by the early nineteenth century much of the area had been brought under some form of agricultural land use.

The systems by which the land was occupied and used reflected the contrasts in geology and soil. The areas of limestone and chalk were predominantly cultivated in open-fields, made up of intermixed strips. The peasants lived in nucleated villages, and tilled their fields according to a common cropping pattern. The clay lands, by contrast, were mostly enclosed. The *Intendant*'s report of 1695 commented on the small fields, enclosed by hedges, of the Pays d'Auge. Farther west the landscape had a wooded aspect, except near the coast, where the wind restricted tree growth. This was the *bocage normand*, which contemporary writers contrasted with the treeless, open-field *champagne* both of the limestone area around Caen and of the Pays de Caux.[173]

A change gradually came over the whole of Normandy in the course of the nineteenth century. Improvements in transport intensified local specialisations and led to the elimination of crops which had become uneconomic. Wine-growing, never of very great importance as far north as this, gradually disappeared. The cultivation of apples and pears, traditional in the *bocage normand*, increased in importance, and good cider replaced bad wine as the local drink. Flax and hemp had formerly been important on the damper, heavier soils throughout the region. But the domestic linen industry was in decline during the first half of the century and with it the cultivation of its principal raw material. The cultivation of woad disappeared, and colza, grown for the sake of its oil, was abandoned as other illuminants became available.

The most important change by far, however, was the increase in pastoral farming. In this way western, or Basse Normandie capitalised on its natural assets. The manufacture and export of cheese – Pont l'Évêque from the Pays d'Auge and Camembert and Port Salut from the Pays de Bray – were important early in the century. The coming of the railway in the 1850–60s allowed liquid milk to be moved efficiently and expeditiously to the cities, and Normandy became the source of dairy produce for the growing population of Paris. Overall the area under arable fell to less than a half in the course of a century and that

Table 6.6. *Land use in Basse Normandie (in hectares)*

	Campagne de Caen		Valleé d'Auge	
	1873	1910	1873	1910
Cereals	11,042	8,600	3,930	3,355
Artificial grasses	5,652	5,501	1,664	1,269
Meadow (*près*)	1,618	3,935	13,884	22,768
Other crops	9,548	8,203	2,907	2,907

under grazing increased more than fourfold (table 6.5).[174] The change intensified after about 1880, as grain prices continued to fall and transport to Paris, Rouen and other cities was improved.[175] Nor was it limited to the clay lands. On the dry soils of the Pays de Caux, fodder-crops, grown increasingly on the fallow, were used to feed dairy cattle. The spread of pastoral farming changed the landscape. By the beginning of the present century it had 'almost wholly displaced arable from the Pays de Bray and greatly reduced its extent on the chalk plateau'.[176] In Basse Normandie the area under cereals fell on both the clay lands of Auge and even the limestone of the Pays de Caen (table 6.6).[177] The somewhat unexpected diminution of the area under 'artificial grasses' was due to the reduction in the area under cereals, with which they were grown in rotation.

Changes in agriculture in Normandy during the nineteenth century were not limited to the replacement of cropland by permanent grass. The crops themselves underwent a change. The elimination of fallow left more space for root crops. The potato was introduced in the late eighteenth century, and met with strong resistance from the peasants. It was adopted only because it was found to crop well at times when cereals failed, and was thus of value in relieving famine.[178] Cultivation was then reduced by the potato blight, and it was not until the last third of the century that the acreage under potatoes began at last to grow significantly. It was really important, however, only where an open-field, three-course rotation continued to be practised.[179] In the Pays de Caux it was the sugar-beet rather than the potato which tended to replace the fallow. Normandy lay outside the principal sugar-beet growing area of France, and Sion attributed the importance of beet in Caux only to an agreement between influential landowners to support a sugar factory at Étrépagny.[180]

Throughout Normandy domestic crafts had formerly supplemented the income from agriculture. But there was an important difference both in their character and their fortunes between eastern or Upper Normandy and Western, or Lower. At the beginning of the century cotton-spinning and weaving were common throughout the villages of Caux. Spinning ended near the beginning of the century as mills, water-powered in the first instance, were established in ther narrow valleys which dropped from the plateau to the Seine. Domestic weaving lasted longer, and had not wholly disappeared from the villages of Caux before the twentieth century. It differed, however, from many branches of domestic industry in that it ceased at an early date to be *un métier d'hiver*, and became the full-time occupation of professional weavers. By contrast, the domestic crafts both of Picardy to the north-east and of Lower Normandy to the west remained predominantly the part-time supplements to farming. This was attributed by Sion, as far as Picardy was concerned, to the fact that the agricultural calendar, with its heavy emphasis on sugar-beet, was acutely unbalanced with a great deal of underemployed labour in winter.[181] This does not explain the persistence of craft industries in western Normandy, where, it might be supposed, pastoral farming would have provided sufficient year-round work. The reason probably lies in the growing density of population, at least until the mid-nineteenth century, in this region of indifferent or poor soils.[182] Spinning, weaving, lace-making and embroidery, tanning, even pottery were intimately bound up with agriculture,[183] and in consequence survived far longer than the domestic industries of Upper Normandy.

The Limousin

The north-western part of the Central Massif consists of a monotonous plateau which rises inconspicuously at its highest point almost to 1,000 m. It is built of pre-Cambrian rocks, widely intruded with granite and planated to form a rolling upland. This is the Limousin, corresponding closely with the *départements* of Creuse, Corrèze and Haute Vienne. It is a region of high rainfall and poor acid soil, and at its higher levels, in the Plateau of Millevaches, is partially covered with peat.[184] Though crossed by one of the main routes from Paris to Bordeaux, most of the Limousin was remote from trunk roads, and even at the end of the century there were parts in the east and south-east which lay 20 km from the nearest railway. Lack of access to markets was not the only reason for the poverty and backwardness of the region. Its poor agriculture and sparse population had failed to attract either good roads or railways.

Apart from Limoges, there were no large towns; all had a considerable agricultural component, and many were functionally little more than overgrown villages. There could thus be little urban demand even within the region to tempt the peasant into commercial production. Apart from the porcelain industry at Limoges there was little manufacturing industry to attract population away from the land. A few factories, most of them on the lower ground to the south-west, made bricks and tiles and small metal goods. Not even domestic craft industries, which relieved the poverty of so many rural areas, were to be found here on any significant scale.[185] The family spun and wove flax, hemp and wool, but almost entirely for its own consumption; rarely did it produce for the market.

The population, very heavily rural, increased slowly during the century, reached a peak about 1890, and thereafter declined sharply. There had been some degree of rural overpopulation, relieved by the seasonal migration of workers. The whole region supplied masons and construction workers to the cities of much of France, but most came from Creuse, the poorest, least urbanised and least developed of the *départements* of the Limousin. Seasonal migration reached its highest level under the Second Empire, when population was still growing and the demand for labour in urban construction was greatest. During the last decades of the century the practice of going to Paris, Lyons or other large cities

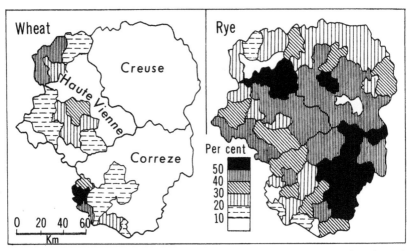

6.14 Distribution of the cultivation of wheat and rye in the Limousin. Figures indicate the percentage of the total cropland

for the summer half-year declined. Many of the migrants ceased to return for the winter, and population in the Limousin fell. The men of the Limousin migrated to the cities, as those of more northerly areas turned to their looms.[186]

There was throughout the century a remarkable stability within the Limousin. The pattern of agriculture changed little. There were very few estates or well-to-do farmers who might have pioneered new methods, and the peasantry as a whole was too poor to invest and too ignorant to experiment. It was almost as if technological advance had passed them by. But not quite: there were minute changes in land use, and after the 1870s, when fallowing had almost disappeared from northern France, the peasants began to plough and sow the fallow here also.[187] At the same time the potato was adopted widely and was often grown as a catch-crop on the fallow. In 1840 farm equipment was rudimentary in the extreme. The *araire* continued in use, and it was not until the end of the century that a modern plough with steel share was adopted. By tradition a two-course rotation was used, with field crops alternating with fallow. Only on the better soil, of which there was very little, did a three-field system prevail. Peasant holdings were small and severely fragmented. In one *commune* 79 per cent had less than 10 hectares, and only 3.8 more than 50; this in a region of low fertility where one might expect larger units to prevail.[188]

Less than 1 per cent of the whole region was said in 1851 to have soil of high quality, and overall only 38 per cent was cultivated, as against about 53.5 per cent for France as a whole. The rest was made up in not unequal proportions of rough grazing, forest and waste. Cropland was devoted almost entirely to cereals, but in the middle years of the century potatoes, grown as a field crop, were increasingly important. In 1852 they occupied about 6 per cent of the cultivated area. The dominant crop throughout the century was rye (fig. 6.14). Wheat was of minor, but slowly increasing importance, but was grown only on the lower ground to the west and south of the region. Rye retained its

Table 6.7. *Yield (in quintals per hectare), Limousin*

	1840	1852	1862	1872	1882
Wheat					
Corrèze	9.36	10.05	11.30	8.4	15.51
Creuse	8.25	10.41	11.19	9.0	17.85
Hte-Vienne	10.07	11.29	12.37	9.9	17.80
France	12.45	13.64	14.69	12.04	17.98
Rye					
Corrèze	9.02	10.51	11.84	12.0	15.55
Creuse	8.57	9.2	10.37	8.0	15.48
Hte-Vienne	10.38	11.29	11.99	9.2	16.25
France	10.79	11.51	12.91	10.86	16.38

overwhelming preponderance on the high ground which made up most of *dépt* Creuse. The only rival to rye was buckwheat, which continued into the present century to be grown on the worst soils and to provide a staple food for some of the poorest of the population.[189] Oats were also grown for the horses which were used for traction. Yields were low relative to the average for France, but improved considerably during the century (table 6.7).[190] By the late nineteenth century yields were approximating to the French average, especially in rye. It must, however, be remembered that by this date rye was being grown mainly on soils which were no better than those of the Limousin. The improvement in yields is attributed by Corbin[191] primarily to the use of better tools of cultivation, especially the plough, and by a small use of fertiliser.

Pastoralism had always been important in this region so abundant in rough grazing, but until the second half of the century animals were reared mainly for local use rather than the commercial market. Store cattle were sold to farms in surrounding areas, but lack of transport prevented the development of a specialised pastoral industry. Later in the century this began to change. Improvements in transport allowed direct marketing to the larger cities. The quality of the animals, especially of the beef cattle, improved. Good stock was introduced for breeding, and there was a small increase in the cultivation of green fodder. Of some significance also was the increase in the area of water-meadow. But these small advances touched only the margins of the region. The number of cattle increased by only 30 per cent between 1852 and 1882, the period when most of the improvements were made, and that of sheep showed a considerable decline. The biggest increase – of almost two-thirds – was in the number of pigs.[192]

The very slow agricultural advance of the Limousin was typical of that in other such areas: the Auvergne, parts of Brittany and the Alps. Here there was no 'agricultural revolution'; at most a series of very small increments was achieved on the farms of the more well-to-do. But the mass of the peasantry, semi-literate at best, cut off from the mainstream of French life, dominated by a patriarchal and conservative social system, continued at the beginning of the present century to farm as their ancestors had done a century before.[193]

Lower Alsace

Alsace, unlike the Limousin, lay open to the winds of progress. It bordered the Rhine and was crossed by one of the main routes from Paris into Germany. It contained one of the foremost cultural centres of France in the city of Strasbourg, and it had benefits of soil and climate which were denied to the Limousin. Physically Lower Alsace consists of an undulating plain, part of the Rhine Rift Valley, bordered on the west by the steep margin of the Vosges; on the east by the Ried or marshes which line the river. Sediments had long been accumulating over the valley floor. Close to the Rhine and its principal tributaries the recent deposit of alluvium had led to the formation of marshes, most of which were reclaimed during the nineteenth century and used for the intensive cultivation of vegetables. Rather higher above the river level were spreads of sand, improved in parts by the deposit of wind-blown loess during the Quaternary. Close to the Vosges the land became more hilly and soils more varied, but they were good wherever the loess was piled up against the hills. The several types of soil formed belts lying roughly from north to south, parallel with the Rhine, whose floods did so much to create them.

Wherever the loess lay deep there were rich soils, easily tilled, densely settled and intensively used, but where it had been worn away, or had never been laid down, the sterile sands and gravels came to the surface. Little attempt was made to cultivate these areas, and throughout the nineteenth century most remained under forest. The Forest of Haguenau, lying to the north of Strasbourg, occupied the most extensive of these sandy deposits, and remained impervious to human settlement. The loess-covered terraces, the forested sands and the marshes and meadows along the Rhine and its tributaries shared between them the Plain of Alsace in the early nineteenth century. There was little change during the next hundred years. The regulation of the Rhine was followed by the building of levées to protect the low-lying areas from the floods which had previously been severe. These lands were in part used for grazing, in part developed for intensive vegetable growing.

Most of the agriculture of Lower Alsace was on the loess-covered terraces. It was in general a traditional three-course, open-field farming. The principal winter crop was wheat on the best soil, but rye was important wherever the loess was thin and the underlying sands showed through. Spelt was grown in the north, beyond the Haguenau Forest, but declined and disappeared during the nineteenth century.[194] The spring-sown grains were oats, grown mainly on the lighter soils of the Vosges foothills, and barley, found chiefly to the south of Strasbourg. The traditional system began to break down earlier in Alsace than in most parts of France. There were places where fallowing had been abandoned by the mid-eighteenth century, and many where it had disappeared a century later. In its place a variety of catch-crops was introduced. Foremost amongst them was the potato, which before 1800 had become a staple food of the poor.[195] Its rapid spread, it was said, marked the end of subsistence crises in Alsace. Green fodder and roots were also grown, but cattle-rearing either for milk or meat never achieved any great importance. More important were specialised crops, notably hops and tobacco. Hop cultivation was introduced from Bohemia at the

Table 6.8. *Wheat yields for Alsace (quintals per hectare): averages for four cantons in each group*

	1814	1852
Loess area	18	23.25
Sands area	14.25	17.75

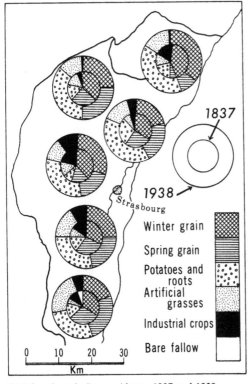

6.15 Land use in Lower Alsace, 1837 and 1938

beginning of the century. At first hops were grown only near Haguenau, but in the second half of the century it spread over much of the northern part of Lower Alsace. It supplied the local brewing industry and was also exported to much of north-western Europe. Tobacco and sugar-beet were introduced later. Tobacco became the chief industrial crop to the south of Strasbourg, and sugar-beet on the good, loess soils to the west, while in a broad belt along the western margin of the plain viticulture and fruit-growing developed (fig. 6.15).

Alsace was noteworthy for its intensive polyculture. It was able, partly on account of the richness and variety of its physical endowment, to accommodate specialised and industrial crops without abandoning its traditional cereal cultivation. This was facilitated by the early abandonment of fallow, the large urban market in the province, and, above all the progressive outlook of its

Table 6.9. *Size of holdings, Alsace* (*in hectares*)

	0.5–2.0	2–5	5–10	10–20	Over 20	Total
1860	c. 33,000	24,000	8,600	3,100	600	c. 69,300
1882	c. 30,000	25,200	11,600		560	c. 67,360
1907	27,500	25,400	12,200		330	65,430

peasantry. Crop-yields, at least on the loess, were amongst the highest in France (table 6.8).[196] On the other hand, farm holdings were on average very small and excessively fragmented, and little progress was made during the century in consolidating them. In 1859 the average size of a parcel of arable land was only 0.12 hectare, and even smaller in the vine-growing area.[197] In the mid-century, almost half the farm holdings were of less than 2 hectares and overall more than 80 per cent had less than 5 (table 6.9).[198]

Rural population was increasing until the middle years of the century, with the most rapid growth in areas such as the Ried where land reclamation was active. At this time the density of rural population was so high that there was a short-lived experiment with rural craft industries. Flax and hemp working had been traditional in the villages, though never practised on a commercial scale. For a decade or two after about 1840 these were expanded into professional occupations. Shoemaking and the production of iron hardware also developed in small rural workshops. But the crisis passed; rural population ceased to increase after about 1860, and there was an actual decline in the number of micro-holdings.

Despite its diversity, the agriculture of Alsace never developed animal husbandry on a significant scale. It was always subordinate and small-scale, with a resulting shortage of manure. Sheep had been numerous as long as there was grazing on the fallow, but when the traditional system was replaced with more complex rotations, fodder crops were of little importance. In the prosperous villages of the loess region farm animals amounted to little more than were necessary for draught purposes. Only in the marshes along the Rhine did the area of grazing increase, and here a small-scale transhumance developed, with stock moving between summer pasture in the Vosges and winter near the river.

Alsace was in the main a region of large villages, each surrounded by its open-fields, and such it has remained. The farms, consisting generally of a courtyard, with house, barn and stables arranged around it, lined the streets. Most were timber-framed buildings of greater size and sophistication than were to be found in much of France. They were very far removed from the simple, stone-built cabins of the Limousin peasant. The question must arise, whether the superiority of agriculture and of the quality of life in Lower Alsace was due entirely to the natural endowment of the region itself. Without it, there could have been no development of diversified agriculture such as actually occurred. But the resources were probably managed better here than in, for example, the Beauce. It is commonly assumed that good agricultural practice was diffused downwards from the estate farmer and rich peasant to the broad mass of the peasantry. This seems to have been what happened in the Beauce and probably also Normandy. The absence of such a prosperous class has been held responsible for the lack of progress in the Limousin. In Alsace, it is claimed,[199] no such lead was given. Neither estate owner nor burgess pioneered in agricultural development. The 'revolution' was the work of the peasant himself, more farsighted, more ingenious, more aware of the market than most others. Juillard has suggested that the settlement here of colonies of industrious Mennonites served to leaven the peasant mass. This may well have been so; but no less important must have been the proximity of urban settlements and the prospect of profit from urban markets. Alsace, furthermore, was a region of France where the work of the Revolution was not undone at the Restoration. The peasants who got possession of their land in the one, succeeded in keeping it through the other. This became a region of freeholders, lacking both estates and sharecropping tenants, one in which a spirit of enterprise was more developed than in most other regions of France.

Conclusion

These examples of agricultural practice and of its development during the nineteenth century have been chosen to represent the great range in farming in Europe during this period. It is impossible to generalise, even for a country or province, least of all for the continent. Variations in cropping patterns and farming techniques, in crop-yields and farm stock were apparent from one *commune* to the next. In some the character of farming changed radically during the century; in others a traveller, could he but come back after the passage of a lifetime, would have been hard put to it to discover any significant alteration.

Nevertheless, one can delimit a number of agricultural regions on the basis of crops and agricultural systems (fig. 6.16). They changed little during the century. Most extensive at its beginning was the region of open-field, three-course agriculture, with wheat-growing dominant in the west, rye in the east. It contracted in the west as, particularly in France, the Low Countries and Denmark, fallowing was gradually abandoned and more complex rotations were introduced. In several parts of the region wheat replaced rye as the dominant cereal. In Mediterranean Europe the two-course system, with wheat alternating with fallow, survived little changed throughout the century.

In some areas the three-course system with its open-fields had already broken down when the century began. In others it had never existed. Here there was a patchwork of small fields, organised and cultivated in various ways. Generally they were under crops for a year or two, and were then sown with grass for a period of several years before again being ploughed and sown. There was an almost infinite number of variations on this theme, deriving from local custom and necessity and the physical constraints of the land itself. Such a system implied an intimate combination of arable with animal farming, and is often called 'mixed' agriculture. The tendency was for such a system to become more widespread during the century. By the end of the period it prevailed throughout the Low Countries, north-western Germany and Denmark. It was replacing the open-fields in southern Sweden and in much of 'Atlantic' Europe, and was common in the hilly country of central Europe. The transition from a pattern

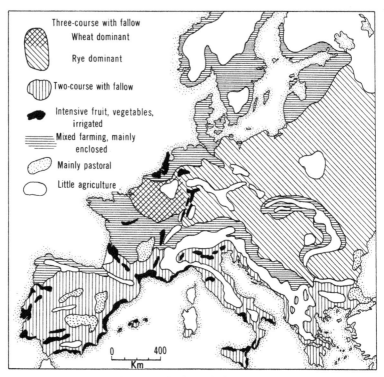

6.16 Agricultural regions in Europe in the mid-nineteenth century

of open-fields to one of enclosed, with no particular system of cropping, was speeded by the coming of the railway and the opportunity which it offered for the swift transport of perishable foodstuffs to the towns.

At the opposite end of the agricultural spectrum from both open-field and mixed farming was the predominantly pastoral farming in the hilly country and over rough terrain. Crop-farming would have been practicable only on the bottomlands, and much of the area had little agricultural value. This category included the hill grazing of Britain and the Dalmatian Karst, and much of the hill country which stretched from the Central Massif of France to the hills of Bohemia, Romania and the Balkans. In general, the frontier of settlement and land use withdrew in such areas during the century, as marginal land was abandoned. In Sweden and other areas near the climatic margin such land was often used only seasonally in a system of transhumance.

A final category of agricultural land consists of land which was intensively cultivated either as vineyard, orchard or market garden, or as irrigated cropland. Land values were high, labour was intensive, and in most instances the capital invested was considerable. Such lands formed at most very small units. They tended to increase in number and area during the century with the extension of viticulture and the cultivation of green vegetables for the urban markets.

Lastly, there was land which could not normally be used agriculturally. It

Agricultural regions

embraced not only the Norwegian Fjeld and Swedish Norrland, but also the forests of eastern Europe and the high mountains of the Alpine system.

Fig. 6.16 is an attempt to represent these regions cartographically. The map should be taken to represent conditions in the later years of the nineteenth century. Small but significant changes would have occurred by the first decade of the twentieth. The map shows the system of agriculture rather than the intensity of cultivation. The level of generalisation is high, and boundaries between agricultural regions are arbitrary. A detailed investigation at the local level would unquestionably show that within each of these over-simplistic regions there was indeed a complex agricultural pattern. The map is intended to show the dominant land-use system, not the exceptions.

7
Manufacturing in the nineteenth century

When the Napoleonic Wars came to an end Great Britain was already an industrial power. A century of innovation had transformed the technology of its textile and metal industries. The steam-engine was being used increasingly to power factories and to operate mines. The greater part, as much as four-fifths, of the world's coal production was in Great Britain, and a smaller fraction, perhaps a half, of the world's iron was smelted and refined there. Most of Europe's cotton spindles were to be found in Lancashire, gathered in large, mechanically powered factories. Everywhere, the traditional, craft-based industries were in retreat.

Little of this industrial technology had yet reached continental Europe. In Great Britain there were laws against the export of industrial equipment and the emigration of skilled craftsmen, and these remained in force until 1843. The law was nevertheless broken on a considerable scale, and in one way and another knowledge of most of the new industrial processes had filtered into continental Europe well before 1815. The latter was, however, unprepared for it. It proved difficult to transplant processes which had been developed empirically in Great Britain into a different environment. Resistance to innovation, combined with the lack of a scientific attitude to industrial processes, made it difficult for the continental entrepreneur to emulate his British counterpart.[1]

This was abundantly demonstrated when attempts began to be made to use coke for smelting iron. Most failed abysmally, and when at last success was achieved in Upper Silesia it was only with the help of workers who had already learned their skills in Great Britain. There is no known instance of the actual building of a steam-engine outside Great Britain before 1815, and up to 1789 only nine Boulton and Watt engines had been ordered by foreign customers. By 1815 the cumulative total had risen to 53,[2] only a fraction of the number employed in the tin mines of Cornwall alone at that time.

Evidence given to a Parliamentary Selection Committee in 1825 showed the kind of difficulty faced by those continental manufacturers who sought to adopt British technology.[3] It was not difficult for them to secure drawings of the most advanced machines. A witness asserted that Cockerill at Liège had obtained such plans from 'a person who was in England for some time and who has rolls of drawings of all the machinery in this country'. Implementing these plans was more difficult. The British laws against the export of machinery had, it was said, 'a tendency to force the French to become machine-makers themselves'. Henry Maudslay, who would have been well aware of the situation, nevertheless added that at Seraing, one of the most advanced centres of manufacture, 'they have not the facility of making tools for themselves'. Lack of machine tools was clearly one of the biggest obstacles to industrial progress in continental Europe.

It must not, however, be assumed that the demand for the new British industrial technology was overwhelming in continental Europe. In some branches of production there was little point in rapidly expanding the output of goods for which there was little effective demand. In others there was vigorous opposition to the new methods from practitioners of the older technology, and in some branches of industry – iron-smelting and refining, for example – it was at first difficult to achieve with new methods and materials the same quality that had been possible with the old. These factors provide a partial explanation of the slow spread of industrial technology in continental Europe. There were other reasons, including the scarcity of investment capital, the slow growth of mass markets, and the long survival of pre-industrial attitudes to society and economic progress.

One cannot, however, generalise for continental Europe as a whole. There were areas, notably Belgium and Switzerland, where British technology was adopted at an early date and where a rapid expansion of industrial production was achieved. There were others in which manufacturing was still carried on in the early twentieth century on a domestic basis and with much the same technology as in the eighteenth.

The diffusion of the new technology was, furthermore, severely hindered by the Revolutionary and Napoleonic Wars.[4] Both commerce and the flow of ideas were interrupted. For a period continental Europe was blockaded and the import of such goods as cast-steel prohibited. Attempts by continental entrepreneurs to fill the gaps thus left received considerable encouragement from the Napoleonic government, but their achievements were minimal.[5] Rewards were offered for the perfection of a method of making cast-steel comparable with that yielded in Great Britain by the Huntsman process, but little progress was made. By 1815 continental Europe was in all probability farther behind Great Britain in industrial achievement than it had been in 1789.[6]

The diffusion of scientific and technical knowledge became easier after 1815 (see p. 5) since social and institutional barriers to innovation had been reduced, if not entirely removed, in the course of the revolutionary changes of the previous quarter of a century. Napoleon had swept away 'a whole deadwood of time-honoured institutions... which had been hampering economic progress'.[7] Nevertheless, gilds with their restrictive practices continued to oppose industrial advance until far into the nineteenth century. It was not until 1868, for example, that the right of certain Rhineland cities to compel all passing ships to discharge their cargoes and to reload (*Umschlagsrecht*) onto different vessels was at last abrogated.[8]

More was needed, however, than the removal of obstacles to the spread of industrial technology. A market had to be created for the products of industry, and capital had to be raised for the building of factories. A commercial infrastructure, including roads and other means of transport and communication, had to be created, and a banking and financial system established. These tasks, which had to some extent been achieved in Great Britain a century earlier, were accomplished only very slowly in continental Europe during the nineteenth century.

There was no local mass market in eighteenth-century Britain ready and waiting for the products of industry as soon as industrial innovations permitted them to be made in abundance.[9] As Hobsbawm has pointed out, a mechanised industry was able to produce far more goods than any local market could absorb. Great Britain, however, had an empire into which to discharge this marginal production. One is, for example, constantly surprised at the volume of eighteenth-century British goods – china and metalware mostly, because these are more likely to have survived – in modern American museums. Colonies and trading bases in Africa and Asia were no less important and together absorbed a large part of Britain's industrial production. The countries of continental Europe had no such advantage. Dutch and French empires provided markets for some products of their homelands though never on the scale offered by Britain's colonial dependencies. At the same time their domestic markets were too small to support any great industrial expansion. The eighteenth-century solution to this problem had been to develop the manufacture of exotic goods for which there could only be a market among the rich and fashionable. The making of silks and porcelain are cases in point. It is significant that after the first attempt at iron-working at Le Creusot had proved unsuccessful and had been abandoned, the buildings were adapted to the production of jewellery.[10] By contrast, when, some 40 years later, the Cockerills found vacant space in the old archiepiscopal palace at Seraing, which they were using for their ironworks, they set up the spinning of coarse linen thread.

During the first half of the nineteenth century a mass market was slowly evolving for the simpler and cheaper forms of consumer goods. Foremost amongst these were textiles and tools and equipment made of metal. Arthur Young had remarked, when he saw the barefooted peasants of Aquitaine, that 'a large consumption among the poor [was] of more consequence than among the rich',[11] because their aggregate demand was far greater. The purchase of clothing and other consumer goods has always varied inversely with the price of food, and any increase in the latter was accompanied by a decline in the demand for the former.[12] During the first half of the nineteenth century the purchase of foodstuffs began to absorb somewhat less of peasant and working-class incomes.[13] The half-century following Waterloo was for much of Europe one of unaccustomed peace. After the shortages of 1816–17, food remained relatively cheap and abundant until scarcities reappeared after the poor harvests of 1846 and 1847. During the ensuing crisis the market for products of craft industries, especially those of Flanders, collapsed, with disastrous consequences for the peasants.[14] But the rest of the century was, in general, a period of increasing effective demand for consumer goods. The railway, with its faster transport and lower freight rates, widened local markets into national ones and put a premium on large-scale manufacture. By this time the marginal demand created by a captive overseas market was no longer necessary, and Italy, Spain and, in some degree, the countries of eastern and south-eastern Europe were able to expand industrial production on the basis of their internal markets. Poland, meanwhile, profited from the Russian market. The growth of markets was nevertheless slow. There remained until late in the century a vicious circle 'in which the narrowness of urban markets...discouraged agricultural improvements, and the sluggishness of rural demand slowed down industrial growth'.[15] A small increment in urban demand, from whatever cause, was enough to break this cycle, and the growth later in the century of large administrative and commercial centres such as Antwerp, Cologne, Paris and Berlin, was an important factor in stimulating demand in all branches of production. It has been argued that the rural and agricultural market in the early phases of the Industrial Revolution extended far beyond the demand for clothing and footwear; that metal goods were increasingly important, and that without this market the demand for ironware would have been insufficient to sustain a growing iron industry.[16] Unquestionably, better farm equipment called for more iron in its manufacture, but the volume of metal used must have remained small, and it is doubtful whether its role could have been decisive.[17] The general improvement in rural living standards did however, contribute to the growing demand for metal goods of all kinds. The Polish peasant, Jan Slomka, described how cast-iron stoves and other domestic hardware began to appear in the cottages of southern Poland in the middle third of the century.[18] If they were at all widespread they must presuppose the growth of a not inconsiderable iron industry.

An extension of this argument claims that capital in the form of tools and equipment displaced labour from agriculture, thus freeing it for work in the factory (see p. 231). It is doubtful whether the new agriculture was any less labour-intensive than the old, though it did increase yields per unit area. Migration from the countryside was due primarily to high birth-rates and rising rural population (see p. 79). Nor can it be shown that agriculture was a significant source of industrial capital. The profits from successful farming tended to go back either into the land or into various forms of conspicuous consumption.[19]

Capital was scarce everywhere in Europe during the early phases of industrialisation, and from the 1840s onwards the demands of the railway companies gravely reduced funds available for other forms of industrial development. The most important source of capital was commerce. Repeatedly we find close links between shipping and trade on the one hand, and industrial development on the other. This was especially the case in the Rhineland, where not a few magnates in the iron and steel industry got their start in Rhine shipping.[20] In this respect Great Britain had a unique advantage. Her extensive empire and large volume of trade provided not only capital for industrial development but also a market for a significant part of its output.

In every region of Europe in which modern industry was established, progress was generally made on a broad front, but there may have been – indeed, there generally was – a 'leading sector', in which innovations were first made and mass production first developed.[21] It had to be a branch of the consumer-goods industries, because it was only here that a large and growing demand could be found. In Great Britain cotton textiles formed the leading sector, and the textile industries in general constituted the leading sector in much of western and central Europe. It could also have been the leather and footwear industries, but in these it was not possible to achieve a large expansion in the supply of the essential raw material, and the industry, at least in the early days of industrialisation, did not lend itself to mechanisation.[22] The food industries could, and sometimes did, constitute a leading sector. Flour-milling, it has been claimed, served this role in Hungary. But nowhere, except perhaps in Sweden, did metal goods lead the movement towards industrial growth. The significance of a leading sector lay, in part at least, in the spin-off effect which it had on other branches of the economy, such as building, transport and engineering and the food and consumer-goods industries.

The textile industries had a great advantage as leader in the process of industrialisation. There was a large but generally hidden demand for their products. Their manufacture presented fewer technical problems and called for a smaller capital investment than most others, and the basic skills in spinning, weaving and cloth-finishing were already widespread.[23] During the last third of the century growth slackened in the textile industries, and in some branches production actually declined. At the same time, newer consumer-goods industries developed to satisfy new patterns of demand, in paper and printing, china and glass, dye-stuffs and paint. In spite of this, the consumer-goods industries tended to lose their position of primacy and the production of capital goods gained in importance as the century progressed. Foremost amongst the latter were smelting and refining of iron, mechanical engineering, and the chemical and non-ferrous metal industries. To these should be added building construction, which in many European cities came to constitute the largest source of employment and the biggest capital investment. Capital-goods industries were clearly important from the earliest phases of industrial growth; without them steam-engines for factories and mines could not have been built. But steam-engines did not become really numerous before the 1840s, and it was not until railway-building gained momentum at about the same time that iron-working came into its own.

Innovations in the iron and steel industries had come early in the Industrial Revolution in Great Britain. But their diffusion was very slow, largely because there was no pressing need for them. The function of innovation was primarily to relieve shortages or to remove a 'bottle-neck' in the process of production. It is symptomatic that coke-smelting was not adopted in the Ruhr – later to become the foremost European iron and steel-working centre – until 1849. The chronology of industrial innovation varied greatly from one country to another, but as a general rule the sharp expansion of the iron, steel and engineering industries – the leading branches of the capital good industries – came a generation or so later than that of textiles, the foremost branch of consumer-goods production.

Technical progress during the 'first' Industrial Revolution was made on a comparatively narrow front. Very few branches of industry were involved; in the first instance only the textile industries, and then primarily cottons. The metal industries and engineering followed. But little advance was made in a broad range of consumer goods such as the production of footwear and leather goods,

Manufacturing

glass and ceramics, wood-working and food-processing. If technology did not advance rapidly in these branches of manufacturing, there was nevertheless significant change in the organisation of most of them. The division of labour, by which an individual workman performed only a single, narrow task, as in Adam Smith's pin factory, became very much more widespread, and the workshop developed into a factory, which may have used mechanical power, but certainly employed a division of labour.

This chapter examines first the transition – slow and very local, from workshop to factory, from the stage of proto-industry to that of fully fledged modern manufacturing. It then takes up the main branches of industrial production. First to be examined are the textile industries, the traditional textiles, woollens and linen, and then the newer fabrics, cottons and silks. Discussion then turns to the iron and steel industries, in which technological change was more significant than in any other branch of manufacturing except chemicals. Lastly there is a short discussion of the newer industries of the later years of the nineteenth century, engineering, chemicals and the vast range of the food-processing and consumer-goods industries, which, at the beginning of the present century, were increasing almost daily in range and complexity.

Workshop and factory

One tends to assume that a revolutionary change must have been accomplished quickly and suddenly.[24] The European Industrial Revolution was, however, a long, slow process in which traditional methods of production yielded very gradually to mechanical methods of mass production. The increase of gross national product, the most effective measure of economic progress, was slow everywhere. Even in Germany, where it was most rapid, it never exceeded 2 per cent a year until after 1880. In France growth was a great deal slower.[25]

Proto-industry

The industrial development of Europe was preceded by a thinly scattered form of manufacturing which has come to be called 'proto-industry'. Its characteristics were its widespread nature, the participation of the peasantry in handicraft production, and the sale of the products through an intermediate merchant to distant markets. In its simplest form it was often only a part-time occupation, and was sometimes carried on only in winter. Units of production were invariably small, and mechanical energy was restricted to that provided by a water-wheel. Manufacturing called for little capital and was thus within the limits of a peasant society. Working capital was commonly provided by the entrepreneur who furnished the raw material and collected the product and sometimes processed it further before marketing it.[26] There could be no standardisation of goods, though sizes and qualities were usually more or less uniform within each proto-industrial region.

The nature of proto-industrial manufacture is well known. It is the *Hausindustrie* of German writers, and 'domestic industry' of British. Only its name is a recent innovation. Domestic crafts were to be found almost everywhere in pre-industrial Europe, but not all pre-industrial crafts constituted a proto-industry. Many of them served only the needs of their immediate neighbours, as did the blacksmith and the tanner. The essence of proto-industry was that it had a regional basis, but supplied a larger and usually distant market through the medium of the merchant.[27]

Its existence raises two fundamental questions: how far did it contribute to population increase, and how far was it a basis on which factory industry could be built? The traditional argument is that the practice of a domestic industry allowed a peasant family to subsist on a very small holding; that it encouraged earlier marriage than might otherwise have been the case and thus larger completed families. 'The regions of proto-industrialization', wrote Hans Medick, 'were also those where demographic growth was most rapid.'[28] There can be little doubt but that this actually happened in some areas, notably Flanders, where Mendels made his initial study. It can also be seen in regions as far apart as Brittany and Bohemia, central Sweden and Saxony. But there were also areas of rapidly increasing population where there was no significant proto-industry, amongst them eastern Germany, southern Italy and parts of Spain. Clearly there were other factors in rapid population increase, amongst them the practice of partible inheritance and the availability of potential farmland. Conversely there were areas with a prominent proto-industry, notably northern and north-western France, where there was only a moderate increase in population. Clearly the relationship between proto-industry and population growth is far from clear-cut.

Nor can it be said with any degree of certainty that regions in which proto-industry was strongly developed had a marked advantage in the development of factory manufactures. Proto-industry was largely engaged in the spinning and weaving branches. To a lesser extent it produced light metal goods. The factory production of cotton, woollen and linen fabrics did, in fact, develop in some regions where the corresponding proto-industry had previously been important. The Münsterland, the Lille district of northern France, the lower Rhineland and the Zwickau-Chemnitz-Plauen area of Saxony are cases in point. But as many examples can be cited where a significant textile industry developed without the advantage of a vigorous proto-industry – northern Italy, Catalonia and Łódź amongst them. Conversely many areas with a strongly developed proto-industry failed to develop factory manufactures. They gradually 'de-industrialised' as the century progressed. Amongst them were much of northern France from Brittany to Picardy, with the exception only of the Seine valley, and also the Flanders Plain, where the linen industry almost disappeared.

The connection between the iron and steel industry on the one hand and proto-industrial metal-working on the other is even more tenuous. The modern industry was, almost without exception, founded on virgin sites, and its most significant manifestation, the Ruhr, was almost wholly devoid of metal-working before the early nineteenth century. The proto-industrial development had been to the south, in the Sauerland and Siegerland, and here the modern industry never really established itself.

Two further aspects of proto-industry deserve consideration. It developed most vigorously, it has been said, in mountainous and other harsh environments.

In a few instances it did, but its scale in such regions was small and its significance marginal. It was almost invariably a winter occupation, and the goods produced, wood-carvings and clocks, for example, did not usually lend themselves to factory manufacture. Some proto-industrial crafts, lace-making, for example, were practised almost exclusively by women, and these rarely moved to the factory.

Proto-industry decayed slowly during the century. In some areas it was replaced by factory industry, but more often it dwindled away as younger men migrated and others grew old on the job and left no successors. The last practitioners of a decaying proto-industry were not infrequently ageing women who made lace or turned the handle of a stocking-knitter. The contribution of proto-industry to later industrial growth has to be looked for in the entrepreneurs who managed the system and, in some instances, accumulated the wealth which they later employed in factory development.

There were many reasons for this long survival of proto-industrial patterns and forms of manufacturing into a period when we tend to think of them as obsolete. To some extent it was due to the lack of capital to replace them with newer forms; to some extent also to institutional and psychological barriers to change. Above all, however, the proto-industrial system survived because it continued to provide a viable alternative to mechanically-powered factory industry. There were, of course, many branches of production which offered few economies of scale and could be carried on only at a small and local scale. Many of the food-processing industries – the bakers and butchers, for example – found their industrial organisation little changed in the course of the century. Many crafts, especially those such as cabinet-making, engraving, glass-cutting and some branches of leather-working, were little affected by the factory system, though mass production at the lower end of the market became important as the century wore on. In other branches of production, such as tanning, and the leather industries, the economies to be derived from scale were small, and in all industries which used local materials to supply goods for local markets there was little advantage in increasing the scale of production. Factory production was predicated on a developed means of transport. In no part of Europe had this been created before the second half of the century, and in some, notably the east and parts of the south, it had not really been established before the outbreak of the First World War. There were areas which still lived in a state of near self-sufficiency.

The proto-industrial system also lingered on in the textile and metal industries which did indeed cater for a mass market and in which the factory system did offer real economies. The process of transition from manual spinning and handloom weaving was long and gradual. The inefficiency of labour was balanced by its cheapness, and the peasant was often able to meet the competition of factory industry by working for a lower reward.

The spinning industry, once the most widespread of domestic industries, had largely passed to the factory by the middle of the nineteenth century. Cotton spinning led the way. It had no long tradition in much of Europe, and there was thus no strong vested interest in maintaining it as a domestic craft. Furthermore, it was relatively easy to mechanise, and could use small units of water-power. The mechanical spinning of wool spread more slowly, but it too had been generally adopted by the mid-century. The mechanisation of weaving was more difficult and, once the flying shuttle had been added to the handloom, the power-loom offered no overwhelming cost advantages.[29] Nevertheless, cottons were generally being woven by the power-loom in western and central Europe soon after 1850, and woollen fabrics by the end of the century. The handloom, sometimes of the most primitive design, continued to be used in the east and south. Late in the nineteenth century Vlaykov found in the villages of Bulgaria 'a loom in nearly every other house...[and] all the widows and quite a few of the married women too, earning their living by the spinningwheel'.[30]

Flax-spinning and linen-weaving resisted the process of mechanisation more successfully, and in some parts of Europe never ceased to be domestic crafts until they died out early in the twentieth century. For this there were two important reasons. In the first place, linen-weaving was a marginal, domestic occupation, carried on chiefly in winter when work in the fields failed to occupy the agricultural worker's day. The English traveller, J. E. Tennent, urged his readers to visit the urban linen markets in Belgium in winter, because that was when rural workers had most to sell.[31] The fact is that where domestic weaving remained important, farm holdings were generally small and the supplementary income from weaving was necessary to support the peasant household (see p. 218).[32] In Belgium the contraction in the demand for *toiles*, or linen cloth, in 1846–7, was disastrous for the welfare of the rural population. The peasant had hitherto been able to compete with the factory because he had few overheads and was prepared, if necessary, to work for a very small reward;[33] thereafter his market slipped away.

The second factor in the survival of the domestic linen industry was that it was more sheltered from competition than other branches of textile production. Linen was of declining importance relative to other fabrics and for many reasons it was being replaced by cotton.[34] Few entrepreneurs were prepared to invest heavily in its production. Flax, furthermore, was less suited than cotton or wool to mechanical spinning and weaving, and it required many years of experiment before machines that had been developed for cotton could be used for flax. Nevertheless, the relatively small-scale production of linen in factories was a real threat to the domestic industry, especially in densely populated areas such as Flanders and the hills of Saxony and Silesia, where it had become an essential complement to an unrewarding system of agriculture. Tennent described the new flax mills in the Belgian cities about 1840, and shared the Belgian government's concern over the resulting distress amongst the traditional rural hand-spinners.[35] At this time a fifth of the population of Flanders, almost all of it rural, was engaged in flax-spinning and linen-weaving and in all some two-thirds of the population was said to be dependent in one way or another on the industry.[36] Those employed in the linen industry were most numerous on the poor, sandy soils of interior Flanders (fig. 6.3). They were fewer on the richer clay soils of the coastal tract, and were almost non-existent over the loess plateau of central Belgium (see p. 264). Domestic clothworkers, though not necessarily in linen, became numerous again in the barren heaths of North Brabant and the Twente district of Overijssel.[37]

Another area similarly noted for the high proportion of its population engaged in domestic weaving was the mountainous belt which separated Bohemia from the German provinces of Saxony and Silesia. The plight of the linen-weavers was emphasised in Gerhart Hauptmann's drama, *Die Weber*. With unusual clarity, Hauptmann showed that the misery which he depicted sprang from declining markets, the competition of steam-powered linen mills, and the harsh and grasping merchants who put out the linen thread for the domestic weavers to manufacture into cloth.[38]

Much has been written about conditions in the factories which were at the heart of the alternative industrial system. The oft-cited instances[39] of insanitary conditions, long hours of work and child-labour must, however, be regarded as extreme rather than typical cases.[40] The conditions of domestic labour could be at least as bad as those prevailing in the factory. In 1848 Audiganne wrote[41] of 'le dénûment le plus absolu, la misère la plus profonde' among domestic workers, adding that hours of work were less among factory employees than with 'those who work freely at home in their families, as at Lyons, Saint-Étienne and Saint-Chamond'. The virtual extinction of domestic work in the textile industries, as far as western and central Europe were concerned, was no great social loss.

The other significant branch of proto-industrial production which lingered on through the nineteenth century was iron-working and metal fabrication. These called for a much larger capital investment than the domestic textile industry, and for this reason had to be practised on a larger scale and on a full-time basis. When the term 'proto-industry' is applied to these forms of iron-working in the nineteenth century it is on account of their older and more primitive technology, rather than the smaller scale of operation. The 'direct' method of iron-smelting on the hearth was still practised in a few areas, such as the eastern Pyrenees, during the century,[42] but had died out before its end. Its scale was very small indeed, and it could be, and often was, carried on as a family business. To this extent it was a domestic industry. Most of the iron was, however, produced in the blast-furnace, using either charcoal or coke as fuel, and was refined first to bar or malleable iron and later directly to steel (see p. 332). The charcoal-fired blast-furnace may have been proto-industrial in its technology, but it was a complex operation, closer to the factory than to domestic industry. The same is true of the refining industry. Although a few refining hearths continued in use in the nineteenth century, the industry quickly came to be dominated by the puddling furnace, supplemented by the cementation furnace which produced steel by a kind of case-hardening process. In the middle years of the century, however, these methods began to be superseded by those which employed the Bessemer converter or the open-hearth. The scale of operations was at once greatly expanded. The size of the capital investment put such undertakings beyond the reach of the traditional ironmaster, and iron- and steelworking entered its modern or factory stage.

It was in the fabrication of iron goods that the proto-industrial structure survived longest. Many such branches of industry could be carried on at home, or at least in small workshops. The nailer was typical of domestic ironworkers. In 1839 Briavoinne described their countless workshops scattered through the hills of Liège, Namur and southern Hainault. They were active only in winter; in summer the craftsmen were busy in the fields.[43] The nailer in Liège was strictly analogous to the linen-weaver in Flanders.[44] The factory made inroads in the course of the century into the sphere of the domestic nail-maker, but in 1896 there were still 549 who worked in their small workshops at home as against only 682 who were employed in factories.[45]

The Belgian nailers were a conspicuous example of the survival of proto-industry in the sphere of metal goods production. There were, however, many others. Any product which demanded a high level of skill and relatively little raw material might be carried on profitably on a domestic basis, though demands of safety and sometimes also of hygiene often required that a separate workshop be established adjoining or close to the artisan's cottage. Amongst such craft industries were the manufacture of sporting guns and other hand weapons in neighbourhood of Liège and also around Suhl in the Thuringian Forest. At Liège there was minute local specialisation. Gun-barrels were drawn and polished in the Vesdre valley; breach actions made in Hervé; the guns assembled in Liège, all by craftsmen at home in their own workshops.[46] At Solingen in the Sauerland there was a similar division of function, reminiscent of Adam Smith's pin-makers, in the manufacture of cutlery and other cutting tools.[47]

The long twilight of craft and proto-industrial production is one of the most curious features of the history of modern manufacturing. That it would survive the coming of the factory system was to be expected because it was well established in rural tradition. That it survived so long is to be attributed to the fact that it formed an essential part of an agricultural system which would have collapsed without it. In time, of course, the rural population which depended upon it drifted away to the towns and to urban industry. In the late nineteenth century it survived mainly in remote and less fertile areas where there was little alternative employment to supplement the income from agriculture.

Some branches of domestic manufacture, such as woodcarving, clockmaking, embroidery and lace, even grew in importance, as their products began to acquire a sophisticated appeal outside the rural community which made them.[48] Nevertheless, there was overall a sharp decline in domestic and craft industries during the last third of the nineteenth century, with a resulting high unemployment. In Germany a very careful survey was made of most branches of *Hausindustrie* towards the end of the century.[49] Little more than 1 per cent of the total German population was fully dependent on domestic industry at this time, but the proportion was very much higher in parts of Silesia and Saxony and in the hills of Thuringia and the Lower Rhineland. Textiles were still the chief household industry, employing somewhat less than a third of the total domestic workforce, but light metal goods were important in the Rhineland.[50] It is commonly assumed that the domestic workforce was mainly female. 'Spinsters' had, it is true, once formed its largest segment, but in weaving and metalworking employment was heavily male. One of the reports published by the *Verein für Sozialpolitik* commented on the high average age of domestic workers. Clearly theirs was a diminishing race, and few of the younger generation were attracted to rural crafts.[51]

The factory system

The factory system is indissolubly linked with the Industrial Revolution, whose essential feature was the application of mechanical power to manufacturing processes. It replaced domestic industry and the workshop, and accounted for a large and growing, though not precisely determinable, proportion of total industrial production. The factory and domestic systems are commonly conceived as in sharp contrast with one another, the 'dark satanic mills' on the one hand, and the craftsman on the other happily working in the bosom of his family. This, as has already been noted, is sentimental nonsense. The line between factory and craft industry is, in fact, a difficult one to draw. Few craftsmen worked entirely alone. Most, especially those in the weaving and metal-fabricating industries, had helpers or companions. They often had a workshop or weaving-shed separate from their homes, and occasionally, like the miller, they made use of water-power.

Outwardly the smallest factories differed little from the largest workshops of the *Hausarbeiter*. Some factories in the textile industry were minute, and many were the instances where, even at the end of the nineteenth century, an entrepreneur supplemented the labour of a small workforce in his factory with that of a more numerous body of domestic workers to whom he supplied the essential materials. The difference between factory and domestic workers lay chiefly in their organisation. The factory worker was employed for a wage and worked during hours which were unquestionably long, but also defined. He worked, furthermore, under the supervision of the factory owner or manager. He commonly used some form of mechanical power, and there were usually as many machines as the local source of energy was capable of activating.

Intermediate between the craftsman's workshop and the factory were those buildings in which handworkers, sometimes in large numbers, exercised their traditional crafts under supervision and at a wage. In this way William Stumpe, clothier, had in the 1540s used the buildings of the dissolved Malmesbury Abbey. Such also were the first woollen 'factories' at Verviers in Belgium.[52] But such controlled labour made no use of water-power and can hardly be said to have constituted a factory in any modern sense.

The earliest factories were based upon water-power, and were in consequence located in hilly regions where streams and rivers could be most easily harnessed for this purpose. Water-power had long been used in the iron-industry, to produce the blast in the furnace and to turn the rolls of a rolling mill, and its availability had been an important factor in the location of the industry. Not infrequently it enforced a spatial separation between the two branches, since refining and rolling made much heavier demands on energy than smelting (see p. 340). The earliest textile mills, which were engaged mainly in cotton-spinning, also used water-power, and its availability helped to locate such manufacturing centres as Chemnitz (Karl-Marx-Stadt), Elberfeld and Barmen (Wuppertal) as well as a multitude of smaller undertakings in the towns of Saxony, Bohemia, the Rhineland and central Belgium. Łódź was chosen as the site of the textile industry in Russian-occupied Poland because of the abundance of small streams capable of providing the motive force for many very small factories.

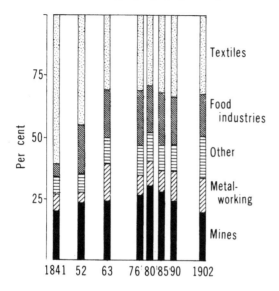

7.1 The use of the steam-engine in Bohemia and Moravia, 1841–1902. This graph shows the diminishing *relative* importance of its employment in mining and the manufacture of textiles, and the increasing diversification of its uses

But the energy that could be derived directly from a stream was severely limited. Many a factory outgrew this source of power early in its history, and turned to steam. For some, however, conversion to steam-power had to await the building of railways for the transport of coal. Others were so unfavourably sited that the supply of fuel was never a practicable possibility, and their importance declined until they were closed.

Steam-power was first used on an important scale in the mines, primarily for pumping water from the shafts, but also for hauling coal, ores and men. Without it, there could have been little increase in the output of coal and metalliferous ores. The first effective steam-engines to be set to work in continental Europe were from Boulton and Watt's Soho works, near Birmingham.[53] They were transported, often with great difficulty, to mining centres where they were assembled and put to work. There is a considerable body of folklore on the subject of the untutored peasant who intuitively solved the problems which their operation presented. It was not long, however, before steam-engines on the Boulton and Watt model were being turned out from continental workshops.[54] In this Cockerill of Seraing, near Liège, was first in the field, but was quickly followed by other manufacturers in Hanover, Berlin, the Ruhr and elsewhere.

Despite the continued importance of water-power, the spread of steam-engines provides the most convenient measure of the progress of modern industry. At first they were employed mainly in the mines and textile mills, but in the second half of the century they were used increasingly in the many branches of industry that made up the 'second' Industrial Revolution. Fig. 7.1[55] shows the changes in the purposes for which steam-engines were used in the Czech lands.

Particularly significant was the increase of steam-power in the food-processing industries, notably flour-milling and brewing.[56]

The diffusion of the steam-engine for industrial, as distinct from transport purposes is an unwritten chapter in the economic history of Europe, and it is impossible here to trace statistically the growth in the use of steam-power and the parallel decline, at first relative and then absolute, in the use of water-power. It is probably true to say that by the mid-nineteenth century steam-power, aside from its use on railways and in ships, was more important than water-power. It is doubtful whether at the end of the century water-power contributed more than 10 per cent of industrial energy. By this time however, electrical energy, derived increasingly from hydroelectric stations, was beginning to cut into the field hitherto dominated by coal. At about the same time the internal combustion engine made its first appearance, but remained of negligible importance in manufacturing until after the First World War.

The branches of manufacturing

Until the second half of the nineteenth century industrial advance was largely in the textile and iron and steel branches of manufacturing. These employed the greater part of the industrial workers and absorbed much of the investment capital that did not go into transport and other social overheads. Indeed, most of the investments in railways was, in fact, spent on metal goods. Not until the latter half – or even third – of the century was the industrial basis of western civilisation significantly broadened, with the development of chemical, engineering and a wide range of consumer good industries. This does not mean that other branches of manufacturing did not progress. The near self-sufficiency of the rural community was broken down, and craft industries grew to satisfy the needs of urban workers. Bread-making ceased in many places to be a family pursuit, and passed to the specialised baker. Tanning and the making of shoes and clothing expanded as urban crafts to satisfy an increasing urban demand for the products. In some instances these branches of production continued to be organised on craft lines, without even a significant division of labour. With the slowly increasing sophistication of life came a growth in the number of small, generally urban, pottery, brick, tile, glass and furniture works. In no respect were they innovative – not even in organisation – until late in the century. But their growth paralleled that of the dominant industries, and, reacting upon one another, they brought about an advance in manufacturing on a broad front.

The location of all branches of manufacturing was strongly influenced by the available means of transport. In most of continental Europe – in contrast with Great Britain – the most important industrial developments came *after* the construction of at least a primary railway net. They could thus use the railways to transport both materials and products, and were thus less constrained in their location than earlier developments in Great Britain had been. There was in continental Europe a very much less conspicuous movement of manufacturing towards the coalfields just because fuel and other raw materials could be distributed relatively cheaply.

There was nevertheless a strong tendency for the several branches of manufacturing to concentrate in particular areas, even though these had in some cases little to offer by way of fuel, energy or physical resources. Many branches of the textile industry inherited their locations from the traditional manufactures which had preceded them. The presence of a trained labour force and a commercial infrastructure marked the transition from pre-industrial to industrial or factory production. The role of the coalfields, so important in the industrial development of Great Britain, must not, however, be underrated. It was of great importance in those branches of industry, such as iron-smelting and refining and non-ferrous metal working, which made extravagant demands on fuel. Every major European coalfield and some of only minor importance became the scene of metal-working during the first half of the century. There was, however, a tendency during the latter half for the coalfields to relax their grip on industry, and this was, by and large, due to an increasingly efficient use of fuel.

There was a corresponding tendency, especially late in the century, for manufacturing to be located as near as practicable to its largest market, irrespective of whether this involved transport of fuel and materials. The industrial development of the larger capital cities reflected the fact that within or near them lay a significant part of their nation's potential market. A third type of location which became important during the century may be called the break-of-bulk site. The loading and unloading of freight, whether onto or from sea-going ship, river barge or rail car, was always expensive. Careful planning of location – and this in general was not common before the later years of the century – would attempt to minimise the number of times freight had to be shifted in this way. Sites thus tended to be chosen where railway met river or canal, or where a navigable river reached the sea. The chemical and metal-using industries of the Rhineland typify the former; the coastal ironworks and the complex range of 'port' industries, the latter.

In the following pages the location and diffusion of, first, the textile, then the metal, and lastly the 'new' industries is examined.

The textile industries

These are amongst the oldest European manufactures and were the first to be brought into the factory system. Broadly speaking, there were two traditional branches of the textile industry: woollens and linen; and two others, cotton and silk, which were introduced during the Middle Ages, and received their greatest impetus during the eighteenth century. Lastly, the synthetic or man-made fibres, products of modern technology, made their appearance towards the end of the nineteenth.

The woollen industry

The manufacture of woollens was the oldest and most widespread of the textile industries. With that of linen, it faced and in varying degrees succumbed to the competition of cotton. To some extent woollens and linen were complementary. Both derived from locally produced raw materials, but areas which were physically suited for flax-growing were rarely the best to support sheep, so that there was little direct competition between the two fibres. Most of continental Europe produced only a coarse wool, suited for making rough cloth. The most significant exception was the Spanish Meseta, where the high-quality merino wool supported an important export trade rather than an indigenous cloth industry. Though there were a few centres where a concentrated cloth industry served distant markets, most woollen spinning and weaving was practised as a cottage industry to satisfy local demand.

The chief centres of the manufacture of woollen textiles at the beginning of the nineteenth century were in northern and southern France, in Belgium, the lower Rhineland and Saxony, in each of which there was a local supply of wool and also a labour supply both skilled in weaving and in need of a craft industry to supplement its agriculture.

The more important concentrations in France were in Languedoc, Champagne and Normandy. In the seventeenth century Colbert did much to stimulate the cloth industry in the Midi, and a significant export to the Levant resulted. The industry was widely scattered through the countryside of Languedoc, though focused in Lodève, Mazamet and Carcassonne.[57] It was however, in decline. Its overseas markets were being lost to English competition. Its workers resisted technical innovations,[58] and by the second half of the century the industry was virtually extinct,[59] and its workers absorbed into the spreading wine industry of the region.[60]

The cloth industry fared better in northern France. It remained active in Champagne, where Reims was still a significant centre of the industry late in the nineteenth century.[61] About 1860 there were still some 40 water-powered spinning mills,[62] though hand-loom weaving continued to be practised in the cottages. Along the Lower Seine valley a traditional woollen industry was able to resist the competition of cottons, and remained in the early twentieth century firmly established at Elbeuf and on the left bank of the river. Machine-spinning was adopted relatively early, though domestic weaving was not wholly displaced by the factory until late in the century.

The factory production of woollen cloth expanded at a small number of well-favoured centres to satisfy a growing mass market. By the last quarter of the century these had, in effect, been reduced to Sedan and Reims in Champagne, Elbeuf, and, on a much smaller scale, Castres and Mazamet which, alone amongst the southern centres of the woollen industry, had been able to stage a recovery.[63] But dominating them all was the industry which had grown up in Lille and its surrounding villages. Woollen cloth had been woven for export in Flanders and northern France since the Middle Ages. In Flanders the industry had very largely succumbed to warfare and the competition of linen. But it survived in the Lille district as well as in Abbeville, Amiens and Saint-Quentin. From the Flanders plain it was carried by refugees southwards to the valleys of the Ardennes.[64] Here it established itself during the eighteenth century, spinning being carried on mainly in the rural areas, and much of the weaving in the small towns. Growth was encouraged by the abundance of water power and the general receptivity of the region to technical innovation.[65]

Verviers became the first major centre of the Ardennes woollen industry. It adopted machinery at an early date, and carding and spinning by water-power spread through the area soon after 1800.[66] In 1816 the first steam-engine was set up; by 1850 almost the whole industry was mechanised,[67] and the Belgian woollen industry had become 'one of the technologically most advanced'[68] in Europe. Verviers constitutes one of the clearest examples of industrial concentration to be found in nineteenth-century Europe. Not only did almost the whole of the Belgian woollen industry establish itself here, but the small cotton industry which Briavoinne found,[69] was squeezed out of existence. Important though it was, the Belgian woollen industry came, in the number of its employees and in the volume and value of its output, far behind the linen industry and not far ahead of the cotton.[70]

The Verviers industry did not develop in total isolation. To the east lay the clothing district of Aachen, Eupen and Burtscheid; to the North, that of the southern Netherlands. The latter survived as a domestic industry in an area of poor agriculture until late in the century with its chief centres in Tilburg and Eindhoven.[71]

The development of the woollen industry in Germany followed a course similar to that in France. It was a widespread domestic industry, which gradually concentrated in a few well-favoured centres of production.[72] It owed much to the example of the Low Countries, and even to the immigration of Flemish and French weavers.[73]

The Aachen district, with its small towns of Montschau, Eupen and Burtscheid, was from the first a leading source of woollen cloth. At first they merely focused the activity of a scattered domestic industry,[74] but steam-powered factories began to be built soon after 1815,[75] and the industry grew with the formation of the Zollverein. The woollen industry of the Aachen district spread to Düren, where it remained important throughout the century,[76] and to Cologne,[77] Düsseldorf[78] and the Ruhr,[79], where, however, it was eclipsed by other industries at an early date. The woollen industry never really established itself in south Germany, where linen and mixed linen and cotton fabrics had long dominated. Factories were established at Nuremberg[80] and in Baden,[81] where Swiss influence was strong, but none ever achieved any lasting success. North Germany was pre-eminently the domain of flax-growing and linen-weaving, and, except in two areas the woollen industry never gained any real importance. The exceptions were the Saxon Ore Mountains and Greater Berlin. The former bore a significant resemblance to the Ardennes and Eifel with their hill-grazing for sheep, abundant small streams and soft water. A traditional cloth industry grew up in the valleys of the Vogtland, centring in the small towns of Gluchau, Zwickau and Plauen.[82] Though threatened by the cotton industry spreading out from Chemnitz, it survived largely because it was able to absorb labour which became available as the metalliferous mines closed (see p. 396).

The second major centre of the woollen cloth industry was greater Berlin. Cloth manufacture had been encouraged by the Prussian government during the eighteenth century, and in 1805 there were said to have been no less than 10,000 active looms in the city.[83] The woollen industry of Berlin, nevertheless, declined during the century with the expansion of manufacturing in Prussia's Rhineland provinces,[84] but in the 1880s there were still, it was claimed, some 6,000 looms

Table 7.1. *Woollen-weaving in Saxony*

	Handlooms	Power-looms
1846	47,290	1,372
1861	67,485	6,247
1875	58,214	29,341

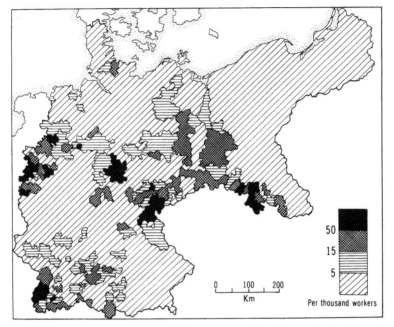

7.2 The distribution of weavers in Germany, about 1885. The principal cloth-making areas – the lower Rhineland, Baden and South Germany, Saxony and Silesia – are readily apparent

of all kinds.[85] The weaving industry disappeared soon afterwards, and its heir was the rapidly growing clothing industry of Berlin (fig. 7.2).

The German woollen industry grew steadily through the middle years of the century in its major centres, the Lower Rhineland and Saxony. The number of fine spindles in Saxony increased between 1840 and 1861 from 33,873 to 104,622. Handlooms also increased in number but formed an ever-diminishing fraction of the total. Power-looms increased nearly fivefold between 1861 and 1875 (table 7.1).[86]

Eastwards from Germany, through Poland to Russia, there was a small and scattered production of coarse woollen cloth for peasant wear. Attempts were made to organise this into a significant weaving industry. The Zamojskis made such an effort early in the century,[87] and for a short time Janów Lubelski was a significant source of cloth. The wool came from the Zamojski estates, and the entrepreneurs from Saxony and Bohemia, but the local labour, much of it supplied by the *corvée*, lacked the necessary skills, and the industry collapsed.

Greater success attended efforts to establish a woollen industry in Central Poland.[88] It was better organised and possessed from an early date the necessary transport and commercial infrastructure. Indeed, it might be said that at Pabianice, Zgierz and Tomaszów, as well as at Łódź itself, the woollen industry rode to prosperity on the coat-tails of the burgeoning cotton manufacture (see p. 412).[89]

Woollen-weaving was of only minor importance in the Alpine lands of Switzerland[90] and Austria.[91] But it assumed great importance within the Habsburg Empire, in Bohemia and Moravia. In the course of the eighteenth century some of the local magnates tried to capitalise on their considerable production of wool and the labour obligations of their peasantry in order to develop a woollen industry on their estates. Clothworkers from Verviers were settled at Liberec and Jihlava in Bohemia and at Brno in Moravia. Mechanical spinning was introduced, and by 1826 there were numerous power-driven mills in both Liberec and Brno,[92] and power-looms began to be used in the 1840s. With the expansion of woollen manufacture at Brno and Liberec the rural industry decayed.[93] By the end of the century not only did Bohemia and Moravia embrace some 80 per cent of the woollen industry in the Habsburg monarchy,[94] but most of this was concentrated near these two towns.

A smaller woollen industry developed in Slovakia, and here Hungarian entrepreneurs established manufactories, partly domestic, partly in small factories, at Halič, Bratislava, Trnava, Prešov and elsewhere.[95] Most of these undertakings, however, succumbed in the second half of the century to the competition of the larger and more efficient factories in the Czech lands.[96]

Southern Europe, with its dry summers and abundant rough grazing, was well suited to sheep-rearing, and the wool produced in Spain was of the highest quality. Nevertheless, the Spanish cloth industry had not at the beginning of the century risen above the level of a domestic craft. In the course of the century, however, in both the Spanish and the Italian peninsulas, a factory-based industry grew up, extinguishing as it did so the traditional cottage industry.[97] In Spain it was located mainly in Catalonia, where its rise followed the success of cotton manufacturing.[98] Its chief centres – Sabadell, Tarrasa, Olot – lay in the immediate hinterland of Barcelona, where they could profit from the infrastructure already built up for the cotton industry.[99]

The woollen industry of Italy was always more important than that of Spain, and was well developed in the Alpine foothills. Woollens belonged to the mountains, it was said, as cottons did to the plain.[100] In the course of the nineteenth century the woollen industry concentrated in Biella and neighbouring small towns, where it could use the water-power of the Alpine streams. Factories were invariably small, and a British consular report mentioned 'four separate small firms carrying on their industries in one building'. Altogether there were, in 1885, 126 factories employing over 9,000 workers.[101]

A second centre of the woollen industry lay in Tuscany, chiefly in the town of Prato,[102] and a third in Naples and other towns in southern Italy. The industry in peninsular Italy showed some promise before the unification of Italy, but after 1860 it was exposed to the full blast of northern competition and quickly withered.[103] The north Italian woollen industry developed with some vigour during the middle years of the century but thereafter stagnated,[104] leaving the Biella region as the only significant centre.[105]

The Balkan peninsula, the least industrialised part of southern Europe, was, paradoxically, the scene of one of the more successful attempts to develop a woollen industry. Domestic spinning and weaving were to be found everywhere, and remained important until the end of the century. In Romania, even at the beginning of the twentieth century, it was estimated that two-thirds of the households still had a loom.[106] In Bulgaria this domestic industry, carried on in the foothills of the Balkan and Rhodope Mountains, produced a coarse cloth, the *aba*, which was widely marketed in the Ottoman Empire. Under these circumstances there was no great market for the product of a factory industry. Nevertheless, it was here, at Sliven, that a broadcloth industry was established in 1836 to supply uniforms for the Turkish army.[107] Other 'factories' followed at Plovdiv, Stara Zagora, Gabrovo and Samokov.[108] Most lasted through the century, though their limited markets combined with competition from central Europe, to restrict their growth.

Linen

If the woollen industry stagnated or at best grew slowly in the course of the nineteenth century, linen production underwent a severe decline. At the beginning of the century flax was grown and linen cloth woven throughout northern and central Europe, and in some parts also of southern Europe. It was essentially a peasant industry. A small patch of flax could supply material enough to occupy a peasant family during the winter months. It could be retted and scutched by the men, spun by the women and woven by both into linen for domestic use or for sale to the merchant. It was the ideal complement to a small-holding with poor soil, and it contributed to the dense rural populations to be found in Brittany, Flanders and parts of Silesia. Its decline in the later nineteenth century was an important factor in the rural depopulation of these regions.

Its rival was cotton, a more flexible, more easily worked and, in the end, a cheaper product. The processing of flax and the spinning of linen thread were, in fact, difficult processes to mechanise, and until technical problems had been solved, the factory production of linen made little progress,[109] while cottons steadily increased their share of the market. Machine-spinning of linen thread did not begin in France until the 1830s, more than a generation later than the corresponding process in cotton production. Hand-spinning had almost disappeared here by the 1860s but there was little evidence even at this date for the use of the power loom in linen-weaving.[110]

Flax was grown in most parts of France, but was of the greatest significance from Brittany to Flanders.[111] Here it was the almost universal supplement to peasant agriculture, and nowhere more so that in Brittany and western Normandy. Without it, wrote the *sous-préfet* of Mayenne in 1832, 'men were reduced to idleness and families deprived of their means of subsistence'.[112] Indeed the volume of linen thread produced was actually increasing in the middle years of the century, a normal peasant reaction to conditions of over-production and falling prices.[113] In some parts of this region the linen industry supported a large part of the population. At Damigny, near Alençon, for example, it gave employment – much of it doubtless part-time – to 810 in a population of only 1,239.[114] The industry here remained relatively prosperous until about 1865, when cottons began to cut into the market for *toiles*. By 1880 production was negligible, and by 1900 the industry was almost extinct.

The domestic linen manufacture gave rise to a factory industry nowhere west of the Seine.[115] The situation was different to the east, where the linen-weavers drifted into factories. The most important were at Douai, and Cambrai, which specialised in fine linen, at Armentières, Hazebrouck and above all at Lille, where coarse fabrics were woven.[116] By 1912, 91 per cent of France's flax spinning was located in *dépt* Nord, though weaving, some of it still by hand, was more widespread.

Nowhere else in Europe did the linen industry achieve the dominance which it possessed in Belgian Flanders. Except for the newly established cotton industry of Ghent, it completely replaced the traditional Flemish cloth industry.[117] It was almost wholly a rural industry, providing a part-time occupation for as much as 40 per cent of the rural population of Flanders. In 1843 287,500 out of a population of 1,440,900 were engaged in spinning, weaving and finishing linen cloth, and as many more in growing and preparing the raw flax.[118] As the average holding was only about 2.5 hectares, the linen industry was of vital importance to the peasantry. It combined with the practice of partible inheritance to produce the densest rural population to be found in western Europe.

In the 1840s the domestic industry was in decline. Its market, both local and foreign, was being attacked by cottons. The mechanical spinning of flax was making progress and factories had been established in Ghent, Audenaarde, Avelghem and Renaix.[119] Add to this the growing import of Russian flax to supply the factories, and the plight of the rural craftsman became acute. The final destruction of the industry, however, came from external causes, the potato blight and the poor harvests of 1846–8. This struck away the other support – agriculture – of the Flemish peasant-artisan. The result was severe famine and migration from Flanders to other parts of Belgium. Thereafter the domestic linen industry of Flanders gradually withered away.

Flax was an important crop throughout the North German Plain from the Netherlands to Prussia and Silesia. Linen was mostly produced to satisfy local needs, but in a few places it was woven for a wider market. Amongst the linen-producing areas were the left bank of the lower Rhine, Münsterland, Hanover, eastern Saxony and Silesia. In the lower Rhineland Viersen, Rheydt and Gladbach were linen-weaving centres,[120] but in the early nineteenth century their economies became more diversified, and the emphasis passed to other textiles. In *Kreis* Gladbach the number of looms sank from 191 in 1822 to 132 in 1849 and to 48 only by 1880. Soon the industry was extinct west of the Rhine. The linen industry flourished for a time in the Wupper valley, where the pure water of the streams was said to be well suited for bleaching. But it was displaced by cotton and silk, and when dyeing was established the waters became too polluted for the bleacher to use.

The plain of Westphalia was noted for the quality of its flax.[121] The crop was, wrote Jacob, 'grown in every part' of it, and either in the form of yarn or of linen cloth 'is almost the sole commodity which creates any external trade'.[122] Münster, Osnabrück and Bielefeld became the centres of this commerce, and machine-spinning was adopted at Bielefeld, though weaving continued to be a predominantly peasant occupation.[123]

The other major centre of the German linen industry lay in eastern Saxony and Silesia.[124] A British consular report noted the large production both of yarn and of linen cloth from Silesia. Much of it was sold by way of the east European fairs into the Russian Empire.[125] But here too machine-spinning was introduced in 1840s, leading to serious disturbances by the domestic craftsmen.[126] In Saxony and Lausitz the linen industry succumbed to the competition of cotton;[127] but survived in Silesia where, at the end of the century, a third of Germany's linen-weavers were to be found.[128] A small factory industry also grew up at Andrychów, in southern Poland.[129]

In southern Germany the linen industry declined and disappeared in the course of the century, driven from foreign markets and undersold at home by the cheaper Silesian linen. In the Allgäu, where it had once been an important craft, it was replaced as a major source of peasant income by cheese.[130] In neighbouring Switzerland, where field-work was difficult in winter, domestic crafts were particularly significant. Linen-weaving was at first widely practised on the Swiss Plateau,[131] but cotton began to compete, especially in the towns, and the rural linen industry slowly declined.[132] By the end of the century it was restricted to a few remote areas in the mountains.

Within the Habsburg lands the linen industry was located chiefly in Bohemia. As in neighbouring Silesia and Saxony, it was carried on in the mountain valleys where it provided a supplementary income 'for hundreds of thousands of people...though alone it did not suffice for even a minimum livelihood'.[133] So large was the manufacture that linen thread was imported from Silesia to supply the weavers. The chief centres of the industry were at Náchod and Děčín in the Giant Mountains. Machine-spinning of flax was introduced in the 1850s, and by 1865 there were several linen mills. Production of linen increased sharply during the first half of the century, but thereafter stabilised as the domestic industry gradually disappeared and demand turned to cotton.[134]

Few parts of southern Europe were suitable for the cultivation of flax, and the linen industry was of negligible importance in Spain[135] and in much of Italy. In northern Italy, however, domestic spinning and weaving were widespread,[136] and a few linen mills were established. Flax-growing and linen-weaving were never of great importance in Hungary and Romania.[137] They were more significant in Scandinavia and Finland, But here too linen was gradually replaced by cotton.[138]

The coarse fibres

In some parts of Europe, especially the coastal districts of the west and north, hemp was commonly associated with flax-growing. Though it was sometimes woven into a coarse fabric, it was most often used for ropes and cordage and supplied the countless small rope works which were common in sea ports.[139] Hemp was, however, of declining importance through the century, and for many purposes was replaced by jute.

The processing of jute was introduced into Great Britain, at Dundee, in 1832. In 1845 a jute mill was established on the Somme near Amiens, and a second a few years later at Dunkirk.[140] The raw material came almost exclusively from South Asia, and the opening of the Suez Canal in 1869 made it cheaper and more abundant. Jute mills were established in southern Italy,[141] and also in Austria, which obtained the raw jute through the port of Trieste. But its biggest development was in Germany, where there were in 1895 no less than 220 jute mills, though many of them employed only a handful of people. They were concentrated in Hanover, Brunswick and Saxony, the hinterlands of the ports of Hamburg and Bremen.[142]

The cotton industry

The chief reason for the stagnation of the woollen industry and the decline in linen production – at a time of increasing population and rising living standards – was the development of cottons. The growth of the cotton industry was indeed phenomenal. The increase in raw cotton consumption in Europe, Great Britain excluded, was of the order of thirtyfold between about 1830 and 1913. The number of spindles increased less spectacularly – about eightfold – but most of the spindles in 1913 were of a much more efficient design. However computed, the total value of cotton products greatly exceeded that of all other textile products by the time of the First World War.

The reasons for the triumphant progress of the cotton industry are clear. Cotton fabrics were versatile and attractive, capable of being both dyed and printed with coloured designs. They could be laundered easily, and their widespread adoption was itself a contribution of no small importance to personal hygiene. These advantages made cotton clothing fashionable long before factory production made it cheap. Cotton fibre became less expensive after the invention in 1793 of a gin to remove the seeds from the cotton boll. Supplies had come at first from the Middle East and South Asia; now a limitless supply was made available from the southern states of America. By 1860 no less than 69 per cent of the raw cotton entering world trade was from the United States.[143] The disastrous consequences of the short interruption of supply during the American Civil War[144] showed how great this dependence had become.

Although at first spun and woven by hand, cotton proved to be much more suited to mechanical spinning and weaving than either wool or flax. In this lay the secret of its success; it could be mass-produced in the factory and sold more cheaply than linen, which for most purposes it gradually supplanted.

For these reasons cotton manufacture became the leading sector in the industrialisation of almost every country in western and central Europe. Innovation in the cotton industry had begun in Great Britain in the 1730s. The rapid expansion of demand encouraged further technical advances. In the half-century before 1815 the British consumption of raw cotton had increased more than twentyfold, and 'the cotton mill [had become] the symbol of Britain's industrial greatness'.[145] It was not long before these technical innovations were introduced into France. The obstacle to continental emulation was not Great Britain's efforts to preserve her monopoly, but the lack of a mass market within the continental countries themselves. Some progress was made before the outbreak of the Revolutionary Wars, notably in France, where Holker imported British technology, and in the Austrian Low Countries (Belgium), where the Cockerill family had begun its brilliant technological career in 1799. But a demand sufficient to justify large-scale, factory production, did not arise until after 1815. Its existence became apparent when, after the Peace of Paris, British goods flooded into continental Europe to satisfy a market long starved of such goods.

The period from 1815 to the revolutions of 1848 saw the emergence of the factory production of cotton goods in France, Belgium, Germany, Switzerland, the Habsburg Empire and Russian-held Poland. The development of the factory industry occurred in two stages. As in Great Britain, the earliest technical innovations were in the combing, carding and spinning branches, and the earliest factories were in every instance spinning mills. This was not only because the mechanisation of spinning was relatively simple, but also because the hand-spinners had considerable difficulty in meeting the demands even of the handloom weavers. As a general rule the mechanisation of weaving came a decade or two later, but once introduced it quickly ended the rule of the handloom weaver in the cotton industry.[146] After 1850 the manually operated handloom remained in significant use only in a few backward areas of central and eastern Europe.

The cotton industry developed primarily in areas already noted for the production of more traditional textiles. This was necessary in order to use the existing body of labour skilled in handling textile materials. It was not difficult for flax and woollen workers to adapt to cotton. But cotton was an imported material; supplies were less certain than in the case of locally grown flax, and the manufacturer was dependent on the merchant. This, in turn, influenced the location and scale of the industry. In the pre-railway age it was usually at no great distance from the ports, through which raw cotton was imported. About 1815 the chief centres of cotton-spinning and weaving were Lille and Ghent in the Flanders plain and Cologne, Elberfeld, Barmen, Gladbach and Rheydt in the Rhineland. In most areas where the cotton industry became established there was a period when raw cotton simply replaced wool or flax in the hands of the spinner. Even after cotton-spinning had been taken over by the factory, weaving remained for a number of years in the hands of craftsmen who had learned their trade on the older fibres. The mills, both spinning and weaving, used mechanical power. In the first instance this was usually water-power; there are even a few instances of the use of a horse-whim. In no case was the industry consciously located with reference to fossil fuels. In a few instances – northern France and the Lower Rhineland are obvious examples – the industry found itself in close proximity to coal, but this was fortuitous. More often, when a site for a factory was chosen, as for example in the Łódź district of Poland, it was only after careful examination of the local resources in water-power.

Once established, a cotton-manufacturing region grew by attracting new units of production. This would have been in part because of the availability of skilled labour, but more importantly perhaps because the cotton industry, being based on an imported raw material, called for a developed infrastructure in transport and commerce.

France At the beginning of the nineteenth century there was already a small but widespread cotton industry in France.[147] It was carried on mainly by traditional methods, and in few places had even begun to use machine-spinning. Already, however, the future centres of the industry were emerging. More than half the yarn was spun in *dépt* Nord and neighbouring *départements*. The Seine valley between Paris and Le Havre produced about 10 per cent, but Alsace, which was later to dominate the fine spinning industry, was still relatively unimportant.

The Seine valley was the avenue through which the modern cotton industry entered France. It was here that Holker had introduced the improved spinning machines. Soon after 1815 the little rivers that flow to the Seine were lined with spinning mills and steam-power was used in Rouen, where English coal was readily available.[148] Weaving, however, continued to be mainly by hand and was carried on intensively in and around Elbeuf, Louviers and Darnétal. In *dépt* Seine-Inférieure, in 1863, almost a third of the population was said to be dependent on the cotton industry, but the part-time rural weaver was giving way to the professional urban artisan, and he too, in the second half of the century, was abandoning his workshop for the factory.[149]

The cotton industry of *dépt* Seine-Inférieure remained relatively backward. The chief growth point was in Alsace, and its focus Mulhouse. The city had been part of the Swiss Confederation until 1798, and Swiss entrepreneurship remained important. It was Swiss craftsmen who introduced cotton-spinning towards the end of the eighteenth century, and were responsible for many mechanical innovations. The first machine-driven spinnery was established in 1802 at Wesseling, and by 1812 there were 11 in *dépt* Haut-Rhin.[150] In the same year the first steam-mill was built at Mulhouse, and by the mid-century hand-spinning was virtually extinct. Hand-weaving continued to be important, though experiments with power-looms were being made – appropriately – by a Swiss entrepreneur from Appenzell. By the 1870s Alsace had overtaken Normandy and was closely rivalling the north in the number both of spindles and of looms. There was a sharp decline in production after 1871, when Alsace was incorporated into the German *Reich*, and many of the cotton workers migrated across the Vosges to Épinal and Remiremont, which remained French.[151] There they established a rival industry. By 1900 this industry was larger in both spinning and weaving capacity than the Alsatian industry from which it derived. A smaller cotton textile industry, centred in Mulhouse, survived the war of 1870, and found itself in competition with the German industry in Saxony and the Lower Rhineland.

Alsace and the Vosges seem at first glance to have had little to recommend them as centres of the cotton industry. The abundance of small units of water-power quickly ceased to be a significant factor. Labour was appreciably cheaper in the early stages of growth than in either Normandy or the north, but this was more than offset by the high cost of transporting the essential raw material from the French port of Le Havre.[152] Until railways had been built this was done by means of convoys of wagons. Yet the 'East', with Alsace, became 'the most considerable cotton-spinning, weaving and printing district on the continent of Europe'.[153] The reason can only lie in the enterprise and powers of organisation of the Swiss entrepreneurs who established and developed the industry.[154] Units of production were on average larger than in the rest of France, and mechanical power, first water- and then steam-power, were used earlier than elsewhere.

The third important centre of the French cotton industry was the north, approximately the *départements* of Nord, Pas de Calais and Somme. Traditions of textile-working were stronger here than elsewhere in France, and an indigenous wool and flax industry survived to share its assets with cotton. A feature of the region in general was its lack of specialisation. Although some towns concentrated on a particular type of fabric, all three branches of the textile industry continued to exist side by side.[155] Another feature, in contrast with the steady adherence of Alsace and the Vosges to cottons, was the readiness with which a factory would change its manufacture from wool to linen, to cotton and vice versa, according to changes in the economic fortunes of particular textiles. The cotton famine of the 1860s, for example, led to widespread abandonment of calicos in favour of linen. Lille became a cotton-spinning centre without, however, abandoning linen. Tourcoing, meanwhile, shifted from cottons to woollens, but showed a marked predilection for mixed cotton and wool fabrics.[156]

It is easy to explain why such a varied textile industry should grow up in the Lille region: local traditions and labour, ease of transport, the large market of north-western Europe, but more difficult to understand the local specialisations that developed, and the not infrequent changes that took place in them. It seems likely that pre-industrial rivalries and vested interests, the chance adoption of innovations, prescriptive rights and grants of privilege continued in the nineteenth century to influence local specialisation.[157] Extraneous factors, such as the cotton famine in the 1860s, led to the closure of some works and locally tipped the balance against cotton.

Mechnical power was adopted fairly early in northern France, and coal was brought by canal to the mills even before the first railways were built in the area. A feature of the cotton industry of northern France was the disproportion between the spinning and weaving branches, much of the thread being sent to centres of weaving, lace-making and embroidery in other parts of France.

The Low Countries The Belgian cotton industry, like that of northern France, was heir to the ancient clothing industry of Flanders. A small machine-spinning industry had been established by 1804, but after a period of prosperity under Napoleon, it declined with the influx of cheap British fabrics after 1815.[158] When the industry revived a decade later the smaller centres had succumbed, and more than two-thirds of the spindles were in Ghent itself.[159] Ghent had important advantages. It was itself the port through which the raw cotton was imported and coal for its mills could be brought by water from the Belgian coalfield. The city continued throughout the century to be the dominant centre of the Belgian cotton industry.[160]

Attempts to establish a cotton industry in the northern Netherlands followed quickly on the success of the Ghent undertaking. Mills were built at Eindhoven and then in the Twente district of the eastern Netherlands. This was an area of poor soil, where the peasantry had long supplemented their income with domestic spinning and weaving.[161] Cotton began to replace flax early in the century, but the conversion of this primitive craft into a modern industry was due to the stimulus provided by the Nederlandsche Handel Maatschappij (Dutch Trading Company).[162] Schools were founded to teach the newest methods. A steam-powered mill was built at Almelo, but hand-spinning long remained important, and power-looms were not important before the 1860s. 'In view of the low wages...in Twente', wrote an adviser to the Dutch company, 'it would be better to equip the weaving factories with hand looms and flying shuttles'.[163] The industry was not fully mechanised before 1870,[164] and by this time the railway net had been developed to bring fuel to the area.

The example of Dutch enterprise in Twente prompted a similar development on the other side of the German boundary.[165] Its chief centres – Rheine, Gronau, Bocholt, Nordhorn – lay mostly in the Ems valley, easily accessible from the cotton port of Bremen. The German industry suffered severely during the cotton 'famine' of the 1860s, but subsequently revived.[166] It was heavily dependent on Dutch labour and always maintained close links with the Twente industry, with which it can be said to have formed a single textile 'region'.[167]

Germany Apart from this development in Westphalia, there were three areas where the cotton industry developed significantly during the century. In each, cottons shared the field with other textiles, and in all of them the industry was developed from an earlier linen or woollen manufacture.

The first was the Lower Rhineland from approximately Cologne northwards to the Dutch border. Here cotton insinuated itself as a domestic industry in areas where flax-spinning and linen-weaving had long been the chief rural crafts. It first became established late in the eighteenth century in the hills of Berg, where abundant streams made the mechanisation of spinning a fairly simple matter. Elberfeld and Barmen, strung out along the valley of the Wupper became the first important centre of the cotton industry, and from here it spread to the left-bank towns of Gladbach, Rheydt and Viersen.

The second major cotton-manufacturing region was 'royal' as distinct from Prussian Saxony. Cotton-working had been introduced to this area at the end of the Middle Ages, and early in the nineteenth century it began to displace linen-weaving as the chief *Hausindustrie*. The spinning branch quickly became mechanised and centred in the towns of Chemnitz, Zwickau and Plauen. Weaving continued to be largely by hand until the 1880s, and when Dehn wrote in 1912 the handloom still provided a winter occupation in remote areas of the Ore Mountains. The cotton industry of Saxony extended across the boundary into Bohemia, where it was widely practised within the mountains.

The Saxon cotton industry derived its inspiration from southern Germany,

Table 7.2. *The cotton industry in Germany*

	Spindles 1887		Spindles 1905		Looms 1905	
	No.	%	No.	%	No.	%
North-west Germany						
Rhineland	435,802		1,051,362		} 50,137	
Westphalia	285,828		1,172,222			
Total	721,630	14.3	2,223,584	25.2	50,137	21.7
South Germany						
Alsace	1,375,000		1,511,586		39,919	
Baden	398,172		468,784		16,744	
Bavaria	924,312		1,578,084		32,781	
Württemberg	362,048		714,905		20,133	
Total	3,059,532	60.5	4,273,359	48.4	109,577	47.4
Central and eastern Germany						
Saxony	1,001,569		1,949,313		39,236	
Silesia	75,064		109,320		16,540	
Total	1,076,633	21.3	2,058,633	23.3	55,776	24.1
Rest of Germany	197,000	3.9	274,440	3.1	15,709	6.8
Total, whole of Germany	5,054,795	100	8,830,016	100	231,199	100

as that of the Rhineland did from the Low Countries. It is all the more remarkable that a modern industry was so slow in developing from the traditional crafts in the south German states of Baden, Württemberg and Bavaria. Early in the nineteenth century, however, a small cotton industry, with calico printing works, was established at Lörrach in southern Baden, mainly by Swiss entrepreneurs, similar to those who had established the Alsatian industry on the opposite bank of the Rhine. The industry grew after the formation of the Zollverein, as the Swiss sought to benefit from the widening German market.[168] This was in turn followed by the foundation of a cotton industry in Augsburg and Nuremberg. Spinning and calico printing were carried on in small factories, but much of the weaving was domestic.[169] Again Swiss initiative was significant. Indeed, the territory from the Vosges to the border of Bohemia owed its development as a textiles region largely to Swiss enterprise. By the end of the century south Germany had emerged as the leading cotton textile region in Germany in both spinning and weaving (table 7.2).[170]

The Swiss cotton industry, which did so much to inspire developments in both Alsace and south Germany, had grown up during the later Middle Ages. In 1800 spinning machines were introduced at Saint Gallen, and at about the same date at Zurich.[171] Others followed, mainly within the triangle Zurich-Winterthur-Baden (Aargau). By 1827 there were 106 mechanical spinneries in Canton Zurich, all of them worked by water-power.[172] From here the spinning mills spread up the valleys of the Oberland, into Glarus, Saint-Gallen and Appenzell. Attempts to mechanise calico-weaving were checked by a violent outbreak of Luddism, and as late as 1866–7 handlooms still outnumbered power-looms by almost four to one.[173] In 1900, in Canton Appenzell there were still only 280 power-looms as against 3,731 handlooms.[174] This reluctance to modernise is the probable cause of the sluggish growth of the industry in the latter half of the century. Besso claimed in 1910 that in some rural areas the mills were deserted in summer while the operatives helped with the harvest.[175]

In Austria cotton workshops were established before the end of the eighteenth century at Ebreichsdorf and Schwechat (both near Vienna) and at Linz and Graz.[176] They were mostly engaged in finishing and printing, spinning and weaving being carried on domestically. Mechanical spinning followed in Vorarlberg,[177] a clear extension of the industry of eastern Switzerland, and in Upper and Lower Austria. From here it spread into Tyrol and Styria.[178] From Austria the industry also spread to the mountains of Upper Hungary, or Slovakia, and mills, some of them short-lived, were established at Nitra. A far greater success attended the cotton industry of Bohemia and Moravia, first at Znojmo and Jihlava, and then through the hill country of Moravia and the mountains of north-eastern Bohemia.[179] By 1828 there were 69 spinning mills and about 118,000 spindles in Bohemia; in 1847, 86 mills with 461,000 spindles; and by 1871 there were over 700,000 spindles.[180] According to Purs, the mechanisation of spinning was complete by 1830, and that of weaving between 1851 and 1872,[181] though this may exaggerate the speed of Czech development.[182] In fact the handloom weavers of Bohemia put up a long struggle, like their fellow workers in Silesia, against the power-loom. The industry continued to expand, though more slowly, during the second half of the century, increasing the Bohemian share of the total production of cotton goods within the Habsburg Empire to about 75 per cent.[183] The industry was located mainly in northern Bohemia, in and around the towns of Liberec, Litoměřice and Boleslav, but finishing processes were carried on in Prague. Geographically the Bohemian industry was part of the Saxon-Silesian textile region, with which it had close economic and technical associations.[184]

The chain of development which, basing itself on British initiative, had originated in the Low Countries and Switzerland, came to an end in eastern Poland. In the 1820s a textile industry was established in Łódź and its surrounding villages. It was mainly a factory industry, but spread to Białystok in north-eastern Poland largely as a domesitic industry. From Silesia the domestic cotton industry was carried along the Carpathian foothills to Bielsko-Biała and Andrychów and to the Zamojski estates in south-eastern Poland.[185] Throughout, it was a domestic industry carried on in the cottages of the peasants and vulnerable, as soon as transport had been developed, to the factory products of Łódź.

Southern Europe Modern industry was late in coming to southern Europe. The region as a whole lacked solid fuels, and its water-power resources, except in the mountains, suffered the handicap of a long summer drought. Yet its backwardness cannot be explained wholly in physical terms. There was a long tradition of craft industries, but their market was restricted by poor communications and, in Italy, by political division.

In Spain the cotton textile industry emerged in the middle years of the nineteenth century in the province of Catalonia, 'an industrial civilization...surrounded by a conservative peasantry'.[186] It is usual for historians to contrast the enterprise and progressiveness of the Catalans with the conservative backwardness of the rest of Spain. As the British consul wrote in 1872, Catalonia was 'free from those legitimist and sacerdotal sympathies and traditions which make the strength of Carlism... The Catalan of the upper classes is a liberal, of the lower classes, a republican. He is industrial, commercial and much under the influence of the political literature of France.'[187]

There had long been a small traditional cotton manufacture in Catalonia. It was almost destroyed during the Peninsular War, but revived after 1815. In 1846 the first factory was founded.[188] Almost immediately the recorded import of raw cotton into Spain doubled, and continued to increase, though at a less rapid pace, into the twentieth century. In 1835 only about 3 per cent of the spindles were mechanical; by 1850 more than three-quarters. Mechanisation of weaving came later.[189] In 1850 only a fifth of the looms were power-driven, but by the end of the century the handloom had ceased to be significant.[190] An important part of Catalonia's market was in the surviving parts of the Spanish empire. Their loss in 1898 was severely felt, and thereafter there was no significant growth in the cotton industry.

The industry was located primarily near Barcelona, though there was for a time a secondary centre in Malaga.[191] Factories were few within the city of Barcelona,[192] and most of the industrial development took place, in fact, in Barcelona's hilly hinterland, where water-power was still available for those works that needed it.[193]

The development of the Italian cotton industry was contemporary with that of the Spanish. Not until the 1830s, and then only with tariff protection, was it able to achieve any importance. Mechanical spinning was introduced, with equipment largely from Switzerland, but weaving continued to be by hand in the rural communities. The progress of the industry was held back by the import of fabrics from Austria, where the factory industry had established itself rather earlier.[194] The chief centres of the Italian industry lay along the Alpine foothills, especially in the Habsburg province of Lombardy, where the many Alpine rivers provided power.[195] The chief centres of cottonspinning were near Monza, Gallarate and Busto Arsizio, where much of the skilled labour came from Switzerland and France. The industry also grew up in Piedmont at Cuneo, Novara and Turin, and there was also an outlier of the north Italian industry at Udine in the Venezia.

The Unification of Italy in 1860 created a larger market for struggling Italian industries,[196] just as the Zollverein did for the German. Industry grew more rapidly, especially after the adoption of a protectionist policy in 1878.[197] The spinning industry was completely mechanised, though handloom weaving continued into the twentieth century. In 1904 the British consul reported that a quarter of the looms were still hand-operated.[198] The number both of spindles and of mechanical looms increased sharply in the late nineteenth and early twentieth centuries (table 7.3[199]). About 85 per cent of the spindles were to be found in the northern plain, more than half of them in the Province of Lombardy, and the rest in Campania, Liguria and Tuscany.

From the middle years of the century cottons began to assume a dominant

Table 7.3. *Cotton textile industry in Italy*

	Spindles	Power-looms
1868	450,000	—
1876	764,862	15,517
1900	2,111,000	70,600
1914	4,620,000	120,000

role among Europe's textile industries. Whether measured by the number of spindles and looms, or by the volume of raw cotton used, the industry expanded very rapidly. In continental Europe, from Spain to Poland, there were about 9,000,000 spindles of all kinds about 1850. By 1913 this total had increased to at least 33,000,000 spindles, most of a greatly improved variety. Within the same period consumption of raw cotton increased from less than 175,000 tonnes to about 1,350,000 tonnes, an almost eightfold increase. The only other textile that came close to this rate of increase was jute, and this only because jute-working started from a very low base. This rate of increase was fastest in Germany, Italy and the Habsburg Empire; slowest in France and Belgium. In 1850 France had by far the largest cotton industry; by 1913, it was only about half the size of the German.

Silk

The silk industry differed from other textile industries in that it catered for a luxury market. Though there was a marked increase in the output of ribbons in the course of the century, there was never any question of mass production, and manual processes continued to dominate production. As in the linen industry, the primary material was a product of peasant agriculture, and the first stages in processing were normally accomplished in the village. The weaving of silk fabric, however, was rarely left to the rough hands of the peasant, but was a craft, sometimes a factory industry. The chief – almost the only – European source of raw silk was Italy. Here silk-reeling and silk-throwing were 'widely and increasingly diffused through the countryside as the part-time occupation of the peasants'.[200]

Reeling consisted of unwinding the silk thread from the cocoons; throwing was the twisting of these strands to produce a thread strong enough for weaving. Throughout the nineteenth century most of the thrown silk was exported to supply the silk-weaving industries of France, Switzerland and Germany. Some was re-imported as dyed silk for weaving in Italy, though the Italian silk-weaving industry remained small until late in the nineteenth century. The production of thrown silk was widespread in northern Italy, and was also found in central and southern Italy. It was 'by far the most important textile industry', and at the beginning of the present century was described as 'progressing rapidly'.[201]

The most important consumer of Italy's thrown silk was the silk-weaving industry of Lyons.[202] Here the ancient craft, long practised in the Croix Rousse district of the city, was spreading into the surrounding countryside (see p. 413).

Amongst the places which benefited from the dispersal of the Lyons industry was Saint-Étienne, where ribbon-weaving rivalled the traditional metal industry. Other centres of the French silk industry were small and unimportant. They included Tours, Ganges (Cevennes) and Paris itself. All ceased production during the century.

In its early stages the industry had drawn most of its raw material from native producers in the south of France. This source soon proved to be inadequate, and was, furthermore, reduced by the silkworm disease. By the middle of the century most of the raw and thrown silk was from Italy and the Middle East.[203]

The Swiss silk industry was established in the eighteenth century in the Zurich region. It spread after 1815 to Basel, Solothurn and the small towns in Aargau and Appenzell. Basel developed an important ribbon manufacture,[204] but Zurich remained the chief centre of the Swiss industry. In 1836 it was estimated that there were 10,000 silk looms, employing 12,000–13,000 rural artisans, most of them on a part-time basis.[205]

The only other important European centre of the silk industry was the lower Rhineland. It had been established in Krefeld by the van der Leyens in the 1720s,[206] but spread to other towns of the lower Rhineland. By 1822 more than half the silk looms were in or near the twin towns of Elberfeld and Barmen,[207] but by 1849 silk-weaving had spread through much of the Düsseldorf region.[208]

In the course of the eighteenth century a silk industry had been established at a number of other places. In every case it had been encouraged by courtly patronage. The largest centre was Berlin itself, but the industry had also a short life at Potsdam and Frankfurt-on-Oder.[209] None of these manufactures survived into the harsher and more competitive conditions of the mid-nineteenth century. Nevertheless, the growth of the German silk industry was particularly rapid. Statistics are scanty and unreliable, but a German source, in 1879, credited Germany with 45 per cent by value of Europe's (including Great Britain's) silk production.[210]

The derivative textile industries

The decay of rural textile crafts during the nineteenth century left a legacy of manufactures which used textile materials and relied upon the skills which had been developed in spinning and weaving. This could be paralleled in the metal industries, in which a variety of iron-using crafts survived the disappearance of the primary smelting and refining industries. The secondary textile industries included lace-making, embroidery and stocking-knitting. Most of these occupations called for little capital, and they could thus be taken up as the primary crafts declined. They could – and did – use all kinds of textile materials and became a winter occupation of the peasants in much of western and central Europe.

The use of traditional patterns and designs did much to popularise the products of these cottage industries even to the point at which they were imitated in factories. In no country were these textile crafts more strongly developed than in Switzerland. Hand-knitting and embroidery were widespread, especially in eastern Switzerland and in the Jura Mountains.[211] Hand-knitting was sufficiently important in canton Zurich to support a considerable export. Machine embroidery and knitting were, however, tending to replace the domestic crafts, and the Cantons of Saint Gallen, Thurgau and Appenzell came before the end of the century to be dotted with small workshops, as these residual handicrafts in their turn came to be mechanised and absorbed into the factory system. In the 1890s 10,000–11,000 were said to have been employed in such workshops.[212]

In France employment to lace-making, embroidery and hand-knitting reached almost a quarter of a million about 1860,[213] though most were not fully employed. Lace-making was particularly well developed in western Normandy, where early in the century 6,000 women and children were employed in the town of Caen alone and 20,000 in the neighbouring countryside. The lace of Alençon and Luxeuil was particularly noted,[214] but lace-making was also important in Brittany, the Auvergne and Vosges, indeed almost everywhere where agricultural conditions were poor. The machine production of lace began in 1824 at Calais. It proved to be successful and spread to several of the textile towns.[215] This ended the traditional lace industry in much of France, as it had done in Switzerland. At the end of the century the craft continued to be practised in Normandy, so it was said, only by a few old women, but was somewhat more prosperous in the Auvergne.[216]

The lace and hand-knitting industry of Belgium followed the same pattern of development.[217] Traditional designs were produced in the villages of Flanders and Brabant, and marketed in Brussels, Malines and Bruges, but were gradually overtaken by factory production.

Statistics of employment and production in the lace, embroidery and hand-knitting industries are almost wholly lacking. It would appear, however, that in the middle years of the century, almost half was to be found in France, and that France, Belgium and Switzerland together accounted for three-quarters. In Germany it seems to have been of small importance, and barely merited a mention in the voluminous reports on domestic crafts published by the *Verein für Sozialpolitik*.

Conclusion

By the middle years of the nineteenth century the textile industries had assumed a geographical pattern which they were to retain until the First World War. They were located in a number of distinct regions, which had been formed by, as it were, a process of crystallisation, as larger centres of production gradually separated out from the broad mass of handicraft industries. This process of concentration first showed itself in the finishing branches – bleaching, dyeing, calico-printing – then in spinning and last in weaving. As mechanisation progressed in each of these branches of the industry, so more and more of the productive processes came to be concentrated in a small number of centres. By the early twentieth century the whole industry with the exception of a small amount of handloom weaving, was carried on in factories. Many were very small; some still used water-power and were located in a rural environment, but with very few exceptions they formed clusters within each of the textile regions.

A feature of these regions is that they were not constrained by national boundaries. Like conditions on each side of a political boundary tended to produce a similar economic development. Within each region there was a movement of entrepreneurs and artisans, of capital and technology. One can, in very general terms, plot the flow of technical innovation within each region and from one region to the next.

A second feature of these regions is that each embraced one or more foci which served as central-places for the industry. Rouen and Lyons; Lille and Mulhouse in France; Aachen, Düsseldorf and Chemnitz in Germany; Zurich and Milan, Turin and Barcelona; Liberec, Brno and Łódź. As a general rule these had once been important for the actual manufacture of textiles, and some in fact remained so, but they also served as bases for the merchant capitalists who put out materials to the domestic artisans. In the course of time the latter function became more important and the former less so, until the role of these towns became that of commercial and business capitals of their respective regions. They were the Leeds and Manchesters of continental Europe.

A third feature of these textile regions is that, in contrast with those of Great Britain, none of them was really specialised. In each there was commonly a dominant branch, whether cotton, woollens, linen or silk, but most other branches were also to be found. Later in the century the older manufactures were in many instances joined by 'mixed' fabrics and by knitwear, hosiery and related industries. There was even, in some regions, a tendency for the mix of textiles to vary through the century with the changing profitability of the different branches.

The most important factor in the location of these regions, fourthly, was historical; the modern industry developed out of a proto-industrial craft structure. Since the latter made no use of mechanical power, the location of the later industry cannot be said to have been determined by the availability of fuel resources, or even of water-power. This does not mean that within the broad regional setting suitable sites for water-power were not chosen when the industry came to be mechanised. But, with the possible exception of the Łódź area of Poland, power resources were not significant in the initial formation of the textile regions of Europe. Some indeed – the Plain of Alsace, the Lille district, the Netherlands, and north-western Germany – were conspicuously lacking in sources of energy. Certainly proximity to a coalfield played no part in determining the location of the industry, though once a manufacture had been established the availability of coal greatly influenced its profitability and future success.

Lastly, as the proto-industrial forms of production decayed and the industry gradually concentrated at a relatively small number of sites, it left behind a scatter of derivative industries. These made use of the products of the major textile industries, notably yarn, and of the infrastructure which that industry had built up.

The iron and steel industries

The textile industries had established a pattern by 1850 which was to last with only minor changes until the First World War. The geography of iron and steel, on the other hand, continued to alter throughout the period. This section will trace the changes that took place in the spatial pattern of this group of industries in relation to an evolving technology, a changing raw material base, and a fluctuating market for its products.

Technology of iron and steel

The technology of the textile industries underwent little fundamental change in the second half of the nineteenth century, a fact which helps to account for their geographical stability. That of iron and steel, however, continued to develop. Not only was charcoal smelting almost wholly eliminated, but the furnaces themselves became incomparably larger and more efficient in the use of fuel. The refining processes evolved from the simple hearth to the puddling furnace, and steelmaking from the cementation furnace and crucible to the converter, open-hearth and electric furnace.

In 1815 all iron-smelting in continental Europe, with the possible exception of a furnace in Upper Silesia, was still with charcoal. Everywhere problems with fuel, with furnace design, and with the blast had combined to defeat the many attempts made to use coal or coke. The problem was becoming urgent; demand for iron was beginning to increase, while in many parts of Europe the forests were being depleted and the price of charcoal was rising. In 1819 Héron de Villefosse described in more careful and scientific terms than those used hitherto the qualities required in the coal and the best methods of coking it,[218] and at the same time the German resources were examined by C. J. B. Karsten.[219]

Progress was being made in the use of mineral fuel. In 1821 coke was used with success at Couvin in the southern Low Countries, and two years later by Cockerill at his Seraing works near Liège.[220] Stumm followed at his works at Dillingen and Neunkirchen in the Saar,[221] and in 1849 coke was first used successfully in the Ruhr at the Friedrich-Wilhelmshütte at Mühlheim. It was simply a matter of choosing the right coal, coking it carefully and using a large furnace with a powerful blast.[222] It may seem surprising that the adoption of coke smelting was not more rapid. Indeed, charcoal furnaces continued to be built and the output of charcoal iron to increase until after the mid-century. There were good economic and technical reasons for this, apart from the natural conservatism of furnace-masters. Coke-smelted iron, owing to its high sulphur content, did not refine well on the hearth or in the puddling furnace. Existing furnaces could not be used effectively with coke without extensive rebuilding and a much higher operating temperature, and this in turn called for a more powerful blast. Early attempts to make coke were costly and inefficient,[223] but after the introduction of the closed retort the quality of the coke improved, but only when the by-products began to be collected and used did coke become cheap and readily available. After about 1850, when many new furnaces were building, coke began to take over from charcoal, and the map of European ironworking changed profoundly.

The refining of high-carbon iron also underwent radical changes. First, the refining hearth was gradually abandoned, and from the 1820s until the 1860s the puddling furnace ruled supreme. It accomplished the task of eliminating carbon and producing a mass of wrought iron quickly, and furthermore, had the immense advantage of being able to use coal, since fuel and metal did not come into direct contact. The process was introduced into France soon after 1815, and by 1827 there were said to have been 149 puddling furnaces in operation.[224] Its use spread quickly – to Belgium, Aachen, the Saar, the Ruhr, and on to Moravia and Silesia. Charcoal-smelted iron from the Siegerland was brought to the Ruhr to be refined with coal. Coal was taken in the opposite direction to fuel the puddling furnaces at Siegen. In general, however, the demand for coal was so extravagant that it proved more profitable to take the pig-iron from the smelting furnaces to the coal for refining. It was the puddling furnace that first brought the iron industry to the coalfields, and it was there that iron for most of the early railways was refined and rolled.

The puddling furnace was always associated with the rolling mill. The rolls squeezed out the slag which remained in the metal; they also produced the shapes most needed by both industry and railways. Though the earliest rolls were operated by water-wheels, the energy produced in this way soon proved inadequate, and they were replaced by the steam-engine. In fact, a high proportion of stationary steam-engines in use in the mid-nineteenth century were employed in ironworks.

Steel was made by 'recarburating' the wrought or bar-iron produced on the hearth or in the puddling furnace. Bar-iron was recarbonised in the cementation furnace, and since the absorption of carbon was uneven the metal was melted down in a crucible to secure an even quality before being poured into an ingot mould. This process was perfected by Krupp at Essen, but remained very costly, could produce only small pieces of steel, and was economically prohibitive for making rails. Clearly, a less complex and expensive method of reducing the carbon content of iron to the range required for steel was one of the most urgent requirements of the industry, now that a large and growing market for steel had developed in engineering, shipbuilding and the railways.

Attempts were made to puddle steel, by the simple process of stopping the operation before the carbon had been wholly burned out. This method was far from satisfactory, and was little used. The Bessemer and open-hearth processes were then introduced, and together accounted for most of the steel made up to the First World War. The former, developed in the 1850s, consisted in passing a stream of air through the hot metal. The carbon was oxydised, and passed off in a shower of sparks, leaving an almost pure iron which could be recarburated according to the type of steel required. The 'converter' was used successfully in England in 1860 and soon after was introduced into France.[225] Its advantages were immense. The process was very rapid; the heat required was generated by the 'combustion' of the carbon, and ingots of any size could be poured.

The Bessemer process quickly spread – more rapidly, in fact, than any previous innovation in the industry. Within ten years it had been adopted at Le Creusot, Montluçon and Saint-Étienne in France; at Seraing in Belgium, in the Ruhr, Upper Silesia and Austria, Hungary and Sweden.

It was not long, however, before imperfections showed themselves in Bessemer iron. The air which passed through the metal left microscopic bubbles which became a source of weakness. More serious was the fact that the stream of air failed to eliminate phosphorus that had passed into the metal from the ore. This in turn made it brittle, or 'cold-short' in the language of the industry. Immediately low-phosphorus ores acquired a new value. They were far from abundant and most deposits were quite small. Amongst them were the ores of Siegen, the *fer fort* of France, and many of those of Styria and Carinthia, but the best hope lay in the deposits of northern Spain and central Sweden.

At about the same time Pierre Martin, a French ironmaster at Sireuil, in *dépt* Charente, adapted the regenerative furnace, which had previously been developed by Werner Siemens, to burn pre-heated gases above a mass of iron. The great heat of combustion caused the oxidation and removal of carbon and of most other impurities either as waste gases or in the slag. The Siemens-Martin, or open-hearth process, came into widespread use in western Europe. Though a great deal slower than the Bessemer converter, it was in many ways superior. It allowed a much closer control over the process, which could be halted at any point, and the resulting metal was free of the minute vesicles which had impaired the quality of Bessemer steel. The open-hearth required a gas supply which was supplied most easily from the coke-ovens – an important factor in the growth of integrated iron and steelworks. It could use scrap as readily – indeed more so – than new metal from the furnace, and could thus be established independently of a blast furnace works, wherever, in fact, pig-iron or scrap metal were available together with a gas supply.

But the Siemens-Martin process was no solution to the fundamental problem of eliminating phosphorus from the metal. This exercised metallurgists and chemists for almost two decades before, in 1879, success was achieved by Gilchrist Thomas. His method was to line the converter with basic firebricks instead of the siliceous bricks used hitherto. This in turn allowed the addition of lime, which combined with the phosphorus without at the same time reacting with and destroying the converter lining. Thomas's first successful 'blow' was at Middlesbrough, and it was at once heralded as a most significant development in modern steel-making. It changed abruptly the relative values of the ore deposits of Europe and opened up unlimited prospects for the industry. The low-phosphorus ores lost much of their importance, though they continued to be used, for not all works were adapted to use the Thomas process. The vast deposits of high-phosphorus ores, which made up the greater part of Europe's reserves – 60 per cent at a rough estimate – at once acquired an immense value. The rush to obtain the patent to use the Thomas process was paralleled by the scramble to gain concessions to work the high-phosphorus ores of Lorraine and Luxembourg.[226]

France and Germany were the chief beneficiaries of the new process. Within weeks of the publication of Thomas's experiment it was in use in the Ruhr, and shortly thereafter in Lorraine, Luxembourg and Austria. The Thomas process was quickly adapted to the open-hearth, which now became the chief source of quality steel until replaced by the electric furnace. The proportion of steel made by the basic process, either in the converter or the open-hearth, rose in Germany from 2.4 per cent in 1880 to 64.3 in 1913, and in France from none to 65.4.

Manufacturing

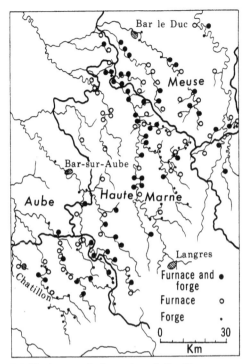

7.3 The iron-producing region of *dépt* Marne

The raw-material base

The iron and steel industry was based upon only two essential raw materials: iron ore and fuel. The latter was entirely consumed in the process, with the exception only of the by-products of the coking ovens. The former was reduced in weight to an extent which varied with the grade of the ore. These considerations, together with variations in the coking quality of coal, profoundly influenced the location and fortunes of the iron and steel industry after the mid-nineteenth century.

In the proto-industrial conditions which prevailed in the early years of the century small, stone-built furnaces were used to smelt ores, often of indifferent quality, with charcoal fuel. The furnaces themselves were commonly sited within a very few kilometres of the ore and not much farther from the forests which provided the charcoal. The fact that these resources were quickly depleted mattered little. The furnace represented no great capital investment, and it could be abandoned and a fresh start made elsewhere. Refineries were similarly insubstantial. Such conditions were repeated hundreds of times in western and central Europe.

In a traditional ironworks the metal from the furnace would have been taken to a refinery where, with bellows operated by a water-wheel, it would have been reduced to soft or bar-iron. The first significant breach in this traditional pattern was made by the puddling furnace which gradually replaced the refinery. It called for a different fuel, coal, of which it was inordinately extravagant. The quality of coal was of no great significance; even lignite is known to have been used. But the puddling furnaces had to be built on, or within easy reach of a coalfield. In fact, almost all that were operating by mid-century were within a few kilometres of a coal mine.

The spreading use of mineral fuel in the blast furnace marked a second phase in the transfer of the iron and steel industry to the coalfields. The early coke-fired furnaces were no less demanding of fuel than the puddling furnace. Early in the century about 2.5 tonnes of coke were required in an efficient furnace to smelt a tonne of pig-iron, and this ratio would have been considerably higher with low-grade ores.[227] In the course of the century furnaces became very much larger and their operating temperatures higher; the ratio of coke to metal fell, until by its end a tonne of iron could be produced with one tonne of coke. This, combined with the decline in the use of the puddling process after about 1880, meant that a coalfield site became less attractive to the industry by the end of the century.

Whereas the puddling process was able to use any quality of coal, the smelting industry was more discriminating. It required a firm coke, able to resist the crushing pressure inside the furnace, and the larger the furnace the more important this quality became. The coke had also to be free of sulphur and other impurities which might impart a weakness to the metal. Some of the early attempts to use coke had failed because the coke was too soft. Very few varieties of coal fitted these exacting requirements. French coal was notoriously poor, and France came ultimately to depend heavily on Germany for supplies of metallurgical coke. Saar coal was better, but Belgium had little good coking-coal. The best in Europe was the *Fettkohle* of the Ruhr, which came from intermediate seams in the coal series.[228] In Upper Silesia also good coking-coal was found in the so-called *Sattelflöze*.[229] Unquestionably the presence of coking-coal influenced the movement of the iron-smelting industry to the Ruhr, where several of the leading industrialists, among them Krupp, Thyssen and Hoesch, acquired mining concessions in the coking-coal area of the central Ruhr. The price of coal varied greatly, and coking-coal was commonly amongst the more expensive. Its pithead price was lowest in Upper Silesia; highest in Liège and Hainault, where the coal furthermore was poorly suited for metallurgical coke. The Saar contained some coking-coal, but only Ruhr coal combined good coking quality with relatively low price and abundant reserves.

The extent and quality of Europe's resources in iron ore have already been examined (pp. 49–53). The history of their exploitation is that of the gradual abandonment of the small deposits as they became exhausted or ceased to be economic, and of a growing concentration on the larger. By the 1870s there was a growing shortage of good ores. Belgium, once declared to be 'the most richly endowed of our continent' in this respect, was obliged to import.[230] In France ironmasters were scraping the last crumbs of ore from their scattered deposits.[231] In the Saar they were forced to add *minette* from Lorraine to the furnace charge, and in the Ruhr the local blackband had to be supplemented with ore from the Sieg-Lahn-Dill region as well as from Spain. In Upper Silesia also the local deposits were ceasing in the 1860s to satisfy demand and ore had to be obtained from Hungary, Sweden and even Spain.[232] In Germany as a whole ore production was static despite the growing demand for steel.

The ultimate solution to the problem of the growing shortage of iron ore was to develop the bedded ores of Jurassic age, which occurred in such abundance in Lorraine, Luxembourg and the North German Plain. Their existence had long been known and for centuries attempts had been made to smelt them. But the ores were high in phosphorus; and *minette*, as this ore was contemptuously called, was fit only for castings. This was changed by Gilchrist Thomas's invention. The looming ore shortage disappeared. High-quality, non-phosphoric ores could be reserved for high-grade and special steels; the *minette* and similar ores would do for the basic process. Steel-production increased almost without interruption up to the First World War, and a growing proportion of it derived from the low-grade, phosphoric ores (fig. 7.5).

A more thorough exploitation of the Jurassic ores began at once. In German-held Lorraine there was a scramble for concessions, and several were acquired by the steel barons who managed the Ruhr industry. Development was for several years almost restricted to German-held Lorraine. The only *known* reserves on the French side of the boundary were near Longwy in the north and Nancy in the south, and it was generally supposed that the ore-bearing beds died out close to the French boundary. In 1882, however, bores were made near Briey and iron ore was reached.[233] Subsequent bores revealed the vast extent of the Briey 'Basin', but they also showed how difficult it would be to exploit. The ores lay below the water-table, and the problem of sinking a shaft through wet and porous rocks defeated the mining engineers for several years. It was not until 1899 that the first shaft – at Auboué – came into use. During the following years French Lorraine became the most important source of iron ore in France. Its ores were more calcareous than the siliceous ores from German-held Lorraine and were very useful to blend with them. For this reason the German steel concerns also acquired concessions in French Lorraine (fig. 8.7).[234]

The Jurassic ores of north Germany had long been worked on a small scale. In the 1850s, smelting works were built near Osnabrück and Ilsede and a few years later at Peine. The metal produced was not used for steel-making, but in the 1880s the Thomas process was introduced and mining began to be pursued on a more serious scale. In Bavaria also the use of phosphoric ores saved the Maxhütte from closure, and in Poland a basic steelworks was built at Częstochowa.

By the end of the nineteenth century iron production had come to be dominated by the low-grade phosphoric ores and the basic process. It is difficult to be precise but it is unlikely that these ores contributed less than 60 per cent of the iron smelted in Europe, and probable that half of this came from the *minette* of Lorraine and Luxembourg.

In the 1880s and 1890s the links between the Ruhr and Lorraine, between the coal and the iron ore, became closer, though a complete mutual dependence was not and could not be achieved. The idea that a shuttle service of freight cars would carry coal or coke in one direction and ore in the other never worked out in practice, and, though Lorraine obtained at most a little more than half its fuel from the Ruhr, the latter seems never to have taken more than a fifth at most of its ore from Lorraine. The Belgian and French smelting industries were, however, much more dependent on Lorraine ore, and the Saar obtained most of its supply from this source.[235]

The problem was in part one of transport. The Moselle and Rhine, it might have been supposed, would have formed an ideal link between Lorraine and both the Ruhr and the Saar. The Rhine presented no problems, but the Moselle was shallow, meandering and unsuitable for any but the smallest craft. All proposals to improve it were abortive and both coal and ore continued to travel by rail (see p. 392).

Towards the end of the nineteenth century ore mined in western and central Europe began to be supplemented with that brought in from overseas. The movement of so bulky a material was a great deal cheaper by water than overland, and it was thus practicable to import it from quite distant sources. Such a movement of iron ore had indeed become significant after the adoption of the Bessemer process, when demand was high for low-phosphorus ores. Ore from northern Spain was imported by sea into France, Belgium and Germany, and – the first from a non-European source – Algerian ore was imported into southern France and taken up the Rhône to the ironworks near Saint-Étienne.

The expectation that imports of ore by sea would continue to be important led to the establishment of ironworks at points on the coast where fuel was readily available. These included the French Mediterranean coast, Bayonne and Bordeaux, the Loire estuary and the north French coast. In Germany coastal ironworks were built at Bremen, Emden, Lübeck and Stettin (Szczeczin).[236]

Furnaces built on the north German coast were clearly intended to smelt Swedish ore, which was much cheaper here than in the Ruhr or Silesia. In the early years of the present century Sweden became a major source of ore for the ironworks of western and central Europe. About 1890 the reserves of high-grade magnetite in northern Sweden were opened up (see p. 53). At the same time the realisation in Germany that Swedish ores were of increasing importance led to the creation of a water connection between the Ruhr and the North Sea ports. In 1899 the Dortmund–Ems Canal was opened (see p. 376), to be continued westwards to the Rhine by the Herne Canal. In 1913 the Ruhr is estimated to have obtained its ore supply from:[237]

Lorraine and Luxembourg	24 per cent
Other German sources	10 per cent
France	7 per cent
Overseas (including Sweden)	59 per cent

Of the overseas ore perhaps three-quarters would have come from Sweden.

The market

It is trite to observe that the market for the products of the iron and steel industry was expanding both in size and in range during the century. Indeed the increase in the volume of iron and steel products must have been of the order of seventyfold between 1815–20 and 1912–13. The market was characterised by certain general features. In the first place iron-castings declined relatively and even absolutely. High-carbon iron had been used for many purposes only because the cost of wrought iron or steel was high, and its usefulness diminished as the price differential between them contracted. There was, however, a late revival in the demand for cast-iron pipes to serve the spreading network of urban water, gas and sewage lines. One important works – Pont-à-Mousson in Lorraine – was developed exclusively for the production of cast-iron pipes.[238]

Until 1870 or even later most of the increasing volume of iron went to the making of bar- or wrought iron, either on the refining hearth or in the puddling furnace. The largest single market for this metal was railway construction. Railway track became heavier during the century as the weights which it was required to support increased. In the 1830s, a kilometre of track incorporated about 35 tonnes of iron.[239] By the mid-century this had increased to about 85 tonnes. At this time track was being laid at the rate of about 3,300 km a year, and one may fairly estimate that the total consumption of iron in laying *new* track was of the order of 300,000 tonnes a year. To this should be added the iron used in replacing existing track, since the older and thinner rails wore out fast, and in building locomotives and freight cars. In England in 1848 the average weight of a goods train, including locomotive and tender, was about 175 tonnes. One does not know the rate at which track was replaced and locomotives and trains built, but it would seem unlikely that the railways during the period of most active construction consumed less than half a million tons of puddled iron a year. This was, in fact, more than was produced in continental Europe at this time, and demand could be satisfied only by importing metal from Great Britain and Russia.

A small though increasing proportion of the bar-iron was converted to steel by the cumbersome and costly methods summarised above. Demand was sufficient to warrant a number of specialised steel-makers, among them Krupp's *Gussstahlwerk* at Essen, Jakob Mayer at Bochum and Terrenoire at Saint-Étienne. Some of the steel was used by craftsmen in the tool and cutlery industries at Remscheid and Solingen, and by the gunsmiths of Liège and Suhl and elsewhere. Most, however, was absorbed by heavy engineering: steam-engines, locomotives and machine tools.

After the introduction of the Bessemer and open-hearth processes, steel began to replace puddled iron for most purposes. Between 1880 and 1913 there was a fifteenfold increase in steel production, the greater part of it basic or Thomas steel. The first generation of railways had been made of wrought iron; the second was of steel. A vast new market was opened in the shipbuilding industry. The first iron ships were also of wrought iron, but this was quickly replaced by steel – generally open-hearth steel for reasons of safety. Early in the twentieth century the shipbuilding industry was taking almost a million tonnes of steel a year in continental Europe, as well as two millions in Great Britain. Chemical engineering was another new market for steel, and older markets in mechanical engineering and durable consumer goods expanded greatly.

The spatial pattern of iron- and steel-making

The factors which have been discussed – technology, raw materials and market – together brought about a changing geographical pattern in the iron and steel industry. For the first decade or two the industry continued to be dominated by its proto-industrial technology and structure, with a multitude of small, charcoal-fired units. In some parts of Europe these lingered on until the end of the century, insulated by distance from competition and protected by the conservatism of peasant demand.

Iron production in two areas will serve to illustrate the spatial pattern of this proto-industrial pattern. Amongst the more important iron-producing regions of pre-industrial France was Champagne, very roughly the *dépt* of Marne.

Haute Marne

In the mid-nineteenth century *dépt* Haute Marne was, and had long been, the chief French source of pig- and bar-iron. In 1826 it produced about 18 per cent of the iron smelted together with 16 per cent of the refined metal.[240] The industry of *dépt* Marne extended across departmental boundaries, so that the region probably produced at least a quarter of the French output (fig. 7.3). It retained this primacy until the 1850s. Its prosperity was founded upon limited reserves of good-quality, high-grade ore and an abundance of charcoal. The ore itself occurred mainly in cavities in the Jurassic limestone.[241] It was found close to the surface and was easy to mine. The metal obtained from it was relatively free of phosphorus and could be refined to a good bar- or forge-iron. At first charcoal was used exclusively both in the furnace and on the refining hearth.[242] But the forests were becoming depleted and, soon after the primary railway net had been completed, coal began to be brought in. By the 1860s some 12 per cent of active furnaces were using coke, and about a third a mixture of coke and charcoal.

Throughout its history, the Marne iron industry was dependent on water-power to work both the bellows and the rolls.[243] This governed the location of works. They lay, without exception, along the rivers which went to make the Marne, Aube and Seine. Refining-hearths, together with their rolling-mills, were commonly at some little distance from the furnaces, since the available water-power could not support both at any single point. There was never any attempt to use steam-power.

Soon after 1850 the industry began to experience difficulties.[244] The raw materials on which it was based were becoming scarcer and more expensive, at a time when iron produced near the coalfields was becoming cheaper. The problem of transport from the smelting or refining works to the main highway or to a navigable river had always been difficult, and as the market became more competitive, so transport costs grew more burdensome. In a few instances attempts were made to modernise the industry by increasing the size of furnaces and adopting puddling, but this merely increased the charges for transporting coal.[245] One by one the less economic furnaces and refineries closed,[246] and the number of active furnaces fell from 75 in 1864 to 33 in 1880, ten in 1890 and only four in 1910.[247]

Central Sweden

To this picture of a declining charcoal-iron industry central Sweden offers a partial exception. Its traditional industry had been during the eighteenth century the most important in Europe, owing in large measure to the excellence of its ores and the abundance of its forests. Late in the century it had to face strong competition from the Russian Urals industry, but nevertheless continued to provide much of the bar-iron for the Sheffield steel industry.[248] During this period bar-iron made up some 60 per cent by value of Swedish exports.

Smelting was carried on close to the ore deposits, but tended to be 'decentralized into as many small units as possible... evenly distributed over the forest area'[249] to reduce the cost of transporting charcoal. The government furthermore restricted the growth of the industry in order to conserve the forests. The ironworkers fitted into a complex system. Many were at the same time small-scale farmers.[250] To the east and south of the metal-working lay a zone in which the farmers supplemented their income by working as carters, taking supplementary food to the mining areas and returning with loads of bar-iron. Lastly, an outer zone of exclusively agricultural land supplied food to the mining area (fig. 6.10).[251] Within the iron-working area itself refining was practised away from the smelting, in order not to put too great a pressure on charcoal resources.

In the 1830s and 1840s, however, the Swedish export began to meet with competition from cheaper puddled iron made in the West. The high cost of fuel prevented Sweden from adopting the puddling process. Instead, the Swedish ironmasters chose to change the basis of their industry and to concentrate on producing high-quality steel which would be above the competition of puddled iron. Charcoal-smelting continued, but the traditional refineries were gradually phased out and 'the last remnants of the old system were swept away about... 1860'.[252] In its place the Lancashire Hearth was adopted.[253]

There was a gradual concentration of the industry.[254] Small furnaces disappeared, and the average yield of those that survived rose from less than a thousand tonnes before 1860 to 6,000 tones half a century later.[255] The Lancashire method proved to be extravagant of fuel; the Bessemer method was ill-suited to produce the high-quality steel in which Sweden was specialising; from the 1890s the open-hearth became important, but the ultimate solution to Sweden's problem was the electric furnace, which came into widespread use after 1910.[256] The use of these processes effectively excluded the small and traditional ironworks. Sweden thus became, in the words of Söderlund, 'the only country in the world without coal deposits, where a virtually unprotected iron industry has been able not only to survive but to keep its position as one of the major export industries'.[257]

Sweden never became a large producer. The export of bar-iron declined, and most of the iron smelted went to the production of high-quality steel, output of which just exceeded half a million tonnes before the War.[258] Much of it went into the manufacture of steel goods for which Sweden became noted, such as dairy machinery, ball-bearings, telephone equipment – all of it calling for small quantities of the highest-quality steel and a large input of skilled labour.[259]

Change in the nineteenth century

The proto-industrial pattern of ironworking lingered on through the century. It contracted in France and Germany, but even in the early twentieth century charcoal ironworks were still to be found remote from coalfields and commercial centres. From the 1820s, however, a new industrial pattern began to develop, superimposed like a palimpsest on the old. It was characterised by the movement of the industry to the coalfields. The puddling process was in the forefront, closely followed by the smelting branch. The extent of the coal resources and the qualities of the coal did not seem to matter greatly. After all, *any* coal could be used in the puddling furnace, and the *small* blast furnaces were not too discriminating in the quality of the coke they used.

In France, between 1840 and 1860, the small coal basins that ringed the Central Massif became a kind of Mecca to the industrialist. The Saint-Étienne coalfield in *dépt* Haute Loire was first developed. A blast furnace was built in 1818 and was quickly followed by puddling and steelworks as well as by other smelting works (see p. 416). The old and decaying works at Le Creusot, beside the Le Creusot-Blanzy coalfield, were redeveloped by Schneider Frères after 1836.[260] On the opposite side of the Massif the Duc de Cazes founded the works which bore his name on the Carmaux coalfield.[261] At the same time the Tamaris and Bessèges works were founded on the Alès coalfield in Provence, and to the north, in the Bourbonnais and Berry, ironworks were built at Montluçon, Commentry, Fourchambault and Nevers, each of them on or close to one or more of the small coal basins of this region.

Another group of furnaces and puddling works arose in the Sambre valley near Maubeuge, within reach by navigable waterway of the Belgian coalfield. Others grew up on the French extension of the same field near Valenciennes. Within Belgium works were established on the Liège coalfield in the 1820s, first by Orban at Grivegnée, then by John Cockerill in the old archiepiscopal palace at Seraing. Others quickly followed on both the Liège and the Charleroi divisions of the field. By 1842 there were no less than 42 blast furnaces on the Belgian coalfield and a proportionate number of puddling furnaces.

In Germany the story was the same: first charcoal-iron from the scattered furnaces was taken to the coalfield for puddling; then furnaces fuelled with coke were established nearby. In the Aachen region charcoal-iron from the Eifel as well as coke-smelted iron from Liège were puddled at Eschweiler and Lendersdorf, and in 1857 blast furnaces were built. But Aachen must count as a failed coalfield. Local ores ran out; Belgian pig ceased, partly for tariff reasons, to be imported, and the local coke proved to be unsuitable for the more modern furnaces. One by one the ironworks near Aachen closed, and the owners of two at least amongst them transferred their assets to the Ruhr.

Development on the Ruhr coalfield, the richest and, in terms of the quality of coal, the most varied in Europe, followed the same course. Puddling was established at Wetter-on-Ruhr in 1826; then at Dortmund, Oberhausen, Essen and elsewhere. Steel-making developed at Essen and Bochum. The metal was at first obtained from the furnaces of the Siegerland. Then blackband ore was discovered in the coal measures and at first smelted, paradoxically, with charcoal, but after 1849 increasingly with coke. Coke-fuelled blast furnaces were built during the next 20 years in a broad belt from Duisburg in the west to Dortmund and Hörde in the east, just where the best coking-coal was obtained. Within the span of two decades the Ruhr had become the largest concentration of iron-smelting and refining works in Europe and by far its largest single source of metal.

The Saar coalfield contained smaller and less varied resources. The greater part lay in Germany, but a westward extension dipped beneath the Triassic rocks of Lorraine. Small ironworks in the surrounding hills at first supplied metal to puddling works, first at Neunkirchen, then at Dillingen, Saarbrücken and elsewhere on the coalfield. The first smelting works on the coalfield were built by de Wendel at Styring, in the French sector, quickly followed by works at Burbach, Halberg and Völklingen.

The only other coalfield that became significant for its iron industry was the Upper Silesian. Here the familiar model was repeated, but with a significant variation; iron-smelting came before puddling. This was due to the initiative of the Prussian government, which introduced British ironmasters to instruct the local workers. Works were established first at Gleiwitz, then at Königshütte. Development was slow, but by 1830 there were about ten ironworks using coke for smelting. But charcoal furnaces long continued to be used, even on the coalfield. The landowners were as interested in using charcoal from their vast forests as they were in opening up coal mines on their estates. Much of the coal, furthermore, was not well suited to blast-furnace use, and throughout the century furnaces remained a great deal smaller and less economic than those in western Germany. Charcoal smelting did not disappear from Upper Silesia until the 1880s.

Puddling was introduced in 1828, and the furnaces came to be used primarily for making railway lines. The Upper Silesian coalfield extended into northern Moravia, where the coal measures came to the surface at Ostrava and Karvinná. Here, at Vítkovice a puddling works was founded in 1826 to refine iron from charcoal furnaces in the surrounding hills. During the next decade blast furnaces were built to use locally produced coke, and another works was built nearby at Třinec.

By 1880 this second phase in the development of the spatial pattern of iron-working was complete. The industry had effectively moved to the coalfields, leaving only a scattering of small charcoal-fired furnaces and refineries. The third phase was characterised by significant developments in technology and by a radical change in the raw-material base. The introduction of converter and open-hearth not only led to a sharp increase in the volume of production, but also made it desirable to integrate steel-making with smelting. A 'heat economy' was thus achieved by using hot metal direct from the furnace and by burning coke-oven gas in the open-hearth. At the same time, the Thomas process permitted a much greater use to be made of low-grade, highly phosphoric ores. The latter were of so low a grade that it ceased to be economic to move them any great distance. Instead, the ratio of the gross tonnages of ore and fuel dictated that the metal be smelted close to its source. Iron- and steelworks, many of them fully integrated, began to be built on or very cloe to the Jurassic orefields

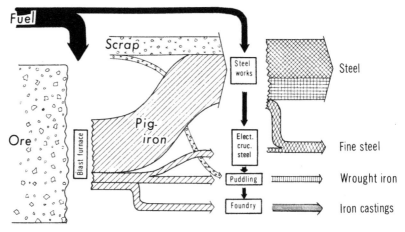

7.4 The flow of materials in the European iron and steel industry, early twentieth century. The scale is only approximate

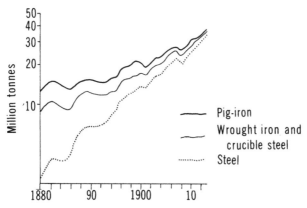

7.5 The increase in the production of iron and steel in Europe during the later nineteenth century (1900 = 100)

(fig. 7.4). Luxembourg led the way, in part because a law required that ore from newly conceded fields be smelted within the Grand Duchy. A succession of smelting and refining works was quickly established below the limestone scarp. German-held Lorraine followed, and then French. Some plants produced at first only cast iron, which was despatched by rail to refineries in northern France, Belgium or the Ruhr. But gradually the economies of integrating steel-making with smelting led to the addition of a basic steelworks.

The phosphoric ores in the North German Plain were more modest in their reserves, and attracted less attention than those of Lorraine and Luxembourg. But the introduction of the basic process stimulated developments near Hanover.

The last phase in the development of the pattern of ironworking during the nineteenth century was marked by the movement of works to the coast or at least to break-of-bulk points, where fuel or ore would be transhipped. This trend resulted from the growing importance of ore imported from overseas. Coastal iron- and steel-works were established from southern Italy to the Baltic. At Bilbao in northern Spain it was the fuel which was imported to smelt ores which occurred close to the sea.

The second Industrial Revolution

The first Industrial Revolution took place in the textile and metal industries, and it was in these that most of the technical advances were made and the vast increase in industrial production achieved. But in the middle years of the century a second and more diverse wave of invention and innovations occurred. Many of the developments sprang from the contemporary scientific revolution; others were due to a rapidly growing population with a steadily rising standard of living. Some of these advances arose from the needs of the textile and metal industries; others were developed to use their products. They ranged from basic chemicals to pharmaceuticals and dye-stuffs; from instruments and machine tools to steel-built ships and automobiles; from paper to footwear. They included capital and consumer goods. Most were cumulative, in the sense that any one technical advance was likely to make farther advances possible. Lastly, the separate branches of industry tended to interact and to promote the progress of one another, so that progress tended to accelerate on a broad front.

Chemical industries

These features are best exemplified in the chemical industries. Of course there had been a chemical industry before the nineteenth century, but its scientific basis was little understood and a deep gulf separated the 'natural philosopher' and experimental scientist on the one hand from the manufacturer of simple chemical products on the other. This changed in the course of the century.[262] During the first third the industry continued to be dominated by the need to produce the basic chemicals used in making soap, glass and other simple products. The chief requirement was an abundant supply of soda, and this was made possible by the development of the Leblanc process. During the first half of the nineteenth century the chemical industry focused on the production of soda and common acids. The chief centres were at first in Marseilles, Alès and Lyons in south-eastern France. Belgium followed. The manufacture of the common acids, mordants and bleach grew up in Switzerland to supply the needs of the textile industry. It spread through the Rhineland, and works were established during the middle years of the century at Heilbronn, Mannheim, Darmstadt, Frankfurt and Duisburg.[263] At Stolberg a plant was built to extract sulphuric acid from zinc-smelter fumes.

In this early phase of the modern chemical industry the pioneers were Great Britain and France, and until about 1860 both Switzerland and Germany followed in their wake.[264] Then a series of revolutionary changes occurred within the industry itself, altering radically the regional balance and pattern of specialisation. In the first place the production of the basic chemicals was immensely improved by the introduction, in 1863, of the Solvay process for making soda ash. Within two or three decades it had completely displaced the

Leblanc process. More important, Germany, where the basic chemicals had achieved little importance by 1860, went direct to the Solvay process, which became the basis of its growing heavy chemicals industry. France, on the other hand, tended to cling to the older method, on which its earlier fortunes had been based.

A second aspect of the industry in the second half of the century was the emergence of the organic chemicals manufacture. This was based on the fractional distillation of coal-tar, itself a product of the gas and coke industry. Coal-tar was at first burned off in the coke-oven. Not until about 1850 were the by-products of the coking process retrieved – first in England – and used. These ranged from ammoniacal liquor at the top of the scale of distillates, through naphtha and benzol to creosote and tar at the bottom. Until about 1880 Great Britain was the chief producer – and exporter – of coal-tar, because it was there that gas production for domestic and industrial use was most developed. The German organic chemicals industry was largely based upon imported coal-tar. Only when tar from the gasworks ceased to suffice – in the 1870s – was a serious attempt made to harness the immense production of waste gases from the coke-ovens.[265]

Coal-tar was the basis of an immense range of chemical products. Most prestigious were the aniline dyes, beginning with the production of mauveine in 1856 and leading on to the fabrication of synthetic indigo at the end of the century. Closely linked with them were drugs and pharmaceuticals. Next came the fertilisers which derived from the ammoniacal liquor from the distillation process; then fuel, solvents and explosives.[266] Crucial to the development of the industry in Switzerland and Germany was the Fuchsine affair of 1863. Fuchsine was a dyestuff, developed by a French chemicals company at Lyons. In 1863 the courts held that the patent of monopoly granted to the company related to the end product, not to the process by which it was obtained, notwithstanding the fact that other and cheaper methods were known. The effect was to kill off the manufacture of organic dyestuffs in France and to lead to the emigration of many of the most skilled chemists.[267] Many of these went to Switzerland where they could pursue their work free from the restrictions imposed by the French patent laws. 'Everyone of the Basel firms traces its existence as a chemical firm to French emigrants.'[268]

Most branches of the chemicals industry were resource-intensive, in that they processed large amounts of raw materials and required large quantities of water and energy. As a general rule they were not labour-intensive, though the organic chemicals required a large investment in research. Most branches were capital-intensive – some very heavily so, in that they required elaborate plant and equipment. Most chemical works, especially those producing the basic chemicals, were carefully located with regard to raw materials and means of transport. As the German industry developed it established itself close to the Rhine, which provided its most important means of transport for fuel and materials, much of which came from the Lower Rhineland. A number of works were established along the river in the 1850s–1860s. Amongst them were the Griesheim works at Frankfurt, Merck at Darmstadt and the Verein Chemischer Fabriken at Mannheim. To these were added the Rhenania company, which derived from

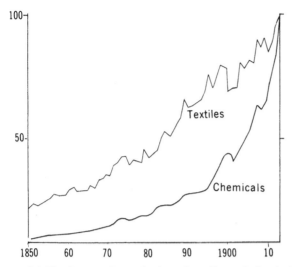

7.6 The increase in production of textiles and chemicals in Germany, 1850–1913 (1913 = 100). Note the contrast between the linear expansion of the textile industry and the exponential growth in the manufacture of chemicals

the Stolberg sulphuric acid works; the Badische Anilin- und Soda-fabrik at Ludwigshafen; Bayer at Leverkusen, and Höchst at Frankfurt.[269] At the southern end of the chain were the Ciba and Geigy works at Basel, which specialised in dyestuffs and pharmaceuticals.

The German chemicals industry received a powerful stimulus from the discovery of the potash deposits at Stassfurt, in Saxony, and of the related salt deposits.[270] These contributed to the growth of the giant basic chemicals and fertiliser industry of central Germany. This was, in part, a response to the needs of German agriculture (see p. 231); in part to the availability of natural materials and by-products which could be used in this way.[271] First imported phosphates were treated with acid to produce superphosphates. Then nitrogenous fertilisers became available from the fractional distillation of coal-tar. Lastly the phosphorus-rich slag from the basic steel works began to be used from the 1880s (fig. 7.6).

Alongside these large-scale and heavily capitalised branches of the industry were the countless factories, most of them small, which produced specialised chemical products. These included bleach for the textile industries, paints and lacquers, inks, soap and other cleansing materials. Their number proliferated during the century. Most were consumer-oriented, and sprang up near the large centres of population and industry. Some cities, Paris and Lyons, Berlin and Hamburg for example, were notable for their number. In fact, they are sometimes called 'Lyons'-type industries.[272] Amongst these manufactures, that of photographic materials had begun to assume an important role even before the First World War. There were many manufacturers, but one of the most important was the A. G. für Anilinfabrikation (AGFA), established first in Berlin and then at Wolfen in Saxony.[273]

The chemicals industry remained small in the Habsburg Empire and scarcely developed at all elsewhere in continental Europe. A basic chemicals industry was established at Ústí (Aussig) in northern Bohemia, where it could be supplied with raw materials from across the border in Saxony, and in Prague.[274]

The metal-using industries

These also were present in pre-industrial Europe, mainly as domestic or craft industries. They survived through the nineteenth century, though shifting gradually from a domestic to a factory basis. Typical of the industries which thus emerged were the gunsmiths of Liège[275] and of Suhl in Thuringia,[276] the cutlery industry of Solingen[277] and the manufacture of tools, saws and files in neighbouring Remscheid. All tended by the end of the century to be carried on in small workshops, using steam or occasionally electric power. There was a scatter of small workshops through the hills to the south of the Ruhr, where ironware goods like screws, nails, springs, hinges and small castings were produced by the million. Similar conditions prevailed in the Thuringian Forest and some other areas where there had once been a flourishing iron-smelting and refining industry.

Beside this small-scale and almost traditional manufacture there grew up in the course of the century larger and more specialised metal-using industries. These can, for convenience, be grouped as (1) mechanical equipment and machine tools, (2) steam-engines, railway locomotives and rolling stock, (3) shipbuilding, and (4) automobile construction. Though some of these manufactures were carried on in units which may seem very small, these were a great deal larger than the workshop industries already described. Nevertheless most of the products were 'custom made'. These was little attempt before the twentieth century to standardise production and almost none to adopt an American-style production line.[278]

Most producers of mechanical equipment and machine tools were strongly market-oriented. Textile machinery was made from the first in textile centres such as Chemnitz and Mulhouse; machinery for the rolling-mills and steelworks at places such as Liège and Wetter, close to the major centres of iron production. The earliest steam-engines in use in continental Europe were imported from Britain, but within a relatively short period of time they began to be imitated by continental craftsmen. Most were built where the demand for them was greatest, in the mining and developing industrial areas such as central Belgium, and the Ruhr. By the middle years of the century the market for steam-engines had broadened. Important centres of production developed in Liège, where Cockerill became the largest builder of locomotives in Europe, and in Ghent. In northern Germany, Hannomag and Borsig developed the manufacture of engines of all kinds. In south Germany, Augsburg and Nuremberg similarly developed heavy engineering, their respective works subsequently merging to form the Maschinfabrik Augsburg-Nürnberg (MAN).[279]

Late in the century two highly significant developments took place in the broad spectrum of the engineering industries: the emergence of electrical engineering and the invention of the automobile. Electricity for industrial and domestic use first became significant in the 1870s. Demand for generators, electric motors, transformers and cables increased throughout the rest of the century, and received further impetus at the beginning of the twentieth from the large number of hydroelectric projects then under construction. From the first Germany dominated the industry. It was a highly capitalised industry, offering little scope for small units of production, and within a short time two firms, those associated respectively with Emil Rathenau and Werner von Siemens, dominated not only the German but also the European industries.[280] Both firms were based on Berlin, where Rathenau's Allgemeine Elektricitäts Gesellschaft (AEG) was the largest electrical engineering company outside the United States. The Siemens firm then merged with Schuckert to form an even larger firm which was able to create the industrial suburb of Siemensstadt for itself and its workers.[281]

Although France and Italy lagged far behind Germany in electrical engineering, they both leapt ahead in the provision of hydroelectric power and in the development of the electro-chemical and electro-metallurgical industries based on it. In Italy hydroelectric power was welcomed as an alternative to expensive imported coal. The first generating station was built on the River Adda in 1898. Others followed. Many of the textile mills of Lombardy were adapted to electric power, and Turin developed electrical engineering industries to service them.

The industry progressed even more rapidly in France. Its centre was Savoy and the northern French Alps, where rainfall was heaviest and the rivers most vigorous. Power was first used for illumination in 1882 and then for small industrial undertakings. In the 1890s electro-chemical works began to appear, and, shortly afterwards, electro-metallurgical.[282] At first certain chemicals were produced electrolitically; then power began to be used to smelt aluminium, of which France became the world's biggest producer after the United States.[283] Lastly, the electric-arc furnace, invented in 1900, was used for refining a particularly high grade of steel.

The motor car originated in Germany where the first was built by Daimler and Maybach at Cannstadt in Württemberg in 1886. Their factory at Untertürkheim, near Stuttgart, became one of the largest in Germany.[284] The automobile industry then proliferated, and there were more than 50 works before 1914. Most developed from the manufacture of much humbler machines; Opel, for example, moved up-market, like a certain English firm, from the manufacture of bicycles.

Although a German invention, the car came to be associated particularly with France. Its manufacture followed quickly on the innovation of Daimler, and France soon exceeded Germany in the number of cars built.[285] It was, of course a luxury industry, for the early motor cars were very expensive. There was a larger market for them in France, and Paris, focus of taste and fashion, seemed to be the natural place at which to manufacture such a commodity.[286] At a relatively early stage in its development the French industry began to delegate the manufacture of components to other firms, so that the car factories themselves began to approximate to the modern concept of assembly plants. The automobile industry was in fact the only branch of engineering in which France achieved and maintained a technical superiority over Germany during the pre-war years, and cars formed during this period a significant part of French exports.

Once it became possible to purchase some of the more sophisticated components from specialised producers, the manufacture of cars could be established almost anywhere. Hundreds of small firms came into being and many lasted for only a short while. Only a fraction ever achieved the scale and organisation of Daimler and Opel, Renault and Peugeot, which, in fact, absorbed many of the firms which failed. The location of these motor manufacturers was, aside from Paris, almost random. No industry was more footloose; a branch was likely to develop wherever an entrepreneur had the idea of developing one. The only other important car industry in Europe before 1914 was the Italian. The first car was built at Turin in 1898, and in the following year the Agnelli workshop, also of Turin and later to become FIAT, started production.

The only other important branch of the engineering industry was shipbuilding. When the century began all ships were built of wood and propelled by sail. The largest did not exceed 1,000 tonnes displacement or draw more than two or three metres. They could be, and indeed were, built on almost any stretch of sheltered water where the materials could be assembled. Shipbuilding materials came mainly from the Baltic region and North America, and the industry was located chiefly on the Baltic coast and round the North Sea.[287] Iron and then steel at first supplemented wood and then replaced it, but so gradually that there was little immediate change in the location of shipbuilding. In any case wooden, sail-driven ships continued to ply around Europe's coasts into the twentieth century. The steamship was slow in taking over on the long intercontinental voyages because it had difficulty in carrying the large quantity of fuel required by its inefficient engines, and for many years sail continued to supplement steam.

Metal began to be used in ship-construction in the 1840s, but did not become really important until, with the converter and open-hearth, the price of steel fell sharply. At the same time shipbuilding began to concentrate in a small number of shipyards of over increasing size: Danzig, Stettin and Lübeck on the Baltic; Hamburg, Vegesack and Bremerhaven,[288] Amsterdam and Rotterdam on the North Sea coast, and Le Havre. A consideration in the location of shipyards in the late nineteenth century was ease of access to steel-producing regions. In the case of German shipyards steel was obtained mainly from the Ruhr and to a small extent from Upper Silesia. Indeed Krupp actually acquired its own shipyard, the Germania Werft at Kiel, and Schneider-Le Creusot owned yards at Le Havre.[289]

Other countries which developed a modern shipbuilding industry included Italy, where a yard was opened at Genoa; Norway, Belgium and Denmark, but the most important after Germany and France was the Netherlands. The Dutch had a long history of shipbuilding to match their long experience as a trading nation. Ships had been built with timber from the Baltic at dozens of small ports in Groningen, Holland, Zeeland and around the Zuider Zee. As shipping began to revive on the Rhine (see p. 436) after Napoleon had removed most of the impediments to trade, the Dutch were well placed to profit from it. The earliest steamships and most of the barges which carried the coal and ore were built on the lower river in the Netherlands. The Dutch established a clear-cut superiority in building tugs and river craft, which they retained throughout the period. But their yards at Rotterdam and Amsterdam never relinquished the construction of the larger, ocean-going vessels which sailed in the Dutch East Indian trade.

Clothing and footwear industries

These were amongst the last industries to pass from the artisan's workshop to the factory, a process which had barely begun over much of Europe in 1914. These were customer-oriented industries. Much of the product was made to order and manufacture was necessarily carried on close to the market. Rural areas would have been, by and large, self-sufficing in clothing. In some areas itinerant tailors made up the coarse local cloth into garments; only quality clothing being purchased in the urban market. The larger towns developed a broader spectrum of production, catering both for working-class demands and the refined taste of the wealthy.

In this respect Paris established its pre-eminent role at an early date. To the decaying cloth industry (see p. 164) there succeeded a clothing industry which drew its varied fabrics from the textile centres of northern France. Developments in Berlin followed a similar course, and the workforce, formerly engaged in textiles, turned to the production of ready-made garments. The *Bekleidungsindustrie* became in terms of employment one of the largest in this rapidly growing city.[290] In Germany as a whole, clothing was second only to textiles among domestic crafts, employing towards the end of the century more than a tenth of all who worked at home.[291]

The manufacture of footwear was also in the main a craft industry, widely distributed throughout Europe. But, like the clothing industry, it showed an exceptional concentration in certain towns. This appears to have been related originally to a local tanning industry. Breslau, for example, was noted for its tanning, and this was reflected in the number of shoemaking businesses. In 1825 there were 615, and in 1895, 1,782 in a city of fewer than 500,000 inhabitants.[292] There were also large numbers of shoemakers in Dresden and Leipzig.[293] but this concentration was insignificant beside that met with in the small towns of Holstein.[294] At Heide in the Ditmärschen, a town of only about 6,000 inhabitants, there were 350 shoemakers in 1854. At the nearby town of Plön there were 350 in a population of 5,000. But soon after this the industry began to pass from the small workshop to the factory. Indeed many of the small footwear factories evolved from domestic workshops, as at Groitzsch near Leipzig, Rosswein in the Ore Mountains and Pirmasens in the Palatinate. Fougères in Brittany was also the unlikely location of a highly concentrated boot and shoe industry, which was said to employ 'fully half the population'.[295] It derived from a pre-industrial manufacture of linen and canvas. At first these fabrics were used to make a kind of canvas slipper. Then, with the decline of the linen industry, felt came to be used, and about 1850, leather was introduced. The industry grew rapidly with the introduction of machinery after 1870. At the time of the Board of Trade's report there were no less than 33 factories, but in spite of this half the workforce was said to have been still employed at home.

Conclusion

The increase in the volume of industrial production was the most noteworthy achievement of the century between Waterloo and the First World War. It is, however, difficult to measure because some goods were substituted for others in the course of the century. The increase in steel production, for example, was from a few hundred tonnes to about 30 millions, but this increase was in some measure offset by the reduction in the output of puddled iron and the virtual extinction of bar-iron production. The decline in linen production was similarly more than balanced by the increase in that of cottons.

Industrial production in France increased, it has been claimed, more than sixfold during the century, and that of Germany more than ten times between 1850 and the outbreak of the First World War. In Sweden the rate of growth was almost as great, but started in this instance from a very low base. But an expansion of industrial production of this magnitude was high even for Europe. It was very much slower in all of southern and eastern Europe and was negligible in most of the Balkans until after 1900.[296]

Rates of growth varied. In general they were slow during the first half of the century or even longer. In France, despite the gains made under the Second Empire, the rate of growth remained low until the 1800s.[297] In Germany expansion did not become rapid until after 1890.[298] In Italy and Sweden this phase was reached even later. In Belgium industrial growth came early, slowed during the middle years of the century, and again accelerated towards its end.

The growth of industrial production was accompanied in very many instances by its relocation. The older branches of manufacturing, essentially textiles and the ferrous metals, tended to concentrate in a relatively small number of industrial regions, the most important of which are discussed in the next chapter. At the centre of some of these regions were coalfields, and the reasons for their growth included the availability of cheap fuel. In a few instances the movement of manufacturing from more traditional sites to the newly industrialising regions was sudden and abrupt. Ironworks in the Aachen district were closed, and their movable assets transferred to other areas where factor costs were lower. More often the shift was slower and more subtle. The older industrial centres were starved of investment capital and slowly passed into obsolescence and decay. New growth was in the new regions. In this way the iron industry slowly declined in the Siegerland, Saxony and central France, and was restricted to steel-making and steel-fabrication in Saint-Étienne, Le Creusot and elsewhere.

The textile industry shifted its location less readily than the iron and steel industry, largely because it was growing less rapidly and attracting less investment. The smallest change was in the linen industry which was virtually starved of capital. Nevertheless, an increasing share of the industry came to be located in its leading centres, notably the Lille district, the Lower Rhineland, Saxony and south Germany. At the same time textiles declined and almost disappeared from the workshops of Berlin and Paris.

Until late in the nineteenth century industrialisation was largely a matter of iron, steel and textiles. In the last third or quarter of the century a new wave of industrialisation blurred the simple spatial pattern that was beginning to emerge. Some of the 'new' industries were merely traditional manufactures transferred to a factory setting. Amongst them were the leather and footwear industries, printing and the manufacture of paper, ceramics and small metal goods. Others were new in so far as they used a new technology to produce a new range of products. Such industries included the chemical, glass and

Manufacturing

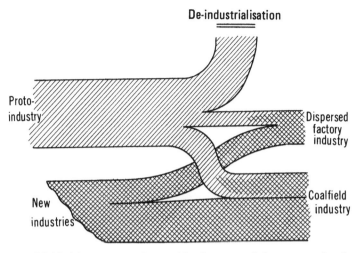

7.7 Model to represent the transition from proto-industry to modern factory industry

electro-metallurgical manufactures. The location was less constrained than that of the older industries. By the time they began to emerge as large-scale factory industries the network of railways had by and large been completed, and the more important waterways had been enlarged and improved. The movement of fuel was easier than it had been previously. The heavy chemicals industry, it is true, was attracted in the Rhineland and Saxony to the proximity of the solid fuel which it used in large quantities, but others were more footloose. Some were drawn to the largest markets, and this effectively meant the large cities; others were drawn to the ports, and there were some, ceramics for example, which moved to the open country where both labour and land were as a general rule cheaper.

In this way the tendency for manufacturing to concentrate into a few highly industrialised regions began late in the nineteenth century to be countered by that of the newer industries to disperse. The resultant of these two contrary forces is difficult to measure. A recent study of Belgium[299] has shown that almost every province had in 1910 a more diversified industrial structure than in 1846. Concentration was least in the basic consumer goods and highest, the extractive industries aside, in the manufacture of capital goods.

Fig. 7.7 is an attempt to reduce these changes to a simple model. Proto-industry either decayed (de-industrialisation), was transformed into factory industry more or less *in situ*, or gave way to factory industry on or in close proximity to the coalfields. 'New' industries, which came into being in the second half of the nineteenth century, were attracted to some extent to the vicinity of the coalfields; but, by and large, they were established close to the markets for their products or at sites convenient for the transport of their raw materials and finished products, such as break-of-bulk points.

8
The growth of industrial regions

There was a tendency, as has been seen, for manufacturing industries to cluster during the nineteenth century, and thus to form industrial regions. The manufactures in question might have been attracted to a particular area by a mutual dependence or by a reliance on a localised resource such as coal or a metalliferous ore. An industrial region might embrace only the many facets of a single industry, such as the textile industry with its related processing and manufacturing activities at Łódź in Poland. Most, however, were larger in terms of area and more diverse in their industrial structure; it is difficult, for example, to think of a branch of manufacturing that was not present in the industrial region of north-west Europe.

It was emphasised in the previous chapter that industrial development was no respecter of political boundaries. The early textile manufacturing regions spread across the Franco-Belgian, the Belgo-German and the Dutch-German borders. There was no essential difference between iron-working in the Belgian or French Ardennes and the German Eifel. Europe was, as Wrigley has emphasised, a single economic community 'within which like circumstances might give rise to similar results'.[1] It has been the usual practice to regard economic development as taking place within national boundaries, as if shaped by national policy or national temperament. But industrial growth was 'essentially a local rather than a national affair...[and] each country was made up of a number of regional economies'. The Industrial Revolution 'spread across Europe like an epidemic, ignoring frontiers and alighting on areas susceptible to the disease, but taking no apparent notice of governmental encouragement'.[2] Within each country there was a wide range between the most developed and technically the most advanced regions, on the one hand, and those, on the other, which remained relatively backward. The advanced regions in all countries had far more in common with one another than each had with the less developed areas of its own country.

If we pursue Pollard's analogy we may ask what were the antibodies that gave immunity to the disease of industrialism? Absence of resources, especially coal, to be sure, but we must not forget that not all coalfields were industrialised, nor that industries developed in areas far removed from coal and most other sources of energy. Perhaps it might be said that some areas had a predisposition to the contagion in the form of an earlier commercial development or a developed

Hausindustrie.³ Much depended, however, on local attitudes to innovation, and these in turn might have been shaped by the presence or otherwise of institutions such as gilds or the 'feudal' ownership of land and resources.

The nature of industrial regions

The emphasis of this chapter is on the formation of regions, most of which reached across political boundaries. The writer accepts fully the arguments of Wrigley and Pollard in this regard, that one is dealing with 'regions operating in a European context', and that in the early phases of industrial development 'governments were at best irrelevant, and frequently took a negative part in a development which drew its main driving force from outside the political and governmental sphere'.⁴

An industrial region grew from a nucleus, sometimes from more than one, like 'a series of small dots on the map'.⁵ Each must have had – or have been thought to have – some advantage for manufacturing. It is rarely possible to isolate the individuals who first located an industry in a particular place; totally impossible to reconstruct their thought-processes as they made the vital decisions. We in retrospect may condemn some of these decisions as unwise or wrong, but we are in no position to know how those who made the decisions perceived their physical environment. The reasons commonly advanced for a particular industrial location, as distinct from industrial growth, are at best plausible rationalisations, at worst merely fictions. Where one can narrow the focus and examine in some detail the early phases of industrial growth, one finds that there was always an element of chance in industrial location; that it 'should be regarded as a stochastic process'.⁶ The factors were always permissive rather than compelling, like the spread of an epidemic.

These industrial nuclei grew in population relatively rapidly, and most became densely peopled, by European standards, before the end of the century. They also became highly urbanised. Some, notably the north-west European region, had a developed urban network before industrial development began. It is noteworthy, however, that in comparatively few instances were factories established within or even close to the older towns. Most often they were established in predominantly rural areas and themselves became the foci around which small towns grew. Nowhere is this process better illustrated than in Upper Silesia, much of which was a forested wilderness when industrial growth began. In all developing industrial regions the population grew mainly by immigration. Agriculture declined in importance, but never entirely disappeared from the interstices which lay between the units in each industrial cluster. On the contrary, as agricultural land diminished the use made of it became more intensive, as market-gardening and dairy-farming developed to supply the towns.

At the same time the provision of public services became more necessary and more complex. Most branches of manufacturing were heavy consumers of water, and it was often for their convenience, rather than that of the domestic users, that a piped water supply was first made available (see p. 146). This in turn necessitated the construction of holding reservoirs, sometimes well outside the industrial region. The converse of water supply is the disposal of sewage and industrial waste. For this there was little provision; for most purposes the nearest river sufficed. In 1848 Banfield described the polluted nature of the Wupper, 'an open receptacle for all sewers, disguising the various tinctures contributed from the dyeing establishments in one murky, impenetrable hue, that makes the stranger shudder on beholding'.[7] The Wupper was, it is true, one of the worst rivers in western Europe owing to the bleaching and dyeing works that had gathered along its banks. But every river that flowed through an industrial region was beginning to acquire the same fetid and lifeless character, for which there was no satisfactory remedy before 1914.

The supply of energy and illumination were hardly less important than that of water. Until the development very late in the century of electricity supply-lines, energy had to be supplied in a *gross* form, as coal or, in rare instances, oil, to be consumed on the site. Gas illumination was first developed and used in Belgium. By the mid-century, a network of gaslines had been laid down in some of the developing regions, and this public utility continued to broaden its sphere until checked by the comparable spread of an electricity net. The latter did not take place until the end of the century even in the most advanced regions.

The growth of industrial regions was fed by a constant inflow of labour, men forced off the land by agricultural improvements or the falling price of agricultural products; women, with little scope for employment or marriage in rural areas, seeking employment in the towns. The growing industrial regions were 'the promised land'[8] where hopes were dashed by long hours of work in dark, insanitary factories. Death-rates continued to be higher in industrial regions than in most other areas and birth-rates to be lower. Although this may be taken to reflect the insanitary and polluted conditions which prevailed in some such areas, it was also due in part to the skewed age structure. Expectation of life must be shorter in a population made up in part of adult immigrants, and birth-rates must be lower where the consequent age at marriage was relatively high.

The spread of working-class housing lagged behind the growth in population, and there was in most industrial regions a severe overcrowding which appears to have worsened through the century. Some employers, such as Krupp in the Ruhr, Siemens in Berlin, and the magnates in Upper Silesia constructed either small houses or tenement blocks for their employees. Both these and the housing erected by speculative builders were poorly planned, crudely built and ill-lit, offering in many instances a floor area that was grotesquely inadequate for family needs. Only where space was not at a premium, and that was so only in a few regions like Lorraine-Luxembourg-Saar, did housing development include the provision of cottages with gardens.[9]

The delimitation of the industrial regions examined in this chapter is particularly difficult. The regions, in the first place, tended to grow by expanding at their periphery and encroaching on rural areas. They were all larger in area in 1914 than they had been half a century earlier. Whatever the date that is being considered, there appear to be two methods of defining the region, in administrative and in functional terms. The former consists in making the regional boundary follow the pre-existing limits of *Kreis, canton* or *commune*.

This is the method adopted by Wrigley.[10] It has the advantage of conforming with the unit areas used for the collection of population data and sometimes also of industrial statistics, and this in turn greatly simplifies analysis. At the same time, however, it may include within the industrial region large areas which are in no sense industrially developed. The alternative method is merely to draw a line around areas which appear on the map to have been industrialised and urbanised. Though this second method is used all too often, it is subjective and unscientific. The first is used wherever possible in this chapter, and the industrial regions analysed are in each case made up of those administrative units which can be regarded as wholly or partly industrialised.

The industrial region of north-west Europe

In this chapter the development is examined of eight industrial regions. The largest and most complex is that which, for lack of a convenient name, is here called the North-west European region. It can, however, be broken down into distinct but adjacent units: northern France, central Belgium, Aachen, the lower Rhineland and the Ruhr, which nevertheless showed a high degree of mutual dependence. Others were smaller in area, population and production, and apart from Saxony, were very much more highly specialised. Most narrowly based of all was Łódź, which with its surrounding industrialised villages, concentrated on textile production alone.

The most important industrialised region in Europe at the end of the nineteenth century was that which extended from the Pas de Calais in northern France, through central Belgium, to Aachen in Germany. Then, after a short interruption it was continued in the industrialised Rhineland and Ruhr. The region ended near Hamm, 400 km to the east. It had at the time of its fullest pre-war development an area of about 5,000 sq km, and in 1910–11 a population of 9.14 million. It contained at least 85 towns of over 10,000, and carried on somewhere within its compass every significant branch of manufacturing. In it was employed about half of the industrial population in France, Belgium and Germany, the countries between which it was divided.

This region had been one of relatively dense population long before the development of modern industry. Much of it was good agricultural land, part of the *limon*-covered belt that formed the southern margin of the North European Plain. Important commercial towns had grown up, linked by routes which followed this belt of open country. Not only was the region well endowed with roads; it was crossed by two of the best navigable rivers in Europe, the Meuse and Rhine, which converged on their common delta in the Low Countries. They linked parts of the region with the ocean ports of Antwerp, Rotterdam and Amsterdam, and also with eastern France and South Germany.

However one may explain the location of certain facets of industrialisation within this region, the basic reason for its growth was coal. It was coal that attracted the metallurgical industries and powered the chemical and textile. Wrigley has demonstrated how closely population growth was related to the increase in the production of coal.[11] Yet the region originated in several nuclei,

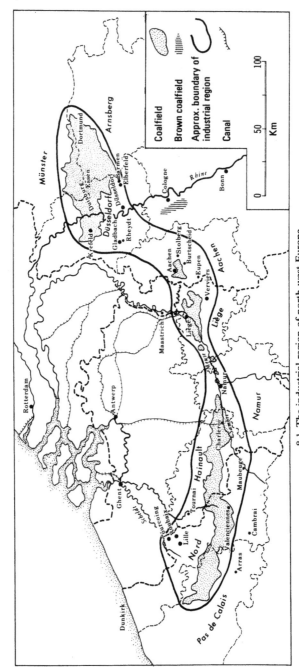

8.1 The industrial region of north-west Europe

'a series of small dots on the map roughly distributed along...the Rhine [and] Scheldt'.[12] These nuclei grew and eventually many of them coalesced to form the highly urbanised and industrialised region that we now know (fig. 8.1).

The coalfields

If the growth of any industrial region can be regarded as inevitable, this would be so of north-western Europe. Its coal resources embraced some two-thirds of those known in Europe. It had reserves of ferrous and non-ferrous ores; it was well served with navigable waterways and ocean ports, and it had a tradition of both craft industries and commerce. But the basic factor was coal, even though the true extent of the reserves was revealed only gradually. Not until about 1860 were the limits of the coalfield in northern France demonstrated. The northern extension of the Ruhr coalfield became known even later and the Campine basin of Belgium was not fully explored until 1906. Clearly those who first used the coal from these fields to refine their metals and power their machines had no conception of the extent of the resources they were beginning to tap.

The coalfields of north-western Europe – for there were at least six separate coal basins – are part of a coal-bearing zone which reaches from Great Britain to the Ukraine. The coal occurs on the northern flanks of an uplift made up of rocks older than the coal series. The latter forms a series of basins, underlain by older rocks and partially covered with younger. In the case of the Ruhr, the coal series dips northward and in this direction its limit has not been determined. The basins lie in an arc, concave towards the north, which stretches from Fléchinelle in Pas de Calais to Hamm, near Münster in Germany, a distance of 450 km. To the west lies the minute coalfield of Hardinghen; to the east the small fields of Ibbenbüren and Hanover.

Within these coal basins occurred every variety of coal. In northern France and Hainault the coalfield is narrow – at no point more than 15 km across – and tapers westwards, pinched between the hidden folds of the Carboniferous series. Though the coal measures come to the surface along the Sambre valley they are hidden from Mons westwards by later deposits. The coalfield ends near Namur, and over the next 40 km is represented only by a few minute occurrences. Then it reappears in the small but historically important Liège coalfield. Another short, barren area intervenes before the Aachen field, itself made up of the small Inde and Würm basins, is reached.

The whole of this western part of the coalfield was intensely folded and faulted in the course of the earth-movements which ended the Palaeozoic era. This has greatly increased the difficulty and hence the cost of mining. Furthermore, the seams were gassy and relatively thin, and almost completely lacking in good quality coking-coal. And yet the Liège basin, one of the worst affected, was highly praised in the mid-eighteenth century as 'in all Europe, the area most richly endowed with coal'.[13] Mining has now virtually ceased.

The decline of the central Belgian field was balanced by the discovery first of the Limburg field, in reality a western extension of the Aachen basins, and then of the Campine field. These were more easily worked than the central Belgian field, and, furthermore, yielded coking-coal. Their industrial importance, however, was slight before 1914.

Table 8.1. *Coal resources of north-western Europe*

	Probable reserves (million tonnes)	Percentage of total	Of which coking-coal (million tonnes)	Percentage of total
Nord and Pas de Calais	3,500	1.5	800	0.5
Central Belgium	3,000	1.2	480	0.3
Campine (Belgium)	8,000	3.3	3,300	2.1
Limburg (Netherlands)	4,402	1.8	2,100	1.4
Aachen	10,500	4.3	3,200	2.1
Ruhr and Ibbenbüren	213,600	87.9	145,000	93.6
Total	243,002	100	154,880	100

A wider area of barren ground separated the Aachen coalfields from the Ruhr, the richest and most extensive coalfield in Europe. The coal seams outcrop in the hills lying to the south of the River Ruhr, which gives its name to this part of the region. From here they dipped northwards. After distances of up to 15 km they passed beneath the Secondary beds, and near the present northern limit of mining lie at a depth of about 300 m. One of the most significant advantages of the Ruhr coalfield, apart from its thick and little folded seams, was the varieties of coal which it had to offer. The lowest and oldest seams were 'lean', approximating to anthracite in quality. Above them were semi-anthracite and *Fettkohle* from which the best coking-coal was obtained, and, highest in the series, gas and long-flame coals.

The coal measures outcrop very approximately along the valley of the Ruhr. Here they could be worked in open cuts on the valley side or in shallow pits. The most easily worked seams were of lean coal; comparatively little of the good-quality coking-coal came to the surface, a fact which helps to explain why coke was so late in being used for smelting in the Ruhr. With the increase in coal production during the nineteenth century, mining spread northwards from the exposed to the hidden part of the coalfield. By 1850 the whole of the exposed coalfield had been conceded and was being worked, and in a few places pits had been sunk onto the hidden area of the coalfield. The total resources of the coal basins of north-western Europe were immense (table 8.1).[14]

The Liège and Ruhr fields had been worked by means of very shallow pits since the Middle Ages. It is, however, very difficult to evaluate their output before the middle years of the nineteenth century. During the early years of the century the most productive field by far was the Belgian, especially the Liège basin. In 1815 coal production in the province of Liège was about 331,000 tonnes. This had doubled by 1828, and had reached 853,000 by 1840. In 1850 output was some 1,250,000 tonnes.[15] Production in Hainault was at first smaller than in *prov.* Liège, but by the mid-century had moved ahead, and continued to increase until the outbreak of the First World War, when production was:

        Hainault basin   16,700,000 tonnes
        Liège basin   6,500,000 tonnes

Table 8.2. *Coal production per year in northern France (thousand tonnes)*

	Nord	Pas de Calais
1858–62	1,627	679
1888–92	4,561	8,528
1908–12	6,589	19,102

The more easterly part of the northern French coalfield had been explored and opened up during the eighteenth century, and here the two huge concessions of Anzin and Aniche were yielding an increasing volume of coal. In 1814 a report in the official *Journal des Mines* credited *dépt* Nord with an output of more than 200,000 tonnes, exceeded only by that of *dépt* Loire.[16]. Thirty years later production was 893,325[17] tonnes, and in 1851, 1.6 million tonnes.[18] At this time the rising demand for coal led to a more intensive search for the westward extension of the seams in this fractured and contorted field. Although extensions were found along the margins of the Anzin-Aniche concessions, it was not until after 1850 that the limits of the coal basin to the west of Douai came to be known. Concessions granted between 1850 and 1864 took the exploitable field westwards almost to Saint-Omer. Subsequent prospecting has only extended the Pas de Calais field a kilometre or two towards the south.[19]

By 1910 output from the Pas de Calais coalfield was almost 20 million tonnes, and that from all of northern France more than 25 million (table 8.2).

The small coal basins near Aachen resembled the Liège field in their contorted, waterlogged seams, their early exploitation and above all, in their failure to live up to earlier expectations. Output in 1816 was of the order of 150,000 tonnes.[20] By about 1850 this had risen to more than 400,000, and output reached a million tonnes about 1870. Thereafter the rate of growth was much slower than elsewhere in north-west Europpe, and by 1910 had reached only about 2.6 millions. The mines lay in two groups, the larger in the Würm valley to the north of Aachen, and the rest to the east, near Eschweiler.[21]

The Ruhr was at the beginning of the century the best endowed and one of the least productive of Europe's coalfields. In 1815 output only just turned 400,000 tonnes. The 150 mines along the Ruhr valley each had an average employment of only 20 and an output of no more than 2,500 tonnes. This grew slowly during the next 30 years, not reaching a million tonnes until 1840–4, and 1.7 million by 1850. Thereafter growth was much more rapid in each of the administrative areas which spanned the Ruhr (table 8.3).[22]

Total production from the *whole* field must have risen from about a million tonnes in 1815–20, to some 9,110,000 tonnes in 1850. Until the coming of railways in the 1840s much of this coal was consumed near the mines. In Belgium the demands of the metallurgical industries were heavy, and the Aachen field supplied the lead and zinc smelters at Stolberg and near Liège. Coal travelled far from the mines only where water transport was available. Ruhr coal was used in ironworks along the Rhine, as for example at Sayn and Rasselstein, near Coblenz, even before it was used on the coalfield itself. Coal from the more

Table 8.3. *Coal production in the Ruhr (thousand tonnes)*

	Düsseldorf	Arnsberg	Recklinghausen	Total
1850	1,666	—	—	1,666
1888–92	10,348	21,691	3,472	35,511
1908–12	27,850	43,495	16,972	88,317

westerly parts of the known field, in Hainault and Nord, supplied Paris in growing quantities after the Saint-Quentin canal had been completed in 1824.[23] Indeed, the correspondence between the early canal networks and the Hainault-Nord coalfield was, as Gendarme noted, remarkable, even if entirely fortuitous.[24]

The only other fossil available in the region was the brown coal of the Ville region near Cologne. Much of it lay on the estates of the Archbishop of Cologne, who allowed it to be dug from shallow pits at least as early as the seventeenth century.[25] By 1812 there were no fewer than 42 separate workings, and about 1821 the coal began to be used commercially to make bricks and burn lime. Later the uses of brown coal became more diversified. In 1877 the first briquetting plant was established, and by 1900 the field was producing over 1.25 million tonnes of briquettes a year.[26] The total brown-coal production was about six million tonnes.[27]

The industrial region until 1850

When peace returned to Europe in 1815 there was already a well developed proto-industry in the future industrial region of north-west Europe. It was based upon the smelting and fabricating of iron; on textiles, especially linen; and on the smelting of non-ferrous metals. A widely scattered iron industry smelted ores from the Ardennes, Eifel, Sauerland and Siegerland with charcoal from the local forests. Furnaces deep in the hills supplied iron to refineries lower down the valleys towards the north, and the latter sent bar-iron to nailers, wire-drawers and a host of craftsmen whose workshops lined the small streams which flowed down to the Sambre, Meuse and Ruhr.

In the early years of the century, furnaces were as a general rule attracted to the source of their ore. A few were built out on the plain in order to use the small quantities of bog-ore that were available, but most lay within the forested hills where both ore and charcoal were available. Most important were the smelting works which clustered round Siegen, and smelted the local haematite, but furnaces were also scattered through the Eifel and Ardennes and their hilly northern and western margins.[28] Iron-refining was distributed more widely, the pig-iron in this case tending to move towards areas where there was less competition for charcoal and there were streams to power the bellows. Iron-refineries and the splitting and rolling mills which produced the metal for the iron trades thus tended to lie in a belt which extended from *dépt* Nord, through the northern Ardennes to Aachen and Eschweiler, and again from the Rhine eastwards through the Sauerland, where the refineries lay along the Wupper and the tributaries of the Ruhr. An iron-refining and fabricating region was thus clearly distinguished from the smelting region to the south.

Refining and metal-working required water-power. The smallest streams were enough to work the bellows of a refinery; a rolling or splitting mill required a more powerful current. But this was a precarious source of power. The hot summer of 1846, wrote Banfield, 'put a stop to nearly all the operations that depended on water-power.... The long winter that succeeded condemned all water-wheels to inaction until late in the spring.'[29] In 1822 no less than 30 water-powered works along the small Hoyoux river, which discharged into the Meuse at Huy, were destroyed by floods.[30] Most were rebuilt and many continued in use until the end of the century, but not all for refining and processing iron. A feature of these early works was the ease with which the water-powered mechanism could be adopted to other purposes. Some ironworks even ended their careers as flour mills.

There was thus throughout the region a northward progression from the smelting to the refining and fabricating industries. The fact that this movement was *towards* the coalfield was entirely coincidental. It nevertheless made the transition from refining on the hearth to puddling with coal an easy one. Coal did not enter into iron-working until well into the nineteenth century, but it was available when a more advanced technology made its use practicable.

The smelting industry remained prosperous through the first half of the century. Some furnaces closed with the exhaustion of their local ore supply, but demand continued high for charcoal pig-iron. In 1811 the Belgian *départements* of the French Empire produced together about 37,500 tonnes of pig-iron.[31] There was also a considerable but indeterminate output from the Eifel[32] and a large production – perhaps 10,000 tonnes – in the Siegerland.[33]

The production of pig-iron increased steadily during the next half-century. Most of it continued to be smelted with charcoal, sometimes mixed with coke. Timber resources were being depleted and the need to perfect a method of using mineral fuel in the furnace was becoming urgent.

The modern iron and steel industry spread outwards from two local areas in north-western Europe, central Belgium and the Ruhr. The first successful attempt to develop a modern iron industry, using coal both to smelt and refine iron, was made near Liège in central Belgium. William Cockerill had established himself in Verviers as a maker of textile machinery. In 1817 his son, John, set up an ironworks in some obsolescent buildings at Seraing, on the Meuse, a short distance above Liège, for the primary purpose of making components for the family's textile machines. These were mainly castings, for which he used pig-iron from furnaces in the Ardennes.[34] So far he had merely established an iron-processing works like dozens of others in the Ardennes foothills. But three years later he added a puddling furnace, the first in this region. Cockerill's works were described by Roentgen, an emissary of the government of the Low Countries, as 'by far the best in the whole land'.[35] In 1823 he added a blast furnace to smelt ore from the nearby hills with coke from the local coalfield, but in this was anticipated by Orban's furnace at nearby Grivegnée. Other furnaces and puddling works followed, and by 1842 there were 45 coke-fired furnaces in central Belgium as against 75 charcoal furnaces in the hills to the south.[36] The former lay in two groups the older was beside the navigable Meuse near Liège, and consisted of Seraing, Grivegnée and Espérance. The other lay on the River Sambre and the Hainault coalfield, close to Charleroi.[37] Here were the works of Couillet, Monceau, Montigny and Châtelineau. Iron production in the Sambre and Meuse valley increased to 90,000 tonnes in 1831 and to 222,000 in 1847, more than half of it from the furnaces around Charleroi.[38] Not all the pig-iron was puddled to make wrought iron. There was a sizeable export to ironworks in both northern France and in the Aachen-Eschweiler district of Germany. Thus early did different sectors of the region become integrated with one another.

Belgian initiative spilled over into both northern France and the Aachen district of Germany. The small-scale production in the western Ardennes now developed into a major iron industry. Small furnaces near Maubeuge were rebuilt to use mineral fuel brought up the Sambre from central Belgium, and iron-mining in the Ardennes was intensified. By 1845 there were nine smelting furnaces in the Maubeuge area. In 1839 blast furnaces, together with puddling and rolling works, were established at Denain, near Valenciennes, on the French extension of the coalfield.[39]

To the east of Liège lay the developing ironworks of the Aachen district. At its centre was Schleiden, where charcoal in the furnaces was slowly yielding to coke from the Würm-Inde field.[40] The Schleidental differed from other traditional iron-working areas in having a group of particularly enterprising and aggressive ironmasters. Amongst them was Albert Poensgen, who built up a highly specialised branch of the industry at Gemünd.[41] The manufacture of welded tubes was first developed in England in the 1820s. In 1843 it was adopted at Eschweiler and two years later Poensgen established a plant to make them at Mauel bei Gemünd. At this time the demand for welded iron pipes was increasing with the development of municipal gas systems. Pig-iron from the Eifel soon ceased to be adequate, and Poensgen and fellow manufacturers of tubes, boilers and metal sheet were obliged to import iron from Belgium. At the same time, Hoesch was operating a puddling works at Lendersdorf, near Aachen, again using imported Belgian pig.[42] In 1854 the German duty on imported pig-iron was raised. This was a severe blow to the refining industry, which had relied on Belgian metal. Within a few years coke-fired furnaces – the *Rote Erde* works – were built near Aachen to supply the puddling furnaces of the area. Lack of local ore was a serious disadvantage, and in the end the Aachen industry proved unable to compete with more favourably located works. Poensgen moved to Düsseldorf; Hoesch to Dortmund; and Rote Erde to Luxembourg.

The second focus of the iron-working industry lay along and to the north of the valley of the Ruhr (fig. 8.2). During the first half of the century there was little evidence of its future industrial pre-eminence. In the 1840s the English traveller Banfield could describe the environs of Essen as 'poetically agricultural'[43] but there were signs of change in the Ruhr valley to the south, where 'castles, ruins and factories rapidly succeed each other'. Here, on the exposed area of the coalfield, the processing and fabrication of iron from the Siegerland were well established. Puddling was introduced by Harkort at Witten and spread to the

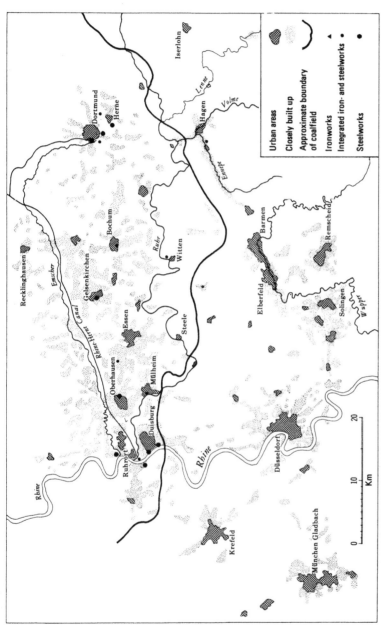

8.2 **The Ruhr industrial region, 1850–70.** About half the smelting capacity was built for the purpose of using coal-measures ore. The rest aimed to use ore imported into the area

vicinity of Dortmund, Bochum and Essen. Numerous attempts had been made to smelt iron, using Ruhr coke, but were uniformly unsuccessful despite the fact that the Ruhr contained ample reserves of the best coking-coal in Europe. The reason probably lay in a lack of understanding of the qualities of coal available and of the geology of the coalfield. There were repeated complaints of the coal and of the total inadequacy of the coke made from it. Nevertheless Ruhr coal was taken up the Rhine and used with apparent success in a Siegen furnace. At last, in 1849, iron was successfully smelted with coke at the Friedrich-Wilhelm works at Mülheim-on-Ruhr.

In the mean time important developments had taken place in the refining of iron. One of the most urgent needs in the field of ferrous metallurgy was a method of making steel from bar-iron. Such a process had been developed by Huntsman in England, but it was imperfectly understood on the continent. The Napoleonic government offered a reward to anyone who could make Huntsman, or 'crucible' steel. Among the claimants was Friedrich Krupp of Essen. He evidently produced steel, but whether of Huntsman quality is not clear. More likely it was his son Alfred, founder of the Gussstahlfabrik at Essen, who perfected his father's process. At about the same time Jacob Mayer, like Huntsman a watchmaker by trade and thus much aware of the need for steel of uniform quality, established another cast-steel works at Bochum. This was the beginning of the Bochumer Verein. Both Mayer and Krupp made steel for the moving parts of machines, and wheels and axles for locomotives were amongst their early specialisations. Steel-making, like puddling, was a very heavy user of coal, and was attracted to a coalfield site. At the mid-century output in all branches of the iron industry remained small. The production of pig-iron had risen only slowly. Within the Ruhr area itself it could only have been a few hundred tonnes at the beginning of the century, and Héron de Villefosse reported an output of only 142,000 quintals – about 14,000 tonnes – of pig-iron and 100,000 quintals of bar-iron in the whole Rhineland.[44] As late as 1860 the Ruhr itself yielded no more than 90,000 tonnes of pig-iron, though the amount of puddled and refined metal was a great deal larger. In 1849 the whole metal industry of the Ruhr area employed only 17 per cent of the working population.[45]

At this time the Ruhr seemed important far more for its coal production, which was increasing steadily, than for its iron and steel. Metal-working was still concentrated mainly in the hills to the south. Banfield described the intense activity along the rivers:

the great number of small works of wire-drawers, coppersmiths, steel-hammers, nail-makers, and countless other trades, which have clustered together on the brooks, like their rivals in the streets of Sheffield. On one stream [near Iserlohn] twenty-four little works are placed. Twenty are counted on another falling into the Lenne, near Altena. The little River Volme is occupied whereever its fall measures but a few feet.[46]

The highest reputation, however, attached to the works near Solingen and Remscheid. Solingen was traditionally a centre for the manufacture of swords. Now that swords were less in demand the craftsmen of Solingen turned to cutlery and cutting tools, in the production of which they very nearly achieved a German monopoly.[47] Banfield wrote:

The Wupper and all the dells that fall into it are peopled with small cutlery establishments. Thirty grinding-mills are driven by the Wupper alone, and the little River Dhün turns the wheels of 59 more, besides four iron-hammers and two steel-refining hammers...Every artisan lives in his own house...and undertakes work by the dozen.[48]

Remscheid was initially less specialised, but later concentrated on high-carbon steel goods, such as files and saws. Important as it was, this *Kleineisenindustrie* of Berg was technically backward, surviving by virtue of the skill of its craftsmen.[49] The less sophisticated crafts were indeed beginning to suffer from the competition of factory production. The traditional refineries in particular were disappearing, as the role of refining pig-iron passed increasingly to puddling-furnaces located on the coalfield.

The textile industries The weaving of textiles was the other prevailing industry in north-western Europe before the Industrial Revolution. Spinning and weaving were carried on almost everywhere; in the poorest and most backward areas because these were largely self-sufficing; in the more advanced because here a higher-quality fabric found a ready market. The whole of the Low Countries in particular was, in Coornaert's phrase, *saturé en draperie*, and had been since the Middle Ages.

There was a series of foci, all of them urban, around which branches of the industry had concentrated. They were the seats of the entrepreneurs who put out the raw fibre to the spinners, the thread to the weavers, and commonly finished the cloth and marketed it themselves. They had little fixed capital, and could readily shift the scene of their activities in response to cheaper or more abundant labour or more readily available materials.

Two features characterised the development of the textile industries: their failure to challenge the supremacy of the iron industry and their surprising flexibility and mobility. The textile industries disappeared from central Belgium and the Ruhr, which became pre-eminently the sphere of coal, iron and steel. They declined in Valenciennes-Maubeuge with the rise of the metal industries, and if they survived in the Aachen district, this was only because here it was the iron industry which failed to hold its own. In the end the textiles dominated in only three areas within the industrial region (i) the district which centred in Lille, (ii) the Verviers-Aachen district and, lastly, (iii) the Wupper valley and the district to the west of the Lower Rhine. The traditional fabrics were linen and woollens, but to these were added in the eighteenth century silk and cotton. Woollen fabrics, which had once been the mainstay of the textile industry in the southern Low Countries and northern France, tended to give place to linen, a local raw material and low-priced product replacing an imported material and a more expensive fabric. But woollen manufacture never disappeared. It remained important in the Lille district, and here was fought a three-way struggle between woollens, cottons and linen. It was one in which fortunes varied, and towns, even individual factories, alternated between one material and another. At first, cottons, in response both to the rising mass demand and to the greater ease with which manufacturing could be mechanised, spread at the expense of the more traditional manufactures. Then, as if by a simple dialectical process, woollens revived.[50] The town of Lille, like most towns in Artois, remained wedded to linen, but in the surrounding villages the spinning and weaving of wool regained the ascendancy. At the spreading industrial village of Tourcoing, cotton was dominant in 1839, but only four years later three-quarters of the fabric woven was either of wool or of mixed wool and cotton.[51] By the mid-century Tourcoing was essentially a woollens centre.[52]

In the neighbouring village of Roubaix a woollen mill was established in 1831,[53] and soon afterwards cotton spinneries were being converted to wool.[54] Roubaix soon became the most important cloth-weaving centre in France with, towards the end of the century, nearly half of the French productive capacity. 'The success of Roubaix-Tourcoing', wrote Landes, 'is the more impressive because no special advantages will account for it.'[55] There was, it is true, a local tradition of cloth-working, but the twin towns were poorly placed to receive coal and raw materials, and their chief advantage for the woollen industry probably lay in the fact that other places were more attractive to cotton. Not until 1846 was a railway link constructed between Lille and the mining district, and only thereafter did the industry have a satisfactory supply of fuel.[56] As late as 1838 11 out of 39 cotton mills in Roubaix 'were still running on horse gins'.[57] Such primitive devices were, however, gradually displaced by the steam-engine, though only 54 were employed in the whole textile industry in 1834.[58] Textile manufactures developed at a considerable distance from the coalfield, and the westward extension of the latter to within 15 km of Lille was not even suspected during the early days of the cotton industry. 'Coal was not', in Landes' words, 'a decisive factor.'[59] Nevertheless, coal was brought by a combination of canal and road to the infant factories.

The woollen industry survived also at Avesnes, Fourmies and other places on the margin of the Ardennes, but elsewhere it was in decline.[60] At Abbeville and Saint-Quentin it yielded place to cotton; around Saint-Amand, Douai and Valenciennes to coal-mining and the metal industries.[61]

The French woollen industry suffered from its failure to adopt quickly and effectively the technical innovations made in England. Mechanical spinning was not important until after 1830, and was far from general as late as 1860. Wool-combing by machine was not introduced until the 1840s, and the handloom remained common until late in the century.[62] Output nevertheless grew, though in a much less spectacular fashion than that of cottons.

The cloth industry was of little importance in central Belgium, where it was overtaken by coal-mining and eventually metal-working. But in the valley of the Vesdre, to the south-east of Liège, a woollen industry had been established by refugees from Flanders. It was successful, as refugee industries so often are, despite the almost complete absence of raw materials.[63] Here the small streams descending from the Ardennes plateau provided soft water for dyeing and finishing, and power for fulling. Wool from the Ardennes was early replaced by merino wool from Spain. By 1850 the industry around Verviers was dependent almost entirely on imported wool. At the same time the quality of the cloth improved, and an English observer grudgingly admitted that the kerseys from Verviers were as good as those woven in England.[64] No planned development of resources would have located such an industry near Verviers, yet in terms of eighteenth-century transport and labour supply the location was not unreasonable and was certainly successful. Success depended during the early phases of industrialisation more on careful management than on resources, and, once established, an industry acquired a momentum which often triumphed over environmental obstacles. The woollen industry was also well developed in Aachen and the hilly margin of the Eifel. It owed much to entrepreneurs from Verviers and was carried on mainly as a domestic craft.[65] Montschau, Eupen, Burtscheid and Düren emerged as local industrial centres,[66] and from here the industry spread to Cologne, Düsseldorf and the Ruhr. After 1815 steam-powered factories, fuelled from the Aachen coalfield, began to take over from the domestic industry.

The plain of the Lower Rhine had long been important for flax-growing and linen-weaving. In the early nineteenth century its economy became diversified. Silk manufacture, introduced by the van der Leyens in the early eighteenth century, dominated the town of Krefeld, and from there spread across the Rhine to Elberfeld and Barmen. At the mid-century, about a third of the silk-looms were still to be found in Krefeld.

Cotton-spinning and weaving were established in the hills of Berg in the eighteenth century. Settlements along the Wupper valley began to specialise in quality cottons, while the manufacture of poorer-quality fabrics spread across the Rhine and was thus included within Napoleon's France and benefited from the large French market.[67]

The history of the Weerth firm of Bonn illustrates the development of the left-bank industry. In 1804 its founder acquired the cloister of a dissolved Capuchin house in Bonn and in it established a small spinning mill.[68] Soon afterwards he established three more such mills. But he put his yarn out to domestic weavers, and continued to do so until 1846. His dependent weavers were then spread over a large area, the largest group being in Cologne, but considerable numbers were to be found in the Eifel.

The first mechanical spinning had in fact been at Düsseldorf in 1783–4, a large water-powered factory on the English model. This was followed by the establishment of spinning mills in nearby Gladbach and Rheydt, which supplied yarn to the predominantly rural weavers, but Krefeld, with its well-established silk industry (see p. 329), resisted the less remunerative cotton-spinning. The steam-engine was introduced for spinning at both Gladbach and Rheydt in the 1820s, and hand-spinning was eliminated over the next 30 years. Handloom-weaving lasted longer, and in 1861 only a fifth of the looms were power-operated. Banfield, in the 1840s, wrote that 'the whole district is peopled with weavers of cotton, linens and silk, Gladbach, Kempen, Dahlen, Rheydt, Viersen, are all manufacturing towns; and in numerous intervening villages the loom is a frequent inmate of the houses'.[69] Even in some of the towns of the Ruhr district, future cradle of the German steel industry, weaving was more important than metal-working, and in 1849 textiles employed more than 40 per cent of the workforce in the heart of the industrial region.[70] In the course of the century the cotton industry increased its grip on the left-bank region, while in Elberfeld and Barmen and the small towns of Berg the finishing branches – dyeing and

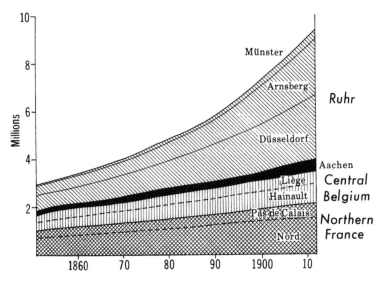

8.3 Growth of population in the north-west European region

printing – already important when Banfield visited the towns, strengthened their hold along the Wuppper 'and dye its waters a different colour every hour of the day'.[71]

The Lower Rhineland thus developed into a textile region as varied in its processes and products as the Lille region in northern France.[72] Linen and cotton weaving were still mainly rural, but woollen and silk-weaving had largely passed to the small urban mills.[73]

Population and urban development About 1850 the population of the whole industrialised region from the Pas de Calais to the eastern end of the Ruhr was about 2,700,000. A generation earlier it may have been about two million (fig. 8.3). A relatively small part of it was urban, and there were few large towns. Amongst them was Lille, a squalid and crowded town of 55,000 at the beginning of the century and of 76,000 half a century later. By the latter date Roubaix and Tourcoing had grown from mere villages to towns of 35,000 and 28,000 respectively. No other place in the French sector of the region approached this size, though Valenciennes, Douai and the port of Dunkirk had reached 20,000, and Maubeuge, Armentières and Halluin had turned 10,000.

In central Belgium only Liège, with a population of about 55,000, had reached any considerable size at the beginning of the century,[74] but Mons grew to be a sprawling mining settlement of 20,000, and the weaving centre of Verviers had reached 12,000. Older towns, like Namur and Huy, remained small, and the future steel centre of Charleroi was only a large industrial village. Until the mid-nineteenth century growth was more in the villages than the towns, because industrialisation was rather a rural than an urban phenomenon.

The eastern or German sector of the region, which was to become the most highly urbanised and densely populated, had few towns of any significance during the first half of the century. About 1815, the largest was the twin city

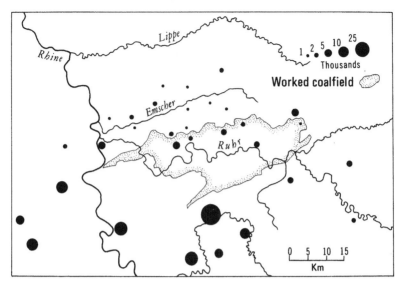

8.4 Urban development in relation to the Ruhr coal resources, about 1800. Based on K. Olbricht, 'Die Städte des Rheinisch-Westfälischen Industriebezirks', *Pet Mitt.*, 57 (1910)

of Elberfeld and Barmen with 45,000 (fig. 8.4). The other textile centres, Krefeld and Gladbach, had respectively 22,000 and 15,000, and Düsseldorf, capital of the Duchy of Berg, 21,000. On the coalfield there were no towns of more than a few thousand inhabitants. Mülheim may have had 7,000, but Essen and Dortmund were appreciably smaller.[75] Amongst the towns of the iron-working district, only Solingen with about 20,000 had reached any considerable size.

By 1855 there were still no large towns. The biggest were the textile centres of Elberfeld-Barmen, Krefeld and Gladbach, and, with Solingen and Düsseldorf, these were the only towns which exceeded 15,000. The low level of urban development is not a true measure of industrial growth. The coal mines, which provided the largest single source of employment, were not opened near to towns; they tended, rather, to be rural and to lead to the formation of mining villages. Most of the industrial region was therefore still rural, and a high proportion of its population, perhaps between a third and a half, still cultivated the soil, if only on a part-time basis.

The period of rapid growth, 1850–1914

The mid-century marked a turning-point in the development of the north-western industrial region. Before, there had been a period of steady but unspectacular growth. After it, every aspect of production in almost every part of the region grew at an unprecedented rate. The output of coal, in some respects a key indicator of growth, increased fifteenfold between 1850 and the First World War; pig-iron, some 25 times, and the manufacture of textiles at least ten times. Population increased rapidly, and, within the industrialised region itself, was three times greater in 1910 than it had been in the middle of the previous century.

Table 8.4. *Size of coal mines in the Ruhr*

Annual output (tonnes)	Number of mines	
	1850	1900
1,000–4,999	107	11
5,000–24,999	51	10
25,000–99,999	32	14
100,000–499,999	—	155
Over 500,000	—	24

During the same period manufacturing had become more diversified. Engineering grew into a major branch of industry; chemicals, scarcely represented in 1850, were of major importance by 1910, and food-processing and consumer-goods industries, of trifling importance before 1850, became in terms of employment as significant as steel or textiles.

Four features characterised this period of rapid growth. In the first place, domestic or family-scale manufacturing declined and disappeared and many, though not all, the small water-powered workshops were replaced by factories worked by steam. In the textile industry, hand-spinning at last ended, and the handloom had by 1900 effectively disappeared from the cotton and woollen industries. It lasted longer in linen, and, for quite different reasons, in the silk-weaving industry. A survey of the surviving domestic industries in Belgium, made in 1896, showed that only in needle-making, firearms and linen-weaving was household manufacture still significant.[76]

A second feature was the increasing size of units of production. This was especially the case with coal-mining. In 1850 most pits in the Ruhr produced less than 5,000 tonnes: by 1900 this had increased to more than 400,000 tonnes (table 8.4).[77] In iron-smelting and refining the new technologies made it necessary for plants to become very much larger in order to achieve the maximum economies of scale.

Thirdly, there was an increasing degree of integration between industrial undertakings both within the region and between it and other industrialised areas. This showed itself not only in the formation of cartels – particularly in coal and the metal industries – but also in the creation of vertical linkages between companies both for the supply of material and the disposal of their products.

Lastly, the effective completion of the railway net in western and central Europe gave a greater freedom than ever before to the entrepreneur in locating a factory. It also went far towards establishing a single market in which coal and steel from Upper Silesia could compete with that from the Ruhr, central Belgium and northern France, and textiles from Lille with those woven in Saxony or Württemberg. Under these circumstances marginal differences in factor costs could make a great deal of difference to success or failure in a particular branch of industry. No producing unit any longer had a secure market, insulated from competition; hence, the contraction and failure of plants in which factor costs were high, and the growing concentration of manufacturing in areas where these were most favourable.

Iron and steel If textiles constituted the foremost growth sector in the first half of the century, this role was unquestionably assumed by iron and steel in the second. The broad economic and technological circumstances of its expansion have been discussed. There was only one exception to the general picture of rapid growth: central Belgium, where the industry continued to expand, but at a significantly slower pace. However measured, the productivity of Belgian industry was relatively a great deal smaller in 1910 than it had been in 1850. About 1850 it contained 20 per cent of the population in the whole industrial region; by 1910 this had fallen to 17 per cent. In 1850 it produced two-thirds of the coal; in 1910, only 17 per cent, and its output of pig-iron fell from more than a half to only a quarter.

To some extent Belgium was paying the price of its early success. Mines were developed and factories built during the early phases of the Industrial Revolution. Its rivals, which developed their resources at a later date, could take advantage of a more advanced technology and thus build on a larger and more economic scale. The blast furnaces in central Belgium were in the later nineteenth century older and smaller than those in northern France or the Ruhr, and steel and rolling mills were less advanced.[78] Belgium had been the first country in continental Europe to adopt the puddling process; it was almost the last to retain it, and in other ways also Belgium was slow to adopt the newer steel technology.[79]

This relative backwardness in the Belgian industry was, of course, due in part to insufficient capital investment, and this in turn reflected a lack of confidence in the future profitability of the industry. For this there were two important reasons: iron ores were exhausted and coal-mining was meeting with increasing difficulties. The ore reserves of the northern Ardennes had never been large, and the prevailing methods of exploiting them were calculated to reduce their value yet more.[80] Except in *prov*. Namur, their extraction was left to anyone who could come to an agreement with the local landowner. Mines, in consequence, were too numerous to count and were so irregularly worked that rational and scientific mining was impossible.[81] By 1910 only one iron mine was left in production within the region.[82] The furnaces in central Belgium had thus to be supplied increasingly with imported ore. This came overwhelmingly from the *minette* field of Lorraine, and, in the absence of water transport, had to be conveyed by rail.

The Belgian coalfields, so highly praised earlier in the century, ceased gradually to satisfy the needs of industry. The thin and contorted seams were increasingly costly to mine. Water, fire-damp, a very high accident rate, and the rapid exhaustion of the better coking-coal added to the problems of the industry. Output of coal per miner rose comparatively little, and at the beginning of the present century was lower in the Liège and Hainault fields than anywhere else in the whole region.[83]

There were other factors in the relative decline of the central Belgian industrial area, amongst them the inadequacy of water transport to the port-city of

Antwerp – not rectified until the cutting of the Albert Canal in the 1930s, and the loss of the German market for Belgian pig-iron. The location of an industry which is no longer expanding is not likely to change, and the Belgian industry remained concentrated around Liège and Charleroi. Several of the Belgian iron and steel companies actually secured rights to ore or formed links with firms in Luxembourg and Lorraine, as if guaranteeing themselves against the ultimate run-down of the industry in central Belgium.[84]

The iron industry that had grown up to the west, in what is sometimes called French Hainault, shared many of the characteristics of the Belgian.[85] The charcoal-iron industry of the Ardennes was at first supplemented with puddling works in the Sambre valley, and eventually replaced by coke-fired furnaces. The latter were built near Maubeuge, and by 1867 there were no less than eleven of them.[86] But the Maubeuge region shared the problems of Belgium itself; the furnaces were gradually blown out, and only four were left in 1914. In the mean time, however, most of the companies active in the Maubeuge-Aulnoye district had also established links with ironworks in Luxembourg and Lorraine. The latter supplied pig-iron and crude steel for the numerous refining and fabricating works in French Hainault.

The coalfield of northern France appeared to have greater advantages than either central Belgium or the Maubeuge district. Coal was more abundant and of better quality, and ore, akin to that of Lorraine, occurred to the west, in Boulonnais. Furnaces were built close to the ore at Marquise and Outreau, but – another failure to appreciate the scale of the resources – the ores were soon worked out and the furnaces closed. At the end of the century the industry revived on the coalfield.[87] The reason was that it was becoming increasingly necessary to provide steel of a different – and higher – quality than that made from *minette*. This, of course, necessitated the use of imported, non-phosphoric ore, whether from Spain, Sweden or North Africa. The coalfield of northern France, with the port of Dunkirk only some 50 km away, seemed as good a site as any for smelting this ore. Existing works at Denain, Anzin and Trith-Saint-Leger were expanded, and in the whole area pig-iron production increased from 577,000 tonnes in 1910 to 933,700 in 1913.[88]

In contrast with the sluggishness shown in Belgium and northern France, the industry in the Ruhr enjoyed a spectacular growth which was interrupted only by the First World War. Fundamental to its development was the northward spread of coal-mining and the increase in coal production. In 1850 the northern limit of mining lay from west to east very roughly through Oberhausen, Bochum and Dortmund. Little of the 'hidden' field had been exploited, and coking-coal (*Fettkohle*) can have made up only a small fraction of the output. During the next half-century, the frontier of mining not only advanced beyond the Emscher but at some points reached the Lippe and even extended west of the Rhine in the direction of Krefeld. The new mines could penetrate the whole gamut of the coal series, and produce whatever was most in demand, and by careful location of the mines on the anticlinal ridges of the coal beds, the thickness of overlying rocks was minimised.

In 1850 about 1,666,000 tonnes of coal were produced from about 200 mines, an average yield of only about 8,000 tonnes. During the following decades both

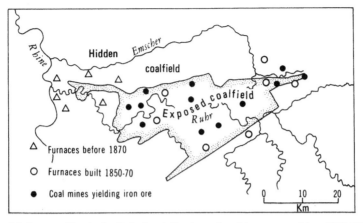

8.5 Iron-working in the Ruhr in relation to coal and iron-ore working, 1850–70

the number and the size of mines increased sharply. Their solid, brick towers, commonly called Malakoff Towers, began to form a prominent feature of the landscape of the central Ruhr. They provided 'apartments for the overseer and captain, with the engine and lifting shafts all under the same roof', the whole forming 'a very tasteful building'.[89] By about 1880 the coal output reached more than 22 million. By 1900 it was 54 million, and by 1910, 88,000,000. By this date the Ruhr was producing about 58 per cent of all the coal mined in western Europe, northern France yielding only 14 per cent and central Belgium eleven.[90] The Ruhr's share in Europe's coke production was even higher – 71 per cent, while northern France produced 8 per cent and central Belgium 9 per cent.[91]

The blackband iron ores, found within the coal-measures, were at first used intensively. Furnaces were built close to the mines which thus yielded *all* the raw materials needed. Amongst them were the Dortmund and Hörde works (fig. 8.5). The output of blackband ore continued to increase until 1865, when it reached 364,800 tonnes.[92] Production then declined, and by the end of the century had ceased to be significant. The search for alternative sources of ore was important if the Ruhr was to survive as an iron-producing area. Attention at first turned to the ores of the Sieg-Lahn-Dill region, which were brought down the Rhine and used in furnaces built near Duisburg. Too much reliance was, perhaps, placed on these ores, and no less than six smelting works were built near the mouth of the Ruhr, where Rhine barges could unload their ore. The ore supply fell far short of requirements, and at the beginning of the twentieth century supplied only about 10 per cent. *Minette* from Lorraine was then used, but the high cost of conveying a low-grade ore over so great a distance reduced its importance, and it never amounted to more than about 15 per cent of the total furnace charge. On the other hand, it was quite practicable to transport coal and coke from the Ruhr to Lorraine, and several firms established works on the orefield (see p. 390). If the problem of ore supply to the Ruhr had not been solved they might indeed have transferred much of their activity to the orefield.[93]

In fact, ore began to be imported, first from Spain and North Africa and then, on a massive and growing scale, from northern Sweden. These ores could all be brought by sea to the Rhine-mouth ports and transported by Rhine barges to the Ruhr. The works built near Ruhrort to smelt ore from the Siegerland could as easily handle ore from overseas brought upriver from Rotterdam.

This left the many large works that had been built in the eastern Ruhr without a water-borne means of supply. Industrialists of the Dortmund area demanded a canal link with the Rhine or the North Sea ports. As early as 1877 a proposal was made to the *Reichstag* for cutting a canal from Dortmund to the navigable Ems with a further extension to the Elbe. In 1886 the scheme was adopted, but, out of deference to Upper Silesian coal-mining interests, without the extension to the Elbe. Work was begun in 1890 and completed in 1899. The volume of traffic on the canal, made up entirely of ore and coal, grew quickly. By 1905 it was over 1.5 million tons, and by 1910 over three.[94] The two major ironworks in Dortmund were equipped with dock basins, and other works quickly clustered near the waterway. The link between the Dortmund–Ems Canal and the Rhine – the Herne Canal – was begun soon afterwards, but was not opened until 1914.

In this way the Ruhr was assured of an adequate supply of high-grade ore. Sweden quickly became the largest overseas source, supplying about 22 per cent of the ore in 1901. This continued to rise, and amounted to 40–50 per cent after the First World War.

With unlimited high-quality coking-coal, a secure supply of ore and an expanding internal and overseas market, the Ruhr was set for spectacular growth. Although some of the older works within the area closed, most were modernised and enlarged, and new works were established, such as the Thyssen works at Hamborn, the Phönix works at Laar, the Krupp works at Rheinhausen, all of them beside the Rhine, with their own docks for unloading ore. A short distance to the east the Gütehoffnungshütte, oldest in the Ruhr, was expanded. The Krupp works, which dominated the city of Essen, concentrated on steel products, amongst which steel plate and armaments were increasingly important. Bochum, 18 km to the east, was second only to Krupp for steel, but tended to concentrate on castings and forgings. In the Emscher valley to the north were the Schalke works of Grillo, around which the town of Gelsenkirchen was beginning to sprawl. The other large cluster of works was around Dortmund, where the Hörde plant was joined in 1871 by the Hoesch furnaces and steelworks (Westfalenhütte), and the Union works. The only iron- and steelworks remaining south of the River Ruhr, Hagen-Haspe, continued in production and, indeed, added new furnaces.[95]

The Bessemer process was adopted in most of the Ruhr works, and shortly afterwards the open-hearth. In 1879 the Thomas process was eagerly welcomed by ironmasters who feared that they would soon be starved of ore. The first 'blows' outside England were at Dortmund and Ruhrort. In the end, neither converter nor the basic process established any clear predominance. It was better to leave basic-steel to the works newly established on the orefield in Lorraine and Luxembourg, and for the Ruhr to concentrate on the superior qualities of metal for which its imported ores were more suitable. For this reason it was the open-hearth rather than the converter that replaced the puddling-furnace. The open-hearth, in addition to allowing a close control over the quality of the metal, could also use the growing volume of process scrap and the coke-oven and blast-furnace waste gases. Indeed, a feature of the development of the Ruhr in these years was the network of gas pipelines which conducted such gases around the area.

Beside the dozen large, integrated iron- and steelworks, within the Ruhr area there was a larger number of works which only refined, processed and fabricated steel. These almost invariably used the open-hearth, though the traditional crucible method lasted a long while for the production of special steels. Foremost amongst such steelworks was, of course, Krupp, which obtained its iron from other plants in the area. But there were also works at Witten, Wetter, Hagen and along the Rhine as far as Düsseldorf. They were even to be found in the textile stronghold of Krefeld.

The volume of pig-iron crept upwards with only minor checks from 1850 to 1914. In 1850 it had been about 11,500 tonnes; on the eve of the First World War this reached about 8.2 millions. Steel production rose from a few hundred tonnes to more than ten millions.

The textile industries The textile industries, with the exception only of linen, continued to grow, after 1850, though a great deal less steeply than iron and steel. The latter had an expanding market. The domestic market for textiles on the other hand was already satisfied, and increased productive capacity was needed only to meet requirements imposed by a growing population, a slowly improving standard of living, and a rising export trade. As a result, there was little change in the geographical pattern of production, contraction here, consolidation there, but no new centres of production developed, and no radical change in technology, apart from the elimination of hand-spinning and the virtual disappearance of the handloom.

About 1850 there were three major centres: the Lille district of northern France, the Verviers-Aachen district, straddling the Belgian-German boundary, and the Lower Rhineland from München-Gladbach to the Wupper valley. Around each there was a scatter of small manufacturing centres and a host of domestic textile workers. The trend during the next 60 years was for the latter to disappear, absorbed into the main concentrations of textile production. A corollary of this was the increase in the number and size of factories, which in turn necessitated ever larger units of mechanical power. The relationship of the textile industry to the coalfields thus became closer, and the westward extension of mining in the Pas de Calais was a factor in strengthening the industry in the Lille district of northern France.

The character of the Lille industrial area had never been clearly defined. It was, in Fohlen's words, 'la seule région d'industries textiles "*interchangeables*"'.[96] By 1850 the pattern was beginning to stabilise. Lille itself had become the chief centre for cotton-spinning, with flax-spinning of secondary importance. In neighbouring Roubaix and Tourcoing cottons and woollens were more evenly balanced, with linen coming a poor third. In 1872, spindles were distributed as shown in table 8.5.[97] The number of cotton spindles steadily increased in the Lille area. The cotton famine of the 1860s led to the closure of many small spinneries in Armentières, Halluin and other small towns, but this was followed by renewed growth in the Lille conurbation. From 20 per cent of the total cotton

Table 8.5. *Textile spindles in the Lille district, 1872*

	Cotton	Woollen	flax, hemp, jute	Silk
Roubaix	325,054	289,128	7,424	7,424
Lille	496,968	5,808	251,331	—
Tourcoing	150,000	377,090	12,484	—

spindles in France about 1850, it increased its share of a greatly enlarged total to more than 40 per cent by 1910.[98] The Lille industry concentrated on spinning. There were fewer than 900,000 spindles there in 1852, but by 1910 this figure had increased to nearly three millions. Weaving on the other hand was relatively unimportant. About 1850 it embraced only about 3 per cent of French power-looms. By 1909 it still had little more than 4 per cent.[99] Much of the cotton yarn from Lille was used in the lace and hosiery industries or sent to weaving areas such as the Vosges and Rouen.

By contrast with the cotton industry, which produced in the main for domestic consumption, woollens served a large export market. Production continued to increase into the 1880s; then, with the contraction in overseas sales, it stabilised. Handloom-weaving remained an important though declining branch of the industry until the end of the century.[100] Roubaix and Tourcoing remained the chief woollen centres,[101] but there was a degree of local specialisation within the region. Wool-combing, carding and spinning tended to predominate in Tourcoing; weaving in Roubaix.[102] Through the latter half of the century the two towns intensified their control of the French woollen industry. About 1850 they contained about 20 per cent of the woollen spindles; by 1875 nearly half were in northern France,[103] and in 1910 no less than 35 per cent were in Roubaix-Tourcoing alone, and a further 42 per cent were scattered through the rest of the northern *départements*.[104]

The linen industry continued its slow decline. Spinning became largely mechanised during the later nineteenth century, though much of the weaving still remained in the hands of a diminishing body of domestic craftsmen in the villages and small towns of the Flanders plain.[105] Lille continued to be the centre of flax-spinning, but at other centres, such as Armentières, Wervicq and Halluin, small factories were beginning to attract labour away from the rural workshop.[106] Indeed much of it travelled daily from the overcrowded farmlands on the Belgian side of the boundary. The location of the linen industry was a kind of compromise between 'les necessités du commerce et celles de la main d'oeuvre'.[107]

The textile industry had once been spread through the whole industrial region of northern France. Linen was important at Douai and Cambrai; lace at Valenciennes, cloth at Saint-Amand and Douai, and scattered through French Hainault was the manufacture of linens and woollens.[108] But wherever iron-working came to be established, as it was near Valenciennes, Douai and Maubeuge, the competition for labour drove out the ill-organised manufacture of textiles. Wherever the iron industry succumbed, as at Fourmies, or where it never developed, as at Cambrai, the textile industry survived.[109] Furthermore, as coal-mining developed, the textile industry, much of it domestic, was also driven from the mining area. This mutual exclusiveness of textiles on the one hand and coal and steel on the other is surprising. About 71 per cent of the labour in textiles was female, and rather higher in the spinning branch. Employment in the rival branches of industry was heavily male. The complementary nature of the industries – 'un facteur d'équilibre', Gendarme called it[110] – seems not to have brought them together on any significant scale.

In the Belgian sector of the region a textile industry survived on a significant scale only in the valley of the Vesdre, where almost the whole of the Belgian woollen industry was located. The industrial area straggled along the narrow valley, though by the mid-century only a minority of its total of nearly 200 factories were still operated by water. The area lay within a few kilometres of the productive Liège coalfield, and by 1853 there were 143 steam-engines active in the industry; by 1861, there were 208, and new factories were being built away from the river on more spacious sites.[111]

Only a spur of the Ardennes and a political boundary separated the Verviers district from the German woollen centres of Aachen, Düren and Monschau. The woollen industry was at first overshadowed by iron-production. Then the decline of the latter (see p. 314) left the field free for the renewed growth of cloth-weaving.[112]

The Lower Rhineland disputed with Saxony the role of largest German centre of textile manufacturing. It was indeed widespread, and the towns of Gladbach, Rheydt, Krefeld, Elberfeld and Barmen were only the most important amid a host of lesser centres. At the mid-century the Ruhr, the very core of the developing iron industry, had nearly 30 per cent more textile workers than iron- and metal-workers. The steel boom of the 1870s drove textiles from the Ruhr, but they remained firmly established to south and west.[113] In the Wupper valley they spread at the expense of the small steel and cutlery workshops, and west of the Rhine their supremacy was undisputed. Kollmann's map of the distribution of weavers of all kinds in 1882 (Fig. 7.2) showed the densest concentration in all Germany in the plain to the west of Düsseldorf.[114]

In the late nineteenth and early twentieth century this area was attracting an increasing proportion of the industry's capacity. From 8 per cent of German cotton spindles in 1887, the Lower Rhineland increased its share in 1909 to about 13 per cent.[115] of a capacity that had in the meanwhile more than doubled. In München-Gladbach, Rheydt and the surrounding villages, 'cotton ruled almost unchallenged'. The chief products were quality fabrics, pure cottons, mixed fabrics, blankets and flannel.[116] Hand-spinning had at last disappeared in the 1860s, but at this date there were still almost five times as many handlooms as mechanical, though it must be remembered that a power-loom could weave very much faster. In 1878 there were all together 5,011 power-looms in the towns, as against 176 handlooms. In the rural areas, on the other hand, there were 846 handlooms and only 110 power-looms.[117] During the following years the number of handlooms continued to decline and they were of negligible importance by the end of the century.

The only rivals to Gladbach-Rheydt were the twin towns of Elberfeld and Barmen. As in Gladbach, mills were scattered through the surrounding villages, where production was mainly for the lower end of the market.[118]

The linen industry was declining throughout the period, as flax passed out of cultivation in the German plain and cottons cut ever deeper into its market. But silk, at the opposite end of the scale, was able to maintain its prestigious position. It continued to dominate the industrial structure in Krefeld and overflowed into neighbouring Rheydt and Viersen.[119]

There was a certain similarity between the textile industry of the Lower Rhineland and that at the western extremity of the industrial region, in the neighbourhood of Lille. Both lay off the coalfield, but near enough for the supply of fuel to present no problem. Both formed single industry regions, in which most other manufactures, such as engineering and chemicals, were of small importance, and in any case were related in some way to the dominant textiles. And within each there were local specialisations, not only in silk, cotton, woollens and linen but also in types and qualities of each.

Other manufactures Textiles, iron and steel dominated manufacturing in the north-west European region. Each evolved in and almost monopolised the areas which seemed best suited to them, and competed with others for dominance in the intervening areas.

In the later years of the century the industrial structure of the whole region began to diversify. Many of the new industries were related in some way to those already there. They provided essential materials, like dyestuffs and bleach for the textile manufactures, or, like engineering and metal construction, they used the products of the older industries.

The smelting and working of non-ferrous metals was not quite a new industry, though it grew and diversified in the later years of the century. It derived from the zinc mines of Vieille Montagne[120] and the lead and copper of the Eifel and Sauerland. Ores were nearing exhaustion in the late nineteenth century, but were supplemented by growing imports, and non-ferrous smelting remained an important branch of industry near Liège and in the western Ruhr.[121]

Amongst the first of the really new industries to emerge was glass-making, 'the Belgian speciality *par excellence*',[122] attracted to the Liège coalfield early in the century. From Val Saint-Lambert, the industry spread to the Hainault coalfield near Charleroi. By 1850 central Belgium was the leading centre of glass manufacture in Europe.[123] The industry continued to produce the traditional blown and spun glass, but tended to concentrate on window and plate glass in which a large export trade was developed. The units of production became larger and fewer as the century progressed and more advanced technologies were employed. By 1896 the glass industry was one of the most important manufacturing industries in Belgium, after textiles, mining and the metal industries.[124]

The metal-using industries had always been carried on in much of the region. At first they were restricted to the production of simple articles made from bar-iron, such as wire, nails, domestic hardware and cast-iron pots and stoves. Small-scale manufacturing units continued throughout the period to produce such goods along the southern fringe of the region, from French Hainault in the west to the neighbourhood of Altena and Iserlohn in the east. Foremost amongst the *Kleineisenindustrien* were the cutlery manufacture at Solingen and the making of tools, files and saws in and around Remscheid. Both industries continued until the First World War to be carried on in small domestic workshops.[125] The work was labour-intensive, and the replacement of water-power by small electric motors had the effect of perpetuating the traditional industrial structure. This was especially true of the grinders, an important group within the labour force, most of whom worked at home.[126]

On this pattern of scattered and mainly domestic industry were superimposed the larger-scale engineering and steel construction industries; first, steam-engines and locomotives, followed by machine tools, marine engines and steel constructions of all kinds. The Ruhr evolved into the largest and most varied producer of such goods, with an important boat-building industry at its Rhine ports of Duisburg and Ruhrort. In addition, the manufacture of textile industry, established at Verviers by Cockerill, continued throughout the century.

Most important by far of the newer industries was the manufacture of chemicals (see p. 346). The industry assumed many forms, most of which were in some way integrated with older branches of manufacturing. Foremost amongst them were those which derived from the mining and smelting industries. One of the first to evolve was the production of sulphuric acid, and hence of fertilisers, from the waste gases of the zinc and lead smelters at Vieille Montagne and Stolberg. The by-products of the coking process were at first allowed to burn off, but in the second half of the century the primitive coking ovens began to be replaced by those which allowed the by-products to be recovered. In the last quarter of the century the Carvès oven, specifically designed to conserve the waste products, began to be used and with modifications has since dominated the coking industry. Nevertheless, in 1898 a consular report noted that, in spite of the adoption of Carvès ovens, some beehive ovens were still in use in the Ruhr.[127] During these years the number of workers in the coking and coal-tar industry in the Ruhr area increased from 7,441 in 1886 to 23,485 in 1907.[128] From the 1860s the dyestuffs industry was firmly linked with the distillation of coal-tar from the coke ovens, and achieved a very considerable importance in the Rhineland, especially at Leverkusen.[129]

A second major branch of the chemicals industry was the manufacture of sulphuric acid and of soda, the starting-points of a chain of more sophisticated chemical products, many of which were essential to the textile industries. Location of the newer branches of the chemical industry was dictated in part by their raw materials, in part by the market for their sometimes bulky products. They were related on the one hand to the coking and steel-making industries and on the other to the textile, glass-making, ceramic and fertiliser industries. They were to be found in the Ruhr and along the Lower Rhine, where at the beginning of the twentieth century Leverkusen was beginning to develop into a giant of the chemical industry; around Liège, and near Lille, where the textile industries offered an immense market for their products.

The transport net The evolution of the north-western industrial region was both the consequence and the cause of the development of a complex network of waterways and railways. The region was well endowed by nature before industrial growth began. In the west the headwaters of the Scheldt; in the centre, the Sambre and Meuse, and to the east, the Rhine and Ruhr all provided navigable waterways, though their quality varied greatly and some had to be deepened and supplemented with canals before they acquired much value. Nevertheless, much of the coal, especially during the earlier stages of industrial development moved by water, and without water-borne transport it is unlikely that the mines of central Belgium and the Ruhr would have attained their importance in the early nineteenth century.

The early railway systems were not devised with the future needs of industry in mind. The axial routes in Belgium avoided the Meuse valley, and when the first railway was built through the central Ruhr as part of the *Köln-Mindener Bahn*, it avoided the town of Essen. By 1850 the railway net therefore bore little relationship to future industrial needs. By 1870 this had changed radically, and all parts of the industrial region were interlinked. By the end of the century the railway system was effectively completed.

Despite the development at a relatively early date of a dense railway network, the waterways never lost their significance. In no other industrial region was their role so important, and as manufacturing came to depend more and more on materials imported from overseas, so the linkages by water with the ocean ports grew in importance. It was not wholly coincidental that those parts of the region which declined, relatively at least, in importance – central Belgium, French Hainault, the Aachen district – were those where navigable waterways were least developed.

In northern France the headwaters of the Scheldt, navigable for small barges, were supplemented at an early date by canals which, furthermore, linked the region both with the ports of Dunkirk and Ghent, and with Paris which constituted a major market for the coal. Charleroi was linked with Brussels by a canal, built in 1827–32 and enlarged half a century later, but its capacity was small, and central Belgium suffered throughout the century from the inadequacy of its water links with the port of Antwerp. The Meuse-Scheldt canal from near Maastricht across the Campine to the Lower Scheldt was constructed in 1844–59, but it also was too small and its usefulness was reduced by persistent disputes with the Dutch.

No part of the Aachen sector had the advantage of water-borne transport, but the Ruhr rejoiced in the finest navigable waterway in Europe and was able to supplement it with the axial Herne canal and the heavily used Dortmund–Ems canal. The Ruhr and Lower Rhineland enjoyed the best of connections not only with the Rhine-mouth ports of Rotterdam and Amsterdam but also with Antwerp to the west and Emden and Bremen to the east.

Urban growth By the eve of the First World War the industrial region of north-west Europe had become the nucleus of the most populous and highly urbanised part of the continent. The development of manufacturing industry brought with it a rapid growth in population and the expansion of towns. The region had a population of about 2,945,000 in 1856–61. By 1880–1 this had risen to 4,390,000, and by 1900–1 to 6,907,000. By 1910 the total was close to nine

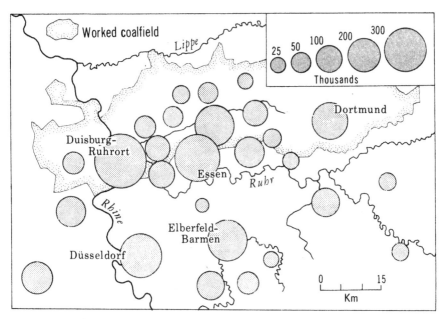

8.6 Urban development in the Ruhr industrial region, 1910

million. At the same time the population became increasingly urbanised. At a rough estimate 20 per cent lived in towns of over 10,000 in the 1850s. By about 1910 this had increased to more than 40 per cent.

Two poles of population growth had emerged in the French sector of the coalfield, the Lille area and the so-called *Pays Noir*, between Béthune and Valenciennes, in which mining and metal-working dominated. Lille itself grew from 75,800 in 1851 to 217,800 60 years later, and its satellites, Roubaix and Tourcoing, from respectively 34,700 and 27,600 to 122,700 and 82,000. Urban development was less vigorous in the 'Black Country'. Instead a large number of small towns and industrial villages grew up around the mines and steelworks, and the area acquired its present aspect of a large, densely settled region without major, urban nuclei, except in Douai, Denain and Valenciennes, none of which greatly exceeded 25,000.

In the Belgian sector of the industrial region the only large city was Liège, which had reached 225,000 by 1910, or, with its industrialised suburbs, 300,000. The textile centre of Verviers was about 75,000, but none of the mining and steel-making centres exceeded 30,000. Within Germany, by contrast, urban growth had continued unabated during the later nineteenth and early twentieth centuries, and the Ruhr industrial region had become a vast conurbation (fig. 8.6). Within the Rhine-Ruhr sector of the region there were no fewer than ten cities of over 100,000, and Essen and Düsseldorf were close to a quarter of a million.

During the second half of the nineteenth century the transport network was completed, and the region came to be criss-crossed by railways which linked every industrial centre not only with the larger cities, but also with the ports, from Calais in the west to Bremen and Hamburg in the east. At the same time water-transport was further developed. On the eve of the First World War the waterways of the region – the Scheldt, Meuse and Rhine – were used intensively for bulk cargoes. They had furthermore been linked by canals with waterway systems to west and east (see pp. 435, 474), so that all parts of the region were in close touch by water as well as by rail with the chief centres of population in western Europe, as well as with its ocean ports.

Luxembourg, Lorraine and the Saar

The Jurassic escarpment, which forms the eastern limit of the Paris Basin, swings to the south of the Ardennes and continues through Lorraine into Burgundy. On the east it rises, generally steeply, from the Triassic plain of eastern Lorraine, forming a barrier which was formerly thought to have great military significance. Its real value, however, lay in the immense reserves of iron ore which it contained, the largest in Europe and one of the most abundant in the world. Farther to the east the Triassic beds lapped against the ancient rocks of the Ardennes, Hunsrück and Vosges. The coal basin of the Saar lay against the flanks of the Hunsrück, and dipped beneath the plain of eastern Lorraine. The twin resources upon which the industry of the region was founded were the coal of the Saar and the iron of Lorraine. But there was another. Triassic beds everywhere are likely to contain deposits of soluble salts. Here, in the vicinity of Nancy, were deposits of rock-salt large enough to support a chemicals industry.

The region was well served with the means of water-transport (fig. 8.7). Beyond its western margin flowed the Meuse. In front of the Jurassic scarp itself flowed the Moselle, its northward course taking it across the Eifel-Hunsrück massif to join the Rhine at Coblenz. Just within the limit of the Hunsrück the Moselle was joined by the Saar, which flowed across the coalfield to which it has given its name. All these rivers were navigable for small craft, though the Moselle below Trier was shallow and little used. The ironmasters of Lorraine persistently demanded that improvements should be made in its channel in order to give them a link with the Rhineland and Ruhr. These had to await the more intimate relations between France and Germany which followed the Second World War. The region was, however, served by a canal which linked the Marne and thus the Paris area with the Meuse and Moselle, and was continued eastwards to the Rhine near Strasbourg.

Coal and iron

The Saar coalfield was exposed in the Saar valley, but dipped to the west beneath the younger rocks of Lorraine. The seams were numerous, but thin and much folded. The coal itself was gassy, and coke made from it was friable and easily crushed in the blast furnace, a factor which greatly restricted the size of the furnaces used. On the other hand the coal was well suited to gas production, and was much used to supply municipal gasworks in south-western Germany.

Until the end of the First World War most of the mines were in the hands

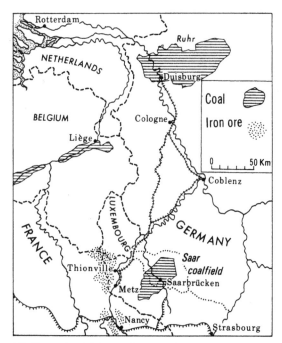

8.7 The Ruhr, Belgian and Saar coalfields and the iron-ore resources of Lorraine and Luxembourg

of the Prussian State, which had inherited them from the Dukes of Nassau-Saarbrücken. The boundary of Germany as determined in 1815 ran parallel to the River Saar and to the south-west of it. The exposed coalfield lay wholly in Germany, but most of the hidden field was in France. Little attempt was made to explore the latter as long as France had possession of much of the exposed field. After 1815 experimental bores were sunk but not until 1846 was a workable seam discovered on the French side of the boundary.[130] A number of mines were opened, but output was still less than 200,000 tonnes when the area was annexed to the German Reich in 1871. Thereafter mining was put on a more rational basis; the number of mines was reduced and output steadily expanded to 3,000,000 tonnes on the eve of the First World War.

Within the Prussian sector of the coalfield mining had been carried on continuously from the Middle Ages, and in 1815 output was of the order of 115,000 tonnes a year. By the mid-century it had increased to about 1.5 million tonnes; by 1890 to 9.4, and in 1913 to 13 million.[131]

The region as a whole was rich in iron ore. Long before the Jurassic ores of Lorraine and Luxembourg had achieved their modern importance, smaller deposits of different quality and geological age were being worked. Pockets of ore in the older rocks of the Eifel and Hunsrück were mined and smelted in charcoal furnaces scattered through the hills. At the opposite end of the geological scale was the *fer fort*, a granular deposit formed in cavities in the surface of the Jurassic limestone. It was easily worked in shallow pits, was free of phosphorus, and yielded an excellent bar-iron. On the other hand individual

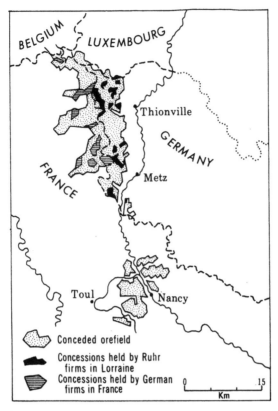

8.8 The conceded orefield in Lorraine, about 1900

deposits were small and total reserves limited. Most lay to the west of the scarp face, where they had since the Middle Ages supported a charcoal-iron industry. To these varieties of ore should be added the coal-measures ironstone which occurred in small amounts in the Saar coalfield itself.

By far the most extensive deposit, however, was the bedded Jurassic ores, or *minette*. These formed thick deposits some of which came to the surface in the face of the scarp and could be worked in open quarries. From here the beds dipped gently to the west. The number of ore-bearing beds varied, but their combined thickness was generally from 25 to 50 m. They were best developed to the west of Thionville. They diminished in number and thickness southward, and disappeared near Nancy. North of Thionville the scarp turned to the west, and lay across southern Luxemburg and the southernmost part of Belgium, where again the ore-bearing beds died out. The western boundary of the orefield was less definite but certainly lay at too great a depth for profitable mining (fig. 8.8).

The ore itself consisted of grains of haematite set in a matrix which could be either calcareous or siliceous. The distribution of the two types of ore was important since neither was by itself suitable for the modern furnace. There had, therefore, to be some movement of ore within the orefield. Ores were in general of a low grade, rarely containing more than 30 per cent metal and usually significantly less. Furthermore they were highly phosphoric. Reserves were, however, immense. Total probable reservers were in 1949 put at over 5,000 million tons of ore. This amounted to about half the workable ore to be found in Europe.[132]

The Lorraine region was the most monolithic of these examined in this chapter. It was solidly based on coal and iron, and scarcely another branch of manufacturing intruded, and no branch of the textile industry was carried on. Even the engineering and metal-using industries were of little consequence. The chemical industry established on the Lorraine saltfield is the only significant exception in this world of coal and steel.

The early nineteenth century

In the early years of the nineteenth century it was a populous, agricultural region. The limestone scarplands formed dry, open country; the Lower Jurassic and Triassic clays which bordered them on west and east were more heavily wooded, and forests spread over the more hilly areas of older rock to north and east. It was not highly urbanised, but contained Nancy, former seat of the Dukes of Lorraine; there also were Saarbrücken and the episcopal cities of Metz and Toul. Lorraine was at this time ringed with ironworks. The forests of Luxembourg and the Ardennes, of the Hunsrück and the Vosges were dotted with furnaces. To the west furnaces used the *minette* as well as the *fer fort*, while to the south lay the ironworks of Franche Comté. It is impossible to be precise but there *may* have been as many as a hundred charcoal ironworks within the limits of the future Lorraine-Luxembourg-Saar industrial region.

The history of industry within this region is divisible into two periods separated by the critical year 1879 which saw the introduction of the basic process of steel-making. It is a mistake to assume that the *minette* had no value and was not used before Thomas devised a method of eradicating its phosphorus. A highly phosphoric iron ore was good for castings, and it could be puddled to make an iron that was adequate for agricultural tools. The calcareous *minette* was, in fact, often mixed with *fer fort* to give a self-fluxing mixture in the furnace. The famous Hayange works of de Wendel were operating throughout the eighteenth century on the Jurassic orefield and unquestionably used *minette*. In 1803 Colchen listed 14 blast furnaces in *dépt* Moselle,[133] all except five of them on the *minette* orefield, which must have supplied part at least of their charge.

After the Napoleonic Wars, higher duties on imported iron encouraged the Lorraine industry. The number of furnaces increased and with them the consumption of *minette*.[134] At the same time the growing scarcity and high price of charcoal compelled the furnace masters in the scarplands to use coke. Links were thus forged between Lorraine and the Saar. Coal was brought by barge along the Rivers Saar and Moselle to Thionville, Metz and Pont-à-Mousson. The de Wendels even built a private railway from their works at Moyeuvre and Hayange to quays built beside the Moselle – surely one of the earliest private mineral railways to be built in continental Europe.

The iron industry suffered severely during the depression of the 1840s, and

Table 8.6. *Iron ore and pig-iron in Lorraine and Luxembourg, 1878 (tonnes)*

	Iron ore	Pig-iron
'French' Lorraine	1,287,452	440,468
'German' Lorraine	822,000	242,531
Luxembourg	1,408,000	248,377

by the time it ended many of the small charcoal-ironworks had ceased production. When the industry revived during the following decade, it was the coke-fired furnaces that made the running.[135] At this stage in its development the iron industry made the greatest progress in the Longwy area, at the north-western end of the orefield. The deposits of *fer fort* were richest in this area, and in 1863 a railway was built along the Chiers valley, below the scarp, to Luxembourg and Saarbrücken. New furnaces were built on the lower ground, beneath Vauban's old fortress, to smelt *fer fort* or *minette* or a mixture of the two with coke from the Saar.

These developments were accomplished by ironmasters who had long worked in the area, but their success attracted outside enterprise. The Belgian firm of La Providence built the Rehon works near Longwy, and companies in northern France, where local ores were also nearing exhaustion, began to turn to Lorraine. Progress was also made in the southern part of the orefield. Works were established at Pont-à-Mousson and near Nancy, and several were sited on the navigable Moselle for the convenience of importing fuel by water.

Developments on the Saar coalfield complemented those in Lorraine and Luxembourg. Early in the century ironworking on the coalfield was stimulated by the introduction of puddling. The first furnaces were built at Dillingen in 1831, and at first refined charcoal-iron from the surrounding forests. Then *minette* began to be smelted on the coalfield. In 1853 de Wendel established a works at Styring-Wendel on the French part of the coalfield to smelt the *minette* which was brought back in the same freight cars that had carried coal to Moyeuvre and Hayange. Other works followed at Saarbrücken, Völklingen, Dillingen and Saint-Ingbert, and the old Neunkirchen works were expanded and converted to use coke.

By the Treaty of Frankfurt of 1871 Alsace and a large part of Lorraine were annexed to Germany. The new boundary ran from Luxembourg southwards to the Moselle, and thence across the plain of Lorraine to Belfort. It was as if the treaty-makers had set out to divide the orefield equally. It is unlikely, however, that the presence of natural resources in any way influenced German claims, which were based wholly on linguistic and tactical considerations.[136] The new boundary left both the Longwy group of ironworks in the north as well as the southern group near Nancy within France. But Germany acquired not only de Wendel's works at Styring-Wendel, but also the Hayanange and Moyeuvre plant and others that had been built near Thionville and Metz. De Wendel retaliated by building the Joeuf plant a few kilometres west of Moyeuvre, but on the other side of the new boundary. Despite its territorial losses, France remained in the

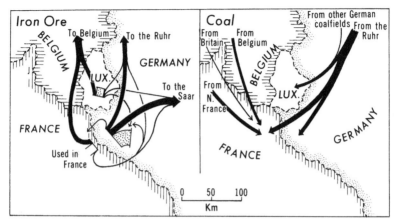

8.9 The movement of ore from the Lorraine–Luxembourg orefield and the import of coal, about 1910

1870s the chief iron producer on the *minette* orefield (table 8.6).[137] The introduction of the basic process was to change this pattern radically within the space of a few years.

Basic iron in Lorraine and Luxembourg

The discovery of the basic process in 1879 revolutionised ironworking in most of Europe, but nowhere as profoundly as in Lorraine and Luxembourg. The *minette* could now be used for steel – as distinct from iron-making – and in 1880 the Thomas process was adopted at Hayange in 'German', and at Joeuf in 'French' Lorraine. The Longwy firms collaborated in order to take a licence to use it in their area. In Luxembourg, where little progress had been made with smelting the *minette*, a series of blast-furnace works was built along the scarp foot: Dudelange, Esch, Rodange, Rumelang, Differdange, Belval. Much of the iron from the furnaces was at first sent to be refined elsewhere, but the economies of integrating steel-making with smelting were too great for this to continue, and most of the Luxembourg works became fully integrated.

In the two sectors of divided Lorraine the industry developed along parallel and closely similar lines. In Luxembourg a law of 1864 required that ore from all future mining concessions be smelted within the country. This explains why so many ironworks were built while, at the same time, outside interests showed so little interest in obtaining concessions. Not so in Lorraine. Almost from the first there was a scramble to acquire rights to the *minette*. The iron and steel companies of the Saar and the Ruhr were foremost. All together they gained possession of about a tenth of the orefield within 'German' Lorraine. Fearing that these would become exhausted in the near future, they then turned their attention to 'French' Lorraine, which was slowly being prospected. Here they gained control of a rather larger fraction.[138]

The German firms established smelting works on the orefield in Luxembourg and 'German' Lorraine, just as French iron and steel firms did in 'French' Lorraine (fig. 8.10). In all, no less than 27 out of a total number of 39 works in

Table 8.7. *Affiliation of* minette *ironworks*

	A total works	B Autonomous	C With outside affiliation	Pig-iron produced (tonnes) A	B
Luxembourg	7	0	7	2,547,861	—
French Lorraine					
Longwy	11	6	5	1,116,151	620,908
Briey-Orne	3	2*	1	572,531	376,522
Nancy	6	2	4	272,089	534,525
German Lorraine	11	2†	9	2,869,866	*c.* 612,500
Belgium	1	0	1	*c.* 100,000	—
Total	39	12	27	*c.* 8,478,498	*c.* 2,144,455

* Includes Joeuf (de Wendel).
† Includes Hayange and Moyeuvre (de Wendel).

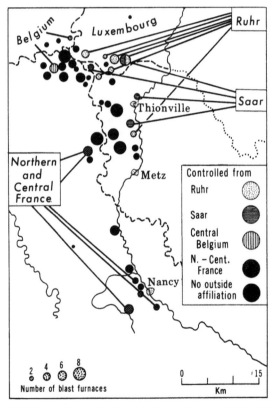

8.10 Ironworks on the Lorraine–Luxembourg orefield and their affiliation with iron and steel companies elsewhere in Western Europe, about 1910

Table 8.8. *Pig-iron and steel production, 1913 (thousand tonnes)*

	Pig-iron	Steel Converter	Open-hearth
Saar	1,307	1,700	300
Luxembourg	2,548	1,275	40
'German' Lorraine	3,819	1,783	150
'French' Lorraine	3,490	2,134	157
Total	11,164	6,892	647

the *minette* area were affiliated with 'outside' firms. These included all seven of the works in Luxembourg and nine out of the 11 in 'German' Lorraine. All four significant Saar firms[139] acquired concessions, or built works, or both, on the orefield. The pattern which became common to all three political sectors of the *minette* area was for the ore to be smelted and refined to steel in integrated works. Most of it was 'converter' steel; the lack of scrap restricting the use of the open-hearth furnace. The steel was exported, some as ingots, most as sheets and 'shapes', to steel-using industries elsewhere.

By the eve of the First World War, pig-iron production in the whole *minette* region had risen to about 8.5 million tonnes. The largest share, about 45 per cent, was made in German Lorraine, followed by Luxembourg with 30 per cent. More significant is the fact that three-quarters was smelted by firms whose bases were elsewhere, in northern France, Belgium, the Saar or the Ruhr (table 8.7). A noteworthy feature of the *minette* industry was the imbalance between iron and steel production. The volume of steel made in 1913 amounted only to two-thirds of that of pig-iron (table 8.8). The rest was shipped out as ingot iron to supply steelworks throughout western Europe (fig. 8.9).

Industry in the Saar

These developments on the orefield might have been expected to threaten the prosperity of the iron industry in the Saar. On the contrary, the Saar industry developed vigorously during the third quarter of the century, as *minette* was imported for smelting and puddling with Saar coal. In time, however, the advantages of the Saar seemed less obvious. Its coal was not of a good coking quality and imposed severe restrictions on furnace design. At the same time, the advantages of smelting on the orefield rather than the coalfield became more obvious. Increasingly pig-iron smelted in Lorraine and Luxembourg was imported to be refined in the Saar, and the smelting industries declined in relative importance, while steelmaking increased.[140]

At the same time a number of steel-fabricating works were attracted to the Saar by the availability of both fuel and crude steel. Mannesmann, for example, established a tube mill near Völklingen, and a number of wire-drawing works, foundries, and mechanical construction and sheet-metal works were founded.[141]

The Lorraine-Luxembourg-Saar region remained predominantly rural and agricultural. There were few towns, and the largest of them, Nancy and Metz, never themselves became centres of the iron and steel industry. Nancy had developed as the capital of Lorraine and its role became that of provincial capital, not of an industrial centre. Saarbrücken and Thionville were the only towns which owed their growth during the century to the steel industry.

Iron- and steelworks were established away from the towns and tended to gather around them sprawling workers' settlements, consisting at this time of little more than rows of cottages with a few shops, fitting incongruously into this attractive and in parts very beautiful rural scene. The local region was quite unable to generate the labour needed by this industrial development, despite immigration both from the Hunsrück-Eifel area and from rural Lorraine. Before the end of the century Lorraine had begun to draw on the pool of cheap, unskilled labour in France's North African possessions, for which barrack-like buildings were constructed, akin to the *Schaffhäuser* in Germany.

The Lorraine-Saar region was more dependent than most on transport. Some ore was exported to northern France, Belgium and the Ruhr; there was a large import of coal and coke, almost exclusively from the Ruhr, and a large movement of pig-iron to refineries on other coalfields as well as of steel 'semis' to metal-using works elsewhere in western and central Europe. Despite the apparent convenience of the waterway system, this was in fact very little used. As barges became larger, so the Moselle became less and less suitable. Some ore and fuel moved by means of the Rhine–Marne canal to Strasbourg, but this necessitated a break of bulk at the Rhine port. And so freight moved almost exclusively by rail. The essential network had been completed before 1880, and the line across the Eifel from Trier to the Ruhr came to be the most used. The rail links between Lorraine–Luxembourg and northern France, central Belgium and the Saar were all well developed before the period of intensive development in Lorraine and were heavily used for the transport of fuel and ore, as well as pig-iron and steel.

Saxony and the Ore Mountains

The industrial region which spread over the kingdom of Saxony and the Ore Mountains was one of the most varied and least concentrated in Europe. At its core lay the forested ridges which reached from the Fichtelgebirge in northern Bavaria north-eastwards to Lausitz and thence south-east into Silesia (fig. 8.11). The mountains themselves resembled the Ardennes and Siegerland in structure and geological age, and were similarly endowed with ores of ferrous and non-ferrous metals. The deposits of iron ore were small, but of good quality. They supported a large number of charcoal furnaces and refineries, but were quite inadequate as a basis for a modern industry.[142] Non-ferrous and precious metals had once been far more important, and occurred mainly in the aptly named Ore Mountains.

The growth of industrial regions

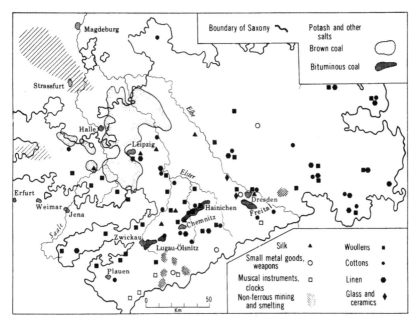

8.11 The industrial region of Saxony and the Ore Mountains

Fuel and mineral resources

The hard rocks of the mountains dipped northward beneath the plain of Saxony, and south under that of Bohemia. The once abundant deposits of coal had been stripped away, leaving only a few small basins. Of these only three acquired any importance during the century. They lay respectively near Plauen, Zwickau and – the Lugau-Ölsnitz field – near Chemnitz.[143] The coal was generally considered to be of poor quality, and that from a deposit near Dresden was said by Jacob to be 'disliked by those who can afford to purchase wood, and... only used by the poor'.[144] The output remained small, and by 1870 amounted to less than 100,000 tonnes in the whole of Saxony. It was used mainly for glass and brick-making. A little was coked and used in the iron industry,[145] but, apart from the despatch of fuel to a few ironworks in Bavaria and Thuringia, none ever moved far from the mines.[146] A greater importance attached to the Lower Silesian coalfield, which straddled the boundary of Silesia and Bohemia. In the early nineteenth century it was the most highly regarded of them all, and, although its resources hardly justified this high opinion, it played an important role in the industrial development of the region.

The vast deposits of brown coal and lignite were far more important for industrial growth than the small reserves of bituminous coal. They occurred in two areas lying respectively to the north and south of the Ore Mountains. To the south they occupied a narrow depression, overlooked by the steep face of the mountains. They stretched with little interruption from Cheb in the west to Ústí nad Labem in the east, and formed thick, level beds, lying close to the surface and easy to work.[147] The brown-coal deposits of Saxony were even more extensive. They covered an area of some 2,000 sq km centring in Leipzig. Other deposits, little developed before 1914, lay to the east, around Cottbus and Frankfurt-on-Oder.[148]

The non-ferrous metals were still important in the early nineteenth century. Freiberg was the chief centre of silver-mining, and Altenberg was the only alternative source of tin to Cornwall.[149]

To these assets should be added the many streams which descended from the Ore Mountains and Thuringian Forest and the dense woodland which covered the higher ground. Without the former, the countless small factories could never have been established, and but for the latter the small ironworking industry would quickly have been extinguished.

Industrial development

At the beginning of the nineteenth century the whole of Saxony and the Giant Mountains was a hive of industry.[150] 'Few of the Kingdoms of Europe', wrote Jacob, 'have so large a proportion of their population employed in manufactures as the Saxon.'[151] He estimated that about 1820 three-fifths of the working population was in manufacturing, though this must have included the part-time textile workers. Only paper, he added, needed to be imported. A few years later the British Commissioner, reporting on the progress of the Zollverein,[152] noted that manufacturing industries were more developed in Saxony than anywhere else in Germany. Even the coal mines and ironworks received a far greater coverage in the technological literature than their real resources warranted. Everything supported the judgement of Dietrich,[153] that, amongst the German states, Saxony was undoubtedly the one in which 'the process of industrialisation began first, progressed most rapidly and achieved the highest level of development'.

During the first half of the nineteenth century development was restricted to textiles, iron and non-ferrous metals. During the second half of the century industry became very much more diversified. The brown-coal deposits were exploited; the salt deposits of the Saxon plain began to support a chemicals industry, and glass, ceramics and other consumer goods industries were greatly expanded.

The non-ferrous metals were, however, a diminishing asset. Silver continued to be mined at Jáchymov on the Bohemian and at Freiberg on the Saxon side of the mountains, but its importance steadily declined.[154] There were 145 active silver mines at Freiberg in 1826,[155] but only 88 some 30 years later, and the decline continued for the rest of the century.[156] Silver-mining increased within Bohemia during the first half of the century, before declining with the exhaustion of deposits.[157] It was from Jáchymov or Joachimsthal, once famous for having minted the first silver 'thalers', that the Curies obtained pitchblende for their experiments.

Iron-mining and the industry based on it proved to be scarcely more durable. According to Jacob some 24,000 tonnes of iron ore a year were mined about 1820,[158] a figure which agrees with de Villefosse's statement that the output of iron was of the order of 8,000 tonnes.[159] This came exclusively from charcoal furnaces in the Ore Mountains. About 1835, it is said, there were some 20 blast furnaces, which, with their refineries, yielded 5,000 tonnes of pig-iron and 2,250 of bar-iron. To this should be added the output, in all probability significantly smaller, from the works in the Thuringian Forest.

By 1850 the charcoal ironworks of the whole region were in deep trouble and many of them had already closed.[160] A writer in a technical journal expressed in unusually forceful terms what he thought were the reasons: 'not the shortage of wood and ore...not the competition of works outside the region...but only the conservatism of the older workers, their failure to keep up with the times, and the lack of competent, knowledgeable and practically educated staffs'.[161] Not all, however, closed. As a writer in the *Berg und Hüttenmännische Zeitung* noted, those ironmasters who had adopted the new technology, which in effect meant coke-smelting and puddling, were enjoying a considerable success. This, however, was dependent on access to coal of suitable quality, and few railways had been built in the hills of Saxony and Thuringia by 1870. In all, no more than half a dozen works ever converted to mineral fuel. On the other hand new works were established close to the Saxon coalfields, at Cainsdorf, near Zwickau, and at Dippoldiswalde, near Dresden.[162] The greatest success, however, was achieved at Unterwellenborn, in the Saale valley to the south of Jena. Here local ores, access to Saxon coal and good management kept the plant in production until far into the next century.[163] With this exception, however, the ironworks all shut down later in the century when their ores became exhausted.

The smelting industry disappeared, but it left behind a legacy of iron-using industries. Foremost amongst them were the cutlers, nailers, makers of agricultural tools and, above all, the gunsmiths of Suhl.[164] This industrial development mirrored that in the Ardennes and Sauerland. But in Saxony it gave rise also to a manufacture of a more sophisticated nature. In 1830 a workshop was founded in Chemnitz for the manufacture of machines for the local textile industry. Within a few years Chemnitz was the only rival in continental Europe to the Cockerill plant at Liège. From this developed the manufacture of machinery in general, and by the end of the century Chemnitz had become one of the foremost centres in Germany for the manufacture of machine-tools.[165] The industry spread to neighbouring small towns within the mountains, as well as to the other large cities of the region, Dresden and Leipzig.

The textile industries

Throughout the century, however, the dominant industry was textiles. To the indigenous linen and woollen industry the spinning and weaving of cotton were added in the sixteenth century, its raw material coming from Italy by way of Augsburg and Hof.[166] As in much of north-western Europe, linen-weaving was at first the most widespread and important branch. It was exclusively a peasant industry, practised throughout the more hilly parts of the region.[167] In the later years of the eighteenth century cotton manufacture began to increase, and the British Consul reported in the 1830s that cotton had 'diverted many of the looms and labourers from...linen'.[168] Spinning and weaving were first established – still on a domestic basis – in and near Chemnitz and Plauen. In countless small towns of the Vogtland and western Erzgebirge cotton gradually took the place of linen-weaving. At first small, water-powered mills were set up along the rivers; then the steam-engine, using coal from the local fields, was used, and by 1846 there were said to have been nearly 200 of them at work in the textile industry.[169] The wider market opened up by the Zollverein was important to Saxony, which probably profited from it more than any other German state.[170] Chemnitz remained the business centre for the industry, while the spinneries were scattered through the small towns of the region: Hof, Zschopau, Flöha, Falkenau and Mittweida. Chemnitz itself came to specialise in hosiery, lace and embroidery, drawing its thread from the surrounding mills, and the lace made at Plauen was declared to be the equal of that made at Saint-Gallen.[171] The advance of the industry was rapid during the middle years of the century, and Chemnitz became the source of the entrepreneurial skills which developed the industry in Łódź. But its units of production remained relatively small and, despite the general cheapness of labour, the competition of the Lower Rhineland and, after 1871, of Alsace became increasingly severe.[172] Much of the weaving, furthermore, remained domestic until the end of the century. Indeed, there were in 1875 no fewer than 12,231 cotton establishments, each with five workers or less, in Oberfranken – the area surrounding Hof. This was described as the last region in Germany where the domestic weaving industry survived, and it was in fact still active there in the 1920s.[173] After about 1870 the Saxon region began to decline relative to other German industrial areas, and in 1893 Saxony held only 14 per cent of the cotton spindles and 15 per cent of the looms.[174]

The traditional woollen industry survived the competition of cottons chiefly where it was less easy to obtain raw cotton. Dresden itself, together with the small towns of eastern Saxony, remained a centre for woollen-weaving.[175] Yet farther to the east, in Lower Silesia, linen-weaving remained the chief industry in both domestic workshops and the small mills. All three branches of the textile industry remained important throughout the century in the villages and small towns of northern Bohemia. Linen-weaving provided a supplementary income, between Děčín and Náchod, 'for hundreds of thousands of people...though alone it did not suffice for even a minimum livelihood'.[176] It was not until the mid-century that the power-loom began to displace the handloom. The woollen industry was scarcely less conservative than the linen. Water-powered spinning mills were established in the 1830s, but the traditional weavers offered a stronger resistance. Not until late in the century was the woollen cloth industry concentrated in factories at Liberec (Reichenberg) and nearby towns.[177]

The cotton industry spread into northern Bohemia in the early nineteenth century. It suffered, as in Saxony, during the years after 1815, and did not expand significantly until the 1830s. With mechanisation the cotton industry concentrated in Liberec, Frýdlant, Litoměřice and Jablonec. By 1850 the cotton industry of northern Bohemia was carried on almost entirely in factories.[178] It grew rapidly during the following decades, and became the chief source of cottons for the protected market of the Habsburg Empire. Wage-rates in Bohemia were marginally lower than in Austria itself.[179]

Textiles and small metal wares were only a few amongst the host of domestic industries that had been pursued in the Thuringian Forest and Ore Mountains.

The growth of industrial regions 397

They were most numerous and most varied to the west, where non-ferrous mining was least important.[180] Most were in decline, and they had almost disappeared from the Fichtelgebirge.[181] But in the Thuringian Forest and Ore Mountains they left a legacy of small, specialised industries, which included the manufacture of musical instruments and toys, footwear, pottery and glass, as well as ironware and textiles.

The new industries

During the second half of the nineteenth century the traditional industries of the region began to face a very powerful rival. The immense reserves of brown coal, lying both to the north and south of the Ore Mountains, were exploited on a serious scale. They had been used much earlier. De Villefosse noted a small production of *bois fossile* in Saxony,[182] and pits had been opened on the Bohemian lignite at Ústí as early as 1760, and at Teplice and Most early in the next century.[183] After 1850 brown coal began to be used as raw material for making coal-tar and its distillates and then for briquetting as furnace fuel. Production increased from half a million tonnes a year soon after 1815 to 2.5 millions in 1848. The industry grew very rapidly after 1860, and output reached three million tonnes by 1869; eight by 1880, and about 20 in 1910.[184] In Saxony its expansion was even more rapid, and by 1903 had reached 30 million tonnes a year.[185]

Potash and rock-salt mining developed alongside that of brown coal. The deposits lay mainly to the north of the largest brown coalfield, and close to Magdeburg. Chemical industries were developed on the southern margin of the field at Borna and Zeitz, and Halle, Merseburg, Leuna and Bitterfeld became major centres for the manufacture of basic chemicals.[186] A similar chemicals industry developed at the same time near the brown coalfields of northern Bohemia, at Ústí, Teplice and Loket. Other chemical raw materials were brought up the Elbe (Labe) from the saltfield of Saxony.

The development of the industry in both sectors of the Saxon-Bohemian region was greatly aided by that of a network of roads and railways. A line had been built by 1850 from both Magdeburg and Berlin to Leipzig and thence up the Elbe valley to Prague. During the next quarter of a century a net of railways was spread over Saxony and northern Bohemia. At the same time navigational conditions were improved on the Elbe, and Ústí became a significant river port.

Saxony and the Ore Mountains formed at the beginning of the century one of the most densely peopled parts of Germany. Over almost the whole province density in 1834 was more, in many areas considerably more, than 100 to the sq km. Landless peasants and smallholders formed about 30 per cent of the total in 1750, rising to 47 per cent in 1843. Only domestic crafts could have supported so dense a rural population.[187] At the same time the population was relatively urbanised. In 1844, about 34 per cent lived in towns, a slight drop from the situation a century earlier owing largely to the growth of rural crafts.[188] Dresden, the largest city, had a population of 66,000, and Leipzig, the second largest, of 45,000. In all, there were five towns of more than 10,000, and 20 with more than 5,000. Saxony was far more highly urbanised than the Ruhr at this time. There were no towns of this size in the Bohemian area of the region, though very high rural densities were to be found in the textile districts of Liberec, Jablonec and Rumberk.[189]

The decline of craft industries in the mountainous areas was accompanied by an outward migration to the growing towns. This was most marked in the Ore Mountains of Saxony, least in the hills of Bohemia and Lower Silesia, where domestic weaving proved more durable. In both Saxony and Bohemia the brown-coal area attracted population and a dense settlement developed, much of it in small, loosely built industrial towns such as Ústí, Chomutov and Most on the one side; Zeitz, Borna and Leuna on the other. By 1910 both Leipzig and Dresden had grown to be cities of over a half million people and were respectively the fourth and five largest in Germany. In addition Chemnitz, 'a factory town of the most pronounced type',[190] had reached more than a quarter of a million, and Plauen 120,000. In all there were more than 30 towns in Saxony of over 10,000 and at least 12 in the industrialised region of northern Bohemia.

Throughout the nineteenth century Saxony had a higher employment in manufacturing than any other province of Germany, with almost twice the national average. Employment remained largest in textiles, though in percentage terms this tended to decline, as that in metal-working and chemicals increased.[191]

Upper Silesia and Moravia

When in 1740 Frederick the Great invaded and annexed the Habsburg province of Silesia, he had no knowledge of the extent of the resources which he thus acquired. Lead had long been mined in the south of the region, and with it, as a kind of by-product, the ores of zinc. For the rest, iron-smelting had grown up in the forests, which covered much of the region, but by and large it was an unprogressive and unimportant industry. Yet beneath the forests of Upper Silesia lay a coal basin second only to the Ruhr in the wealth and variety of its resources.

Between the headwaters of the Oder and of the Vistula lies a plateau (fig. 8.12). To the north are low limestone hills, part of the Jurassic system, containing bedded ores of iron. To the south the land rises to the Beskidy, the foothills of the Tatra. Towards the south-west the plateau narrows between the Beskidy and the mountains which encircle Bohemia. This is the Moravian Gate, the gap in the mountains which opens a low-level route between the norther plain and the basin of the Danube.

The coalfield

This plateau is over much of its area underlain by the coalfield of Upper Silesia. The coal basin forms a vast triangle, almost 150 km from west to east at its base and reaching 50 km to its northern apex. In the west it reached the Oder, and in the east the Vistula, while to the south the steep rise to the Beskidy marked its limit. The coalfield itself forms a saucer-like depression, with coal seams rising around its margin and lying at their greatest depth in the middle. The simplicity of this geological structure is fortunately marred by a 'saddle' or upfold lying

The growth of industrial regions

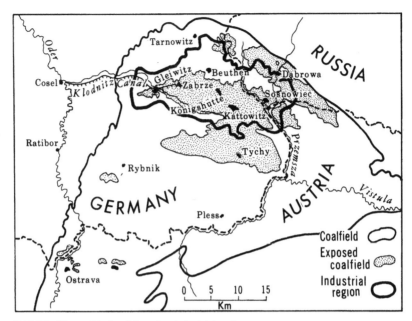

8.12 The Upper Silesian–Moravian industrial region

from west to east across the northern part of the basin, thus bringing the lower and richer seams close to the surface. To the north of this saddle lay the small Beuthen (Bytom) basin; to the south, the very much more extensive main basin over which much of the coal lay too deep for profitable mining. Over most of its extent the coal basin was hidden beneath later deposits. In the north this cover was formed by beds of the Triassic series amongst which the Shelly Limestone (*Muschelkalk*) was the most important because it contained lead and zinc ores, and was also a source of water for the industrial region. The coal beds were exposed in the area of the 'saddle', but from here southwards lay a thick cover of Tertiary sands and gravels, which even today remain forested. As the Moravian Gate was approached, the coal measures again rose to the surface in a number of small outcrops, amongst them beds of coking-coal. The 'saddle' together with this south-western or Moravian sector of the coalfield constituted the twin poles around which industry concentrated.[192]

Industrial growth before 1850

Development around the northern pole was precocious (fig. 8.13). Coke was used for smelting with some success at Małapanew as early as 1789. This was followed by the building of a furnace at Gleiwitz (Gliwice) on the exposed coalfield for the specific purpose of using coke. Other furnaces followed at Königshütte – the present Chorzów. Shortly afterwards some of the noble landowners themselves entered the business, so that by 1840 there were no less than seven smelting works using coke as fuel. In the mean time, in 1828, the first puddling furnace was built, and by the mid-century there were no fewer than six such works. The Upper

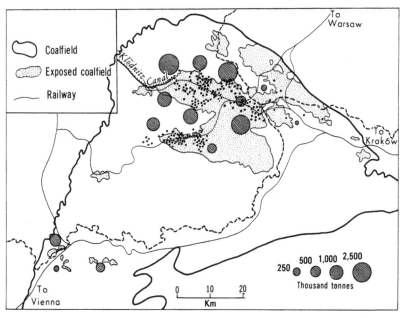

8.13 The spread of coal-mining in Upper Silesia–northern Moravia. The distribution of mines in 1860

Silesian industrial development was described in 1842, with pardonable exaggeration, as 'the equal of England and foremost on the continent of Europe'.[193]

The Upper Silesian region had great advantages; coking-coal was adequate, at least in the early years, and iron ore was available from the coal measures as well as the Trias. The problem was that there was in the early days of the industry little market for the products of the iron industry in this remote and thinly peopled region. The activities of the Prussian government had been primarily for the purpose of casting ordnance and producing military supplies. Only after the railway had come to the region in 1845 did the demand for rails and other transport equipment begin to develop. Even so it remained very small compared with the enormous market which was developing in the west. The iron industry remained unprogressive in many respects. Charcoal-fired furnaces continued active even on the coalfield into the second half of the century, and it was not until the 1850s that the volume of coke-smelted iron began to exceed that of charcoal iron (fig. 8.14).

A further factor restricting growth was the nature of its ownership and control. Apart from the holdings of the Prussian State, resources were owned by a small number of noble families, the richest of whom were the two branches of the Henckel von Donnersmarck family, and the Hohenlohe, Colonna-Renard and Tiele-Winkler families. Their instincts were far from commercial and their interests often dillettante. But change was imminent. Small groups of manufacturers in Breslau and Berlin took concessions and built works to supply their factories with iron. This trend was strengthened after a railway link had been established with central Germany in 1845.[194] Most important of these

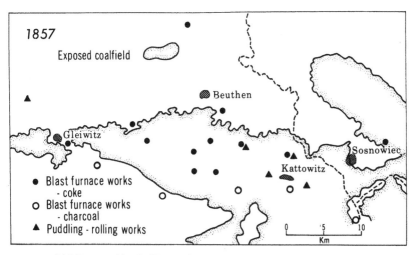

8.14 Iron-working in Upper Silesia, 1857

developments was the building by August Borsig, the engine manufacturer of Berlin, of a blast furnace and puddling works – Borsigwerk – near Zabrze.

The iron industry in Upper Silesia failed to attract other branches of manufacturing that might have been complementary to it. Coal-mining also suffered from the lack of a large local market and the high cost of rail transport. A canal linked the coalfield with the Oder, but its capacity was small and the Oder itself poorly suited to barge traffic. Although mining was relatively easy, with thick, level seams at shallow depths, and the pit-head price of coal consistently lower than in the Ruhr, this could not entirely offset the problem of distance. Silesian coal could rarely be distributed farther afield than Berlin, where it met the full competition of coal from the Ruhr. For this reason growth in coal mining remained very slow until after 1860.

The coal industry had, however, one important market apart from the iron industry, the non-ferrous metal refineries. The deposits of lead and zinc were amongst the largest known. Zinc ores – calamine and blende – were the more extensive, and occurred in a broad belt within the *Muschelkalk*. Zinc was, however, difficult to smelt owing to the low boiling-point of the metal. The technical problems were largely overcome at the beginning of the century, at least as far as calamine was concerned, and production of zinc, largely from works owned by the noble landowners, began to increase from about 1820. By 1821 there were no fewer than 33 zinc smelters. It was many years, however, before a method was devised for smelting blende, which was in fact the more abundant ore, and it was not until the 1870s, when demand for metallic zinc was increasing sharply, that it began to replace calamine in the furnaces.

A feature of zinc smelting was its immense consumption of fuel. The proportions of ore and fuel were so unbalanced that there was never any question of smelting at the source of the metal. Mining was on the *Muschelkalk*, and smelting on the exposed coal measures (fig. 8.15).

Lead was a less abundant but more tractable metal, and was at first the more

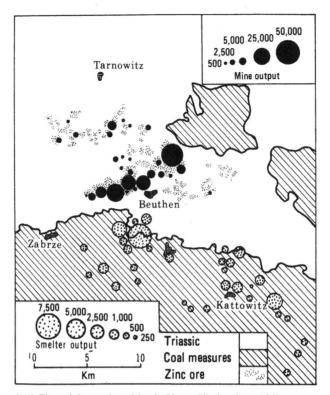

8.15 Zinc-mining and smelting in Upper Silesia, about 1860

sought after. It occurred in association with zinc, and more abundantly to the north at Tarnowitz (Tarnowskie Góry). Lead, unlike zinc, which was treated in mining law as an 'earth', was a regalian mineral, and the Prussian State exercised a monopoly of mining and smelting. There was a lead-smelter, the Friedrichshütte, near Tarnowitz, but a private smelter, owned by the zinc firm, von Giesche's Erben, the Walther Croneck Huta, near Katowice smelted galena as a by-product of its zinc works.

Since the final Partition of Poland in 1796 the area of the coalfield had been partitioned between Prussia, Russia and Austria, so that we have in fact to deal with three contrasted areas of development. The Russian sector, or Dąbrowa, lying to the north-east of the 'Saddle', was by far the smallest in area, but embraced a large part of the exposed coalfield. Coal had been mined here in the eighteenth century, and after 1815 production was increased, but Russian policy was vacillating, and little progress was made until after 1860. In the 1830s the Bank of Poland financed the building of blast-furnace works in Dąbrowa – the Huta Bankowa, but its fortunes were as chequered as those of coal-mining. Its output remained small, and it was not until 1881 that a second works was established.[195]

The ore-bearing Trias, which was so important a factor in the development of Prussian Silesia, extended eastwards across Dąbrowa. Although a number of mines were opened and worked intermittently, the only successful operations were near Olkusz, the ore from which was smelted at Będzin on the exposed coalfield.[196]

The Austrian sector of the coalfield was of roughly the same extent as the Prussian, but over much of its area the coal lay too deep for profitable mining. It was divisible on historic and economic grounds into two parts. The larger, known as Western Galicia, came to the Habsburgs in the First Partition of Poland in 1772. It resembled Dąbrowa in containing part of the exposed coalfield as well as Triassic beds with their zinc deposits. Coal was worked at Jaworzno and Tenczynek, and a zinc-smelter was built at Szersza.

The rest, known as Austria Silesia and Teschen (Cieszyn, Těšín), had been part of the Habsburg Empire since early in the seventeenth century. Coal came to the surface over only a small area at Ostrava and Karvinná. The seams were thinner and more contorted than farther north, but they contained some of the best coking-coal to be found in the whole region. Coal-mining began on a significant scale in the 1840s. By 1850 there were eight mines producing only about 140,000 tonnes a year, but thereafter mining expanded steadily.

In 1826 Archduke Rudolf established a puddling works on the Ostrava coalfield to refine charcoal-iron. Ten years later coke-fired blast furnaces were added, and the Vítkovice works became the most advanced in the Habsburg Empire. Other works followed – at Stefanau, Třinec (Trzyniec), Węgierska Górka and elsewhere, though, as in the Prussian sector, away from the coal mines iron long continued to be smelted with charcoal.

There were basic similarities in the development of the Austrian and Prussian sectors of the coalfield. Despite its considerable potential, however, the Russian-held sector lagged behind the others, on account mainly of the ineffective management of its resources and the feebleness of demand.

When the mineral resources of Upper Silesia began to be exploited the region as a whole was thinly populated and much of it was forested. There were few towns. The largest in Prussian Silesia were Gleiwitz and Beuthen, but neither had more than 2,000 inhabitants. Others, like Pless (Pszczyna) and Auschwitz (Oświęcim), were even smaller. They were the market centres of a poor and backward province. both Gleiwitz and Beuthen grew as their surrounding areas became industrialised, and many a small Polish village found itself overnight the scene of mining or ironworking. The population of the whole of Upper Silesia was in 1820 of the order of half a million, only a small fraction of which actually lived in the future industrial area.[197] By 1867 this had increased to about 1.25 million; by 1890 to 1.58, and in 1910 to 2.2. Most of this increase was in the industrial towns and villages which lay close to the exposed area of the coalfield where industrial development took place.

In 1846 the railway from Breslau reached Kattowitz and was continued to Mysłowice and Kraków. Two years later a link was established with the line from Vienna to Warsaw. In the following years branches were built to Racibórz and Oświęcim and a direct line was laid to Vienna.

Upper Silesia seemed poised in the middle years of the century for a period of rapid growth. Much of the exposed coalfield was being worked, and there were in 1856 no less than 116 active mines in the Prussian sector alone, producing more than two million tonnes of coal.[198] Output at this time was running

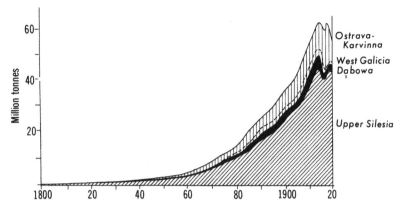

8.16 Graph showing the increase in coal production in Upper Silesia–Dąbrowa–northern Moravia, 1800–1920

marginally ahead of that of the Ruhr. In the smelting and refining of iron Upper Silesia had an immense lead. The region had furthermore an important non-ferrous metals industry, the value of whose product may have amounted to a fifth of the total industrial production of the region. The writer who in 1851 claimed that Upper Silesia was still by far the most important German province for heavy industry was close to the mark[199] and that was without including the Russian and Austrian sectors of the region.

From 1850 to 1914

All branches of industrial activity continued to expand. Nevertheless the region was overtaken by the Ruhr in both coal and iron production (fig. 8.16). The assets of the region favoured a precocious development, but not long-term growth. The good coking-coal was easily mined but quickly exhausted, and towards the end of the century the coke available was so soft and friable that blast-furnaces had to be kept small. At the same time, the local iron ores in both Austrian and Prussian Silesia became scarce and had to be supplemented and then replaced by more expensive ore from as far afield as the Ukraine. Production costs of pig-iron thus became significantly higher in Upper Silesia than in the Ruhr. At the same time the costs of coal-mining were increasing more rapidly, and the differential between them was narrowing. This in turn reduced the area in central Europe within which Uper Silesian coal could compete with both that from the Ruhr and imported coal from Britain.

The fundamental weakness of Upper Silesian industry, however, was its distance from potential markets. Although engineering and metal-using industries developed in eastern Germany, notably in Breslau, Berlin, Posen and on the Baltic coast at Stettin and Lübeck, this was insufficient to maintain the momentum that had been developed in the earlier years of the century. The market in Russia and the Habsburg Empire was partially closed by a tariff barrier to products of Prussian Silesia, and the Russian market itself was too

Table 8.9. *Size of coal mines*

	Active mines	Output (tonnes)	Workers	Output per mine (tonnes)	Output per worker (tonnes)
1856	116	2,102,337	11,802	18,124	178
1875	110	8,252,465	32,193	75,022	256
1890	56	16,862,876	49,708	307,123	339
1902	63	24,470,788	80,038	388,425	306

Table 8.10. *Number and size of ironworks*

	Active works	Output of pig-iron (tonnes)	Workers	Output per worker (tonnes)
1852	63	61,969	1,913	32·4
1875	20	266,836	2,774	96·2
1890	14	508,624	4,212	120·8
1902	11	685,450	3,414	200·8

small and too distant to stimulate much growth even in Dąbrowa. Only Austrian Silesia had a secure though limited market within the Habsburg Empire.

It had been supposed that the River Oder, together with the Klodnitz Canal which linked it with the mines near Zabrze, would give the region access both to Berlin and to the eastern German market. But the river was not continuously navigable even for the small coal barges then in use, and the canal remained inadequate even after extensive work on it in the 1880s. Not until the construction of canals in north-western Germany in the last decade of the century posed a new threat to the market for Upper Silesian coal was a belated attempt made to improve navigation on the Oder. And so Upper Silesia remained overwhelmingly dependent on rail transport at a time when its only significant competitor, the Ruhr, was able to benefit increasingly from cheap water-transport for all bulky commodities.

In all branches of manufacturing there was a marked concentration of production. Mines became larger and fewer,[200] and the contraction in the number of iron-smelting works with their increase in size was even more marked (tables 8.9 and 8.10).

Coal output continued to grow until the First World War, when total production from the whole field reached 60 million tonnes. The greater part of this increase came from Prussian Silesia, but there was a significant change in the spatial pattern of mining. The so-called 'Saddle' between Gleiwitz and Kattowitz remained overwhelmingly important, yielding in general about three-quarters of the total output (fig. 8.17). But mining was spreading outwards from the shallow seams of the 'Saddle' into the Beuthen basin to the north and southwards into the main basin. Towards the west of the main basin mines were opened up in the Rybnik district, where a miniature 'saddle' brought the richer,

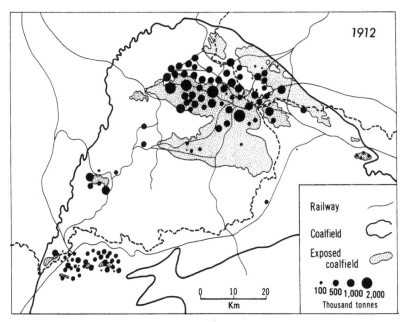

8.17 Coal-mining in the Upper Silesian–Moravian coalfield in 1900

lower seams close to the surface. Coal-mining also increased in the Austrian and Russian sectors of the coalfield, though together they never produced more than 40 per cent of the Prussian output. Mines were always much smaller in Dąbrowa and Austrian Silesia, and proportionately more numerous.

Iron production increased slowly through the second half of the century and reached about a million tons a year at the outbreak of the First World War. Charcoal-smelting had at last disappeared, but the region was slow to adopt the newer technology. Blast furnaces remained small. The Bessemer process achieved little success, and the ores generally used had too much phosphorus for the acid process and too little to be really suited to the basic.

The region as a whole generated little scrap metal so that the open-hearth was not really economic. In consequence, puddling remained the dominant means of refining iron even at the end of the century, when there were still about 300 active puddling furnaces and only about ten open-hearths. The next decade, however, saw a sharp change. The growing importance of engineering and other iron-using industries in Silesia meant an increased production of process scrap, and this in turn made it possible to increase the open-hearths. At the same time the number of puddling furnaces began to decline (fig. 8.18).

There was little change in the geographical pattern of the iron and steel industry.[201] Some new works were founded and several of the early works ceased production. A few, including the royal works at Gleiwitz, the oldest of them all, suspended smelting and concentrated on open-hearth steel and rolling. In the Dąbrowa area the antiquated Bankowa works took on a new lease of life; the Huta Katarzyna was founded and other steelworks established largely by German enterprise from Prussian Silesia.

Table 8.11. *Value of industrial production in Upper Silesia*

	Coal and iron ore (%)	Iron and steel (%)	Lead and zinc (%)
1880–4	34.4	41.7	29.3
1903–13	43.7	37.6	18.7

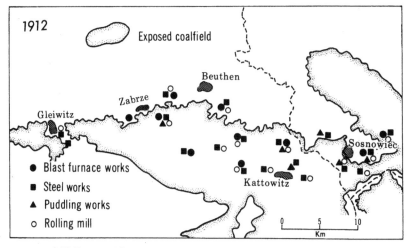

8.18 Iron-working in Upper Silesia, 1912

Expansion in Austrian Silesia was slower, despite the good quality of its metallurgical fuel. The exhaustion of local iron ores placed a burden on the industry which met, furthermore, with competition from works in Bohemia and even from the old industry of Styria and Upper Hungary.

The production of zinc and lead, chiefly from Prussian Silesia, continued to increase until the outbreak of the First World War. The more easily smelted calamine was nearing exhaustion, and by 1914 output was almost entirely of blende. The share of lead and zinc in the total industrial production of Upper Silesia, however, was falling (table 8.11).[202]

Transport

All industrial regions are heavily dependent on transport facilities, and none more so than the Upper Silesian-Moravian region because of the restricted nature of its local market. One of the gravest disadvantages from which the region suffered was its lack of an effective system of water-borne transport. Its rivers were not really navigable, even though coal was sometimes sent down the Przemsza to the Vistula for the supply of Kraków. The Klodnitz Canal could take only the smallest barges, and navigation on the Oder itself was difficult and seasonal.[203] No more than 60,000 tonnes of coal were ever shipped by canal *and* river in a single year, before the coming of the railway took away what little traffic there was. Although the Klodnitz Canal was enlarged later in the century and improvements were made progressively in the Oder,[204] traffic remained with the railways. Downstream traffic from Kosel (Koźle) in 1910 amounted only to 2.75 million tonnes.

Upper Silesia was first linked by rail with the rest of Germany in 1845 by way of Kosel and the Oder valley. Shortly afterwards the area was joined to the Vienna–Warsaw line. By the 1870s a network of standard-gauge lines covered the area, and was supplemented by a vastly more complex net of narrow-gauge lines which served the mines and factories.[205] The Moravian sector of the region was served from 1847–8 by the *Ferdinands Nordbahn*, which continued through Dąbrowa to Warsaw. A connection between the northern industrial area and Ostrava was not completed until the 1850s.

Urban growth and economic development

Industrial growth inevitably brought with it an expansion, very rapid after about 1890, of the towns of the region. The only two cities of medieval origin, Beuthen and Gleiwitz, grew to about 68,000 and 67,000 respectively. They retained something of their pre-industrial atmosphere, though architecturally they were wholly without distinction. Their medieval walls disappeared and their tightly built cores were enveloped by factories, smelters, mines and all the debris of nineteenth-century industrial growth. It was, however, the 'new towns' which grew most rapidly. The largest municipality in Prussian Silesia was Königshütte, which had grown up to serve the blast furnaces from which it was named.[206] Close behind it were those other mushroom growths, Zabrze and Kattowitz, the latter to become the administrative and economic capital of the region. In northern Moravia the town of Mährisch Ostrau (Ostrava) had grown from a village into a sprawling, characterless and ill-built town, at its heart and focus the iron- and steelworks of Vítkovice. But the largest town in the whole region was Sosnowiec, in Dąbrowa, which grew from 57,000 in 1904 to more than twice that total only ten years later, when the local coal-mining industry was said to employ 20,000.[207] It was the unquestioned capital of the Russian sector of the industrial region, and, like many Russian towns of intermediate size, it attracted a significant Jewish community which occupied a compact area near the town centre, spoke Yiddish and retained its Jewish culture and customs. This was, in fact, the only significant Jewish settlement in the whole industrial region.[208]

Population growth was too rapid for the towns to assimilate. Housing was inadequate and living conditions were squalid and congested in the extreme. In Königshütte the working class, which made up most of the population, lived 'in tenements of one or two rooms'.[209]

The towns of the Prussian and Russian sectors of the region were closely spaced within an elongated area, 40 km in length from Gleiwitz in the west to Dąbrowa Górnicza in the east and no more than 10 km from Kattowitz on its southern margin to Beuthen. Despite their uncontrolled spread there still remained areas of open country, encroached upon by the growing industrial villages and spreading slag and waste heaps and undermined by coal-workings which led to subsidence and the waterlogging of the surface.

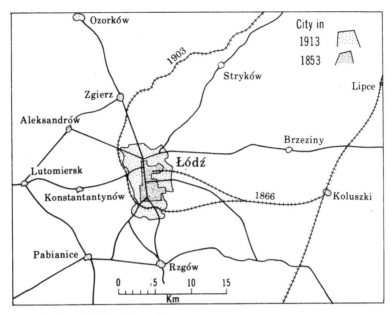

8.19 The city of Łódź and its region

The Łódź region

The Łódź industrial region developed in the Russian-held Kingdom of Poland. Its focus was the small town of Łódź, lying about 120 km to the west-south-west of Warsaw, and a few kilometres within the boundary of the Kingdom. The site straddled the watershed between the Oder and the Vistula (fig. 8.19). It had no major river, and its many small streams were adequate only to turn the wheels of its early mills. The region was from its inception given over to a single industry: textiles. It was founded at a known date for this specific purpose, and never attracted any industries that were not related to its main preoccupation. It is an example of that rare phenomenon, an industry consciously planned and located at a site which seemed to offer the physical advantages which it required. Indeed, Dylik has argued that it was the very backwardness of the region, the complete lack of any proto-industrial antecedents, that attracted the German entrepreneurs to this industrially virgin site. Even a low level of development would have 'reacted negatively against adaptation to a new form of economic structure'.[210]

The Łódź textile industry was founded in the 1820s by entrepreneurs from Chemnitz in Saxony. The region had no clothworking tradition, though the local population was not unfamiliar with the processes of the textile industry. Two factors attracted the entrepreneurs. The first was the policy of the administrators of the semi-autonomous Kingdom of Poland, actively encouraging the establishment of manufacturing enterprises. They sponsored the development of ironworking in the so-called Old Polish Basin – *Staropolskie Zagłębie*[211] – and of a textile industry at Łódź. The second advantage was that in 1820 the customs barrier which had hitherto separated 'Congress' Poland from Russia was abolished and henceforward Polish manufactures could command the vast Russian market. The transfer of manufacturing from Saxony to almost any site within Poland thus seemed to offer immense advantages. The first industrialists set up business in 1821; by 1824 some of the well known figures in the textile industry had built workshops.[212] A report of 1825[213] described Łódź, which then had barely 2,000 inhabitants, as a town of wooden buildings with an abundance of small, easily regulated streams. Each 'master' was thus able to find the energy that he needed. There were, it continued, 31 *drapiers* with 59 cotton-weavers and 27 master-spinners with 47 employees, as well as some linen-weavers.[214] The industry was still very small, and might indeed have perished after the Rising of 1830. In 1831 the Russians retaliated against the Poles by re-imposing a tariff barrier between the Kingdom and Russia. The confidence of the German industrialists was shaken. Geyer, one of the foremost amongst them, contemplated shifting his activities to Riga.[215] Others did indeed make the move. They capitalised on a feebly developed domestic industry in the vicinity of Białystok, and there established a cotton industry. Białystok then lay within the Russian province of Grodno, but is now in north-eastern Poland.[216] It was never more than a small, domestic industry scattered through a miserably poor countryside; later in the century the offending tariff was again removed and the Białystok industry lost the principal reason for its existence.

In the mean time the Łódź industry continued to expand, despite the greater difficulty of access to the Russian market. It underwent a crisis after 1851 due to over-expansion in the previous years and again ten years later owing to the cotton famine.[217] But the industrialists were able to alternate between cotton, wool and flax in order to make the best of the market. There were even factories in which all three materials were being used at the same time.

In 1847 there were three spinning mills in Łódź, and three others in the surrounding villages. Amongst them was one founded by the Frenchman, Philippe de Girard, at Żyrardów, which despite its spelling, bore his name.[218] Mills were then established even farther afield, at Ozorków, Zgierz, Aleksandrów and Pabianice, which all grew into small mill-towns. At Pabianice there were some 25 textile mills, some of them physically divided between cottons and woollens. Everywhere there were large, mechanised factories, powered by steam, and small mills which were nothing more than large workshops, where manual labour was helped out by a water-wheel.[219]

Much of the expansion of the later nineteenth century was achieved without the Russian market. The Polish land reform of 1864 was followed by a rise in the standard of living of the masses and a consequent increase in commercial activity.[220] The steam-engine was adopted after a branch line had been completed in 1866 from the Warsaw–Upper Silesia railway, thus making it possible to bring in coal, and large factories began to be built. Many of these survive; some were vast brick-buildings, similar to those being built in Oldham and Rochdale. Others hid behind over-elaborate classical façades.

A labour force was drawn from the surrounding countryside, where land reform combined with a high birth-rate to produce a large body of surplus labour. It moved to the 'promised land' of the emerging city which appeared from the distance to offer so much more than the crowded Mazowian farms (fig.

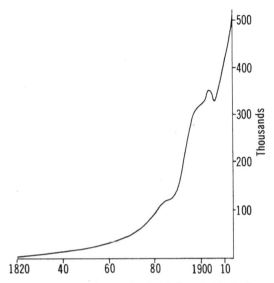

8.20 The growth of population in Łódź, probably the fastest growing city in nineteenth-century Europe

8.20). Population increased rapidly, perhaps more rapidly than that of any other industrial city at this time. From about 16,000 in 1850 it rose to 321,000 in 1900 and over half a million by 1913.[221] It grew far faster than the provision of housing, utilities and amenities, and Łódź became as crowded and insanitary as the Manchester and Glasgow described by Friedrich Engels nearly half a century earlier. Nor did Łódź lack literary descriptions. The Polish novelist, Władysław Reymont, lived at Lipce, only 30 km away, and used Łódź as the setting of his novel *Ziemia Obiecana*,[222] 'The Promised Land', to which the ignorant and impoverished peasants crowded in the hope – how bitterly disappointed – of a better life. Through his pages one can see the unlit, unpaved streets, deep in mud; the alleys and courts so similar to those described by Engels, around which most of the workers lived; the factories, unventillated, ill-lit and unhealthy, and the entrepreneur who resided nearby in vulgar ostentation, overawing his workers by his continual presence and always driving them to greater efforts.[223]

Conditions in the smaller textile centres were less squalid than those in Łódź itself, if only because they were closer to the open countryside. But everywhere one found evidence of conditions of work comparable with those in western Europe half a century earlier. But there were signs of change. A gas supply was made available in 1869 and electric lighting began to be installed in 1886.[224] The chief beneficiaries, apart from the merchants and entrepreneurs, were the factories, where night-work could now be carried on more easily. But even on the eve of the First World War only a minute proportion of the houses had a sewer or a piped water supply. The town had no definite centre. It was grouped along a central, north–south axis, the *ulica Piotrkowska*, from which narrow alleys branched to form a regular net. An electric tramway system was built in the central city in 1898 and was extended to the suburbs and surrounding industrialised villages ten years later.

The number of factories in the Łódź complex was immense. Soon after the end of the First World War there were more than 1,100 separate undertakings, and the number cannot have been significantly different before the war. Nearly three-quarters of them employed fewer than 50 workers, and only 33 had more than 500, but the latter employed far more than half the total.[225] Although all branches of the textile industry were carried on, more than two-thirds of the output was cotton goods, and most of the remainder woollens. There was a small production of linen, jute, hemp and even silk fabrics. The industry concentrated on coarse fabrics and brightly coloured prints, such as sold most readily in the slowly expanding peasant markets of eastern Europe.

The Lyons-Saint-Étienne region

In the mid-nineteenth century the Lyons-Saint-Étienne region was regarded as the most important and most promising industrial region in France. But, as in central Belgium, the high expectations of the first half of the century were not fulfilled during the second. The region was heir to a long history of commercial and industrial development. Lyons itself had been in turn the foremost city of Roman Gaul, an important medieval town, the site of an international fair, and the focus of routes which radiated across much of western Europe.[226] The River Rhône, difficult though it was to navigate, provided from the earliest times an avenue southwards to the Mediterranean and northwards by way of its tributary the Saône into Burgundy. To the east, the Isère and Upper Rhône had created routes into the Alps and thence to the Italian plain. On the west the Rhône valley was overlooked by a succession of mountain masses, from the Montagnes du Charollais in the north to the Cevennes which looked down on to the plain of Languedoc. To the south-west of Lyons a gap between the Montagnes du Lyonnais and the Montagnes du Vivarais broke their continuity and provided a routeway to the valley of the Upper Loire. This gap was drained to the Rhône by the Giers, while the Ondaine and Furan flowed westward to the Loire. It thus linked the south-flowing Rhône with the north-flowing Loire, and in the pre-railway age was a much used route from Paris to the Mediterranean. In the midst of this gap lay the ancient town of Saint-Étienne (fig. 8.21).

Routeways alone would not have made this an important industrial region had the depression linking the Rhône with the Loire not been underlain by coal. It was, in fact, a down-faulted trough, bordered by older rocks. The coalfield was larger in area and richer in resources than all except the northern coal-basin. It embraced the whole range of bituminous coals, with the exception only of anthracite. Its seams, furthermore, came to the surface over a large area, and had been worked at least since the later Middle Ages. The coal of Saint-Étienne was considered to be especially suitable for the forge, and the region acquired an early reputation for the manufacture of iron goods.

Lyons lay some 12 km from the nearest point on the coalfield, and was in its economic development little influenced by it. The modern city grew up as a route centre and fair town. In the fifteenth century it received by royal charter a

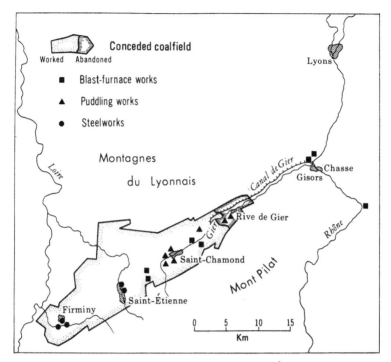

8.21 The Loire coalfield and the industrial region of Saint-Étienne

monopoly of silk-weaving in France. Although it subsequently lost its monopolistic role, it remained in the early nineteenth century by far the most important centre of the industry. Silk-weaving was at this time a domestic craft. Though concentrated in the weavers' quarter of the city, it was beginning to spread into the surrounding countryside, even as far as Saint-Étienne, where it competed uneasily with coal-mining and iron-working. Lyons and Saint-Étienne emerged as the twin poles around which the contrasted industries of this region developed: Lyons, lying between the confluence of Rhône and Saône, with its industrial structure based upon a luxury product, and Saint-Étienne, 50 km to the south-west, thriving on coal and the industries based on it.

Lyons and its industries

Silk-weaving had come to Lyons in the later Middle Ages, and as the city's role as a fair town declined, so its importance in the silk trade grew. Its raw material was produced in the Rhône valley to the south, where the silkworm dominated the life of the peasant as the mulberry tree did the landscape. In the course of the nineteenth century, however, the local supply had to be supplemented by imports from Italy and ultimately from the Far East. In the early nineteenth century most of the silk-weavers lived in the Croix Rousse district, which covered the high ground between the Saône and the Rhône, to the north of the city centre.[227] In 1820 the city had a population of about 25,500, among whom there

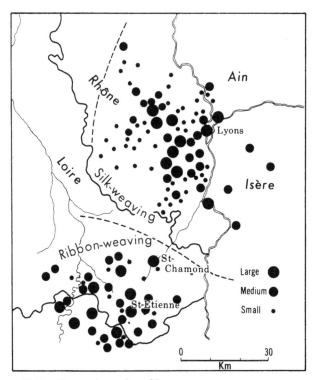

8.22 The silk-weaving region of Lyons

were over 19,000 silk-looms. Many more were active in the small towns of the surrounding countryside, in *dépts*. Ain, Isère and Rhône.[228] Indeed, 300 merchants controlled as many as 20,000 looms, putting out the raw silk to the weavers and collecting and marketing the finished fabrics. The industry had previously been tightly organised, with a gild structure which maintained the quality of the product. But in the early years of the nineteenth century this organisation was beginning to crumble. About 1825 the Jacquard loom, capable of weaving patterns into the silk, was introduced. It was a heavy and expensive machine, beyond the capacity of independent weavers to purchase or of the floors of their ill-built houses to support. A consequence was unemployment and unrest amongst the poorer weavers,[229] and many migrated to newly built houses and workshops in suburban Lyons and its rural hinterland. By 1830, only half the silk-looms were in the industry's traditional quarters, but within the city itself a third of the population was still employed either in silk-weaving or in related industries (fig. 8.22).

Silk-weaving was an export-oriented industry. In 1811, three-quarters of the silk fabrics were sold abroad; by 1832 this had increased to more than four-fifths.[230] It was also a luxury industry. Its products were subject to a highly elastic demand, and in times of economic recession, such as the later 1840s, the weavers suffered severely. but with rising real incomes during the following decades the industry revived, and the Lyons district became the foremost centre of the silk industry in Europe, exceeding even the Krefeld district in the volume of its production. This development was achieved at the expense of changes within the industry. It became increasingly mechanised. The number of power-looms grew from 7,000 in 1875 to more than 40,000 early in the present century, while that of the traditional handlooms declined from about 80,000 to less than a quarter of this total.[231] At the same time the industry ceased to concentrate on highly priced luxury goods and began to appeal to a mass market.

But silk was only one of several textile industries that had emerged in the Lyons-Saint-Étienne region. Textiles were made throughout the mountainous area which lay to the west of the Rhône. Most were based on cotton. Thread spun at Roanne or Amplepuis was woven into tulle, muslin, cretonnes and 'vichys' on countless looms in the cottages of the highlanders. Every farm, every household, it was said in 1860, turned itself into a workshop in winter.[232] Each local area was in some degree specialised, but the most marked specialisation was in silk ribbons and velours.

In the course of the sixteenth century the silk industry of Lyons spread south-westwards into the Giers Valley, where it earned the jealous enmity of the merchants of Lyons. Saint-Chamond, where ribbon-weaving was first established, lay within the spheres of influence of Lyons, and weavers had to conform with the regulations of its gilds. For this reason the focus of manufacture then moved south-westwards to Saint-Étienne, which lay beyond the control of Lyons. Here a rival silk-weaving centre developed, with its specialisation in ribbon-weaving.[233] Its structure and organisation resembled that of Lyons. *Fabricants* maintained their offices and held their store of silk near the centre of the town. Weavers, scattered through Saint-Étienne and over its surrounding countryside, wove it into ribbons. These were at first marketed at Lyons, but the Saint-Étienne merchants then established their own outlets, and acted independently of their rivals.

The silk-weaving industry continued to grow in Saint-Étienne and its vicinity. In 1839 there were more than 10,000 ribbon-looms, almost a third of them in the town, and another third was to be found in the neighbouring *bourg* of St Jean Bonnefond.[234] The industry continued to be carried on in small workshops, each with at most a few Jacquard looms. It disappeared from the east of the region, displaced by mining and iron-working, and at Saint-Chamond, the earliest seat of the industry in the Saint-Étienne district, there was only a handful of looms left at the mid-century.

The most curious feature of the silk ribbon manufacture of Saint-Étienne is not that it survived the hostility and formidable competition of Lyons, but that it held out, and even throve, alongside coal-mining and the most powerful concentration of iron and steel in France at this time. Here, unlike the north-west European region, a balance was achieved between light and heavy industry, between silk ribbons and coal and steel.

One reason lay, perhaps, in the fact that only a few steps from the town one was outside the coalfield and in the mountains of Forez where the products of an ungrateful soil had to be supplemented by the profits from domestic industry. Another reason was that silk-weaving provided female employment in an area of male-dominated heavy industry. But, one may reply, female labour was no less abundant in the Ruhr, and there its most obvious employment was allowed to decay.

Coal-mining

A strong demand arose in the eighteenth century for the coal of the Saint-Étienne basin. It was used by the forge-masters and was shipped by way of the Loire and Rhône to brick-kilns, glassworks and lime-burners. It went down the Loire as far as Nantes and on the Rhône it was taken upstream to Lyons. Development came first on the north-eastern part of the coalfield along the River Giers, where the Givors Canal was cut to carry the coal to the Rhône port of Givors.[235] In 1812, 190,000 tonnes of coal were produced from the Rive-de-Gier area, but only about 100,000 from Saint-Étienne.[236] Output continued to increase. In 1836 the coal basin was described as 'the most important in the country both for the quality and the quantity of its coal'.[237] The output of the whole field rose to half a million tonnes in 1824, and a million by 1835. By 1856 production had reached 2,240,000 tonnes. Then output received a check; rose again to 3,500,000 tonnes in 1880 and fluctuated about this level until the First World War and the destruction of the northern mines gave it a renewed importance.

During the period of their greatest prosperity, the 1820s–1830s, the mines of Saint-Étienne were the chief source of supply both to Paris and the whole Loire valley. The three earliest railways to be built in France were constructed first to convey coal to the Loire at Andrézieux (1827), then to Roanne (1835), and to Lyons (1832). Why, then, was this early promise not fulfilled? The answer lies partly in the successful competition of the northern coalfield for the market in Paris and northern France; in part in high mining costs and the exhaustion of the eastern part of the field; in the growing import of English coal, and, lastly, in the failure of local industries to expand significantly in the second half of the century.[238]

Iron and steel

Most important of these local uses for coal was iron-smelting and the production of puddled iron and steel. Saint-Étienne had been a centre of iron-working since the later Middle Ages. Iron, smelted from local ores, was worked up into arms and armour, domestic ironware and, above all, nails. The local coal was excellent for the forge, and there was abundant water-power to operate the bellows. The making of small iron goods, such as nails, was, like ribbon-weaving, an occupation for the rural population. In the seventeenth century Colbert established a royal weapons manufacture here, thus beginning that close association of Saint-Étienne with the military, which has continued until today. At the end of the eighteenth century the small rivers were lined with water-wheels, which operated hundreds of forges, hammer-works and splitting mills. Indeed this was the only area in all Europe where a traditional iron industry was vigorously pursued *on* a significant coalfield. Not surprisingly it was chosen early in the nineteenth century as the scene of a more modern industry. In 1820 blast furnaces were built at Saint-Chamond, together with puddling and rolling mills.[239] This was closely followed by the introduction of steel-making by the cementation and crucible processes. These methods, perfected in England, had not been used in France when the government invited James Jackson to establish a steelworks.

He chose Trablaine, near Saint-Étienne, as the site of his plant.[240] This was quickly followed by others – smelting, puddling, rolling, steel-making – at Terrenoire, Saint-Julien, Izieux, Lorette. Within little more than a decade the whole region, from Rive-de-Gier to Le Chambon, was transformed into one of the busiest and fastest developing industrial areas in Europe. Indeed, its only rival was central Belgium. Amongst the works established at this time were smelting plants at Givors, where the Gier flowed into the Rhône, and at Chasse on the other side of the river. The location of these works reflected a basic weakness of the Saint-Étienne industry: it had scarcely any iron ore, and was dependent on imports from the Alps and Languedoc. A riverside site allowed both coal from Rive-de-Gier and ore to be brought in by water, while the pig-iron could be despatched by rail to the puddling and steel-making works on the coalfield. It was a familiar spatial pattern.

It is not surprising that the Saint-Étienne industry concentrated on iron-refining and steel-making, nor that it emphasised high-quality products rather than common steel. Iron continued to be smelted both on the coalfield and beside the Rhône up to the First World War, but the cost of transporting ore became a factor of increasing importance.[241] The Bessemer and open-hearth processes were adopted, but the discovery of the basic process was a crippling blow to the industry. The Loire could not compete with the cheap steel produced in Lorraine and Luxembourg; it was even less competitive than central Belgium. The production of heavy rolled goods, such as rails, girders and plate, passed to Lorraine. The production of pig-iron, which had reached 61,000 tonnes in 1880, sank to 20,000 in 1895.[242] Puddling gradually fell out of use, though the crucible process, important for high-quality steels, was still used in 1913, and the electric furnace was introduced. Total steel production declined from 116,000 tonnes in 1880 to 61,000 only 15 years later.

The decline in the volume of steel production disguised the fact that the Saint-Étienne region had become an important source of quality and alloy steels and a centre for the manufacture of steel-castings, forgings and precision pieces. This was part of the process of moving 'up-market'. Gun manufacture, not unlike that which had long been carried on in the vicinity of Liège, gained rather than lost in importance, and to it was added the construction of bicycles and sewing machines. *Dépt* Loire continued to be a major source of artillery and armour-plate for the military and of steel forgings for ships, turbines and generators.

A not dissimilar fate was overtaking Lyons at the other end of the industrial axis. Its basic industry, silk-weaving, did not decline; it deserted the city for rural locations, chiefly in Bas-Dauphiné, where the factors of production were cheaper.[243] By 1870 only a quarter of the looms remained in Lyons and its suburbs; by 1898 this had fallen to less than a tenth, though the *marchant-fabricant*, based in the city, continued to control the rural industry up to the First World War.[244] After about 1880 the fortunes of Lyons ceased to hinge on those of the silk industry. Foremost amongst the manufactures which then began to dominate the city's economic activities was chemicals. The chemical industry had first made its appearance early in the century, when it provided dyestuffs and other materials for the textile industries. It then began to produce the basic acids and alkalis, and from this point it became increasingly diversified.[245] Hydroelectric power from the Alps was then transmitted to the Lyons region, encouraging the growth of electro-chemical and electro-metallurgical industries.

Second in importance only to the chemical industries was engineering and mechanical construction. The development of river navigation led to boat-building, and the opening of the railway from Paris to the Mediterranean was followed by the establishment of engine workshops. This was in turn followed by automobile engineering, and Lyons became the home of several of the early French car manufacturers.

In these developments Lyons had the advantage, which Saint-Étienne lacked, of excellent means of transport and communication. In the 1850s the railway was completed along the Rhône valley, putting Lyons into direct contact with both Paris and the Mediterranean ports. Materials used in Lyons' industries were generally of relatively low bulk and high value, and tended to move by rail. The river, which had been important in the early stages of development became almost deserted. Even ore for the furnaces at Givors and Chasse came in the end to be transported by rail.

Population and urban growth

The industrial region formed from the beginning of the nineteenth century an island of relatively dense population amid the thinly peopled mountains of Dauphiné and the Central Massif. From Saint-Étienne to Rive-de-Gier it derived its income primarily from coal-mining, metal-working and the weaving of ribbons. To the east, between the coalfield and the Rhône, population was more sparse, but a belt of denser population followed the Rhône northwards to Lyons.[246]

On the coalfield in 1806 there was a population of about 56,000;[247] Lyons had about 25,000 inhabitants, and the whole region about 100,000. Within it were only two towns of significance, Saint-Étienne with about 20,000, and Lyons itself.[248] Between them lay Saint-Chamond, Rive-de-Gier and Gisors, none of which had more than 5,000 inhabitants. No other settlement, even within the area of the coalfield, amounted to more than a village.

Population grew in all parts of the region until the middle of the century. That on the coalfield increased threefold to more than 160,000. Saint-Étienne itself became a town of 90,000, its medieval core enveloped within a pattern of straight streets laid out at right angles, bordered by high tenements and girdled by a ring of factories and steelworks. Saint-Chamond grew to 16,000, and Rive-de-Gier to 13,000. Industrial communities like Terrenoire, L'Horme, Lorette and Firminy evolved from villages into sprawling industrial towns.

The first check to this growth came in the 1850s. The coal resources in the east of the basin, which had been worked most intensively, were running out. Mines were closed, and the population of Rive-de-Gier actually contracted. In the west of the region growth continued into the 1880s, when the region was struck by a second crisis, resulting this time from the successful competition of the newly developed steel industry in Luxembourg and Lorraine. There was a small drop in the population of the region as a whole, followed by an increase at a significantly slower rate than hitherto. The population of Saint-Étienne itself ceased to grow towards the end of the nineteenth century, and in 1911 had 146,800, and the coalfield region about 300,000.

The Lyons region, on the other hand, continued to expand with the diversification of its industrial structure and improvement of its transport net. The city grew rapidly, from about 177,000 in 1851, to 376,600 in 1881 and 472,100 in 1911, and consistently maintained its position as third largest city in France.

Population growth in the Lyons-Saint-Étienne region was maintained through the century in part by natural increase, but very largely by immigration. In Saint-Étienne births continued to exceed deaths until about 1900, but there was a steady immigration from the surrounding hills.[249] In 1911 little more than half the urban population was born there, and more than 26 per cent had been born outside *dépt* Loire.[250] The chief source of immigrants to the Saint-Étienne area was the mountainous *départements* of the Central Massif, especially Haute-Loire. but there was also a large movement of foreign workers, mainly Italian, to the coal-mines and steelworks. As early as 1844 complaint was made of their excessive numbers.[251] Their immigration continued, however, on such a scale that there was at Rive-de-Gier one of the few violent expressions in France of anti-foreign feeling.

The growth of Lyons was yet more dependent on immigrant labour from the surrounding mountains. They came both from the Alps of Dauphiné and the Central Massif. During the period of the city's most rapid growth there were said to be some 6,000 masons from Creuse working there.[252] The mountains, from the Alps in the east to Forez, Livradois and Velay in the west, were being depopulated, except when the surviving textile industries or the newly developing electrical served to retain their population.[253]

Southern Europe

The development of manufacturing industries south of the Alpine system was on a very much smaller scale than in north-western Europe. Southern Europe was poorly endowed with those energy resources upon which northern manufacturing had been based. There was little fossil fuel, and most of it was of poor quality. Much of the coal consumed in Italy and the Iberian peninsula was, in fact, imported by sea from Great Britain. Deposits of iron ore were few and small, and, apart from those found in northern Spain and on the island of Elba, were of little significance for modern industry. Water-borne transport, so important in northern Europe, was severely hindered by the long summer drought and was, in fact, scarcely used at all. The railway net was well developed only in the plain of northern Italy, and over much of southern Europe it was built with an eye more for military defence than for the needs of a developing economy.[254] Only coastwise shipping provided the means of cheap and regular movement of goods, and in much of southern Europe economic development was, by and large, coastal. The domestic market for manufactured goods, especially in Greece and the Iberian peninsula was small, but there was, on the other hand, a long tradition of craftsmanship, and in some parts, notably northern Italy, a deep familiarity with the processes and methods of textile production. The potential for hydroelectric power was large in the Pyrenees and the Alps, but in 1913 was little developed. Units of production were almost invariably small, and manual and domestic methods of manufacture remained more important than in western and central Europe.

There were in consequence no concentrations of industrial production, such as have been described above. Instead, there were at most loose clusters of small, manufacturing towns, still maintaining close links with the domestic handicrafts of the surrounding villages. A symbiotic relationship between small urban factory and rural workshop, like that which characterised western Europe during the earlier phases of the Industrial Revolution, was still to be found at the end of the nineteenth century. Two such industrialised areas emerged: Catalonia and the North Italian Plain, with a third – not examined in this chapter – in the Basque region of northern Spain.[255]

Catalonia

Catalan industry focused in Barcelona, but was scattered through the small towns of the hinterland. It was dominated by textiles, of which cotton was by far the most important. The cotton industry became the outstanding, indeed the only successful, manufacturing industry in Spain.[256] The foundations of the Catalan industry lay in the traditional cottage industry – chiefly woollens – and the commercial and progressive outlook within the province, contrasting with the conservatism of most of Spain (fig. 8.23). Cotton manufacture also developed on a domestic basis during the eighteenth century, and Catalan cottons were shipped to the Spanish Empire in the New World. A modern industry began to develop after about 1830, when steam-power was introduced in the spinning industry.[257] A local ordinance excluded steam-mills from the city of Barcelona,[258] but they were established during the following decades in its immediate hinterland. Coal, imported through the port of Barcelona, was relatively expensive and its transport inland difficult. In consequence, most mills at any great distance from the Catalan capital used water-power. This restricted their size, and Catalonia at the mid-century was noteworthy for its immense number of very small mills. There were said at this time to have been more than 4,500 mills and workshops.[259] Linen-weaving was widespread in Catalonia at this time, but the manufacture of woollens was still a cottage industry, pursued to satisfy only local needs.

The cotton industry, as measured by the import of raw cotton, continued to grow during the 1840s and 1850s. Then, in the 1860s, it was hit by the cotton 'famine'. Attempts to grow cotton in southern Spain were in the main unsuccessful, and many mills were obliged to close. The Catalan industry emerged from the crisis slimmed down but more efficient. The following years were in the main prosperous for the cotton industry. The import of raw cotton

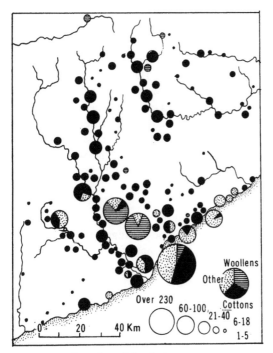

8.23 The textile region of Catalonia

– almost all of it through Barcelona – rose from an annual average of about 24,000 tonnes in 1869–73 to about 81,000 tonnes in 1904–8.[260] Demand in the domestic market was increasing, and the Catalan industry was able to command that of the remaining Spanish colonial dependencies. These, with their markets, were lost after the war of 1898, but domestic demand continued to increase and the industry to expand up to the First World War. A British consular report of 1893–4 noted the increase in the number of steam-driven mills, and the almost complete disappearance of hand-weaving.[261] The industry, in fact, spread up the valleys of the Llobregat, Ter and their tributaries right up to the Pyrenees.[262]. The British consul reported that steam had almost entirely replaced water-power; it had, indeed, in the vicinity of Barcelona, but if the consul had travelled into the interior he would have found that most of the small mills 'amid the terraced vineyards of the valley of the Llobregat', were still operated by water-wheel.

The woollen industry of Catalonia grew more slowly. It was once more widespread than cotton, but not until the middle years of the century did woollen manufacturers, following the example of those in cotton, begin to build large mills. They were helped by the lower tariff rates introduced in 1869 allowing the free import of Australian wool.[263] The industry which developed in the last third of the century was located almost exclusively in the towns of Tarrasa and Sabadell, which together accounted for about 80 per cent of Spain's woollen industry.[264] Most of the remainder lay in Barcelona itself. Elsewhere there was only a relict cottage industry.

Other branches of the textile industry, especially silk and linen, were also scattered through the region, together with derivative crafts such as lace and embroidery, but in terms of employment and production they were of small importance beside cotton and woollens. Few other branches of manufacturing intruded into the textile domain of Catalonia. Barcelona itself had a more diversified industrial structure; but in its hinterland only tanning was unrelated to the weaving and finishing of cotton and woollen fabrics.

In the 1920s the Catalan textile industries employed nearly 200,000; the total could not have been significantly different a decade earlier. At this time 60 per cent were engaged in the cotton industry and the average factory employment was about 125. Only 14 per cent were employed in the woollen industry, but 16 per cent were in the derivative industries, especially hosiery.[265]

Northern Italy

Despite its early promise, Italy was, at the beginning of the nineteenth century, a backward country. Its once distinguished manufactures had either ceased or had degenerated into rustic crafts. Even the raw silk, which Italy produced in abundance, was largely exported to be woven elsewhere. The political fragmentation of the country and the long survival of restrictive gilds hindered development. It was not until the latter had been abolished in the eighteenth century and the country temporarily united under Napoleonic rule that prospects for industrial growth began to improve. Mechanical spinning was introduced in 1808, and the first steam-powered textile mill was established in 1819. Little progress was, however, made before the mid-century. There was little market for manufactured goods and local advantages were slight. Although some progress was made in Piedmont and Lombardy, Italy remained on the eve of the *Risorgimento* an undeveloped country.[266]

Such development as had occurred was largely in the textile industries, and was located mainly in the North Italian Plain (fig. 8.24). An industry located in Naples showed some promise only as long as it was protected by the tariff barriers of the Sicilian Kingdom. After 1861 it withered in the face of northern competition.[267]

The preliminary stages of silk manufacture were already well established when the century began. They were rural crafts, practised for only a short period each year, and for that reason intimately bound up with the life of the countryside. Small mills, most of them run on a family basis, reeled the raw silk from the cocoons, and then twisted it to make thread strong enough for weaving. Although silk was woven on the handloom at Milan and Como, most of the thrown silk was exported to France and Germany.[268]

The woollen industry was only a shadow of its former self. It was carried on early in the century as a cottage industry in Tuscany and in the Alpine foothills, and its product was small in amount and of poor quality. Flax- and hemp-weaving were traditional crafts in the northern Plain, and to these was added during the century a small manufacture of jute fabrics.

Only the cotton industry grew in any significant fashion before 1861.[269] A small manufacture had been carried on since the Middle Ages. Early in the nineteenth century it received a considerable impetus from the introduction of

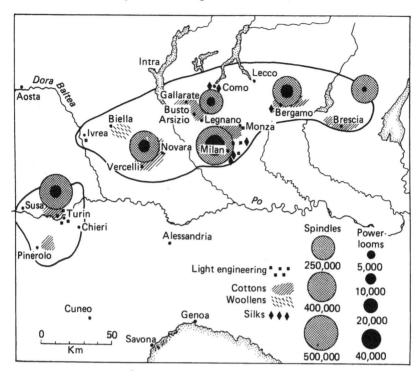

8.24 The North Italian industrial region

British technology, and small spinning mills were established well before the unification of Italy. There were no powerful factors influencing their location, other than a railway to the port of Genoa, through which coal and raw cotton were imported. Most, however, used water-power, and were built beside the Alpine rivers where these entered the Plain. Labour was drawn from the local peasantry, already familiar with the processes of textile manufacture, and their market was chiefly in the urbanised region of northern Italy, though an export market developed later in the century.

The import of raw cotton grew steadily after 1861. After some 15 years it had increased tenfold, and by 1900 was more than 30 times its volume in 1862.[270] The mills were mostly established in the small towns and large villages lying between Milan and the mountains, where the demand for labour, water-power and transport were most easily met. The greatest number both of spindles and of looms came to be located in Lombardy, especially in the districts of Milan, Bergamo and Brescia, and, in, Piedmont, in the districts of Turin and Novara.[271] Small mill towns, closely similar to those in Catalonia, developed, notably Gallarate, Busto Arsizio Monza and Novara. Most had some local specialisation. Besso wrote that 'the district between Milan and Lago Maggiore...with its numerous mill villages linking up the manufacturing towns, bears a striking resemblance to Lancashire'.[272] This resemblance, alas, was entirely in their function; visually they were worlds apart.

Most mills were small by the standards of north-western Europe. Spinning mills had commonly 30,000–60,000 spindles, and important weaving sheds might have as few as 400 looms. Indeed, handloom weaving continued to be practised, even in the cotton industry, at the beginning of the present century.[273] There was a tendency for the industry to concentrate in larger units. Average employment in 1876 was about 80; this had doubled by 1893, and the number of both spindles and looms had increased by about 70 per cent during the period.[274]

The woollen industry made no such progress. It continued to be carried on in part as a cottage industry throughout the century. The chief centre of the factory industry was Biella in Piedmont. A consular report of 1887 showed that there were no less than 145 mills, but that many were very small. 'I saw at least four separate small firms carrying on their industries in one building', wrote the consul, adding, however, that there was one mill employing as many as 600.[275] The woollen industry was slower to modernise than the cotton. Its market, both domestic and overseas, was small, and it was not until late in the century, when the industry gained tariff protection, that the finer fabrics began to be made.[276]

The silk industry was the least progressive. Reeling and throwing continued to be by hand, and weaving was mainly by the handloom until the First World War.

The textile region of northern Italy formed a narrow belt from Venezia in the east to the mountains of Savoy. Here it profited from the many small units of water-power and from the abundance of labour which had moved from the Alps to the Plain. Yet in its organisation it was archaic, and much of it remained integrated with rural life until the end of the century.[277] Very broadly, the woollen industry characterised the mountains, the cotton the nearby plains, while the processes of preparing silk were to be found wherever mulberry trees were grown and silkworms raised (see p. 259).[278]

Backward as it was in so many respects, the textile industry nevertheless accounted at the beginning of the present century for almost a third of Italy's industrial production by value, most of it in the northern region.[279]

Second in importance to textiles were the food industries, but chemicals accounted for only about 1 per cent of industrial output, and metallurgy and engineering together for only about 10 per cent.

Throughout the pre-war period Italy's iron industry remained small and backward. There had formerly been a traditional iron industry in the Alpine foothills of Lombardy and in the Val d'Aosta.[280]. It was already in decline in the mid-century, when the reduction of duties on imported iron goods exposed it to the full weight of competition from north-western Europe. As late as 1859 there were 20 charcoal furnaces active in Lombardy and nine near Aosta. There was little iron ore and even less coal in the Italian Alps, and the smelting industry declined and disappeared except in the Val d'Aosta.[281] Italy's small iron and steel industry came in fact to be established in peninsular Italy, where it was more accessible to imported fuel, ore and scrap. Meanwhile the northern industry had given rise to a small metal-fabricating industry, located chiefly in the Val d'Aosta and at Dongo, Bergamo, Brescia, Lecco and Milan. Some works were poorly sited for the supply of fuel and scrap in which they were increasingly dependent, and closed. The industry thus tended to concentrate at a few centres, of which the most important were, apart from Genoa and the Ligurian coast, Milan and Turin. Belatedly, a modern iron-smelting and steel industry was established near Aosta to use the local ore. But most of the north Italian works came to depend on metal from furnaces located in peninsular Italy.

The metal refined in northern Italy was at first largely used in railway engineering. Manufacturing then turned to producing agricultural, electrical and mechanical equipment. During the last two decades of the nineteenth century the range of metal-using and mechanical industries greatly increased, and to them was added at the turn of the century the most important of them all, the automobile industry. At least 40 companies, it is said, were formed to build cars, many of them in Turin and Milan. Construction was by hand, and the number built relatively small. Some of the firms were short-lived, but one of them, established in Turin in 1899 by Giovanni Agnelli, grew into the Fabbrica Italiana Automobili Torino (FIAT).[282] Other firms which had a lasting importance were Lancia in Turin and Romeo, Bianchi and Bugatti in Milan. Italian pre-eminence in fine engineering was demonstrated at the same time by the foundation at Ivrea in 1898 of the Olivetti typewriter works.

It is doubtful whether there could have been so vigorous an industrial expansion in the last two or three decades before the First World War without the concurrent development of hydroelectric power. The direct use of water-power was restrictive and the cost of imported British coal was high. On the other hand resources in hydroelectric power seemed unlimited, and electric power furthermore was well suited to the small-scale and mainly manual industries for which Italy had become noted. There had previously been experiments with power generation from imported coal. This proved costly, and in 1898 the first major hydroelectric generator was built on the River Adda at Paderno.[283] By the outbreak of the First World War there were some 50 generators supplying the manufacturing industries of northern Italy.[284] Without them the rapid industrial growth of Turin and Milan and of numerous smaller towns in the years before the First World War would have been impossible.

The industrial region which had emerged in northern Italy by 1914 resembled that in Catalonia rather than the coalfield regions of north-western Europe. Apart from a few large urban centres, manufacturing was widely scattered through the small towns and industrialised villages of the region. The reason lies, of course, in the fact that factory industry here emerged directly from rural proto-industry, and, once established, had no reason to change its location. Many of the 'new' industries developed only after hydroelectric power had become available, and were located wherever power lines could reach them.

Conclusion

In the preceding pages the growth of Europe's major manufacturing regions, together with that of some of lesser importance, has been discussed. This conclusion examines features that have been common to them all. In the first place industrial growth was generally preceded by a system of proto-industry. Only in Upper Silesia and the Łódź district was there no significant 'industrialisation before industrialisation', and in these instances manufacturing industry was consciously founded by entrepreneurs from without. Elsewhere entrepreneurs and in some measure the workforce derived from the earlier system of domestic industry.

The labour force in industrial regions was recruited in large measure by immigration. It does not appear that domestic or cottage workers transferred to the factory on any considerable scale. It seems rather that the domestic crafts ceased gradually to attract the younger workers, who then turned to the factories for employment. Much of the migration was over relatively short distances, the only conspicuous exception being that from the German East to Berlin, Westphalia and the Lower Rhineland.

Although most industrialised regions contained coalfields, industrial growth was initiated in almost all of them on the basis of water-power. Only gradually was this replaced by steam-power generated by burning coal. If coal was unnecessary as a source of energy during the early days of industrial growth, it was nevertheless essential for its fullest development. Even the manufacturing industries of Catalonia, Saxony and Łódź were by the end of the century based mainly on steam-power, since the small units of water-power that were available could not support a large factory.

If the presence of coal was an advantage only slowly revealed and used, a means of transport was essential from the earliest days of industrial growth. Concentrated industrial production was dependent on transport since its market was necessarily widely distributed. In most instances reliance was at first placed on water-borne transport, and the presence of easily navigated rivers was unquestionably a factor of great importance in the development of the north-west European industrial region. Water-borne transport was first supplemented and then replaced, except for bulky goods of low value, by the railways. Regions which lacked a system of navigable waterways had, in effect, to await the coming of the railways before industrial growth became significant.

Industrial regions, lastly, tended to become more diversified in the course of the century. This was due primarily to the growing complexity and mutual dependence of industrial processes, one manufacture serving as a raw material or component in another. Textile industries attracted the manufacture of dyestuffs; coal-mining and coke-ovens helped to support a chemicals industry, and the iron and steel industries contributed to the engineering and steel construction industries located in their vicinity. The tendency was for industrial regions to expand around their peripheries, as new factories were attracted to virgin land at no great distance from their cores. At the same time, older industrial sites in the centres of the regions were in some instances abandoned, leading to the appearance of dereliction which many such areas acquired.

9
Transport and trade

Economic development in the nineteenth century was marked by the growth of local specialisations in agriculture and manufacturing. A limited number of points became the source of supply of particular goods for wide areas, from fresh produce and liquid milk to steel wire and chemical fertilisers. The closed local community, autonomous and autarchic, may have been a figment of the imagination of economists, but the scale of dependence on external markets and sources of supply was small in pre-industrial Europe. This restricted degree of self-sufficiency was gradually eroded by market forces, and the local market became merely a stage in the distribution of goods between the local community and the distant city, factory and farm.

This independence of the local community had begun to break down long before the nineteenth century. The process was, however, conditioned by the available means of transport. Of these there were only two before the railway age: pack animal and wagon on the roads, and boat on the navigable waterways and along the coast. The one was costly, the other restricted to those routes where waterways were present, and both were excessively slow and exposed freight to damage and loss. Movement over more than short distances added greatly to the cost of commodities, and only the more valuable of them could bear the additional cost of transport to a distant market. This is one of the reasons why expensive goods, such as silk and china, were prominent amongst manufactures during the early phases of industrialisation. Cheap and bulky commodities such as grain, timber, coal and metalliferous ores entered into long-distance trade only when they could be transported by water.

The development of manufacturing industries catering for a mass market was dependent on an extension of the network of transport and an improvement in its quality. The first half of the nineteenth century was marked by attempts to improve the road system in western and central Europe. This was accompanied by the construction of new canals and the improvement of river navigation. The volume of freight transported increased many times, though it is impossible to produce quantitative evidence except for the movement of coal from the coalfields of north-western Europe.

The first railway was opened in 1835. It was quickly followed by others, though it was several years before they began to carry freight on a significant scale. By 1870 the primary railway net in western and central Europe was complete. It

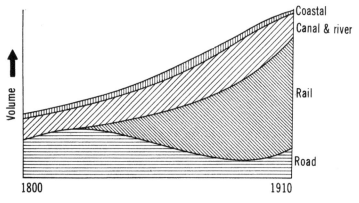

9.1 A simple model of the development of transport during the nineteenth century

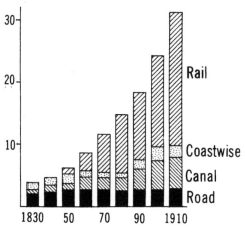

9.2 Internal transport in France

was far more closely adjusted to the needs of developing industry than that of rivers and canals could possibly have been. Traffic began to desert the waterways and the roads for the railways. Even traffic between the cities of the Rhineland began to move by rail. Little further improvement was made in the road system because it was supposed that such improvements were no longer necessary (fig. 9.1).

Late in the century there was a return to the waterways. The construction of wider and deeper canals with larger locks encouraged the building of bigger barges. The volume of coal, iron ore, timber and other bulk commodities again increased on the waterways, especially in north-western Europe. Although many of the lesser rivers and canals which had played so important a role in the early phases of industrialisation were abandoned, the role of the few waterways capable of taking the largest barges increased sharply and attracted manufacturing industries to their banks. In the meantime railways strengthened their grip on the transport of light and perishable goods. It was not until the end of the century that the roads began again to challenge the supremacy of rail and waterway.

This was dependent on the internal combustion engine, and although there were motor vehicles on the roads as early as the 1890s (see p. 349), it was not until the eve of the First World War that road transport became significant (fig. 9.2).

The road system

At the beginning of the nineteenth century the road system of Europe was made up of countless trackways which had done service for centuries. The more important of them were broad ribbons of territory along which horses galloped and wagons and coaches picked their tortuous way. The quality of roads varied. It was consistently bad where they crossed belts of clay or the floodplains of rivers. In *dépt* Mayene, for example, the general backwardness of the area was attributed in part to the fact that many of the roads, built on heavy clay, were impassable for much of the year.[1] Dry, limestone uplands, like those of Burgundy and Champagne made the going easier, and many a traveller took the more hilly route in order to avoid the marshy lowlands. The loess belt often provided a drier route, though wheeled vehicles were liable to sink into its friable surface. In mountainous areas the route often became a track hewn from the steep slopes, and many of the Alpine crossings were perilous in the extreme. The narratives of those who made the Grand Tour often described the discomforts of the route and the pleasure experienced when, for a short distance, the coach could be exchanged for a boat on a Swiss or Italian lake. Guide books were written for the convenience of such travellers, warning them of the discomforts and dangers which lay ahead. One of the last of these old-style route books was published in 1833 by John Murray, the veteran publisher of travel guides, and provided a valuable picture of the road system of at least part of Europe on the eve of the railway age.[2]

A change began to come over road-building near the beginning of the century. Napoleon ordered the construction of a number of improved roads, *routes impériales*, most of them radiating from Paris. Their purpose was primarily military, but they also served the needs of commerce. The construction of national roads continued after the Restoration. By 1824 there were 14,000 km of such routes. The programme was particularly active under Louis-Philippe, and by 1855 the network embraced 35,000 km.[3] Thereafter road-building was pursued less vigorously. A primary network had been completed by then and progress made in railway construction appeared to make trunk roads less necessary.

This was probably a false economy. Many parts of France could be reached only by roads, many of which were quite unsuited to the traffic which was developing. Most of those in the Limousin, for example, were quite unfit for wheeled vehicles.[4] The quality of the roads, wrote Price, was 'still a fundamental restraining factor on economic activity'.[5] Roads were, in fact, essential to the prosperity of the railways since they served as feeders to them. At the mid-century nearly half the freight transported in France still went by road.[6] The railways, in particular, were quite incapable of handling the traffic which radiated from all the large urban centres. In Paris, for example, there were in 1845 more than 4,500 transport companies, operating nearly 30,000 carts and wagons, and on the routes between Lyons and the Midi it is estimated that there were some 35,000 carts and 86,000 horses.[7] A consequence of the improvements in the roads was the increasing speed of travel, from on average of about 3.4 km an hour in 1800 to 9.5 km in 1848. Nevertheless very long journeys with heavy freight were sometimes made. Raw cotton for the growing Alsatian industry was carried by horse-drawn wagons from Le Havre to Mulhouse, a journey of more than 600 km which occupied at least six days and sometimes as much as a month.[8]

In 1851 roads were far more intensively used than the railways and internal waterways together. In tonnes per kilometre the roads carried 2,400 as against 1,670 on the canals and rivers and 485 on the railways.[9] The intensity of road use continued to increase until the mid-1860s, though its share of the total traffic gradually declined. A service of horse-drawn diligences was maintained on the main roads far into the railway age. They could carry from nine to 18 passengers and made stops at places inaccessible to both railway and water-borne traffic.[10] The volume of road traffic remained stable until the beginning of the present century, when the increasing use of the automobile led to a significant growth. The heaviest use of the roads was always to the north of a line from Nantes to Strasbourg, though there were areas of relatively intense traffic around Marseilles and near Lyons-Saint-Étienne.

The only other countries in which the government played an active role in road-building were Belgium and Germany. In the former the road network increased from 3,400 km in 1830 to 9,750 in 1910. It was well developed in the north, but poorly in the Ardennes and Luxembourg.[11] In the Netherlands the efficient use made of waterways precluded a policy of road-building and even postponed that of railways. German roads appear to have been even worse than French at the beginning of the century, and their condition was reported by Jacob to have been so bad that trade was inhibited wherever water transport was not available.[12] In 1816, however, the State of Prussia began to build roads to link Berlin with the chief towns. This was followed by a similar policy in Bavaria. The formation of the Zollverein stimulated road-building by encouraging agreements between states to build and maintain roads.[13]

Elsewhere in Europe little was done to improve the roads before the period of railway development. Within the Habsburg Empire the roads were deplorable; the government showed no initiative, and the best roads were built and maintained privately.[14] A contributor to the *Quarterly Review* noted that even in the vicinity of Vienna roads were so bad that they contrasted 'unfavourably even with Russia, where the principal approaches to the capital have been solidly constructed and are kept in excellent repair'.[15] In the Hungarian Plain the vast extent of marsh and of soft, yielding sand made road construction almost impossible.

In the Balkan peninsula there could scarcely be said to be any roads. A report to the British Foreign Office as late as 1897 claimed that in Serbia communications were so bad that 'peasants [were] quite ignorant of the country 10 or 15 miles distant'.[16]

In Spain a system of 'royal highways' had been begun in the eighteenth century, but was still unfinished in 1840. Most other roads were mere mule tracks along which it was impossible for wheeled vehicles to pass.[17] In the populous plain of Andalusia conditions were even worse than in the mountains, for the soft and yielding soil provided no firm basis for roads.[18] Even at the end of the century Spain had only a rudimentary road net, and in 1907 a United States consul reported that 'Spanish highways are less numerous and distinctly inferior to those of any other country of Europe'.[19]

A road-building programme in Italy had to await the unification of the country in 1860. Even then railway construction was given priority. An adequate road net was built in the northern plain, but throughout central and southern Italy, as well as in Sicily and Sardinia, it was totally insufficient for economic development.[20] Some of the roads in the South were said to be quite impassable in winter.[21]

Water-borne transport

Transport by river and canal played an essential role in industrial development, especially during the first half of the nineteenth century. Although roads handled in the aggregate more freight than internal waterways, only the latter could handle heavy and bulky goods, notably coal, lumber and metalliferous ores. Factory development was effectively limited to the coalfields and to those areas where the coal could be distributed by barge. Water-borne transport was not restricted to bulk cargoes. Grain, sugar-beet and other foodstuffs were often distributed by boat. The major ports were served by a network of water communications, and in some parts of Europe there was a regular service of passenger transport by river and canal.

River transport had, however, severe limitations. Throughout Mediterranean Europe rivers were swift, shallow and in varying degrees seasonal. Only rivers, such as the Po, Rhône and Ebro, which had their source outside the region, could maintain a sufficient flow for navigation. North of the Alpine system obstacles were more likely to be ice in winter and floods at almost any season of the year. Although sails could be used for small craft, especially on broader rivers like the Rhine, most barges had to be pulled by human or animal power, at least during the first part of the century. It was thus necessary to maintain a towpath, a matter of some difficulty if floods were frequent or severe.

Downstream movement was always easier than upstream, and on the upper courses of some rivers only downstream traffic was possible. This was accomplished on the Loire, Vistula and other rivers by using boats, roughly made from timber cut near their headwaters, which were broken up at their destination and sold for construction or firing. On some rivers only rafts could be floated downstream. So difficult was movement overland that the smallest rivers were used for navigation until the building of railways gradually took away their role.

Attempts had been made long before the nineteenth century to regulate and improve rivers for navigational purposes and to link them with one another by canals. Canals were built where the need for transport was greatest. 'The map of English canals', wrote Hartwell, 'is the map of industrial England',[22] and the same, with reservations, could be said of continental Europe. Canals served more than one purpose. In the Netherlands they were maintained primarily in order to carry water from the polders to the sea; in Spain, Provence and northern Italy,

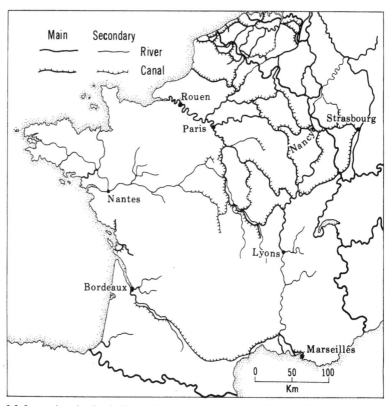

9.3 Internal navigation in France, late nineteenth century. The classification into 'main' and 'secondary' waterways is that adopted in the Freycinet Plan

to bring water from the mountain streams to the land; in Belgium, France and Germany, to transport coal and other bulky commodities.

Inland waterways continued to increase in importance until the middle years of the century or later. Then the railways began to draw traffic away from rivers and canals, until the need for largescale transport of bulky commodities restored a degree of importance to *some* waterways.[23]

Inland waterways of France

France had unusual advantages for the development of water-borne transport: a roughly radial pattern of rivers which could without great difficulty be interlinked by canals, a relatively steady discharge, without severe droughts and floods or prolonged periods of frost, and a large demand for transport.

Most of the larger French rivers had been used in the eighteenth century, some of them only for downstream traffic. Paris received most of its fuel and building materials and much of its food supply by river,[24] and all the larger French cities carried on much of their trade by water. A number of canals had been built before the Revolution to link the river systems (fig. 9.3). The well known Languedoc Canal from the Garonne to the Mediterranean had little importance in the nineteenth century, but those linking the Loire with the Seine system, the Canals d'Orléans and de Briare, were of great value. In addition, there were links between the Seine system of waterways and the rivers of the Low Countries. These were already gaining importance for the movement of Belgian coal to Paris. They were further augmented under Napoleon by the building of the Saint-Quentin Canal, which tapped the expanding coalfield of northern France. At the same time the Canal de Bourgogne was cut from the Paris Basin to the Saône near Dijon, and further connections were built during the century to join Paris by way of the Marne with the Meuse, Moselle and ultimately the Rhine. Lastly, a canal was completed between the Rhine and the Rhône by way of the Belfort Gap.

An expansion of the canal network was foreseen in Becquey's Plan of 1821–2, and the years until 1840 saw much activity, but the grandiose plans of earlier years succumbed to the railway fever which followed.[25] The traffic on French rivers and canals declined and on the Loire and Saône-Rhône it almost disappeared. A problem confronting internal navigation was the varying age and capacity of canals and their locks. Only the smallest and least economic barges could make a long voyage which made use of several separate waterways. In 1879 de Freycinet, the Minister for Public Works, introduced a plan to modernise and standardise the system. Barges of two standard sizes were adopted, and canals adjusted to take either the smaller or both.[26] New canals were to be constructed and river navigation improved. The plans were never wholly realised, but they gave fresh life to the French waterways. Freycinet himself foresaw a division of the traffic between rail and waterway, with the railways taking high-value, low-bulk goods and those requiring fast transport, leaving to the waterways all 'heavy, low-value goods, which yielded little revenue to the railways and choked rather than assisting them'.[27] This was in large measure realised. There was an appreciable increase in the freight on French rivers and canals, made up mainly of coal and building materials.[28] Indeed, it was claimed that the continued prosperity of French coalfields was dependent on the improvement of the canals which served them, and that the decline of the Saint-Étienne coal basin was due in part to the fact that it was not really served by a canal.[29] The freight carried by waterways more than doubled between about 1860 and 1900, though its percentage of total freight increased only marginally.[30]

The three river-canal systems whose varying fortunes hold the greatest interest, were the Saône-Rhône, the Loire and the Seine.

The Rhône had been used for navigation since classical times, but it was by French standards a difficult river.[31] Its current was always swift, especially during the summer, and its floods were sometimes violent. There were shifting sandbanks, and it was difficult to maintain the towpath. Whereas the downstream journey could be accomplished from Lyons to Arles in three days, the return voyage took 20 days or more.[32] Above Lyons the Rhône itself was barely navigable and little used, but its major tributary, the Saône, flowed deep and slow for 250 km from as far upstream as Gray.

The Rhône valley route was much used. The roads were bad and there was every inducement to improve the conditions of navigation. A passenger traffic developed on the Saône, and in 1827 a steamboat was introduced between Lyons and Chalon. Two years later steam navigation commenced on the Rhône itself.[33] Despite continuing navigational difficulties the 1830s and 1840s saw a growing volume of river traffic. About half a million tonnes were handled at the quays of Lyons. About 45 per cent of the upstream traffic was by steamer, though most of the much larger volume of downstream shipping moved more cheaply with the swift current. The competition of the railways after 1850 greatly reduced the traffic on both Saône and Rhône, and it virtually disappeared under the Second Empire.

The Freycinet programme provided for a large expenditure on regulating the Rhône,[34] deepening the channel and removing the many obstacles in its course. The result was to reduce greatly the number of days each year when shipping was held up, and the Rhône became in this respect a better-managed river than the Rhine.[35] Traffic, however, remained small except on the deltaic tract of the river, where the newly built Canal Saint-Louis provided a link with the port of Marseilles.

The Loire, like the Rhône, was obstructed by shifting sands and swift currents, and was described as the most *défectueuse* of all the major rivers of France.[36] It was used to distribute goods from the port of Nantes and the fact that it flowed westward allowed small, sail-driven boats to move upstream with the wind.[37] The river remained important as long as roads were bad, but, as on the Rhône, traffic declined in the second half of century, and here no radical improvements in the river were called for by the Freycinet Plan.[38] The volume of traffic below Orléans fell from over 600,000 tonnes in 1850 to 32,000 in 1892, and has since ceased altogether.

The upper river, however, presented a very different picture. It was a turbulent stream, and upstream traffic was difficult if not impossible. It had long been used for transporting forest products from the mountains to the Paris Basin, and it was to assist this traffic that canals were built from the Loire to the Seine. Coal from the Saint-Étienne field was brought by mule to the banks of the upper Loire and taken downstream by barge. In 1827 a horse-railway was built to Saint-Andrésieux (see p. 416), and the volume of coal shipped by the Loire and by the Canal de Briare at once increased. Despite the great distance, about 450 km, Saint-Étienne became the chief source of coal for Paris, a role which it retained until waterways improved and coal-mining expanded in northern France.

The rivers of the Seine system were the most easily navigated, and the gentle terrain through which they flowed made the construction of canals relatively easy. Paris, lying at the hub of this system, was the destination of much of the freight carried. The lower river, below Paris, required little regulation. The tolls which had hindered traffic were abolished at the Revolution,[39] and the volume of traffic increased through the century. Much of the commerce of the ports of Le Havre and Rouen used the river, and the canals and navigable rivers extended their hinterland as far as Lorraine.[40]

The river Oise, which joined the Seine some 70 km below Paris, was improved and linked by canal both with the Somme and with the network of waterways of Nord and Flanders. In 1836-8 the Oise was joined with the Sambre near Maubeuge and thus with the Meuse. Coal from first the Borinage and later the French coalfields of Nord and Pas de Calais was distributed over all northern France and through Flanders to the Ports of Antwerp and Dunkirk.[41] Although there was some decline in river and canal traffic after the building of the primary railway net, it increased again with the implementation of the Freycinet programme. The capacity of the Saint-Quentin Canal which carried much of the coal was increased, and in 1901 the Sensée Canal was constructed from the western part of the coalfield to the Oise.[42] At the end of the century almost half of the water-borne freight of France was carried on these northern rivers and canals.[43]

The Rhine

The Rhine is outstanding among the rivers of Europe not only for its natural advantages, but also for its role in the development of trade and industry. It rises in the Alps of Switzerland, where its headwaters lead up to several Alpine crossings. Until the great bend at Basel is reached the river is interrupted by rapids and is navigable only for short stretches.[44] At Basel it swings to the north and for 360 km flows across the plain of Alsace and Baden. In its primitive state it was a shallow, meandering river, changing its course with each flood and fringed by a marshy plain. It was joined by two important tributaries, the Neckar and Main, which gave access to south Germany. Below the confluence of the latter the plain ended, and the Rhine traversed a deeply incised and highly picturesque gorge for 110 km. The Moselle joined from the west at Coblenz, and the Lahn and Sieg from the east. Then, near Remagen, the hills drew back and the Rhine entered its plain tract. It cut across the Ruhr coalfield, where it was joined by the small but important rivers Ruhr, Emscher and Lippe, and entered the Netherlands and the last section of its course, the delta. Here it divided and subdivided. Many of its watercourses were narrow, shallow and of little commercial importance. In the early nineteenth century the most used branches of the river were the Lek, which gave direct access to the port of Rotterdam, and the Waal, which lay to the south and was the more easily navigated branch. A number of rivers and canals of varying degrees of difficulty gave access to the Zuider Zee and the port of Amsterdam.

River navigation is always conditioned by the river's regime, the regular and seasonal variations in discharge. The Rhine flows most strongly in summer when the Alpine ice-melt is most vigorous, and its level is lowest in winter.[45] The tributaries, however, carry most water in winter when they are swollen by rains. The river's discharge thus tended to even out downstream, though floods might occur at almost any season, and ice was a hazard everywhere in the winter.[46] Freight rates varied from month to month, reflecting the expected conditions on the river. They were highest in winter, when navigation was liable to interruption by ice and barges could not be fully loaded owing to the shallow draught in parts of the upper river.[47] In the six years from 1896 to 1901 shipping was held up for a maximum of 199 days in 1899, and a minimum of 53 in 1896, either by high or by very low water. In 1899 there were no fewer than 192 days with less than 2 m of water at Cologne.

The depth of water greatly influenced the size of both steamboat and barge.

Early in the century they were very small. Although boats of 500 tonnes were used on the lower river, the maximum size of a vessel sailing above Strasbourg was only 75 tonnes.[48] Barges increased in size through the century as larger tugboats were built to sail against the fierce summer current. At the end of the century the majority were still of less than 200 tonnes, but the number of large barges of over 1,000 tonnes was increasing, and accounted for a rapidly growing proportion of freight carried. In 1905 the largest were of 1,800 tonnes.[49]

A great deal of work was, however, necessary to permit the river to take vessels of this size. The chief problem areas were the upper river, above Mainz; the Bingerloch at the southern end of the gorge tract, and parts of the plain where at high water the river was liable to spread out over the bordering countryside. Everywhere there were shallows on which barges were liable to ground. The problem in regulating the river lay in the fact that if the speed of the current was used to scour a deeper channel, the problems of upstream navigation were increased. This, however, proved to be the only course. In 1809 Johann Gottfried Tulla, an engineer in the employ of the Baden government, prepared a plan for the regulation of the river.[50] It involved cutting a channel through the complex and shifting meanders of the upper river and securing it with dykes. The gradient and thus the speed of the current were increased, and its scour used to keep the channel clear. In the face of opposition, Tulla's plan was implemented, 1824–32, and was extended after his death. Despite this work, navigation remained difficult above Mannheim; without it, it would have been impossible.[51]

The next obstacle lay in the gorge tract, where a series of rocky bars produced rapids. Most significant of these was the Bingerloch. Small vessels had been able to negotiate the shallows, but for larger barges it was necessary to quarry away the rock in the river bed. At the same time, dykes were constructed *in* the river to confine the main stream and thus increase its speed and scour. Lastly, along the lower river levées were built all the way to the delta in order to reduce the ever-present danger of flooding.

All the engineering work accomplished on the river would have been of little value if it had not been accompanied by a change in its legal status. Until the Napoleonic conquest the use of the river had been obstructed by tolls and prescriptive rights which dated from the Middle Ages. These included the rights of certain cities, including Cologne and Mainz, to compel passing vessels to unload the goods they carried and offer them for sale (*Stapelrecht*). No less burdensome was the obligation at some of them to reload on to different ships, regarded as more suitable for the next section of the river (*Umschlagsrecht*). These, combined with tolls levied at more than 30 stations, had effectively extinguished long-distance river traffic by the eighteenth century.

In 1792 the French Revolutionary government proclaimed the freedom of navigation on all international rivers. This freedom was reasserted in the Final Act of the Congress of Vienna, which provided for the creation of a Commission to regulate navigation. Tolls were abolished and the only dues chargeable were those necessary to maintain services on the river. It was not, however, until 1831 that the Rhine Commission was established by the Mainz Convention, and the last obstacles to freedom of navigation, such as the staple rights of certain cities, were abolished. The work of the Mainz Convention was extended and amplified

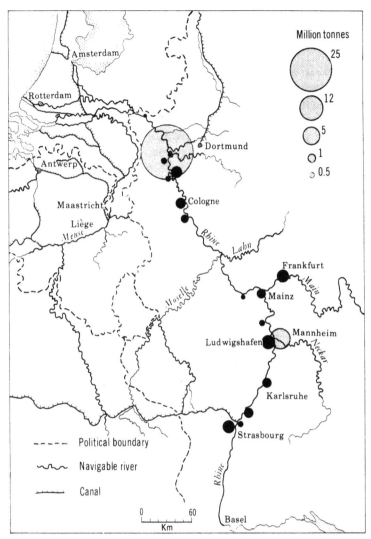

9.4 Rhine navigation about 1910. Symbols indicate the volume of freight loaded and unloaded

by that of Mannheim in 1868 in the light of the greatly increased volume of traffic which the river then carried.[52]

Navigation on the Rhine The traffic traditionally carried by the Rhine had been of two kinds. The more important was the long-distance movement of goods, generally of high value, from Italy by way of the Alpine passes. This traffic had received a crippling blow with the development of the sea route from the Mediterranean to north-western Europe, and finally succumbed to the tolls and other hindrances to navigation. The other traffic was a local one, which centred in each of the cities of the Rhineland. A service of small vessels plied between riverine villages and their nearest city, carrying passengers and produce for the market. This service, the *Mess-* and *Marktshiffahrt*, was particularly developed in the highly urbanised tract of the middle Rhine, and remained active into the nineteenth century,[53] when it was gradually superseded by railway and road.

Long-distance river traffic had gradually to be rebuilt during the nineteenth century, but there was no revival of transalpine commerce. The new focus of trade was the lower Rhineland, the Ruhr coalfield and the ports of the Rhine delta. Throughout the century coal was the most important commodity. It was carried both upstream to Mannheim and downstream to the Dutch ports. Most was loaded in Duisburg and Ruhrort, at the confluence of the Ruhr with the Rhine.[54] About 1831, 300,000 tonnes of coal were being shipped a year. By 1850 this had increased to 850,000 tonnes; to two millions in 1880 and seven by 1900. Later in the century iron ore, imported through the Dutch ports for the furnaces in the Ruhr, began to rival the movement of coal.[55] Iron and iron products, timber, stone and other building materials, and cereals were also amongst the cargoes.[56]

The heaviest traffic was always on the lower river between Cologne and the Dutch ocean ports. From Cologne upstream to Mainz the volume was less, and diminished yet more between Mainz and Mannheim. The port of Mannheim was long the effective limit of navigation. The traffic upstream to Strasbourg was diminishing when the completion of the railway through the plain of Alsace led to the closure of the port. The cessation of navigation above Mannheim was not unrelated to the fact that here the river was the boundary between France and Germany, between which co-operation to improve the river was minimal. After 1871 both banks lay within Germany. Work was at once recommenced, and gradually traffic revived at Strasbourg. But still there was no regular movement above Strasbourg. The port of Basel, which had closed in 1846, was not reopened until 1904, when the first cargo of coal arrived from Duisburg.

Rhine navigation was highly organised. As early as 1817 the Stinnes brothers acquired a fleet of boats which sailed regularly between Cologne and the Netherlands. By 1820 they had 66 barges in the coal trade. Other entrepreneurs followed their example, amongst them Haniel of Duisburg, who, like Stinnes, later used his profits to promote heavy industry in the Ruhr. Shipping could not have been developed on this scale without the use of steamboats and steam-tugs. The first steamship on the Rhine sailed in 1817 from London to Coblenz. This was followed by the formation of a Dutch steamship company. By 1830 several companies were operating steamship services between Mannheim and the Netherlands, with occasionally trips to Strasbourg. These vessels carried only passengers and light freight, and were thus particularly vulnerable to competition from the railways. The companies failed on the upper Rhine, though on the lower river and in the Netherlands, where the competition of railways was less severe, the steamboat service remained profitable. In particular the voyage through the gorge tract had a romantic appeal which kept the passenger boats in business.

Some of the boats carrying heavy freight continued as late as 1860 to be driven by sail, although the number of steam-powered tugs steadily increased after 1838. By 1860 there were 16 steam-tugs based on Ruhrort, which were used to haul coal to Mannheim. Thereafter Rhine navigation was dominated by trains of barges, towed by a smoking tug-boat; there was no longer room for the sailing boat to tack against the wind within the narrow confines of the river.

Traffic on the Rhine was at first between the long-established river ports. Although quantitive data are scanty, it appears that, apart from the ocean ports of the Rhine delta, the most used were Cologne, Mainz and, 40 km away on the lower Main, Frankfurt. The section of the river between them undoubtedly carried the heaviest traffic. With the reeopening of the river after 1815 the older ports took on a new lease of life. Quays which bordered the river were extended, and handled a growing range of goods.[57] There were in all some 40 river ports between Strasbourg and the sea. Some handled chiefly the specialised products of their immediate hinterlands: Wesel, coal from the northern Ruhr; Weseling, the brown coal of the Ville field; Andernach, andesite from the volcanic hills to the west; and ports near Coblenz the iron ore from the Lahn region. Coke was imported for the smelting furnaces which lay close to the river, but most merely imported coal for their local needs and shipped out cereals, wine, timber or whatever agricultural goods were available. By 1860 the railway had been built along the Rhine, linking the small ports which had previously relied upon the river for transport. Their volume of river traffic declined, in some cases never again to become significant. In the second half of the century manufacturing began to develop along the upper river. Some industrial sites were chosen in part because the river provided cheap and convenient transport for coal. Docks were built and these in turn attracted other manufactures, and the volume of traffic carried on the river began to increase again after about 1890 and continued to grow until the First World War.[58]

A number of major river ports had emerged by the early years of the present century. First to develop was the port of Duisburg-Ruhrort. Duisburg had long served as a transhipment point where coal brought down the River Ruhr was loaded on to the larger Rhine barges. This traffic increased after 1815 and in 1832–40 a short canal, promoted by the Ruhr industrialist, Friedrich Harkort, was cut from the Rhine near the town to the Ruhr. There was, however, little scope to extend the docks to the south of the Ruhr, and in 1860 work was begun on dock basins to the north, in the area now known as Duisburg-Ruhrort. The excavation of dock basins in the soft river silt continued until the First World War.[59]

Meanwhile, however, shipments of coal down the River Ruhr declined as the focus of mining shifted northwards to the Emscher valley (fig. 9.6). Plans were made for a canal to link the newer mines with Duisburg-Ruhrort, but it was not until 1908 that work began on the Rhine-Herne Canal. Other docks were built nearby: Wesel, 40 km downstream in 1890, to serve the northern part of the coalfield; Ürdingen on the opposite bank, as the port of Krefeld; Alsum (1884), Schwelgern (1904), Rheinhausen (1895) and Walsum-Süd (1905) to handle ore for ironworks.

Less than 40 km upstream from Duisburg lay Düsseldorf. It had carried on a small traffic since the Middle Ages, and about 1600 a small dock basin was constructed. Facilities were expanded in the 1830s, and in the 1890s four modern basins were built, followed in 1908 by the Heerdt basin on the opposite bank of the Rhine. Cologne was one of the oldest and at one time the most important

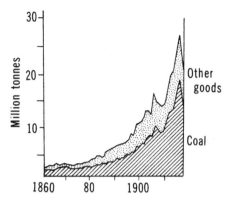

9.5 Freight handled in the port of Duisburg–Ruhrort

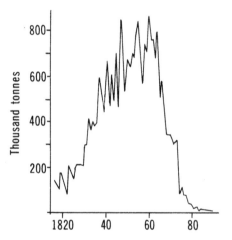

9.6 Coal shipped on the River Ruhr. The abrupt decline about 1870 marks the closure of the more southerly mines of the Ruhr coalfield

river port. It was overtaken by Duisburg-Ruhrort and by other ports handling bulk cargoes, and it was not until the 1880s that the riverside wharfs were supplemented by the first basin, built close to the city centre. This was quickly followed by basins at Deutz, Mülheim (near Leverkusen) and Niehl, built to serve the manufacturing industries which were springing up along the Rhine between Cologne and Düsseldorf.

Mainz, like Cologne, was a medieval staple port. It lay on the left bank of the Rhine opposite the confluence of the Main, and was the limit of navigation for the larger river craft. The docks were expanded in the later nineteenth century near the mouth of the Main and close to the developing industries of the Rheingau.

For much of the century Mannheim served as the limit of commercial navigation. The town had been founded near the beginning of the seventeenth century, at a time when traffic on the Rhine was in decline. It lay on the peninsula which separated the Rhine from its affluent, the Neckar. In 1804 it was incorporated into the State of Baden and, during the following years Tulla so improved the river's course that by 1825 steamships were able to reach the town. The new port developed quickly, and became an emporium for the upper Rhineland, the developing railways serving as feeders to the river. In particular, it shipped cereals and timber to the lower Rhine. The first wharfs, built along the Rhine, were superseded by a basin excavated in the alluvium between the Rhine and Neckar. In the last third of the century this triangle, bounded by the city and the two rivers, was developed as an extensive dockland, and beyond the Neckar a bend of the Old Rhine, eliminated by Tulla's 'correction', became the Industrial Docks. Upstream from the city another complex of docks was constructed at Mannheim-Rheinau in 1896–1907.

In the meantime a system of docks and warehouses had grown up on the opposite bank of the Rhine at Ludwigshafen. It originated in an attempt by the Bavarian Palatinate to rival the commercial achievements of Baden. A small wharf was built early in the century. Then King Ludwig I of Bavaria not only gave his name to the settlement, but encouraged the cutting of the first dock basin in 1843. The port, however, owed most to the establishment here of the chemical industry. In 1865 the Bädische Anilin- und Sodafabrik was founded, followed by that of Gebr. Giulini. The choice of site was dictated by the existence of a port less intensively used than Mannheim and the relative cheapness of land west of the river. Thereafter the growth of the port of Ludwigshafen paralleled closely that of Mannheim. In volume of traffic they were together second only to the Rhine-Ruhr ports.

The rectification of the upper river had paradoxically the effect of making navigation – upstream at least – more difficult. A port was, however, developed at Karlsruhe, but never achieved the importance of Mannheim-Ludwigshafen. It made use of an old meander of the Rhine in which first the docks at Maxau and then the larger basins at Karlsruhe-Mühlburg were built. Like the closely comparable docks at Ludwigshafen they attracted a range of manufactures, prominent amongst them chemicals.

Strasbourg suffered severely from the increasing speed of the current, and regular commercial traffic ended in 1853. It did not revive until the 1890s, by which time tugboats powerful enough to breast the current with a train of barges had been developed. Docks were constructed in a cut-off on the Strasbourg side of the river, and from 1900 even larger basins at Kehl on the eastern or Baden bank. There was a rapid growth of trade in both. That of Strasbourg benefited from the opening of the canal between the Rhine and the Marne and of the Rhine-Rhône Canal, which, by way of the Belfort Gap, linked it with the Saône.

The port of Basel was idle for a longer period than that of Strasbourg, and was not reopened to river traffic until 1904. The old problems, however, remained – shallow water in winter and the swift current in summer. Traffic, effectively restricted to the summer months, remained small until after the First World War.

The traffic on the Rhine owed much to its tributaries and canals to west and east and, above all, by the ocean ports of the Low Countries. In earlier centuries the shifting, silting waterways of the Rhine delta had discouraged the use of this section of the river. The chief ports were Amsterdam, Dordrecht and Antwerp, none of which enjoyed easy communications with the Rhine itself. Rotterdam, later to become the chief port of the Rhine mouth, was of little consequence until the later nineteenth century, and then only after much work to improve the waterway. Navigation was, furthermore, hindered by the obstructive policies of the Dutch. At first they insisted that upstream navigation from Dutch ports should be in Dutch ships and, after conceding in 1831 that all ships registered in Rhineland states might sail down to the sea, continued until 1863 to levy tolls.[60]

Right-bank affluents of the Rhine The more easily navigated of the Rhine's tributaries were from the east and together they opened up routeways into south Germany and the northern plain. The most important of them were the Neckar, Main and Lahn and the three rivers of the Ruhr industrial area: Lippe, Emscher and the Ruhr itself.

The importance of the Neckar lay in the fact that it provided a low-level route across the Black Forest and Odenwald between the plains of Bavaria and the Rhine. This contributed to the expansion of the port of Mannheim at the confluence of Neckar and Rhine. As early as 1782 the states of Bavaria and Württemberg had agreed to improve conditions on the river from Mannheim up to Cannstadt, close to Stuttgart. But the river could carry only small craft – the largest was of no more than 40 tonnes – and subsequently suffered from the competition of the railway. There was a revival with the adoption in the 1870s of steam-towing, and it was even proposed to cut a canal from the navigable Neckar, near Stuttgart, to the Danube.[61] Traffic nevertheless remained very small until the First World War.

The Main had a far greater potential than the Neckar. It was a larger and a longer river, and even before the nineteenth century was navigable to Nuremberg by small craft. Indeed, in 1836–45 the Ludwigskanal was cut from near Nuremberg to the Danube near Regensburg, but was too small to have permanent value. Traffic in the port of Frankfurt grew through the century. In 1883–6 the river was canalised from Mainz up to Frankfurt, a distance of 35 km, so that the largest Rhine barges could reach the city. In 1897–1900 a dock was built at Offenbach and plans were made to extend the canalised waterway to Aschaffenburg and beyond, but the vision of water-borne trade between the Rhineland and the Danube valley was not to be realised until late in the twentieth century.

The Lahn was important in the nineteenth century because it flowed through a region richly endowed with iron and other metalliferous ores.[62] Its traffic was made up largely of the products of the mines, and as their output declined in the later years of the century so the river faded into insignificance.

The rivers of the Ruhr area were the shortest and least navigable, but economically the most important of all tributaries of the Rhine. The Ruhr itself carried most of the coal exported until the 1860s (fig. 9.6), but by the end of the century shipping had almost ceased on the Ruhr. The decline of mining in the Ruhr valley was compensated by its expansion in the valley of the Emscher. The Emscher itself was too small to be of any navigational significance, and the problem of distributing coal and of bringing in iron ore to the more easterly part of the coalfield was one of increasing importance. The remedy was to construct a canal from Dortmund, in the east of the industrial belt, to the navigable river Ems, thus establishing a route to the port of Emden. This waterway was opened in 1899 and a decade later was supplemented by the Herne Canal, cut from Dortmund along the marshy Emscher valley to the dock complex at Ruhrort. The third river of the region, the Lippe, was navigable for small craft, but lay too far north to be useful for the coal and iron-ore traffic before 1914.

The Low Countries and northern France

No part of Europe offered greater opportunities for the development of navigable waterways or more serious obstacles to their use. The area was drained by the Scheldt, Meuse and Rhine, which shared a common delta and made it technically possible for vessels to pass from one to the other. Political boundaries had, however, been drawn without any reference to the requirements of commerce and navigation. Both the Scheldt and Meuse were cut by the Belgian-Dutch boundary, and the Rhine itself flowed from Germany across the Netherlands to the sea. At the beginning of the century there was a complex system of waterways in Flanders and in the western Netherlands (fig. 9.17). Most had been constructed in the course of land-drainage but the larger amongst them were regularly used for conveying freight and even passengers. In Flanders the tributaries of the Scheldt formed a basic network. The problem was to give it access to the sea and to link it with the Meuse. The former was achieved through the ports of Ghent and Antwerp (see pp. 474–8). The latter was more difficult. As industry developed in central Belgium the need grew for a water link with the port of Antwerp. The natural route by way of the river Meuse, the Zeeland archipelago and the Scheldt estuary was far from direct and, furthermore, lay mostly in Dutch territory.

The Dutch did nothing to facilitate access to Antwerp and attempted, rather, to draw traffic to their own port of Rotterdam. A struggle for the traffic of the industrial hinterland was waged upon the waterways of the Low Countries by the two countries. Before the separation of Belgium from the Netherlands in 1831 a canal – the Zuid-Willems Canal – had been cut within Dutch territory between Maastricht and 's Hertogenbosch. Its purpose was simple: to avoid the great bend of the Meuse at Venloo and put central Belgium in direct contact with the Hollandsch Diep and the sea. This was followed by another canal between the Meuse near Liège and the Scheldt, which kept to the Belgian side of the boundary. This waterway was completed in 1859 and the branch through Turnhout in 1866. The Netherlands at once complained that the abstraction of water from the Meuse to supply the Liège–Antwerp canal lowered the level and made navigation hazardous on the river. This, and numerous other disputes regarding the waterways, helped to embitter relations between the two countries during much of the century. The Dutch reply to Belgium was to improve navigation on the Maas (Meuse) and to cut the Wilhelmina Canal (1906) to entice traffic to the Dutch waterways.

In 1827–32 the Scheldt was joined, by way of Brussels, with the Meuse at Charleroi in order to facilitate the distribution of coal. At the same time there was great activity near the border of Belgium with France, aimed at linking the Scheldt system with the Oise and thus facilitating the movement of coal to Paris. Traffic on the Belgian canals continued to increase late in the century. A heavy barge traffic developed based on Antwerp, which served as a distribution centre for imported wheat. Much of this was sent on to the Rhineland, by way of the Zeeland waterways if quick delivery was needed, but through the French canal system to Strasbourg if there was no urgency.[63] It was cheaper, so it was said, to send it by this long and circuitous journey than to hold it in storage at Antwerp.

Within the Netherlands there was a network of waterways spanning all except the south-east of the country. The capacity of most of them was small and they were used chiefly for local farm and market traffic. A greater importance attached to the rivers and canals by which the Rhine and Meuse could be reached from the sea. For much of the century Amsterdam was the leading ocean port of the Rhineland, but its linkages with the Lek were described as 'narrow, very crooked, shallow'.[64] In 1892 this tortuous waterway was replaced by the Merwede Canal. Both Amsterdam and Rotterdam had been severely hindered, at a time of steadily increasing trade, by their poor connections with the sea. These were greatly improved by the construction of ship canals, details of which are postponed to the discussion of the ocean ports of north-west Europe (pp. 474–7).

The north European rivers

Four important rivers, the Weser, Elbe, Oder and Vistula, flow obliquely across the North European Plain to the North or Baltic Seas. All four were navigable and all had been used for many centuries. Furthermore they were linked one with another by marshy depressions formed during the closing phases of the Ice Age. These north European rivers shared a common regime. All tended to have low water in summer and long, hard winters when ice obstructed navigation. The number of days when on average they were open tended to diminish eastwards:[65]

Rhine	304
Elbe	307
Oder	284
Vistula	261

The size of river craft increased during the century as physical conditions were improved, but they were never as large as on the Rhine (table 9.1). River transport on the north German rivers was clearly far less economic than on the Rhine and suffered even more severely in competition with railways. In general, traffic tended to disappear from the smaller rivers, but to maintain and even increase its importance on the large and more easily navigated, like the Elbe.

The Ems, the most westerly of the north German rivers, was of little significance until the Dortmund–Ems Canal was built to link it with the Ruhr region. The Weser was a larger river, and was used regularly despite its physical shortcomings. It was, wrote William Jacob, 'very unfavourable, especially

Table 9.1. *Size of ship in regular use* (*in tonnes*)

	Rhine	Weser	Elbe	Oder	Vistula
1840	400	n.d.	150	75	n.d.
1880	800	300	600	150	150
1900	2,000	450	800	450	350

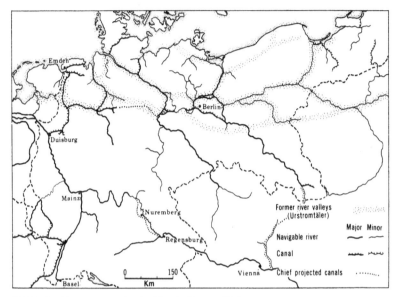

9.7 Navigable waterways in the North German Plain

during...summer. When I was [there] a vast number of loaded craft had been detained a long time, for want of sufficient depth of water to ascend the stream; and near two months later...at Münden...many barges were waiting for sufficient water to descend to Bremen.'[66] The growth of the port of Bremen ensured that the lower Weser remained well used but navigation gradually declined above Minden.

The Elbe has always been an important river for navigation. It was larger and less impeded than others in north Germany, and furthermore it transversed the industrial region of Saxony, was linked with Berlin and served the port of Hamburg. It carried a great variety of goods: timber and brown coal, grain, flax and salt from Saxony; colonial goods from the docks of Hamburg and manufactured goods from Berlin.[67] Traffic increased during the first half of the century, then declined between 1850 and 1870 as railways took over the business of moving light goods and foodstuffs. Then, after about 1870, it expanded with increasing shipments of fuel and industrial raw materials.

The Bohemian section of the Elbe (Labe) and its tributary, the Vltava, were less useful, and for the larger river craft Ústí, near the German boundary, was an effective limit. Only small craft were able to ascend to Prague. Of the Elbe tributaries the most important were the Saale, which flowed through the brown coal and potash-producing areas of Saxony, and the Havel and Spree, which linked Berlin with the Elbe. These rivers were supplemented at an early date by a network of canals as dense as that to be found in northern France. Most had been built before the railway age; they were small and in the second half of the century were falling into disuse. A debate developed on the 'canal question', whether to rebuild them to contemporary standards or to abandon them entirely.[68] In the end it was decided to modernise those canals that were likely to carry most low-value freight, and to extend the system where necessary.[69] Little was done for the many short canals in the Baltic provinces, which had been built primarily to assist the export of grain. Their usefulness was ended. But the network which linked the Elbe with the Oder by way of Berlin (fig. 9.7) was modernised and extended, and came to play an important role in the industrial development of the capital.[70]

Late in the nineteenth century the plan, first mooted by Napoleon,[71] to cut a canal from the Rhine eastwards to the Elbe and Oder was revived.[72] The Dortmund–Ems Canal was the first segment of it to be completed. It was opened in 1896, and its extension westwards to the Rhine, the Herne Canal, was opened in 1914. By this time, also, the waterways between the Elbe and the Oder had been improved; the Havel and Spree had been canalised; the Oder–Spree and Finow Canals rebuilt, and the Teltow Canal constructed (1901–6) to circumvent Berlin. It only remained to build a canal from the Ems eastward to the Elbe. The section friom the Ems to the Weser was authorised in 1905, but the link between the Weser and the Elbe was not commenced until after the First World War.

The effect of this canal-building programme was to link the Ruhr directly with the North Sea at Emden and the Berlin district with both the Elbe and the Oder.[73] The Oder, however, was a river of very limited value, 'an unsatisfactory stream for navigation', wrote the British consul, 'in spite of all the improvements that have been made'.[74] The channel was shallow and shifting and could be used only by small barges. The chief cargo handled was Upper Silesian coal destined for Berlin, but even this travelled more often by rail. In addition a small amount of grain and other agricultural products went by river.[75]

East of the Oder, inland navigation was of little economic value. Timber was floated down the Warthe (Warta) from Russia; some grain travelled from Poland to Berlin. But nothing was done in Russian Poland to improve, or even maintain, the waterways, and there was, in any case, no bulky freight of low value, apart from timber, for them to handle.[76]

The Danube

The Danube was a longer and incomparably more complex river than any other in Europe. Its regime varied throughout its course and it was, in its natural state, obstructed by rapids and shallows. In its upper course, above Linz, the current was swift during the whole summer season of snow-melt. Across the Hungarian Plain its meandering course was strewn with shifting sandbanks, and, wrote Quin, 'our boat rubbed upon 'he natural bed of the river'.[77] Then, where the river left the plain, there lay the biggest single obstacle in the river's turbulent course, the Iron Gate. This was merely the most dangerous part of a particularly difficult section of the river. Here the Danube had cut a gorge, 120 km long, across the Carpathian Mountains. Its course was constricted and the speed of the current was increased. At the lower end of the gorge a rocky bar gave rise to rapids, 'an insuperable obstacle to the enormous flat-bottomed barges'.[78] If these difficulties, wrote Urquhart, 'were removed...the whole of the East would communicate' with Germany.[79] Undoubtedly the Iron Gate, in the minds of merchants, presented an absolute barrier to transport, and turned the trade of Walachia and northern Bulgaria eastwards to the Black Sea, and that of Hungary towards central Europe.

The lower river was not entirely free of obstacles. While there was no great difficulty in navigating the main channel, its northern or Romanian bank was fringed with marshes which hindered access to its wheat-growing hinterland. Below Galaţi the river entered its delta, splitting up into numerous and mostly unnavigable channels. Of these only three could be used: the Kilia Channel in the north, the Sulina which flowed across the middle of the delta and the St George's Channel in the south. All channels were shallow, and shifting sands made navigation hazardous. The Kilia Channel was the most easily navigated, but its mouth was obstructed by a bar on which many vessels had been wrecked. For this reason ships tended to use the Sulina Channel (fig. 9.8)[80]

The Final Act of the Treaty of Vienna of 1815 included the Danube amongst international rivers, and prescribed the setting up of an international commission to assist and supervise navigation. Little, however, was done. There were no tolls, because commerce was too small to justify them. The Habsburgs evaded their responsibilities on the upper river, and below the Iron Gate the matter lay between Russia and the Ottoman Empire, which showed no greater desire to improve conditions. By the Treaty of Adrianople of 1829 the St George's Channel was declared to be open to the shipping of both countries. The lands bordering the lower Danube had long served as the granary of Istanbul, but the grain was carried in very small ships, able to navigate the shallow waterways of the delta. In the 1830s, however, the grain trade expanded, especially from Walachia, and Western European ships sailed by way of the Turkish Straits to the Danube in increasing numbers (see p. 29).[81] But there were both legal and practical difficulties. Freedom to use the river, announced at Vienna, had not been recognised locally[82] and absolutely nothing had been done to safeguard shipping from the natural hazards of the delta. The masts of wrecked ships, standing above the water, were said to be the only available navigational aid.[83] The Kilia Channel, under Russian control, was allowed to silt up to the point at which it could no longer be used, and it is probable that the Russians were intent to divert traffic from the Danube ports to their own port of Odessa.[84] It was even proposed in the west to circumvent the delta altogether by cutting a canal from the Danube across the Dobrogea to the Black Sea coast.[85] In the mean time, the growing traffic concentrated on the Sulina Channel, which, in the 1840s, allowed ships of 300 tonnes to ascend the river to Galaţi.[86]

The first opportunity to impose an international regime on the maritime Danube arose in 1856 with the end of the Crimean War. The deteriorating

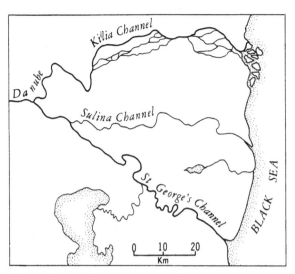

9.8 The Danube delta and its navigable channels

waterway, combined with the growing volume of shipping using it, at last compelled the Powers to act. A commission with extensive powers was established for the maritime Danube, defined as the river below Galaţi and its deltaic branches. In 1883 the jurisdiction of the Commission was extended upstream to Brăila. Work began almost at once on the Sulina Channel. Dykes were constructed along its course in order to narrow the channel and increase the scour. In this way the channel was deepened to 15 feet. The bar at its mouth was kept clear, and steps were taken to establish a system of pilotage.[87] These improvements led to a greater use of the river. In 1856 2,240 ships visited the delta, with a combined tonnage of 340,810 tonnes; by 1907, the number of ships had fallen to 1,258 but their total tonnage had risen to 2,200,000 tonnes.[88]

The fluvial Danube, which extended upstream from Galaţi to Ulm in Bavaria, had no international commission before 1921, since the Austrian government managed to evade the conditions of both the Vienna Final Act and the Treaty of Paris of 1856.[89] Yet no river called as desperately for some form of international regulation. It was, in fact, managed by Austria-Hungary in the interests of Hungarian wheat-growers and Austrian shipping companies. It was necessary only to hinder improvements at the Iron Gate in order to maintain the fluvial and the maritime sections of the Danube as two independent systems. It even became the practice for agricultural exports from the Hungarian Plain to move down the Danube to Beograd, then up the Sava and overland to Rijeka. 'While the difficulties of the rapids at the Iron Gates remain', wrote an anonymous British commentator, 'little or no produce will be brought from above that point.'[90]

The barrier of the Iron Gate was not, however, absolute. Small vessels occasionally passed the rapids downstream, and after 1834 a steamer occasionally made the passage.[91] A problem, however, was the supply of fuel, and not until the opening up of the small coalfield near Orsova was steam navigation practicable below the Iron Gate.[92] Attempts were made to create a deep channel within the Iron Gate, but progress was very slow. The Paris Treaty of 1856 called for the creation of a navigable waterway, as did the Treaty of Berlin of 1878. But it was not until 1883–98 that a canal was constructed within the river's course, and first a tugboat and then a railway locomotive was used to pull ships against the current.[93]

Difficulties within the Hungarian Plain were of a different order. Both the Danube and its tributaries meandered across the plain, shifting their courses with each flood, dividing around marshy islands and then reuniting. Much of the work on these rivers aimed to prevent flooding by straightening their courses. Work began on the Tisza and Sava in the eighteenth century, and was continued in the nineteenth. The Bega Canal was cut from the wheat-growing Banat to the Danube in order to bring grain to Budapest.[94] The Balkan tributaries of the Sava were of little use for navigation, but served to float lumber down to the almost treeless plain.[95]

Above Budapest traffic on the Danube increased slowly. The current was so swift in summer that only a steamboat could make headway against it, and in winter ice obstructed navigation. The first steamboat on the Danube, in 1830, ran a passenger service between Vienna and Budapest. Seven years later a service opened between Vienna and Linz and was later extended to Regensburg and then Ulm. Traffic was never heavy on the upper river.[96] In 1845 iron for the Szechenyi bridge in Budapest was brought from England by way of Rotterdam, the Rhine, Main and Ludwigskanal, but there were few freights like this. Indeed, without a waterway capable of taking large barges between the Rhine and the Danube, there was little demand for transport on the upper river.[97]

There was little scope for water-borne transport in southern Europe. Most rivers were short and steep, torrents rather than rivers, and, with only a few exceptions, water levels in summer were too low for navigation. Amongst these exceptions were the major rivers of Spain and Portugal, the Duero (Douro), Tagus, Quadalquivir and Ebro. All were used for fairly short distances in their lower or plain tracts, and the Duero was of some importance for the transport of wine to Oporto. In the late eighteenth century an ambitious programme of canal construction was initiated, but left unfinished. In 1840 only two short sections of canal, in the Ebro valley near Zarazoza and beside the Duero in Old Castile, were in use,[98] and their economic significance was negligible.

In Italy only the river Po and its tributaries were of any practical importance, but navigation was difficult and river tolls were onerous. Such canals as were built, like the Cavour Canal between the Dora Baltea and the Ticino, were intended primarily for irrigation, though they were also used by small craft.[99]

Development of the railway system

The building of a primary railway net in western and central Europe was accomplished in a remarkably short period of time. The first lines with steam-drawn transport were constructed in Belgium and south Germany in 1835, but had been preceded by a number of horse-drawn tramways, used mainly for the movement of coal. Almost at once plans were formulated, either by the

Table 9.2. *Railway development*

	Kilometres of railway per 100 sq. km		Kilometres of railway per 100,000 of population	
	1850	1896–7	1850	1880
Austria, with Bohemia	0.5	5.8	7.8	70.0
Belgium	2.8	15	19.4	83.2
France	0.6	7.6	7.6	106.0
Germany	1.1	8.8	16.6	91.0
Great Britain	3.4	10.8	39.3	86.0
Hungary	0.1	1.7	4.8	86.6
Italy	0.2	5.6	2.5	32.7
Netherlands	0.5	8.1	5.8	114.4
Polish Kingdom	0	2.9	0	29.3
Spain	0.006	2.6	0.2	75.9
Sweden	0	2.4	0	229.9
Switzerland	0.6	8.9	1.0	130.0

government or by private individuals, in Belgium, France and Germany. In Belgium it was proposed as early as 1834 to build two lines running approximately from Ostend to Liège and the German boundary and from Antwerp southwards by way of Brussels to the French border near Maubeuge, intersecting at Malines. This basic network was in fact completed by 1842.[100]

In France the legislature provided in 1833 for the building of a 'railway system', consisting of lines from Paris to the chief provincial cities.[101] Progress was, however, a great deal slower than in Belgium, and the greater part of this net was not built until after 1850. No government in Germany was in a position to formulate a general plan for railway development, but between 1833 and 1837 Friedrich List campaigned vigorously in books and pamphlets for an integrated system for the whole of Germany, and sketched a series of seven lines which he thought adequate to span the whole country.[102] His plan was politically unacceptable, and the German system began as a series of short disconnected sections, authorised by the individual states. Indeed, there was a wasteful competition between railway promoters, so that few lines became profitable or really served the public needs.[103] Nevertheless, it is not a little surprising that the immense potential of an integrated system was realised so early.

Important as the increase in the gross mileage of railways was during the nineteenth century, it is the ratio of the length of line to the area and population of each country which gives a truer measure of their development. Table 9.2 presents these data for the years 1850 and 1880.

By 1860 a primary network, linking the chief centres of population and industry in western and central Europe had been completed (fig. 9.10). Progress was a great deal slower in Spain and Italy, where there were both political and economic reasons for the delay. In eastern and south-eastern Europe, again largely for political reasons, no progress was made before the last third of the century. Railway-building in continental Europe benefited greatly from the British example, and in its initial stages used equipment imported from Great Britain. This had the effect of imposing on the whole of Europe, with the

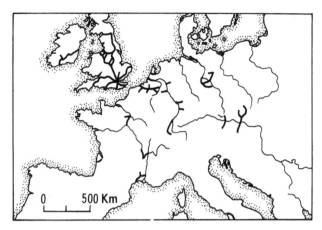

9.9 Railway development in Europe, 1840

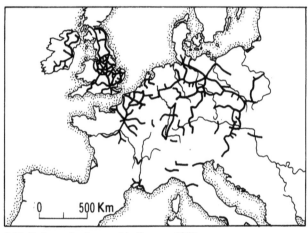

9.10 Railway development in Europe, 1850

exception of Russia, the standard gauge of 4 feet $8\frac{1}{2}$ inches which prevailed in much of Britain. The few attempts to use other, and perhaps more rational, gauges were abandoned, and it thus became practicable technically if not also politically to run trains from one end of the continent to the other at a very early date.

By 1840 1,481 km of track had been completed and were in use in continental Europe, excluding Russia (fig. 9.9). Ten years later the total had increased to 12,132 km (fig. 9.10), and by 1860 to 33,405 km. Thereafter the expansion of the network was even more rapid, as the primary network came to be filled in with local and linking lines (fig. 9.11). The *rate* of growth was greatest between 1860 and 1880. Growth came earliest and ceased first in north-western Europe. In Belgium for example, there was little expansion after 1880, and in France the net was virtually completed in the 1890s. In Germany, Switzerland, Sweden and the Austro-Hungarian Empire there was little railway building after about 1900, when construction was still in full swing in the Russian Empire and the Balkans.

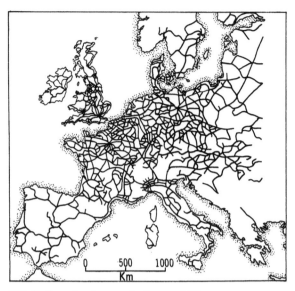

9.11 Railway development in Europe, 1880

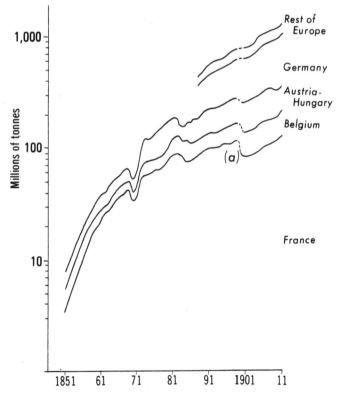

9.12 Growth in the use of railways for freight, 1851–1911; 'a' indicates a change in the method of calculating the volume of freight

In retrospect the primary reason for building railways appears to have been commercial, to carry freight for which the roads and waterways were either unsuitable or inconvenient. Yet there were other, and to contemporaries more important, reasons. Passengers, not freight, figured most prominently on the earliest railways, and the purpose of some suburban lines was, so it was claimed, to allow town-dwellers to enjoy the delights of the countryside. Only gradually did the conveyance of freight assert itself as the primary function of railways (fig. 9.12). Even so, early railways were seen as subordinate to waterways, traditionally the chief movers of heavy freight. The first railways in west Germany were, for example, built as feeders to the river Rhine.

Building a railway was a political act. Some form of legislative enactment was necessary before land could be acquired and track laid. In some European countries the government itself operated the railways from the start, and in others, took into public ownership lines which had been built privately.[104] Some government-built lines were never expected to be profitable commercially. They had been built for social or, more often, military purposes, and their cost could in part be set against the social net profit which resulted. Military considerations are said to have been paramount when Bismarck acquired for the state rights over the many private and local lines in Germany.[105] Within Russia the railway net was conceived in terms of troop movements, and the location of stations, far from the centres of towns, was designed to protect them from the urban mob. In France, despite the existence of a national plan, there were in 1846 no fewer than 33 private railway companies.[106] This number was reduced by amalgamations and eventually the companies were bought out by the state under Napoleon III. The lines were then 'organically grouped in a few coherent units, and ever since, rail transport has dominated French communications'.[107] Beginning in 1879, the Freycinet Plan completed the rationalisation of the French transport system. In Italy, early railway development was fragmented, like the country itself. After unification in 1861 a unified system became an urgent necessity, and 'huge sums [were] spent in the construction of lines which were never expected to pay, but the existence of which was a political necessity, and without which the unification of the country could never have been accomplished'.[108] It was in eastern Europe and the Balkans that railways assumed the greatest political importance. Even the gauge of the track became a political issue. The Russian lines were built on a gauge of 7 feet, not because this was more economical, which it probably was, but so that locomotives and wagons from the west might not run over them. It was only with the greatest reluctance that the Tsar permitted the line which had been built from Vienna to the Polish border, the Ferdinands Nordbahn, to be continued to Warsaw on the western gauge. All other lines in Russian-occupied Poland were built on the broader gauge.

National networks

The maps on pp. 451–2 show the development of the railway network during the first half-century of the railway age, during which its total length, excluding Russia, increased from a few horse-drawn tramways to some 115,000 km. It is impossible to represent the growing complexity of the system after 1880 on a single map. The European net was made up of a number of national systems each responding to national needs and ambitions, and only gradually, and in some instances reluctantly, coalescing into a system of continent-wide proportions.

France The railway age in France was preceded by the construction of a number of tramways over which horses pulled coal-wagons from the mines to the nearest navigable waterway. The longest and most important of these lines were on the coalfields of Burgundy and Saint-Étienne.[109] The first line using mechanical traction was built in 1837, from Paris to Saint-Germain. It was the signal for a logical and ambitious plan of railway construction. Unfortunately the implementation of the plan was left to private promoters, each of whom was interested in only a relatively short segment of line.[110] By 1846 there were no fewer than 33 separate railway companies, over which the state exercised only a remote and ineffectual control.[111] This is reflected in the railway map for 1850, where no fewer than seven discrete systems are shown, only one of which, that which radiated from Paris, could have had any great economic significance.

The great railway-building period in France was under the Second Empire. Not only was the network increased from 2,915 km to 15,544 between 1850 and 1870, but the several segments were integrated into a system.[112] The private lines were bought up by the state and organised into six *grands réseaux*, a process which somewhat resembled the British reorganisation of 1923. Trunk lines from Paris to the major provincial cities were completed. When the Second Republic began there were still many gaps in the network, and large areas, such as the Pyrenees, the French Alps and the Central Massif were still virtually without railways. A new campaign, initiated by Freycinet in 1879, almost doubled the length of railway by 1910 and led to the building of secondary and branch lines between the trunk routes.[113]

The Low Countries Belgium was as precocious in its railway development as in other branches of its economy. A primary network which spanned the northern plain was projected in 1834 and completed within ten years.[114] After this early burst of activity railway-building languished. The state permitted private enterprise to take over the work, and a confusing state of affairs emerged with two conflicting systems. After 1870 the state again asserted itself, acquired possession of most private lines, and extended and by the early 1880s had almost completed the network. The resulting system was dense to the north of the Meuse, sparse to the south. It was, indeed, not until 1859 that a line – the Guilleaume–Luxembourg railway – was built across the Ardennes.[115]

Railway-building was as late in the Netherlands as it was early in Belgium.[116] The reason lay in the availability of navigable waterways and of the organisation of water-borne traffic. A railway linking Amsterdam with Haarlem was opened in 1839, and was later extended to The Hague and Rotterdam in the one direction and Utrecht and the German border in the other.[117] But here the matter rested until the 1860s, when the government belatedly sponsored railway-building in the rest of the country. By 1890 a loose net was completed.[118]

The slowness with which the Dutch developed their railway network had a

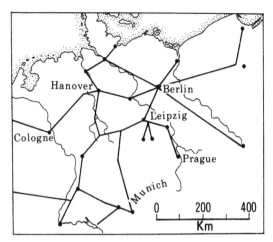

9.13 Friedrich List's plan for a unified German railway system

curious influence on German developments. Industrial growth in the Rhineland was dependent on access to the sea. Although heavy freight generally moved by water, the need nevertheless arose for a railway linkage with the ocean ports. Amongst the early German railway projects were those designed to circumvent the Netherlands by joining the German Rhineland with Bremen and Hamburg and with the Belgian port of Antwerp. Indeed, the Cologne–Liège line was the first international connection to be established in Europe.

Germany In Germany, even before the first rails had been laid, Friedrich List had published his plan for a railway network (fig. 9.13).[119] It was remarkably prophetic or, perhaps, self-fulfilling. The first railway to be built, however, was the rather comic *Ludwigsbahn*, between Nuremberg and Fürth, but this was quickly followed, in 1839, by more significant lines between Leipzig and Dresden; Berlin and Potsdam, and Brunswick and Wolfenbüttel. These became the nuclei of the railway system of their respective states, and by 1845 the three had been linked together and extended to the Baltic at Stettin. German railways proved to be remarkably profitable. During the middle years of the century manufacturing, especially of iron and steel, was developing rapidly. There was a rising demand for transport, much of it in areas ill-provided with navigable waterways. The Zollverein encouraged the building of lines across the boundaries between the German states; above all, the widely scattered lands of the Prussian State called for railways to link them together. By 1850 the eastern provinces of Prussia had been joined, through Hanover, with the Rhineland and Belgium. A few years later this northern system was linked with the lines already built in Bavaria and Württemberg and along the Rhine valley to the Swiss border. The essential network was complete before the unification of the German Empire in 1871,[120] and subsequent building consisted largely of doubling the track and filling in the interstices in the system.

The early completion of an integrated railway system was of incalculable importance for the economic development of Germany. It completed, in the words of Treitschke, what the Zollverein had merely begun. It permitted manufacturing to be more widely dispersed than, for example, in England, where its concentration close to the coalfields had antedated the building of the railways. Nevertheless, the network as it took shape by the end of the century was particularly dense in three areas: the Ruhr, the middle Rhineland and Württemberg, and Saxony, reflecting the higher level of industrial development in these regions.

Switzerland and the Habsburg lands The first railway within the Habsburg lands was a horse tramway from Linz on the Danube to the Elbe system at Budejovice in Bohemia. It was built in 1822–32 primarily to convey salt from the Salzkammergut; it was of little commercial importance and, in fact, retained its horse traction until 1854.[121] In the mean time, in 1835, it was proposed to construct a line from Vienna through Olomouc to the Moravian coalfield, with branches to Brno, Prague and the Saxon border. In the opposite direction it was designed to build a line from Vienna through Graz to the Adriatic port of Trieste, but engineering problems delayed the completion of the latter until the 1860s.[122] Meanwhile the first Hungarian line had been built in 1846, and by the mid-1850s it was possible to travel by rail from Vienna to Budapest and thence across the Great Alföld to Szeged. This simple net clearly delimited what the Habsburgs regarded as the core of their Empire.[123] Lines were built probing towards the Balkans. The Moravian line was linked with the Russian line to Warsaw; the Trieste line was pushing through the mountains of Slovenia ever closer to the Adriatic, but of any deliberate attempt to link with the German railway system there was not a sign.[124]

During the last quarter of the nineteenth century there was intense activity within the Habsburg lands. Vienna and Budapest each became the focus of railway routes, and Prague, at one time the terminus of a line from Vienna, became the centre of a Bohemian network. In the course of this development the rivalry of Austria and Hungary within the dual monarchy came to the fore. After the *Ausgleich* of 1867 each controlled railway development within its own territory. Austria impeded the export of Hungarian agricultural produce to Germany; Hungary obstructed Austria's access to the Dalmatian coast, and Austria retaliated by holding up Hungary's request for a line across Moravia to Prussian Silesia and Berlin.[125]

The intense rivalry between river and railway, such as developed in western Germany, was never apparent in the Austro-Hungarian lands, chiefly because the opportunities for inland navigation were severely limited. Greater use could certainly have been made of the Danube, but this would have called for closer collaboration between Austria and Hungary on the one hand and Germany on the other than was in fact possible.

The Alps and their continuation, the Dinaric Mountains of Dalmatia, presented a formidable obstacle to railway construction, and in Switzerland this was intensified by the conflicting policies of the cantons.[126] In 1848 the Federal government took over authority for railway building, but construction and operation were, until 1898, left to a number of small private companies. Until 1870 there was little more than an open network on the Swiss Plateau between

Lake Constance and Lake Geneva, with a few branch lines thrusting into the mountains. This was intensified by 1900 but there was still only one crossing of the Alpine range, the Gotthardbahn. The first Alpine tunnel had been the so-called Mont-Cenis, completed in 1871 between Modane in the French Alps and the Dora Riparia in Piedmont. At the same time a railway was built up the Reuss valley from Lake Lucerne towards the Saint Gotthard massif. It was a more elaborate and costly piece of railway engineering than had hitherto been attempted in Europe, and it was not until 1882 that it was opened. Not only did it give access from the Rhineland to the plains of Italy; it also joined the Italian-speaking canton of Ticino, whose tenuous link had hitherto been the road over the pass, to the rest of the Confederation. Twenty years later the Rhône Valley line was continued through the Simplon tunnel, opened in 1905, to Italy.

The Austrian government was no less eager than the Swiss to build a railway across the Alps to the Italian plains, where lay its provinces of Venezia and Lombardy. The line across the Brenner Pass was not built until 1864–7, by which time the Habsburgs had lost their Italian possessions with the exception of South Tyrol. A line was subsequently built from the Brenner route eastwards by way of the Drau (Drava) to Klagenfurt and the Hungarian Plain.

Poland It is curious that Poland's earliest external links by rail should have been with Austria rather than Germany. Even more strange is the fact that this was because the Tsar did not wish to become dependent on the German Baltic ports, preferring to channel Russian trade by way of the Ferdinands-Nordbahn to Vienna and thence to the Adriatic. Economic nonsense though this was, Tsar Nicholas nevertheless signed a decree in 1839 authorising the construction of the line from Warsaw by way of Częstochowa to the Austrian boundary where it joined the Nordbahn. Furthermore, he conceded that it might be built on the western or standard gauge. Work on the line was entrusted to a Polish company, and after delays due to financial difficulties, it was completed in 1848.[127] The line from Warsaw to St Petersburg by way of Vilna was begun in 1862, and that to Moscow soon afterwards. The only other railways built during these years were short lines in Dąbrowa and the line which followed the Vistula valley to Toruń (Thorn) where it was later joined with the Berlin railway. Łódź was not linked with this rudimentary system until 1866.[128] A number of short lines were built later in the century, and financed by the Banka Polski, but the Polish Kingdom had still in 1914 one of the lowest densities of railway development to be found in Europe.

The Balkans The last parts of Europe to be penetrated by the railway were the Balkans and Scandinavia. Both lay on the economic frontier of nineteenth-century Europe, but whereas railway construction in Scandinavia proceeded peacefully and methodically, in line with the economic needs of the region, that in the Balkans became a pawn in the imperial designs and national ambitions of south-eastern Europe. After the mid-nineteenth century Hungarian railways began to reach out towards the boundary of the Empire, into eastern Galicia, Transylvania and Croatia. Beyond lay the Romanian principalities, Serbia and the territory of the Ottoman Empire. But until 1860 not a mile of railway had been built south of the Sava and Danube. In that year a short line was opened across the base of the Dobrudja from Cernavoda to Constanţa. Six years later a parallel line was built from Ruse on the Bulgarian bank of the Danube to the Black Sea port of Varna.[129] These were the first railways to be built within the Ottoman Empire, and their purpose was to allow the agricultural produce of Walachia and northern Bulgaria to be exported by way of the Black Sea and Turkish Straits.

In the mean time the Sultan had in 1855 proposed the construction of a line from Constantinople to Beograd by way of Sofia. Nothing came of the proposal for fourteen years. Then, in 1869 he granted a concession to Baron Hirsch, a Bavarian financier, to build a railway from the Bosporus to link up with the Austrian system. The route assigned to Hirsch was a curious one. It was to follow the Marica valley to Sofia, then to strike westwards to Priština, and so to the north-west, through Novi Pazar, to Sarajevo and the river Sava. The purpose was to avoid crossing Serbian teritory and to keep the line wholly within the Ottoman Empire until it reached the Hungarian border. This was, indeed, one of the problems in planning a railway in the Balkans; 'each state wanted to control the lines within its territory and each had plans for connections and new through lines which would serve commerce and make armies mobile'.[130] Baron Hirsch made very slow progress with his railway. He was paid by the kilometre by the Turkish authorities, and allegedly made his route exceptionally circuitous, avoiding expensive cuttings and embankments. From Constantinople he reached Sarembey, between Plovdiv and Sofia, and then gave up his contract 'at the point where real difficulties began'.[131] At this point the construction of the Orient Railway, as we may now call it, was interrupted by the Balkan War of 1875–6. The Ottoman Empire emerged humiliated. Serbia acquired territory which lay in the path of Hirsch's railway, and the Habsburgs occupied Bosnia and Hercegovina.

There was no longer any question of building the railway on Ottoman territory, and no reason for not taking the obvious and from an engineering point of view the most practicable route from Niš along the Morava valley to Beograd. The railway company was reconstituted under Austrian rather than Turkish sponsorship. By 1881 the route was completed across Bulgaria, and shortly afterwards from the Bulgarian border to the Danube at Beograd where it linked with the Hungarian system. In 1888 the first 'Orient Express' ran from Vienna to Constantinople.

In the mean time, however, Baron Hirsch's original route through the Sanjak of Novi Pazar had not been forgotten. Amongst the earlier Turkish projects was a route from Thessaloníka up the Vardar valley to Skoplje (Üsküb). This was completed in the early 1870s, and was soon afterwards continued to Mitrovica in the Kosovo basin. At the same time the northern section of Baron Hirsch's railway, from the Hungarian border to Banjaluka and Sarajevo was built, and, despite immense engineering problems, was continued to the northern border of the Sanjak.[132] These lines were built on a narrow gauge of 2 feet 6 inches, and their commercial value was minimal. A distance of about 160 km separated the railhead at Uvac in Bosnia from the terminus of the Thessaloníka line at Mitrovica. To the north-east lay Serbia, ambitious to expand into the Sanjak, and thence through Montenegro to the sea.[133] To the north-west lay Bosnia, occupied jointly by Austria and Hungary, waiting for the opportunity to extend through the Sanjak to Macedonia and thus to stifle the ambitions of Serbia. The completion of the Sarajevo–Mitrovica railway was part of Austria-Hungary's plan, and on it hinged much of Balkan diplomacy during the first decade of the twentieth century.[134] The line was never built, and its construction, at least by the Austro-Hungarians, would have been precluded by the Balkan War of 1912 and the resulting partition of the Sànjak between Serbia and Montenegro.

The simple rail net within the Ottoman Empire was completed by linking the Thessaloníka line at Skoplje with the main 'Orient' line at Niš and with Constantinople by way of the plain of Thrace. A line was also laid from Thessaloníka through the mountains to Bitolj. It was never pretended that these lines had any significant commercial value. They were overtly strategic. The Thracian line kept sufficiently far inland to be immune at that time from naval bombardment, and loop lines were built so that traffic might, if necessary, avoid the coastal ports.

Austria-Hungary was interested before 1914 not only in preventing Serbia from gaining access to the Adriatic Sea, but also in improving her own routes from the Danube basin to the Mediterranean. The obstacles were formidable, a high plateau of karst limestone dissected by gorge-like valleys. There were no natural routeways across this region, and it was a cultural as well as commercial divide. The Dinaric System could be crossed with the least difficulty in the north-west where it was narrowest, but the Vienna–Trieste line was not completed until 1857, and the more difficult line from Zagreb to Fiume (Rijeka) not until 1873. In 1875 work began on a line from Metković on the coast, up the Naretva valley and over the mountains to Sarajevo. It was not completed until 1890, and required the assistance of a curious rack-and-pinion device to assist trains in ascending to the karst plateau. The inadequacy of Metković as a port led soon afterwards to the building of a line parallel with the coast to Gruž, the port of Dubrovnik, and on to the Gulf of Kotor in Montenegro. A line farther to the south-east linking Beograd directly with the southern Adriatic coast was a Serbian dream in the 1890s, and so it remains.[135]

The railway net in Greece was, until after the First World War, independent of that in the rest of the Balkan Peninsula. The first line was the short one, built in 1867–9, between Athens and Piraieus. It was not until the 1880s, following the acquisition of Thessaly from the Ottoman Empire, that the railway system was extended. By the early twentieth century it consisted of a line from the south-western Peloponnesos to Athens, and another from Athens northwards to Thessaly. This, with a few branch lines, made up the Greek rail network.

Late nineteenth-century railway development in the Balkans is generally – and probably correctly – regarded as an example of the economic imperialism of Germany and her Austro-Hungarian ally. It was German capital that financed the Orient Railway, and it was an agreement between Germany and the Ottoman Empire in 1903 which provided for the continuation of the railway across Asia Minor. In 1912 the so-called Baghdad Railway reached the Euphrates.[136] The purpose of German – and also Austrian – participation in the enterprise was political. Its construction was a means, at first, of preventing expansion by

Serbia, backed by Russia, and later of holding Turkey fast in the web of German alliances.

The economic significance of the long-distance railways in the Balkans and Middle East remained negligible up to the First World War. The Orient Railway was opened in 1888, but 'a decade later [it] still carried little more than express mail and passengers; the staples of trade moved by sea'.[137] German commercial interests in the Balkans were small. Between 1891 and 1913 less than 2 per cent of Germany's foreign trade was with Romania, and this was carried mainly by the Hungarian railways. Trade with the rest of the Balkans, excluding territory under Austro-Hungarian rule, never amounted to as much as 1 per cent of the total, quite insufficient in itself to justify Germany's railway-building activities.

The Italian and Iberian peninsulas Both Italy and Spain were slow to develop their railway networks, the one on account of its political fragmentation, the other because of its slow economic growth. By 1850 there were only a few short segments of line (fig. 9.10) in the whole of Italy, and in Spain there were none. The first serious attempts to develop a network in Italy were made in the 1850s in Piedmont and in the Austrian-held provinces of Lombardy and Venezia.[138] After the political unification of the country in 1860, the integration of these scattered networks into a national system became an urgent necessity. Southern Italy was almost cut off from the more developed north. Road transport was bad, and travellers were obliged to take a ship even for short journeys like that from Genoa to Leghorn and Tuscany.[139] After about 1861 railway-building began in earnest, much of it by private enterprise, but with heavy governmental subventions. Between 1870 and 1910 the length of the network was increased threefold, and there was a period when the *rate* of construction was one of the fastest in Europe. By the end of the century there was a dense network in the northern plain, where the economic need was greatest, and as full a development in the peninsula as the terrain permitted. Needless to say, lines built primarily for political reasons were not fully used and had to be subsidised.[140] This was the price of unification in a land such as Italy.

In Spain a short line from Barcelona to Mataró was opened in 1848, but another decade passed before real progress began to be made. Railway building then proceeded in fits and starts. A large number of private companies was formed to build and operate relatively short railways. In the course of time many of these amalgamated and their respective tracks were knit together.[141] This, combined with the desire to pick up as much freight as possible from the small towns, explains the circuitous routes followed by some of the lines. There was, for example, no direct line as late as 1914 from Madrid to the most important Pyrenean border crossing at Irun. Nevertheless the railway net which emerged from this chaotic development showed a pattern of lines radiating from Madrid, with a somewhat denser network in Catalonia and Valencia and in Andalusia, where a number of branch lines had been built to tap the mineral wealth of the region.

The Portuguese railway system was built later than the Spanish, and consisted essentially of a line which followed the coast, with branches which penetrated the interior to link up with the Spanish lines. That the whole was never conceived as a single system is apparent from the discrepancies in gauge. Spain adopted a broad gauge of 5 feet 6 inches, but also built a number of standard- and narrow-gauge lines. Portugal also adopted a broad gauge, but made it 5 feet $5\frac{1}{2}$ inches, as well as standard and narrow gauges. In no part of Europe was the pattern of gauges more confusing or the difficulty of running through trains greater.

Scandinavia Railways came late to northern Europe as they did to southern Europe. But whereas there were political reasons for the delay in the latter, Scandinavia was late in developing any need for railways. Only in Denmark was there any development by 1850, and much of this was in the Duchies ceded to Prussia in 1864. The first railway in Denmark proper was that from Copenhagen westward across the island of Sjoelland to Korsör. This was followed by a line across Fyn and one through southern Jutland to Slesvig. By 1880 a broad network enveloped the whole country. Lines of particular importance converged on the new port of Esbjerg, which was of growing importance for the export of agricultural produce.

A railway system for Sweden was first planned in 1856-8, and by 1880 a primary network had been completed in central and southern Sweden. Its chief features were lines from Stockholm to Göteborg and Malmö. In addition links were established with Oslo and a number of lines, many of them narrow-gauge, were laid in the Bergslagen mining district of central Sweden. It was not until the last years of the century that a railway was built into the vast, sparsely populated Norrland, and not before the early twentieth that it was extended around the head of the Gulf of Bothnia into Finland.[142] In 1902 a mineral line was opened from the iron-mining district of Kiruna, across the mountains to the Norwegian port of Narvik, but this was the only railway in Norrland to be intensively used.

The Norwegian system began in the 1860s with a number of short lines in the Christiania (Oslo) district. These were then linked with the Swedish system. The extension of the Norwegian railway net to the west and the north of the country met with enormous difficulties from both the rugged terrain and the severe climate. In 1875 a commission, set up by the government to report on railway-building, recommended that a system should be established to bring the country together, whether or not it could show any profit. A result of this report was the building of the Oslo-Trondheim railway (1877) and of the technically more difficult line across the high Fjeld to Bergen, which was not opened until 1907.[143]

Ports and shipping

In the pre-industrial age sea-borne traffic was almost as important as that carried by river and canal. No country with an extended coastline failed to make use of coastal shipping, and the shores of all such countries were studded with small ports between which frail coasting craft shuttled with passengers and local products. Scarcely a river reached the coast without sheltering near its mouth a small port where the commerce of its hinterland met that of the sea. Where,

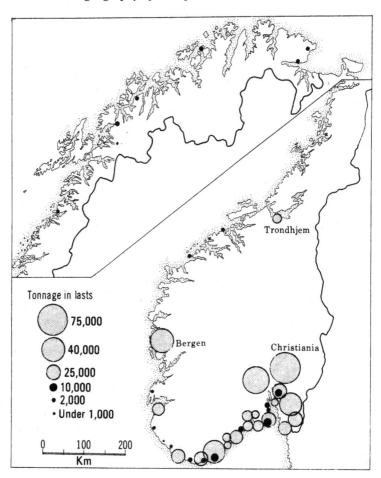

9.14 Freight handled in Norwegian ports, about 1860

as in Italy and Spain, internal communications were poorly developed, the sea assumed a proportionately greater importance. Even today, in Greece and its islands small craft carry much of the internal commerce from port to port.

It is impossible to list the hundreds of small ports which served the commercial needs of Europe. In most no record was kept of entrances and clearances and of the tonnage of freight handled. For Norway, however, a country within which communications were notoriously difficult, a list is available for 1861.[144] It records only trade with foreign countries, and if sea-borne domestic trade had also been recorded the list would have been very much longer (fig. 9.14). Christiania (Oslo) carried on the greatest volume of trade, but the difference between it and ports far down the table was not large, and there was a curious equality amongst most Norwegian ports.

In the course of the nineteenth century the great majority of small ports declined in importance, and trade completely disappeared from many of them. At the same time a small number of particularly well-favoured ports increased their traffic, so that by the early twentieth century a dozen ports handled between

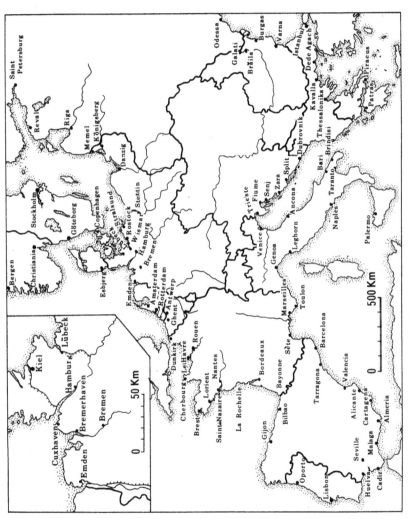

9.15 Ports named in chapter 9

them at least three-quarters of the sea-borne commerce of continental Europe. There were many reasons for the decay of the small ports. Some silted, at a time when the draught of ships was increasing. Above all, the improvement of overland transport was detrimental to them. Their hinterlands were served increasingly by railways, and goods which had formerly been moved by barge and coasting vessel now tended to travel overland. The advent of motorised road transport in the twentieth century was the last blow to the commerce of many of them.

Typical of decaying ports was Papenburg near the mouth of the river Ems.[145] In 1815, 50 sea-going ships were based on Papenburg, as well as 40 river craft. There was also a small ship-building industry, and no fewer than 307 ships, all of them sail-driven, were built there between 1815 and 1840. There were altogether about 180 ships in 1866, when Papenburg, along with the Kingdom of Hanover, passed to Prussia. Thereafter the number declined steadily, from 106 in 1885 and 67 in 1890 to 23 in 1900 and only nine on the eve of the First World War. The largest of them was of only about 300 tonnes. Another port which shared the same fate was Brake, in Oldenburg, which lay on the Weser estuary.[146] In the first half of the century it carried on trade with Asia and the Americas and was a small whaling port. It was the chief port of the Grand Duchy of Oldenburg and a significant rival to Bremen and Bremerhaven. But it was the latter, aided by the railway, which captured the growing traffic in the second half of the century.

From the hundreds of ports at the beginning of the nineteenth century, a small number – no more than a dozen – had achieved the status of world-ports a century later. They included Hamburg and Bremen-Bremerhaven; the three great ports of the Rhine delta – Amsterdam, Rotterdam and Antwerp; Rouen and Le Havre, Barcelona, Marseilles and Genoa. To this list should be added Göteborg, Trieste and Athens-Piraeus, because of their importance in the trade of their respective countries. All had great initial advantages, and, with the exception of Rotterdam, a long history of maritime activity. Each had developed the necessary infrastructure: quays and warehouses; chandlers, brokers and merchants. They figured in the earliest attempts to plan a railway network, and when the railways reached them their further growth became inevitable (fig. 9.15).

But a long history of maritime trade was not in itself a guarantee of continued prosperity in the nineteenth century. Lübeck, foremost amongst the cities of the German Hanse, grew very little in absolute terms and, relative to the great ports, declined disastrously. Nantes[147] and Bordeaux, both highly important in the eighteenth century, declined relatively after the mid-nineteenth century. In many other instances early promise was not fulfilled in the more fiercely competitive world of the later nineteenth century.

The factors which most influenced the growth of a great port appear to have been threefold. It needed to have favourable conditions for the creation of port facilities; it required a productive, and that in the nineteenth century meant an industrialised, hinterland, and lastly it needed excellent means of communication by both land and water with that hinterland. It was because one or more of these factors was lacking that so many important ports of the eighteenth century failed to maintain the momentum of their earlier growth.

It was once sufficient if ships could tie up at quays built along the waterfront. During the nineteenth century the growing volume of shipping, the increasing size of ships and the need to insulate them from the rise and fall of the tide made it necessary to excavate dock basins, cut off by lock gates from river or sea.[148] At Antwerp, for example, ships had merely tied up along the Scheldt until in 1803 Napoleon ordered the construction of the first two dock basins, the Bonaparte and Willem Docks, capable, as he said, of accommodating 52 ships of the line.[149] This could not have been accomplished if the Tertiary beds which make up most of Flanders had not yielded close to the city to soft alluvium in which the basins could be dug. In all the great ports of northern Europe the construction of docks was relatively easy. In Mediterranean Europe it was the reverse. Most of the ports were located on rocky coasts where docks had to be created at the expense of the sea. A saving feature, however, was the fact that in this almost tideless sea there was no need for floating basins. A sufficient depth of water for ships of increasing draught and a relative freedom from silting were also desirable, though in most ports the regular dredging of the channels became a necessary and expensive obligation.

Every port needed a hinterland, and the great ports which emerged in the nineteenth century were each dependent upon a large and well-developed hinterland. That of Hamburg embraced much of central and eastern Germany and reached into Bohemia. The great ports of the Rhine mouth competed for the trade of the whole Rhineland as far as the Swiss mountain passes. Rouen and Le Havre served Paris and the Paris basin; Barcelona served the industrialised region of Catalonia, and Genoa, despite certain difficulties in transport, the most developed parts of northern Italy.

Hinterlands of individual ports varied during the century; they were extended as canals were constructed or railways built. Improvement in navigation on the Main or Neckar, for example, broadened the hinterland of the Rhine-mouth ports in southern Germany. They were influenced by national policy, operating through subsidised dock construction or preferential freight rates. The Prussian government, for example, successfully diverted part of the traffic of the Ruhr from its 'natural' outlet in the Low Countries to the ports of Emden and Bremen. The Austrians supported the port of Trieste, thus extending its hinterland deeper into central Europe than might otherwise have been possible, while, after the *Ausgleich* of 1867, Hungary set out to make Fiume the exclusive port of the Hungarian monarchy. It was the lack of an industrialised hinterland which caused the relative decline of the ports of Lisbon, Bordeaux, Nantes, as well as many on the Mediterranean coast.

A great port, lastly, had to have cheap and easy access to its hinterland. The most convenient, at least in the first half of the nineteenth century, was by river. Without exception the ports which grew rapidly during this period were located near the mouths of large, navigable rivers. The most obvious examples were Hamburg, the Rhine-mouth ports and Rouen-Le Havre. The attempts made at this time (see p. 433) to improve navigation on the Rhône were not unrelated to the expansion of the port of Marseilles, and the failure of Nantes and Bordeaux can be attributed in part to the difficulties presented by their respective rivers. Within the Baltic, Stettin overtook Lübeck because the former was served by the navigable Oder, whereas Lübeck had only a canal which led directly to its triumphant rival, Hamburg.

The advent of railways further reinforced the advantages of the great ports. By 1850 the most important of them, in particular Stettin, Hamburg, Bremen, Antwerp and Rouen-Le Havre had been linked by rail with their hinterlands, which were extended into areas not served directly by river. As the railway net intensified so the proportion of the freight brought to the port by rail increased at the expense of that conveyed by river and canal. Antwerp was thus linked by rail with the German Rhineland, and Bremen, where the Weser hardly ranked as one of the more easily navigable rivers, became very largely a railway port.

In southern Europe port development was dependent overwhelmingly on the railways. Trieste and Fiume did not grow significantly until they were reached by the railway.[150] The development of Barcelona accompanied the building of the Catalan railways, the only significant network in nineteenth-century Spain, and that of Marseilles, given the none too successful attempts to improve navigation on the Rhône, was dependent on the completion of the railway northwards to the Paris Basin and the Rhineland. The ports of the Danube mouth and Black Sea coast, which handled vast quantities of wheat in the first half of the century, were linked with their hinterlands by roads and later by railways; little use was made of river transport.[151] Elsewhere in the Balkans the growth of ports was overwhelmingly dependent on road and rail. The only significant exception was Constantinople, which inevitably relied heavily on shipping for both local and long-distance commerce.

In addition to the great ports, handling general cargo, there were two types of port, smaller in scale and more restricted in scope, which grew up during the century. These are the outport and the specialised port. The former supplemented the great port; the latter was created to handle a specialised traffic. Except in the Mediterranean, none of the great ports was located on the coast; all grew up on a navigable estuary or at the head of a bay where there was natural protection from storms. However convenient in earlier centuries when ships were smaller and more vulnerable, a river or estuarine site came in the nineteenth century to have grave disadvantages. It restricted the size of ships that could reach it; it called for the maintenance of a channel through banks of silt, and time was lost in taking a ship up the estuary sometimes only to take on a few passengers or to discharge a part cargo. An 'outport', created near the entrance to the estuary could assume these functions.[152]

The medieval port of Bremen lay some 75 km from the open sea, at the head of an estuary which even in the Middle Ages had proved difficult to navigate. As early as 1618 a harbour was established at Vegesack, 20 km below Bremen. This did not suffice, and in 1827 Bremen acquired a site from the state of Hanover and on it founded the port of Bremerhaven. Attempts by Bremen itself to improve navigation on the Weser were consistently obstructed by Hanover and Oldenburg until 1867. In the 1880s the river was improved; docks were constructed close to the city, and Bremen began to regain trade which had been lost to its subsidiary Bremerhaven.

Hamburg lay even farther from the sea than Bremen, and, though the Elbe was more navigable than the Weser, it too developed an outport. Cuxhaven, as a port, was founded in 1857. It lay nearly 100 km to the north-west of Hamburg, and even farther by the twisting river channel.[153] It became the chief port of the Hamburg-Amerika Line and, during the later nineteenth century, developed in contrast with Bremerhaven, more as a passenger and liner port than as a cargo port.

The relationship of port and outport is best illustrated by the paired ports of Bremen-Bremerhaven and Hamburg-Cuxhaven, but it can be seen in the Low Countries, where Hook of Holland, founded in 1890, stood in a similar relationship to Rotterdam, and Flushing to both Rotterdam and Antwerp. Both Hook of Holland and Flushing were important chiefly on account of the passenger traffic with Great Britain. More typical of the function of port and outport is the combination of Rouen, Le Havre and Cherbourg. Rouen, about 125 km from the sea, had been for centuries the chief port of the Seine basin and by far the most significant maritime outlet for Paris. Le Havre-de-Grace was founded by Francis I in 1517, but achieved little importance before the nineteenth century. Then the increasing size of ships, coupled with the slowness of the voyage up to Rouen, led it to take over many of the functions performed hitherto by the latter. Cherbourg evolved as a kind of outport to Le Havre. It had originated in the eighteenth century as a strategic base for French warships. Its military role continued through the nineteenth century, but from about 1870 it began to be visited by Atlantic liners for the purpose of taking on and dropping passengers.

Saint-Nazaire developed, like Le Havre and Bremerhaven, because the navigation of the Loire estuary up to the port of Nantes was becoming increasingly difficult. The first dock basin at Saint-Nazaire was built in 1856, but later in the century a lateral canal was cut through the mud of the estuary, and the supremacy of Nantes re-established.

La Rochelle, like Nantes, was an important commercial port of pre-revolutionary France. Its growth in the nineteenth century was hindered by its small size, and in 1873 an outport was established at La Pallice. Bordeaux, at the head of the Gironde and 98 km from the sea, was one of the least accessible of Europe's ocean ports. It had, nevertheless, played an important role in France's colonial trade, and its commerce grew during the second half of the century. Nevertheless, the increasing size of ships, combined with the difficulties of navigation, led in 1894 to the establishment of an outport at Pauillac.

The second category of ports consisted of those created to handle a particular and sometimes temporary trade. No port had had a more restricted commercial basis than Narvik. It was built in 1902 for the sole purpose of providing an outlet for Swedish iron ore. It was linked with the orefield by rail, and its highly mechanised operations were from the start geared to the regular arrival of the ore trains. Few other commodities were handled, only a small export of timber and an import of coal.

The construction of the Danish port of Esbjerg was a consequence of the development of agricultural exports to Britain. The west coast of Jutland was fringed with dunes to the north, marshes to the south, but between the two, at Esbjerg, firm rock reached the coast and provided access for road and rail as well as a suitable site for a town. In 1868 the Danish government authorised the building of Esbjerg, and ten years later the dock was completed and was linked by rail with the rest of the country. The agricultural revolution of the 1880s (see p. 267) increased its trade by adding butter, cheese and bacon to the earlier agricultural exports. In a few instances the prosperity of these specialised ports was short-lived. Bayonne, for example, rose to temporary prominence in the later years of the century because of iron-working and chemical industries established in its hinterland.[154] Many small ports of the Baltic coast of Germany throve on the export of linen from the eastern provinces and declined with the decay of the industry.[155] Several ports developed late in the century to handle the export of iron-ore, such as Portoferraio on the island of Elba, Luleå in Sweden and Bilbao in northern Spain.

To these special categories of port must be added the free port or entrepôt. It was outside the customs of the country in which it lay. Goods could be discharged on to its quays and re-exported without customs examination or payment. A free port became the focus of routes and a place of exchange. Hamburg and Bremen were free ports until their inclusion within the German customs system in 1881, and even after this date Hamburg retained a large free zone on the left bank of the Elbe. In 1894 Copenhagen established a free zone, and Stockholm became a free port in 1919. Another port which acquired a significant free zone was Thessaloníka, but in this case it was instituted to facilitate the trade of Serbia (see p. 486).

Baltic sea ports

The ports of the Baltic Sea had long served as intermediaries in the trade between their relatively undeveloped hinterlands and western Europe. In the past they had been organised under German hegemony in the Hanseatic League. But by the early nineteenth century this organisation had long been defunct, and the basis of Baltic trade had changed. Most of the 70 or more ports which had once sent their delegates to the *Hansatag* had utterly decayed, and such trade as was still carried on was concentrated in little more than a dozen ports. This trade consisted essentially of an export of grain and timber, together with diminishing amounts of German linen and Swedish bar-iron and, late in the century, a growing export of iron ore. These were requited by miscellaneous goods from the west, many of them of a luxury nature.

The Baltic export of grain – most of it rye – and of forest products passed through some ten ports, of which three – Riga, Reval and St Petersburg – lay in Russia. Linen, mainly from eastern Germany, was one of the few remaining commodities traded in the small ports, and Swedish iron came chiefly from Stockholm. Lübeck, once the premier city of the German Hanse, had sadly decayed. It lay too far west to share significantly in the grain and timber trade, and Hamburg, thanks to the River Elbe, had tapped its local hinterland. At the same time the small river Trave, which gave access to the city from the sea, restricted the size of vessels which used it. Nevertheless, trade increased late in the century when navigation on the Trave was improved and a new canal cut from Lübeck southward to the Elbe.

The once important ports of Wismar, Stralsund and Rostock carried on very little trade until railways broadened their hinterlands and the accession of Mecklenburg to the Zollverein in 1867 gave them a wider market. Stettin was an exception to the general picture of stagnation. Unlike most others, it was a river port, even though the Oder was not one of the most easily navigated rivers. It attracted both capital and trade, and in 1843 the railway from Berlin was completed, earlier than that to any other Baltic port. The exports were at first dominated by agricultural and forest products, but from the early nineteenth century coal from Upper Silesia began to constitute an important cargo. Stettin began to rival Hamburg as the port of Berlin, and it handled much of the overseas trade of the Silesian industrial region.[156] It attracted manufacturing industries, including, late in the century, iron-smelting, steel-working and shipbuilding, and its commerce, unlike that of other German ports of the Baltic increased through the century. By the end of the century Stettin handled nearly a quarter of Germany's Baltic trade and some 8 per cent of all German sea-borne commerce.[157]

Danzig (Gdańsk) seemed to share the advantages which had made Stettin the third most active port in Germany. It lay on one of the distributaries of the Vistula which drained much of Poland. It had once handled the immense export of timber and grain from the Vistula basin, and the granaries and the floating basins, where the timber rafts were held before export, still lined the lower river. But the Vistula was a poor river for navigation; there was little upstream traffic, and, above all, the Second Partition of Poland had allocated the city to Prussia, but had left most of its hinterland in Russia. Not until 1852 was a railway built to Danzig, and then only a branch from the East Prussian line at Bromberg (Bydgoszcz). Commerce failed to grow, but the city became instead an administrative and above all a military centre. It was 'transformed from a city of traders and shippers to the more or less temporary abode of a host of German civil servants and soldiers'.[158] Not until after 1900 were steps taken to improve the docks and access to the sea by way of the 'Dead' Vistula (Martwa Wisła).[159]

Königsberg (Kaliningrad) suffered under similar disadvantages and shared the same fortunes. It lay some 37 km from the sea, on the banks of the Pregel. Access to the Baltic was across the shallow Frisches Haff. In the opposite direction, the Pregel was navigable only for small craft and its nineteenth-century traffic was negligible. Königsberg remained throughout the century a port for grain and timber, but the Russian boundary was only 100 km away, and the German hinterland was smaller even than that of Danzig. Access by rail to the port from the rest of Germany was possible only after 1857. In 1861 the German system was linked with the Russian, and Königsberg developed in the later nineteenth century a port for goods in transit from Russia. It is estimated that three-quarters of the exports of Königsberg consisted of Russian grain and lumber. The Niemen (Nemunas) was a longer and more navigable river than the Pregel, and it tapped a broader Russian hinterland. Its comercial outlet was through the small

German port of Memel (Klaipeda), and the chief commodity handled was timber floated down from Lithuania and White Russia. The ports of Riga, Reval (Tallinn) and St Petersburg not only lay within Russia but handled exclusively Russian trade.

Sweden carried on an important trade in bar-iron and timber. Its chief ports were Stockholm and Göteborg; indeed the port of Göteborg had been developed to further this trade. In the course of the century the timber, pulp and paper trade spread to Norrland, and was largely handled by small ports which developed along the Baltic coast.[160] Göteborg, however, maintained its role as the leading Swedish port through its close links, including the Göta Canal, with central Sweden. Stockholm assumed a secondary role because of its greater distance from the entrance to the Baltic Sea.

The Danish Straits and access to the Baltic Sea

The Baltic is a sheltered and almost tideless sea, but the price which has to be paid for these advantages is a narrow entrance through the territorial waters of Denmark and Sweden. There were, in fact, three natural routes, two of them – the Little and the Great Belts – lay between the Danish islands, and the third – the Sound – separated Denmark from Sweden. The Little Belt was narrow and tortuous and has never been much used. The Great Belt is wider, but in its commercial importance has always yielded place to the Sound, through which most of the recorded traffic has always passed. Denmark's control of the Sound dates from the time when both shores were under Danish sovereignty. The Danes were thus able to impose a toll on all ships as they passed the narrows – less than 5 km wide – between Helsingör and Halsingborg. In 1677 Sweden gained possession of the eastern shore of the Sound, but Denmark continued to claim jurisdiction over the whole breadth of the waterway and to levy tolls. In 1848, the United States, whose ships in fact made only a minimal use of the Sound, denied the right of the Danes 'under the public law of nations' to levy tolls. The controversy thus stimulated led ultimately to the Conference of Copenhagen, 1857, and their abrogation in return for an indemnity for Denmark's rights over the waterway. This indemnity was contributed by the countries whose ships used the Sound in proportion to their use. Seventy per cent of it was paid by Great Britain, Russia and Prussia.[161]

In the mean time attempts had been made to circumvent the Danish Straits. This could be done only by cutting a canal across either the base of the Danish peninsula or across southern Sweden. In the last years of the fourteenth century the citizens of Lübeck constructed the Stecknitz Canal from their city to the Elbe. It fell into decay and disuse, and its place was taken after a long interval by the Elbe–Trave Canal, which was opened along a similar route in 1900.

A much more important waterway was the Kaiser-Wilhelm or Kiel Canal. It was constructed in 1887–95 from the mouth of the Elbe, across Holstein to Kiel Fjord on the Baltic Coast. It was a sea-level canal, capable of taking the largest German warships. Its purpose was primarily strategic, to permit naval craft to move between the North Sea and the Baltic without close observation by the Danes. It was, however, also used by commercial shipping because for most vessels sailing between the Baltic and North Sea it saved a distance of over 600 km as well as a sometimes perilous voyage round the Skaw. It was less easy to cut a canal across Sweden. Nevertheless the Göta Canal was completed in 1832 to join up some of the lakes of Central Sweden and to connect Stockholm with Göteborg. Its purpose was more to give access to the scattered industrial centres of this region than to circumvent the Danish tolls. It had some success in its primary objective, but declined in importance with the coming of the railways later in the century.[162]

The North Sea ports

A number of factors combined before the end of the nineteenth century to make the North Sea the most important focus of the world's shipping. Around it lay the most advanced industrial nations with a dense population and a high level of consumption. To these must be added the rivers which discharged into it, the Meuse, Rhine, Weser and Elbe, which brought a great deal of central and western Europe within the hinterlands of the North Sea ports. No feature in the geography of rivers was of greater importance than the oblique course of the River Elbe, which, with its system of canals, put not only Berlin and the cities of Saxony, but even Bohemia and Silesia within reach of Hamburg.

The North Sea ports are here taken to include all from Norway to northern France. In Norway the large number of small ports (see p. 462 and fig. 9.14) compensated for the difficulties of overland travel. Though the proportion of the trade passing through Christiania (Oslo) and Bergen steadily increased, dozens of very small ports and landing-places right up to the North Cape remained active up to the First World War.

In contrast with the east coast of Denmark and the Danish islands, the straight west coast offered no natural haven, and the only port of significance, Esbjerg, was created after 1868 to meet the very special needs of the growing agricultural commerce with England (see p. 268).

Along the south coast of the North Sea, the process of concentrating trade in a few giant ports was carried to an extreme. Nowhere else have small ports decayed so conspicuously, and nowhere else in Europe was the growth of large ports more rapid. By 1914 most of the sea-borne trade of the Low Countries and western Germany was carried on through only five ports: Hamburg, Bremen and the Lower Weser ports, Amsterdam, Rotterdam and Antwerp. Most of the rest passed through Emden and Ghent. All were, in the first instance, river-ports, dependent on rivers for transporting freight to and from their hinterlands. Difficulties of navigation gradually reduced Bremen's dependence on the Weser, and rail transport assumed some importance in all of them. An intense rivalry developed between these ports. The building of railways and the extension of the network of waterways by the construction of linking canals meant that their hinterlands overlapped. From any one point there was a choice of routes to the sea, and the one chosen for a particular commodity was likely to depend on a variety of factors, of which time, distance and cost were not necessarily the most important. National policy was always an overriding consideration, since all countries were eager to increase the volume of trade passing through their own

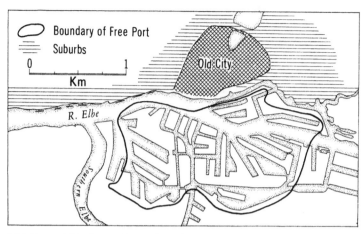

9.16 Port of Hamburg, about 1910, showing dock basins excavated in the alluvial deposits

ports. Both Germany and Belgium attempted with considerable success to divert traffic away from the Rhine-mouth ports of Amsterdam and Rotterdam, while in France efforts were made to use Dunkirk or even Rouen-Le Havre instead of the ports of the Low Countries. Most intense of these port rivalries was that between the three principal ports of the Low Countries: Amsterdam, Rotterdam and Antwerp.

Hamburg, most easterly of the great ports of north-western Europe, had a less developed hinterland than its rivals, and was in consequence slower to develop (fig. 9.16). The port, furthermore, faced more serious political problems. Although it had inherited the status of Free City and had been an important member of the Hanse, it was hemmed in by rivals in Denmark, Holstein and Hanover. Upstream trade was hindered by tolls, and on the lower Elbe a particularly burdensome toll continued to be exacted at Stade by the State of Hanover.[163] The British, always the most important trading partner of Hamburg, protested vigorously, and, after the termination of the Danish Sound dues in 1857, actually sent coal to Berlin by way of Stettin. The tolls on the upper river were removed with the progress of the Zollverein, but the Stade tolls continued to be levied until 1861.

It was not until all river tolls had been removed that Hamburg began to develop a modern dock system. Hitherto ships had tied up to posts set in the river bed, where they were off-loaded into lighters. In 1866 the first dock basin was excavated on the edge of the Old City. This was quickly followed by docks of increasing size on both banks of the river. The traditional commercial policy of Hamburg had been one of free trade. For this reason the city did not join the Zollverein, and, even after 1871, remained outside the German customs system. This situation, however, could not last, especially when the *Reich* was becoming increasingly protectionist. In 1881 Hamburg became part of the German customs area, but retained a free port located on the south bank opposite the Old City. At the same time Hamburg's links with its hinterland by both rail and waterway were improved. The rapid growth of Berlin in the last third of the century and the expansion of its manufacturing industries contributed powerfully to the increase of traffic in the port of Hamburg.[164]

The tonnage of ships entering the port increased from some 600,000 tonnes at the mid-century to 1,328,000 tonnes in 1866, when the first dock basin was opened, to 2,178,000 tonnes in 1876 and to 14,185,000 in 1913. Cargoes increased proportionately, but it was a severely unbalanced trade, imports greatly exceeding exports owing to the large import of coal, lumber, grain and raw materials. From the first the port relied heavily on river-borne traffic. in the 1850s up to three-quarters of the goods reaching Hamburg for export came by river and almost as high a proportion of imports was distributed in the same way. The ratio of rail traffic to and from the port to that carried by river varied. In the 1870s rail traffic exceeded river-borne by a small margin; in 1913, more than 60 per cent went by river, thanks mainly to the large consignments of coal to Berlin.

Bremen had been a Free Imperial City and a member of the Hanseatic League, and, like Hamburg, was not integrated into the German Reich until 1884. Like Hamburg, it acquired a free port, built between 1885 and 1888 and consisting of two dock basins to the north of the Old Town. There, however, similarities with Hamburg ended. The River Weser ceased to be significant for the movement of freight to and from the port. The railway reached Bremen from Hanover in 1847; other linkages followed, and Bremen became almost exclusively a railway port. Below the city the estuarine tract of the river also presented difficulties, solved by the creation of the outport of Bremerhaven in 1827. Nevertheless, the port developed rapidly during the middle years of the century and was for a time far more important than Hamburg. It was the home port of the North German Lloyd and a major passenger port from which the majority of German emigrants sailed in the nineteenth century. Although Bremen was the major port for the import of tobacco and later developed an important trade in raw cotton, its commerce did not expand rapidly in the later years of the century, and it was soon overtaken by Hamburg. By the end of the century the volume of shipping using the port of Bremen was little more than a tenth of that which sailed into Hamburg.[165] Bremen was overtaken even by its own outport of Bremerhaven in the volume of shipping using it, because many ships, including ocean liners, called only at Bremerhaven.

The third important German North Sea port was Emden. It lay on the northern shore of the shallow estuary, known as the Dollart, by which the Ems reached the sea. It had been a small port engaged mainly in coasting trade until, in the last decade of the century, it became a major port for the Ruhr industrial area. In 1887 a canal, the Ems–Jade, was completed between Emden and the developing naval base of Wilhelmshaven, 70 km to the east. More important was the Dortmund–Ems Canal, opened in 1892. Its purpose was to divert to Emden as much as possible of the traffic of the Ruhr which until then passed through the ports of the Rhine delta. Dock basins were excavated in the silty plain which lay between the town and the Dollart for the export of Ruhr coal and import of iron-ore.

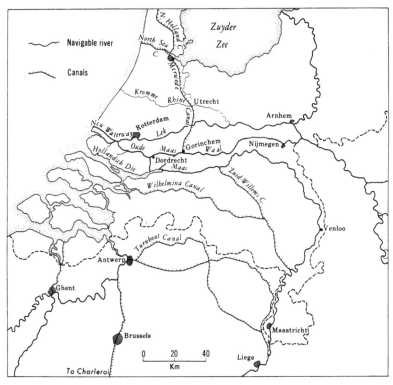

9.17 Ports of the Rhine delta

The ports of the Rhine Delta

Near the Netherlands boundary the Rhine began to divide into the IJssel, Kromme (Crooked) Rhine, Old Rhine, Lek, Waal and the many streams which mingled their waters with those of the Maas (Meuse). The direction of the Rhine itself was continued in the Waal, which divided near Dordrecht to make the Oude Maas and Merwede which, with the Maas itself, flowed into the Hollandsch Diep. To the south, beyond the shifting islands of Zeeland, lay the Honte or Western Scheldt, the estuary of the River Scheldt. The geography of the Rhine delta was confusing at any time, but confusion was confounded by continuing changes in the courses of the waterways and in the names by which they were known (fig. 9.17). Superficially regarded, the many waterways might have been expected to produce an abundance of navigable routes between the coast and the Rhineland. In reality, however, most of them were too shallow and their sandbanks too shifting and uncertain for them to have provided a regular means of transport. Indeed, before the nineteenth century these waterways were little used. Traffic from the lower Rhineland in general made its way to the sea at Antwerp or to the Flemish coast by one of several overland routes. Dordrecht, lying where the Waal merged into the Oude Maas, was the hub of the delta, handling most of the river-borne trade between the Rhineland and the sea. Its traffic declined in the eighteenth century and by the nineteenth, the waterways near Dordrecht had silted and larger vessels could no longer pass.

With industrial development in the Rhineland, the need for direct access to an ocean port became overwhelming. Amsterdam was the first contender for this role. It had grown up near the western shore of the Zuyder Zee primarily as an intermediary in the trade between the Baltic and western Europe. It had inherited the role formerly played by Antwerp and had developed an important trade with the Indies. But Amsterdam's links with the Rhine were tenuous. The IJssel and Kromme Rhine were used, but routes were long and tortuous and open only to small craft. Throughout the century it was the ambition of the merchants of Amsterdam to shorten and improve the waterway.

In the mean time, Rotterdam, heir to Dordrecht, had developed as the chief outlet from the Rhineland. It became the starting-point of the first steamship service on the Rhine, the River Waal providing a fairly direct route. The chief problem facing the port was the improvement of its outlet to the sea.

The third great port of the Rhine delta was Antwerp, lying on the east bank of the Scheldt and within the boundary of Belgium. Its access to the sea, however, was through Dutch waters. Its hinterland lay in Belgium, but road links with Cologne also allowed it to serve in some measure as an outlet for the Rhineland. It was also linked by way of the waterways of Zeeland with the Hollandsch Diep, and thus with the Waal. The route was circuitous, though no more so than that from Amsterdam to the Rhine, and was liable to interruption where it passed between the islands of the Netherlands, a source of continuous recrimination between the Belgians and the Dutch.

In the long run it was Rotterdam which, benefiting from its shorter and easier access to the Rhine, became the most important maritime outlet for north-western Europe. Though Amsterdam and Antwerp continued to share in the trade of the Rhineland, their role was increasingly restricted to handling the sea-borne commerce of their immediate hinterlands, Holland and Belgium respectively.

Amsterdam At first Amsterdam enjoyed the advantage of a sheltered harbour where the little River IJ entered the Zuyder Zee. This, however, turned to an acute liability as the waterway shallowed and ships grew larger, and the greatest problem facing the port became that of access to the sea. The first attempt to solve it consisted in building, 1819–24, the tortuous North Holland Canal from the IJ through the lakes and polders to Den Helder at the entrance to the Zuider Zee. But large ships could not use it, and commerce failed to grow. The ultimate solution was the opening of the North Sea Canal in 1876. It was cut from the IJ westwards across the polders and the coastal dunes to the sea at IJmuiden. It was a sea-level canal, though protected by lock gates at each end, and was large enough to take all vessels likely to use the port. The port of Amsterdam turned its back on the Zuyder Zee and dock basins were built to the west, along the new canal, which became a giant floating basin. Trade revived, and from the 1880s Amsterdam grew to be the leading port in the domestic commerce of the Netherlands.[166]

It did not, however, cease to be a Rhineland port, and one of its objectives in the continuing competition with Rotterdam was to improve the waterways which led to the Rhine. A variety of routes was used to gain access to the Lek, but eventually that which followed the Vecht came to be the most used. The Lek, however, was not the most navigable of rivers, and the waterway was extended to the Waal at Gorinchem by a canal which came to be known appropriately as the Keulsche Vaart (Cologne Canal). Even so, the route was inadequate, and was replaced in 1883–92 by the Merwede Canal.[167]

Rotterdam This port, like Amsterdam, faced an acute problem of keeping open its channel to the sea. The Brielsche (Brill) Maas which had given access to the port silted rapidly in the eighteenth century and had to be abandoned. Ships then took the circuitous route to the sea around the island called the Hoeksche Waard and so to the Hollandsche Diep. But this was little better, and after 1815 prospects for Rotterdam looked no better than those for the moribund port of Dordrecht. In 1830 Rotterdam followed the example of Amsterdam and opened up an artificial waterway to the sea. A canal was cut across the island of Voorne to Hellevoetsluis. The respite was, however, only temporary, and the increasing size of ships quickly made the Voorne Canal obsolete. The final solution, adopted in 1863, was to cut a canal westwards from Rotterdam to the sea to the north of the silted Brielsche Maas. This became the New Waterway, opened in 1872 and progressively widened and deepened until it could accommodate the largest ships. The commerce of Rotterdam expanded with the growth of manufacturing in the Rhineland. The Lek, which provided the most direct route to the Rhine, proved to be too constricted, and a cut, known as the Noord, was made from the Lek to the Waal.

The earliest docks were close to the city on the northern bank of the New Maas, but with the opening of the New Waterway larger docks were established on the south, and were extended westwards towards the sea. The volume of trade in the port of Rotterdam continued to increase; in the 1860s it began to exceed that of Amsterdam, and by 1870 half the sea-borne commerce of the Netherlands passed through Rotterdam. It differed sharply, however, from the trade of Amsterdam. In the first place much of its trade was in transit to or from the Rhineland; secondly, it was mainly a river port, only a small part of its freight being received or distributed by rail, and lastly the goods handled consisted mainly of bulky raw materials. With the development of the Ruhr industrial region iron ore became the largest single commodity passing through the port, followed by grain and lumber. Foremost amongst the exports was coal. The opening of the Dortmund–Ems Canal and the creation of the port of Emden dented, but did not seriously impair, the traffic, which, up to the First World War rode high on the commerce of a rapidly developing Ruhr and Rhineland.

Antwerp This, the third of the trinity of great ports at the Rhine mouth, also had problems of access to the sea, but they were political rather than technical. Twenty-five kilometres below the city the River Scheldt entered the Netherlands. From 1609, when the United Netherlands gained *de facto* independence, until 1792 when the armies of revolutionary France overran the Low Countries, the River Scheldt and the port of Antwerp were closed to ocean shipping. Trade did not cease, but sea-going ships were obliged to off-load at the entrance to the Western Scheldt, and their freight was carried up to the city in lighters.[168] Such a procedure was cumbersome and costly, and the volume of trade which survived such restrictions was small. Napoleon opened the river, ordered the building of new docks and encouraged the renewed growth of the port. The Final Act of the Congress of Vienna confirmed that navigation of the Scheldt up to the city should be unrestricted. Until 1831 there could be no question of Dutch interference with Antwerp's trade, since both Belgium and the Netherlands were united in a single kingdom. In 1831 Belgium asserted its independence, and the Dutch reimposed constraints on Belgian commerce. It was not until 1839 that the Netherlands agreed that, by the terms of the Vienna treaty, navigation was free. The Dutch, however, imposed a toll on all vessels except their own sailing through the Scheldt to Antwerp. This continued to be levied until, in 1863, it was bought out by Belgium in collaboration with the maritime countries which used the port.[169] Thereafter the commerce of the port of Antwerp increased rapidly.

The natural hinterland of Antwerp was the basin of the Scheldt and the province of Brabant. The rivers were navigable, at least for small craft, and the region was well served by railways from a relatively early date. But Antwerp looked for a wider hinterland in western Germany and eastern France, and in this its ambitions came into conflict with those of the Dutch ports. A railway to the Rhine by way of Liège was completed in 1850 and provided a means of conveying light freight to and from the port. The need, however, was for a water link both with the Rhineland and with the industrialised Meuse valley of central Belgium, and the satisfaction of this requirement became a source of friction between Belgium and the Netherlands throughout the century. The first canal, cut before 1831, remained in Dutch territory after the separation of the two states, and later attempts to forge a water connection with the central industrial belt were in varying ways inadequate.[170] The question of Antwerp's access to the Meuse and, *a fortiori*, to the Waal and Rhine, remained unresolved when the First World War suspended all negotiations.[171]

In the mean time, Belgium had been developing her own routes from Antwerp to its industrial hinterland in the middle valley of the Meuse. First, Brussels was made accessible to Antwerp by a short canal from the Rupel. Then in 1827–32 the canal was extended from Brussels across Hainault to the Meuse near Charleroi. Later in the century the canal was enlarged, and a branch was constructed westwards to the coalfield at Mons and on to the French boundary.

Antwerp suffered in the competition with Rotterdam for the trade of the Rhineland, but came to dominate the domestic trade of Belgium, as Amsterdam did that of the Netherlands. An interlocking system of dock basins was built to the north of the original Napoleonic docks between 1860 and 1913, and Antwerp became by 1913 the most important continental European port for handling general cargo.

Ghent This port lay 90 km up the Scheldt from Antwerp and a similar distance from the Flemish coast near Zeebrugge. Like Antwerp, it was cut off from the sea by the closure of the Scheldt, but, unlike Antwerp, was able to gain access to the sea at Ostend without passing through Dutch waters. Early in the nineteenth century Ghent became a centre of the cotton industry (see p. 323). With its increasing commerce its maritime connections by way of either Ostend or the Scheldt became inadequate, and in 1824–7 a canal 34 km in length was cut from the city to Terneuzen on the Western Scheldt and docks capable of receiving sea-going ships were built on the outskirts of Ghent.

Dunkirk The only other North Sea port of major importance in the nineteenth century was Dunkirk. Although used commercially before the nineteenth century, its importance was military and strategic rather than economic. Its modern growth dates from the industrial expansion of northern France. The first dock was built in 1845. The canals of Flanders were extended to Dunkirk, but they were small and provided no satisfactory connection with the coalfield and industrial centres of northern France. In 1848 the railway reached Dunkirk, which developed primarily as a railway port. Industrial raw materials, especially raw cotton, wool and iron ore, were the most important imports; coal and manufactured goods, the chief exports. Initially the hinterland of Dunkirk was the industrial region of northern France. The port lay too far west to be able to share significantly in the traffic of the Rhineland, but in the later nineteenth century the port developed as an outlet for eastern France and Luxembourg. In this, however, it faced the competition of Antwerp, which not only lay nearer but could use the Meuse to ship bulky goods.

The Atlantic ports

The coast west of the Straits of Dover has many excellent harbours, but few great ports. This is due to the relative lack of development in the hinterlands which they serve. Only Rouen and Le Havre, joint ports for Paris and the Paris Basin, can compare with those of the Rhine mouth and North Sea. Channel ports developed during the century as packet stations for the traffic with England. Regular sailing began from Calais in 1820. The Boulogne packet began in 1843, and that from Dieppe in 1856. The rise of the Channel ports was dependent on speedy transport to Paris, and their expansion was closely linked with the growth of the railways. They commonly handled a small amount of general cargo, but freight was in general not important.

This, however, was not so with the ports which grew up on the Lower Seine. Rouen, the most important of them, lay 125 km from the sea, reached by a shifting and difficult channel. Despite the difficulties of navigation, the trade of Rouen grew steadily with the growth of Paris itself and the expansion of the textile industry of the Seine valley (see p. 322). By the mid-nineteenth century the increasing size of ships made it difficult for them to negotiate the mudbanks of the estuary and the twisting channel of the river. From 1848 extensive works were undertaken to improve navigation. The channel was deepened and stabilised by means of training walls; the Tancarville Canal was cut through the silt of the estuary, but larger ships could still reach Rouen only on the tide, and the port remained the most difficult of access of all the great ports of Europe.

The trade of Rouen remained, up to the First World War, heavily unbalanced. Imports were dominated by coal, much of it transhipped to river barges for Paris. Rouen, before 1914, was handling up to a fifth of all coal imported into France. Textile raw materials, some of them sent by rail to Lorraine and Alsace, came second. Exports consisted mainly of light and manufactured goods.

Le Havre developed late in the eighteenth century as an outport for Rouen, when the growth of trade made it necessary to build the first dock basin. Growth became more rapid after the mid-nineteenth century, and after 1845 a complex system of docks was built.[172] Le Havre developed the true functions of an outport, being used by larger vessels, for which the voyage up to Rouen was impracticable. It handled part cargoes and was the chief French port for ocean liners.

The coast of Brittany was dotted with ports, all of them small, and most engaged only in the Atlantic fisheries or the domestic coasting trade. Their number and importance reflected in part the fact that it was easier in this region of inadequate roads and poorly developed railways to move goods along the coast than overland. Brest was an exception. It has the best harbour on this coast, and was chosen in the seventeenth century as a French naval base. But its hinterland was relatively undeveloped, and despite a small commercial traffic, its function has remained largely military. The port of Lorient, farther to the south-east, was founded by the French Compagnie des Indes Orientales for trade, as its name suggests, with the East. The project was ill-conceived. Its hinterland was undeveloped and routes into the interior almost non-existent. The loss of the French empire in Asia left the port without any real function.

The twin ports of Saint-Nazaire and Nantes offer an interesting parallel with Le Havre and Rouen. They lie respectively at the mouth and head of a shallow and sandy estuary, while to the east the Loire stretches into the heart of France. The port of Nantes had before the Revolution been important in France's colonial trade. Goods were carried up the Loire and by canal to the Seine. But commerce failed to expand in the nineteenth century. The river was difficult to navigate and was largely replaced by the railway. Faced with a declining trade the port authorities tried to improve conditions of navigation by a combination of training walls and a lateral canal through the silt. They met with some success and before the end of the century the trade of Nantes began to recover. As in so many French ports, it was an unbalanced trade, with imports, amongst which coal occupied a dominant place, greatly exceeding exports. In the mean time, in 1856, the port of Saint-Nazaire had been established at the river mouth. It developed a highly important shipbuilding industry during the second half of the nineteenth century, but was never a serious rival to Nantes as a port.

La Rochelle, 120 km to the south of Nantes, had once rivalled the latter in the colonial trade. Its trade again expanded in the early nineteenth century but the docks were cramped and inadequate, and between 1881 and 1891 a deep-water port was constructed at La Pallice, 5 km to the west. Unlike Nantes, La Rochelle was unable to regain its lost trade, and by 1914 cargo was handled mainly at La Pallice.

The dominant port of south-west France was Bordeaux. It lay 98 km from the sea, at the head of a wide, shallow estuary, the Gironde. Its growth during the nineteenth century was slow at first, like the development of its hinterland, and the first dock basin was not built until 1869. Thereafter trade expanded more rapidly. As in other ports of western France, coal imports were significant though a trade in colonial goods, especially sugar, survived from the eighteenth century, when Bordeaux shared with Nantes the trade with the French empire. Exports were made up largely of wine and softwood from the Landes. Bordeaux suffered, like all the estuarine ports of western France, from difficulties in navigating the channel which led from the ocean, and was obliged to establish outports late in the nineteenth century.

Bayonne, the only other port of significance on the Atlantic coast of France, is one of a group lying at the head of the Bay of Biscay. It was at first a fishing and whaling port; then a commercial port handling timber from the Landes. Later on, the founding of ironworks at Le Boucau, on the Adour below the town, in 1882 (see p. 344) led to a sharp increase in freight, especially of iron ore and fuel.

The long coastline of the Iberian peninsula is characterised by few large ports, but a plenitude of small ones. This is in part due to the indented coastline, but in part also to the fact that population and economic activity tended to be peripheral and to be concentrated within 100 km of the sea. The hilly terrain, the lack of railways and navigable rivers and the poor quality of the roads forced domestic transport to take to the sea. In few parts of Europe did coastal shipping play a bigger role in domestic commerce. Most of the overseas trade, on the other hand, was carried on through a small number of large ports, which handled chiefly mineral ores and fuel.

The only important ports between the French border and the Straits of Gibraltar were Bilbao, Gijon and Huelva in Spain, and Oporto and Lisbon in Portugal. Others, whose names are far more familiar, had decayed with the decline of empire, and retained but a shadow of their former importance. Bilbao lay on the River Nervión some 10 km above the point where it entered the sheltered Bay of Bilbao. To the south lay one of the few gaps in the Cantabrian-Pyrenean mountains, and this had allowed it to develop as the port for central Spain through which much of the wool produced by the Mesta had been exported to northern Europe. Along the banks of the Nervión many of the ships of Spain's imperial navy had been built. But by the early nineteenth century these activities were in decline, and Bilbao was just another decaying Spanish port, until the development of the iron industry and the iron-ore trade. About 1870 the traditional iron industry took a new lease on life. The ships which took the iron ore to Britain and Germany brought back fuel for the Spanish furnaces. By the early years of the present century Bilbao was a significant though highly specialised port. Gijón, 220 km to the west, was no less narrowly based. The chief commodity passing through its docks was coal from the Oviedo coalfield, much of it to be distributed by coasting vessels along the coast of Iberia.[173]

The ports of north-western Spain – El Ferrol, Corunna, Vigo – resembled those of Brittany. They had excellent harbours, poor communications with their hinterlands and little traffic. The only other port of significance on the Atlantic coast of Spain was Huelva. It too was highly specialised. Though of some importance during the century of Spain's greatness – Columbus sailed from the port of Palos de la Frontera, close to Huelva – it had decayed utterly when the copper mines of the interior were reopened about 1870. Thereafter Huelva became the port for the mining district.

The plain of Andalusia had formerly been one of the most developed parts of Spain and the ships of the *conquistadores*, though built on the north coast of Spain, sailed from the ports at the mouth of the Guadalquivir. It was at Seville, 90 km up the twisting and silting Guadalquivir, that the treasure of the Indies was unloaded at the Casa de Contratación. Difficulties of navigation led to the decline of the port, and not until after 1900 was any serious effort made to dredge the river and cut a channel through its meanders. In consequence trade was beginning to revive on the eve of the First World War. One might have expected that Cádiz would profit from the difficulties of Seville. It lay on a headland, linked with the mainland by a long peninsula behind which was one of the largest and most sheltered harbours in western Europe. Cádiz became the chief commercial port serving the Spanish Empire, but with the loss of the latter, nothing remained for Cádiz, and despite its illustrious history, it joined the long list of decayed Iberian ports.

Between the ports of northern and of southern Spain lay those of Portugal, foremost amongst them Oporto and Lisbon. Oporto was the port for northern Portugal. It imported coal and exported wine – chiefly to England – as well as other agricultural products. Its harbour was merely the incised River Douro whose steep banks precluded any dock development. For this reason a wholly artificial port was created late in the nineteenth century at Leixões, on the open coast 12 km away to the north-west.

Lisbon had no such disadvantage. Its harbour, formed by the estuary of the Tejo (Tagus) was one of the largest and most sheltered in Europe. Its eastern and southern shores were flat and silting, but along the north-west the rock of the Lisbon peninsula rose steeply from the water, and here, protected by the narrow entrance from the sea, docks were established. Lisbon has always carried on most of the foreign trade of Portugal, much of it general cargo, but including an import of coal, important for a country which had none, and an export of cork and other agricultural products.

The ports of Mediterranean Europe

The ports which lay within the Straits differed in many respects from those which fronted the Atlantic and North Sea, differences which sprang in part from the lower level of economic development in southern Europe; in part also from contrasted physical conditions. The Mediterranean Sea is, for practical purposes, tideless, and though storms occur, they never approach the degree of severity experienced in the North Atlantic. This meant that dock basins, closed by lock gates, were quite unnecessary for loading and discharging cargo, and that a system of moles and breakwaters, partially enclosing a bay, together with quays or wharfs, is all that was required of a port.

A second characteristic of the Mediterranean was the absence of long estuaries, such as occurred in north-western Europe. Instead, rivers entered the sea by deltas, rarely navigable, except for the smallest craft, while the rivers themselves, even if navigable, were little used. There was thus no need in the Mediterranean for the familiar system of port and outport. Indeed, Mediterranean ports tended to be located away from river mouths, protected, like Marseilles and Thessaloníka, from the silt which they discharged.

In the absence of navigable rivers opening up their hinterlands, Mediterranean ports faced serious problems of access. They were primarily road and railway ports, and their growth in the nineteenth century was heavily dependent on progress in railway construction. Lastly, the frequency of coves and sheltered anchorages had encouraged the proliferation of small ports, handling the restricted traffic of their limited hinterlands. Coastal trade was particularly important and in some degree made up for the deficiencies of overland travel.

Mediterranean Spain The Mediterranean coastlands were at the beginning of the nineteenth century the most prosperous parts of Spain. A succession of fertile, populous and well-cultivated lowlands were separated from one another by barren ranges which reached out from the Meseta to the Mediterranean Sea. Each had its port: Málaga, Almería, Cartagena, Alicante, Valencia, Tarragona and Barcelona. All developed port facilities during the century; exported the produce from their *vegas* and occasionally minerals from the highlands behind, and imported fuel and industrial raw materials, such as cotton, timber and iron. But only Barcelona became a port of national importance, reflecting the industrial development of Catalonia.[174]

Southern France The ports of Mediterranean France were more important than those of Spain because they had behind them a more highly developed hinterland with which, furthermore, they had easier commercial links. West of the Rhône delta lay Sète, to the east Marseilles and its satellite ports of the Rhône delta. Marseilles was the oldest port on this coast. It was founded beside a small cove, now the Port-Vieux, overlooked by limestone mountains. It had shelter and deep water, but its links with its hinterland were poor since the hills which enclosed it cut it off from the Rhône valley. As a port, Marseilles remained small and of mainly local importance until the nineteenth century. Then the French conquest of Algeria and, nearly 40 years later, the opening of the Suez Canal transformed the port. Trade grew; the Old Harbour ceased to be adequate, and the first basin, enclosed by jetties and a protective mole, was built on the coast to the north. Other basins were added progressively until by 1900 they stretched along the shore for 4 km.

To the west of Marseilles lay the mouth of the principal branch of the Rhône delta, reached by a voyage of 40 km across the Golfe de Fos. Earlier ports of the delta, like Arles and Aigues Mortes, had silted and had long been abandoned. During the nineteenth century attempts were made to reopen a route through the delta. In 1834 a canal was cut from the head of the delta to Bouc on its eastern edge. It proved inadequate, and improvements were then made on the river's main distributary. These failed to secure an adequate depth of water, and in the 1860s a canal – the Canal St Louis – was cut from the Rhône eastwards to the Golfe de Fos, and on it the Port-Saint-Louis-du-Rhône was established. These ports and waterways were created in the first instance to provide a means whereby Marseilles could use the River Rhône. But it did not long remain a river port. The volume of shipping on the river declined during the century, and Marseilles became a railway port.[175] The railway from Paris and Lyons along the east bank of the Rhône was completed by 1855, and that on the western bank followed later in the century. Lines were built eastwards to Toulon, despite the difficult terrain, and westwards through Languedoc to Toulouse and Aquitaine. Marseilles became not merely the regional port serving the whole south of France, but incomparably the most important French port in Mediterranean and Asiatic trade. This was reflected in the nature of the commodities handled. The largest categories of goods imported were fuel, especially coal from Great Britain, and tropical agricultural products: raw sugar, vegetable oils, fruit and cereals. The large volume of coal made it, as in all French ports, an unbalanced trade. Exports were made up largely of manufactured goods and the products of French agriculture.

The mountainous coast of Provence offered little scope for commerce. The excellent harbour of Toulon became the chief Mediterranean base for the French fleet in the eighteenth century, and this it remained up to the First World War. The commercial port, which always existed alongside the naval, was small and its trade unimportant. The only other significant French port on this coast was Nice. Its small natural harbour had long been of local importance when in 1860 the city passed to France. Its facilities were then increased, but its hinterland, consisting of the sparsely populated Alpes Maritimes, offered little scope for growth.

Italy Italy, with its long and indented coastline and mountainous interior, has always depended heavily on sea-borne commerce. Ports were numerous, even if their trade was small. The unification of most of Italy in 1861, followed by the opening of the Suez Canal and industrial growth in northern Italy, brought about a rapid growth of the better placed and more adequately equipped ports. Foremost amongst Italian ports was Genoa. Disputing its domination of the sea-borne trade of northern Italy was Venice, with Naples the most important port of the south.

Genoa bears some resemblance to Marseilles. Both developed at a very early date around small natural harbours which sufficed for their needs until the nineteenth century. Both were enclosed by mountains which greatly hindered their connections, by land, and both came to rely almost exclusively on railway links with their hinterlands. Until 1815 Genoa had been an independent republic, but was then attached to the Kingdom of Sardinia. Regret at the loss of its independence was moderated by the prosperity which came to it as the chief port of the most progressive state in Italy. In 1853 the railway from Turin was completed through the Ligurian Apennines to Genoa. During the following years the rail net spread through Piedmont, linking its small but growing industrial centres with the coast.[176] The original harbour of Genoa was larger than that of Marseilles, and for much of the nineteenth century port development consisted of building jetties radially into the bay. By the early twentieth century the available space had been used, and jetties, protected by a mole parallel with the coast, began to be built, as at Marseilles, along the shore to the west (fig. 9.18).[177]

The nature of Genoese trade was determined in large measure by its industrial hinterland. It was heavily unbalanced, with large imports of coal and materials for the textile and metal industries of Lombardy and Piedmont. Exports were made up largely of products of the farms and factories of northern Italy.

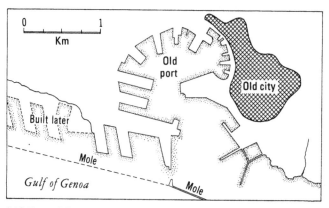

9.18 Port of Genoa, about 1910, showing a port constructed on a rocky coast

Venice rivalled Genoa as the port of northern Italy. Its advantage was that no natural barrier separated it from its hinterland; its disadvantage, that the shifting sands of its lagoons made navigation difficult. The modern growth of the port started later than that of Genoa. Until 1866 it was included within the empire of the Habsburgs, who favoured Trieste as their Adriatic port rather than Venice. Significant growth did not begin until after 1880, even though the railway had reached the shores of the lagoon at Porto Marghera as early as 1842, and was extended by a causeway to Venice itself four years later. During the period of its commercial greatness the port of Venice lay at the eastern end of the principal island and along its southern shore. This became inadequate during the nineteenth century and was first supplemented by docks dug in the silt at the western end of the city and then largely replaced in the late nineteenth century by a larger and more complex system of docks on the mainland at Porto Marghera. The trade carried on from the island of Venice was limited to light goods and general cargo. Heavy goods were handled at Porto Marghera, where, early in the present century, modern industry began to develop.

The chief port of peninsular Italy was Naples, the largest city with the most populous hinterland. The early port lay at the head of the Bay. During the late nineteenth century this grew in classic Mediterranean fashion. Jetties were built at right angles from the shore behind the protective barrier of a mole. In the late nineteenth century Naples grew into the most important passenger port of the Italian peninsula, on account of the heavy migration (see p. 85) from southern Italy to the New World. The opening of the Suez Canal put Naples very close to a major shipping route. It became a very important port of call, handling a greater shipping tonnage than any other Italian port in the pre-war era. The volume of cargo handled, however, never approached that passing through the northern ports. The chief port of central Italy was Leghorn (Livorno). Though built in the Middle Ages, it remained small until late in the century. Its docks – unusually for a Mediterranean port – were largely excavated in the soft silt of the Maremma marshes, where they attracted shipbuilding, chemical and other heavy industries.

There were dozens of ports around the Italian coast; some like the iron-ore port of Piombino and the Sicilian sulphur port of Porto Empedocle, were specialised. Others, like Palermo, Taranto, Brindisi, Bari and Ancona, became general ports for their regions. But most in 1914 were under-equipped and greatly in need of dredging and modernisation. Some imported coal and exported agricultural products but, by and large, they carried on only a local and coastwise trade.

Dalmatia and the Balkans The mountainous Dalmatian coast, from the silted and abandoned Roman port of Aquileia to the Gulf of Kotor, where the crags of Montenegro look down upon the sea, was under Habsburg control from 1815 to the First World War. It was natural, therefore, that both Austria and Hungary should attempt to develop ports along this coast, so rich in natural harbours. The problem, however, lay in the mountains which formed a broad belt of rugged country across which no rivers had created any obvious or natural route (see p. 459). The ports which had grown up on this coast before the nineteenth century – Senj, Zara, Split, Dubrovnik – mostly tapped the produce of their immediate hinterlands, and despatched it by sea to Venice or some other Italian port. Their importance was small, and grew little during the century. Even the completion of a narrow-gauge railway to Metković and Dubrovnik brought little trade to the ports.[178]

Only in the north-west where the Dinaric system was both narrower and lower was any significant trade developed between the interior and the Dalmatian coast. Here it was possible to cross the range at several points without rising much more than 500 m above sea-level. Even in the eighteenth century the Austrians tried to develop trade with the Turks through the north Adriatic ports of Trieste, Fiume and Senj.[179] Trieste and Fiume lay at the head of the gulfs which lay respectively on the west and east of the Istrian peninsula. Senj lay to the south-east, overlooked by the rugged Velebit. The easiest routes from the interior converged on Trieste and Fiume, and the prosperity of Senj was short-lived. Trieste had been a port for Austrian trade before 1815, exporting not only iron, agricultural produce and timber, but also glass from Bohemia and even linen from Silesia.[180] It is not surprising, then, that the first railway planned within the Habsburg Empire was that between Bohemia and Moravia in the north and the port of Trieste.[181] It was not completed until 1857. In the mean time, however, the Austrian Lloyd Shipping Company had been established in 1832, with its headquarters at Trieste, and by the mid-century had become the 'biggest steamship company in the Mediterranean'.[182] When at last the railway reached Trieste all was set for a vigorous commercial expansion. Both the docks and the volume of shipping using them grew rapidly. The hinterland of Trieste remained essentially Austria and the middle Danube basin, and the port derived little from northern Italy.

The railway from Vienna to Trieste ran through Graz and Ljubljana (Laibach), avoiding Hungarian territory. Although Croatia, with its ports of Fiume and Senj, were part of Hungary, little use was made of them. It was cheaper, wrote a contributor to *The Quarterly Review*, to send a cask of Hungarian wine to the west by way of the Danube and Constantinople than over the hills to Fiume or Trieste.[183] When, in 1867, Hungary was separated administratively from Austria, the port accommodation at Fiume was primitive, and was linked by execrable roads with its hinterland. Thereafter, however, its development was rapid.[184] In 1873 the railway from Budapest was completed. The Hungarian government spent large sums on improving the port and establishing Hungarian-owned shipping lines. Docks spread westwards along the coast in typically Mediterranean fashion, and the value of trade passing through the port increased more than fivefold between 1871 and 1895. Fiume was, in fact, one of the fastest growing ports in Europe, and there can be no doubt that it drew a great deal of traffic away from its Austrian neighbour, Trieste.[185]

Despite the projects formulated at the end of the century for a railway from Serbia to an unspecified port on the Dalmatian or Albanian coast, nothing was in fact done, and between Fiume and the southern headlands of Greece there was only one port of significance, the Greek port of Patras. It was destroyed by the Turks during the Greek war of independence, and it was long before town and port recovered. The harbour, in fact, was not built until after 1880. Patras then developed as the chief port of southern and western Greece, and became the focus of a network of coastal shipping routes. Agricultural products, especially currants and wine, were collected at Patras for export, and imported goods were distributed from the port.

Piraeus, on the Aegean coast, became in the course of the century by far the most important Greek port. In classical times it had been the port of Athens, but was destroyed by the Romans and never rebuilt until the nineteenth century, when Athens was again the capital of an independent state. After it had been refounded in 1834 it grew rapidly with the expansion of Athens itself. As in classical times the port occupied the deep, narrow and sheltered bay to the north of the town. It imported fuel, foodstuffs and industrial raw materials for the capital, and, in a very unbalanced trade, exported mainly agricultural products. It developed also as an entrepôt for the trade of the Aegean. The opening of the Corinth Canal in 1893 extended the area served by the local boat traffic of Piraeus to the Gulf of Corinth and beyond, thus encroaching on the sphere of Patras.

The port of Thessaloníka had as great a potential as any in the Mediterranean. Under Ottoman rule it was as a commercial outlet for the Balkans second only to Constantinople. It lay to the east of the delta of the Vardar, free from the river's silt. Though the Vardar was not really a navigable river, its valley opened up a route through the mountains into Serbia, and during the middle years of the nineteenth century there were those who envisaged Thessaloníka as a port for the Danube basin. After 1870 a rail net began to be developed in Macedonia. It was not an ambitious net, and it tapped an impoverished and undeveloped region. Not until the beginning of the present century was it found necessary to construct a modern harbour. In 1912 the city passed from Ottoman to Greek rule, and much of its hinterland was annexed by Serbia. The potential of the port was at once increased, as Serbia looked to it as its principal access to seaborne commerce. Early in 1914 the Greek and Serbian governments reached an agreement whereby part of the port of Thessaloníka would become a free zone for use of the Serbs, but the war broke out before it could be implemented.[186]

The northern shore of the Aegean, like the western, has a number of small ports, none of which would deserve mention were it not for Bulgaria's ambitions to occupy and develop them, and thus to gain direct access to the Mediterranean. At the end of the First Balkan War, in 1912, the coast of Thrace, including the ports of Kavalla and Dede Agach, was annexed by Bulgaria (see p. 29). A year later, after the Second Balkan War, Kavalla was lost to Greece, and at the end of the First World War Greece also took the rest of Thrace, thus ending Bulgaria's short-lived corridor to the sea before there had been time to develop any trade through it.[187]

Constantinople and the Black Sea

Throughout the nineteenth century Constantinople was the hub of the Ottoman Empire. The city spread over the hills which bordered the Bosporus. Opening from the Bosporus was the creek called the Golden Horn. From classical times it had served to shelter ships, which could tie up at the wharfs which fringed the waterfront. To the west of the Bosporus a dry grassland, a kind of European steppe, reached into Bulgaria. There were few towns, and population was sparse. In effect, Constantinople had a very restricted hinterland; most of its traffic came by water and left by water. The phenomenon, seen at Patras and Piraeus, of an entrepôt which received and despatched its freight by sea, was here carried to its extreme. Sailing caiques brought in foodstuffs, agricultural produce, timber and other goods and carried back to the coastal villages on both the European and the Asiatic shores manufactured and consumer goods. At the same time large ships carried a surplus to the west and requited it with western products.

The Bosporus formed the only exit from the Black Sea. Ships sailing to and from the Russian ports and the Lower Danube had to pass beneath the minarets of Constantinople and the guns of the Ottomans. The Russians had gained certain rights by the Treaty of Kuchuk-Kainarji of 1774, including the freedom of navigation through the Straits for their shipping. By the Treaty of Adrianople of 1829 this freedom of navigation was extended to the ships of all countries at peace with the Porte.[188] Henceforward the Black Sea was open to the trade of the western powers, and the volume of shipping passing the Straits began to increase. Western trade with the Black Sea grew rapidly because of the grain of Walachia, Moldavia and Russia. First to develop was the grain trade with the Danubian ports of Galați and Brăila. Galați lay on a low bluff well above the Delta, where the north-flowing Danube turned eastwards to the sea. Sea-going ships could sail up to the town (see p. 447) and tie up along the river's banks. It was, wrote McCulloch, following a British consular report,[189] 'ill-built and dirty'. The trade was chiefly in the hands of Greek merchants, though recently foreign trading houses had been established.[190] 'It may', he added, 'be said to be the port of the Danube' and would, if political difficulties could be overcome, 'become a first-rate emporium.' Brăila, 20 km upstream, developed later. Both were essentially river ports, the grain being brought down to them by an immense number of lighters from loading-places farther up the Danube.[191]

Other ports on the Black Sea coast also shipped grain to western Europe. There was a small export from the Bulgarian ports of Varna and Burgas,[192] but the dominant port in the Black Sea grain trade was Odessa. It had been founded as recently as 1794, but quickly developed the export of wheat and rye from the western parts of the Russian Steppe.[193] Grain was brought to Odessa overland from Bessarabia and even from Podolia and Volhynia on the borders of Poland. By 1850 Odessa was 'the greatest wheat export center in the Russian Empire', and a major factor in the Russo-Turkish war of 1828–9 was Turkish interference with this trade as it passed through the Straits.[194] The Russian grain trade reached its maximum in the 1890s.

Internal trade

Only a fraction of Europe's commerce passed through its hundred or more ports. Most of the domestic trade and much of that between countries was overland, between one market centre and another, one country and its neighbour. In the first half of the nineteenth century there was still a high degree of regional self-sufficiency, and most goods changed hands in the local market. Only a small fraction of total local production – agricultural or industrial – ever reached an urban or regional market, and an even smaller fraction passed through the latter to a national or international market.

These conditions varied. The condition of near self-sufficiency – the 'economy of no markets' – was more marked in the east and south-east than in the west. But everywhere the relationship between town and country, regional capital and rural market, central-place and scattered community was in flux. The chief factors in this changing spatial pattern of trade can be summarised as improving means of transport, increasing specialisation, and rising demand resulting from higher real incomes. Other factors were the increase in the tertiary, or service sector of the economy; standardisation, at least within individual countries, of weights, measures and currency; reduction in tariffs, the elimination of tolls, and the emergence of a banking system which made low-interest commercial credit more readily available. These were interrelated, improvement in one leading to an increase in the others in a reciprocal fashion.

But the most important of these factors was unquestionably the increasing ease with which goods could be moved. Improvements in the means of transport had begun in the pre-railway age with the construction of canals and the building of new roads. Travel became faster, and wagons larger. Heavy freight of low value began to be transported by road over very considerable distances. One of the most remarkable instances of the transport of heavy freight by horse-drawn wagons was that of raw cotton, already mentioned, from Rouen to Alsace. It was, however, the railway which did most to break down the near self-sufficiency and commercial isolation of the local community. Most railway nets, as they developed, showed a radial pattern around the larger cities and regional capitals. Inevitably they drew to themselves much of the trade of their surrounding regions.[195] Small towns often stagnated; the larger usually grew rapidly. The impact of railways began to be felt in western Europe from about 1850, but not until considerably later in central Europe, while it was not before the twentieth century that modern modes of transport began to have much impact in eastern and south-eastern Europe. In France the increase in the volume of merchandise transported between 1830 and the outbreak of the First World War was accounted for almost entirely by the railways (fig. 9.2).[196]

The increasing importance of local specialisation would have been impossible without a comparable increase in effective demand, and this in turn sprang from a slowly rising standard of living. A great deal of controversy has surrounded this matter, the pessimists claiming that there was very little improvement generally until late in the century. It is, in fact, difficult to generalise. It is doubtful whether the standards of the rural population increased appreciably. The real wages of agricultural workers remained low, principally because there was a surplus of rural labour. In urban areas, where industrialisation was taking place, there seems to have been a sharp contrast between the standards of the skilled artisan, which were rising, and those of the unskilled or semi-skilled day-labourer who bore the brunt of cyclical changes in business prosperity and food supply. Overall, however, there is evidence for a very slowly improving diet, an increasing standard of housing and, above all, a growing demand for consumer goods. This increasing per capita consumption became marked during the second half of the century, first in western Europe; then in central, and lastly in eastern Europe. The Polish peasant, Jan Slomka, wrote in his autobiography that 'about 1870...the peasants [began] to build proper brick chimneys, when the iron cooking stoves came in...'[197] It was by such small increments that demand was increased in the course of the century.

A final factor in the slow transition from an 'economy of no markets' to one in which a large part of the goods produced actually changed hands and were transported over distances more or less great before being used or consumed, was the expansion of the tertiary sector of the economy. There had always been a small section of the population which was engaged not in production or manufacture but in providing intangible goods or services. This section increased greatly during the nineteenth century, and its growth was to be related to the expansion of industry and other productive activities. 'It has been held', wrote Colin Clark, 'that a high level of income per head is always associated with a high proportion of the working population engaged in tertiary industries.'[198] The tertiary sector was made up in varying proportions of five different types of service:[199]

(1) Social overhead services, consisting of transport and communications, provision of public utilities, education, public order and defence.
(2) Business services, including wholesale, retail and financial services.
(3) Personal services, including domestic and legal.
(4) Community services, including religious and social.
(5) Cultural and recreational services.

Not all these were important in the nineteenth century but in the aggregate they tended to absorb an increasing proportion of the total population.

By 1870 the tertiary sector had absorbed the following (approximate) proportions of the total population:

    35 per cent: Great Britain, Belgium
    25 per cent: Germany, Netherlands, Norway
    12–20 per cent: France, Italy, Spain, Denmark, Austria-Hungary and Portugal.

By 1900–10, the proportions (again, approximate) had risen appreciably:

45 per cent: Great Britain
30–35 per cent: Germany, Belgium, Norway, Netherlands
16–25 per cent: France, Denmark, Austria-Hungary, Italy, Spain and Portugal.[200]

Since the tertiary sector produced no tangible goods, it had to be supplied with all foodstuffs and consumer goods through the market mechanism. At the same time part of this tertiary sector was itself engaged in transport and trade, wholesale and retail, in these commodities.

Yet other factors were of great, though sometimes only local importance. They included the abolition of tolls on river and canal, and the reduction and in some instances the elimination of tariffs on the movement of goods between two or more countries. The consequences of the formation of a tariff-free area, the Zollverein, in central Europe, were in this respect immense. Also of importance was the gradual standardisation of weights and measures and the elimination of countless purely local standards. The long-distance sale of goods required some generally accepted unit in which to express their volume and value. The metric system, adopted by the French government in 1791, came into general use very slowly even in France, and it was long before it replaced older measures in the rest of Europe. In 1875 an international bureau was established, with its seat near Paris, for the purpose of encouraging its use, and by the end of the century the metric system was generally used in all except small-scale and local transactions.

The mechanics of marketing

At the beginning of the nineteenth century trade within Europe was carried on at three different levels. At the lowest was the petty huckstering of the village or small town market. Peasants sold or bartered their agricultural surpluses, always in very small quantities, for equally small amounts of consumer goods. The market was held frequently – once or twice a week – because many of the goods traded were perishable and few of those who attended were able to deal in large quantities. Most of the goods which were bought and sold went no farther; they were used or consumed by those who thus acquired them, and the whole transaction was within the compass of a very nearly self-sufficing community.

On a higher plane was the business transacted through urban shops. They were open most days of the week; they carried an inventory more or less large, for which they had to be adequately capitalised. In some instances they served as outlets for local craftsmen, whose products they sold. Many were supplied by merchants, serving as wholesalers, who acquired goods from local markets, from manufacturing centres or from importers.

Lastly, there were fairs. These differed from markets in their lesser frequency and their concentration on non-perishable goods and long-distance trade. Some were specialised, dealing mainly in cloth, animals – especially horses – or exotic goods imported into Europe. Most lasted for several days, at the end of which stalls were dismantled and the traders dispersed until fair time next year brought them together again. There were, it is said, 26,000 fairs in France in 1836, but most were in decline, and many had probably degenerated into 'fun fairs'.[201] The fairs of Beaucaire, amongst the last to retain their vigour, almost disappeared after the railway had been built in 1839 to the nearby town of Tarascon.

Fairs retained their importance somewhat longer in central Europe, and some remained important throughout the century, though their function changed.[202] The Frankfurt fairs were of great importance in the cloth trade, especially in silks and high quality woollens. Leather and metal goods were also brought to the fairs in quantity.[203] The Leipzig fairs were even more important, and, indeed, their business was increasing during the middle years of the century.[204] Leipzig lay near the frontier of developed, industrialised Europe. Beyond was the vast area of eastern and south-eastern Europe, with towns small and infrequent and craft industries only feebly developed. The products of the west, especially silk and woollen cloth, lace and metal goods, passed through the Leipzig fairs to be disseminated to the markets of the Russian and Ottoman Empires.[205]

Fairs were most important and survived longest in areas where town life was least developed. Nowhere was this more conspicuously the case than in the Balkans, where towns were too few and too undeveloped to serve as a medium of trade. Indeed, the number of fairs is said to have increased in spectacular fashion during the first half of the nineteenth century.[206] The goods traded consisted mainly of cloth and such small fabricated and luxury goods as could be transported easily, and, of course, animals. Fairs were especially numerous and important in the zones of contact between mountain and lowland, on the edge of the plains of Hungary, Macedonia, Thrace or Walachia. In the nineteenth century, it is said, most of the trade was carried on at fairs. They represented little by way of fixed assets, and destruction in the frequent wars could be repaired without great difficulty. Amid the hundreds that were active two were of especial importance throughout most of the century.[207] Most famous was that of Uzundžovo, held in Thrace between Plovdiv and Edirne (Adrianople). The other was that of Eski Džumaja, which took place below the Balkan Mountains, between Šumen and Ruse. These and other fairs in the Balkans remained very active until late in the century. Then changes in political boundaries combined with urban growth and the building of railways to break down the linkages on which the Balkan fairs had been based.

In the rest of Europe fairs had declined long before this, and their business had passed to the towns. Wholesale trade became much more organised. It ceased to depend on the systematised randomness of the fairs, and came gradually to be managed by means of advertising, the issue of catalogues and price-lists and even the periodic visits to retail shops of itinerant salesmen. Ordering by post and delivery by rail became increasingly common. Goods became more standardised. It ceased to be necessary for the seller to carry around with him a large stock of the commodity which he had to sell; a sample became adequate. This revolution in the ways in which goods were bought and sold was due to increasing real incomes and rising standards of living, but it would not have been possible without means of rapid communication by post and of quick delivery by rail.

Retail outlets increased in number and size. In Prussia, for example, there were said to be three times as many shops per thousand of population in 1900 as there had been 50 years earlier. Furthermore, shops were changing their character. Few now served as outlets for the goods made in workshops lying behind their façades. Instead their wares came from larger and more distant factories.[208] Even food shops became detached from their sources of supply. Fruit and vegetables continued to be brought to market almost daily, even in the largest towns, but those foodstuffs with a longer 'life' were increasingly handled by a wholesaler who supplied a number of retail shops. In the last third of the century the grocery 'store' began to appear in the larger towns. Some were even organised into 'chains', which, by means of bulk-buying from wholesalers and the economies of scale, were able to reduce their prices appreciably. A further advance in retailing was the creation of the first department stores, with the foundation of Bon Marché in Paris in the 1860s.

These developments in retailing took place first in France, Belgium and western Germany where the growth in urban population (see p. 149) was most rapid and the railway system most developed. They spread very slowly to central Europe, but their impact on eastern and south-eastern Europe remained slight. Here, the market and fair yielded up their role to the wholesaler and retail shop only very slowly. Indeed in these areas of low effective demand the pedlar continued into the twentieth century to hawk from village to village the goods he had acquired at the fairs.

Conclusion

The development of a system of transport and communication lay at the heart of nineteenth-century industrial and urban growth. Without it, there could have been no regions of concentrated manufacturing; large cities could not have been supplied with their basic materials and foodstuffs, and the economic life of the small ones would have remained integrated only with that of their surrounding rural areas.

When the century began there was a system of roads, supplemented, especially in north-west Europe, by one of navigable waterways. Apart from the small amount of freight which moved by sea, these systems carried everything that moved. Roads were bad at all times, and some were almost impassable in winter. Attempts to improve their quality were in general frustrated by the lack both of engineering skills and of the necessary raw materials. Travellers moved by coach, on horseback and by foot. Freight was carried slowly and laboriously by wagon and pack-animal. The cost was formidable, and the prominence of luxury goods in eighteenth-century trade was due in part to the fact that they could most easily bear the cost of transport.

Water transport was used wherever possible, and it alone proved capable of conveying bulky and heavy goods relatively cheaply. But transport by river and canal faced severe problems. Boats were propelled by wind or traction, both highly unreliable, and there were few waterways which did not cry out for major engineering works. Nevertheless the role of navigable waterways in early industrial development is incalculable. Every one of the early industrial regions had the benefit of transport by water, and without it their course of development would have been very different. It would probably not be unfair to say that the early development of the north-west European region owed more to its navigable rivers than to its coal resources.

After about 1840, at least in western Europe, the railway began to draw traffic from the rivers and canals and to displace the roads in all except local and short-distance movement. Although the earliest railways were merely tramways, it was the steam locomotive which gave to the railways their spectacular importance. Of the three major uses to which the steam-engine had been put since its improvement in the eighteenth century – mines, factories and railways – the last was by far the most significant, because without a developed transport system the others could not have achieved their fullest development. Surely one of the most remarkable aspects of the change in Europe's geography during the century was the speed with which a primary railway net was constructed and brought into use. Of all the indices of growth, that of freight carried by the railways is the most impressive (fig. 9.12).

Only for France is it possible to follow the changes in the mode of transport during the century (fig. 9.2). The railway gained at the expense of both its rivals. Water transport recovered somewhat late in the century, as canals were improved and extended and the volume of bulk goods transported, notably coal and iron ore, increased. But a recovery in long-distance road transport had to await the internal combustion engine and the improved road-building which it demanded.

Trade which passed through Europe's ports grew steadily if not spectacularly during the century. France's foreign trade had roughly doubled in volume by 1850, and by 1913 had increased fourfold in the previous 60 years. That of Germany grew only threefold between 1880 and 1913, a time when Germany's manufacturing underwent a very rapid expansion. The foreign trade of the Low Countries increased about 25 times between about 1846, when statistics were first compiled, and 1913, but much of this growth occurred in the last 25 years and was due to a large extent to the growing trade with the Rhineland. In other countries growth was much slower, and in some, Portugal and the Balkan countries, for example, foreign trade virtually stagnated throughout the century. Domestic trade, as measured by movement on the rivers, canals and railways, grew at a very much faster rate than foreign and overseas commerce.

A feature of commercial growth during the century was its tendency to concentrate in a small number of giant ports. By 1913 some 12 of them handled at least three-quarters of Europe's sea-borne trade. The reasons for this development lay both in the increasing size of ships, preventing them from entering many smaller ports, and the infrastructure of docks, warehouses, cranes and railways, which could not be duplicated at more than a small number of them.

Domestic trade, lastly, ceased to be dominated by the market and fair, and was transferred to retail shops and commodity markets located in the larger cities. The growing importance of factory production was leading to a greater uniformity or consistency in the products. It became possible to buy by sample and to order from a catalogue. This, together with the slowly increasing purchasing power of the mass of the people, led to the growth late in the century of the 'store', which served as an urban outlet for a wide range of sometimes unrelated goods, deriving from factories in different parts of the country or continent. The department store was the complement to factory production, as the small-town shop had once been to the domestic craftsman and proto-industrial worker.

10
Europe in 1914

A previous volume ended with a survey of Europe on the eve of the Industrial Revolution. This one will finish with a description of Europe's geography almost a century later, when a long period of steady and barely interrupted growth was cut short by the guns of the First World War. In the course of that century population had more than doubled; gross national product had increased many times – in the more advanced countries up to ten times. A continent which had once been on balance self-sufficing in foodstuffs had become dependent on the rest of the world, with which it was linked in a trading network of growing complexity. Urban population, which had been no more than 15 per cent of the total at the earlier date, had increased to 45 per cent of a very much larger total. In the early nineteenth century the largest city, with the possible exception of Constantinople, was Paris, with about 550,000 inhabitants. Fewer than 20 cities exceeded 100,000, and there were only about 200 with more than 20,000. By 1914 this had changed radically. There were four cities of more than a million, and 150 of over 100,000. Factories, which in 1815 could have been counted on the fingers of one hand, were now scattered in their hundreds from one end of Europe to the other, and their raw materials were drawn from the ends of the earth.

What were the essential conditions of this transformation? It has been suggested that there were two, without which it could not have taken place: mechanical power and bulk transport. The two were connected, in so far as both were dependent on steam-power. Without the steam-engine factories could have been nothing more than efficient workshops, and the railways only a system of horse-drawn tramways. Mass-production, specialisation and exchange, essential features of the Industrial Revolution, could have been developed only on the feeblest scale.

The transport net

It is impossible to map with any pretension to accuracy the use of the steam-engine and other forms of mechanical power in the Europe of 1914. The number of such installations had grown too large, and in most countries it was no longer recorded. But the transport net – the other condition of economic growth – was both apparent and intensively used. By 1914 the railway system had been completed except in parts of eastern and south-eastern Europe. In some areas

Table 10.1. *European railway track 1914*

	Length of track (km)	Track (km) per 1,000 sq. km
Austria-Hungary	22,981	36.8
Belgium	4,676	153.3
Bulgaria	2,109	24.2
Denmark	3,868	89.7
Finland	3,560	10.9
France	40,770	76.0
Germany	63,378	117.2
Greece	1,584	24.4
Italy	18,873	65.9
Netherlands	3,305	97.8
Norway	3,085	9.5
Poland	2,796	22.0
Portugal	2,958	32.6
Romania	3,549	27.0
Serbia	1,598	33.1
Spain	15,088	29.8
Sweden	14,377	33.6
Switzerland	4,832	114.2
Great Britain	32,623	139.5

it was even over-developed. Table 10.1 presents the total track in each country and its length per thousand square kilometres of territory.

The networks in Germany, the Low Countries, northern France and the Swiss Plateau were most developed. Belgium in fact had a denser rail net than Great Britain. Least developed were Scandinavia, southern Europe and, above all, the Balkans. No indicator emphasises more strongly than the density of the railway net and the intensity of its use the contrast between the developed, industrialised core of north-western Europe and the remainder of the continent.

Most of the European system was built on the standard gauge of 4 feet $8\frac{1}{2}$ inches. Narrower gauges were used for short distances in mountainous areas, notably the Dinaric region of the Balkans; the broad or 5-foot gauge was employed in Russian-held Poland, with the exception of the line from Warsaw to the Austrian border, and in the Iberian peninsula, where a strange assortment of gauges had been allowed to develop. Despite the tendency for lines to terminate as they approached an international frontier, there were enough cross-boundary routes for freight to move with relative ease from one part of Europe to another.

The intensity with which the railways were used varied greatly. Estimates of the tonne-kilometres of freight carried have been published by B. R. Mitchell.[1] From these it is possible to calculate an index representing the use made of the lines (table 10.2 and 3).[2] Data are not available for Portugal, Serbia and most of the Balkans, but it may be presumed that the index of utilisation of their railways would have been comparable with those of Bulgaria and Greece.

A similar table can be compiled, using Mitchell's data for passenger traffic (table 10.3). The correlation between the two tables is high.[3] In both, the intensity of use is greatest in the Low Countries and Germany, that in the

Table 10.2. *Intensity of use of European railways*

	km/tonnes of freight (millions)	Index
Belgium	5,729	1.224
Germany	67,700	1.106
France	25,200	0.622
Netherlands	1,802	0.560
Romania	1,443	0.420
Austria-Hungary	17,287	0.399
Italy	7,070	0.391
Switzerland	1,458	0.309
Sweden	3,184	0.230
Spain	3,179	0.217
Finland	649	0.178
Denmark	578	0.168
Norway	401	0.135
Bulgaria	176	0.091
Greece	50	0.032

Table 10.3. *Intensity of use of European railways (passengers)*

	km/passengers (millions)	Index
Belgium	6,242	1.334
Germany	41,400	0.676
Switzerland	2,685	0.569
France	19,300	0.477
Netherlands	1,433	0.446
Denmark	950	0.276
Italy	5,000	0.276
Romania	871	0.253
Austria-Hungary	8,321	0.193
Finland	704	0.193
Greece	297	0.188
Norway	462	0.155
Spain	2,139	0.146
Sweden	1,848	0.134
Bulgaria	136	0.070

Netherlands, however, being reduced by the considerable use made of water-borne transport. These were followed by France, with Italy, Austria-Hungary and Switzerland (except in passenger traffic) in the middle bracket, and Scandinavia and the rest of southern and south-eastern Europe much lower in the scale. The greatest intensity of use by both passengers and freight was clearly in the most industrialised countries. What is surprising, however, is the very considerable differential between the most heavily used railway systems, in Belgium and Germany, and the least used in Bulgaria and Greece. It follows that the cost of maintaining a railway system was disproportionately high for the less developed countries, which were least able to support it. The remarkable showing of Romania must be attributed to the traffic in cereals and oil.

Railways had at first been tributary to the canals and navigable rivers. By 1914 the position was reversed. Most of the rivers and many of the canals had passed out of use, and where inland navigation remained important it was restricted to bulk commodities of low value. Coal was the most important, followed by iron ore, timber and building materials. The size of barges and other craft had been increasing for many years, and smaller rivers and canals could no longer accommodate them. The French canal system had been modernised, and was in part well used since it provided a cheap means of transporting coal. It could, indeed, be said that the prosperity of the mines was in large measure founded on interior navigation.[4] But away from the coalfields the canals were little used, and traffic was disappearing from most rivers. By 1914 only those rivers, such as the Lower Seine, the Scheldt, Meuse, Rhine and Elbe, which provided a highway from the great ocean ports into their hinterlands, continued to be heavily used. And the only canals which remained active were those which either served to distribute coal from the coal basins or, like the Merwede and the Dortmund–Ems, extended the hinterlands of the great ports of north-west Europe. Inland navigation remained important in just those areas which already possessed the densest railway net. Elsewhere, from the Loire to the Vistula, horse-drawn barges and small sailing ships had almost disappeared from the rivers. Inland navigation remained significant only in an area which reached from Paris and the Lower Seine north-eastwards to the Low Countries and the Lower and Middle Rhineland. A canal linked this system with the port of Emden, but proposals to extend it eastwards to the Elbe and Danube, canvassed for decades, had still made no real progress. Not only were they strongly opposed by the railway interests, but the expected traffic appeared to be too small to justify the very large investment.[5] There continued to be traffic on the Lower Elbe, much of it in imported coal, and on the waterways which encircled the city of Berlin,[6] but there was little traffic on the Oder and farther east canals served little economic purpose beyond floating lumber.[7]

The high hopes that had once been entertained of developing shipping on the Danube had foundered on the physical difficulties of navigation and the political problems of the region. The Danube Steamship Company in 1913 handled only about 2.3 million tonnes of freight on the whole river, less than that carried on the Lower Elbe and only a fraction of the Rhine traffic.

Only on the Rhine was there any great volume of water-borne traffic. The work of regulation and correction carried on over the previous century had made it navigable to the Swiss border, and over most of this distance it could be used by the largest barges. It flowed through the most populous and industrialised area of continental Europe, and by its distributaries was linked with three of the greatest ports. Yet the Rhine's traffic was narrowly based. About 54.5 million tonnes of freight were carried on the German section of the river in 1913 or 10,470 million tonne-kilometres.[8] By contrast the railways carried about 63,000 million tonne-kilometres. Almost 70 per cent of the river freight was international, in so far as it crossed the boundary at Emmerich, and either came from or was destined for the Low Countries. The focus of this traffic was the Rhine-Ruhr ports, and coal from the Ruhr and iron ore for its furnaces made up about 85 per cent of the total traffic.[9]

The importance of road transport had been eclipsed over much of Europe by the development of the railways. Short-distance wagon traffic – to the nearest market, canal or river – was still significant, but horse-drawn coaches had largely disappeared from the roads, and the long-distance traveller, as well as freight, had turned to the railways. The roads, in consequence, had been neglected and their condition was generally bad in the early twentieth century, when new forms of road transport made their appearance. Attempts to develop a steam-powered vehicle which could use the highways met with little success.[10] The next attempt to bring life back to the roads was by the development of the bicycle. The modern bicycle took shape in the 1880s, but it clearly could not be used for freight and was valuable only for short distances. The motor vehicle, which followed soon afterwards (see p. 349), was destined to bring back to the roads much of the traffic they had lost earlier in the century. The number of motor vehicles increased rapidly, and by 1914 there were at least 200,000 and possibly 250,000 on the roads of Europe.[11] Apart from Great Britain, the greatest numbers were in France and Germany, where the automobile had first been developed. Few, however, could have been regarded as commercial vehicles; most were private limousines, bought and maintained more as status symbols than for any economic benefit they might offer.

The population map

The area of intensive railway development and of vigorous waterway traffic was also one of dense population and strong industrial growth. In 1914 the contrast between the most densely populated areas and the least populous was greater than ever before. The population map (fig. 10.3) suggests a fourfold division of the continent. There were in the first place highly urbanised and heavily industrialised regions in which only a small proportion of the population – generally less than 20 per cent – was employed on the land (fig. 10.1). The most important of these regions was that which reached from the Pas de Calais eastwards to the Ruhr, with an extension up the Rhine valley to Stuttgart and Strasbourg. To the east lay the urbanised and industrialised areas of Hanover, Saxony, northern Bohemia and greater Berlin. Yet farther afield lay a number of lesser concentrations of population and industry, developed, as at Paris and Vienna, around a political capital, or, like Upper Silesia and the Lyons-Saint-Étienne region, on important mineral resources, or, as in the hinterlands of Barcelona, Marseilles and Genoa, where port facilities encouraged the growth of manufacturing industries.

A second populous region consisted of areas of dense rural and agricultural population. Here, as a result of a high birth-rate and, often enough, of a system of partible inheritance, a dense population lived by cultivating small and poorly equipped family holdings. Such areas were southern Italy and Sicily, where rural density was probably greatest,[12] and both Spanish and Austrian Galicia. The contrast between dense rural and dense urban-industrial population is sometimes equated with that between eastern and western Europe (fig. 10.2).[13] This, however, ignores the fact that areas of very dense rural population were also to be found in peripheral areas of western Europe, as in Brittany, Sicily and parts

Table 10.4. *The population of Europe, 1910*

	Total	Per sq. km
Belgium	7,424	651
Bulgaria*	4,338	129
Denmark	2,757	177
Finland	2,943	9
France	39,192	189
Germany	64,926	311
Greece (1907)	2,632	41
Habsburg Empire		
Austria	28,572	247
Hungary	20,886	166
Bosnia-Hercegovina	1,898	96
Italy	34,671	314
Montenegro*	286	32
Netherlands	5,858	465
Norway	2,392	19
Ottoman Empire	6,000	90
Poland	13,056	103
Portugal	5,958	65
Romania	7,235	143
Serbia*	2,912	60
Spain	19,927	102
Sweden	5,522	32
Switzerland	3,753	235
Total	283,138	

* Data relate to areas before the Balkan Wars.

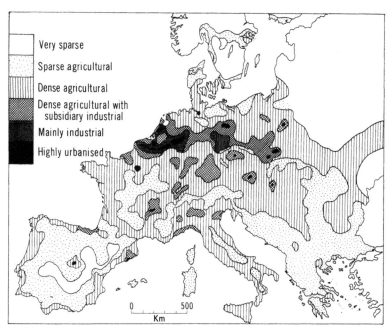

10.1 Types of population density: a simple model

Europe in 1914

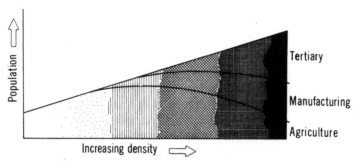

10.2 Diagram showing the relationship of population density to the economy. For scale of shading, see Fig. 10.1

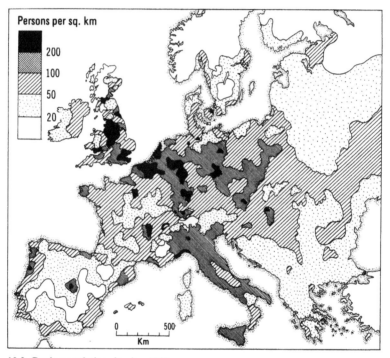

10.3 Crude population density, 1910

of the Iberian peninsula. Curiously, the quality of the soil was not a regular factor in the formation of such areas of dense rural population. Austrian Galicia, it is true, was characterised by a loess soil of high fertility, but in southern Italy, the Asturias and Spanish Galicia the dense population arose more from social than from physical factors.

These regions of almost Malthusian population pressure passed gradually into others with a more moderate density (fig. 10.3). These covered much of the plains of western and central Europe. They were in general good agricultural land where, for reasons as much social as environmental, urban and industrial growth had not taken place. They were characterised rather by small towns and craft industries, still linked by mutual dependence with their surrounding countryside.

Such were much of France and the North European Plain, of northern Italy and the Danube valley. They had densities of from 50 to 100 to the sq km, lower where the soil was poorer and the terrain more rugged; higher in areas of greater fertility and easier transport.

Peripheral Europe – consisting of the north and much of the east, south-east and south – formed a fourth population region. Density was low, as a general rule less than 50 to the sq km. Most of Norway and much of Sweden and Finland were in this category. The scanty population lived close to the coast, where alone the soil and climate were conducive to agriculture. The Baltic region and much of eastern Europe were also sparsely peopled. To the north were forested glacial deposits, and, in the south, the undrained marshes of the Prypeć region. In south-east Europe the Carpathian, Balkan, Rhodope and Dinaric mountains, with their continuation, the Píndhos and Grammos mountains of Greece, offered little scope for settlement, and population was chiefly to be found on the alluvial soil of the small basins like Ohrid, Bitolj and Prizren, ringed by bare, limestone mountains. To the east the Bulgarian platform bordering the Danube, the Marica valley, and the plains of Macedonia and Thrace offered more scope for settlement. But here the long period of misrule and civil strife had prevented development, and the density of population was far less than the land was capable of supporting.

The southern peninsulas presented curious contrasts and contradictions. In Greece, as in the Balkans, populous plains were separated by almost uninhabited mountains. Population lived mostly around the coast, on the narrow plains which separated the mountains from the sea. The situation in Spain was similar. The vast empty heart of Spain was much as it had been a century earlier. Transhumant flocks no longer traced their seasonal way between south and north, but, apart from the growth of the agglomeration of Madrid, there had been little increase in population. The situation along the coastline, between the Meseta and the sea, was very different. Here, in Asturias and the Basque country, in Catalonia, Valencia, Andalusia, and in northern Portugal and Galicia, its growth had been rapid, more like that of Italy than of the coastal regions of Greece. The Mediterranean islands were thinly peopled, with the exception of Sicily, which shared the high birth-rate and population pressure of central and southern Italy.

These contrasts within the peninsulas and islands of southern Europe are not easy to explain. The Italian birth-rate had long been high and in recent decades the death-rate, and in particular the infant death-rate, had fallen sharply.[14] In Spain, by contrast, the ravages both of civil war and of frequent and catastrophic epidemics held back population growth, even though birth-rates did not differ greatly from the Italian.

Birth-rates had been falling for several decades. They were lowest in Scandinavia and western Europe, and highest in Italy and Spain. The lowest, less than 20 live births per thousand of population, were to be found in France, with those of French-speaking Belgium and Switzerland very little higher. Death-rates had also been declining in much of Europe for half a century, and in 1914 were falling steeply. They remained highest in Spain and Italy. Data are lacking for eastern and south-eastern Europe; but birth- and death-rates were

Table 10.5. *Birth- and death-rates*

	Births per thousand	Deaths per thousand
Austria	31.3	20.5
Belgium	22.5	14.5
Bulgaria	41.7	20.6
Denmark	25.6	12.5
France	18.8	17.7
Germany	27.5	15.0
Hungary	34.3	23.2
Italy	31.7	18.7
Norway	25.1	13.3
Netherlands	28.2	12.3
Portugal	32.5	20.6
Romania	42.1	26.1
Spain	31.5	23.4
Sweden	24.0	13.8
Switzerland	23.2	14.3

Table 10.6. *Employment of economically active population*

	Agriculture, forestry, fishing (%)	Manufacturing (%)	Commerce, finance and services (%)
Austria	56.9	21.4	21.7
Belgium	23.2	39.8	37.0
Bulgaria	81.9	8.0	10.1
Denmark	41.7	24.1	34.2
Finland	69.3	10.6	20.1
France	40.9	31.9	27.2
Germany	36.8	36.3	26.9
Hungary	64.0	16.7	19.3
Italy	55.4	25.9	18.7
Norway	39.6	25.1	35.3
Netherlands	28.3	31.8	39.9
Portugal	57.4	21.5	21.1
Romania	79.6	7.8	12.6
Spain	56.3	13.8	29.9
Sweden	46.2	25.1	28.7
Switzerland	26.7	45.4	27.9

certainly high, in this traditional society. The 'vital revolution' was only beginning to show itself in the remote valleys of the Dinaric and Rhodope. Table 10.5 presents the crude birth-and death-rates for most of Europe at this time.[15]

The population of Europe was becoming increasingly divorced from the soil. Estimates of agricultural population must be subject to a wide margin of error because of its confusion with total rural population.[16] The latter is comparatively easy to estimate. The difficulty is to know how much to allow for the rural but non-farming population. Table 10.6 shows the percentage of the economically active population engaged in each of the major branches of activity.[17]

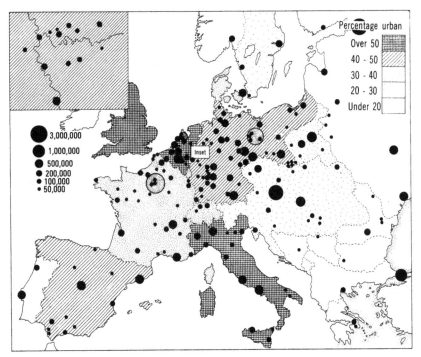

10.4 An urban map of Europe, 1910. Only cities of more than 50,000 are shown

The urban map

A consequence of changes in the occupational structure of the population of Europe was the rapid growth of cities and towns, in which most of the secondary and tertiary activities were carried on. No great change had taken place in the vast number of small towns, but the larger towns were in 1914 increasing rapidly in size and in the diversity of their functions (fig. 10.4). In 1914, about 45 per cent of the population of Europe was classed as urban, though definitions of 'urban' were not consistent (see p. 127). There is no evidence for the urban population of some countries. Those for which there is firm evidence are listed in table 10.7.[18] The urban population of England and Wales was, by comparison, about 75 per cent. Evidence is very uncertain for south-eastern Europe, but the proportion is unlikely to have been more than 10 to 15 per cent. There was a fairly high correlation between the level of urbanisation and the degree of economic development, as represented by the gross national product per head.[19] In this respect the fairly high levels of urban growth in Italy and Spain are misleading, since the very large villages and agricultural towns, especially numerous in southern Italy and Sicily, were counted as urban. On the other hand, a significant part of Sweden's industrial activity was carried on outside the towns and the workers were classed as rural.

There were in 1914 about 150 cities and towns of more than 100,000 inhabitants, representing about 43 per cent of total urban population. About 25 of these, with almost a quarter of the urban population, each had more than

Table 10.7. *Percentage of population living in towns, 1910*

Italy	62.4
Belgium	56.6
Netherlands	53.0
Germany	48.8
Spain	42.0
France	38.5
Switzerland	36.6
Denmark	35.9
Hungary	30.0
Austria	27.3
Sweden	22.6
Romania	16.0
Portugal	15.6

half a million. Fig. 10.4 shows the distribution of these giant cities, most of which were concentrated in north-western Europe. These great cities served many functions. Amongst the largest of them were the political capitals, the primate cities, with scarcely an exception, of their respective countries. Berlin and Paris; Brussels and Rome; Madrid, Vienna and Budapest, all of them cities of more than half a million in 1914, were widely diversified in function. All were, in the first instance, centres of public administration. Each contained a large civil service, together with a military garrison of varying size. Capital cities always attract a wide range of business, commercial and service functions, from banking and commodity markets to library and museum services. Some were attracted by the advantages of close association with government; others by the centrality and ease of transport and communication which capitals possessed. Most capitals were the foci of their respective railway systems, if only because the earliest lines were made to radiate from them. Capital cities, furthermore, attracted a considerable population merely because they were the centres of national life in their respective countries. Although they tended to be centres of consumption rather than of production, they became without exception the seats of manufacturing industry. As a general rule, however, the manufactures attracted to capital cities were consumer-oriented. They profited from the large market which the capital itself provided, and from the ease of transport to all parts of the country. They were, with few exceptions, light industries, labour-intensive, and requiring relatively little energy and raw materials. There were, however, some partial exceptions. The electrical engineering industry was established in Berlin, largely because it was here that Werner von Siemens founded its humble predecessor. The automobile industry was similarly established in Paris. Nevertheless, their dominant manufactures were typified by clothing, furniture and luxury goods, all of which required a large shop-window and a relatively wealthy clientele.

Not dissimilar in many respects to the national capitals were the larger regional centres, such as Prague, Munich, Cologne and Milan. As regional capitals they reflected on a humbler scale the services and functions of the primate cities themselves. As commercial, financial and transport centres they were less important, but in many instances made up for this by a relatively larger employment in manufacturing.

The great port-cities formed a second category. Administrative and cultural functions were relatively unimportant. Dock work, much of it unmechanised, demanded a large labour force. Transport services to and from the docks were essential, and in most port-cities a particular range of industries had grown up. These included, in addition to shipbuilding and repairing, the processing of imported materials. The manufacture of soap and vegetable oils from imported oilseeds, the smelting and refining of ores and concentrates, and the preparation of imported foodstuffs had all become significant at the major ports by 1914.

Lastly there were the industrial towns, those in which manufacturing provided the basic occupations. Service industries were no larger than was necessary for the town itself and its immediate area. This narrower range of functions in most manufacturing towns meant that as a general rule they were smaller than the capital cities, and regional centres; smaller also than many of the great ports. Typical of these manufacturing towns were Lille, Saint-Étienne, Essen, Chemnitz, Kattowitz and Łódź. Most of them were relatively specialised. Amongst those cited, Lille, Chemnitz and Łódź were dominantly centres of the textile industry; Saint-Étienne, Essen and Kattowitz of the iron and steel-working industries. This is true also of industrial towns lower in the scale of size, such as Liège, Zwickau, Dortmund and Longwy, and the textile towns of Catalonia and northern Italy. All had developed a relatively narrow range of industrial functions. A broad spectrum of industrial activity was paradoxically to be found only in those towns which were primarily national and regional capitals. The lower one goes in the scale of industrial towns the narrower becomes their specialisation, until in Commentry or Decazeville, Salzgitter, Recklinghausen or Königshütte, and the mill towns of Catalonia and the North Italian Plain one finds towns developed around a single industry or even a single factory.

The industrial map

By 1914 the continent of Europe and in particular the north-western part of it, had become the dominant industrial region in the world. It had far outstripped Great Britain, which had pioneered most of the industrial processes carried on there, and was still well ahead of that young industrial giant which had recently arisen in the New World. The basis of this industrial development was still, as it had been when growth began a century earlier, the energy derived from burning coal, and much of the manufacturing industry was still being attracted to sites easily accessible from the coalfields. Nevertheless, the supremacy of coal was being challenged by the development of both the oil industry and hydroelectric power. The latter, in particular, was beginning to stimulate growth in areas remote from the coalfields.

Continental Europe possessed less than 10 per cent of the world's known coal resources, but its output amounted to about 30 per cent of world production on the eve of the First World War. If Great Britain were included the proportion would have risen to more than a half. In 1913 continental Europe, including Russian Poland, produced about 285,185,000 tonnes of bituminous coal. Of this

Table 10.8. *Sources of energy, 1913*

	10^{15} metric calories	Percentage
Solid fuel	3,997.0	90.53
Wood used as fuel	207.0	4.69
Oil and petroleum	66.5	1.51
Natural gas	0.5	0.01
Hydroelectric power	144.0	3.26
Total	4,415.0	100.00

Table 10.9. *Production of bituminous and brown coal, 1913*

	Bituminous		Brown	
	Total	Per cent	Total	Per cent
Austria	16,460	5.77	27,378	21.83
[of which Bohemia-Moravia]	[14,087]	—	[23,137]	—
Belgium	24,371	8.55	—	—
Bulgaria	11	—	358	—
France	40,844	141.32	—	—
Germany	190,109	66.66	87,233	69.99
Greece	0.2	—	—	—
Hungary	1,320	0.46	8,954	7.14
Italy	1	—	700	—
Norway	33	—	—	—
Poland	5,770*	2.02	—	—
Portugal	25	—	—	—
Serbia	32*	—	273	—
Sweden	364	—	—	—
Spain	3,971	1.39	277	—
Romania	—	—	250	—
Total	283,311		125,423	

* Figures are for 1911.

Germany yielded almost precisely two-thirds.[20] There was also an output of about 125,423,000 tonnes of lignite or brown coal, of which Germany produced nearly 70 per cent, and Bohemia more than 18 per cent. Solid fuel supplied almost all the energy used in Europe at this time (table 10.8).[21] It is not possible to calculate the *direct* use of water-power, since this could be used only in thousands of very small units. Europe was on balance more than self-sufficing in energy, with a net export of about 4 per cent. The export of coal was partially balanced by an import of oil, mainly from Russia.

The distribution of coal production is shown in table 10.9 and fig. 10.5. There was also a small but unrecorded production of brown coal in the Balkans. Brown coal was mostly consumed close to where it was extracted and scarcely entered into long-distance trade. But the predominance of Germany, particularly the Ruhr, in bituminous coal production ensured that there was a considerable long-distance trade in coal. The largest export was unquestionably from Great

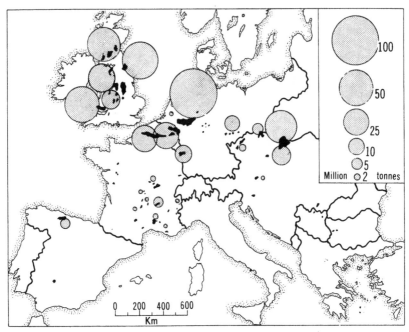

10.5 Coal production in Europe, 1912. The overwhelming importance of Great Britain and Germany is apparent

Britain, and British coal competed successfully with that from continental mines along the whole littoral of northern, western and even southern Europe. But there was also a significant movement of coal from Moravia to Vienna and other parts of the Habsburg Empire; from Upper Silesia to the Oder valley and Berlin; from the Saar to Lorraine and south Germany, and from the Ruhr to Luxembourg, Lorraine and all parts of the Rhineland.

The most important factors in the coal trade were its pithead price and the available means of transport. The former depended not only on the technical efficiency of the mines, but also on the geological conditions under which the coal was won. In these respects the small coal basins of France and the Liège basin of Belgium were highly uneconomic, and many mines had been closed. Most efficient were the new mines along the expanding northern margin of the Ruhr and in the Belgian Campine. The lowest pit-head prices were to be found in Upper Silesia, where favourable geological conditions combined with low wages to produce the cheapest coal in Europe.[22] Output per man/shift was highest in Poland, lowest in Belgium, where it was less than half the level of production in Upper Silesia–Dąbrowa.

For most countries in Europe the year 1913 marked the culmination of a long period of growth in the iron and steel industries (table 10.10). In that year the production of pig-iron was about 31,173,000 tonnes, and that of steel, 30,560,000 (fig. 10.6). About a quarter of the metal drawn from the smelting furnaces went into castings or was used in the puddling furnace to make 'soft' or wrought iron. It is doubtful whether the latter accounted for more than 2 or 3 per cent of the

Europe in 1914

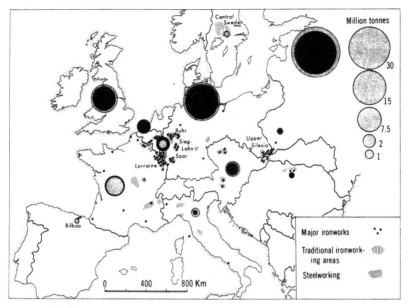

10.6 Iron and steel production, 1912

Table 10.10. *Production of pig-iron and steel, 1913 (thousand tonnes)*

	Pig-iron	Steel
Austria (including Bohemia and Moravia)	1,758	2,611
Belgium	2,485	2,403
Finland	9	7
France	5,207	4,687
Germany	16,761	17,609
Hungary	623	—
Italy	427	934
Luxembourg	2,548	1,326
Spain	425	242
Sweden	730*	591
Poland	418	n.d.
Total	31,391	30,410

* A significant part of the iron was refined to wrought or 'soft' iron.

metal produced. The rest was used for steel-making. Since the introduction of the open-hearth the practice had developed of using scrap metal in steelmaking, so that it was possible for the volume of steel made to be considerably in excess of that of pig-iron. This was the case in Germany, Austria and Italy. Only in Luxembourg, Lorraine and Spain was pig-iron production significantly in excess of steel production (see p. 343).

The iron industry was concentrated to the extent of about 72 per cent in just six areas; of these five, namely northern France, central Belgium, the Ruhr, the Saar and Upper Ssilesia, were coalfield sites. The other – Lorraine and

Table 10.11. *Pig-iron and steel production in north-western and central Europe* (*thousand tonnes*)

	Pig-iron	Steel
France		
Northern France	933	1,131
'French' Lorraine	3,560	2,291
Rest of France	514	1,265
Belgium	2,301	2,403
Luxembourg	2,548	1,326
Germany		
Rhineland-Westphalia	7,605	10,112
Saar	1,301	2,080
'German' Lorraine	3,819	2,233
Upper Silesia	1,048	1,407
Rest of Germany	1,827	1,777

Table 10.12. *Steel production in France and Germany, 1913* (*million tonnes*)

	France	Germany
Converter steel		
Basic	2,806.5	10,630
Acid	272.7	155
Open-hearth steel, basic and acid	1,582.5	7,613
Crucible steel	24.1	85
Electric steel	21.1	89
Total	4,686.9	18,572

Luxembourg – was an orefield site. Steel-making had been as a general rule linked with iron-smelting owing to the heat-economy that could be achieved. Nevertheless, steel-furnaces, mainly open-hearth and electric, were now becoming divorced from the smelting branch. The reason was that, owing to the cost of transporting low-grade ore, it was most profitable to smelt it near the orefield or at some convenient break-of-bulk point. Pig-iron could then be sent more cheaply to steel-making centres. Smelting had in fact ceased at a number of once important works, leaving only a residual steel industry, which used scrap-metal and imported pig-iron. Saint-Étienne was the foremost example, but there were several others in western and central Europe. Table 10.11 presents a more detailed breakdown of iron and steel production in the major producing countries.[23]

A feature of the iron and steel industry on the eve of the First World War was the growing importance of the basic-process. Approximately two-thirds of the ore smelted was high in phosphorus, and non-phosphoric ores were available in significant quantities only in Spain, the Rhineland, Austria and central Sweden. The vast consumption in Luxembourg and Lorraine was entirely of phosphoric ore, and most of that used in northern France, Belgium and the Ruhr was also high in phosphorus. In consequence more than 90 per cent of the steel made in north-western Europe was basic, most of it Thomas steel (table 10.12).[24] Acid steel continued to be made chiefly at coastal sites, where imported ores could be obtained most easily and in Austria, northern Spain and central Sweden. The puddling process was in rapid retreat, as the usefulness of 'soft' or wrought iron continued to decline. Within Germany only about 4 per cent of the pig-iron went into the puddling furnaces, more than half of it in Upper Silesia.[25] It was relatively more important in areas such as central France, Styria and Sweden, where traditional methods had survived more vigorously.

The vast expansion of the iron and steel industry of recent years would have been impossible without the commensurate growth of the iron-using industries. Shipbuilding, heavy engineering and the newly developing automobile industry were heavy consumers of steel, more than replacing the demand which had been created half a century earlier by the railways. These iron- and steel-using industries were in the main located near their sources of material. Shipbuilding was established mainly in the ports of north-western Europe, from Saint-Nazaire to Hamburg, with the heviest concentration near the Rhine mouth. Heavy engineering and steel construction were to be found close to the steelworks, with a particular concentration in cities, such as Düsseldorf, Aachen, Liège and Breslau, which lay near the primary iron- and steel-making regions. In some instances smelting and steel-making had actually retreated, leaving a legacy of steel-using industries. Amongst such centres were Saint-Étienne, Nevers, Commentry-Montluçon and Aachen. Some branches of the metal industry were more strongly market-oriented. These were in general more labour-intensive than the capital-goods production. Typical of such products were automobiles, precision goods and typewriters. Amongst the earlier centres for the first were Paris, Lyons and Stuttgart, none of which had any particular advantage in the supply of materials. Other branches of light and labour-intensive engineering were established in Berlin, Hanover, Bavaria and Württemberg, and in northern Italy.

A feature of the iron, steel and engineering industries early in the present century was their increasing integration not only between industrial regions but across political boundaries. No longer, except in parts of eastern Europe, did a works use local materials to produce goods for local consumption. Every significant iron- and steelworks was dependent on a distant source for one or more of its materials, and its market might span much of the continent. This contributed to two forms of integration. The more important in the present context was vertical integration, by which the steel producer sought to control not merely the source of iron but also of iron ore and fuel, and reached forward even to metal-fabricating and marketing. Many examples have been cited: the Ruhr, Belgian and French industrialists acquiring ore concessions and building smelting works on the Lorraine orefield; the Ruhr steel magnates owning coal-mines, coking ovens and fleets of Rhine barges; industrialists in Berlin and Breslau acquiring works in Upper Silesia to supply their steel. The list is almost endless, even though in some instances control was masked by the niceties of

Table 10.13. *Consumption of textile raw materials (thousand tonnes)*

	1900	1913	Percentage increase
Cotton	870.4	1,505.5	73
Wool	457.7	707.3	54
Flax and hemp	382	452	18
Synthetic fibres	none	7.3	—

commercial law. Nor did the mechanism of control stop at international boundaries. The long arm of a Krupp, a Stinnes or a Thyssen reached not merely into neighbouring France and Luxembourg, but into Austria, Sweden and Spain.

The other form of integration was horizontal. Two or more firms producing the same type of goods agreed to associate or to merge, usually in order to make a fuller use of resources or to share a market. Countless small works were in this way taken over and, in one way or another, absorbed into a larger operation. The formation of cartels was the culmination of this process. Without actually accepting common ownership and control, firms agreed on production and price levels and in some instances acquired a monopolistic position within their own countries. There was a cartel in the German coal industry and a series of cartel agreements within the steel industry. The cartel was, in effect, a German invention and up to 1914 almost a German monopoly.

Europe had once accounted for the greater part of the world's production of non-ferrous metals, especially of copper, lead and zinc, with most of the tin coming from Great Britain. On balance, Europe had maintained its output during the previous half-century, and in some areas, notably Upper Silesia and southern Spain, had actually increased it. But in 1913 European production formed a diminished fraction both of world output and of European demand. This change brought about a shift in the location of the non-ferrous smelting industries, which tended to leave the mining areas, such as Vieille Montagne, Harz and the Ore Mountains, and to become established in the ports and on the coalfields. Only in Upper Silesia, where the metalliferous ores and coal occurred in close proximity, did the older pattern of industry survive.

Coal-mining and the iron and steel industries were expanding rapidly when the First World War began. By contrast, the other dominant industry of nineteenth-century Europe, textiles, showed no such vitality. Although statistics are far from complete and reliable, it appears that the linen branch had been in decline for several decades. It had virtually disappeared as a cottage industry, and production, mainly of relatively high-quality fabric, was in factories in northern France and north-western Germany. At the lower end of the market it had been replaced by the cheaper cottons.

The woollen industry, as measured by the consumption of raw wool, was still expanding, though in its major centres, Verviers and the lower Rhineland, growth was slow. Only the cotton industry had been able to maintain its earlier momentum, though even this was just beginning to be clouded by two developments: the competition of synthetic fibres and of cottons woven in the 'new' countries (table 10.13).[26] The former did not become really significant until

Table 10.14. *European manufacturing, 1913 (percentages)*

	France	Germany	United Kingdom	Other
Bituminous coal	7	34	52	7
Steel	12	46	20	22
Machinery	5	48	27	20
Chemicals	14	41	19	26
Raw cotton consumption	11	19	43	27
Total manufacturing capacity	13	32	27	28

after 1925, and, in 1913, the number of spindles in Japan was only about half of that to be found in Italy.[27] In 1914 continental Europe is estimated to have had about 25 per cent of the world's cotton spindles, with a further 39 per cent in Great Britain.[28] Continental Europe had only 22 per cent of looms, though power-looms were probably more numerous relatively in Europe than elsewhere. There had been no significant change in the location of the textile industries during recent years, though there was a tendency for the smaller and older producing units, especially in Saxony, Bohemia and Silesia, to close.

The chemical industry, lastly, was a European invention, and had remained very nearly a European monopoly. Outside Europe it was important only in the United States, though the manufacture of basic chemicals and fertilisers was being developed in European Russia and in Japan. It is difficult to measure its rate of growth and its size in 1913 because of the diversity of its products. Its rapid expansion in the years immediately prior to that is, however, apparent. The production of sulphuric acid, for example, increased by 142 per cent between 1900 and 1913, and the growth in output of other basic chemicals as well as of fertilisers, pharmaceuticals and photographic materials must have been at least as rapid. The industry was located chiefly in France, Switzerland and, above all, Germany (see p. 346), where its heavy use of solid fuel had led to a concentration along the Rhine and in Saxony.

In 1914 a long period of economic and, in particular, of industrial growth came to an end, cut short by the weapons which it had produced. By this date continental Europe had displaced Britain as 'workshop of the world'. The latter remained the leading producer of bituminous coal and cotton textiles, but in most other branches of industrial production the British contribution was in fact smaller than that of Germany and in the aggregate amounted to no more than a quarter of the total European production. Over 70 per cent of Europe's manufacturing capacity was to be found in Germany, France and Great Britain (table 10.14).[29] If war had not come, the expansion of recent years would certainly have slowed and its direction would have changed. There was a preponderance of capital-goods industries, owing, in part, to the expectation that the recent heavy spending on investment would continue; in part to the large-scale expenditure on armaments. But it was made possible by the low wage rates and consequently low purchasing power of the mass of the population. Industry, furthermore, was oriented, at least in the more developed countries, towards the export market. This also would have changed, as overseas countries began to develop their own manufacture of goods for mass consumption. The war and its aftermath merely hastened changes which were inevitable.

Much of the manufacturing capacity of Europe had become concentrated within a relatively narrow belt which extended from northern France, through Belgium to the Rhineland and north-west Germany. From this 'core-area' manufacturing spread eastwards and southwards, 'into soil that was less and less well prepared to receive' it.[30] Even so, much of the continent was in 1914 still untouched by the technological advances which had been made within the core-area.

Around both the core-area and the other and smaller concentrations of manufacturing activity lay a peripheral region marked by a relatively dense population, an intensive agriculture geared to the provisioning of the industrialised areas, and a scattering of factories and other manufacturing centres. The latter were of two kinds. There were relict industries which had in varying degrees been superseded by the technologically more advanced manufactures within the core and nuclear areas. They included the manufacture of small iron goods in the Sauerland, Siegerland and Ardennes; linen-weaving in Flanders and relict textile industries at Arras, Cambrai and Saint-Quentin in the west, and at Bielefeld and Münster in the east. The second category of manufacturing consisted of new factories and new industries which marked, as it were, the expanding fringe of the industrial regions themselves. These included chemical industries in the Rhineland; glass and china in Belgium, Saxony and Bohemia, and a vast range of consumer-goods industries.

Agriculture

Agricultural output was in most parts of Europe larger in 1913 than it had ever been, and a greater number of people was fed with a smaller input of labour. Statistical data are uneven, incomplete and not wholly reliable. Nevertheless one can present a quantitative picture of farming in Europe when the Great War began. The increasing productivity which had characterised recent years was not uniform across the agricultural field. It was most marked in the new crops, such as potatoes, sugar-beet and maize; least in such traditional cereals as rye and oats. Overall there had been a decline in the less palatable bread-crops like rye, barley and buckwheat, and an increase in the more desirable wheat. Olive-growing was in decline in Mediterranean Europe owing to the competition of animal fats and, though viticulture was increasing in southern France, Italy and Spain, it was contracting in marginal areas near the climatic limit of the vine.

Animal farming

The most conspicuous changes of recent years were in animal-rearing. There was in many parts of Europe a swing towards mixed farming and an increase in the cultivation of fodder-crops. The number of cattle increased and, more significantly, their quality was greatly improved (fig. 10.7). Pig-rearing was expanding, and the only significant decline had been in the number of sheep.

Cattle had become the most numerous farm animal throughout western and

Europe in 1914

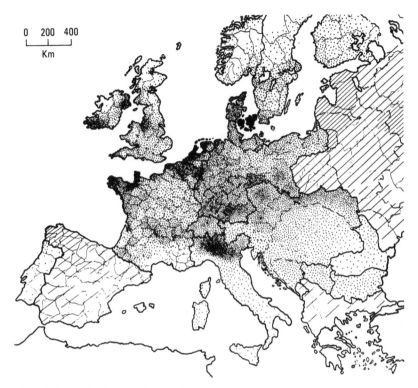

10.7 Dairy cattle about 1913. Each dot represents 5,000 head

central Europe. They had always been multiple-purpose animals, providing both meat and milk and serving also to pull the plough and the wagon. There was, however, a movement away from breeding cattle for beef and towards dairy-farming. In some countries, notably Denmark, the Netherlands and north-western France, farming was being directed increasingly towards producing milk, butter and cheese for the urban market. The *Geography of the World's Agriculture*,[31] based mainly on statistics gathered shortly before the war began, can be taken as representative of conditions of 1914. The map, fig. 10.7, shows the distribution of cattle, dense in 'Atlantic' Europe, south Germany and northern Italy; sparse in southern Europe, the area of high-farming in France, and throughout eastern Europe. Cattle kept specifically for dairying were very much fewer, probably no more overall than a third of the total. But areas of concentration were clearly shown: Brittany and Normandy, the Netherlands, and Denmark, especially the Danish islands (table 10.15).

By contrast, there were few countries in which the sheep population had not declined, and in some the fall in numbers was catastrophic. Overall their decline in western and central Europe was of the order of 45 per cent. In Austria and Hungary they were reduced by a half, and in Germany by 80 per cent. Only in Italy and perhaps the Balkans had there been any significant increase in the size of flocks. The reason lay in the fact that sheep had been reared primarily for wool; meat and milk were of only minor importance. Not only had the woollen

Table 10.15. *Numbers of cattle (in thousands)*

	Cattle	Cattle per thousand of population
Austria	9,160	320.6
Belgium	1,880	253.2
Bulgaria	1,606	370.2
Denmark	2,254	817.6
Finland	1,199	407.4
France	14,532	370.8
Germany	19,332	297.8
Greece	298	113.2
Hungary	6,184	296.1
Italy	6,337	182.8
Netherlands	2,027	346.0
Norway	1,088	454.8
Romania	2,667	368.6
Spain	2,369	118.9
Sweden	2,748	497.6
Switzerland	1,443	384.5
Serbia	957	328.6

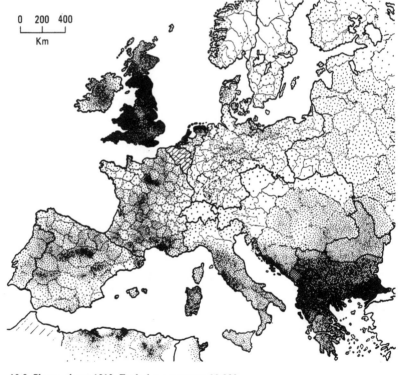

10.8 Sheep, about 1913. Each dot represents 10,000

Table 10.16. *Numbers of farm animals in western, central and northern Europe*

	1860	1910
Cattle	48,062	64,371
Sheep	86,325*	47,594
Pigs	23,369†	46,941

* Estimated for Sweden.
† Estimated for Austria-Hungary.

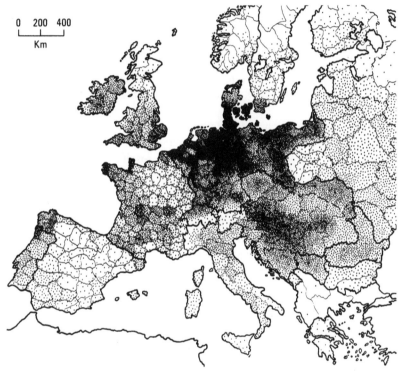

10.9 Pigs, about 1913. Each dot represents 5,000

textile industry failed to expand significantly, but it was supplied increasingly with a superior wool from overseas. There were, however, other reasons. Sheep had once grazed the fallow, but by 1913 little fallow remained in western and central Europe. The map (fig. 10.8) shows that, apart from Great Britain and the polders of the Low Countries, only Mediterranean Europe had a really dense sheep population. They were numerous on the Spanish Meseta, Mediterranean France and peninsular Italy, but their density was greatest in the Balkan peninsula, where, with a low human population, sheep provided the only effective means of using marginal land.

Pig-rearing (fig. 10.9), on the other hand, had increased greatly in importance, and the growth in numbers approximately balanced the decline in sheep.

Table 10.17. *Land use in Europe*

	Area (ha)	Percentage of cultivated area	
Wheat	25,718	16.8	
Rye	15,427	10.1	
Barley	8,227	5.4	
Oats	15,892	10.4	49.0
Maize	8,085	5.3	
Other grains, including buckwheat	1,567	1.0	
Potatoes	9,051	5.9	

Keeping pigs required little space, but an abundance of feed, which was available increasingly not only as coarse cereals, but also as waste from the dairy and the city. Swine were especially numerous where they could be fed on potatoes, as in Germany, or the by-products of the dairy, as in Denmark and the Low Countries. Broadly speaking, the distribution of pigs was the inverse of that of sheep.

By 1913, the overall numbers of farm animals in western and central Europe were as shown in table 10.16. The trend in livestock farming was clearly towards dairy farming, with which was integrated the rearing of bacon pigs.

Arable farming

Despite the growing diversification of agriculture, bread-grains still in 1914 dominated the farming system in most of Europe. Overall, about 152.6 million hectares were regularly cultivated, and of this about a half were under cereals. It is not easy to estimate the area sown with fodder and root crops; it may have been of the order of 10 per cent. In addition some 70 million hectares were used as pasture and rough grazing (table 10.17).

Wheat had been gaining in importance for many years. In much of France it had largely displaced rye and coarse grains in human diet, and was being eaten increasingly in Germany, Austria and Poland. The distribution of wheat cultivation is shown in fig. 10.10. Wheat occupied a high proportion of cultivated land in northern and in several favoured areas in central and southern France; on the plains of Old Castile, between Burgos and Salamanca; the lowlands of northern and central Italy; Saxony; the Hungarian Plain and Romania.

The cultivation of wheat was more strongly influenced than that of most crops by climate and soil. It was grown really successfully only in areas with deep, loamy soil and warm winters. It was grown on only a restricted scale in Germany and scarcely at all in Scandinavia and eastern Europe. The map of wheat *production* (fig. 10.11) contrasts with that of area sown. In much of southern Europe, yields per hectare were very low, though in the northern Meseta there was, near Palencia, a concentrated area of high productivity, the aptly named *Tierra del Pan*. In Germany, the province of Saxony, by no means notable for the area under wheat, appears as one of the heaviest producers in the continent.

Europe in 1914 519

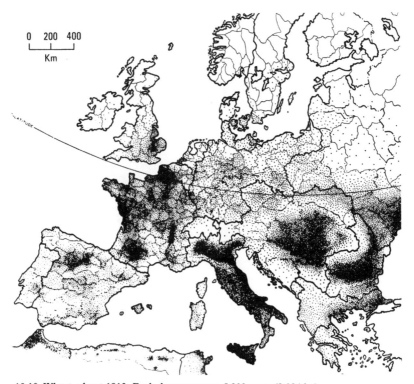

10.10 Wheat, about 1913. Each dot represents 5,000 acres (2,024 ha)

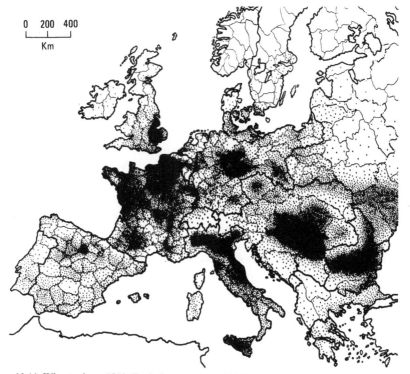

10.11 Wheat, about 1913. Each dot represents 100,000 bushels

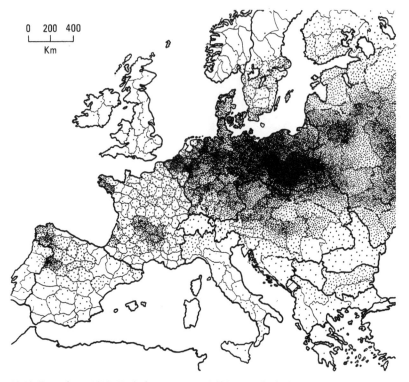

10.12 Rye, about 1913. Each dot represents 5,000 acres (2,024 ha)

The map of heavy wheat production is very close indeed to that of the good loam and *limon* soils.

The distribution of rye was in many respects the converse of that of wheat (fig. 10.12). It was a crop of poor soils and damp, cool climates. In southern Europe it was significant only in the north-west of the Iberian peninsula; in France it was widely grown only in Brittany and the Central Massif. But in Germany, especially the east, it came into its own, and rye cultivation spread eastwards through Poland to Russia, and was also important in Scandinavia.

Other cereals are classed as coarse grains and it is assumed, not altogether correctly, that they were grown as animal feed. Most significant of them was oats, like rye, a crop which did well in cool and moist areas and on soils of indifferent quality (fig. 10.13). Oats formed the traditional feed of horses, and its distribution bore some relationship to the use of horses for traction. Oats, furthermore, fitted well into the traditional three-course rotation. Barley was used, like oats, as a human and animal feed, but it had one particular use; it served for malting, and was widely grown wherever brewing was important. Elsewhere, however, it was tending to be abandoned as more flexible rotations came to be adopted (fig. 10.14). Mixed corn, or *méteil*, made up of oats and barley, was occasionally grown. It was a means of insuring against the failure of one of them, and also of holding up the weak stems of the oats during summer storms. The resulting grain could only be used as animal feed.

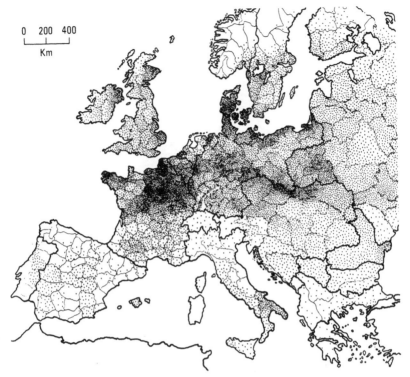

10.13 Oats, about 1913. Each dot represents 5,000 acres (2,024 ha)

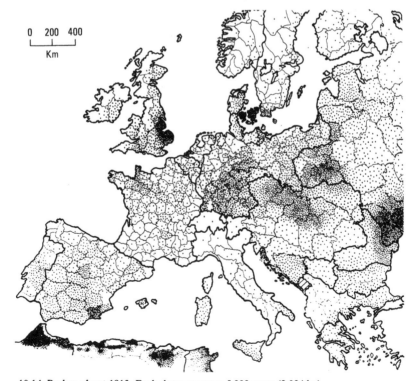

10.14 Barley, about 1913. Each dot represents 5,000 acres (2,024 ha)

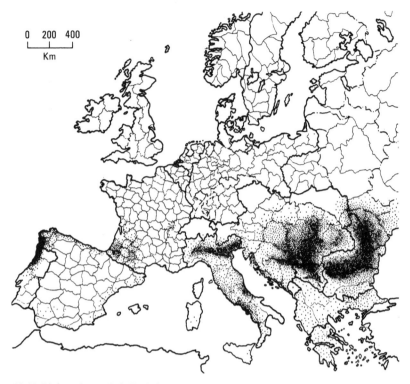

10.15 Maize, about 1913. Each dot represents 5,000 acres (2,024 ha)

Maize played no role in the traditional rotations. It was a recent introduction, and assumed its current distribution pattern only during the nineteenth century. It demanded a hot and moist summer, and, provided this requirement was met, it cropped very heavily. For sheer volume of return it had no rival, and for this reason alone would be attractive to a poor and hungry peasantry. It was widely grown in the southern half of Europe, but was really abundant only where there was a significant summer rainfall (fig. 10.15). These conditions were satisfied in five areas: northern Portugal and Spanish Galicia, south-western France, northern Italy and certain parts of the Italian peninsula, the Hungarian plain, and, lastly, the plains of Romania and neighbouring Bulgaria and Bessarabia. Maize was a multi-purpose cereal. It was fed to stock, though animals were not of great importance in the chief maize-growing areas, and was at the same time an important food. It served as a human food in Romania, thus allowing the wheat to be exported, and it was the basis of the *polenta* of the Italian peasant.

Buckwheat, which, despite its name, is not a cereal,[32] was still grown in 1914 on the poorest and roughest of cultivable soils. It is a herbaceous plant, whose seed can be used both for animal and human food. It is, however, less palatable than the bread-grains, but had the immense advantage that it grows and matures quickly and could be sown in late spring if the bread-grains should fail. It was of rapidly diminishing importance in 1914, and was significant only on the poorer soils of eastern Europe.

Europe in 1914

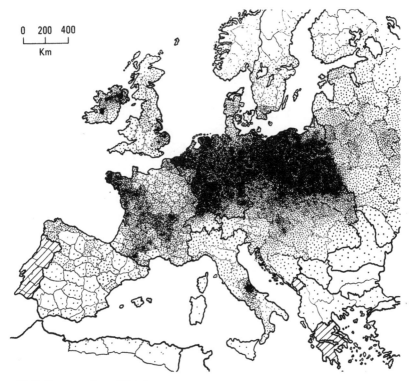

10.16 Potatoes, about 1913. Each dot represents 2,000 acres (810 ha)

Europe had long ceased to be self-sufficing in bread-grains and other cereals. This was due in part to the increase in population, in part to the decline of arable husbandry in favour of fodder crops and permanent grass. Wheat, being the most valuable cereal crop, entered most into international and intercontinental trade. If one excludes the British Isles, by far the largest net importer, and Russia, a considerable exporter, one finds that the whole of eastern Europe, from Poland southwards to the Balkans, had a wheat surplus. This was greatest in Hungary, which supplied the industrialised areas of Austria, and in Romania and Bulgaria, which continued to export to the West (see p. 233). The rest of Europe was a net importer of wheat. In the five years from 1909 to 1914 Europe imported about a fifth of its total consumption.[33] This ranged, on a country-by-country basis, from less than 2 per cent of total consumption in Spain and 4 per cent in France to more than a fifth in Germany, Portugal and Italy; a third in Sweden, Denmark and Greece; three-quarters in Belgium and Switzerland, and almost the total consumption in Norway and Finland. One may assume that the wheat surplus of eastern Europe by and large went to make good the deficit in western Europe. This left almost 10 per cent of total consumption, again excluding the requirements of the British Isles, to be supplied from overseas.

Trade in other cereals was small compared with that in wheat. Some countries in which there was a notable dairy industry, such as Denmark and the Netherlands, imported feed-grains, but rye, the only other bread-grain, entered

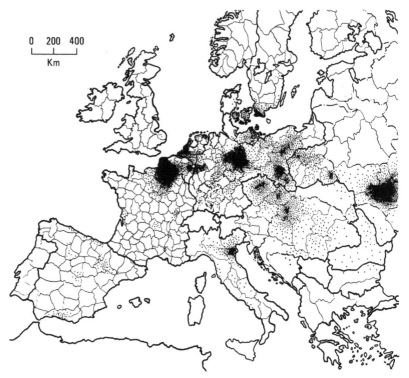

10.17 Sugar beet, about 1913. Each dot represents 1,000 acres (405 ha)

into international trade only to a very small extent. If cereals had to be transported over great distances, then the more desirable wheat would have been preferred to rye.

The two root crops, potatoes and sugar-beet, had by 1914 come to play an important role in crop-rotations. Both were relatively new crops (see pp. 235–7); both were tolerant of cool wet weather, though not of frost, and could generally be relied upon to crop heavily. Neither was important in southern Europe, which was too dry, nor in Scandinavia where late spring frosts were liable to ruin the crop. In France they had little importance in areas of traditional open-field cultivation, but were beginning to appear with the abandonment of fallowing. The most intensive cultivation of roots was in the Low Countries, Germany and Poland. The distribution of the potato (fig. 10.16) almost demarcates the German culture-area, and here, with rye, it formed the staple diet of the mass of the population. Sugar-beet (fig. 10.17) was grown intensively in a number of restricted areas for reasons which have already been examined (see p. 237).

The productivity of European agriculture had been increasing slowly during the early years of the present century. Yields of any particular crop varied from year to year, and there were some for which the harvest of 1913 was by no means the largest. Nevertheless production in the aggregate was as high as it had ever been. Output per unit area varied greatly from one part of Europe to another, with variations in soil quality, in the use of manure and fertiliser, and in farming

Table 10.18. *Crop production and yield per hectare, 1911*

(a) WHEAT	Area (thousand hectares)	Production (thousand tonnes) (1911)	Yield per hectare (thousand tonnes)
Austria	1,226	1,623	1.32
Belgium	156	402	2.58
Bulgaria	1,015	1,184	1.17
Denmark	57	182	3.19
Finland	5	4	0.75
France	6,141	8,690	1.42
Germany	2,126	5,094	2.40
Hungary	3,245	4,119	1.27
Italy	4,649	5,690	1.22
Netherlands	59	136	2.31
Norway	10	12	1.2
Romania	1,893	2,291	1.21
Spain	4,041	3,059	0.76
Sweden	124	259	2.09
Switzerland	70	97	1.39
Poland	508	620	1.24
Portugal	393	191 (1914)	0.49

(b) RYE	Area (thousand hectares)	Production (thousand tonnes)	Yield per hectare (thousand tonnes)
Austria	2,025	2,709	1.34
Belgium	253	571	2.26
Bulgaria	205	205	1.00
Denmark	214	432	2.02
Finland	239	271	1.13
France	1,107	1,270	1.15
Germany	6,408	12,222	1.91
Hungary	1,058*	1,327	1.25
Italy	118	160	1.36
Netherlands	206	446	2.17
Norway	14	22	1.57
Poland	2,099	2,014	0.96
Portugal	188	—	—
Romania	126	95	0.75
Spain	763	709	0.93
Sweden	379	585	1.54
Switzerland	25	45	1.80

(c) POTATOES	Area (thousand hectares)	Production (thousand tonnes)	Yield per hectare (thousand tonnes)
Austria	1,257	11,552	9.19
Belgium	161	3,201	19.88
Denmark	70	1,070	15.29
Finland	78	484	6.21
France	1,495	13,590	9.09
Germany	3,209	54,121	16.87
Italy	294	2,958	10.06
Spain	295	2,534 (1912)	8.59
Sweden	155	1,969	12.70
Switzerland	52	725	13.94
Hungary	602	4,875	8.10
Poland†	1,013	9,726	9.60

* Inclusive of mixed grain.
† From *Handbuch von Polen*, ed. E. Wunderlich, Berlin, 1918, 376.

Table 10.19. *Average yields of chief crops*

	Average yield (unweighted, quintals per hectare)	Standard deviation
Wheat	1.53	0.66
Rye	1.45	0.46
Potatoes	11.63	3.93

techniques. Table 10.18[34] presents the production of the more important crops, together with the yield per hectare. The wide range in yields per hectare is apparent. This is especially the case with wheat, which was clearly grown in some countries under conditions for which it was quite unsuited (table 10.19). This is the probable reason for the poor showing of France. On the other hand, the relatively high yield of rye in Germany is probably due to the fact that this crop, so often grown on the worst soils, was there also sown on the better. The most consistent yields were for potatoes, with low yields only in Finland, Spain and Hungary, none of which could be regarded as truly suitable for the crop. The consistently high yields in Denmark and the Low Countries were probably due to the intimate relationship between arable and dairy farming, and the large volume of manure thus made available.

International trade

By 1914 Europe had become the focus of a complex network of international trade. At this date about half the world's international commerce was still with European countries, though Europe's share had for many decades been declining.[35] Comparisons with earlier periods are difficult, but European overseas trade was in 1913 about twice as large as it had been only 25 years earlier, and almost three times that of half a century before. In 1913 the total value of the exports and imports of the countries of continental Europe amounted to 16,052 million U.S. dollars (table 10.20).[36] By contrast the total overseas commerce of the United Kingdom, by far the most important commercial country, was 5,806 million dollars. Of the total for continental Europe 9,802 million dollars – about 60 per cent – was with other European countries. Only in the United Kingdom did extra-European trade make up a really large proportion – about 70 per cent – of the whole. In no country of continental Europe did trade with other continents amount to more than 45 per cent. It was largest in France, the Netherlands and Germany, all of which had significant colonial dependencies.

The export trade of the more developed countries was dominated by manufactured goods, particularly metal goods and textiles. Foodstuffs were relatively unimportant, but industrial raw materials, notably coal in Germany and iron-ore in France and Sweden, were significant. Foodstuffs and certain industrial raw materials, in particular raw cotton and metalliferous ores, were prominent amongst these country's imports. In the less developed countries the

Table 10.20. *The foreign trade of European countries, 1913 (in millions of dollars)*

	Imports		Exports			
	Total	From other European countries	Total	To other European countries	Total trade	Trade per capita
Industrial north-western Europe						
France	1,618	752	1,323	881	2,941	63.2
Belgium-Luxembourg	875	513	682	540	1,557	202.6
Netherlands	824	387	489	319	1,313	224.1
Germany	2,565	1,045	2,405	1,594	4,970	151.1
Switzerland	358	283	264	183	622	165.7
Scandinavia						
Sweden	227	185	219	183	446	112.6
Norway	148	125	115	85	263	109.9
Denmark	230	169	173	165	403	146.2
Finland	95	68	77	53	172	58.4
Eastern and south-eastern Europe						
Austria-Hungary	688	453	563	452	1,251	24.4
Romania	114	102	131	114	245	33.9
Bulgaria	36	29	18	15	54	12.4
Southern Europe						
Portugal	84	63	32	19	116	20.5
Spain	253	149	206	143	459	23.0
Italy	704	410	482	283	1,186	34.2
Greece	31	22	23	18	54	20.4

balance swung in the opposite direction. Foodstuffs and other primary goods were the biggest exports. Wheat, for example, made up 80 per cent by value of Romania's exports and almost 30 per cent of those of Bulgaria. Foodstuffs and wine constituted 49 per cent of those of Greece, 43 per cent of those of Spain, and ranked high in the trade of Italy. The imports of these countries were dominated by manufactures.

National income

In the first chapter estimates were presented of the gross national product per head of population near the beginning of the century. By 1914 the statistical data had become more abundant and reliable, and table 10.21[37] and figs. 10.18 and 10.19 are offered with a greater pretension to accuracy. A feature of the previous century had been the growing differential between the richest and the poorest countries. Near the beginning of the century the Mediterranean countries were not significantly below the rest of Europe, but by 1913 they formed a group at the bottom of the table of gross national products (fig. 10.18). Below them came only the Balkans, for which the data are inadequate. At the same time, the G.N.P. of the Scandinavian countries, which had, at the beginning of the century

Table 10.21. *Gross National Product and G.N.P. per head, 1913*

	G.N.P. (in million U.S. dollars at 1960 prices)	G.N.P. per head (in U.S. dollars)
Austria-Hungary	26,050	498
Belgium	6,794	894
Bulgaria	1,260	263
Denmark	2,421	862
Finland	1,670	520
France	27,401	689
Germany	49,760	743
Greece	1,540	322
Italy	15,624	441
Netherlands	4,660	754
Norway	1,834	749
Portugal	1,800	292
Romania	2,450	336
Serbia	725	284
Spain	7,450	367
Sweden	3,824	680
Switzerland	3,700	964
[United Kingdom]	[44,074]	[965]

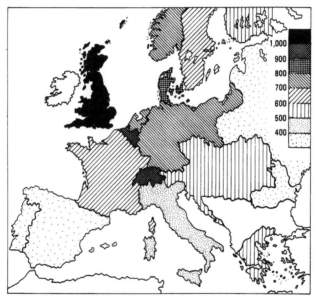

10.18 Gross National Product per head, 1913 (in U.S. dollars)

been closely comparable with that of southern Europe, was growing steadily, and by the early twentieth century had reached the level of north-western Europe.

The per capita G.N.P. for the Balkans would, if the data were available, have been seen to be the lowest in Europe. Bairoch estimates that in Bulgaria and Serbia it was lower even than in Portugal, and it must have been lower still in

Europe in 1914

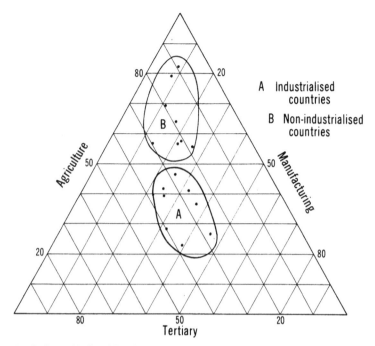

10.19 Gross National Product per head by sector for (a) industrialised countries, (b) non-industrialised countries. Data for 1913

Albania, Montenegro and what remained of the Ottoman Empire. It is doubtful whether there had ever been a time in the history of Europe when the economic gradient between the richest and the poorest was steeper than it was when a Balkan skirmish precipitated the First World War.

It was pointed out earlier that most statistics were published for countries and major administrative areas, and that these rarely coincided with economic regions. But even with crude national estimates of production and wealth, the contrast between the two Europes, the developed and the less developed areas of Europe, is apparent, with a gross national product per head in the Low Countries and Scandinavia more than three times the level in southern Europe and, in all probability four times that in the Balkans (fig. 10.19). A more refined analysis of the data shows that even in the most developed countries there were areas of backwardness. They can be defined in various ways: in terms of lack of manufacturing and of transport facilities, of low yields of agriculture, of a high level of illiteracy and, as a general rule, of a high birth-rate and vigorous out-migration. If one compares region with region rather than country with country, the contrast is even greater. Because of the lack of statistics, however, one can only suggest how much below their national averages was the level of wealth in the villages of southern Spain, the mountains of Basilicata, the Auvergne or even of the fertile but crowded plains of Württemberg. A more detailed map of wealth and welfare must therefore be subjective and impressionistic, a summation of all the factors – demographic, agricultural, industrial and commercial – that have been examined in this book.

A final question should be asked: why did this differential develop during the nineteenth century? Why did eastern, south-eastern and southern Europe remain relatively undeveloped? Their backwardness was not particularly apparent at the beginning of the century, and their economic level not significantly below the European average. A partial answer is to be found in the less developed areas within the developed countries themselves. Economic development, at least in the early phases of growth, is spurred by small local advantages in physical resources and social organisation. Reserves of solid fuel and metalliferous ores are most often cited as a necessary and even a sufficient cause. It was, indeed, a very significant factor, but never, in the early stages of growth, was industrial development prompted only by the presence of these naturally occurring resources. There had first to be entrepreneurs and a demand for the goods which they produced and sold, and this in turn called for the commercial infrastructure of markets, means of transport and monetary transactions.

In early modern times a social climate was developing in north-western Europe which encouraged and rewarded the entrepreneur. At the same time the mass of the peasantry began to acquire what amounted to economic freedom and some kind of title to the land which they cultivated. The strength of tradition began to weaken, and they became freer to innovate and experiment. The sharp change in the economic climate of Poland after 1864 illustrates the broader implications of such factors. By and large, however, such changes came slowly in east-central, south-eastern and southern Europe. In most of this area the system of great estates began to break down only in the later nineteenth century, and in some parts it survived until the time of the First World War. Until a significant measure of land reform had been achieved there could not be much growth in peasant demand.

At the same time, however, physical influences must not be minimised. Most of this area of slow development was poorly endowed with fuel and mineral resources. That this was not an insuperable barrier to growth, however, is shown by the industrial development of Catalonia and northern Italy. But in these areas the social conditions were very much more favourable than in other countries which developed slowly. The role of transport and communications in development has already been stressed. In most of the less developed areas of south-eastern and southern Europe these were poor. Water transport was impossible over much of the area, and road transport was hindered by the broken terain. Indeed, much of the area was fragmented into an immense number of small basins and valleys, between which communication was always difficult and sometimes impossible.

Contrast this with the ease of movement both by water and overland in north-western Europe. Commercial routes and trading towns were already part of the infrastructure when modern industrial development began, and they were strengthened yet more as modern development progressed. The core-area of modern economic growth lay across north-western Europe, from the Paris Basin to Berlin and Saxony, forming a zone of intensive urban, industrial and above all transport development. Even without its reserves of fossil fuel, north-western Europe had immense advantages. It was one of nature's strangest quirks to endow this region yet further with coal, iron ore and non-ferrous metals to a degree not met with anywhere else.

Conclusion

The outbreak of war in August 1914 brought to an end a century of economic growth. During the previous three decades this development had been rapid, and despite growth in the United States, Russia and Japan, western Europe remained by far the most important centre of manufacturing in the world. Western and central Europe produced in 1913 the following percentages of the world's output of:

Bituminous coal	50
Steel	57
Sulphuric acid	70
Synthetic fibres	93
Cotton spindles	69

One may ask whether this trend of development would have continued during the coming years if the First World War had not intervened; whether production of key commodities would have continued to expand at the same rate as before. European industries were increasingly dependent on foreign and generally overseas markets, but already in 1913 a threat to Europe's markets could be seen in the growing manufactures of Japan and the United States. The effect of the war was to encourage industrial development in other non-European countries, notably in Latin America, India and Australia.

Europe's share in the production of key commodities dropped sharply during the war years, and, though pre-war levels had in general been regained by 1925, they constituted a significantly smaller fraction of world output than they had been ten years before:

Bituminous coal	45
Steel	46
Sulphuric acid	52
Synthetic fibres	68
Cotton spindles	62

In Svennilson's words, 'Europe had to face a trend in world economy whereby its traditional position as the foremost industrial workshop of the world was becoming undermined by industrial progress in overseas countries'. The First World War was something more than an interruption in a process of growth. It is impossible to regard the European economy 'as a clock which has been set back a number of hours and which then resumes its movement at the same speed as before'. The checks to long-term economic growth were too complex for this.

When the First World War ended the economy of Europe was in ruins; half the continent was in revolt, and five million lay dead on Europe's battlefields. By the mid-1920s the material damage had been largely repaired, but the human losses could not be so easily replaced. For the next generation and more there remained in much of Europe a distorted age and sex structure. The age-group that would have provided the leaders in the 1930s and 1940s had been decimated in the war.

In 1913 no less than 36 per cent of the land surface of the globe was dominated and in varying degrees controlled by a small number of European states. This control was severely weakened as a consequence of the war. The German Empire was terminated, and a new concept of the obligations of empire, faintly anticipated in the Congo Act of 1885, was widely accepted, if not generally applied. The train of events which led to the political independence of the Indian subcontinent and of most of Africa was initiated when the European powers called upon their subject peoples overseas to help to fight their battles for them.

By the same token, the economic value of colonies to their parent country was diminished, and the gloomy prognostications, from the imperialist point of view, of A. R. J. Turgot and J. A. Hobson were at last fulfilled. The cycle of colonialism was nearing its end, and colonies, like ripe fruit, were preparing to drop from the branches which had sustained and nurtured them.

The decades before the First World War had been characterised by rapid technological advances in manufacturing and agriculture. Most of these innovations had been made in Europe, and Europe was their chief beneficiary. Development was yet more rapid in the post-war years, but advances were more significant outside Europe than within. The United States and Japan, in particular, were in the strongest position to benefit from the newest developments, for many of which the United States was responsible. At the same time industrial production was tipped increasingly towards consumer goods, and the 'consumer society' came into being. An immense range of goods, many of them pioneered in America, came on to the market, responding to the increasing productivity and real incomes of labour.

All these changes brought about severe dislocations in production and supply of resources and of manufactured goods. Before the war, markets were in general certain and expanding; afterwards they became fluctuating and uncertain. This in turn led to short-term or cyclical changes in the economy, obscuring whatever long-term trends there may have been. In the final analysis the Great Depression can be viewed as a consequence of the dislocations resulting from the First World War.

Some of these developments would have occurred eventually with or without the First World War. A 'consumer society' would have emerged; technological innovation would have increased the productivity of labour; factor costs would have changed, and the location of manufacturing would slowly have adjusted to the new circumstances. As it was, the armistice of 11 November 1918 disclosed a radically different and more rapidly changing geographical scene.

Notes

Chapter 1. From Waterloo to the First World War

1 Paul Bairoch, 'Europe's Gross National Product: 1800–1975', *J. Eur. Ec. Hist.*, 5 (1976), 273–340.
2 A. R. J. P. Ubbelohde, 'The Beginnings of the Change from Craft Mystery to Science as a Basis for Technology', *A History of Technology*, ed. Charles Singer et al., vol. IV (Oxford, 1958), 663–81.
3 Eric (Lord) Ashby, 'Education for an Age of Technology', ibid., vol. V (Oxford, 1958), 776–98.
4 'Reports of the Royal Commission on Technical Instruction', *Parl. Pap.*, 1884.
5 Gabriel Jars, *Voyages métallurgiques*, 3 vols. (Paris, 1774–81).
6 H. L. Duhamel de Monceau, ed., *Description des arts et des métiers*, 27 vols. (Paris, 1761–88).
7 E. Swedenborg, *Circa Ferrum* (Amsterdam, 1721); a French translation appeared in *Description des arts*, vol. XIX (Paris, 1762).
8 Notably *Mémoires d'agriculture, d'économie rurale et domestique* (Paris, 1761–1872). The Bath and West and Southern Counties Agricultural Society began its publications on agriculture in 1780.
9 Jean Vial, *L'Avènement de la civilisation industrielle de 1815 à nos jours* (Paris, 1973), 49ff.
10 Jeffrey Kieve, *The Electric Telegraph: A Social and Economic History* (Newton Abbot, 1973), 13–28.
11 Eugen Weber, *Peasants into Frenchmen* (London, 1979).
12 I. N. Lambi, 'Free Trade and Protection in Germany 1868–1879', *Viert. Soz. Wirtgesch.*, Beiheft 44 (1963), 22.
13 W. O. Henderson, 'A Nineteenth-Century Approach to a West European Common Market', *Kyklos*, 10 (1957), 448–59.
14 Oscar Jaszi, *The Dissolution of the Habsburg Monarchy* (Chicago, 1929), 190–3.
15 Joseph A. Schumpeter, *Business Cycles* (New York, 1939), vol. I, 258–380.
16 *The French Colonial Empire*, Roy. Inst. Int. Aff. (London, 1940).
17 J. A. Hobson, *Imperialism*, edn of 1938 (London), esp. chap. 2.
18 E. J. Hobsbawm, *Industry and Empire* (Harmondsworth, 1969), 69.
19 J. S. Furnivall, *Netherlands India* (Cambridge, 1939), 115–47.
20 Quoted in *Germany's Claim to Colonies*, Roy. Inst. Int. Aff. (London, 1939), 8.
21 Ivan T. Berend and György Ránki, *The European Periphery and Industrialization 1780–1914* (Cambridge, 1982), 142–60.
22 Herbert Feis, *Europe the World's Banker 1870–1914* (New Haven, Conn., 1930), 342–60, 379–81.
23 Grover Clark, *The Balance Sheet of Imperialism* (New York, 1936), 5.
24 The Spanish Empire in Central and South America became independent of Spain in 1806–25, and Brazil from Portugal in 1822–3. Canada is included in this list because by the British North America Act of 1867 the *Dominion* of Canada acquired a *de facto* independence. Australia and New Zealand did not acquire this status until respectively 1901 and 1907.
25 Paul Bairoch, 'Geographical Structure and Trade Balance of European Foreign Trade from 1800 to 1970', *Jl Eur. Ec. Hist.*, 3 (1974), 557–608.
26 Patrick O'Brien, 'European Economic Development: The Contribution of the Periphery', *Ec. Hist. Rv.*, 35 (1982), 1–18.
27 Scott M. Eddie, 'The Terms of Trade as a Tax on Agriculture: Hungary's Trade with Austria, 1883–1913', *Jl Ec. Hist.*, 32 (1972), 298–315.

28 Angelo Tamborra, 'The Rise of Italian Industry and the Balkans (1900–1914)', *Jl Eur. Ec. Hist.*, 3 (1974), 87–121.
29 N. J. G. Pounds, 'France and "les limites naturelles" from the Seventeenth to the Twentieth Centuries', *Ann. A.A.G.*, 44 (1954), 51–62.
30 Roderick Geikie and Isabel A. Montgomery, *The Dutch Barrier 1705–1719* (Cambridge, 1930).
31 J. P. D. van Banning, 'Belgium and Dutch Limburg 1830–1839', *Symbolae Verzijl* (The Hague, 1958), 22–36.
32 Shepard B. Clough, *A History of the Flemish Movement in Belgium* (New York), 1930.
33 W. O. Henderson, *The Zollverein* (Cambridge, 1939), 29–69.
34 Vaclav L. Benes, 'The Slovaks in the Habsburg Empire: A Struggle for Existence', *Aust. Hist. Jb.*, 3 (1967), 335–64.
35 L. S. Stavrianos, *The Balkans Since 1453* (New York, 1958), 138–42; Bruce McGowan, *Economic Life in Ottoman Europe* (Cambridge, 1981).
36 Sir Edward Hertslet, *The Map of Europe by Treaty*, vol. IV (London, 1891), 2997–3009.
37 Stavro Skendi, *The Albanian National Awakening 1878–1912* (Princeton, N.J.), 88–108.
38 From the Phanar suburb of Constantinople, where the Sultans were accustomed to recruit many of their servants.
39 On regional variations of G.N.P. within particular countries see: Herman Freudenberger and Gerhard Mensch, 'Regional Differences, Differential Development and Generative Regional Growth', *Disparities in Economic Development since the Industrial Revolution*, ed. Paul Bairoch and Maurice Lévy-Leboyer (New York, 1981), 199–209; A. Graziani, 'Regional Inequalities in Italy', ibid., 319–30; Jan de Vries, 'Regional Inequality in the Netherlands since 1600', ibid., 189–98.
40 P. Bairoch, op. cit.
41 Id., 'Le mythe de la croissance économique rapide du XIXe siècle', *Rv. Inst. Soc.*, 35 (1962), 307–31.
42 Simon Kuznets, *Modern Economic Growth* (New Haven, Conn., 1966), 88–93; Colin Clark, *Conditions of Economic Progress* (London, 1940).
43 Tihomir J. Markovitch, 'The Dominant Sectors of French Industry', *Essays in French Economic History*, ed. R. Cameron (Homewood, Ill., 1970), 226–44; M. Lévy-Leboyer, 'La croissance économique en France au XIXe siècle', *Ann. E.S.C.*, 23 (1968), 788–807.
44 W. G. Hoffmann, *Das Wachstum der deutschen wirtschaft seit der Mitte des 19. Jahrhunderts* (Berlin, 1965).
45 G. Otruba, 'Wachstumsverschiebungen in der wirtschaftssektoren Österreichs 1869–1961', *Viert. Soz. Wirtgesch.*, 42 (1975), 40–61. Peter Hanak, 'Hungary in the Austro-Hungarian Monarchy: Preponderancy or Dependency?', *Aust. Hist. Jb.*, 3 (1967), 260–302; Ervin Pamlenyi, ed., *A History of Hungary* (London, 1975), 346–58.

Chapter 2. The resource pattern of Europe

1 John Maynard Keynes, *The Economic Consequences of the Peace* (London, 1919), 75.
2 N. J. G. Pounds, 'Geographical Factors in the Exploitation of Minerals', *London Essays in Geography*, ed. L. Dudley Stamp and S. W. Wooldridge (London, 1951), 241–54.
3 Quoted in *Ann. Bret.*, 37 (1925–6), 404–6.
4 E. A. Wrigley, 'The Supply of Raw Materials in the Industrial Revolution', *Ec. Hist. Rv.*, 15 (1962), 1–16.
5 Id., *Industrial Growth and Population Change* (Cambridge, 1961), 1–9.
6 L. Cordier, 'Sur les mines de houille de France', *Jl Mines*, 36 (1814), 321–94.
7 M. Lefebvre, 'Aperçu général des mines de houille exploitées en France, de leur produits, et des moyens de circulation de ces produits', *Jl Mines*, 12 (An. 10), 325–458.
8 *Berg. Hütt. Ztg*, 2 (1843), 201–8, 220–9, 244–53; 9 (1850), 261.
9 Notably Morand, 'Du charbon de terre et de ses mines', *Description des arts et des métiers*, vol. II (1768), 83–8, 133–6.
10 Édouard Grar, *Histoire de la recherche, de la découverte et de l'exploitation de la houille dans le Hainaut française et dans l'Artois*, 3 vols. (Valenciennes, 1847).
11 N. J. G. Pounds, *The Ruhr* (London and Bloomington, Ind., 1952); id., *The Upper Silesian Industrial Region*, Slavic and East European Series, 11 (Bloomington, Ind., 1958).
12 Gerhard Jacobi, 'Der Steinkohlenbergbau in den Grafschaften Tecklenburg und Lingen im ersten Jahrhundert preussischer Herrschaft', *Münst. Btr. Gesch.*, 23 (1909); Paul Kukuk, *Geologie des niederrheinisch-westfälischen Steinkohlengebietes* (Berlin, 1938), 348–62.
13 A. M. Héron de Villefosse, *De la richesse minerale*, vol. I (Paris, 1819), 17–18.

14 Ibid., vol. I, 376.
15 Ibid., vol. I, 388; J. F. Daubuisson, 'Notice sur l'exploitation des houillères de Waldenburg en Silésie', *Jl Mines*, 15 (An. 12), 88–103.
16 Michel Chevalier, 'Note sur les richesses de la Bohême en combustibles fossiles et sur le bassin de Radnitz en particulier', *Ann. Mines*, 4e série, 1 (1842), 575–602; E. Gruner and G. Bousquet, *Atlas général des houillères* (Paris, 1911); N. J. G. Pounds, 'The Spread of Mining in the Coal Basin of Upper Silesia and Northern Moravia', *Ann. A.A.G.*, 48 (1958), 149–63; Franz Schwackhofer, *Die Kohlen Osterreich-Ungarns und Preuss.-Schlesiens* (Vienna, 1901), 50—7.
17 Melrad Mellen and Victor H. Winston, *The Coal Resources of Yugoslavia*, Mid-European Studies Center (New York, 1956).
18 P. Nicou and C. Schlumberger, 'L'Industrie minière et métallurgique dans les Asturies', *Ann. Mines*, 10e série, 7 (1905), 203–57.
19 F. J. Monkhouse, 'The South Limburg Coalfield', *Ec. Geog.*, 31 (1955), 126–37.
20 'Report on the Coal Mining Industry of Belgium,' For. Off. Misc. Ser., 664, *Parl. Pap.*, 1908, vol. CVII, 761–807.
21 F. J. Monkhouse, *The Belgian Kempenland* (Liverpool, 1949), 107–19.
22 'Report of the Board of Trade on Railways in Belgium, France and Italy', Cd 5106, *Parl. Pap.*, 1910, vol. LVII, 137.
23 *Das Östliche Deutschland*, Göttinger Arbeitskreis (Würzburg, 1959), 675–82.
24 Francois Crouzet, 'Le charbon anglais en France au XIXe siècle', *Charbon et Sciences Humaines*, Éc. Prat. Htes. Ét., Ind. et Art., 2 (1966), 173–206.
25 W. Stanley Jevons, *The Coal Question* (London, 1861), 282–3.
26 Fritz Wundisch, 'Zur Geschichte des rheinischen Braunkohlenbergbaus', *Rhein. Vbl.*, 17 (1952), 197–221.
27 S. Schneider, 'Das Braunkohlenrevier im Westen Kölns', *Köln und die Rheinlande* (Wiesbaden, 1961), 341–52.
28 T. H. Elkins, 'The Brown Coal Industry of Germany', *Geog.*, 38 (1953), 18–29.
29 H. Bartel, *Braunkohlenbergbau und Landschaftdynamik*, *Pet. Mitt. Erg.*, 270 (1962), 29–31.
30 Antoni Wrzosek, *Czechoslowacja* (Warsaw, 1960), 144–8.
31 'Report on Brown Coal Mining in the Rhineland', For. Off. Misc. Ser., 497, *Parl. Pap.*, 1899, vol. XCVII.
32 E. Pape, 'Der deutsche Braunkohlenhandel unter dem Einfluss der Kartelle', *Zt Ges. Staatsw.*, 62 (1906), 234–71.
33 Konrad Fuchs, 'Die Bergwerks- und Hüttenproduktion im Herzogtum Nassau', *Nass. Ann.*, 79 (1968), 368–77.
34 Otto Busch, 'Industrialisierung und Gewerbe im Raum Berlin/Brandenburg 1800–1850', *Veröff. Hist. Kom. Berlin*, 9 (1971), 59.
35 'Reports of Commissioners', no. 7, *Parl. Pap.*, 1864, vol. LXI.
36 R. L. Rudolph, *Banking and Industrialization in Austria-Hungary* (Cambridge, 1976), 41.
37 V. Sandor, 'Die Hauptmerkmale der Industriellen Entwicklung in Ungarn zur Zeit des Absolutismus (1849–1867)', *St. Hist. Ac. Sci. Hung.*, 28 (1960), 15.
38 'Report on the Mining Industries in Bosnia and Herzegovina', For. Off. Misc. Ser., 354, *Parl. Pap.*, 1895, CII; P. F. Sugar, *Industrialization of Bosnia-Hercegovina 1878–1918* (Seattle, 1963), 15–16.
39 'Report on the Mineral Resources of Greece', For. Off. Misc. Ser., 576, *Parl. Pap.*, 1902, vol. CIII.
40 Constantin C. Giurescu, 'Contributions to the History of Romanian Science and Technology from the 15th to the early 19th Century', *Bibl. Hist. Rom.*, 50 (1974), 108–45.
41 A. Birembaut, 'L'industrie du pétrole au XIXe siècle (1780–1900)', *Cah. Hist. Mond.*, 5 (1959), 149–81.
42 R. J. Forbes, 'Petroleum', *A History of Technology*, ed. Charles Singer et al., vol. V (Oxford, 1958), 102–23.
43 Miron Nicolesco, 'Description de la distribution géographique du pétrole en Roumanie', 9e Congr. Int. Géog. (1911), vol. III, 147–57; id., 'Distribution géographique du pétrole en Roumanie', Atti X Congr. Int. Géog. (1915), vol. II, 1194–1207.
44 'Report on the Petroleum Industry in Roumania', For. Off. Misc. Ser., 411, *Parl. Pap.*, 1897, vol. LXXXVIII.
45 Germaine Veyret-Verner, *L'Industrie des Alpes françaises* (Grenoble, 1948), 144–210.
46 Raoul Blanchard, *Les Alpes Occidentales*, vol. II (Paris, 1938), 433–6; id., 'L'industrie de la houille blanche dans les Alpes françaises', *Ann. Géog.*, 26 (1917), 15–41.
47 *Rapport général sur l'industrie française: sa situation, son avenir*, Min. de Commerce, Impr. Nat. (1919), vol. I, 18–19.
48 H. Cavaillès, 'La houille blanche dans les Pyrénées françaises', *Ann. Géog.*, 28 (1919), 425–68.

49 J.-F. Bergier, *Naissance et croissance de la Suisse industrielle* (Bern, 1974), 148–9.
50 M. Mussy, 'Description de la constitution géologique et des ressources minérales du Canton de Vicdessos et spécialement de la mine de Rancié', *Ann. Mines*, 6ᵉ série, 14 (1868), 57–112, 193–299; 15 (1869), 327–404.
51 A. Habets, 'Note sur l'état actuel des mines de fer de Bilbao', *Rv. Univ. Mines*, 3ᵉ série, 4 (1888), 1–15; M. Baills, 'Note sur les mines de fer de Bilbao', *Ann. Mines*, 7ᵉ série, 15 (1879), 209–33.
52 'Ausfuhr von Bilbaoer Eisenerz', *St. u. Eisen*, 14 (1894), part 2, 1145–6; M. W. Flinn, 'British Steel and Spanish Ore', *Ec. Hist. Rv.*, 8 (1955), 84–90.
53 Celso Capacci, 'The Iron Mines of the Island of Elba', *Jl I.S. Inst.*, 84 (1911), 412–50.
54 L. Puzenat, *La Sidérurgie armoricaine*, *Mém. Soc. Géol. Bret.*, 4 (1939); C.-E. Heurteau, 'Note sur le minerai de fer Silurien de Basse-Normandie', *Ann. Mines*, 10ᵉ série (1907), 613–68; H. V. Janau, 'The Iron Ores of Normandy', *Sc. Geog. Mag.*, 41 (1925), 266–84; Jean de Maulde, *Les Mines de fer et l'industrie métallurgique dans le département de Calvados* (Caen, 1916).
55 Pierre Léon, 'Deux siècles d'activité minière et métallurgique en Dauphiné: l'usine d'Allevard', *Rv. Géog. Alp.*, 36 (1948), 215–58.
56 M. de Grossouvre, 'Étude sur les gisements de minerai de fer du centre de la France', *Ann. Mines*, 8ᵉ série, 10 (1886), 311–418; M. F. Rigaud, 'Notice sur les minières de la Haute-Marne', *Ann. Mines*, 7ᵉ série, 14 (1878), 9–62; M. Lieffroy, 'L'industrie métallurgique en Franche Comté', *Ac. Sci. Bes.* (1982), 220–33.
57 A. Noggereth, 'Die Grube Stahlberg bei Müsen', *Zt Berg. Hütt. Sal.*, 11 (1863), 63–93; G. Einecke and W. Kohler, 'Die Eisenerzvorrate des Deutschen Reiches', *The Iron Ore Resources of the World*, Int. Geol. Congr., Stockholm (1911), 671–719; S. Jordan, 'État actuel de la métallurgie du fer dans le pays de Siegen', *Rv. Univ. Mines*, 16 (1864), 425–555; Ernst Frohwein, *Beschreibung des Bergreviers Dillenburg* (Bonn, 1885), 2–11; 'Fortschritte beim Eisenerzbergbau im Siegeneschen seit 1874', *St. u. Eisen*, 11 (1891), part 2, 611–33.
58 'Italie', *L'Industrie du fer en 1867* (Paris, 1867), 626–40.
59 Karl A. Redlich, 'Die Geologie der innerösterreichischen Eisenerzlagerstätten', *Beitr. Gesch. Öst. Eis.* (1931), 99–110; Dr Ahlburg, 'Der Erzbergbau in Steiermark, Kärnten und Krain', *Zt Berg. Hütt. Sal.*, 55 (1907), 463–521.
60 For a map of metalliferous mines in the Habsburg Empire, see *Mont. Rdsch.*, 11 (1919).
61 Ivan Avsenek, *Yugoslav Metallurgical Industry*, Mid-European Studies Center (New York, 1955), 11–12.
62 *Bulgaria*, ed. A. D. Dellin (New York, 1957), 47.
63 Gunnar Lowegren, *Swedish Iron and Steel: A Historical Survey* (Stockholm, 1948), 6–8; A. F. Rickman, *Swedish Iron Ore* (London, 1939), 92–7.
64 N. J. G. Pounds, *The Ruhr* (London, 1953), 73.
65 P. Angot, *Le Bassin ferrifère de Lorraine* (Nancy, 1939); L. Cayeux, *Le Minerai de fer de Lorraine*, Travaux de Comité d'Études, Section Géologique, IPN (Paris, 1919).
66 Estimates differ; see *European Steel Trends in the Setting of the World Market*, Steel Division, Economic Commission for Europe (Geneva, 1949); *World Iron Ore Resources and their Utilization*, Dept. Econ. Aff.., U.N. (New York, 1950).
67 A. F. Rickman, op. cit., 24; G. Lowegren, op. cit., 8–17.
68 R. Chadwick, 'New Extraction Processes for Metals', *History of Technology*, 5 (1958), 72–101.
69 Adolf Soetbeer, 'Edelmetall-Produktion', *Pet. Mitt. Erg.*, 57 (1879).
70 G. Otruba and Rudolf Kropf, 'Bergbau und Industrie Böhmens in der Epoche der Fruhindustrialisierung (1820–1848)', *Böh.*, 12 (1972), 53–232.
71 Rudolf Magula, 'Z dejín tažby a výroby striebra v Spišsko-Germerskom Rudohori v 19. storoči', *Hist. Carp.*, 10 (1979), 163–75.
72 Ernest A. Smith, *The Zinc Industry* (London, 1918), 11–39; F. Krantz, *Die Entwicklung der oberschlesischen Zinkindustrie* (Kattowitz, 1911), 5–6.
73 Fernand Bezy, 'Les évolutions longues de l'industrie du zinc dans l'Ouest européen 1840–1939', *Bul. Inst. Rech. Éc. Soc.*, 16 (1950), no. 1, 3–56.
74 Christopher J. Schmitz, *World Non-Ferrous Metal Production and Prices, 1700–1976* (London, 1979), passim; Rondo E. Cameron, *France and the Economic Development of Europe 1800–1914* (Princeton, N.J., 1961), 353–7; id., 'Some French Contributions to the Industrial Development of Germany 1840–1870', *Jl Ec. Hist.*, 16 (1956), 281–321.
75 R. Chadwick, op. cit.
76 Jordi Nadal, *El Fracaso de la Revolucion Industrial en España 1814–1913* (Barcelona, 1975), 98–101; id., 'Industrialisation et désindustrialisation du Sud-est espagnol 1820–1890', *L'Industrialisation en Europe au XIXᵉ siècle*, Coll. Int. Cent. Nat. Rech. Sci. (Paris, 1972), 201–12.
77 Jan Novotny, 'Zur Problematik des Beginns der industriellen Revolution in der Slowakei', *Hist.(P)*, 4 (1962), 129–89.

78 'Report on the Mines of Servia', For. Off. Misc. Ser., 350, *Parl. Pap.* (1895), CIII; Fritz Behrend, 'Die Kupfer- und Schwefelerze von Osteuropa', Öst. Inst. *Qu. u. St.*, 3 (1921), 8–16.
79 T. C. Banfield, 'The Progress of the Prussian Nation, 1805, 1831, 1842', *Jl Roy. Stat. Soc.*, 11 (1848), 25–37; W. P. Jervis, 'The Mansfeld Copper-Slate Mines in Prussian Saxony', *Jl Soc. Arts*, 9 (1860–1), 592–98, 603–9, 616–23, 627–32; R. Dietrich, 'Zur industriellen Produktion, technischen Entwicklung und zum Unternehmertum in Mitteldeutschland, speciell in Sachsen im Zeitalter der Industrialisierung', *Jb. Gesch. Mitt. Ostd.*, 28 (1979), 221–72; Christopher J. Schmitz, 'German Non-Ferrous Metal Production in the Early Nineteenth Century', *Jl Eur. Ec. Hist.*, 3 (1974), 129–48.
80 L. Gallois, 'La production de la bauxite en France', *Ann. Géog.*, 26 (1917), 386–8.
81 A. M. Héron de Villefosse, op. cit., vol. I, 189–98.
82 L. F. Haber, *The Chemical Industry during the Nineteenth Century* (Oxford, 1958), 49–50.
83 H. Ormsby, *France* (London, 1950), 449–50.
84 'Report on the Sulphur Industry of Sicily', For. Off. Misc. Ser., 297, *Parl. Pap.*, 1893–4, vol. XCI, 69–87.
85 Hugh D. Clout, 'Reclamation of Coastal Marshland', *Themes in the Historical Geography of France* (London, 1977), 185–213; J. Huguet, 'Un polder du Marais Poitevin', *Nor.*, 2 (1955), 19–39.
86 R. Nasse, 'Über Landentwasserung in Holland', *Pet. Mitt.*, 30 (1884), 9–14.
87 J. Kuyper, 'Die Trockenlegungen der Zuiderzee', *Pet. Mitt.*, 22 (1876), 284–9.
88 'Report on the Draining of the Zuider Zee', For. Off. Misc. Ser., 565, *Parl. Pap.*, 1902, vol. CIII, 167–204. This report contains the text of the Bill.
89 Wilhelm Ehlers, 'Die Besiedlung der Mooregebiete in den Niederungen der Wumme, Worpe, Hamme und der mittleren Oste', *Zt Hist. Ver. Niedersachs.*, 79 (1914), 1–105.
90 R. Hansen, 'Die Besiedlung der Marsch zwischen Elbe- und Eidermundung', *Pet. Mitt.*, 37 (1891), 105–8.
91 John D. Post, 'A Study in Meteorological and Trade Cycle History: The Economic Crisis Following the Napoleonic Wars', *Jl Ec. Hist.*, 34 (1974), 315–49.
92 Joseph B. Hoyt, 'The Cold Summer of 1816', *Ann. A.A.G.*, 48 (1958), 118–31.
93 M. L. Parry, *Climatic Change, Agriculture and Settlement* (Folkestone, 1978), 169–71.
94 Emmanuel Le Roy Ladurie, *Time of Feast, Time of Famine* (New York, 1971), 214–16.
95 Jan de Vries, 'Histoire du climat et économie: des faits nouveaux, une interpretation différente', *Ann. E.S.C.*, 32 (1977), 198–226.
96 Gosta H. Liljequist, 'The Severity of Winters at Stockholm 1757–1942', *Geog. Ann.*, 25 (1943), 81–104.
97 M. L. Parry, op. cit., 58.
98 K. W. Butzer, 'Climatic Change', *The New Encyclopaedia Britannica*, vol. IV (Chicago, 1983).
99 E. Le Roy Ladurie, op. cit., 120.
100 Mack Walker, *Germany and the Emigration 1816–1885* (Cambridge, Mass., 1964), 1–41, 44–5.
101 C. Easton, *Les Hivers dans l'Europe Occidentale* (Leyden, 1928), 158.
102 An exception is M. L. Parry, op. cit., especially 70–1.
103 John D. Post, 'Meterological Historiography', *Jl Intdis. Hist.*, 3 (1972), 721–32.
104 W. S. Jevons, *The Coal Question* (London, 1865).
105 Clive Trebilcock, *The Industrialization of the Continential Powers 1780–1914* (London, 1981), 395–6.

Chapter 3. The population of nineteenth-century Europe

1 A. M. Carr-Saunders, *World Population: Past Growth and Present Trends* (Oxford, 1936), 8–9.
2 B. R. Mitchell, *European Historical Statistics, 1750–1970* (London, 1975).
3 Colin Clark, *Population Growth and Land Use* (London, 1967), 50.
4 Ibid., 179.
5 Marcel R. Reinhard, André Armengaud and Jacques Dupaquier, *Histoire générale de la population mondiale* (Paris, 1968), 320.
6 Kurt B. Mayer, *The Population of Switzerland* (New York, 1952), 53.
7 M. R. Reinhard *et al.*, op. cit., 333–8.
8 John D. Post, *The Last Great Subsistence Crisis in the Western World* (Baltimore, Md, 1977).
9 Gunther Meinhardt, 'Die Auswirkungen der Hungerjahre in der ersten Hälfte des 19. Jahrhunderts auf Göttingen', *Gött. Jb.* (1966), 211–19.
10 Gabriel Désert, 'La population de la Plaine de Caen et la crise de 1846–1847', *Ann. Norm.*, 1 (1951), 252–65; M. Baudot, 'La crise alimentaire des années 1846 et 1847 dans le Département de l'Eure', *Ann. Norm.*, 2 (1952), 51–6.

11 Michael Drake, 'The Growth of Population in Norway 1735–1855', *Sc. Ec. Hist. Rv.*, 13 (1965), 97–142; id., *Population and Society in Norway 1735–1865* (Cambridge, 1969), 54–65.
12 E. F. Heckscher, *An Economic History of Sweden* (Cambridge, Mass., 1954), 150–2.
13 G. Jacquemyns, *Histoire de la crise économique des Flandres (1845–1850)*, *Ac. Roy. Belg.*, *Mém.*, 2ᵉ série, 26 (1929), 15; Jurgen Bohmbach, 'Die Hungerjahre 1846/47 in Oberhessen', *Hess. Jb. Lgesch.*, 23 (1973), 333–65; M. Baudot, op. cit.
14 Gabriel Désert, 'Vivande et poisson dans l'alimentation français au milieu du XIXᵉ siècle', *Ann. E.S.C.*, 30 (1975), 519–36.
15 Tadeusz Sobczak, 'Wrost ludności a wrost produkcji rolnej w Królestwie Polskim w XIX wieku', *Kw. Hist. Kult. Mat.*, 13 (1965), 101–20.
16 Miriam Halpern Pereira, 'Niveaux de consommation, niveau de vie au Portugal (1874–1922)', *Ann. E.S.C.*, 30 (1975), 610–31.
17 Guy Thuillier, 'L'alimentation en Nivernais au XIXᵉ siècle', *Ann. E.S.C.*, 20 (1965), 1163–84.
18 Christian Vandenbroeke, 'L'alimentation à Gand pendant la première moitié du XIXᵉ siècle', *Ann. E.S.C.*, 30 (1975), 584–91.
19 Gunther E. Rothenberg, 'The Austrian Sanitary Cordon and the Control of the Bubonic Plague: 1710–1871', *Jl Hist. Med. All. Sci.*, 28 (1973), 15–23; J. D. Post, 'Famine Mortality and Epidemic Disease in the Process of Modernisation', *Ec. Hist. Rv.*, 29 (1976), 14–37.
20 Zbigniew Kuchowicz, 'Choroby i przyczyny zgonów ludności dóbr wilanowskonie porękich w latach dwudziestych XIX wieku', *Kw. Hist. Kult. Mat.*, 18 (1970), 261–75.
21 Angelo Celli, *The History of Malaria in the Roman Campagna from Ancient Times* (London, 1933), 102–23, 131.
22 L. Lemaire, 'Inondations et paludisme en Flandre Maritime (1622–1922)', *Rv. Nord.*, 8 (1922), 173–209.
23 J. Callot, 'La régression du paludisme en France', *Ann. E.S.C.*, 2 (1947), 328–35.
24 M. Drake, in *Sc. Ec. Hist. Rv.*, 13 (1965), 97–142.
25 G. Jacquemyns, op. cit., 342–48.
26 J. D. Post in *Ec. Hist. Rv.*, 29 (1976), 14–37.
27 William H. McNeill, *Plagues and Peoples* (New York, 1976), 261–4.
28 Roderick E. McGrew, *Russia and the Cholera 1823–1832* (Madison, Wis., 1965), 100.
29 Louis Chevalier, ed., *Le Choléra: la première épidémique du XIXᵉ siècle*, Bibl. Hist. Rv., 20 (1958), 13–15; Patrice Bourdelais, Jean-Yves Raulot and Michel Demonet, 'La marche du choléra en France: 1832–1854', *Ann. E.S.C.*, 33 (1978), 125–42.
30 Monique Dineur and Charles Engrand, 'Lille', in L. Chevalier, op. cit., 47–95.
31 Jean Vidalenc, 'Les départements normands', in L. Chevalier, op. cit., 99–108.
32 Dr Freour, 'Bordeaux', in L. Chevalier, op. cit., 109–20; Pierre Guiral, 'Marseille', in L. Chevalier, op. cit., 121–40.
33 A. Suheyl Unver, 'Les épidemies de choléra dans les terres balkaniques aux XVIIIᵉ et XIXᵉ siècles', *Ét. Balk.* (1973), part 4, 89–97.
34 G. Melvyn Howe, *Man, Environment and Disease in Britain* (Harmondsworth, 1976), 179–88.
35 G. Thuillier, 'Pour une histoire du quotidien au XIXᵉ siècle en Nivernais', *Éc. Prat. Htes. Ét., Civ. et Soc.*, 55 (1977), 12–20.
36 Charles E. Rosenberg, 'Cholera in Nineteenth-century Europe: A Tool for Social and Economic Analysis', *Comp. St. Soc. Hist.*, 8 (1965–6), 452–63.
37 Gabriel Désert, 'Les Archives hospitalières', *Ann. Norm. Cah.*, 10 (1977), 232–5.
38 P. Gonnet, 'Épidémies et société au XIXᵉ siècle', *Ann. Bourg.*, 31 (1959), 111–20.
39 Edmonde de Vedrenne-Villeneuve, 'L'inégalité sociale devant la mort dans la première moitié du XIXᵉ siècle', *Pop.*, 16 (1961), 665–98.
40 A. Armengaud, 'Industrialisation et démographie dans la France du XIXᵉ siècle', *L'Industrialisation en Europe au XIXᵉ siècle* (Paris, 1972), 187–200.
41 A. Armengaud, *La Population française au XIXᵉ siècle* (Paris, 1971), 17–18.
42 Wilhelm Abel, 'Der Pauperismus in Deutschland', *Wirtschaft, Geschichte und Wirtschaftsgeschichte: Festschrift Fr. Lütge*, ed. Wilhelm Abel (Stuttgart, 1966), 284–98.
43 A. Armengaud, *La Population française*, 28–31.
44 Jacques Houdaille, 'La population de Boulay (Moselle) avant 1850', *Pop.*, 22 (1967), 1055–84; Marcel Lachiver, *La Population de Meulan du XVIIᵉ au XIXᵉ siècle*, Éc. Prat. Htes, Ét., Dém. et Soc., 13 (1969), 210.
45 Yves Tugault, *Fécondité et urbanisation*, Inst. Nat. Ét. Dém., Trav. Doc., 74 (1975), 4.
46 Dudley Kirk, 'Population and Population Trends in Modern France', *Population Movements in Modern European History*, ed. H. Moller (New York, 1964), 92–100.
47 Frank Thistlethwaite, 'Migration from Europe Overseas in the Nineteenth and Twentieth Centuries', ibid., 73–92.

48 J.-C. Toutain, *La Population de France de 1700 à 1959*, *Hist. Quant. Éc. Fr.*, 3 (1963), 47–55.
49 Philippe Pinchemel, *Structures sociales et dépopulation rurale dans les campagnes picardes de 1836 à 1936*, Cent. Ért. Éc., Ét. et Mém., 34 (1957), 60–3.
50 Raymond Rousseau, *La Population de la Savoie jusqu'en 1861*, Éc. Prat. Htes. Ét., Dém et Soc., 1 (1960), 180–221.
51 Abel Chatelain, 'La formation de la population lyonnaise: l'apport d'origine montagnarde (XVIIIe–XXe siècles)', *Rev. Géog. Lyon.*, 19 (1954), 91–115.
52 André Armengaud, 'Les débuts de la dépopulation dans les campagnes toulousaines', *Ann. E.S.C.*, 6 (1951), 172–8.
53 Gabriel Désert, 'Aperçu sur l'exode rural en Basse Normandie à la fin du XIXe siècle', *Rv. Hist.*, 250 (1973), 107–18.
54 A. Corbin, 'Migrations temporaires et société rurale au XIXe siècle: le cas du Limousin', *Rv. Hist.*, 246 (1971), 293–334.
55 Abel Chatelain, 'Les migrations temporaires françaises au XIXe siècle', *Ann. Dém. Hist.* (1967), 9–28.
56 Jean-Charles Bonnet, 'Les travailleurs étrangers dans la Loire sous la IIIe République', *Cah. Hist.*, 16 (1971), 67–80.
57 Jacques Grandjonc, 'Élements statistiques pour une étude de l'immigration étrangère en France de 1830 à 1851', *Arch. Sozgesch.*, 15 (1975), 211–300.
58 A. Armengaud, *La Population française*, 95–101.
59 Attilio Mori, 'Les Italiens en France', *Ann. Géog.*, 13 (1904), 420–6.
60 Louis Chevalier, 'L'émigration française au XIXe siècle', *Ét. Hist. Mod. Cont.*, 1 (1947), 127–71.
61 Abel Chatelain, *Les migrations françaises vers le Nouveau Monde*', *Ann. E.S.C.*, 2 (1947), 53–70.
62 Wolfgang Kollmann, 'The Population of Germany', *Population Movements* (New York, 1964), 100–8.
63 Marcus L. Hansen, 'The Revolutions of 1848 and German Emigration', *Jl Ec. Bus. Hist.*, 2 (1929–30), 630–58.
64 Karl Kiel, 'Grunde und Folgen der Auswanderung aus dem Osnabrücker Regierungsbezirk', *Mitt. Ver. Gesch. Osnbr.*, 61 (1941), 85–176.
65 Ibid. See also Mack Walker, *Germany and the Emigration 1816–1885* (Cambridge, Mass., 1964), 1–41, 44–5.
66 F. Burgdorfer, 'Migration across the Frontiers of Germany', *International Migrations*, ed. W. F. Willcox, National Bureau of Economic Research (New York, 1931), II, 313–89.
67 Wolfgang Kollmann, 'Die Bevölkerung Rheinland-Westfalens in der Hochindustrialisierungsperiode', *Viert. Soz. Wirtgesch.*, 58 (1971), 359–88.
68 Klaus J. Bade, 'Massenwanderung und Arbeitsmarkt in Deutschen Nordosten von 1880 bis zum ersten Weltkrieg', *Arch. Sozgesch.*, 20 (1980), 265–323.
69 Stanisław Borowski, 'Emigration from the Polish Territories under German Rule 1815–1914', *St. Hist. Oec.*, 2 (1967), 151–83.
69 Hans-Ulrich Wehler, 'Die Polen im Ruhrgebiet bis 1918', *Viert. Soz. Wirtgesch.*, 48 (1961), 203–35.
70 Heinrich Rubner, 'En forêt de Bohême: immigration et émigration 1500–1960', *Ann. Dém. Hist.* (1970), 135–42; Ladislav Tajtak, 'Slovak Emigration and Migration in the Years 1900–1914', *St. Hist. Slov.*, 10 (1978), 43–86.
71 Felix Klezl, 'Austria', *International Migrations* (New York, 1931), vol. I, 390–410.
72 Paul Mitrovic, 'Aspects du mouvement ouvrier Yougoslave au début du XXe siècle', *Rv. Hist. Mod. Cont.*, 11 (1964), 203–28.
73 Robert F. Foerster, *The Italian Emigration of our Times* (Cambridge, Mass., 1919), 23–60.
74 Renée Rochefort, 'Un pays du Latifondo sicilien: Corleone', *Ann. E.S.C.*, 14 (1959), 441–60.
75 G. Yver, 'L'émigration italienne', *Ann. Géog.*, 6 (1897), 123–32.
76 Jaime Vicens Vives, *An Economic History of Spain* (Princeton, N.J., 1969), 621–2.
77 Jorge Nadal, *La poblacion Española (siglos XVI a XX)* (Barcelona, 1966), 158–62.
78 Jovan Cvijić, *La Péninsule balkanique* (Paris, 1918), 135.
79 Chrysos Evelpidis, 'L'exode rural en Grèce', *Contributions to Mediterranean Sociology*, ed. J.-G. Peristiany (Paris, 1968), 201–5.
80 B. J. Hovde, 'Notes on the Effects of Emigration upon Scandinavia', *J. Mod. Hist.*, 6 (1934), 253–79.
81 Adolph Jensen, 'Migration Statistics of Denmark, Norway, and Sweden', *International Migrations*, vol. II, 283–312.
82 E. Jutikkala, 'Migration in Finland in Historical Perspective', 3e Conf. Int. Hist. Éc. (1972), vol. IV, 123–36.

83 Max Engman, 'Migration from Finland to Russia during the Nineteenth Century', *Sc. Jl Hist.*, 3 (1978), 155–77.
84 Friedrich Edding, 'Intra-European Migration and the Prospect of Integration', *Economics of International Migration*, ed. Brinley Thomas (London, 1958), 238–44.
85 Imre Ferenczi, 'A Historical Study of Migration Statistics', *Int. Lab. Rv.*, 20 (1929), 356–84.
86 F. Edding, op. cit.
87 Arthur Ruppin, *The Jews in the Modern World* (London, 1934), 22–64.
88 Jakob Lestschinsky, 'Jüdische Wanderungen im letzten Jahrhundert', *Weltw. Arch.*, 25 (1927), part 2, 69–86.
89 Israel Cohen, 'The Economic Activities of Modern Jewry', *Ec. Jl*, 24 (1914), 41–56.
90 H. G. Wanklyn, 'Geographical Aspects of Jewish Settlement East of Germany', *Geog. Jl*, 95 (1940), 175.
91 Salo W. Baron, 'The Jewish Question in the Nineteenth Century', *Jl Mod. Hist.*, 10 (1938), 51–65.
92 I. Cohen, op. cit.
93 Jacob Lestschinsky, 'Jewish Migrations, 1840–1946', *The Jews: Their History, Culture and Religion*, ed. L. Finkelstein (New York, 1949), vol. II, 1198–1238.
94 Statistics mainly from Liebmann Hersch, 'International migration of the Jews', *International Migrations*, vol. II, 471–520.
95 Published as *Die Grundlagen des neunzehnten Jahrhunderts* (Munich, 1899); the English translation appeared in 1911.
96 Ruth Benedict, *Race and Racism* (London, 1942), 131.
97 Carleton S. Coon, *The Races of Europe* (New York 1939), 241–96.
98 H. J. Fleure, *The Peoples of Europe* (London, 1922), 42.
99 C. S. Coon, op. cit., 298–306.
100 G. M. Morant, *The Races of Central Europe* (London, 1939), 19.
101 Hans Kohn, *The Idea of Nationalism* (New York, 1945), 429.
102 Aristide R. Zolberg, 'The Making of Flemings and Walloons; Belgium 1830–1914', *J. Intdis. Hist.*, 5 (1974–5), 179–235.
103 Paul M. G. Lévy, 'Quelques problèmes de statistique linguistique à la lumière de l'expérience belge', *Rv. Inst. soc.*, 37 (1964), 251–73.
104 A. R. Zolberg, op. cit.
105 H. M. Kendall, 'A Survey of the Population changes in Belgium', *Ann. A.A.G.*, 28 (1938), 145–64.
106 *Statistisches Jahrbuch der Schweiz* (1949); K. B. Mayer, op. cit., 105.
107 K. B. Mayer, op. cit., 105.
108 Ibid., 181.
109 A. J. P. Taylor, *The Habsburg Monarchy 1815–1918* (London, 1941), 302.
110 Ibid., 303.
111 Paul Teleki, *The Evolution of Hungary and its Place in European History* (New York, 1923), 107–12; Isaiah Bowman, *The New World* (New York, 1928), 360.
112 In P. Teleki, op. cit., reproduced in C. A. Macartney, *Hungary and Her Successors 1919–1937* (London, 1937).
113 C. A. Macartney, op. cit., 33.
114 J. Cvijić, op. cit., 162–4. A much higher estimate is given by A. J. B. Wace and M. S. Thompson, *The Nomads of the Balkans* (London, 1914), 10; Wilhelm Giese, 'Zur jetzigen situation der Aromunen auf dem Balkan', *Acta Hist. Dac.*, 4 (1965), 5–12.
115 A. J. B. Wace and M. S. Thompson, op. cit., 7–8.
116 Wayne S. Vucinich, 'The Nature of Balkan Society under Ottoman Rule', *Slav. Rv.*, 21 (1962), 597–616.
117 Stavro Skendi, *The Albanian National Awakening 1872–1912* (Princeton, N.J., 1967), 3–27.
118 Elisabeth Barker, *Macedonia: Its Place in Balkan Power Politics*, Roy. Inst. Int. Aff. (1950), 12–20.
119 H. R. Wilkinson, *Maps and Politics: A Review of the Ethnographic Cartography of Macedonia* (Liverpool, 1951), 1–7.
120 Bruce Mcgowan, *Economic Life in Ottoman Europe* (Cambridge, 1981), 95–103.
121 Huey Louis Kostanick, *Turkish Resettlement of Bulgarian Turks, 1950–1953*, Univ. Calif. Pubns in Geog., 8, no. 2 (1957), 71–80.
122 J. Cvijić, op. cit., 133–52.
123 Jacques Dupaquier, 'Problèmes démographiques de la France Napoléonienne', *Rv. Hist. Mod. Cont.*, 17 (1970), 339–58.
124 Jean Bourgeois-Pichat, 'Évolution générale de la population française depuis le XVIIIe siècle', *Pop.*, 6 (1951), 635–62.

125 Catherine Rollet, 'L'effet des crises économiques du début du XIXe siècle sur la population', *Rv. Hist. Mod. Cont.*, 17 (1970), 391–410.
126 Abel Chatelain, 'Valeur des recensements de la population française au XIXe siècle', *Rv. Géog. Lyon.*, 29 (1954), 273–80.
127 David Grigg, *Population Growth and Agrarian Change* (Cambridge, 1980), 190–206.
128 Y. Tugault, op. cit., 15.
129 P. Pinchemel, op. cit., 60–3.
130 Gabriel Désert, in *Rv. Hist.*, 250 (1973), 107–18.
131 J. -Cl. Perrot, 'Documents sur la population du Calvados pendant la Révolution et l'Empire', *Ann. Norm.*, 15 (1965), 77–128.
132 Lucien Gachon, 'Dans les massifs cristallins d'Auvergne: la ruine du paysage rural et ses causes', *Ann. E.S.C.*, 5 (1950), 448–60.
133 P. Pinchemel, op. cit., 85, 208.
134 *Histoire économique et sociale de la France*, ed. F. Braudel and E. Labrousse, vol. IV, part 1 (Paris, 1979), 98–9; Charles H. Pouthas, *La Population française pendant la première moitié du XIXe siècle*, Inst. Nat. Ét. Dém., Trav. Doc., 25 (1956), 71–4.
135 Fernand Maurette, 'La population de la France au début du XXe siècle', *Ann. Géog.*, 18 (1909), 125–40.
136 C. H. Pouthas, 'L'évolution démographique du département du Nord dans la première moitié du XIXe siècle', *Rv. Nord*, 36 (1954), 331–8.
137 Jacques Grandjonc, 'Élements statistiques pour une étude de l'immigration étrangère en France de 1830 à 1851', *Arch. Sozgesch.*, 15 (1975), 211–300.
138 *Histoire économique et sociale de la France*, vol. IV, part 1, 104–6.
139 A. d'Angeville, *Essai sur la statistique de la population française* (Bourg, 1836), edn of E. Le Roy Ladurie (Paris, 1969).
140 Jean-Paul Aron, Paul Dumont and Emmanuel Le Roy Ladurie, *Anthropologie du conscrit français*, Éc. Prat. Htes. Ét., Civ. et Soc., 28 (1972), especially maps 9–20.
141 Placide Rambaud, *Économie et sociologie de la Montagne: Albiez-le-Vieux en Maurienne*, Éc. Prat. Htes. Ét., Ét. et Mém., 50 (1962), 169–79.
142 E. W. Hofstee, 'Population Increase in the Netherlands', *Acta Hist. Neer.*, 3 (1968), 43–125.
143 H. M. Kendall, op. cit.
144 Friedrich Leyden, 'Die Volksdichte in Belgien, Luxemburg und den Niederlanden', *Pet. Mitt. Erg.*, 204 (1929).
145 F. J. Monkhouse, *The Belgian Kempenland* (Liverpool, 1949), 201–16.
146 E. W. Hofstee, *Korte Demografische Geschiedenis van Nederland van 1800 tot Heden* (Haarlem, 1981), esp. 9–12. B. H. Slicher van Bath, 'Report on the Study of Historical Demography in the Netherlands', *Actes Coll. Int. Dém. Hist.* (1965), 185–98.
147 J. A. Faber et al., – Population Changes and Economic Development in the Netherlands: A Historical Survey', *A.A.G. Bij.*, 12 (1965), 47–114.
148 H. K. Roessingh, 'Het Veluwse inwonertal, 1526–1947', *A.A.G. Bij.*, 11 (1964), 79–150.
149 K. B. Mayer, op. cit., 101–11.
150 Klaus J. Bade, 'Massenwanderung und Arbeitsmarkt in deutschen Nordosten von 1880 bis zum Ersten Weltkrieg', *Arch. Sozgesch.*, 20 (1980), 265–323.
151 W. Kollmann, 'Industrialisierung, binnenwanderung und "SozialeFrage"', *Viert. Soz. Wirtgesch.*, 46 (1959), 45–70.
152 Id., 'Demographische Konsequenzen der Industrialisierung in Preussen', *L'Industrialisation en Europe au XIXe siècle* (Paris, 1972), 267–84.
153 Stanisław Borowski, 'Demographic Development and Malthusian Problem in the Polish Territories under German Rule 1807–1914', *St. Hist. Oec.*, 3 (1968), 159–79.
154 Witold Russ, 'Społeczenstwo Królestwa Polskiego w XIX i początkach XX wieku', *Prz. Hist.*, 68 (1977), 259–88; Irena Gieysztor, 'Recherches sur la démographie historique et en particulier rurale en Pologne', *Kw. Hist. Kult. Mat.*, 12 (1964), 509–28.
155 Tadeusz Sobczak, 'Wzrost ludnosci a wzrost produkcji rolnej w Królestwie Polskim w XIX wieku', *Kw. Hist. Kult. Mat.*, 13 (1965), 101–20.
156 *Handbuch von Polen*, ed. E. Wunderlich (Berlin, 1918), 334–41.
157 W. Russ, op. cit.
158 Axel Nielsen, *Dänische Wirtschaftsgeschichte* (Jena, 1933), 341.
159 Sima Liberman, 'Norwegian Population Growth in the 19th Century', *Ec. Hist.*, 11 (1968), 52–66.
160 Samuel Laing, *Journal of a Residence in Norway* (London, 1836), 403–4.
161 Id., *A Tour in Sweden in 1838* (London, 1839), 183–5, 232.
162 Ibid., 232, 357.

163 S. Laing, *Observations on the Social and Political State of Denmark* (London, 1852), 42, 119, 134–40.
164 Michael Drake, 'The Growth of Population in Norway 1735–1855', *Sc. Ec. Hist. Rv.*, 13 (1965), 97–142; id., *Population and Society in Norway 1735–1865* (Cambridge, 1969), 47–62.
165 E. F. Heckscher, 'Swedish Population Trends before the Industrial Revolution', *Ec. Hist. Rv.*, 2 (1950), 266–77.
166 Christer Winberg, 'Population Growth and Proletarianization', *Chance and Change* (Odense, 1978), 170–84.
167 Dorothy Swaine Thomas, 'Internal Migrations in Sweden: A Note on their Extensiveness as Compared with Net Migration Gain or Loss', *Am. Jl Soc.*, 42 (1936–7), 345–57.
168 Id., 'Economic and Social Aspects of Internal Migrations: an Exploratory Study of Selected Swedish Communities', *Economic Essays in Honor of Wesley Clair Mitchell* (New York, 1935), 447–76; Gunnar Myrdal, 'Industrialization and Population', *Economic Essays in Honour of Gustav Cassel* (London, 1933), 435–57.
169 Adolf Jensen, 'Migration Statistics of Denmark, Norway and Sweden', *International Migrations*, vol. II, 283–312.
170 B. R. Mitchell, 'Statistical Appendix 1700–1914', *Fontana Economic History of Europe* (London, 1973), vol. IV, part 2, 738–42.
171 *Raum und Bevölkerung in der Weltgeschichte*, ed. E. Kirsten, E. W. Buchholz and W. Kollmann (Wurzburg, 1956), vol. II, 163–4.
172 Ivan T. Berend and György Ránki, *Economic Development in East-Central Europe in the 19th and 20th Centuries* (New York, 1974), 20.
173 Kurt Witthauer, 'Die Bevölkerung der Erde', *Pet. Mitt. Erg.*, 265 (1957), 195.
174 Marcel R. Reinhard, André Armengaud and Jacques Dupaquier, *Histoire générale de la population mondiale* (Paris, 1968), 377–82.
175 J. Kovacsics, 'The Population of Hungary in the Eighteenth Century', 3ᵉ Conf. Int. Hist. Éc., 4 (1968), 137–45; id., 'An Account of Research Work in Historical Demography in Hungary', *Actes Coll. Int. Dém. Hist.*, 249–72; L. Katus, 'Economic Growth in Hungary during the Age of Dualism (1867–1913)', *Sozial- Ökonomische Forschungen zur Geschichte von Ost-Mitteleuropa*, ed. E. Pamlenyi (Budapest, 1970), 35–127.
176 Eugene Csocsan de Varallja, 'La population de la Hongrie au XXᵉ siècle', *Rv. Est.*, 5 (1974), 133–77; 8 (1977), 75–170.
177 Gustav Thirring, 'Hungarian Migration of Modern Times', *International Migrations*, vol. II, 411–39; J. Puskas, 'Emigration from Hungary to the United States before 1914', *St. Hist.*, 113 (1975); *Raum und Bevölkerung*, vol. II, 166.
178 Jorge Nadal Oller, 'La contribution des historiens catalans à l'histoire de la démographie développement économique', *Stud. Fanfani*, 2 (1962), 491–510.
179 J. V. Vives, op. cit., 618.
180 S. E. Widdrington, *Spain and the Spaniards in 1843* (London, 1844), 406.
181 Frederick Hendriks, 'A Review of the Statistics of Spain down to the Years 1857 and 1858', *Jl Roy. Stat. Soc.*, 23 (1860), 147–200; J. Nadal, op. cit., 20; J.-G. da Silva, op. cit., *Raum und Bevölkerung*, vol. II, 154; conflicting totals are given by K. Witthauer, op. cit., 205.
182 J. V. Vives, op. cit., 618–19.
183 J. Nadal, op. cit., 181–97; K. Witthauer, op. cit., 207.
184 Jorge Nadal Oller, 'La contribution des historiens catalans à l'histoire de la démographie générale', *Pop.*, 16 (1961), 91–101.
185 *Raum und Bevölkerung*, vol. II, 154. Colin Clark's estimate of 19 millions in *Population Growth and Land Use* (London, 1967), 64, appears to be too high.
186 Carlo Cipolla, 'Four Centuries of Italian Demographic Development', *Population in History*, ed. D. V. Glass and D. E. C. Eversley (London, 1965), 570–87.
187 *Raum und Bevölkerung*, vol. II, 214; M. R. Reinhard, A. Armengaud and J. Dupaquier, op. cit., 385.
188 R. F. Foerster, op. cit., 59–60.
189 Samuel Laing, *Notes of a Traveller on the Social and Political State of France, Prussia, Switzerland, Italy and other Parts of Europe* (London, 1842), 396.
190 G. Yver, 'L'émigration italienne', *Ann. Géog.*, 6 (1897), 123–32.
191 R. F. Foerster, op. cit., 7, 38; Anna Maria Ratti, 'Italian Migration Movements 1876 to 1927', *International Migrations*, vol. II, 440–70.
192 Antonio Gallenga, *Country Life in Piedmont* (London, 1858), 31–2.
193 Ibid., 86.

Notes to pp. 119–37

Chapter 4. Urban development in the nineteenth century

1 Gideon Sjoberg, *The Preindustrial City* (New York, 1960), 182–219.
2 G. Albers, 'Das Stadt-Land-Problem im Städtebau der letzten Hundert Jahre', *Stud. Gen.*, 16 (1963), 565–75.
3 Étienne Juillard, 'L'urbanisation des campagnes en Europe occidentale', *Ét. Rur.*, 1 (1961), 18–33.
4 Mack Walker, *German Home Towns: Community, State and General Estate, 1648–1871* (Ithaca, N.Y., 1971), 19–32.
5 Jean Vidalenc, 'Les sociétés urbaines et les villes dans les arrondissements littoraux de la Seine-Inferieure sous le Premier Empire', 'Villes de l'Europe Méditerranéenne et de l'Europe occidentale du moyen age au XIXe siècle, *Ann. Fac. Nice*, 9–10 (1969), 291–314.
6 Honoré de Balzac, *Béatrix*, edn of 1941, 319–23. This passage is quoted in N. J. G. Pounds, *Historical Geography of Europe*, vol. II (Cambridge, 1979), 147–8.
7 Alain Corbin, *Archaisme et Modernité en Limousin au XIXe siècle 1845–1880* (Paris, 1975), vol. I, 19.
8 Based on Michel Rochefort, *L'Organisation urbaine de l'Alsace*, Pubns Fac. Strasbg, 139 (Paris, 1960), 195.
9 Ibid., 194.
10 The province of Białystok was excluded from the Congress Kingdom, but included in the Poland of 1919, and remained Polish in 1945.
11 Leszek Kosiński, *Miasta Województwa Białostockiego*, Polish Academy of Sciences, Prace Geograficzne, 32 (Warsaw, 1962).
12 For a similar study of towns in central Poland see I. F. Tłoczek, *Miasteczka rolnicze w Wielkopolsce* (Warsaw, 1955), 30–51; Andrzej Jezierski, 'Extensive Development of Towns in the Agricultural Region of Poland', *Stud. Hist. Oec.*, 13 (1978), 139–52.
13 Konrad Olbricht, 'Die deutschen Grossstädte', *Pet. Mitt.*, 59 (1913), part 2, 57–67.
14 Gerald Brause, 'Entwicklungsprobleme von Grosstädtzentren unter besonderes Berücksichtigung Leipzigs', *Jb. Reggesch.*, 3 (1968), 184–203.
15 Louis Jean Laurent, 'Mutation économique et développement urbain dans les Alpes Maritimes de 1860 à 1914', *Villes du Littoral, Ann. Fac. Nice*, 25 (1975), 51–74.
16 Except the Grande Corniche, which was built by Napoleon to provide a road which could not be seen from the sea.
17 Alain Ruggiero, 'Aspects de l'économie niçoise 1814–1860', *Villes du Littoral*, 29–49.
18 Pierre Lavedan, *Histoire de l'urbanisme: époque contemporaine* (Paris, 1952), 198.
19 For a discussion of the definitions used in official statistics see Paul Bairoch, 'Population urbaine et taille des villes en europe de 1600 à 1970', *Rv. Hist. Éc. Soc.*, 54 (1976), 304–35.
20 Hugh D. Clout, 'Urban Growth, 1500–1900', *Themes in the Historical Geography of France* (London, 1977), 483–540.
21 Adna Ferrin Weber, *The Growth of Cities in the Nineteenth Century* (New York, 1899), 8.
22 Paul Meuriot, *Des Agglomérations urbaines dans l'Europe contemporaine* (Paris, 1897), 38–42.
23 Oskar Buchner, 'Der Einfluss der Wirtschaftskrise auf die Wanderungsbewegung in den deutschen Städte', *Rv. Int. Stat.*, 4 (1936), 1–26.
24 Mark Jefferson, 'The Law of the Primate City', *Geog. Rv.*, 29 (1939), 226–32.
25 Paul Bairoch, 'Urbanisation and Economic Development in the Western World: Some Provisional Conclusions of an Empirical Study', *Patterns of European Urbanisation Since 1500*, ed. H. Schmal (London, 1981), 61–75.
26 Kurt B. Mayer, *The Population of Switzerland* (New York, 1952), 250–1.
27 Nikolaj Todorov, 'The Balkan Town in the Second Half of the 19th Century', *Ét. Balk.*, 5 (1969), part 2, 31–50.
28 This can be expressed as: $P_n = P_i(N)^{-1}$ where P_i is the primate town and N the rank of a particular town.
29 Bavaria, Württemberg and Baden, approximately the area covered by Christaller's study.
30 Walter Christaller, *Die Zentralen Orte in Suddeutschland* (Jena, 1933), translated as *Central Places in Southern Germany* (Englewood Cliffs, N.J., 1966).
31 A good exposition of central-place theory in the present context is Robert E. Dickinson, *The City Region in Western Europe* (London, 1967), 30–59.
32 Alsace was included in Christaller's original study.
33 For an analysis of the linear development of central places see W. K. D. Davies, 'Centrality and the Central Place Hierarchy', *Urban Studies*, 4 (1967), 61–79.
34 John W. Alexander, 'The Basic–Non-Basic Concept of Urban Economic Functions', *Ec. Geog.*, 30 (1954), 246–61.

35 Adam Szczypiorski, 'Struktura zawodowa i społeczna Warszawy w pierwszym okresie epoki kapitalistycznej (1864–1882)', *Kw. Hist. Kult. Mat.*, 8 (1960), 75–102. J.-P. Viennot, 'La population de Dijon d'après le recensement de 1851', *Ann. Dém. Hist.* (1969), 241–60.
36 André-E. Sayous, 'L'évolution de Strasbourg entre les deux guerres (1871–1914)', *Ann. E.S.C.*, 6 (1934), 1–19, 122–32.
37 P. Meuriot, op. cit., 38–55.
38 A. Szczypiorski, op. cit.
39 J.-P. Viennot, op. cit.
40 A.-E. Sayous, op. cit.
41 Andrzej Stasiak, *Miasto Królewska Huta* (Warsaw, 1962), 32.
42 Irwin T. Sanders, 'An Approach to Studying Rural–Urban Relationships in Southeastern Europe', *Congr. Int. Ét. SE. Eur.*, 2 (1972), 561–67.
43 N. Todorov, op. cit.
44 The name of W. Reymont's novel, *Ziemia Obiecana*, dealing with the growth of Łódź.
45 Louis Chevalier, *La Formation de la population parisienne au XIXe siècle*, Inst. Nat. Ét. Dém. Ét. Doc., 10 (1950), 174–217; David Pinkney, 'Migrations to Paris during the Second Empire', *Jl Mod. Hist.*, 25 (1953), 1–12.
46 Rudolf Heberle and Fritz Meyer, *Die Grossstädte im Ströme der Binnenwanderung* (Leipzig, 1937), 23–24.
47 Wolfgang Kollmann, 'Industrialisierung, Binnenwanderung und "Soziale Frage"', *Viert. Soz. Wirtgesch.*, 46 (1959), 45–70.
48 R. Heberle and F. Meyer, op. cit., 155.
49 Wilhelm Brepohl, *Der Aufbau des Ruhrvolkes im Zuge der Ost-West Wanderung* (Recklinghausen, 1948), 244; W. Kollmann, 'Binnenwanderung und Bevölkerungsstrukturen der Ruhrgebietsgrossstädte im Jahre 1907', *Soz. Welt.*, 9 (1958), 219–33.
50 Hans-Ulrich Wehler, 'Die Polen im Ruhrgebiet bis 1918', *Viert. Soz. Wirtgesch.*, 48 (1961), 203–35.
51 W. Kollmann, 'The Population of Germany in the Age of Industrialism', *Population Movements in Modern European History*, ed. Herbert Moller (New York, 1964), 100–8.
52 Calculated from tables in W. Brepohl, op. cit., 249; see also W. Kollmann, 'Die Bevölkerung rheinland-Westfalens in der Hochindustrialisierungsperiode', *Viert. Soz. Wirtgesch.*, 58 (1971), 359–88.
53 Paul Hohenberg, 'Migrations et fluctuations démographiques dans la France rurale 1836–1901', *Ann. E.S.C.*, 29 (1974), 461–97.
54 Abel Chatelain, 'L'attraction des trois plus grandes agglomérations françaises: Paris–Lyon–Marseille en 1891', *Ann. Dém. Hist.* (1975), 27–41; id., *Les Migrants temporaires en France de 1800 à 1914*, Pubns. Univ. Lille (n.d.), passim.
55 L. Chevalier, op. cit.
56 A. Chatelain, 'L'attraction des trois...agglomérations...'
57 Georges Dupeaux, 'Immigration urbaine et secteurs économiques: l'exemple de Bordeaux au début du XXe siècle', *Ann. Midi*, 85 (1973), 209–20.
58 A. F. Weber, op. cit., 285–92.
59 J.-P. Viennot, op. cit.
60 Marie-Thérèse Plégat, 'L'évolution démographique d'une ville française au XIXe siècle: l'example de Toulouse', *Ann. Midi*, 54 (1952), 227–48.
61 André Cornette, 'Arras et sa banlieue', *Rv. Nord*, XLII, Livr. Géog. IX (1960), 45.
62 A. W. Flux, 'Urban Vital Statistics in England and Germany', *Jl Roy. Stat. Soc.*, 73 (1910), 207–53.
63 P. Pierrard, 'Habitat ouvrier et démographie à Lille au XIXe siècle et particulièrement sous le Second Empire', *Ann. Dém. Hist.* (1975), 37–48.
64 Guy Pourcher, *Le Peuplement de Paris*, Inst. Nat. Ét. Dém. Trav. Doc., 43 (1964), 11.
65 Louis Chevalier, *Labouring Classes and Dangerous Classes* (London, 1973), 187–8.
66 P. Lavedan, op. cit., 63.
67 Gerard Jacquemet, 'Urbanisme parisienne: la bataille du tout-à-l'égout à la fin du XIXe siècle', *Rv. Hist. Mod. Cont.*, 26 (1979), 505–48.
68 L. Chevalier, *Labouring Classes*, 202–10.
69 David Pinkney, *Napoleon III and the Rebuilding of Paris* (Princeton, N.J., 1958), 127.
70 André Armengaud, 'Quelques aspects de l'hygiène publique à Toulouse au début du XXe siècle', *Ann. Dém. Hist.* (1975), 131–8.
71 Eberhard Schmieder, 'Wirtschaft und Bevölkerung', *Berlin und die Provinz Brandenburg im 19. und 20. Jahrhundert*, ed. Hans Herzfeld and Gerd Heinrich, Veröff. Hist. Kom. Berlin (1968), 382–3.

72 Ernst Busse, 'Die Gemeindebetriebe Münchens', *Schr. Ver. Sozpk*, 129 (1908); this volume also contains studies of other German cities.
73 F. W. Robins, *The Story of Water Supply* (London, 1946), 199–200.
74 Pierre Lavedan, *Géographie des villes* (Paris, 1936), 184.
75 D. H. Pinkney, op. cit., 105.
76 J.-P. Goubert, 'Eaux publiques et démographie historique dans la France du XIXe siècle', *Ann. Dém. Hist.* (1975), 115–21.
77 René Dons, 'Un aspect de l'alimentation en eau de la ville de Bruxelles', *Cah. Brux.*, 19 (1975), 14–45.
78 A. Armengaud, op. cit.
79 Hans Spethmann, *Das Ruhrgebiet* (Berlin, 1933), vol. II, 568–83.
80 N. J. G. Pounds, *The Ruhr* (London, 1952), 228.
81 P. Lavedan, op. cit., 185.
82 E. Schmieder, op. cit.
83 Inge von Karolyi, 'Verhandlungen über den Bau einer Gasanstalt in Göttingen 1836–1861', *Gött. Jb.* (1969), 109–26. It is claimed that a gasworks was built in Düsseldorf in 1826: Otto Most, 'Die Gemeindebetriebe der Stadt Düsseldorf', *Schr. Ver. Sozpk*, 129 (1908), part 2, 1–13.
84 R. H. Parsons, *The Early Days of the Power Station Industry* (Cambridge, 1940), 1–3.
85 'Die Gemeindebetriebe', *Schr. Ver. Sozpk*, 129 (1908), passim.
86 E. Schmieder, op. cit., 383.
87 Raoul Blanchard, 'L'industrie de la houille blanche dans les Alpes françaises', *Ann. Géog.*, 26 (1917), 15–41.
88 Louis Sebastien Mercier, *Tableau de Paris* (Amsterdam, 1782), vol. I, 117; P. Lavedan, op. cit., 97.
89 H. D. Clout, op. cit.; John P. McKay, *Tramways and Trolleys: The Rise of Urban Mass Transport in Europe* (Princeton, N.J., 1976), 10–11.
90 'Report on the Tramway System of Paris and the Department of the Seine', *For. Off. Misc. Ser.*, 362, *Parl. Pap.*, 1895, vol. CII.
91 'Die Gemeindebetriebe', passim.
92 E. Schmieder, op. cit.
93 Peter Hall, *The World Cities* (London, 1966), 73.
94 Maurice Block, *Statistique de la France* (Paris, 1860), vol. II, 414.
95 Victor Hugo published *Les Misérables* in 1862, but had begun work on it before 1850.
96 'Report of an Enquiry by the Board of Trade into Working Class Rents, Housing and Retail Prices...in the Principal Industrial Towns of the German Empire', Cd 4032, *Parl. Pap.*, 1908, vol. CVIII, 253.
97 'Report of an Enquiry by the Board of Trade into Working Class Rents, Housing and Retail Prices...in the Principal Industrial Towns of Belgium', Cd 5065, *Parl. Pap.*, 1910, vol. XVC, 43–306.
98 Friedrich Mielke, 'Studie über den Berliner Wohnungsbau zwischen den Kriegen 1870/71 und 1914/18', *Jb. Gesch. Mitt. Ostd.*, 20 (1971), 202–38.
99 'Report...into Working Class Rents...German Empire', 14, 85.
100 Ibid., passim.
101 Ibid., 179; this estimate is certainly too high.
102 Ibid., 354.
103 Hans Herzfeld, 'Berlin als Kaiserstadt und Reichshauptstadt', *Jb. Gesch. Deutsch. Ost.*, 1 (1952), 141–70.
104 'Wohnungsnoth der armeren Klassen in deutschen Grossstädten', *Schr. Ver. Sozpk*, 30 (1886), passim.
105 'Report of an Enquiry by the Board of Trade into Working Class Rents...Towns of France', Cd 4512, *Parl. Pap.*, 1909, Vol. XCI, 1–486.
106 Ibid., 218.
107 P. Pierrard, op. cit.
108 F. Raefler, 'Das Schlafhauswesen im Oberschlesischen Industriegebiet', *Berg. Hütt. Rd.*, 14 (1917–18), 1–6, 9–14, 17–21, 25–9, 33–7; N. J. G. Pounds, *The Upper Silesian Industrial Region* (Bloomington, Ind., 1958), 148–9.
109 'Report of an Enquiry...into Working Class Rents...German Empire', passim.
110 Krzystof Baronowski, 'Udział budownictwa drewnianego w zabudowie Łódzi w XIX i XX wieku', *Kw. Hist. Kult. Mat.*, 21 (1973), 225–39.
111 P. Lavedan, *Géographie des villes*, 127–37.
112 Othmar Birkner, 'Die Bedeutung der Bauordnung im Städtebau des 19. Jahrhunderts', *Zt Stadtgesch.*, 3 (1976), 26–37.

113 'Report of an Enquiry...into Working Class Rents...German Empire', 354.
114 For city plans see R. E. Dickinson, op. cit., passim.
115 E. W. Burgess, 'The Growth of the City', *The City*, ed. R. E. Park, E. W. Burgess and R. D. McKenzie (Chicago, 1925), 47–62.
116 Homer Hoyt, *The Structure and Growth of Residential Neighborhoods in American Cities*, U.S. Federal Housing Administration (Washington, D.C., 1939).
117 J. Ziołkowski, *Sosnowiec* (Katowice, 1960), is one of the very few historical studies of cities to use the urban models discussed here.
118 Ibid., 202.
119 F. Delrange-Vancomerbeke, op. cit.
120 Chauncy D. Harris and E. L. Ullmann, 'The Nature of Cities', *Ann. Am. Acad. Pol. Soc. Sci.*, 242 (1945), 7–17.
121 P. Lavedan, *Géographie des villes*, 110; id., *Histoire de l'urbanisme*, 25, 376.
122 D. H. Pinkney, op. cit., 15.
123 William L. Langer, 'The Pattern of Urban Revolution in 1848', *French Society and Culture since the Old Regime*, ed. E. M. Acomb and M. L. Brown (New York, 1966), 90–118.
124 L. Chevalier, *La Formation de la population parisienne*, maps.
125 Adeline Daumard, *La Bourgeoisie parisienne de 1815 à 1848*, Cent. Rech. Hist. Dém. Soc., 8 (1963), 190–9; maps, 182–3.
126 'Verfassung und Verwaltungsorganisation der Städte', *Schr. Ver. Sozpk* 117, (1906).
127 Louis Bergeron, 'Approvisionnement et consommation à Paris sous le Premier Empire', *Paris I. de F.*, 14 (1963), 197–232.
128 Jean-Marc Leri, 'Aspects administratifs de la construction des marchés de la ville de Paris 1800–1850', *Bul. Soc. Hist. Paris*, 103–4 (1976–7), 171.
129 Jean Bastié, *La Croissance de la banlieue parisienne*, Pub. Fac. Let. Paris, 17 (1964), 89.
130 Helmuth Croon, 'Die Versorgung der Grossstädte des Ruhrgebietes im 19. und 20. Jahrhundert', 3e Conf. Int. Hist. Éc. (1968), 130–46.
131 N. Todorov, op. cit.
132 Id., 'Sur quelques aspects du passage du féodalisme au capitalisme dans les territoires balkaniques de l'Empire Ottoman', *Rv. Ét. SE Eur.*, 1 (1963), 103–36.
133 Jan Havranek, 'Social Classes, National Ratios and Demographic Trends in Prague 1880–1900', *Hist.* (P), 13 (1966), 171–208.
134 W. Brepohl, op. cit.
135 F. Delgrange-Vancomerbeke, op. cit.
136 H. G. Wanklyn, 'Geographical Aspects of Jewish Settlement East of Germany', *Geog. Jl*, 95 (1940), 175–90.
137 Jacob Lestchinsky, 'Jewish Migrations, 1840–1946', *The Jews: Their History, Culture and Religion*, ed. L. Finkelstein (New York, 1949), vol. II, 1198–1238.
138 Israel Cohen, 'The Economic Activities of Modern Jewry', *Ec. Jl*, 24 (1914), 41–56.
139 Adam Szczypiorski, op. cit.
140 Marian Marek Drozdowski, 'The Urbanization in Poland in the Years 1870–1970', *Stud. Hist. Oec.*, 9 (1974), 223–44.
141 N. Todorov, op. cit.
142 P. Bairoch, op. cit.
143 Françoise Carrière and Philippe Pinchemel, *Le Fait urbain en France*, Ec. Prat. Htes. Ét., Ét et Mém., 57 (1963), 104; Alex Supan, 'Die Verschiebung der Bevölkerung in den industriellen Grossstädten Westeuropas im letzten Jahrzehnt (1881–91)', *Pet. Mitt.*, 38 (1892), 59–66.
144 A. Cornette, op. cit.
145 C. Arnéodo-Frangville, 'La population de Poitiers des origines à 1954', *Nor.*, 7 (1960), 273–315.
146 M.-T. Plégat, op. cit.
147 J. Vidalenc, op. cit.; J.-P. Viennot, op. cit.
148 George Fasel, 'Urban Workers in Provincial France, February–June 1848', *Int. Rv. Soc. Hist.*, 17 (1972), 660–74.
149 J. Vidalenc, *La Société française de 1815 à 1848: le peuple des villes et des bourgs* (Paris, 1973), passim.
150 G. Fasel, op. cit.
151 D. H. Pinkney, 'Migrations to Paris...'
152 G. R. Crone, 'The Site and Growth of Paris', *Geog. Jl*, 98 (1941), 35–47.
153 Guy Pourcher, op. cit.
154 M. Block, op. cit., vol. II, 414–15.
155 'Report of an Enquiry...into Working Class Rents...France', 72.

156 Andre Cuchot, 'Mouvement de la population de Paris', *Rv. Deux Mondes*, 9 (1845), 718–36, as quoted in *Metternich's Europe*, ed. Mack Walker (London, 1968), 248–65.
157 Louis Chevalier, op. cit., 215.
158 L. S. Mercier, op. cit., vol. I, 14–18.
159 A. Daumard, op. cit., 182–95.
160 D. H. Pinkney, op. cit., 20–4.
161 Jean Bastié, op. cit., 53.
162 D. H. Pinkney, 'Napoleon III's Transformation of Paris: the Origins and Development of the Idea', *Jl Mod. Hist.*, 27 (1955), 125–34.
163 Ibid.
164 P. Pierrard, op. cit.
165 Roger Levy, Le Havre entre trois révolutions (1789–1818)', Bibl. Hist. Rv., 4 (1912), 151.
166 P. Guillaume, 'La population d'une grande cité de province au XIXe siècle', *Ann. Dém. Hist.* (1966), 23–35.
167 'Report of an Enquiry...into Working Class Rents...Belgium.'
168 Mina Martens, *Histoire de Bruxelles* (Toulouse, 1976), 273–4.
169 Aristide R. Zolberg, 'The Making of Flemings and Walloons: Belgium 1830–1914', *Jl Intdis. Hist.*, 5 (1974–5), 179–235.
170 A. Demangeon, 'Anvers', *Ann. Géog.*, 27 (1927), 307–39.
171 'Report of an Enquiry...into Working Class Rents...Belgium.'
172 Samuel Brown, 'On the Statistics of the Kingdom of the Netherlands', *Jl Roy. Stat. Soc.*, 32 (1869), 192–214.
173 P. Hall, op. cit., 95–101; R. van Engelsdorp Gastelaars and M. Wagenaar, 'The Rise of the "Randstad", 1815–1930', *Patterns of European Urbanisation since 1500*, ed. H. Schmal (London, 1981).
174 Hans Mauersperg, *Wirtschafts- und Sozialgeschichte zentraleuropäischer Städte in neuerer Zeit* (Gottingen, 1960), 31–2.
175 Erich Keyser, *Bevölkerungsgeschichte Deutschlands* (Leipzig, 1941), 374–6.
176 Heinzpeter Thummler, 'Zur regionalen Bevölkerungsentwicklung in Deutschland', *Jb. Wirtgesch.* (1977), part 1, 55–72.
177 H. Wichmann, 'Orte des deutschen Reichs mit mehr als 25000 Einwohnern', *Pet. Mitt.*, 57 (1911), part 1, 131.
178 Oskar Buchner, 'Der Einfluss der Wirtschaftskrise auf die Wanderungsbewegung in den deutschen Städte', *Rv. Inst. Int. Stat.*, 4 (1936), 1–26.
179 Konrad Olbricht, 'Die deutschen Grossstädte', *Pet. Mitt.*, 59 (1913), 57–67.
180 Karin Weimann, 'Bevölkerungssentwicklung und Frühindustrialisierung in Berlin 1800–1850', *Untersuchungen zur Geschichte der frühen Industrialisierung vornehmlich im Wirtschaftsraum Berlin-Brandenburg*, Hist. Kom. Berlin, 6 (1971), 150–90.
181 Ingrid Thienel, 'Industrialisierung und Städtewachstum', ibid., 106–41.
182 Maurice Halbwachs, 'Gross Berlin: grande agglomération ou grande ville?', *Ann. E.S.C.*, 6 (1934), 547–70.
183 E. Schmieder, op. cit.
184 'Report of an Enquiry...into Working Class Rents...Germany.'
185 'Die Wohnungsnoth der armeren Klassen...'; F. Mielke, op. cit.
186 R. E. Dickinson, op. cit., 240.
187 Richard Dietrich, 'Von der Residenzstadt zur Weltstadt', *Das Hauptstädtproblem in der Geschichte*, *Jb. Gesch. Mitt. Ostdeutsch.*, 1 (1952), 111–39; Ernst Kaeber, 'Beiträge zur Berliner Geschichte', *Veröff. Hist. Kom. Berlin*, 14 (1964), 180–9.
188 Studia Geograficzne nad Aktywizacja Małych Miast, Prace Geograficzne, vol. IX, Polish Acad. Sci. (1957), 37–60; I. F. Tłoczek, op. cit., 30–51.
189 *Slownik Geografii Turystycznej Polski* (Warsaw, 1959), vol. II, sub 'Warszawa'.
190 Eino Jutikkala, 'Town Planning in Sweden and Finland until the Middle of the Nineteenth Century', *Sc. Ec. Hist. Rv.*, 16 (1968), 19–46.
191 Frederick Hendriks, 'On the Vital Statistics of Sweden from 1749 to 1855', *Jl Roy. Stat. Soc.*, 25 (1862), 111–74.
192 S. Laing, *A Tour in Sweden in 1838* (London, 1839), 73–6, 197, 200–1, 212–14.
193 B. Ohngren, 'Urbanisation in Sweden, 1840–1920', in H. Schmal, op. cit., 181–227.
194 P. Bairoch, op. cit.; Lennart Jörberg, 'Structural Change and Economic Growth: Sweden in the Nineteenth Century', *Essays in European Economic History*, 259–80.
195 A. Kjar and C. F. Frisch, 'Die Bevölkerung Norwegens nach dem Census von 1865', *Pet. Mitt.*, 12 (1866), 247–50.

196 Axel Nielsen, *Dänische Wirtschaftsgeschichte* (Jena, 1933), 197.
197 K. B. Mayer, op. cit., 243–9.
198 Jean-François Bergier, 'Genève', *Citta Mercanti Dottrine nell' Economia Europea*, ed. A. Fanfani (Milan, 1964), 151–69.
199 *Statistisches Jahrbuch der Schweiz* (Bern, 1949).
200 Wilhelm Bickel, *Bevölkerungsgeschichte und Bevölkerungspolitik der Schweiz* (Zurich, 1947).
201 Rudolf Till, 'Zue Herkunft der Wiener Bevölkerung im 19. Jahrhundert', *Viert. Soz. Wirtgesch.*, 34 (1941), 15–37.
202 Calculated from Gustav Otruba, 'Wachstumsverschiebungen in der Wirtschaftssektoren Österreichs 1869–1961', *Viert. Soz. Wirtgesch.*, 62 (1975), 40–61.
203 Carl E. Schorske, *Fin-de-Siècle Vienna: Politics and Culture* (New York, 1980), 24–110.
204 Manfred Straka, 'Die Bevölkerungsentwicklung im Raume von Gross-Glatz', *Zt Hist. Ver. Steiermk*, 68 (1957), 22–60.
205 S. Harrison Thomson, 'The Germans in Bohemia from Maria Theresa to 1918', *Jl Cent. Eur. Aff.*, 2 (1942), 161–80.
206 J. Havranek, op. cit.
207 H. G. Wanklyn, *The Eastern Marchlands of Europe* (London, 1941), 275–6.
208 Paul Teleki, *The Evolution of Hungary and its Place in European History* (New York, 1923), 107–10.
209 A. N. J. Den Hollander, 'The Great Hungarian Plain; A European Frontier Area', *Comp. St. Soc. Hist.*, 3 (1960), 74–88, 155–69.
210 Henry Marczali, *Hungary in the Eighteenth Century* (Cambridge, 1910), 165.
211 N. J. G. Pounds, 'Land-use on the Hungarian Plain', *Geographical Essays on Eastern Europe*, ed. N. J. G. Pounds (Bloomington, Ind., 1961), 54–74; A. N. J. Den Hollander, *Nederzettingsvormen en-problemen in de Groote Hongaarsche Laagvlakte* (Amsterdam, 1947).
212 Edit Lettrich, 'Urbanization of Hungary in the Light of the Occupational Structure of her Population', *Applied Geography in Hungary* (Budapest, 1964), 164–78.
213 Friedrich Gottas, 'Anmerkungen zum Urbanisierungsprozess der Stadt Budapest von der Vereinigung (1873) bis zue Jahrhundertwende', *Südost. Forsch.*, 32 (1973), 123–67.
214 H. Marczali, op. cit., 165.
215 Márton Pécsi and Béla Sárfalvi, *The Geography of Hungary* (Budapest, 1964), 179–84.
216 N. J. G. Pounds, 'The Urbanization of East-Central and Southeast Europe: An Historical Perspective', *Eastern Europe: Essays in Geographical Problems*, ed. G. W. Hoffman (London, 1970), 45–81.
217 Gaston Gravier, 'Le Sandzak de Novi Pazar', *Ann. Géog.*, 22 (1913), 41–67.
218 Ion Donat and G. Retegan, 'La Valachie en 1838 (d'après une source statistique inédite)', *Rv. Roum. Hist.*, 4 (1965), 925–41.
219 Julius de Hagemeister, *Report on the Commerce of the Ports of New Russia, Moldavia and Wallachia* (London, 1836), 89–91.
220 *Istoria Oraşului Bucuresti*, ed. Florian Georgescu (Bucharest, n.d.), vol. I, 295 and passim.
221 N. Todorov, op. cit.
222 A. Ischirkoff, 'Die Bevölkerung in Bulgarien und ihre Siedlungsverhältnisse', *Pet. Mitt.*, 57 (1911), 117–22, 179–85.
223 P. Meuriot, op. cit., 230.
224 A. F. Weber, op. cit., 450.
225 David R. Ringrose, *Transportation and Economic Stagnation in Spain, 1750–1850* (Durham, N.C., 1970), 5.
226 T. E. Gumprecht, 'Die Städte-Bevölkerung von Spanien', *Pet. Mitt.*, 2 (1856), 303.
227 Richard Ford, *A Handbook for Travellers in Spain* (London, 1845); edn of 1966 notes that Ford's figures may have come from Caballero, *Noticias topografico-estadisticas*.
228 J. M. Houston, 'Urban Geography of Valencia: the Regional Development of a Huerta City', *Trans. I.B.G.*, 15 (1951), 19–35.
229 Charles Dickens, *Pictures from Italy*, edn of 1973 (London), 114.
230 Ibid., 103.
231 Henry Swinburne, *Travels in the Two Sicilies* (London, 1790).
232 C. Dickens, op. cit., 215.
233 Pietro Donazzolo and Mario Saibante, 'Lo sviluppo demografico di Verona e della sua Provincia dalla fine del sec. XV ai giorni nostri', *Metron*, 6 (1974), 56–180.
234 John Goodwin, 'Progress of the Two Sicilies under the Spanish Bourbons from the Year 1734–35 to 1840', *Jl Roy. Stat. Soc.*, 5 (1842), 47–73, 177–207.
235 John Bowring, 'Report on the Statistics of Tuscany, Lucca the Pontifical and the Lombardo-Venetian States', *Parl. Pap.*, 1939, vol. XVI, 421–586.

236 Samuel Brown, 'On the Statistical Progress of the Kingdom of Italy', *Jl Roy. Stat. Soc.*, 29 (1866), 197–225.
237 A. Caracciolo, 'Some Examples of Analyzing the Process of Urbanization: Northern Italy (Eighteenth to Twentieth Century)', *Patterns of European Urbanisation since 1500*, ed. H. Schmal (London, 1981), 131–41.
238 Bolton King, 'Statistics of Italy', *Jl Roy. Stat. Soc.*, 66 (1903), 213–72.
239 'Report on the Economic Progress of the Kingdom of Italy for the past Twenty-five Years', For. Off. Misc. Ser., 195, *Parl. Pap.*, 1890–1, vol. LXXXIV.
240 P. Bairoch, 'Urbanisation and Economic Development in the Western World: Some Provisional Conclusions of an Empirical Study', *Patterns of European Urbanisation since 1500*, ed. H. Schmal (London, 1981), 61–75.

Chapter 5. Agriculture in the nineteenth century

1 I. Chiva, 'Social Organisation, traditional economy and customary law in Corsica', *Mediterranean Countrymen*, ed. Julian Pitt-Rivers (Paris, 1963), 97–112.
2 H. W. Finck von Finckenstein, *Die Entwicklung der Landwirtschaft in Preussen und Deutschland 1800–1930* (Würzburg, 1960), 182–94.
3 A. Babeau, *Le Village sous l'ancien régime*, as quoted in J. Blum, 'The European Village as Community: Origins and Functions', *Agr. Hist.*, 45 (1971), 157–79.
4 Ingvar Svennilson, *Growth and Stagnation in the European Economy*, U.N., E.C.E. (Geneva, 1954), 88; H. W. Finck von Finckenstein, op. cit., 381–2; Wilfrid Malenbaum, *The World Wheat Economy 1885–1939*, Harv. Ec. Stud., 92 (1953), 238–47.
5 H. W. Finck von Finckenstein, op. cit., 331.
6 Based on J. C. Toutain, *Le Produit de l'agriculture française de 1700 à 1958*, *Hist. Quant. Éc. Fr.*, 2 (1961), 16.
7 W. Malenbaum, op. cit.
8 I. Svennilson, op. cit., 89.
9 R. F. Crawford, 'Notes on the Food Supply of the United Kingdom, Belgium, France and Germany', *Jl Roy. Stat. Soc.*, 62 (1899), 597–638.
10 M. Liepmann, *Tariff Levels*, as quoted in M. Tracy, *Agriculture in Western Europe* (London, 1964), 27–31.
11 Ernest Labrousse, Ruggiero Romano and F. G. Dreyfus, *Le Prix de froment en France au temps de la monnaie stable (1726–1913)*, Éc. Prat. Htes. Ét., Monn.-Pr.-Con. (1970).
12 Prices are tabulated by *département* and month in E. Labrousse *et al.*, op. cit.
13 *Histoire économique et sociale de la France*, vol. III, *1789–1880*, part 2 (Paris, 1976), 718.
14 C. Easton, *Les Hivers dans l'Europe occidentale* (Leyden, 1928), 145; E. Vanderlinden, *Chronique des événements météorologiques en Belgique*, Ac. Roy. Belg. Mém., 2ᵉ série, 6 (1924), 304–7.
15 G. Jacquemyns, *Histoire de la crise économique des Flandres (1845–1850)*, Ac. Roy. Belg. Mém., 2ᵉ série, 26 (1939), 248–55; F. P. Codacconi, 'Le textile lillois durant la crise 1846–1851', *Rv. Nord*, 38 (1956), 29–63.
16 G. Jacquemyns, op. cit., 256–7.
17 Étienne Juillard, ed., *Histoire de la France rurale*, vol. III, *Apogée et crise de la civilisation paysanne 1789–1814* (Paris, 1976), 140–1.
18 Alain Corbin, *Archaisme et modernité en Limousin au XIXᵉ siècle 1845–1880*, vol. I (Paris, 1975), 422; André Armengaud, *Les Populations de l'Est-Aquitaine au début de l'époque contemporaine*, Éc. Prat. Htes. Ét., Soc. et Id. (1961), 171–80.
19 A. Corbin, op. cit., 423.
20 Ilie Corfus, *L'Agriculture en Valachie durant la première moitié du XIXᵉ siècle*, Bibl. Hist. Rom., 23 (1969), 174.
21 Jan de Vries, 'Histoire du climat et économie: des faits nouveaux, une interpretation différente', *Ann. E.S.C.*, 32 (1977), 198–226.
22 Cecil Woodham-Smith, *The Great Hunger: Ireland 1845–9* (London, 1962), 33.
23 M. L. Parry, *Climatic Change, Agriculture and Settlement* (Folkestone, 1978), 171.
24 Charles Christians, 'Quelques aspects de la structure agraire et de l'aménagement rural en Ardenne belge', *Ann. Est Mém.*, 21 (1959), 79–89.
25 Émile de Laveleye, 'The Land Systems of Belgium and Holland', *Systems of and Tenure in Various Countries*, ed. J. W. Probyn (London, 1881), 445.
26 Henry Colman, *The Agriculture and Rural Economy of France, Belgium, Holland and Switzerland* (London, 1848), 218; E. de Laveleye, op. cit., 446.

27 H. Hitier, 'La statistique agricole de la France', *Ann. Géog.*, 8 (1899), 350–7; Michel Augé-Laribé, *L'Évolution de la France agricole* (Paris, 1912), 47–9; Keith Sutton, 'Reclamation of the Waste during the Eighteenth and Nineteenth Centuries', *Themes in the Historical Geography of France*, ed. H. D. Clout (London, 1977), 247–300 gives slightly different figures, but the trend is the same.
28 Robert A. Dickler, 'Organization and Change in Productivity in Eastern Prussia', *European Peasants and their Markets*, ed. W. N. Parker and E. L. Jones (Princeton, N.J., 1975), 269–92.
29 August Pflug, 'Die wirtschaftliche Erschliessung der Lüneburger Heide', *Zt. Ges. Staatsw.*, 46 (1890), 288–305.
30 M. Szuhay, 'L'évolution des cultures à charrue en Hongrois de 1867 a 1914', *Nouv. Ét. Hist.*, 1 (1965), 639–66; Imre Wellmann, 'Esquisse d'une histoire rurale de la Hongrie depuis la première moitié du XVIIIe siècle jusqu'au XIXe siècle', *Ann. E.S.C.*, 23 (1968), 1181–1210.
31 G. Bruguera, 'Les communaux et le développement de la production agricole en Espagne (1808–1833)', *Cah. Hist.*, 14 (1969), 141–79.
32 R. J. Thompson, 'The Development of Agriculture in Denmark', *Jl Roy. Stat. Soc.*, 59 (1906), 374–419; F. Skrubbeltrang, *Agricultural Development and Rural Reform in Denmark*, Agricultural Studies, 22, F.A.O. (1953), 202.
33 B. J. Hovde, *The Scandinavian Countries, 1720–1865* (Ithaca, N.Y., 1948), 296.
34 Gerd Enequist, 'Advance and Retreat of Rural Settlement in Northwestern Sweden', *Geog. Ann.*, 42 (1960), 211–19; Ivar Uhnbom, 'Befolkningsutvecklingen i Norrland från äldre tid och till våra dagar', *Ymer*, 62 (1942), 255–70.
35 W. R. Mead, *Farming in Finland* (London, 1953), 43–9; id., 'Land-use in Early Nineteenth-Century Finland', Publications Instituti Geographici Universitatis Turkuensis, 26 (Turku, 1953).
36 David Mitrany, *The Land and the Peasant in Rumania* (London, 1930), 24; John R. Lampe and Marvin R. Jackson, *Balkan Economic History, 1550–1950* (Bloomington, Ind., 1982), 92–5.
37 Ilie Corfus, 'L'agriculture en Valachie depuis la Révolution de 1848 jusqu'a la Réforme de 1864', Bibl. Hist. Rom., 53 (1976), 15–67.
38 N. J. G. Pounds, *An Historical Geography of Europe 1500–1840* (Cambridge, 1979), 163–4.
39 I. Corfus, in Bibl. Hist. Rom., 23 (1969), 200–1, 244–5.
40 Pierre Pédelaborde, 'L'agriculture dans les plaines alluviales de la presqu'île de Saint-Germain-en-Laye', Cent. Ét. Éc., Ét. et Mém., 49 (1961), 47.
41 Étienne Juillard, *La Vie rurale dans la plaine de Basse-Alsace*, Fac. Univ. Strasb., 123 (1953), 224–34; *Hist. Éc. Soc. Fr.*, 3, part 2, 686.
42 Jerome Blum, *Noble Landowners and Agriculture in Austria, 1815–1848*, Johns Hopkins St. Ser., 65, no. 2 (1948), 112.
43 Bolton King, 'Statistics of Italy', *Jl Roy. Stat. Soc.*, 66 (1903), 213–72. Hansjörg Dongus, 'Die Reisbaugemeinschaften des Po-Deltas, eine neue Form Kollektiver Landnutzung', *Zt Aggesch.*, 11 (1963), 201–12; id., 'Gutsbetrieb und Bauernhof in den Marschen der östlichen Po-Ebene', ibid., 17 (1969), 194–214.
44 Audrey M. Lambert, *The Making of the Dutch Landscape* (London, 1971), 263–9; Paul Wagret, *Polderlands* (London, 1968), 104–13.
45 Kent Roberts Greenfield, *Economics and Liberalism in the Risorgimento* (Baltimore, Md., 1965), 41–2; G. Luzzatto, 'The Italian Economy in the First Decade after Unification', *Essays in European Economic History* (Oxford, 1974), 203–25.
46 Bolton King, op. cit.
47 R. Baird Smith, *Italian Irrigation, being a Report on the Agricultural Canals of Piedmont and Lombardy* (Edinburgh, 1855), vol. I, 73–5.
48 Michel Augé-Laribé, *L'Évolution de la France agricole* (Paris, 1912), 65.
49 Raymond Carr, *Spain 1808–1939* (Oxford, 1966), 14, 18–21.
50 J. M. Houston, 'Irrigation as a Solution to Agrarian Problems in Modern Spain', *Geog. Rv.*, 117 (1950), 55–63.
51 G. Bruguera, op. cit.
52 U.S. Consular Report, as quoted in W. Cortada, 'Catalan Politics and Economics, 1906–1911', *Cuad. Hist. Ec. Cat.*, 13 (1975), 129–81.
53 J. M. Houston, op. cit.; J. Vicens Vives, *An Economic History of Spain* (Princeton, N.J., 1969), 652–3.
54 Roland Courtot, 'Irrigation et propriété citadine dans l'Acequia Real de Jucar au milieu du XIXe siècle', *Ét. Rur.*, 45 (1972), 29–47.
55 Bela K. Kiraly, 'The Emancipation of the Serfs of East Central Europe', *Antem.*, 15 (1971), 63–85; Werner Conze, 'The Effects of Nineteenth Century Liberal Agrarian Reforms on Social Structure in Central Europe', *Essays in European Economic History*, ed. F. Crouzet, W. H. Chaloner and W. H. Stern (London, 1969), 53–81.

56 *Hist. Éc. Soc. Fr.*, 3, part 1.
57 J. H. Clapham, *The Economic Development of France and Germany 1815–1914* (Cambridge, 1936), 13–21.
58 B. Seebohm Rowntree, *Land and Labour: Lessons from Belgium* (London, 1911).
59 Frieda Wunderlich, *Farm Labor in Germany 1810–1945* (Princeton, N.J., 1961), 3–5; W. Conze, op. cit.
60 W. Conze, op. cit.
61 Lujo Brentano, 'Agrarian Reform in Prussia', *Ec. Jl*, 7 (1897), 1–20, 165–84; R. B. D. Morier, 'The Agrarian Legislation of Prussia during the Present Century', *Systems of Land Tenure in Various Countries*, ed. J. W. Probyn (London, 1881), 351–441.
62 W. Conze, op. cit.; J. Blum, op. cit., 90, 145–55.
63 'Report on the Peasantry and Peasant Holdings in Poland', For. Off. Misc. Ser., 355, *Parl. Pap.*, 1895, vol. CIII.
64 R. F. Leslie, *Polish Politics and the Revolution of November 1830*, Univ. Lond. Hist. St., 3 (1956), 55–73.
65 B. K. Kiraly, op. cit.
66 Zbigniew Stankiewicz, 'Serwituty w dobrach rządowych Królestwa Polskiego przed reformą uwłaszczeniową', *Prz. Hist.*, 49 (1958), 45–68.
67 W. Conze, op. cit.; J. Blum, op. cit., 45–90.
68 W. Conze, op. cit.
69 Gyula Merei, 'L'essor de l'agriculture capitaliste en Hongrie dans la première moitié du XIXe siècle', *Rv. Hist. Mod. Cont.*, 12 (1962), 51–4; Gyorgy Spira, 'La dernière génération des serfs en Hongrie', *Ann. E.S.C.*, 23 (1968), 353–67.
70 Bruce McGowan, *Economic Life in Ottoman Europe* (Cambridge, 1981), 58–60; Hristo Hristov, 'The Agrarian Problem and the National Liberation Movements in the Balkans', *Ac. Prem. Congr. Balk.*, 4 (1969), 65–70; John R. Lampe and Marvin R. Jackson, op. cit., 33–7.
71 Peter Sugar, *Industrialization of Bosnia-Hercegovina* (Seattle, 1963), 8–12.
72 D. Mitrany, op. cit., 24–38; John R. Lampe and Marvin R. Jackson, op. cit., 183–190.
73 I. Corbus, 'L'agriculture en Valachie depuis en Révolution de 1848 jusqu' à la Réforme de 1864', Bibl. Hist. Rom., 53 (1976), 15–67.
74 B. K. Kiraly, op. cit.
75 D. Mitrany, op. cit., 51–82; Philip Gabriel Eidelberg, *The Great Rumanian Peasant Revolt of 1907* (Leiden, 1974), 23–65.
76 J. Vicens Vives, op. cit., 519–22.
77 Ibid., 639. These figures relate to 1930 and are the earliest available.
78 Aa H. Kampp, *An Agricultural Geography of Denmark* (Budapest, 1975), 12–15; H. Thorpe, 'The Influence of Enclosure on the Form and Pattern of Rural Settlement in Denmark', *Trans. I.B.G.*, 17 (1951), 111–29.
79 Harry Rainals, 'Report upon the Past and Present State of the Agriculture of the Danish Monarchy', *Jl Roy. Stat. Soc.*, 21 (1860), 267–328.
80 Veikko Anttila, 'The Modernisation of Finnish Peasant Farming in the Late Nineteenth and early Twentieth Century', *Sc. Ec. Hist. Rv.*, 24 (1976), 33–44.
81 F. Skrubbeltrang, op. cit., 157–9; P. Ove Christiansen, 'The Household in the local Setting; a Study of Peasant Stratification', *Chance and Change: Social and Economic Studies in Historical Demography in the Baltic Area*, ed. S. Akerman, H. O. Johansen and D. Gaunt, Od. St. Hist. Soc. Sci., 52 (1978), 50–60.
82 Orvar Löfgren, 'The Potato People: Household Economy and Family Patterns among the Rural Proletariat in Nineteenth Century Sweden', ibid., 95–106. Florence Edith Janson, *The Background of Swedish Immigration (1840–1930)* (Chicago, 1931), 84–116.
83 E. de Laveleye, 'The Land Systems of Belgium and Holland', *Systems of Land Tenure* (London, 1881), 443–95.
84 Maurice Block, *Statistique de la France* (Paris, 1860), vol. II, 19; M. Augé-Laribé, op. cit., 47.
85 Reiner Gross, 'Die bürgerliche Agrarreform in Sachsen in der ersten Hälfte des 19. Jahrhunderts', *Schr. Dresd.*, 8 (1968), 146.
86 Robert A. Dickler, 'Organization and Change in Productivity in Eastern Prussia', *European Peasants and their Markets*, ed. W. N. Parker and E. L. Jones (Princeton, N.J., 1975), 269–92.
87 H. W. Finck von Finckenstein, op. cit., 329.
88 R. B. D. Morier, 'The Agrarian Legislation of Prussia during the Present Century', *Systems of Land Tenure*, 351–441.
89 Regina Chomać, *Struktura agrarna Królestwa Polskiego na przełomie XIX i XX w.* (Warsaw, 1970), 107.
90 H. C. Siegfried von Strakosch, *Die Grundlagen der Agrarwirtschaft in Österreich* (Vienna, 1917), 78 and table 1.

91 I. Wellman, op. cit.
92 M. Szuhay, op. cit.; *A History of Hungary*, ed. Ervin Pamlenyi (London, 1975), 351.
93 I. Wellman, op. cit.
94 Barbara Jelavich, 'Servia in 1897: A Report of Sir Charles Eliot', *Jl Cent. Eur. Aff.*, 18 (1958), 183–9.
95 Harry Rainals, 'Report upon the Past and Present State of the Agriculture of the Danish Monarchy', *Jl Roy. Ag. Soc.*, 21 (1860), 267–328.
96 B. J. Hovde, *The Scandinavian Countries 1720–1865* (Ithaca, N.Y., 1948), vol. I, 296.
97 K. T. Eheberg, 'Agrarische Zustände in Italian', *Schr. Ver. Sozpk.*, 29 (1886), 10.
98 J. Vicens Vives, op. cit., 644–9.
99 Ibid., 649–52.
100 Based on table in Colin Heywood, 'The Role of the Peasant in French Industrialisation, 1815–80', *Ec. Hist. Rv.*, 34 (1981), 359–76.
101 H. J. Habakkuk, 'Family Structure and Economic Change in Nineteenth Century Europe', *Jl Ec. Hist.*, 15 (1955), 1–12.
102 T. E. Cliffe Leslie, 'The Land System of France', in J. W. Probyn, op. cit., 291–312.
103 Laurence Wylie, *Chanzeau: A Village in Anjou* (Cambridge, Mass., 1966), 40–2; Gordon Wright, *Rural Revolution in France* (Stanford, Calif., 1964), 5–8.
104 Notably Arthur Young, but see J. S. Mill, *Principles of Political Economy* (1848), book 2, ch. 8, 3–4, who greatly exaggerates the extent of *métayage* in France.
105 Henry Higgs, '"Métayage" in Western France', *Ec. Jl*, 4 (1894), 1–13.
106 Alan Mayhew, *Rural Settlement and Farming in Germany* (London, 1973), 179–81.
107 F. von Reitzenstein and Erwin Nasse, 'Agrarische Zustände in Frankreich und England', *Schr. Ver. Sozak*, 27 (1884), 17–23.
108 F. Braudel and E. Labrousse, *Histoire économique et sociale de la France* (Paris, 1976), vol. III, part 2, 653–4.
109 *Reports from Her Majesty's Representatives abroad on the Position of Peasant Proprietors in the Countries in which they reside*, Cd 6250, *Parl. Pap.*, 1890–1, vol. LXXXIII, 355–428.
110 Bolton King, op. cit.
111 Report on the Agriculture of Tuscany, For. Off. Misc. Ser., 648, *Parl. Pap.*, 1906, Vol. CXXII, 273–88.
112 David I. Kertzer, 'European Peasant Household Structure: Some Implications from a Nineteenth Century Italian Community', *Jl Fam. Hist.*, 2 (1977), 333–49.
113 Renate Baruzzi-Leicher, 'Die landwirtschaftlichen Besitz- und Betriebsverhältnisse Italiens im 19. Jahrhundert', *Zt Aggesch.*, 10 (1962), 195–211.
114 Id., 'Historische Grundlagen der landwirtschaftlichen Besitz- und Betriebsverhältnisse in Italien', *Viert. Soz. Wirtgesch.*, 47 (1960), 145–85.
115 Frieda Wunderlich, *Farm Labor in Germany 1810–1945* (Princeton, N.J., 1961), 3–5.
116 Georg Droege, 'Zur Lage der rheinischen Landwirtschaft in der ersten Hälfte des 19. Jahrhunderts', *Landschaft und Geschichte: Festschrift für Franz Petri* (Bonn, 1970), 143–56; *Report on Agriculture in the Rhenish Province*, For. Off. Misc. Ser., 652, *Parl. Pap.*, 1906, vol. CXXII, 209–58.
117 R. A. Dickler, op. cit.
118 Saalfeld, 'Zur Frage des bauerlichen Landverlustes', *Zt Aggesch.*, 11 (1963), 163–71; A. Mayhew, op. cit., 193–4.
119 United States Senate Document 214, 63rd Congress (1913).
120 A. Sartorius von Waltershausen, *Deutsche Wirtschaftsgeschichte 1815–1914* (Jena, 1923), 464–5.
121 R. J. Thompson, 'The Development of Agriculture in Denmark', *Jl Roy. Stat. Soc.*, 49 (1906), 374–419.
122 H. Rainals, op. cit.
123 P. Ove Christiansen, 'The Household in the Local Setting: A Study of Peasant Stratification', *Chance and Change*, 50–60.
124 B. J. Hovde, op. cit., 282–3.
125 Regina Chomać, op. cit., 78–88.
126 'Report on the Peasantry and Peasant Holdings in Poland', For. Off. Misc. Ser., 355, *Parl. Pap.*, 1895, vol. CIII.
127 B. G. Ivanyi, 'From Feudalism to Capitalism: the Economic Background to Szechenyi's Reform in Hungary', *Jl Cent. Eur. Aff.*, 20 (1960), 270–88.
128 I. Wellmann, op. cit.
129 David Mitrany, *The Land and the Peasant in Rumania* (London, 1930), 232.
130 Hristo Hristov, 'The Agrarian Problem and the National Liberation Movements in the Balkans', *Ac. Prem. Congr. Balk.*, 4 (1969), 65–70.
131 Quoted in B. Jelavich, op. cit.

132 Based on Michael Palairet, 'Fiscal Pressure and Peasant Impoverishment in Serbia before World War I', *Jl Ec. Hist.*, 39 (1979), 719–40; L. S. Stavrianos, op. cit., 260–4.
133 Peter F. Sugar, *Industrialization of Bosnia-Hercegovina* (Seattle, Wash., 1963), 8–12.
134 K. Grünberg, quoted in D. Warriner, ed., *Contrasts in Emerging Societies* (Bloomington, Ind., 1965), 374–87.
135 Roger Price, *The Economic Modernisation of France* (London, 1975), 55.
136 Michel Augé-Laribé, *L'Évolution de la France agricole*, 103–4.
137 E. Labrousse, 'The Evolution of Peasant Society in France from the Eighteenth Century to the Present', *French Society and Culture since the Old Regime*, ed. E. M. Acomb and M. L. Brown (New York, 1966), 44–64.
138 George W. Grantham, 'Scale and Organization in French Farming 1840–1880', *European Peasants and their Markets* (London, 1975), 293–326.
139 Vernon W. Rattan, 'Structural Retardation and the Modernization of French Agriculture: A Skeptical View', *Jl Ec. Hist.*, 38 (1978), 714–21.
140 C. Heywood, op. cit.
141 B. Seebohm Rowntree, *Land and Labour: Lessons from Belgium* (London, 1911), 107.
142 E. de Laveleye, in J. W. Probyn, op. cit.
143 Louis Verhulst, 'Entre Senne et Dendre', *Ac. Roy. Belg. Mém.*, 23 (1926), 96–162; G. Jacquemyns, 'Histoire de la crise économique des Flandres (1845–1850)', ibid., 26 (1929), 230–2. Farm studies are given in B. Seebohm Rowntree, op. cit.
144 H. von Scheel, 'Die Landwirtschaftlichen Betriebe im Deutsche Reiche', *Jb. Ges. Verw. Volksw.*, 11 (1887), 1011–25; T. H. Middleton, 'The Recent development of German Agriculture', *Reports from Commissioners*, 4 (1916).
145 'Reports from Her Majesty's Representatives abroad on the Position of Peasant Proprietors in the Countries in which they Reside', Cd 6250, *Parl. Pap.*, 1890–91, vol. LXXXIII, 355–428.
146 K. T. Eheberg, 'Die Landwirtschaft in Bayern', *Jb. Ges. Verw. Volksw.*, 14 (1890), 1121–41; T. H. Middleton, op. cit.
147 *Parl. Pap.*, 1890–1, vol. LXXXIII, 355–428.
148 R. Chomać, op. cit., 181.
149 Jaroslav Purs, 'Die Entwicklung des Kapitalismus in der Landwirtschaft der Böhmischen Länder in der Zeit von 1849 bis 1879', *Jb. Wirtgesch.* (1963), part 3, 31–96.
150 H. Böker and F. W. von Bulow, *The Rural Exodus in Czechoslovakia*, I.L.O., Studies and Reports, series K, no. 13 (1935).
151 J. Blum, op. cit., 45–90.
152 Ernst Bruckmüller, 'Wirtschaftsentwicklung und politisches Verhalten der agrarischen Bevölkerung in Österreich 1867–1914', *Viert. Soz. Wirtgesch.*, 59 (1972), 489–529.
153 R. Chomać, op. cit., 181.
154 L. Makkai, 'Production et productivité en Hongrie à l'ère de féodalisme tardif (1550–1850)', 3e Conf. Int. Hist. Éc., 28 (1968), 171–80.
155 Scott H. Eddie, 'The Changing Pattern of Landownership in Hungary, 1867–1914', *Ec. Hist. Rv.*, 20 (1967), 293–310.
156 S. M. Eddie, 'Agricultural Production and Output per Worker in Hungary, 1870–1913', *Jl Ec. Hist.*, 28 (1968), 191–222.
157 Gyorgy Spira, 'La dernière génération des serfs de Hongrie', *Ann. E.S.C.*, 23 (1968), 353–67; Gyula Merei, 'L'essor de l'agriculture capitaliste en Hongrie dans la première moitié du XIXe siècle', *Rv. Hist. Mod. Cont.*, 12 (1965), 51–64.
158 H. Böker and F. W. von Bülow, op. cit.
159 Vasile Liveanu, 'La repartition de la propriété foncière en Transylvanie, fin du XIXe siècle – début du XXe', *Rv. Roum. Hist.*, 9 (1970), 273–90.
160 Henry Marczali, *Hungary in the Eighteenth Century* (Cambridge, 1910), 196–246.
161 M. Palairet, in *Jl Ec. Hist.*, 39 (1979), 719–40; Josef Matl, 'Die Agrarreform in Jugoslavien', Öst. Inst. Qu. u. St., 8 (1927), 26.
162 J. Matl, op. cit.
163 Ibid.; L. Katus, op. cit.
164 D. Mitrany, op. cit., 57–60.
165 Philip Eidelberg, 'The Agrarian Policy of the Rumanian National Liberation Party 1900–1916', *Rum. St.*, 1 (1970), 69–93; id., *The Great Rumanian Peasant Revolt of 1907* (Leiden, 1974).
166 Gerald Brenan, *The Spanish Labyrinth* (Cambridge, 1943), 102–10; J. Vicens Vives, op. cit., 625–43.
167 J. Vicens Vives, op. cit., 639.
168 Gerd Zimmermann, 'Die bauerliche Kulturlandschaft in Südgalicien', *Heid. Geog. Arb.*, 23 (1969).
169 G. Brenan, op. cit., 92–7.

170 E. H. G. Dobby, 'Agrarian Problems in Spain', *Geog. Rv.*, 26 (1936), 177–89; G. Brenan, op. cit., 110–14.
171 G. Brenan, op. cit., 114–26.
172 Hansjörg Dongus, 'Gutsbetrieb und Bauernhof in den Marschen der östlichen Po-Ebene', *Zt Aggesch.*, 17 (1969), 194–214.
173 Renate Baruzzi-Leicher, 'Die Landwirtschaftlichen Besitz- und Betriebsverhältnisse Italiens im 19. Jahrhundert', *Zt Aggesch.*, 10 (1962), 195–211.
174 W. N. Beauclerk, *Rural Italy* (London, 1888), 69–91; For. Off. Misc. Ser., 648, *Parl. Pap.*, 1906, vol. CXXII.
175 F. Lillin de Châteauvieux, 'Travels in Italy descriptive of the Rural Manners and Economy of that Country', *New Voyages and Travels* (London, 1819), 3.
176 Renate Leicher, 'Historiche Grundlagen der landwirtschaftlichen Besitz und Betriebsverhältnisse in Italien', *Viert. Soz. Wirtgesch.*, 47 (1960), 145–85.
177 G. A. Montgomery, *The Rise of Modern Industry in Sweden* (London, 1939), 56–7.
178 H. Rainals, op. cit.
179 R. J. Thompson, op. cit.
180 Jean-François Bergier, *Problèmes de l'histoire économique de la Suisse* (Bern, 1968), 65.
181 'Reports...on...Peasant Proprietors', Cd 6250.
182 P. Brunet, 'Problèmes relatifs aux structures agraires de la Basse-Normandie', *Ann. Norm.*, 5 (1955), 115–34.
183 William Jacob, 'A Report respecting the Agriculture and the Trade in Corn in some of the Continental States of Northern Europe', *Pamph.*, 29 (1828), 361–456.
184 C. Garnier-Deschesnes, 'Sir la vaine pâture et les jachères', *Mém. Agr.* 5 (an. 11), 318–32.
185 Henri See, 'La vaine pâture en France sous la monarchie de Juillet d' après l'enquête de 1836–1838', *Rv. Hist. Mod.*, 1 (1926), 198–213.
186 Francois Sigaut, 'Pour une cartographie des assolements en France au début du XIXe siècle', *Ann. E.S.C.*, 31 (1976), 631–43.
187 Władysław Biegajło, 'Szachownica gruntów i gospodarka trójpolowana terenie województwa Białostockiego', *Prz. Geog.*, 25 (1957), 533–59.
188 Jacqueline Moguelet, 'Les pratiques communautaires dans la plaine vendéenne au XIXe siècle', *Ann. E.S.C.*, 18 (1963), 666–76.
189 Jean Gottmann, *Documents pour servir a l'étude de la structure agraire dans la moitié occidentale de la France*, Cent. Ét. Éc., Ét. et Mém., 55 (1964), 47.
190 I. T. Berend and G. Ranki, *Economic Development in East-Central Europe in the Nineteenth and Twentieth Centuries* (New York, 1974), 53–4.
191 *Histoire économique et sociale de la France*, vol. III (1976), 672–3.
192 Irena Kostrowicka, 'Les productions végétales dans le Royaume de Pologne (1815–1864)', *Kw. Hist. Kult. Mat.*, 12 (1964), *Ergon* supplement, 542–7; M. Szuhay, 'Le dévelopement de la production des labours dans l'agriculture de la Hongrie entre 1867 et 1914', 3e Conf. Int. Hist. Ec. (1965), 181–5.
193 Aa H. Kampp, *An Agricultural Geography of Denmark*, 12–13.
194 B. J. Hovde, op. cit., 276–80.
195 Heinz Haushofer, *Die Deutsche Landwirtschaft im technischen Zeitalter* (Stuttgart, 1963), 53–8; A. Mayhew, op. cit., 187–9.
196 A. Mayhew, op. cit., 189.
197 Mechthild Wiswe, 'Veränderungen des Flurgefüges durch die Braunschweigische General-Landes-Vermessung', *Niedersachs. Jb.*, 37 (1965), 147–54. See also Frank Norbert Nagel, 'Historische Verkoppelung und Flurbereinigung der Gegenwart', *Zt Aggesch.*, 26 (1978), 13–41.
198 A. de Foville, *Le Morcellement* (Paris, 1885), 149.
199 Ibid., 139–41.
200 *The Consolidation of Fragmented Agricultural Holdings*, F.A.O. (Washington, D.C., 1950), 78–9.
201 Ibid., 99.
202 *The Land Tenure Systems in Europe*, European Conference on Rural Life, 1939, Technical Documentation, League of Nations (1939), 62; Jovan Cvijić, *La Péninsule Balkanique* (Paris, 1918), 176–7.
203 R. Chomać-Klimek, 'Problems of Investment expenditures in Farms in the Congress Kingdom of Poland on the turn of the 19th and 20th century', *St. Hist. Oec.*, 10 (1975), 177–81. See also instances cited in *Agricultural Cooperation and Rural Credit in Europe*, Senate Doc. 214, 63rd Congress (1913).
204 E. J. T. Collins, 'Labour supply and demand in European Agriculture 1800–1880', *Agrarian Change and Economic Developments*, ed. E. L. Jones and S. J. Woolf (London, 1969), 61–94.

205 Abel Chatelain, 'La lente progression de la faux', *Ann. E.S.C.*, 11 (1956), 495–99.
206 Jean Vidalenc, 'L'agriculture dans les départements normands à la fin du Premier Empire', *Ann. Norm.*, 7 (1957), 179–201.
207 Roger Thabault, *Education and Change in a Village Community: Mazières-en-Gâtine 1848–1914* (London, 1971).
208 *Agricultural Cooperation and Rural Credit*, 270.
209 H. Rainals, op. cit.
210 G. Merei, op. cit.
211 Roger Price, 'The Onset of Labour Shortage in Nineteenth Century French Agriculture', *Ec. Hist. Rv.*, 28 (1975), 260–79.
212 H. Rainals, op. cit.
213 A. G. Haudricourt and J.-B. Delamarre, *L'Homme et la charrue à travers le monde* (Paris, 1955); Paul Leser, *Entstehung und Verbreitung des Pfluges*, Int. Sam. Eth. Mon., 3 (1931).
214 Folke Dovring, 'The Transformation of European Agriculture', *Camb. Econ. Hist. Eur.*, vol. VI, part 2 (Cambridge, 1965), 604–72; Jan Kruse, *Mechanisation, Commercialisation and the Protectionist Movement in Swedish Agriculture, 1860–1910'*, *Sc. Ec. Hist. Rv.*, 19 (1971), 23–44.
215 J. Kruse, op. cit.
216 W. N. Beauclerk, op. cit., 116; For. Off. Misc. Ser., 648.
217 Joanna Macijewska-Pavković, 'Tradycje orki krzyżowej sochą na Suwalszczyźnie w koncu XIX i na początku XX wieku', *Rocz. Biał.*, 7 (1966), 251–60.
218 Stanisław Borowski, 'Esquisse du développement de la mécanisation de l'agriculture dans la Grande Pologne des années 1807–1918', *Kw. Hist. Kult. Mat.*, 12 (1964), Ergon suppl., 535–41.
219 J. Kruse, op. cit.; *Histoire de la France rurale*, ed. E. Juillard, vol. III (Paris, 1976), 200–1.
220 E. J. T. Collins, op. cit.
221 R. Thabault, op. cit., 143.
222 J. H. Clapham, *The Economic Development of France and Germany* (Cambridge, 1945), 170.
223 *Histoire Économique et Sociale de la France*, vol. III, 681.
224 S. Borowski, op. cit.
225 Henry Colman, *The Agriculture and Rural Economy of France, Belgium, Holland and Switzerland* (London, 1848), 71.
226 G. P. H. Chorley, 'The Agricultural Revolution in Northern Europe, 1750–1880: Nitrogen, Legumes and Crop Productivity', *Ec. Hist. Rv.*, 34 (1981), 71–93.
227 P. Pédelaborde, 'L'agriculture dans les plaines alluviales de la presqu'île de Saint-Germain-en-Laye', Cent. Ét. Éc., Ét. et Mém., 49 (1961), 127–8; H. D. Clout and A. D. M. Phillips, 'Fertilisants minéraux en France au XIXe siècle', *Ét. Rur.*, 45 (1972), 9–29.
228 R. Thabault, op. cit.
229 Laurence Wylie, ed., *Chanzeaux: A Village in Anjou* (Cambridge, Mass., 1966), 39.
230 F. Dovring, op. cit. tabulates the use of fertilisers, p. 656.
231 Roger Price, 'The Onset of Labour Shortage in Nineteenth Century French Agriculture', *Ec. Hist. Rv.*, 28 (1875), 260–79; E. J. T. Collins, op. cit.
232 Quoted in A. Soboul, 'The French Rural Community in the Eighteenth and Nineteenth Centuries', *P. & P.*, 10 (1956), 78–95.
233 G. W. Grantham, op. cit.
234 M. Augé-Laribé, *L'Évolution de la France agricole*, 19.
235 Bohdan Baranowski, 'Changes in the Significance of the Cultivation of Buckwheat and Millet in Modern Times', *St. Hist. Oec.*, 9 (1974), 67–76.
236 J.-C. Toutain, *Le Produit de l'agriculture française de 1700 à 1958*, Hist. Quant. Éc. Fr., 2 (1961), part 2, *La Croissance*, 16.
237 H. W. Finck von Finckenstein, op. cit., 313–16.
238 R. J. Thompson, op. cit.
239 B. J. Hovde, op. cit., 298; B. Fullerton, 'The Northern Margin of Grain Production in Sweden in the Twentieth Century', *Trans. I.B.G.*, 20 (1954), 181–91.
240 H. C. Siegfried von Strakosch, op. cit., table 2.
241 'Notes on Landed Estates in Hungary', For. Off. Misc. Ser., 291, Parl. Pap., 1893–4, vol. XC; M. Szuhay, in *Nouv. Ét. Hist.*, 1 (1965), 639–66.
242 Bela Sebestyen, 'Maisanbau in Ungarn', *Giess. Abh. Agr. Forsch.*, ser. 1, 32 (1965), 147–78.
243 Ilie Corfus, 'L'agriculture en Valachie durant la première moitié du XIXe siècle', Bibl. Hist. Rom., 23 (1969), 51–5, 200–45; Ion Donat and G. Retegan, 'La Valachie en 1838', *Rv. Roum. Hist.*, 4 (1965), 925–41.
244 O. S. Morgan, ed., *Agricultural Systems of Middle Europe* (New York, 1933), 54.
245 Slavka Draganova, 'De la production agricole, l'imposition fiscale et la différenciation sociale de la population paysanne en Bulgarie du Nord-Est durant les années 60–70 du XIXe siècle', *Bulg. Hist. Rv.*, 5 (1977), part 2, 70–92.

246 'Report on the Natural Products and Agriculture of the Province of Florence', For. Off. Misc. Ser., 88, *Parl. Pap.*, 1888, vol. XCIX.
247 Bolton King, op. cit.
248 G. Bruguera, 'Les communaux et le développement de la production agricole en Espagne (1808–1833)', *Cah. Hist.*, 14 (1969), 141–79.
249 R. Thaubault, op. cit., 25–9.
250 J. F. R. Phillips, 'Die Agrarstruktur Sudlimburgs in der ersten Hälfte des 19. Jahrhunderts', *Acta Hist. Neer.*, 4 (1970), 84–104.
251 R. N. Salaman, *The History and Social Influence of the Potato* (Cambridge, 1949), 73–100.
252 J. Videlenc, 'L'agriculture dans les départements normands à la fin du Premier Empire', *Ann. Norm.*, 7 (1957), 179–201.
253 Cecil Woodham-Smith, *The Great Hunger: Ireland 1845–9* (London, 1962), 24–32.
254 Jadwiga Pilawska, 'Silesia–The World's Oldest Beet Sugar Producer', *Ann. Sil.*, 2 (1961), 21–37.
255 J. A. Perkins, 'The Agricultural Revolution in Germany 1850–1914', *Jl Eur. Ec. Hist.*, 10 (1981), 71–118.
256 André Dubuc, 'La culture de la pomme de terre en Normandie', *Ann. Norm.*, 3 (1953), 50–68; G. Desert, 'La culture de la pomme de terre en Normandie', *Ann. Norm.*, 5 (1955), 261–70; J. C. Toutain, op. cit., 13.
257 Quoted in M. Bergman, 'The Potato Blight in the Netherlands and its Social Consequences (1845–1847)', *Int. Rv. Soc. Hist.*, 12 (1967), 390–431.
258 G. Jacquemyns, op. cit., 247; B. Seebohm Rowntree, op. cit., 175.
259 'Bauerliche Zustände in Deutschland', *Schr. Ver. Sozpk.*, 22 (1883), 148.
260 I. Kostrowicka, op. cit.
261 R. Chomać, op. cit., 16.
262 H. C. Siegfried von Strakosch, op. cit., 68–70; François Matejek, 'La production agricole dans les pays tchécoslovaques à partir du XVIe siècle jusqu'à la première guerre mondiale', 3e Conf. Int. Hist. Éc. (1968), 205–19.
263 H. Rainals, op. cit.
264 R. J. Thompson, op. cit.
265 B. J. Hovde, op. cit., 298; Orvar Löfgren, 'The Potato People: Household, Economy and Family Patterns among the Rural Proletariat in Nineteenth Century Sweden', *Chance and Change*, 95–106.
266 J. H. Clapham, op. cit., 25.
267 Yves Guyot, 'The Sugar Industry on the Continent', *Jl Roy. Stat. Soc.*, 65 (1902), 419–46.
268 M. Camier, 'La betterave dans le Cambrésis', *Rv. Nord*, 38 (1956), *Livr. Géog.*, 5, 53–74.
269 Maurice Lévy-Leboyer, *Les Banques européennes et l'industrialisation internationale dans la première moitié du XIXe siècle* (Paris, 1964), 264–5; B. Seebohm Rowntree, op. cit., 175.
270 Gerhard B. Hagelberg and Hans-Heinrich Müller, 'Kapitalgesellschaften für Anbau und Verarbeitung von Zuckerrüben in Deutschland im 19. Jahrhundert', *Jb. Wirtsgesch.* (1974), part 4, 113–47; V. vom Berg, D. Hofmann and J. Heisterkamp, 'Der Zuckerrübenanbau unter dem Einfluss der Frühindustrialisierung', *Zt Aggesch.*, 20 (1972), 198–213.
271 Czesław Łuczak, 'Technika cukrownicza w Wielkopolsce w latach 1820–1861', *Kw. Hist. Kult. Mat.*, 4 (1956), 36–61; Marian Eckert, 'Die Rolle der Zuckerindustrie in der wirtschaftlichen Entwicklung der Polnischen Gebiete im 19. Jahrhundert', *St. Hist. Oec.*, 13 (1978), 153–62.
272 Hansjörg Dongus, 'Die Reisbaugemeinschaften des Po-Deltas, eine neue Form Kollektiver Landnutzung', *Zt Aggesch.*, 11 (1963), 201–12.
273 J. A. Perkins, op. cit.
274 Ibid.
275 J. Moguelet, op. cit.
276 George W. Grantham, 'The Diffusion of the New Husbandry in Northern France, 1815–1840', *Jl Ec. Hist.*, 38 (1978), 311–37.
277 L. Hedin, 'Les conditions d'exploitation de la prairie en Normandie dans le XIXe siècle', *Ann. Norm.*, 1 (1951), 45–69; Jules Sion, *Les Paysans de la Normandie Orientale* (Paris, 1909), 384–400.
278 Georg Droege, 'Zur Lage der rheinischen Landwirtschaft in der ersten Hälfte des 19. Jahrhunderts', *Lanschaft und Geschichte – Festschrift für Franz Petri* (Bonn, 1970), 143–56.
279 T. C, Banfield, *Industry of the Rhine* (London, 1846), vol. I, 18.
280 Robert A. Dickler, 'Organization and Change in Productivity in Eastern Prussia', *European Peasants and their Markets*, ed. W. N. Parker and E. L. Jones (Princeton, N.J., 1975), 269–92; H. W. Finck von Finckenstein, op. cit., 34.
281 K. T. Eheberg, 'Die Landwirtschaft in Bayern', *Jb. Ges. Verw. Volksw.*, 14 (1890), 1121–41.

282 T. H. Middleton, op. cit.
283 Étienne Juillard, *La Vie rurale dans la plaine de Basse-Alsace*, Pub. Univ. Strasbg., 123 (1953), 344.
284 Remy Joseph François, 'Agriculture d'hier et aujourd'hui: quelques aspects de son evolution dans un village du Val-de-Saône', *Ann. Bourg.*, 45 (1973), 217–37; Laurent Dechesne, op. cit., 456.
285 Gertrud Schröder-Lembke, 'Die Entwicklung des Raps- und Rubsenanbaus in der deutschen Landwirtschaft', *Zt Aggesch.*, 24 (1976), 145–60.
286 Roger Dion, *Histoire de la vigne et du vin en France* (Paris, 1959).
287 Aimé V. Perpillou, 'Un siècle d'évolution agricole dans les vignobles français du Languedoc et du Roussillon', *Festschrift Leopold G. Scheidl zum 60. Geburtstag* (Vienna, 1965), vol. I, 268–78.
288 J. Vicens Vives, op. cit., 650–1.
289 'Report on Vine Culture in Bosnia and the Herzegovina', For. Off. Misc. Ser., 228, *Parl. Pap.*, 1892, vol. LXXIX, 437–43.
290 Dwight W. Morrow, jun., 'Phylloxera in Portugal', *Agr. Hist.*, 47 (1973), 235–47.
291 For. Off. Misc. Ser., 228, *Parl. Pap.*, 1892, vol. LXXIX, 437–43.
292 *Histoire de la France rurale*, vol. III, 388–91.
293 Michel Morineau, 'Y a-t-il eu une révolution agricole en France au XVIIIe siècle?', *Rev. Hist.*, 239 (1968), 299–326; R. Thabault, op. cit., 30.
294 H. W. Finck von Finckenstein, op. cit., 214; Rudolf Berthold, 'Einige Bemerkungen über den Entwicklungsstand des bäuerlichen Ackerbaus vor den Agrarreformen des 19. Jahrhunderts', *Beiträge zur deutschen Wirtschafts-und Sozialgeschichte des 18. und 19. Jahrhunderts*, Deutsch. Akad. Wiss. (Berlin, 1962), 121.
295 William Jacob, *A View of the Agriculture, Manufactures, Statistics, and State of Society of Germany* (London, 1820), 338–9.
296 Leszek Wiatrowski, 'Wieś Śląska w latach 1840–1847', *Kw. Hist.*, 75 (1968), 125–37.
297 B. H. Slicher van Bath, 'Yield Ratios', *A.A.G. Bij.*, 10 (1963).
298 Zs Kirilly, L. Makkai, I. N. Kiss and V. Zimanyi, 'Production et productivité agricoles en Hongrie à l'époque du féodalisme tardif', *Nouv. Ét. Hist.* (1965), vol. I, 581–638.
299 M. Morineau, op. cit.
300 Alain Corbin, *Archaisme et modernité en Limousin au XIXe siècle* (Paris, 1975), vol. I, 23–4.
301 Heinz Haushofer, 'Die deutsche Landwirtschaft im technischen Zeitalter', *Deutscher Agrargeschichte*, ed. G. Franz (Stuttgart, 1963), 194–7; 'Report on Agriculture in Germany', For. Off. Misc. Ser., 452, *Parl. Pap.*, 1898, vol. XCIII, 125.
302 Andrzej Jezierski, 'The Problems of Economic Growth of Poland in the Nineteenth Century', *St. Hist. Oec.*, 9 (1974), 121–40.
303 F. Matejek, op. cit.; M. Szuhay, op. cit.; H. C. S. von Strakosch, op. cit., table 2.
304 H. W. Finck von Finckenstein, op. cit., 262–3.
305 J.-B. Henry, 'L'industrie laitière en Bretagne et les révolutions techniques du XIXe siècle', *Ann. Bret.*, 73 (1966), 255–81.
306 J. Videlenc, op. cit.
307 Senate Document 214, 63rd Congress, First Session (1913), 471–84.
308 Georges Castellan, 'Fourrages et bovins dans l'économie rurale de la Restauration: L'exemple du département du Rhône', *Rv. Hist. Éc. Soc.*, 38 (1960), 77–97.
309 Fritz Glauser, 'Handel mit Entlebucher Käse und Butter', *Schw. Zt Gesch.*, 21 (1971), 1–63.
310 Louis-Joseph d'Humières, 'Sur l'état de l'agriculture dans le département du Cantal', *Mém. Agric.*, 3 (1801–2), 140–54.
311 E. Estyn Evans, 'Transhumance in Europe', *Geog.*, 25 (1940), 172–80; Elwyn Davies, 'The Patterns of Transhumance in Europe', *Geog.*, 26 (1941), 155–68.
312 A. J. B. Wace and M. S. Thompson, *The Nomads of the Balkans* (London, 1914), 73–96; J. K. Campbell, 'The Kindred in a Greek Mountain Community', *Mediterranean Countrymen*, ed. Julian Pitt-Rivers (Paris, 1963), 73–96.
313 Julius Klein, *The Mesta* (Cambridge, Mass., 1920), 348–9; André Fribourg, 'La transhumance en Espagne', *Ann. Géog.*, 19 (1910), 231–44.
314 *Agricultural Cooperation and Rural Credit in Europe*, Senate Document 214, 63rd Congress, First Session (Washington, D.C., 1913), for an immense volume of detail on European co-operatives.

Chapter 6. Agricultural regions

1. An English translation by Michael Dziewicki was published as *The Peasants* (New York, 1924). All quotations are from this edition.
2. *The Peasants: Autumn*, 5.
3. Ibid., *Spring*, 8.
4. Regina Chomać, *Struktura agrarna Królestwa Polskiego na przełomie XIX i XX w.* (Warsaw, 1970), 107.
5. Krystyna Bielecka, 'Przemiany struktury rolniczego użytkowania ziemi w Królestwie Polskim w latach 1863–1913 w świetle materiałów statystycznych, *Kw. Hist. Kult. Mat.*, 14 (1966), 491–517.
6. In 1866 more than 40 per cent of the land in the Polish Kingdom belonged to the peasants: 'Report on the Peasantry and Peasant Holdings in Poland', For. Off. Misc. Ser., 355, *Parl. Pap.*, 1895, vol. CIII.
7. *The Peasants: Spring*, 120.
8. R. Chomać, op. cit.
9. *The Peasants: Autumn*, 125.
10. R. Chomać, op. cit.
11. For. Off. Misc. Ser., 355.
12. *The Peasants: Winter*, 119, 132, 253.
13. Ibid., 253.
14. For. Off. Misc. Ser., 355.
15. Zbigniew Stankiewicz, 'Serwituty w dobrach rządowych Królestwa Polskiego przed reformą uwłaszczeniową, *Prz. Hist.*, 49 (1958), 45–68.
16. *The Peasants: Autumn*, 44.
17. Ibid., 102.
18. R. F. Leslie, *Polish Politics and the Revolution of November 1830*, Univ. Lond. Hist. Stud., 3 (1956), 64.
19. Czesław Rajca, 'Unowocześnienie chłopskich zabudowań mieszkalnych w połowie XIX wieku na terenie Ordynacji Zamojskiej', *Kw. Hist. Kult. Mat.*, 20 (1972), 103–14.
20. *The Peasants: Autumn*, 19.
21. Ibid., *Autumn*, 7.
22. Regina Chomać-Klimek, 'Problems of Investment Expenditures in Farms in the Congress Kingdom of Poland at the Turn of the 19th and 20th Century', *St. Hist. Oec.*, 10 (1975), 177–81; Stanisław Borowski, 'Esquisse du développement de la mécanisation de l'agriculture dans la Grande Pologne des années 1807–1918', *Kw. Hist. Kult. Mat.*, 12 (1964), Ergon suppl., 535–41; Julian Bartyś, 'Die Anfänge der Mechanisierung in der polnischen Landwirtschaft', *St. Hist. Oec.*, 3 (1968), 135–57.
23. Leszek Wiatrowski, 'Wieś Śląska w latach 1840–1847', *Kw., Hist.*, 75 (1968), 125–37.
24. *Von Thünen's Isolated State*, ed. Peter Hall (Oxford, 1966), xiii–xiv.
25. William Jacob, *Second Report on the Agriculture and Corn of Some of the Continental States of Europe, 1827* (London, 1928), 13.
26. 'Report on Agriculture in Germany', *Parl. Pap.*, 1898, vol. XCIII, 125–209.
27. William Jacob, 'A Report respecting the Agriculture and the Trade in Corn in some of the Continental States of Northern Europe', *Pamphl.*, 29 (London, 1828), 361–456.
28. Albert D. Thaer, *The Principles of Agriculture*, trans. W. Shaw, vol. I (London, 1844), 209–11.
29. Irmgard Brinkmann, 'Die von Thünensche Rentenlehre und die Entwicklung der neuzeitlichen Landwirtschaft', *Zt Ges. Saatsw.*, 108 (1951), 307–56.
30. W. Jacob, in *The Pamphleteer*.
31. Alfred Wielopolski, 'The Economic Regression of Western Pomerania in the Era of Capitalism', *Poland at the XIth Int. Conf. of Historical Sciences* (Warsaw, 1960), 239–63.
32. *Von Thünen's Isolated State*, 62.
33. Ibid., 63.
34. Ibid., 7.
35. Andreas Grotewold, 'Von Thünen in Retrospect', *Econ. Geog.*, 35 (1959), 346–55.
36. H. Paasche, 'Die rechtliche und wirtschaftliche Lage des Bauerstandes in Mecklenburg-Schwerin', *Bauerliche Zustände in Deutschland*, part 3, *Schr. Ver. Sozpk*, 24 (1883), 327–81.
37. 'Report on Agriculture in Germany', *Dip. & Cons. Repts, Parl. Pap.*, 1898, vol. XCIII, 125–209.
38. Ibid., 151.
39. Ibid.
40. H. Paasche, op. cit.
41. *Das Östliche Deutschland*, Göttinger Arbeitskreis (Würzburg, 1959), 661–75.

42 'Report on Agriculture in Germany', 151.
43 H. Paasche, op. cit.
44 Marion I. Newbigin, *Southern Europe* (London, 1949), 156–66.
45 This account has been based mainly on 'Report on the Statistics of Lombardo-Venetian States', *Parl. Pap.*, 1839, vol. XVI, 421–586, and Emile de Laveleye, *La Lombardie et la Suisse: études d'économie rurale* (Paris, 1869), 55–94. See also 'Agricultural Conditions in Lombardy' in *Agricultural Co-operation and Rural Credit in Europe*, Senate Documents, 63rd Congress, First Session (Washington, D.C., 1913), 27–8.
46 'Report on...Lombardo-Venetian States', 99; 'Irrigation in Italy', *Int. Rv. Agric.*, N. S., 19 (1928), 16–22; Senate Document 214, 63rd Congress (1913).
47 'Report on...Lombardo-Venetian States', 99.
48 R. Baird Smith, *Italian Irrigation, Being a Report on the Agricultural Canals of Piedmont and Lombardy*, 2 vols. (Edinburgh, 1855), vol. I, 295ff.
49 Kent Roberts Greenfield, *Economics and Liberalism in the Risorgimento* (Baltimore, Md, 1965), 31.
50 W. N. Beauclerk, *Rural Italy* (London, 1888), 186.
51 V. C. Finch and O. E. Baker, *Geography of the World's Agriculture*, U.S. Dept of Agriculture (Washington, D.C., 1917), 46.
52 K. T. Eheberg, 'Agrarische Zustände in Italien', *Schr. Ver. Sozpk*, 29 (1886), 11.
53 'Report on...Lombardo-Venetian States', 521.
54 Frederick Lillin de Chateauvieux, 'Travels in Italy descriptive of the Rural Manners and Economy of that Country', *New Voyages and Travels* (London, n. d., but about 1819), 14–15.
55 'Report on the Dairy Industry of Italy', For. Off. Misc. Ser., 450, *Parl. Pap.*, 1898, vol. XCIII, 333–41.
56 K. R. Greenfield, op. cit., 41–3.
57 'Report on...the Lombardo-Venetian States', 517.
58 R. Baird Smith, op. cit., vol. I, 295.
59 K. R. Greenfield, op. cit., 83.
60 'Report on the Condition of the Italian Silk Trade and on the Yield of Cocoons in Italy in 1904', For. Off. Misc. Ser., 632, *Parl. Pap.*, 1905, vol. LXXXVI.
61 Sir John Sinclair, *Hints regarding the Agricultural State of the Netherlands* (London, 1815), 5.
62 Jules Michelet, *Histoire de France* (Paris, ed. of 1971), vol. V, 320.
63 Emile de Laveleye, 'The Land System of Belgium and Holland', in J. W. Probyn, op. cit., 443–95.
64 Raoul Blanchard, *La Flandre* (Lille, 1906), 42–59.
65 Ibid., 271–5.
66 Ibid., 342–3. For a regional description see S. W. E. Vince, 'The Agricultural Regions of Belgium', *London Essays in Geography*, ed. L. D. Stamp and S. W. Wooldridge (London, 1951), 255–88.
67 Emile de Laveleye, 'L'agriculture Belge', Congr. Agric. Int. Paris (1878), xxi; id., *Essai sur l'économie rurale de la Belgique* (Paris, 1875), 21–36.
68 Thomas Radcliff, *A Report on the Agriculture of Eastern and Western Flanders* (London, 1819), 15; E. de Laveleye, *Essai*, 33.
69 E. de Laveleye, *Essai*, 33.
70 Bénoit Verhaegen, *Contribution à l'histoire économique des Flandres*, Pub. Univ. Lov. Leop., 8 (1961), vol. I, 107.
71 E. de Laveleye, in J. W. Probyn, op. cit.
72 G. Jacquemyns, *Histoire de la crise économique des Flandres (1845–1850)*, Ac. Roy. Belg. Mém., 2ᵉ série, 26 (1929), 435–8.
73 Louis Verhulst, *Entre Senne et Dendre*, Ac. Roy. Belg. Mém., 2ᵉ série, 23 (1926), 57–8; 196; 207; E. de Laveleye, 'L'Agriculture Belge', xliii–xliv.
74 E. de Laveleye, *Essai*, 39.
75 B. Seebohm Rowntree, *Land and Labour: Lessons from Belgium* (London, 1911), 179–80.
76 B. Seebohm Rowntree, op. cit., 122; B. Verhaegen, op. cit., 180.
77 P. Deprez, 'The Demographic Development of Flanders in the Eighteenth Century', *Population in History*, ed. D. V. Glass and D. E. C. Eversley (London, 1965), 608–30; id., 'Évolution démographique et évolution économique en Flandres au dix-huitième siècle', 3rd Int. Cong. Ec. Hist, 4 (1972), 49–53.
78 Georges Bublot, *La Production agricole Belge: étude économique séculaire 1846–1955*, Inst. Rech. éc. Soc. (Louvain, 1957), 54.
79 B. Seebohm Rowntree, op. cit., 89.
80 Ibid., 96.
81 G. Jacquemyns, op. cit., 129–63; R. Blanchard, op. cit., 376–8.

82 R. Blanchard, op. cit., 382.
83 R. Seebohm Rowntree, op. cit., 195.
84 See sample households described in B. Seebohm Rowntree, op. cit., 341–91.
85 William Jacob, *Second Report of the Agriculture and Corn of some of the Continental States of Europe, 1827* (London, 1828), 21.
86 *The Travel Diaries of Thomas Robert Malthus*, ed. Patricia James (Cambridge, 1966), 52.
87 Samuel Laing, *Observations on the Social and Political State of Denmark* (London, 1852), 18–22.
88 W. Jacob, op. cit., 23.
89 S. Laing, op. cit., 119.
90 H. Thorpe, 'The Influence of Inclosure on the Form and Pattern of Rural Settlement in Denmark', *Trans. I.B.G.*, 17 (1951), 111–29.
91 Harry Rainals, 'Report upon the Past and Present State of the Agriculture of the Danish Monarchy', *Jl Roy. Stat. Soc.*, 21 (1860), 267–328.
92 S. Laing, op. cit., 18.
93 W. Jacob, op. cit., 22.
94 *The Travel Diaries of T. R. Malthus*, 45; F. Skrubbeltrang, *Agricultural Development and Rural Reform in Denmark*, Agric. St., F.A.O., 23 (1953), 102.
95 E. Jensen, *Danish Agriculture: Its Economic Development* (Copenhagen, 1937), passim.
96 H. Rainals, op. cit.
97 'Report on Dairy Farms in Denmark', For. Off. Misc. Ser., 101, *Parl. Pap.*, 1888, vol. XCIX, 57–76.
98 F. Skrubbeltrang, op. cit., 190.
99 R. J. Thompson, 'The Development of Agriculture in Denmark', *Jl Roy. Stat. Soc.*, 69 (1906), 374–419.
100 Palle Ove Christiansen, 'The Household in the Local Setting: A Study of Peasant Stratification', *Chance and Change* (Odense, 1978), 50–60.
101 R. J. Thompson, op. cit. Also 'Denmark', Senate Document 214, 63rd Congress, First Session (Washington, D.C., 1913), 545–77.
102 E. H. Pedersen, 'Modernization of the Danish Peasant and Cottager Agriculture during the first part of the Danish Agricultural Revolution 1860–1880', *St. Hist. Oec.*, 10 (1975), 109–13.
103 S. A. Hansen, *Early Industrialisation in Denmark*, Univ. Copen. Inst. Econ. Hist., 1 (1970), 11.
104 R. J. Thompson, op. cit.
105 Birgit Nuchel Thomsen and Brinley Thomas, *Dansk-Engelsk Samhandel: et historisk rids 1661–1963* (Aarhus, 1966).
106 George Borrow, *The Bible in Spain*, Everyman edn (London, 1906), 192.
107 Richard Ford, *Gatherings from Spain* (London, 1846), 9.
108 S. E. Widdrington, *Spain and the Spaniards in 1843*, 2 vols. (London, 1844), vol. I, 399.
109 Salvador de Madariaga, *Spain* (London, 1942), 110–17; John Naylon, 'Land Consolidation in Spain', *Ann. A.A.G.*, 40 (1959), 361–73.
110 Gerald Brenan, *The Spanish Labyrinth* (Cambridge, 1943), 334–5; Glen A. Waggoner, 'The Black Hand Mystery: Rural Unrest and Social Violence in Southern Spain 1881–1883', *Modern European Social History*, ed. Robert J. Bezucha (Lexington, Mass., 1972), 161–91; Edward Malifakis, 'Peasants, Politics and Civil War in Spain, 1931–1939', ibid., 192–227.
111 Jaime Vicens Vives, *An Economic History of Spain* (Princeton, N.J., 1969), 649–752.
112 Julius Klein, *The Mesta* (Cambridge, Mass., 1920), 347–8; André Fribourg, 'La transhumance en Espagne', *Ann. Géog.*, 19 (1910), 231–44.
113 J. Vicens Vives, op. cit., 656.
114 G. Brenan, op. cit., 114.
115 Richard Ford, *A Handbook for Travellers in Spain and Readers at Home* (London, edn of 1966), vol. I, 462.
116 Ibid., vol. II, 771.
117 Roland Courtot, 'Irrigation et propriété citadine dans l'Acequia Real del Jucar au milieu du XIXe siècle', *Ét. Rur.*, 45 (1972), 29–47.
118 Emili Giralty Raventos, 'Mouvements paysans et problèmes agraires de la fin du XVIIIe siècle à nos jours', *Les Mouvements paysans dans le monde contemporain*, vol. II, *Cah. Int. Hist. Éc. Soc.*, 7 (1970), 96–126.
119 Jose luis Hernandez Marco, 'Evolucion de cultivos y estructura de la propiedad en el Pais Valenciano', *Est. Rv. Hist. Mod.*, 7 (1978), 111–24.
120 S. E. Widdrington, op. cit., vol. I, 400–1.
121 The idea of linking and comparing these two institutions derives from Richard Busch-Zantner, 'Tschiftlikwesen und Zadruga', *Viert. Soz. Wirtgesch.*, 30 (1937), 72–81. See also John

Notes to pp. 273–82

R. Lampe and Marvin R. Jackson, *Balkan Economic History 1550–1950* (Bloomington, Ind., 1982), 33–9.
122 Philip E. Mosely, 'The Distribution of the Zadruga within Southeastern Europe', *The Joshua Starr Memorial Volume, Jewish Social Studies*, 5 (1953), 219–30.
123 Rebecca West, *Black Lamb and Grey Falcon* (London, 1942), vol. I, 502.
124 Arthur John Evans, *Through Bosnia and Herzogovina on Foot* (London, 1877), 57.
125 Georg Tschemschiroff, 'Die bulgarische Hauskommunion (Zadruga)', *Jb. Ges. Verw. Volksw.*, 61 (1937), part 1, 181–220.
126 I. E. Geshov, quoted in *Contrasts in Emerging Societies*, ed. D. Warriner (Bloomington, Ind., 1965), 257–9.
127 E. A. Hammel, 'The Zadruga as Process', *Household and Family in Past Time*, ed. P. Laslett (Cambridge, 1972), 335–73.
128 G. Tschemschiroff, op. cit.
129 Jovan Cvijic, *La Péninsule balkanique* (Paris, 1918), 174–6.
130 Joel Martin Halpern, *A Serbian Village* (New York, 1958), 150.
131 P. E. Mosely, op. cit.
132 K. J. Jireček, as quoted in *Contrasts in Emerging Societies*, 260.
133 Olive Lodge, *Peasant Life in Yugoslavia* (London, 1941), 92–143.
134 J. Cvijic, op. cit., 222–3.
135 L. S. Stavrianos, *The Balkans Since 1453* (New York, 1958), 478.
136 Bruce McGowan, *Economic Life in Ottoman Europe* (Cambridge, 1981), 72; but see also John R. Lampe and Marvin R. Jackson, *Balkan Economic History, 1550–1950* (Bloomington, Ind., 1982), 33–7.
137 Ibid., 72; *Parl. Pap.*, 1861, vol. LXVII, 72.
138 Troian Stoianovich and George C. Haupt, 'Le maïs arrive dans les Balkans', *Ann. E.S.C.*, 17 (1962), 84–93; L. S. Stavrianos, op. cit., 142.
139 Dimitri Kossev, 'Mouvements paysans et problèmes agraires de la fin du XVIIIe siècle à nos jours en Bulgarie', *Mouvements paysans dans le monde contemporaine*, 182–234.
140 K. J. Jireček, as quoted in *Contrasts in Emerging Societies*, 248–55.
141 L. S. Stavrianos, op. cit., 478, 677.
142 G. A. Montgomery, *The Rise of Modern Industry in Sweden* (London, 1939), 50–5.
143 Staffan Helmfrid, 'The *Storskifte, Enskifte* and *Laga Skifte* in Sweden – General Features', *Geog. Ann.*, 43 (1961), 114–29.
144 Ibid.
145 G. A. Montgomery, op. cit., 54.
146 Jan Kruuse, 'Mechanisation, Commercialisation and the Protectionist Movement in Swedish Agriculture, 1860–1910', *Sc. Ec. Hist. Rv.*, 14 (1971), 23–44.
147 Hugo Osvald, *Swedish Agriculture* (Stockholm, 1952), 32.
148 'Report on the Dairy Industry of Sweden', For. Off. Misc. Ser., 439, *Parl. Pap.*, 1898, vol. XCIII, 573–95.
149 Lennart Jorberg, 'Structural Change and Economic Growth: Sweden in the Nineteenth Century', *Essays in European Economic History*, ed. F. Crouzet, W. H. Chaloner and W. M. Stern (London, 1969), 259–80.
150 Orvar Lofgren, 'The Potato People: Household Economy and Family Patterns among the rural Proletariat in Nineteenth Century Sweden', *Chance and Change*, 95–106.
151 G. Apri, 'The Supply with Charcoal of the Swedish Iron Industry from 1830 to 1950', *Geog. Ann.*, 35 (1953), 11–27.
152 David Gaunt, 'Household Typology: Problems, Methods, Results', *Chance and Change*, 69–83.
153 Gosta A. Eriksson, 'The Decline of the Small Blast-furnaces and Forges in Bergslagen after 1850', *Geog. Ann.*, 39 (1957), 257–77.
154 Gerd Enequist, 'Advance and Retreat of Rural Settlement in Northwestern Sweden', *Geog. Ann.*, 42 (1960), 211–19.
155 Hans Aldskogius, 'Changing Land Use and Settlement Development in the Siljan Region', *Geog. Ann.*, 41 (1959), 250–61.
156 B. Fullerton, 'The Northern Margin of Grain Production in Sweden in the Twentieth Century', *Trans. I.B.G.*, 20 (1954), 181–91.
157 Gordon Wright, *Rural Revolution in France* (Stanford, Calif., 1964), passim; Paul Hohenberg, 'Change in Rural France in the Period of Industrialization, 1830–1914', *Jl Ec. Hist.*, 32 (1972), 219–40.
158 Quoted in George Fasel, 'The Wrong Revolution: French Republicanism in 1848', *Fr. Hist. Stud.*, 8 (1973–4), 654–77.

159 Roger Price, *The Economic Modernisation of France* (London, 1975), 45–79; George W. Grantham, 'The Diffusion of the New Husbandry in Northern France, 1815–1840', *Jl Ec. Hist.*, 38 (1978), 311–37; Vernon W. Rattan, 'Structural Retardation and the Modernization of French Agriculture: A Skeptical View', *Jl Ec. Hist.*, 38 (1978), 714–21; Hugh D. Clout, 'Agricultural Change in the Eighteenth and Nineteenth Centuries', *Themes in the Historical Geography of France* (London, 1977), 407–46.
160 Roger Price, 'The Onset of Labour Shortage in Nineteenth Century French Agriculture', *Ec. Hist. Rv.*, 28 (1975), 260–79.
161 J.-C. Toutain, *Le Produit de l'agriculture française de 1700 à 1958*, part 2, *La Croissance*, Hist. Quant. Éc. Fr., 2 (Paris, 1961), 175. See also Hugh D. Clout, *The Land of France 1815–1914*, London Research Series in Geography, 1 (London, 1983), which presents an excellent picture of spatial changes in French agriculture.
162 Michel Morineau, 'Y a-t-il eu une révolution agricole en France au XVIIIe siècle?', *Rv. Hist.*, 239 (1968), 299–326.
163 Hilda Ormsby, *France* (London, 1931), 124–5.
164 Emile Zola, *Terre* (Paris, 1887); excerpts are from the abbreviated English translation, *Earth*, New English Library (London, 1962).
165 The similarity of Rognes to Lipka (pp. 249–52) is not accidental. Reymont admitted that *Terre* provided him with a model.
166 Jean Pautard, *Les Disparités régionales dans la croissance de l' agriculture française*, Techniques Économiques Modernes, vol. II (Paris, 1965), passim.
167 Jean Vidalenc, *La Société française de 1815 à 1848: le peuple des campagnes* (Paris, 1970), 65.
168 'France', *Agricultural Cooperation and Rural Credit in Europe*, Senate Document 214, 63rd Congress, First Session (Washington, D.C., 1913), esp. 719–24.
169 Gregor Dallas, *The Imperfect Peasant Economy: The Loire Country, 1800–1914* (Cambridge, 1982), 205; V. E. Ardouin-Dumazet, *Voyage en France*, 1ère série, *Le Morvan, le Val de Loire, le Perche* (Paris, 1902), 167–9.
170 Jules Sion, *Les Paysans de la Normandie Orientale* (Paris, 1909), 347.
171 Jean Vidalenc, 'L'agriculture dans les départements normands à la fin du Premier Empire', *Ann. Norm.*, 7 (1957), 179–201.
172 G. Dubois, 'La Normandie économique à la fin du XVIIe siècle', *Rv. Hist. Ec. Soc.*, 21 (1933), 337–88.
173 P. Brunet, 'Problèmes relatifs aux structures agraires de la Basse Normandie', *Ann. Norm.*, 5 (1955), 115–34.
174 L. Hedin, 'Les conditions d'exploitation de la prairie en Normandie depuis le XIXe siècle', *Ann. Norm.*, 1 (1951), 45–69.
175 H. D. Clout, 'The Retreat of the Wasteland of the Pays de Bray', *Trans. I.B.G.*, 47 (1969), 171–89.
176 J. Sion, op. cit., 400.
177 Maurice Lévy-Leboyer, *Le Revenu agricole et la rente foncière en Basse-Normandie*, Pub. Univ. Paris-Nanterre, series A, no. 15 (1972), 4–32.
178 André Dubuc, 'La culture de la pomme de terre en Normandie', *Ann. Norm.*, 3 (1953), 50–68.
179 G. Desert, 'La culture de la pomme de terre dans le Calvados au XIXe siècle', *Ann. Norm.*, 5 (1955), 261–70.
180 J. Sion, op. cit., 367–9.
181 Ibid., 305–14, 325; A. Demangeon, *La Picardie et les régions voisines* (Paris, 1905).
182 F. Dornic, 'L'évolution de l'industrie textile aux XVIIIe et XIXe siècles: l'activité de la famille Cohen', *Rv. Hist. Mod. Cont.*, 3 (1956), 38–66.
183 Dietrich, 'Die gegenwartige wirtschaftliche Lage der Spitzenindustrie in Frankreich', *Jb. Ges. Verw. Volksw.*, 24 (1900), 75–142; Paul-M. Bondois, 'Colbert et le développement économique de la Basse-Normandie', *Bul. Soc. Ant. Norm.*, 41 (1933), 41–141.
184 H. Ormsby, op. cit., 57–9.
185 J. Pautard, op. cit., 136.
186 Alain Corbin, *Archaisme et modernité en Limousin au XIXe siècle* (Paris, 1975), vol. I, 177–225.
187 Ibid., vol. I, 429–84.
188 Albert Goursaud, *La Société rurale traditionelle en Limousin*, Contributions au Folklore des Provinces de France, 12 (1976), vol. I, 175.
189 A. Corbin, vol. I, 22–3.
190 Ibid., 25, 450.
191 Ibid., 450.
192 Ibid., 459.

193 Louis-Joseph d'Humières, 'Sur l'état de l'agriculture dans le département du Cantal', *Mém. Agr.*, 3 (1801–2), 140–54.
194 This follows Étienne Juillard, *La Vie rurale dans la plaine de Basse-Alsace*, Pub. Fac. Strasb. (1953), 29–35.
195 E. Juillard, op. cit., 213–15.
196 Ibid., 253.
197 Based on *Das Reichsland Elsass-Lothringen, Landes- und Ortsbeschreibung*, 3 vols. (Strasbourg, 1898–1903).
198 E. Juillard, op. cit., 423.
199 Ibid., 208–13.

Chapter 7. Manufacturing in the nineteenth century

1 Shepard B. Clough, 'The Diffusion of Industry in the Last Century and a Half', *Studi Sapori*, 2 (1955), 1339–57.
2 Jennifer Tann and M. J. Brecken, 'The International Diffusion of the Watt Engine 1775–1825', *Ec. Hist. Rv.*, 31 (1978), 541–64.
3 'Report from the Select Committee on the Laws relating to the Export of Tools and Machinery', *Parl. Pap.*, 1825, vol. V, 115–66.
4 Louis Bergeron, 'Remarques sur les conditions du développement industriel en Europe à l'époque napoléonienne', *Francia*, 1 (1973), 537–56.
5 David S. Landes, 'Technological Change and Development in Western Europe, 1750–1940', *Camb. Econ. Hist. Eur.*, vol. VI, part 1 (Cambridge, 1965), 371–4.
6 F. Crouzet, 'Western Europe and Great Britain: Catching Up in the First Half of the Nineteenth Century', *Economic Development in the Long Run*, ed. A. J. Youngson (London, 1972), 98–125.
7 Ibid.
8 It had not in fact been exercised for many years.
9 E. Hobsbawm, *Industry and Empire* (Harmondsworth, 1968), 32–3.
10 Jean Chevalier, *Le Creusot* (Paris, 1935), 101–20.
11 Arthur Young, *Travels in France During the Years 1787, 1788, 1789* (London, edn of 1915), 27.
12 S. Pollard and C. Holmes, *Documents in European Economic History* (London, 1968), 151–4.
13 Walter Minchinton, 'Patterns of Demand 1750–1914', *Fontana Economic History of Europe*, vol. III (Harmondsworth, 1973), 77–186.
14 G. Jacquemyns, *Histoire de la crise économique des Flandres (1845–1850)*, Ac. Roy. Belg. Mém., 2ᵉ série, 26 (1929), 15ff.
15 F. Crouzet, op. cit.; E. L. Jones, 'Le origini agricole dell'industria', *Stud. Stor.*, 9 (1968), 564–93.
16 P. Bairoch, 'Le rôle de l'agriculture dans la création de la sidérurgie moderne', *Rv. Hist. Ec. Soc.*, 44 (1966), 5–23.
17 F. Crouzet, 'Agriculture et Révolution Industrielle: quelques réflexions', *Cah. Hist.*, 12 (1967), 67–85.
18 J. Slomka, *From Serfdom to Self-Government: Memoirs of a Polish Village Mayor* (London, 1941), 18.
19 F. Crouzet, 'Agriculture et Révolution Industrielle...'
20 Richard Tilly, *Financial Institutions and Industrialization in the Rhineland 1815–1870* (Madison, Wisc., 1966); Maurice Lévy-Leboyer, *Les Banques européennes et l'industrialisation internationale*, Pub. Fac. Paris, 16 (1964).
21 W. W. Rostow, *The Stages of Economic Growth* (Cambridge, 1960), 52–8.
22 For a discussion of Fougères, see p. 351.
23 W. G. Hoffmann, *The Growth of Industrial Economies* (Manchester, 1958), 38–41; id., *Das Wachstum der deutschen Wirtschaft seit der Mitte des 19. Jahrhunderts* (Berlin, 1965).
24 P. Bairoch, 'Le mythe de la croissance économique rapide au XIXᵉ siècle', *Rv. Inst. Soc.*, 35 (1962), 307–31.
25 Id., 'Europe's Gross National Product 1800–1975', *Jl Eur. Ec. Hist.*, 5 (1976), 273–340.
26 D. C. Coleman, 'Proto-Industrialization: A Concept too Many', *Ec. Hist. Rv.*, 36 (1983), 435–48.
27 Franklin Mendels, 'Proto-Industrialization: The First Phase of the Industrialization Process', *Jl Ec. Hist.*, 32 (1972), 241–61.

28 Hans Medick, 'The Proto-Industrial Family Economy', *Soc. Hist.*, 1 (1976), 291–315.
29 Duncan Bythell, *The Handloom Weavers* (Cambridge, 1969), 89ff.
30 T. G. Vlaykov, as quoted in *Contrasts in Emerging Societies*, ed. D. Warriner (Bloomington, Ind., 1965), 241.
31 J. Emerson Tennent, *Belgium* (London, 1842), vol. I, 129.
32 P. Kriedte, H. Medick and J. Schlumbohm, *Industrialisierung vor der Industrialisierung*, Veröff. Max-Planck Inst., 53 (1977), 36ff.
33 G. Jacquemyns, op. cit.
34 See reports cited below in *Schr. Ver. Sozpk.*
35 J. E. Tennent, op. cit., vol. I, 55–9, 62–7.
36 G. Jacquemyns, op. cit., 44–5.
37 Joel Mokyr, *Industrialization in the Low Countries, 1795–1850* (New Haven, Conn., 1976), 99–121.
38 Gerhart Hauptmann, *Die Weber*, published in 1892, but related to events a generation earlier. The play opens in the merchant's office, with the weavers handing in their cloth and having their payments reduced for alleged faulty workmanship.
39 Notably by Friedrich Engels, *Conditions of the Working Class* and by Charles Dickens, *Hard Times*. In the same vein is W. Reymont, *Ziemia Obiecana*, descriptive of the developing textile town of Łódź.
40 W. H. Hutt, 'The Factory System of the Early 19th Century', *Econ.*, 6 (1926), 78–93.
41 A. Audiganne, *De l'Organisation du travail* (Paris, 1848), 66–7.
42 M. François, 'Précis historique sur le traitement du fer dans l'Ariège', *Ann. Mines*, 3e série, 12 (1937), 580–96.
43 M. N. Briavoinne, *De l'Industrie en Belgique* (Brussels, 1939), vol. II, 305.
44 Georges Hansotte, 'La clouterie liègeoise et la question ouvrière au XVIIIe siècle', *Anc. Pays États*, 55 (1972), 1–9.
45 Cl. Heiss, 'Die belgische Industrie- und Gewerbezahlung vom 31 Oktober 1896', *Jb. Ges. Verw. Volksw.*, 27, part 1 (1903), 209–53.
46 'Report... by the Board of Trade into Working Class Rents, Housing... in the Industrial Towns of Belgium', Cmd 5065, *Parl. Pap.*, 1910, 43–306.
47 Frederic Le Play, *Les Ouvriers européens* (Paris, 1855), 152–7.
48 Eberhard Gothein, *Wirtschaftsgeschichte des Schwarzwaldes* (Strassburg, 1892), 818–33.
49 The reports were published in the series *Schriften des Vereins für Sozialpolitik* in the 1880s and 1890s.
50 W. Stieda, 'Die deutsche Hausindustrie', *Schr. Ver. Sozpk*, 39 (1889), 57–9.
51 Ibid., 68ff.
52 Pierre Lebrun, *L'Industrie de la laine à Verviers pendant le XVIIIe et le début du XIXe siècle*, Bibl. Fac. Liège, 114 (1948), 187–219.
53 J. Tann and M. J. Brechin, op. cit. The few 'atmospheric' engines of Newcomen type were of little importance.
54 W. O. Henderson, *Britain and Industrial Europe 1750–1870* (Liverpool, 1954), 149–53.
55 Based on Jaroslav Purš, 'Použití parních strojů v průmyslu v českych zemích v obdobi do nástupu imperialismu', *Česk. Čas. Hist.*, 3 (1955), 254–90.
56 I. T. Berend and G. Ranki, *Economic Development in East-Central Europe in the Nineteenth and Twentieth Centuries* (New York, 1975), 112–13.
57 Le Comte Chaptal, *De l'Industrie française* (Paris, 1819), vol. II, 122; Étienne Baux, 'Les draperies audoises sous le Premier Empire', *Rv. Hist. Éc. Soc.*, 38 (1960), 418–32.
58 F.-E. Manuel, 'La grève des tisserands de Lodève en 1845', *Rv. Hist. Mod.*, 10 (1935), 209–25, 352–72.
59 Claude Fohlen, *L'Industrie textile au temps du Second Empire* (Paris, 1956), 181.
60 Id., 'En Languedoc: vigne contre draperie', *Ann. E.S.C.*, 4 (1949), 290–7.
61 Georges Clause, 'L'industrie lainière rémoise à l'époque napoléonienne', *Rv. Hist. Mod. Cont.*, 17 (1970), 574–95.
62 C. Fohlen, op. cit., 175–8.
63 Ibid., 465.
64 Pierre Lebrun, 'Croissance et industrialisation: l'expérience de l'industrie drapière verviétoise 1750–1850', 1ere Conf. Int. Hist. Éc. (1960), 531–68.
65 J. Mokyr, op. cit., 17–18.
66 J. A. Sporck, 'Le rôle de l'eau dans la localisation de l'industrie lainière dans la région verviétoise', *Bul. Soc. Belge Ét. Géog.*, 17 (1948), 154–72.
67 P. Lebrun, *L'Industrie de la laine...*, 187–233.
68 J. Mokyr, op. cit., 50.

69 N. Briavoinne, op. cit., vol. II, 378.
70 Table in W. Bowden, M. Karpovich and A. P. Usher, *An Economic History of Europe since 1750* (New York, 1937), 459.
71 Ernst Baasch, *Holländische Wirtschaftsgeschichte* (Jena, 1927), 452–3; J. A. Van Houtte, 'Economic Development of Belgium and the Netherlands from the Beginning of the Modern Era', *Jl Eur. Ec. Hist.*, 1 (1972), 100–20; H. R. C. Wright, *Free Trade and Protection in the Netherlands 1816–30* (Cambridge, 1955), 4, 34–42.
72 F. P. Stegmann von Pritzwald, 'Zur Geschichte der deutschen Wöllenindustrie', *Jb. Nat. Stat.*, 6 (1866), 186–254; 7 (1866), 81–153.
73 Pierre Benaerts, *Les Origines de la grande industrie Allemande* (Paris, 1933), 99–100.
74 Clemens Bruckner, 'Zur Wirtschaftsgeschichte des Regierungsbezirks Aachen', *Schr. Rhein-Westf. Wtgesch.*, 16 (1967); Max Barkhausen, 'Die sieben bedeutendsten Fabrikanten der Roerdepartements im Jahre 1810', *Rhein Vbl.*, 25 (1960), 100–13.
75 Gerhard Adelmann, 'Die deutsch-niederländische Grenze als textilindustrieller Standortfakter', *Landschaft und Geschichte: Festschrift für Franz Petri* (Bonn, 1970), 9–43; Alfred Engels, *Die Zollgrenze in der Eifel*, *Schr. Rhein.-Westf. Wtgesch.*, 2 (1959), 35–9.
76 Franz Decker, *Die betriebliche Sozialordnung der Dürener Industrie im 19. Jahrhundert*, *Schr. Rhein.-Westf. Wtgesch.*, 12 (1965), 22–4.
77 Herbert Milz, 'Das Kölner Grossgewerbe von 1750 bis 1935', *Schr. Rhein.-Westf. Wtgesch.*, 7 (1962), 35–7.
78 Bernard Balkenhof, 'Armut und Arbeitslosigkeit in der Industrialisierung – dargestellt am Beispiel Düsseldorfs 1850–1900', *St. Düss. Wirtgesch.*, 3 (1976), 28.
79 See map in Wolfgang Zorn, 'Eine Wirtschaftskarte Deutschlands um 1820 als Spiegel der gewerblichen Entwicklung', *Wirtschaftliche und soziale Probleme der gewerblichen Entwicklung im 15., 16. und 19. Jahrhundert*, ed. Friedrich Lutge, *Forsch. Soz. Wirtgesch.*, 10 (1968), 143–55; id., 'Neues von der historischen Wirtschaftskarte der Rheinlände', *Rhein. Vbl.*, 30 (1965), 334–45.
80 W. Zorn, *Handels- und Industriegeschichte Bayerisch-Schwabens 1648–1870*, *Veröff. Schwab. Forsch.*, ser. 1, vol. VI.
81 Wilhelm Kaiser, 'Die Anfange der fabrikmässig organisierten Industrie in Baden', *Zt Gesch. Oberr.*, 46 (1933), 612–35.
82 Rudolf Forberger, *Die Manufaktur in Sachsen vom Ende des 16. bis zum Anfang des 19. Jahrhunderts*, *Deutsch. Akad. Wiss.* (1958), 152–61.
83 Horst Kruger, *Zur Geschichte der Manufakturen und der Manufakturarbeiter in Perussen*, *Schr. Inst. Allg. Gesch.*, 3 (1958), 262–3.
84 'Report on the Prussian Commercial Union', *Parl. Pap.*, 1840, vol. XXI, 381–672.
85 *Berichte aus der Hausindustrie in Berlin, Osnabrück, im Fichtelgebirge und in Schlesien*, *Schr. Ver. Sozpk*, 92 (1890), 1–24.
86 Hans-Jurgen Teuteberg, 'Das deutsche und britische Wollgewerbe um die Mitte des 19. Jahrhunderts', *Vom Kleingewerbe zur Grossindustrie*, ed. H. Winkel, *Schr. Ver. Sozpk*, new series, 83 (1975), 78–85.
87 Julian Bartyś, 'Sukiennictwo w ordynacji zamojskiej w I połowie XIX w', *Prz. Hist.*, 49 (1958), 486–509.
88 M. L. de Tegoborski, *Commentaries on the Productive Forces of Russia* (London, 1855), vol. I, 8–22.
89 Gryzelda Missalowa, 'Rozwój form wytwórczości przemysłowej w Pabianicach w latach 1820–1865', *Prz. Nauk. Hist. Spol.*, 3 (1953), 270–307; id., 'Les crises dans l'industrie textile au royaume de Pologne à l'époque de la Révolution Industrielle', *St. Hist. Oec.*, 8 (1973), 285–303.
90 Ludwig Schumann, 'Der Fortschritt vom Kleinbetrieb zum Grossbetrieb in der Schweiz', *Jb. Gesch. Verw. Volksw.*, 20 (1896), 246–53.
91 Richard L. Rudolph, *Banking and Industrialization in Austria-Hungary* (Cambridge, 1976), 60.
92 Heinrich Benedikt, 'Die Anfänge der Industrie in Mähren', *Der Don.*, 2 (1957), 38–51; Hermann Kellenbenz, 'Die wirtschaftlichen Beziehungen zwischen Westdeutschland und Böhmen-Mahren in Zeitalter der Industrialisierung', *Boh.*, 3 (1962), 239–59.
93 Arnost Klima, 'Industrial Growth and Entrepreneurship in the early Stages of Industrialisation in the Czech Lands', *Jl Eur. Ec. Hist.*, 6 (1977), 549–74.
94 Jaroslav Purš, 'The Industrial Revolution in the Czech Lands', *Hist. (P)*, 2 (1960), 183–272.
95 R. L. Rudolph, op. cit., 41.
96 Jan Novotny, 'Zur Problematik des Beginns der industriellen Revolution in der Slowakei', *Hist. (P)*, 4 (1962), 129–89.

97 Gyorgy Tolnai, 'Textilná manufaktúra a rolnícka tkáčka a pradiarska výroba na Slovensku (1850–1867)', *Hist. Stud.*, 14 (1969), 17–46.
98 Raymond Carr, *Spain 1808–1939* (Oxford, 1966), 391.
99 'Report on the Textile Industries of Catalonia', For. Off. Misc. Ser., 301, *Parl. Pap.*, 1893–4, vol. XCI; Max Sorre, 'L'industrie espagnole en 1914', *Ann. Géog.*, 26 (1917), 302–6.
100 Pierre Léon, 'L'industrialisation du Piedmont durant le XIXe siècle', *Ann. E.S.C.*, 16 (1961), 1218–24.
101 'Further Notes on the Industries of the District of Biella', For. Off. Misc. Ser., 44, *Parl. Pap.*, 1887, vol. LXXXII.
102 'Report on the Industries of the Province of Florence', For. Off. Misc. Ser., 570, *Parl. Pap.*, 1902, vol. CIII.
103 Domenico Demarco, 'L'économie italienne du Nord et du Sud avant l'unité', *Rv. Hist. Éc. Soc.*, 34 (1956), 369–91.
104 John Goodwin, 'Progress of the Two Sicilies under the Spanish Bourbons from the Year 1734–35 to 1840', *Jl Roy. Stat. Soc.*, 5 (1842), 47–73, 177–207.
105 Bolton King, 'Statistics of Italy', *Jl Roy. Stat. Soc.*, 66 (1903), 213–72; Richard S. Eckaus, 'The North–South Differential in Italian Economic Development', *Jl Ec. Hist.*, 21 (1961), 285–317.
106 G. Zane, 'L'industrie roumaine au cours de la seconde moitié du XIXe siècle', Bibl. Hist. Rom., 43 (1973), 22–53.
107 N. Todorov, 'The Genesis of Capitalism in the Balkan Provinces of the Ottoman Empire in the Nineteenth Century', *Expl. Ec. Hist.*, 7 (1970), 313–24.
108 Traian Stoianovich, *A Study in Balkan Civilization* (New York, 1967), 89.
109 *Histoire économique et sociale de la France*, ed. F. Braudel and E. Labrousse (Paris, 1979), vol. IV, part 1, 293.
110 A. L. Dunham, op. cit., 236–51; C. Ballot, 'Philippe de Girard et l'invention de la filature mécanique du lin', *Rv. Hist. Éc. Soc.*, 7 (1914), 135–95.
111 Comte Chaptal, op. cit., vol. II, 134–41.
112 Quoted in C. Fohlen, op. cit., 164.
113 Michel Philipponneau, 'Le rôle de l'industrie dans le développement de la Bretagne', *Festschrift Leopold G. Scheidl* (Vienna, 1965), vol. I, 279–87; F. Dornic, 'L'évolution de l'industrie textile aux XVIIIe et XIXe siècles', *Rv. Hist. Mod. Cont.*, 3 (1956), 38–66; Jean Vidalenc, 'L'industrie dans les départements normands à la fin du Premier Empire', *Ann. Norm.*, 7 (1957), 281–307.
114 G. Margry, 'L'industrie textile et le blanchissage du fil à Damigny au XIXe siècle', *Ann. Norm.*, 5 (1955), 173–86.
115 Jules Sion, *Les Paysans de la Normandie Orientale* (Paris, 1909), 319.
116 F.-P. Codaccioni, 'Le textil lillois durant la crise 1846–1851', *Rv. Nord*, 38 (1956), 29–53; René Gendarme, *La Région du Nord: essai d'analyse économique*, Cent. Ét. Éc., Ét. et Mém., 20 (1954), 66; *Histoire économique et sociale de la France*, vol. IV, part 1, 294.
117 M. N. Briavoinne, op. cit., 338–46; Jean de Béthume, 'La crise linière et le pauperisme dans le Courtraisis entre 1825 et 1850', *Féd. Arch. Hist. Belg.*, 35e Congrès (Courtrai, 1953), part 2, 71–133; J. Mokyr, op. cit., 12–15.
118 G. Jacquemyns, *Histoire de la crise économique des Flandres (1845–1850)*, Ac. Roy. Belg. Mém., 2e série, 26 (1929), 44–5.
119 Ibid., 68–83.
120 Willy Franken, 'Die Entwicklung des Gewerbes in den Städten Mönchen-Gladbach und Rheydt im 19. Jahrhundert', *Schr. Rhein.-Westf. Wtgesch.*, 19 (1969), 19–24; Gerhard Adelmann, 'Strukturwandlungen der rheinischen Leinen- und Baumwollgewerbe zu Beginn der Industrialisierung', *Viert. Soz. Wirtgesch.*, 53 (1966), 62–84.
121 Edith Schmitz, 'Leinengewerbe und Leinenhandel in Nordwest Deutschland (1650–1850)', *Schr. Rhein.-Westf. Wtgesch.*, 15 (1967), 16–45.
122 W. Jacob, op. cit., 78, 91.
123 Alex J. Warden, *The Linen Trade* (London, 1864; edn of 1967, Cass Reprints), 267–8, 275.
124 Jungen Schlumbohm, 'Der saisonale Rhythmus im Osnabrücker Land', *Arch. Sozgesch.*, 19 (1979), 263–98; 'Die Hausindustrie des Bezirks der Handelskammer Osnabrück', *Schr. Ver. Sozpk*, 42 (1890), 35–44.
125 A. J. Warden, op. cit., 267–8. 'Report on the Prussian Commercial Union', *Reports from Commissioners*, *Parl. Pap.*, 1840, vol. XXI, 381–672.
126 They were the subject of G. Hauptmann's *Die Weber*.
127 R. Forberger, op. cit., 162–6.
128 Heinz Potthoff, 'Die Leinenindustrie', *Schr. Ver. Sozpk.*, 105 (1903), 3–126.
129 Mariusz Kulczykowski, 'Chłopskie tkactwo bawełniane w osrodku Andrychówskim w XIX

wieku', *Prac. Kom. Nauk. Hist.*, 38 (Wrocław, 1976); id. 'En Pologne au XVIIIe siècle: industrie paysanne et formation du marché national, *Ann. E.S.C.*, 24 (1969), 61–9; C. Ballot, op. cit.
130 W. Zorn, 'Handels- und Industriegeschichte Bayerisch-Schwabens 1648–1870', *Veröff. Schwab. Forsch.*, 1961, 71–2; id., 'L'industrialisation de l'Allemagne du Sud au XIXe siècle', *L'Industrialisation en Europe au XIXe siècle*, Cent. Nat. Rech. Sci. (1972), 379–92.
131 'Report on the Commerce and Manufactures of Switzerland', *Parl. Pap.*, 1836, vol. IX, 655–802.
132 W. Bodmer, op. cit., 198–203, 313–15.
133 J. Purš, op. cit.
134 Gustav Otruba and Rudolf Kropf, 'Bergbau und Industrie Böhmens in der Epoche der Frühindustrialisierung (1820–1848)', *Boh.*, 12 (1972), 53–232; J. Purš, 'The Situation of the Working Class in the Czech Lands in the Phase of the Expansion and Completion of the Industrial Revolution', *Hist. (P)*, 6 (1963), 145–237.
135 'Report on the Textile Industries of Catalonia', For. Off. Misc. Ser., 301, *Parl. Pap.*, 1893–4, vol. XCI.
136 Kent Roberts Greenfield, *Economics and Liberalism in the Risorgimento* (Baltimore, Md, 1965), 89; 'Further notes on the Industries of the District of Biella', For. Off. Misc. Ser., 44, *Parl. Pap.*, 1887, vol. LXXXII; 'Report on the Industrial Development of Italy', For. Off. Misc. Ser., 610, *Parl. Pap.*, 1904, vol. XCVI.
137 G. Zarte, 'L'industrie roumaine au cours de la seconde moitié du XIXe siècle', *Bibl. Hist. Rom.*, 43 (1973), 24–30; Gyula Merei, 'L'essor de l'industrie capitaliste en Hongrie au cours de la première moitié du XIXe siècle', *Mélanges Jacquemins* (Brussels, 1968), 507–25.
138 P. Virrankoski, 'Replacement of Flax by Cotton in the Domestic Textile Industry of South-West Finland', *Sc. Ec. Hist. Rv.*, 11 (1963), 27–42.
139 J. Vidalenc, op. cit.
140 Robert Grevey, 'Localisation et structure de l'industrie du jute dans la région du Nord de la France', *Rv. Nord*, 35 (1953), Livr. Géog., 2, 1–28; C. Fohlen, op. cit., 462.
141 Friedrich Vöchting, 'Die Industrialisierung Süditaliens', *Zt Ges. Staatsw.*, 107 (1951), 120–50; 'Report on the Industrial Development of Italy'.
142 'Textilindustrie', *Schr. Ver. Sozpk*, 105 (1903), 11.
143 George McHenry, *The Cotton Trade* (London, 1863), 59.
144 W. O. Henderson, 'The Cotton Famine on the Continent 1861–5', *Ec. Hist. Rv.*, 4 (1932–4), 195–207.
145 David S. Landes, 'Technical Change and Industrial Development in Western Europe, 1750–1914', *Camb. Ec. Hist. Eur.*, vol. VI, part 1, 275.
146 Duncan Bythell, *The Handloom Weavers* (Cambridge, 1969).
147 Comte Chaptal, op. cit., vol. II, 12–15.
148 J. Sion, op. cit., 299–307.
149 J. Levainville, 'Les ouvriers du cotton dans la région de Rouen', *Ann. Géog.*, 20 (1911), 52–64.
150 Paul Leuillot, *L'Alsace au début du XIXe siècle*, Bibl. Gen. Éc. Prat. Htes. Ét., 2 (1959–60), 357–8.
151 R. B. Forrester, *The Cotton Industry in France*, Univ. Manchester Econ. Ser., 15 (1921), 5–6; P. Leuillot, op. cit., 357–427.
152 M. Lévy-Leboyer, *Les Banques européennes et l'industrialisation internationale* (Paris, 1964), 73–9.
153 R. B. Forrester, op. cit., 6.
154 C. Fohlen, op. cit., 207–8.
155 M. Lévy-Leboyer, op. cit., 71–2.
156 C. Fohlen, 'Esquisse d'une évolution industrielle: Roubaix au XIXe siècle', *Rv. Nord*, 33 (1951), 92–102; Georges Franchomme, 'L'évolution démographique et économique de Roubaix dans le dernier tiers du XIXe siècle', *Rv. Nord*, 51 (1969), 201–47; Jacques Toulemonde, 'Notes sur l'industrie roubaisienne', *Rv. Nord*, 48 (1966), 321–36.
157 Louis Trenard, 'Roubaix, ville drapante entre Lille et Tournai', *Rv. Nord*, 51 (1969), 175–99.
158 Jan Craeybeckx, 'Les débuts de la révolution industrielle en Belgique et les statistiques de la fin de l'Empire', *Mélanges Jacquemyns*, 114–44.
159 N. Briavoinne, op. cit., vol. II, 378.
160 J. Mokyr, op. cit., 28–41.
161 Fred Cornelissen, *Les Industries des Pays-Bas* (Paris, 1932), 138–40.
162 R. T. Griffiths, 'Eyewitnesses at the Birth of the Dutch Cotton Industry 1832–1839', *Ec. Soc. Hist. Jb.* 40 (1977), 113–81.
163 Quoted in K. van der Pols, 'The Introduction of the Steam Engine to the Netherlands', *Acta Hist. Neer.*, 12 (1979), 110–25.

164 I. J. Brugmans, 'Die industrielle Revolution in den Niederlanden', *Rhein. Vbl.*, 29 (1964), 124–37.
165 Ernst Baasch, *Holländische Wirtschaftsgeschichte* (Jena, 1927), 456–62; G. Adelmann, 'Die Deutsch-Niederlandische Grenze als textilindustrieller Standort', *Landschaft und Geschichte: Festschrift für Franz Petri* (Bonn, 1970), 9–43.
166 P.-J. Hutter, 'La famine de coton en Westphalie, 1861–1865', *Rv. Hist. Éc. Soc.*, 20 (1932), 392–405.
167 R. M. R. Dehn, *The German Cotton Industry*, Univ. Manchester Econ. Ser., 14 (1913), 15–16.
168 Wolfgang Fischer, 'Ansätze zur Industrialisierung in Baden 1770–1870', *Viert. Soz. Wirtgesch.*, 47 (1960), 186–231.
169 W. Zorn, 'Zu den Anfängen der Industrialisierung Augsburgs im 19. Jahrhundert', *Viert. Soz. Wirtgesch.*, 38 (1951), 155–68.
170 G. von Schulze-Gaevernitz, *The Cotton Trade in England and on the Continent* (London, 1895), 79; R. M. R. Dehn, op. cit., 24–7.
171 'Report on the Commerce and Manufactures of Switzerland', *Parl. Pap.*, 1836, vol. XLV, 655–802; Rudolf Braun, 'The Rise of a Rural Class of Industrial Entrepreneurs', *Cah. Hist. Mond.*, 10 (1966–7), 551–66.
172 R. Braun, *Sozialer und Kultureller Wandel in einem ländlichen Industriegebiet* (Zurich, 1965), 17–19; W. Bodmer, op. cit., 220–37.
173 Beatrice Veyrasset-Herren, 'Les centres de gravité de l'industrialisation en Suisse au XIXe siècle: le rôle de coton', *L'Industrialisation en Europe*, 481–95.
174 S. L. Besso, *The Cotton Industry in Switzerland, Vorarlberg and Italy*, Univ. Manchester Econ. Ser., 13 (1910), 40.
175 Ibid., 65.
176 Herbert Matis, 'Über die sozialen und wirtschaftlichen Verhältnisse österreichischer Fabrik- und Manufakturarbeiter um die Wende vom 18. zum 19. Jahrhundert', *Viert. Soz. Wirtgesch.*, 53 (1966), 433–76.
177 S. L. Besso, op. cit., 113–19.
178 Viktor Hofmann, 'Die Anfänge der österreichischen Baumwollwarenindustrie in den österreichischen Alpenländern im 18. Jahrhundert', *Arch. Öst. Gesch.*, 110 (1926), 415–742.
179 Jerome Blum, 'Transportation and Industry in Austria, 1815–1848', *Jl Mod. Hist.*, 15 (1943), 24–38.
180 Herbert Matis, *Österreichs Wirtschaft 1848–1913* (Berlin, 1972), 181–2.
181 J. Purš, op. cit.
182 As is claimed by Bernard Michel, 'La révolution industrielle dans les pays tchèques au XIXe siècle', *Ann. E.S.C.*, 20 (1965), 984–1005.
183 R. L. Rudolph, op. cit., 41.
184 G. Otruba and R. Kropf, op. cit.
185 M. Kulczykowski, op. cit.; J. Bartys, op. cit.
186 R. Carr, op. cit., 20.
187 Quoted in James W. Cortada, 'British Consular Reports on Economic and Political Developments in Cataluna, 1842–1875', *Cuad. Hist. Ec. Cat.*, 10 (1973), 149–98.
188 Jordi Nadal, 'Industrialisation et désindustrialisation du Sud-est espagnol, 1820–1890', *L'Industrialisation en Europe*, 201–12.
189 Pierre Vilar, 'La Catalogne industrielle: réflexions sur un démarrage et sur un destin', ibid., 421–33.
190 J. Nadal, *El Fracaso de la Revolución en España 1814–1913* (Barcelona, 1975), 196.
191 J. Nadal, 'Industrialisation et désindustrialisation'.
192 Quoted in J. W. Cortada, op. cit.
193 'Report on the Textile Industries of Catalonia'.
194 Pierre Léon, 'L'industrialisation du Piemont durant le XIXe siècle', *Ann. E.S.C.*, 16 (1961), 1218–24.
195 Shepard B. Clough, *The Economic History of Modern Italy* (New York, 1964), 62–2; D. Demarco, op. cit.
196 G. Luzzatto, 'The Italian Economy in the First Decade after Unification', *Essays in European Economic History*, ed. P. Earle (Oxford, 1974), 203–25.
197 Tariffs were further raised in 1887; see S. B. Clough, op. cit., 63.
198 'Report on the Industrial Development of Italy', For. Off. Misc. Ser., 610, *Parl. Pap.*, 1904, vol. XCVI; Bolton King, op. cit.
199 S. L. Besso, op cit.; S. B. Clough, op. cit.; Bolton King, op. cit.
200 S. L. Besso, op. cit., 132–3.
201 K. R. Greenfield, op. cit., 82.

202 'Report on the Industrial Development of Italy'.
203 Pierre Cayez, 'L'industrie lyonnais au XIXe siècle: du grande commerce à la grande industrie', *Cah. Hist.*, 22 (1977), 3–11; 'Silks of Lyons', *Ciba Rv.* (Feb. 1938).
204 Robert J. Bezucha, *The Lyon Uprising of 1834* (Cambridge, Mass., 1974), 25; M. Lévy-Leboyer, op. cit., 131–6; Monique Coulesque, 'La crise americaine et la crise de la soierie lyonnais, 1860–1864', *Cah. Hist.*, 9 (1964), 261–78.
205 W. Bodmer, op. cit., 188; 'Zurich Silks', *Ciba Rv.*, no. 119 (1957).
206 'Report on the Commerce and Manufactures of Switzerland', *Parl. Pap.*, 1836, vol. XLV, 655–802.
207 N. J. G. Pounds, *An Historical Geography of Europe 1500–1840*, 241–2; 'Silk Industries of Crefeld', *Ciba Rv.*, no. 83 (1950).
208 W. Zorn, *Rhein. Vbl.*, 30 (1965), 334–45.
209 G. Adelmann, ibid., 43 (1979), 260–88.
210 H. Kruger, op. cit., 162.
211 Alphons Thun, 'Die Krefelder Seidenindustrie und die Krisis', *Jb. Ges. Verw. Volksw.*, 3 (1879), 113–43.
212 W. Bodmer, op. cit., 216–19.
213 Otto Hintze, 'Die Schweizer Stickereiindustrie und ihre Organisation', *Jb. Ges. Verw. Volksw.*, 18 (1894), 1251–99.
214 M. Block, op. cit., 157–59.
215 J. Vidalenc, op. cit.
216 Dietrich, 'Die gegenwärtige wirtschaftliche Lage der Spitzenindustrie in Frankreich', *Jb. Ges. Verw. Volksw.*, 24 (1900), 75–142; William Felkin, *History of the Machine-wrought Hosiery and Lace Manufactures* (London, 1845), 407–8.
217 'La dentelle à la main en France – sa décadence: efforts tentés pour y remédier', *Rv. Éc. Int.*, 3 (1904), 180–99.
218. Dietrich, 'Die gegenwärtige wirtschaftliche Lage der Spitzenindustrie in Belgien', *Jb. Ges. Verw. Volksw.*, 23 (1899), 1123–54.
219 A. H. Héron de Villefosse, op. cit., vol. III, 428–32.
220 C. J. B. Karsten, *Untersuchungen über die Kohligen Substanzen des Mineralreichs* (Berlin, 1826); summarised in *Ann. Mines*, 1ère série, 13 (1826), 111–74.
221 J. Franquoy, 'Mémoire sur l'histoire des progrès de la fabrication du fer dans le Pays de Liège', *Mém. Soc. Émul. Liège*, 1 (1860), 313–448; E. Mahain, 'Les débuts de l'établissement John Cockerill à Seraing', *Rv. Univ. Mines*, 4e série, 13 (1906), 171–92.
222 E. Weigert, 'Die Grosseisenindustrie des Saargebiets', *Jb. Ges. Verw. Volksw.*, 46 (1922), part 2, 117–61.
223 Hedwig Behrens, 'Der erste Kokshochofen des Rheinisch-westfälischen Industriegebietes auf der Friedrich-Wilhelmshütte in Mülheim a. d. Ruhr', *Rhein. Vbl.*, 25 (1960), 121–5.
224 *The History of Coke Making and of the Coke Oven Managers' Association* (Cambridge, 1936), 59–68.
225 A. M. Héron de Villefosse, in *Ann. Mines*, 2 (1827), 401–620.
226 J. Cournot, 'La fabrication de l'acier au convertisseur', *Rv. Mines*, 20 (1923), 695–711.
227 L. Guillet, 'L'histoire des procédés basiques de fabrication de l'acier et Sidney Gilchrist Thomas', *Rv. Mines, Mém.*, 14 (1917), 2–38; '30 Jahre Thomas Verfahren in Deutschland', *St. u. Eisen*, 29 (1909), 1465–90.
228 A. M. Héron de Villefosse, op. cit., vol. III, 432.
229 N. J. G. Pounds, *The Ruhr* (London, 1952), 135–9.
230 Id., *The Upper Silesian Industrial Region* (Bloomington, Ind., 1958), 42–50.
231 Léon Jacques, 'Étude sur la houille du bassin de Liège', *Rv. Univ. Mines*, 22 (1867), 149–342.
232 De Grossouvre, 'Étude sur les gisements de minerai de fer du centre de la France', *Ann. Mines*, 8e série, 10 (1886), 311–418.
233 Czeslaw Kuzniar, 'Erzbergbau in Polen: I. Geologischer Bau und Vorräte der Erzlagerstätten in Polen', *Zt Oberschles. B.H.V.*, 68 (1929), 460–9, 514–18.
234 D. C. McKay, 'The Pre-War Development of Briey Iron Ores', *Essays in the History of Modern Europe*, ed. D. C. McKay (New York, 1936), 168–84.
235 N. J. G. Pounds, 'Lorraine and the Ruhr', *Econ. Geog.*, 33 (1957), 149–62.
236 Colin Ross, *Die Entstehung von Grosseisenindustrie an der deutschen Seekuste* (Berlin, 1911).
237 Guy Greer, *The Ruhr–Lorraine Industrial Problem* (New York, 1925).
238 The high-phosphorus ores of Lorraine were particularly suitable for making cast-iron.
239 Calculations based on Dionysius Lardner *Railway Economy* (London, 1850), 58–9.
240 A. M. Héron de Villefosse, 'Des métaux en France: rapport au jury central de l'exposition des produits de l'industrie française', *Ann. Mines*, 2e série, 2 (1827), 401–620.

241 S. Jordan, 'Notes on the Resources of the Iron Manufacture in France', *Jl I.S. Inst.* (1878), 316–56.
242 Hector Rigaud, 'Mémoire sur la situation des forges de France et de Belgique', *Ann. Mines*, 4e série, 8 (1845), 371–496.
243 Marcel Bulard, 'L'industrie du fer dans la Haute Marne', *Ann. Géog.*, 13 (1904), 223–42, 310–21.
244 Michel Wittmann, 'L'évolution de la métallurgie du Barrois', *Rv. Hist. Sid.*, 2 (1961), 89–114; Jean-Bernard Silly, 'La disparition de la petite métallurgie rurale', ibid., 2 (1961), 47–61.
245 'Report on the Progress of the Iron and Steel Industries', *Jl I.S. Inst.* (1873), 459–60.
246 'Revue de l'Exposition de 1867', publ. as *Rv. Univ. Mines*, 21, 25, 26.
247 *La Sidérurgie française, 1864–1914* (Paris, 1914).
248 E. F. Soderlund, 'The Impact of the British Industrial Revolution on the Swedish Iron Industry', *Studies in the Industrial Revolution*, ed. L. S. Presnell (London, 1960), 52–65.
249 G. Arpi, 'The Supply with Charcoal of the Swedish Iron Industry from 1830 to 1950', *Geog. Ann.*, 35 (1953), 11–27; Olof Nordström, *Die Beziehungen zwischen Hüttenwerken und ihrem Umland in Sudschweden*, Lund Studies in Geography, series B, no. 8 (1953).
250 F. Le Play, *Les Ouvriers européens* (Paris, 1955), 92–7.
251 David Gaunt, 'Household Typology: Problems, Methods, Results', *Chance and Change: Social and Economic Studies in Historical Demography in the Baltic Area*, Od. St. Hist. Soc. Sci., 52 (1978), 69–83.
252 G. A. Montgomery, *The Rise of Modern Industry in Sweden*, Stockholm Econ. Studies, 8 (London, 1939), 79–81.
253 This was built to operate at so high a temperature that the bottom plate had to be cooled by passing water under it; see H. R. Schubert, *History of the British Iron and Steel Industry from 450 B.C. to A.D. 1775* (London, 1957), 278–9, 288–9.
254 Gosta A. Eriksson, 'Advance and Retreat of Charcoal Iron Industry and Rural Settlement in Bergslagen', *Geog. Ann.*, 41 (1959), 267–84.
255 Id., 'The Decay of Blast-Furnaces and Ironworks in Vaster Bergslagen in Central Sweden 1860–1940', *Geog. Ann.*, 35 (1953), 1–10; id., 'The Decline of the Small Blast-furnaces and Forges in Bergslagen after 1850', ibid., 39 (1957), 257–77.
256 B. Boethius, 'Swedish Iron and Steel 1600–1955', *Sc. Ec. Hist. Rv.*, 6 (1958), 144–75. The history of the Uddeholm works is traced in Helge Nelson, 'The Modern Steelworks of Sweden', *Balt. Scand. C.*, 4 (1938), 47–52.
257 E. F. Soderlund, op. cit.
258 Lennart Jörberg, *Growth and Fluctuations of Swedish Industry 1869–1912*, Univ. Lund Inst. Ec. Hist. (1961), 69.
259 Jan Kuuse, 'Foreign Trade and the Breakthrough of the Engineering Industry in Sweden 1890–1920', *Sc. Ec. Hist. Rv.*, 25 (1977), 1–36.
260 Jean Chevalier, *Le Creusot* (Paris, 1935), 152–4.
261 Jacques Wolff, 'Decazeville: Expansion et déclin d'un pole de croissance', *Rv. Écon.*, 23 (1972), 753–85.
262 A. J. Ihde, 'Chemical Industry, 1780–1900', *Cah. Hist. Mond.*, 4 (1958), 957–84.
263 L. F. Haber, *The Chemical Industry During the Nineteenth Century* (Oxford, 1958), 42–8.
264 Paul M. Hohenberg, *Chemicals in Western Europe 1850–1914* (Chicago, 1967), 35–6.
265 *The History of Coke Making...*, 64–9.
266 Fritz Redlich, 'Die völkswirtschaftliche Bedeutung der deutschen Teerfarbenindustrie', *Staats. Soz. Forsch.*, 180 (1914), 1–41; 'Report on Chemical Instruction in Germany and the Growth and Present Condition of the German Chemical Industries', For. Off. Misc. Ser., 561, *Parl. Pap.*, 1901, vol. LXXX.
267 P. M. Hohenberg, op. cit., 35–6.
268 Ibid., 40.
269 L. F. Haber, op. cit., 121–36; Walter Dau, 'Geographische Verbreitung der Berufsgruppe des Deutschen Reiches', *Pet. Mitt.*, 52 (1906), 193–204.
270 Rudolf Heimann, 'Die neuere Entwicklung der Kaliindustrie und des Kalisyndikates', *Jb. Ges. Verw. Volksw.*, 30 (1906), 1489–1565.
271 Hans-Werner Schutt, 'Anfänge der Agrikulturchemie in der ersten Hälfte des 19. Jahrhunderts', *Zt Aggesch.*, 21 (1973), 83–91.
272 P. M. Hohenberg, op. cit., 90.
273 W. Dau, op. cit.
274 L. F. Haber, op. cit., 50–2.
275 'Report on the Arms Industry of Liège', For. Off. Misc. Ser., 650, *Parl. Pap.*, 1906, vol. CXXII.
276 Udo Rottstadt, 'Besiedlung und Wirtschaftsverfassung des Thuringer Waldes', *Staats. Soz. Forsch.*, 179 (1914), 40.

277 G. I. H. Lloyd, 'Labour Organisation in the Cutlery Trade of Solingen', *Ec. Jl*, 18 (1908), 373–91; F. Le Play, op. cit., 152–7.
278 A. Milward and S. B. Saul, *The Development of the Economies of Continental Europe 1850–1914* (London, 1977), 99–100.
279 P. Steller, 'Maschinenindustrie', *Schr. Ver. Sozpk*, 107 (1903), 7–76.
280 Josef Loewe, 'Die elektro-technische Industrie', *Schr. Ver. Sozpk*, 107 (1903), 77–116; A. Milward and S. B. Saul, op. cit., 35–8.
281 'Auslese und Anpassung der Arbeiterschaft in der Elektroindustrie, Büchdrückerei, Feinmechanik und Maschinenindustrie', *Schr. Ver. Sozpk*, 134 (1910), 12–65.
282 Henri Morsel, 'Les industries électrotechniques dans les Alpes françaises du Nord de 1869 à 1921' *L'Industrialisation en Europe*, 557–92.
283 L. Gallois, 'La production de la bauxite en France', *Ann. Géog.*, 26 (1917), 386–97; G. Veyret-Verner, op. cit., 286–97.
284 'Auslese und Anpassung der Arbeiterschaft in der Automobilindustrie und einer Wiener Maschinenfabrik', *Schr. Ver. Sozpk*, 135, part 3 (1911), 239–40.
285 Patrick Fridenson, 'Une industrie nouvelle: l'automobile en France jusqu'en 1914', *Rv. Hist. Mod. Cont.*, 19 (1972), 557–78.
286 For the French motor industry, see A. Milward and S. B. Saul, op. cit., 97–100.
287 Gunther Leckebusch, 'Die Beziehungen der deutschen Seeschiffswerften zur Eisenindustrie an der Ruhr in der Zeit von 1850 bis 1930', *Schr. Rhein.-Westf. Wirtgesch.*, 8 (1963), 17.
288 Dirk Peters, 'Der Seeschiffbau in Bremerhaven von der Stadtgründung bis zum Ersten Weltkrieg', *Niedersachs. Jb.*, 51 (1979), 25–45.
289 G. Leckebusch, op. cit., 84.
290 Von Stulpnagel, 'Über Hausindustrie in Berlin', *Schr. Ver. Sozpk*, 42 (1890), 1–24.
291 Wilhelm Stieda, 'Die deutsche Hausindustrie', *Schr. Ver. Sozpk*, 39 (1889), 59; Gustav Lange, 'Die Hausindustrie Schlesiens', *Schr. Ver. Sozpk*, 42 (1890), 51.
292 'Die Lohgerberei in Breslau', *Schr. Ver. Sozpk*, 65 (1895), 1–22; 'Die Schuhmacherei in Breslau', ibid., 23–78.
293 'Lage des Handwerks in Deutschland, vol. II, Konigreich Sachsen', *Schr. Ver. Sozpk*, 63 (1895), 169–312.
294 Eugen Fridrichowicz, 'Die Lage des Schumacherhandwerks in Deutschland', *Zt Ges. Staatsw.*, 55 (1899), 120–60, 241–86.
295 'Report of an Enquiry by the Board of Trade into Working Class Rents, Housing and Retail Prices...in the Principal Industrial Towns of France', Cd 4512, *Parl. Pap.*, 1909, vol. XCI.
296 John R. Lampe and Marvin R. Jackson, *Balkan Economic History 1550–1950* (Bloomington, Ind.), 1982, 237–77.
297 T. Markowitch, *L'Industrie française de 1789 à 1964: conclusions générales*, Cah. I.S.E.A. (1966).
298 Walther G. Hoffmann, *Das Wachstum der deutschen Wirtschaft seit der Mitte des 19. Jahrhunderts* (Berlin, 1965), 454–5.
299 G. L. de Brabander, *Regional Specialization, Employment and Economic Growth in Belgium between 1846 and 1970* (New York, 1981), 60–77.

Chapter 8. The growth of industrial regions

1 E. A. Wrigley, 'The Supply of Raw Materials in the Industrial Revolution', *Ec. Hist. Rv.*, 15 (1962), 1–16.
2 Sidney Pollard, in *Industrialisierung und europäische Wirtschaft im 19. Jahrhundert*, ed. O. Busch, W. Fischer and W. Herzfeld, *Veröff, Hist. Komm. Berlin*, 46 (1976), 5; the idea is more fully expressed in S. Pollard, *European Economic Integration 1815–1970* (London, 1974).
3 This point is made by François Crouzet, ibid., 48–51.
4 S. Pollard, *Peaceful Conquest: The Industrialization of Europe 1760–1970* (Oxford, 1981), vii.
5 F. Crouzet, 'Western Europe and Great Britain: Catching Up in the First Half of the Nineteenth Century', *Economic Development in the Long Run*, ed. A. J. Youngson (London, 1972), 49.
6 N. F. R. Crafts, 'Industrial Revolution in England and France: Some Thoughts on the Question, "Why was England First?"', *Ec. Hist. Rv.*, 30 (1977), 429–41.
7 T. C. Banfield, *Industry of the Rhine* (London, 1946–8), vol. II, 123.
8 *Ziemia Obiecana* – 'The Promised Land' – was the name of Władysław Reymont's novel set against the industrial growth of Łódź.
9 'Die Wohnungsnoth der armeren Klassen in deutschen Grossstädten', *Schr. Ver. Sozpk*, 30 (1886).
10 E. A. Wrigley, *Industrial Growth and Population Change* (Cambridge, 1961), 62–8.

11 Ibid., 72–3.
12 F. Crouzet, op. cit.
13 Comte de Buffon, *Histoire naturelle des minéraux* (Paris, 1783), vol. I, 452.
14 *Rapid Survey of Coal Reserves and Coal Production, 1946* (London, 1947); 'Coal Mining in Europe, 1939', *U.S. Bureau of Mines, Bul.*, 414; *European Steel Trends*, Ec. Comm. Eur. (Geneva, 1949); *The Coal Resources of the World: Inquiry of the Executive Committee of the XIIth International Geological Congress* (Toronto, 1913).
15 Robert Demolin, 'Le bassin mosan pendant la première moitié du XIXe siècle', *Bul. Inst. Arch. Liège*, 63 (1939), 141–52; Pierre Lebrun, Marinette Bruwier, Jan Dhondt and Georges Hansolte, *Essai sur la Révolution Industrielle en Belgique 1770–1847, Hist. Quant. Dév. Belg.*, 2, part 1 (1979), 343–466.
16 L. Cordier, 'Sur les mines de houille de France', *Jl Mines*, 36 (1814), 321–94.
17 G. R. Porter, 'A Statistical View of the Recent Progress and Present Amount of Mining Industry in France', *Jl Roy. Stat. Soc.*, 7 (1844), 281–91.
18 E. A. Wrigley, op. cit., 42.
19 Marcel Gillet, *Les Charbonnages du Nord de la France au XIXe siècle*, Éc. Prat. Htes. Ét., Ind. et Art., 8 (1973), 44–5.
20 Gotz Voppel, *Die Aachener Bergbau- und Industrielandschaft: eine wirtschaftsgeographische Studie, Köln, Forsch.*, 3 (1965), 70; Clemens Bruckner, *Zur Wirtschaftsgeschichte des Regierungsbezirk Aachen, Schr. Rhein.-Westf. Wtgesch.*, 16 (1967), 107–8.
21 Wolfgang Zorn, 'Zur historischen Wirtschaftskarte der Rheinlande 1818', *Rhein. Vbl.*, 29 (1964), 106–18.
22 Manfred D. Jankowski, 'Law, Economic Policy and Private Enterprise: the Case of the Early Ruhr Mining Region, 1766–1865', *Jl Eur. Ec. Hist.*, 2 (1973), 688–727.
23 René Gendarme, *La Région du Nord: essai d'analyse économique*, Cent. Ét. Ec., Ét. et Mém., 20 (1954), 42.
24 Ibid., 41.
25 Fritz Wundisch, 'Zur Geschichte des rheinischen Braunkohlenbergbaus', *Rhein. Vbl.*, 17 (1952), 197–221.
26 Sigfrid Schneider, 'Das Braunkohlenrevier im Westen Kölns', *Köln und die Rheinlande* (Wiesbaden, 1961), 341–52.
27 Ernst Pape, 'Der deutsche Braunkohlenhandel unter dem Einfluss der Kartelle', *Zt Ges. Staatswiss.*, 62 (1906), 234–71.
28 Georges Hansotte, 'La sidérurgie belge du XIXe siècle avant l'acier', *Rv. Hist. Sid.*, 7 (1966), 211–32; Marcel Bourguignon, 'La sidérurgie, industrie commune des pays d'Entre Meuse et Rhin', *Anc. Pays États*, 28 (1963), 81–120; Philippe Moureaux, 'La sidérurgie dans la région de Beaumont et de Chimay à la fin de l'Ancien Régime', ibid., 56 (1972), 231–49; K. C. Edwards, 'Historical Geography of the Luxembourg Iron and Steel Industry', *Trans. I.B.G.*, 29 (1961), 1–16.
29 T. C. Banfield, op. cit., vol. II, 234.
30 Fernand Discry, 'L'ancien bassin sidérurgique du Hoyoux', *Anc. Pays États*, 50 (1970), 21–43.
31 Jan Craeybeckx, 'Les débuts de la Révolution Industrielle en Belgique et les statistiques de la fin de l'Empire', *Mélanges Jacquemyns* (Brussels, 1968), 114–44; J. Mokyr, *Industrialization in the Low Countries* (New Haven, Conn., 1976), 309–33.
32 Wilhelm Gunther, 'Zur Geschichte der Eisenindustrie in der Nordeifel', *Rhein. Vbl.*, 30 (1965), 309–33.
33 Fritz Schulte, *Die Entwicklung der gewerblichen Wirtschaft in Rheinland-Westfalen im 18. Jahrhundert, Schr. Rhein.-Westf. Wtgesch.*, 1 (1959), 25–30.
34 W. O. Henderson, *Britain and Industrial Europe 1750–1870* (Liverpool, 1954), 111–33; Richard M. Westebbe, 'State Entrepreneurship: King William I, John Cockerill, and the Seraing Engineering Works, 1815–1840', *Explns Entrepr. Hist.*, 8 (1955–6), 205–32.
35 M. G. de Boer, 'Twee memorien van G. M. Roentgen over den toestand der Britsche en Zuid-Nederlandsche ijzerindustrie uit den jaren 1822 en 1823', *Ec. Soc. Hist. Jb.*, 9 (1923), 3–156.
36 'Das Eisenhüttenwesen in Belgien', *Berg Hütt. Ztg*, 1 (1842), 541–50, 565–72; N. Briavoinne, op. cit., vol. II, 294; P. Lebrun, M. Bruwier, J. Dhondt and G. Hansotte, op. cit.
37 Léopold Genicot, *Histoire de la Wallonie* (Toulouse, 1973), 361.
38 G. Hansotte, op. cit.
39 R. Gendarme, op. cit., 55.
40 W. Gunther, op. cit.
41 Lutz Hatzfeld, *Die Handelsgesellschaft Albert Poensgen Mauel-Düsseldorf, Schr. Rhein.-Westf. Wtgesch.*, 11 (1964).
42 Justus Hashagen, 'Zur Geschichte der Eisenindustrie vernehmlich in der Nordwestlichen Eifel', *Eifel-Festschrift zur 25 Jährigen Jubelfeier des Eifelvereins* (Bonn, 1913), 269–94.

43 T. C. Banfield, op. cit.
44 Guy Thuillier, 'La métallurgie rhénane de 1800 à 1830', *Ann. E.S.C.*, 16 (1961), 877–907.
45 Klaus Tenfelde, 'Arbeiterschaft, Arbeitsmarkt und Kommunikationsstrukturen im Ruhrgebiet in der 50er Jahren des 19. Jahrhunderts', *Arch. Sozgesch.*, 16 (1976), 1–59.
46 T. C. Banfield, op. cit., vol. II, 65–6.
47 Frédéric Le Play, *Les Ouvriers européens*, vol. II (Paris, 1855), 152–7.
48 Ibid., vol. II, 152.
49 G. Thuillier, op. cit.; Alfons Thun, *Die Industrie am Niederrhein und ihre Arbeiter, Staats. Soz. Forsch.*, 2, part 3 (1879), passim.
50 *Histoire d'un métropole: Lille–Roubaix–Tourcoing*, ed. L. Trenard (Toulouse, 1977), 321–2.
51 Jacques Toulemonde, 'Notes sur l'industrie roubaisienne et tourquennoise dans la première moitié du XIXe siècle', *Rv. Nord*, 48 (1966), 321–36; Michel Raman, 'Mesure de la croissance d'un centre textile: Roubaix de 1789 à 1913', *Rv. Hist. Éc. Soc.*, 51 (1973), 470–501; René Fruit, *La Croissance économique du pays de Saint-Amand (Nord) 1668–1914*, Éc. Prat. Htes. Ét., Ét. et Mém., 55 (1963), 177–8; Henri Sée, *La Vie économique de la France sous la monarchie censitaire (1815–1848)* (Paris, 1927), 55.
52 A. Chanut, 'La crise économique à Tourcoing (1846–50)', *Rv. Nord*, 38 (1956), 77–105.
53 David Landes, 'Religion and Enterprise: the Case of the French Textile Industry', *Enterprise and Entrepreneurs in Nineteenth and Twentieth Century France*, ed. E. C. Carter, R. Foster and J. N. Moody (Baltimore, Md, 1976), 41–86.
54 M. Raman, op. cit.; Léon Machu, 'La crise de l'industrie textile à Roubaix au milieu du XIXe siècle', *Rv. Nord*, 38 (1956), 65–75.
55 D. Landes, op. cit.
56 R. Gendarme, op. cit., 44.
57 D. Landes, op. cit.
58 W. O. Henderson, op. cit., 27–9.
59 D. Landes, op. cit.; Maurice Lévy-Leboyer, *Les Banques européennes et l'industrialisation internationale*, Pub. Fac. Let. Paris, 16 (1964), 163–7.
60 *Histoire économique et sociale de la France*, ed. F. Braudel and E. Labrousse, vol. IV, part 1 (Paris, 1979), 295; C. Fohlen, op. cit., 235.
61 R. Fruit, op. cit.
62 A. L. Dunham, *The Anglo-French Treaty of Commerce of 1860 and the Progress of the Industrial Revolution in France*, Univ. Michigan Pubns. Hist. and Pol. Sci., 9 (1930), 218–23.
63 Pierre Lebrun, *L'Industrie de la laine à Verviers pendant le XVIIIe et le début du XIXe siècle*, Bibl. Fac. Liège, 114 (1948), 128, 138–43; id., 'Croissance et industrialisation: l'expérience de l'industrie drapière vervietoise 1750–1850', 1ere Conf. Int. Hist. Éc. (1960), 531–68; P. Lebrun, M. Bruwier, J. Dhondt and G. Hansotte, op. cit., 161–260.
64 'Report of the Select Committee on Manufactures, Commerce and Shipping', *Parl. Pap.*, 1833, vol. VI, 69.
65 Max Barkhausen, 'Die sieben bedeutendsten Fabrikanten der Roerdepartements im Jahre 1810', *Rhein. Vbl.*, 25 (1960), 100–13.
66 Franz Decker, 'Die betriebliche Sozialordnung der Dürener Industrie im 19. Jahrhundert', *Schr. Rhein.-Westf. Wtgesch.*, 12 (1965), 13–40.
67 H. Kisch, 'The Impact of the French Revolution on the Lower Rhine Textile Districts: Some Comments on Economic Development and Social Change', *Ec. Hist. Rv.*, 15 (1962–3), 304–27.
68 Martin Schumacher, 'Wirtschafts- und Sozialverhältnisse der rheinischen Textilindustrie im frühen 19. Jahrhundert', *Rhein. Vbl.*, 35 (1966), 162–84.
69 T. C. Banfield, op. cit., vol. II, 231.
70 K. Tenfelde, op. cit.
71 R. M. R. Dehn, op. cit., 13.
72 Willy Franken, *Die Entwicklung des Gewerbes in den Städten Mönchen-Gladbach und Rheydt im 19. Jahrhundert*, Schr. Rhein.-Westf. Wtgesch., 19 (1969), 25.
73 Based on table in Gerhard Adelmann, 'Die ländlichen Textilgewerbe des Rheinlandes vor der Industrialisierung', *Rhein. Vbl.*, 43 (1979), 260–88.
74 Étienne Helin, *La Démographie de Liège aux XVIIe et XVIIIe siècles*, Ac. Roy. Belg. Mém., Cl. Lettres, 56 (1963), 238–41; A. Quetelet, 'Sur les anciens recensements de la population Belge', *Bul. Com. Cent. Stat.*, 3 (1847), 1–38.
75 K. Olbricht, 'Die Städte des rheinisch-westfälischen Industriebezirks', *Pet. Mitt.*, 57 (1911), 4–8.
76 Cl. Heiss, 'Die belgische Industrie- und Gewerbezahlung vom 31 Oktober 1896', *Jb. Ges. Verw. Volksw.*, 27, part 1 (1903), 209–53.
77 *Glück.*, 45 (1909), 404–14.
78 'Die Berg- und Hüttenindustrie Belgiens', *Zt Berg. Hütt. Sal.*, 55 (1907), 547–74.

79 J. A. van Houtte, *Esquisse d'une histoire économique de la Belgique* (Louvain, 1943), 157; Baron de Laveleye, 'Historical Survey of the Metallurgy of Iron in Belgium', *Jl I.S. Inst.*, 88 (1913), 8–31.
80 Georges Lespineux, 'Note rétrospective sur les Mines de Fer en Belgique', *The Iron Ore Resources of the World*, Int. Geol. Congr. (Stockholm, 1910), vol. II, 649–61.
81 N. Briavoinne, op. cit., vol. I, 283–7.
82 G. Lespineux, op. cit., 659
83 E. A. Wrigley, op. cit., 38–9.
84 At Liège: Seraing, Grivégnée, Ougré-Marihaye, Sclessin, Ésperance, and at Charleroi: Couillet, Châtelineau, Marcinelle.
85 R. Gendarme, op. cit., 60–2.
86 R. Matton, 'L'industrie du fer dans le bassin de Maubeuge', *Ann. Géog.*, 36 (1927), 309–27; 'Revue de l'exposition de 1867', 419.
87 *La Sidérurgie française, 1864–1914*, Comité des Forges (Paris, 1914), 194.
88 *St. u. Eisen*, 31, part 1 (1911), 990–1.
89 T. C. Banfield, op. cit., vol. II, 58.
90 W. N. Parker, in N. J. G. Pounds and W. N. Parker, *Iron and Steel in Western Europe*, (London, 1957), 277.
91 Ibid., 286.
92 Otto Berger, *Mülheim a. d. Ruhr als Industriestadt*, Diss., Köln, 1932.
93 W. N. Parker, op. cit., 288–9; N. J. G. Pounds, 'Lorraine and the Ruhr', *Ec. Geog.*, 33 (1957), 149–62; *St. u. Eisen*, 52 (1932), 1–3.
94 Details of freight carried were published regularly in *Glückauf*.
95 E. Voye, *Geschichte der Industrie im Märkischen Sauerland* (Hagen, 1908–13), vol. I, 211–13.
96 C. Fohlen, op. cit., 228.
97 M. Raman, 'Mesure de la croissance d'un centre textile: Roubaix de 1789 à 1913', *Rv. Hist. Éc. Soc.*, 51 (1973), 470–501; *Histoire d'une métropole: Lille–Roubaix–Tourcoing*, ed. Louis Trenard (Toulouse, 1977), 320–61.
98 R. B. Forrester, *The Cotton Industry in France*, Univ. Manchester Econ. Ser., 15 (1921), 13–14.
99 Ibid.
100 *Histoire économique et sociale de la France*, vol. IV, part 1, 294–5.
101 Georges Franchomme, 'L'évolution démographique et économique de Roubaix dans le dernier tiers du XIXe siècle', *Rv. Nord*, 51 (1969), 201.
102 M. Raman, op. cit.; A. Chanut, op. cit.
103 C. Fohlen, op. cit., 465.
104 *Rapport général sur l'industrie française: sa situation, son avenir*, Ministère de Commerce (Paris, 1919), vol. I, 539.
105 A. L. Dunham, op. cit., 243–51.
106 *Histoire économique et sociale de la France*, vol. III, part 2, 508; C. Fohlen, op. cit., 230–1.
107 R. Gendarme, op. cit., 68–9.
108 R. Fruit, op. cit.
109 R. Gendarme, op. cit., 67–8.
110 Ibid., 53.
111 J. A. Sporck, op. cit.; L. Dechesne, *Histoire économique et sociale de la Belgique depuis les origines jusqu'en 1914* (Liège, 1932), 434–6.
112 F. Decker, op. cit.; C. Bruckner, op. cit., 199.
113 K. Tenfelde, op. cit.
114 Paul Kollmann, 'Die gewerbliche Entfaltung im Deutschen Reiche', *Jb. Ges. Verw. Volksw.*, 12 (1888), 437–528.
115 R. M. R. Dehn, op. cit.; G. von Schulze-Gaevernitz, *The Cotton Trade in England and on the Continent* (London, 1895), 79.
116 R. M. R. Dehn, op. cit., 14.
117 Willy Franken, op. cit., 64–5.
118 R. M. R. Dehn, op. cit., 12–13.
119 W. Franken, op. cit.
120 J. Mokyr, op. cit., 77; L. Genicot, op. cit., 364.
121 Fernand Bezy, 'Les évolutions longues de l'industrie du zinc dans l'Ouest européen 1840–1939', *Bul. Inst. Rech. Éc. Soc.*, 16 (1950), part 1, 3–56.
122 A. S. Milward and S. B. Saul, *The Development of the Economies of Continental Europe 1850–1914* (London, 1977), 167; P. Lebrun, M. Bruwier, J. Dhondt and G. Hansotte, op. cit., 448–51.
123 J. Mokyr, op. cit., 77.

124 Cl. Heiss, op. cit.; T. C. Barker, *Pilkington Brothers and the Glass Industry* (London, 1960), 133–43, 186–7.
125 F. Le Play, op. cit., 152–7.
126 G. I. H. Lloyd, 'Labour Organisation in the Cutlery Trade of Solingen', *Ec. Jl*, 18 (1908), 373–91.
127 'Report on the Coal Industry of the Rhenish-Westphalian Provinces', For. Off. Misc. Ser., 454, *Parl. Pap.*, 1898, vol. XCIII.
128 Fritz Redlich, 'Die volkswirtschaftliche Bedeutung der deutschen Teerfarbenindustrie', *Staats. Soz. Forsch.*, 180 (1914), 15.
129 L. F. Haber, op. cit., 80.
130 E. Jacquot, 'Note sur les recherches quin ont été executées le long de la frontière nord-est du département de la Moselle', *Ann. Mines*, 5e série, 11 (1857), 107–48.
131 W. Zorn, *Rhein. Vbl.*, 29 (1964), 106–18.
132 *European Steel Trends*, Ec. Comm. Eur. (Geneva, 1949), 115.
133 Colchen, *Mémoire statistique du département de la Moselle*, Paris, An. 11 (1803), 148–9.
134 F. Verronais, *Statistique historique, industrielle et commerciale du département de la Moselle* (Metz, 1844), 170–2.
135 Georges Hottenger, *L'Ancienne Industrie du fer en Lorraine* (Nancy, n. d.), 149.
136 R. Hartshorne, 'The Franco-German Boundary of 1871', *World Politics*, 2 (1950), 209–50.
137 E. Greau, *Le Fer en Lorraine* (Paris, 1908).
138 L. Bruneau, *L'Allemagne en France* (Paris, 1915), 32–120.
139 These were ARBED, Roechling, Stumm and Dillingen.
140 L. Gallois, 'Le bassin houiller de la Sarre', *Ann. Géog.*, 28 (1919), 268–79.
141 J. Kollmann, *Die Grosseisenindustrie des Saargebiets* (Stuttgart, 1911).
142 A. M. Héron de Villefosse, op. cit., vol. I, 374.
143 Richard Dietrich, 'Zur industriellen Produktion, technischen Entwicklung und zum Unternehmertum in Mitteldeutschland, speciell in Sachsen im Zeitalter der Industrialisierung', *Jb. Gesch. Mitt. Ostd.*, 28 (1979), 221–72.
144 W. Jacob, op. cit., 280.
145 P. Benaerts, *Les Origines de la grande industrie Allemande* (Paris, 1933), 419–20.
146 'Über den Betrieb der Königlich sächsischen Steinkohlenwerke im Plauenschen Gründe bei Dresden', *Berg. Hütt. Ztg*, 9 (1850), 321–7, 339–43, 357–61, 371; 'Der Steinkohlen-Bergbau im Königreiche Sachsen', ibid., 2 (1843), 201–8, 220–9, 244–53; 'Der Kokshochofenbetrieb in Sachsen und Thuringen', ibid., 9 (1850), 221–2.
147 V. Häufler, J. Korčák and V. Král, *Zeměpis Československa* (Prague, 1960), 134–8; Ludmila Kárníková, *Vývoj uhelného Průmyslu v českých zemích do r. 1880* (Prague, 1960), 236–66.
148 *Wirtschaftsterritorium Deutsche Demokratische Republik*, ed. Gerhard Schmidt-Renner (East Berlin, 1962), 72–81.
149 A. M. Héron de Villefosse, op. cit., vol. I, 360–3.
150 Rudolf Forberger, *Die Manufaktur in Sachsen vom Ende des 16. bis zum Anfang des 19. Jahrhunderts, Deutsch Akad. Wiss.* (1958), 35–119.
151 W. Jacob, op. cit., 314.
152 John Bowring, 'Report on the Prussian Commercial Union', *Reports from Commissioners, Parl. Pap.*, 1840, vol. XXI, 381–672.
153 R. Deitrich, op. cit.
154 Adolf Soetbeer, 'Edelmetallproduktion', *Pet. Mitt. Erg.*, 57 (1879), 1.
155 A. M. Héron de Villefosse, vol. I, 360–4.
156 R. Dietrich, *Jb. Gesch. Mitt. Ostd.*, 28 (1978), 221–72.
157 G. Otruba and R. Kropf, op. cit.
158 W. Jacob, op. cit., 316–17.
159 A. M. Héron de Villefosse, op. cit., vol. I, 234.
160 *Berg. Hütt. Ztg*, 9 (1850), 161–3.
161 Ibid., 257–8.
162 Herbert Pönicke, 'Sachsens Wirtschaft in der ersten Phase der industriellen Revolution', *Mitte*, 2 (1966), 113–52.
163 *St. u. Eisen.*, 2, part 1 (1882), 35–6.
164 Udo Rottstadt, 'Besiedlung und Wirtschaftsverfassung des Thüringer Waldes', *Staats. Soz. Forsch.*, 179 (1914), 40–9.
165 H. Pönicke, op. cit.
166 R. Forberger, op. cit., 170–6.
167 Klaus Guth, 'Kleinbäuerliche Leinenweberei im Sechsämterland (1789–1825)', *Jb. Fränk. Lforsch.*, 40 (1980), 119–53.

168 John Bowring, op. cit.
169 P. Benaerts, op. cit., 373–7.
170 Rudolf Martin, 'Der wirtschaftliche Aufschwung der Baumwollspinnerei im Königreich Sachsen', *Jb. Ges. Ver. Volksw.*, 17 (1893), 639–89.
171 R. M. R. Dehn, op. cit., 22.
172 'Report of the Commission for the Cotton and Linen Industries', *Parl. Pap.*, 1878.
173 W. Zorn, 'Probleme der Industrialisierung Oberfränkens im 19. Jahrhundert', *Jb. Fränk. Lforsch.*, 29 (1969), 295–310.
174 S. J. Chapman, *The Cotton Industry and Trade* (London, 1905), 106–20; G. von Schulze-Gavernitz, op. cit., 79.
175 P. Benaerts, op. cit., 382.
176 Jaroslav Purš, 'The Industrial Revolution in the Czech Lands', *Hist. (P)*, 2 (1960), 183–272.
177 G. Otruba and R. Kropf, op. cit.
178 J. Purš, op. cit.; G. Otruba and R. Kropf, op. cit.; Arnošt Klíma, 'Industrial Growth and Entrepreneurship in the early Stages of Industrialization in the Czech Lands', *Jl Eur. Ec. Hist.*, 6 (1977), 549–74.
179 Richard L. Rudolph, *Banking and Industrialization in Austria-Hungary* (Cambridge, 1976), 46–7.
180 M. Gau and C. Neubert, 'Die Hausindustrie im nördlichen Thüringen', *Schr. Ver. Sozpk*, 40 (1889), 75.
181 'Berichte aus der Hausindustrie in Berlin, Osnabrück, im Fichtelgebirge und in Schlesien', *Schr. Ver. Sozpk*, 42 (1890), 45–9.
182 A. M. Héron de Villefosse, op. cit., vol. I, 21.
183 G. Otruba and R. Kropf, op. cit.
184 L. Kárníková, op. cit., 348–9, 356.
185 Hellmuth Barthel, 'Braunkohlenbergbau und Landschaftdynamik', *Pet. Mitt. Erg.*, 270 (1962), 3034.
186 E. Pape, op. cit.
187 Karlheinz Blaschke, *Bevölkerungsgeschichte von Sachsen bis zur Industriellen Revolution* (Weimar, 1967), 190–1.
188 Ibid., 163.
189 *Atlas Československých Dějin* (Prague, 1965), plate XXVII.
190 'Report of an Enquiry by the Board of Trade into Working Class Rents, Housing...of the German Empire', Cd 4032, *Parl. Pap.*, 1908, vol. CVIII.
191 Frank B. Tipton, *Regional Variations in the Economic Development of Germany during the Nineteenth Century* (Middletown, Conn., 1976), 37.
192 For fuller references to the development of Upper Silesia see N. J. G. Pounds, *The Upper Silesian Industrial Region*, Indiana Univ. Slavic and East European Series, 11 (1958), passim.
193 *Berg. Hütt. Ztg*, 1 (1842), 58.
194 Alfred Hornig, *Komunikacja na Górnym Śląsku*, Górno-sląskie Prace i Materiały Geograficzne (Katowice, 1963), 44–5.
195 Edward Pietraszek, 'Zagłębie Krakowskie w latach 1796–1848', *Kw. Hist. Kult. Mat.*, 9 (1961), 743–69; M. L. de Tegoborski, *Commentaries on the Productive Forces of Russia* (London, 1855), vol. I, 210–11.
196 Kurt Flegel, 'Die wirtschaftliche Bedeutung der Montanindustrie Russlands und Polens', *Ost. Inst., Qu. u. St.*, 1 (1920), 61–4.
197 Pawel Rybicki, 'Rozwój ludnosci Górnego Śląsku', *Górny Śląsk: Prace i Materiały Geograficzne* (Kraków, 1955), 247–343.
198 Kazimierz Popiołek, 'Koncentracja w przemyśle górniczo-hutniczym Górnego Śląska w drugiej połowie XIX wieku', *Studiów i Materiàłów z Dziejów Śląska*, 2 (1947), 63–181.
199 *Berg. Hütt. Ztg*, 10 (1951), 436.
200 K. Popiołek, op. cit., 157–8.
201 Ibid.
202 Franciszek Ryszka, 'Kapital monopolistyczny na Górnym Śląsku i formy jego polityki', *Prz. Zachod.*, 8 (1952), 209–65.
203 Andrzej Grodek, 'Handel odrzański w rozwoju historycznym', *Monographia Odry* (Poznań, 1948), 384–418.
204 Artur Born, 'Regulacja Odry i rozbudowa urządzen technicznych', ibid., 419–553.
205 A. Hornig, op. cit., 44, 60–2.
206 Andrzej Stasiak, *Miasto Królewska Huta*, *Prac. Inst. Bud. Miesz.*, 13 (1962), 21–30.
207 Janusz Ziołkowski, *Sosnowiec* (Katowice, 1960), 200.
208 Ibid., 256.
209 'Report of the Board of Trade into Working Class Rents...the German Empire', 352–7.

210 Jan Dylik, *Rozwój osadnictwa w okolicach Łódzi*, Acta Geographica Univ. Lodziensis, 2 (1948), 40–1.
211 N. J. G. Pounds, *Historical Geography of Europe 1500–1840*, 259–61.
212 Adam Ginsbert, *Łódź* (Łódź, 1961), 15–17.
213 Quoted in G. Kurnatowski, 'Les origines du capitalisme en Pologne', *Rv. Hist. Mod.*, 8 (1933), 236–67.
214 Jadwiga König-Haźdźyńska, 'Geneza rozwoju miasta kapitalistycznego w Łódzkim okręgu przemysłowym', *Prz. Nauk. Hist. Spol.*, 5 (1954), 99–108; Roman Gawinski, 'Studia nad dziejami manufaktury w Zgierzu', ibid., 109–23.
215 E. M. Sigsworth, 'Fosters of Queensbury and Geyer of Lodz, 1848–1862', *Yorks. Bul. Ec. Soc. Res.*, 3, no. 2 (1951), 67–82.
216 Pawel Korzec, 'Genesis und Entwicklung des Białystocker Textilindustriegebietes im 19. Jahrhundert', *St. Hist. Oec.*, 1 (1966), 75–93.
217 Gryzelda Missalowa, 'Les crises dans l'industrie textile au Royaume de Pologne à l'epoque de la Révolution Industrielle', *St. Hist. Oec.*, 8 (1973), 285–303.
218 C. Ballot, 'Philippe de Girard et l'invention de la filature mécanique du lin', *Rv. Hist. Éc. Soc.*, 7 (1914), 135–95.
219 G. Missalowa, op. cit.
220 Ireneusz Ihnatowicz, 'Rynki zbytu przemysłu łódzkiego w drugiej połowie XIX wieku', *Prz. Hist.*, 56 (1965), 413–31.
221 There was no census before 1897, and these data must be treated as only approximate; see A. Ginsbert, op. cit., 23.
222 First published in 1899, and in English translation in 1928.
223 N. J. G. Pounds, 'Fabryka im. Juliana Marchlewskiego: a Textile Plant in Łódź, Poland', *Focus on Geographic Activity*, ed. R. S. Thoman and D. J. Patton (New York, 1964), 154–8. The Poznański factory described in this article appears to be that described by W. Reymont.
224 A. Ginsbert, op. cit., 54–7.
225 Ibid., 84; Jan Fifałek, Wiesław Pruś and Bolesław Pełka, 'La structure des entreprises modernes de l'industrie textile en territoire polonaise à partir de 1850 environ jusqu'à 1913', *St. Hist. Oec.*, 9 (1974), 197–222.
226 Pierre Léon, 'La région lyonnaise dans l'histoire économique et sociale de la France', *Rv. Hist.*, 137 (1967), 31–62.
227 'Silks of Lyons', *Ciba Rv.*, 1 (Feb. 1938).
228 Michel Laferrère, *Lyon ville industrielle* (Paris, 1960), 121–34; Maurice Lévy-Leboyer, op. cit., 131–6; Germaine Veyret-Verner, *L'Industrie dans les Alpes Françaises* (Grenoble, 1948), 98–102.
229 Robert J. Bezucha, *The Lyon Uprising of 1834* (Cambridge, Mass., 1974), 23–5.
230 M. Lévy-Leboyer, op. cit., 145.
231 *Histoire économique et sociale de la France*, vol. IV, part 1, 295.
232 C. Fohlen, op. cit., 185–7.
233 Maxime Perrin, *Saint-Étienne et sa région économique* (Tours, 1937), 269–73.
234 M. Lévy-Leboyer, op. cit., 131–3.
235 L.-J. Gras, *Histoire économique générale des mines de la Loire* (Saint-Étienne, 1922), vol. I, 69–75; A. Meugy, 'Historique des Mines de Rive-de-Gier', *Ann. Mines*, 4e série, 12 (1847), 143–94, 395–541, 543–68.
236 L. Cordier, 'Sur les mines de houille de France', *Jl Mines*, 36 (1814), 321–94; L.-J. Gras, op. cit.
237 A. L. Dunham, 'The Development of Coal Mining in France, 1815–48', *Pap. Mich. Acad. Sci.*, 27 (1941), 567–80.
238 François Crouzet, 'Le charbon anglais en France au XIXe siècle', *Charbon et les Sciences Humaines*, Éc. Prat. Htes. Ét., Ind. et Art., 2 (1966), 173–206.
239 Ludwig Beck, *Geschichte des Eisens* (Brunswick, 1899), vol. IV, 330.
240 L. Babu, 'L'industrie métallurgique dans la région de Saint-Étienne', *Ann. Mines*, 9e série, 15 (1899), 357–460.
241 Ibid.
242 M. Perrin, op. cit., 231–4; L. Babu, op. cit., 357–460.
243 Pierre Cayez, 'L'industrialisation lyonnaise au XIXe siècle: du grande commerce à la grande industrie', *Cah. Hist.*, 22 (1977), 3–11.
244 Pierre Léon, *Géographie de la fortune et structures sociales à Lyon au XIXe siècle (1815–1914)*, Cent. Hist. Éc. Soc. Rég. Lyon., 4 (1974), 14.
245 'Report for 1907 on the Chemical Metal and Mining Industries of the Consular District of Lyons', For. Off. Misc. Ser., 669, *Parl. Pap.*, 1908, vol. CVIII.
246 M. Laferrère, 'La concentration industrielle Lyonnaise', *Rv. Géog. Lyon.*, 36 (1961), 179–87.

247 M. Perrin, op. cit., 356.
248 Ibid., 410.
249 Gilbert Garrier, *Paysans du Beaujolais et du Lyonnais 1800–1970* (Grenoble, 1973), vol. II, maps 9 and 10.
250 M. Perrin, op. cit., 374–8.
251 Jean-Charles Bonnet, 'Les travailleurs étrangers dans la Loire sous la IIIe République', *Cah. Hist.*, 16 (1971), 67–80.
252 Abel Châtelain, 'La formation de la population lyonnaise: l'apport d'origine montagnarde', *Rv. Géog. Lyon.*, 19 (1954), 91–115.
253 Lucien Gachon, 'Dans les massifs crystallins d'Auvergne: la ruine du paysage rural et ses causes', *Ann. E.S.C.*, 5 (1950), 448–60.
254 Sidney Pollard, *Peaceful Conquest: The Industrialization of Europe 1760–1970* (Oxford, 1981), 207–9.
255 Joseph Harrison, 'Heavy Industry, the State, and Economic Development in the Basque Region, 1876–1936', *Ec. Hist. Rv.*, 36 (1983), 535–61.
256 Raymond Carr, *Spain 1808–1939* (Oxford, 1966), 31.
257 J. Vicens Vives, 'La industrialización y el desarrollo económico de España de 1800 à 1936', 1ère Conf. Int. Hist. Éc. (Paris, 1960), 129–36.
258 Quoted sub 1847 in James W. Cortada, 'British Consular Reports on Economic and Political Developments in Cataluña 1842–1875', *Cuad. Hist. Ec. Cat.*, 10 (1973), 149–98.
259 J. Vicens Vives, *An Economic History of Spain* (Princeton, N.J., 1969), 669–70.
260 Jordi Nadal, *El Fracaso de la Revolución Industrial en España 1814–1913* (Barcelona, 1975), 207–22; P. Vilar, 'La Catalogne industrielle: reflexions sur un démarrage et sur un destin', *L'Industrialisation en Europe au XIXe siècle*, Cent. Nat. Rech. Sci. (1972), 421–33.
261 'Report on the Textile Industries of Catalonia', For. Off. Misc. Ser., 301, *Parl. Pap.*, 1893–4, vol. XCI.
262 Pierre Vilar, 'La vie industrielle dans la région de Barcelone', *Ann. Géog.*, 38 (1929), 339–65.
263 J. Vicens Vives, op. cit., 672.
264 P. Vilar, *Ann. Géog.*, 38, 339–65.
265 Ibid.
266 Shepard B. Clough, *The Economic History of Modern Italy* (New York, 1964), 15–16; Carlo De Cugis, 'Lo sviluppo industriale in Italia dal 1861 al 1914', 1ère Conf. Int. Hist. Éc. (Stockholm, 1960), Communications, 227–50.
267 Domenico Demarco, 'L'économie italienne du Nord et du Sud avant l'unité', *Rv. Hist. Éc. Soc.*, 34 (1956), 369–91.
268 Kent Roberts Greenfield, *Economics and Liberalism in the Risorgimento* (Baltimore, Md, 1965), 83–6.
269 'Report on the Industrial Development of Italy', For. Off. Misc. Ser., 610, *Parl. Pap.*, 1904, vol. XCVI, 687.
270 Mario Romani, *Storia Economica d'Italia nel Secolo XIX* (Milan, 1976), vol. II, 238.
271 S. L. Besso, *The Cotton Industry in Switzerland, Vorarlberg and Italy* (Manchester, 1910), 133.
272 Ibid.
273 Alberto Caroncini, 'Die Krisis in der italienischen Baumwollindustrie und der italienische Baumwollhandel auf der Balkanhalbinsel', *Weltw. Arch.*, 2 (1913), 393–417.
274 'Report on the Italian Cotton Industry', For. Off. Misc. Ser., 364, *Parl. Pap.*, 1895, vol. CII, 583; Bolton King, 'Statistics of Italy', *Jl Roy. Stat. Soc.*, 66 (1903), 213–72.
275 'Further Notes on the Industry of the District of Biella', For. Off. Misc. Ser., 44, *Parl. Pap.*, 1887, vol. LXXXII.
276 S. B. Clough, op. cit., 63–5.
277 Paul Guichonnet, 'Vers de nouvelles formes d'industrialisation; le type alpin, l'expérience italienne', *L'Industrialisation en Europe au XIXe siècle*, Cent. Nat. Rech. Sci. (1972), 547–55.
278 Pierre Léon, 'L'industrialisation du Piémont durant le XIXe siècle', *Ann. E.S.C.*, 16 (1961), 1218–24.
279 Gianni Toniolo, ed., *Lo sviluppo economico Italiana 1861–1940* (Rome, 1973), 20. For capital investment see Luciano Cafagna, 'L'industrializzazione italiana: la formazione di una "base industriale" fra il 1896 e il 1914', *Stud. Stor.*, 2 (1961), 690–724.
280 Mario Abrate, 'Tableau schématique de la métallurgie italienne avant l'adoption des procédés modernes', *Rv. Hist. Sid.*, 5 (1964), 173–85; id., 'Lo sviluppo della siderurgia e della meccanica nel Regno di Sardegna dal 1831 al 1861', *Rec. Trav. Hist. Phil.*, 4e série, 21 (1960).
281 'Statistique de l'industrie minérale de l'Italie en 1897', *Ann. Mines*, 9e série, 15 (1899), 244.
282 A. Milward and S. B. Saul, *The Development of the Economies of Continental Europe, 1850–1914* (London, 1977), 262–3; S. B. Clough, op. cit., 93–4.

283 Luigi Bulferetti, 'Le déclin de l'influence des sources energétiques sur l'implantations industrielles dans l'Italie du XIXe siècle', *L'Industrialisation en Europe au XIXe siècle* (Paris, 1972), 233–6.
284 G. Anfossi, 'L'industrie de la houille blanche en Italie', *Ann. Géog.*, 27 (1918), 196–226; Luciano Cafagna, 'The Industrial Revolution in Italy', *Fontana Economic History of Europe* (London, 1963), vol. IV, 277–328.

Chapter 9. Transport and trade

1 J. Suret-Canale, 'L'état économique et social de la Mayenne au milieu du XIXe siècle', *Rv. Hist. Éc. Soc.*, 36 (1958), 294–331.
2 Mariana Starke, *Travels in Europe, for the Use of Travellers on the Continent, and likewise in the Island of Sicily* (London, 1833).
3 Pierre Léon, 'La conquête de l'espace national', *Histoire économique et sociale de la france*, ed. F. Braudel and E. Labrousse, vol. III (Paris, 1976), 241–73.
4 Alain Corbin, *Archaisme et modernité en Limousin au XIXe siècle* (Paris, 1975), vol. I, 119–20.
5 Roger Price, *The Economic Modernisation of France* (London, 1975), 9.
6 J.-C. Toutain, *Les Transports en France de 1830 à 1965* (Paris, 1967), 248.
7 R. Price, op. cit., 9.
8 Ibid., 10.
9 Dominique Renouard, *Les Transports de marchandises par fer, route et eau depuis 1850, Recherches sur l'économie française*, 2 (Paris, 1960), 37.
10 For a case study in Burgundy see Jean-François Minonzio, 'Les transports publics de voyageurs en Côte d'Or à la veille de l'ouverture du chemin de fer (1845–1849)', *Ann. Bourg.*, 42 (1970), 105–52.
11 Gerard Placq, 'Le développement du réseau routier belge de 1830 à 1940', *Bul. Inst. Rech. Éc. Soc.*, 17 (1951), 425–69.
12 William Jacob, *A View of the Agriculture, Manufactures, Statistics and State of Society of Germany* (London, 1820), 89.
13 Paul Thimme, 'Strassenbau und Strassenpolitik in Deutschland zur Zeit der Grundung des Zollvereins 1825–1835', *Viert Soz. Wirtgesch.*, 21 (1931), 4.
14 Jerome Blum, 'Transportation and Industry in Austria 1815–1848', *Jl Mod. Hist.*, 15 (1943), 24–38.
15 'The Resources and Future of Austria', *Quart. Rv.*, 114 (1863), 1–41.
16 Barbara Jelavich, 'Serbia in 1897: A Report of Sir Charles Eliot', *Jl Cent. Eur. Aff.*, 18 (1958), 183–9.
17 David R. Ringrose, *Transportation and Economic Stagnation in Spain, 1750–1850* (Durham, N.C., 1970), 14–15.
18 Pierre Ponsot, 'En Andalusie occidentale: systèmes de transports et développement économique (XVIe–XIXe siècle)', *Ann. E.S.C.*, 31 (1976), 1195–1212.
19 James W. Cortada, 'Catalan Politics and Economy, 1906–1911', *Cuad. Hist. Ec. Cat.*, 13 (1975), 129–81.
20 Richard S. Eckaus, 'The North–South Differential in Italian Economic Development', *Jl Ec. Hist.*, 21 (1961), 285–317.
21 Domenico Demarco, 'L'économie italienne du Nord et du Sud avant l'Unité', *Rv. Hist. Éc. Soc.*, 34 (1956), 369–91.
22 R. M. Hartwell, 'Economic Change in England and Europe, 1780–1830', *New Camb. Mod. Hist.*, 9 (1965), 31–59.
23 Albert Charles, 'L'isolement de Bordeaux et l'insuffisance des voies de communication en Gironde au début du Second Empire', *Ann. Midi*, 72 (1960), 59–73.
24 Marie-Hélène Bourquin, 'L'approvisionnement de Paris en bois de la Régence à la Révolution', *Études d'histoire du droit parisien* (Paris, 1970), 158–228.
25 J.-C. Toutain, op. cit., 11–73; Paul Léon, 'La navigation intérieure en France', *Rv. Éc. Int.*, 3 (1904), 549–92; Pierre Léon, 'La conquête de l'espace national', *Histoire économique et sociale de la France*, ed. F. Braudel and E. Labrousse, vol. III (1976), 241–73.
26 'Report on Canal Traffic in France', For. Off. Misc. Ser., 342, *Parl. Pap.*, 1895, vol. CII.
27 Quoted in Daniel Bellet, 'Mémoire sur la navigation intérieure', 4e Congr. Int. Sc. Géog. (Paris, 1890), 330–53.
28 Maurice Block, *Statistique de la France* (Paris, 1860), vol. II, 227.
29 Pierre Léon, op. cit.
30 J.-C. Toutain, op. cit., 79–81.

31 Daniel Faucher, *L'Homme et le Rhône*, Géographie Humaine, 35 (Paris, 1968), 210–48.
32 Felix Rivet, *La Navigation à vapeur sur la Saône et Rhône (1783–1863)*, Colln Cah. Hist., 5 (Paris, 1962), 25–39.
33 Id., 'American Technique and Steam Navigation on the Saône and Rhône, 1827–1850', *Jl Ec. Hist.*, 16 (1956), 18–33.
34 F. Rivet, *La Navigation*, 158–75.
35 Pierre Clerget, 'La navigation actuelle du Rhône, ses améliorations possibles et leur influence au point de vue du commerce international', 9e Congr. Int. Géog., Compte rendu des travaux (Geneva, 1911), vol. III, 77–89; 'L'aménagement du Rhône', *Ann. Géog.*, 20 (1911), 376–9.
36 L. Gallouedec, 'La Loire navigable', *Ann. Géog.*, 6 (1897), 45–60.
37 G. Gasnier, 'La navigation sur la Loire et ses affluents vers 1785', *Ann. Bret.*, 36 (1924–25), 76–95.
38 Roger Dion, 'Orléans et l'ancienne navigation de la Loire', *Ann. Géog.*, 47 (1938), 128–54; id., *Le Val de Loire: étude de géographie régionale* (Tours, 1934).
39 N. J. G. Pounds, *An Historical Geography of Europe 1500–1840* (Cambridge, 1979), 300.
40 'Report for the Year 1894 on the Fluvial Traffic of Rouen and the Waterways of the Seine Basin', For. Off. Misc. Ser., 366, *Parl. Pap.*, 1895, vol. CII.
41 Maurice Lévy-Leboyer, *Les Banques européennes et l'industrialisation internationale*, Pub. Fac. Let. Paris, 16 (1964), 268–74.
42 'Création d'une grande artère fluviale dans le Nord de la France', *Rv. Éc. Int.*, 1 (1904), 455–63; André Fortin, 'Les chemins de fer et les voies navigables du Pas-de-Calais durant le Second Empire', *Rv. Nord*, 53 (1971), 443–57.
43 'A Census of Inland Navigation in France', *Jl Roy. Stat. Soc.*, 55 (1892), 482–8.
44 A useful account of the physical conditions of navigation is *Der Rhein: Ausbau, Verkehr, Verwaltung* (Duisburg, 1951).
45 Maurice Pardé, *Fleuves et rivières* (Paris, 1933), 114–17.
46 B. Auerbach, 'Étude sur le régime et la navigation du Rhin', *Ann. Géog.*, 2 (1892–3), 212–38.
47 Peter Stubmann, 'Die Rheinschiffahrt', *Die Störungen im deutschen Wirtschaftsleben*, vol. IV, Schr. Ver. Sozpk, 107 (1903), 211–45.
48 Jean Dollfus, *L'Homme et le Rhin*, Géographie Humaine, 32 (Paris, 1960), 92, 142–6.
49 P. Stubmann, op. cit.; Christian Eckert, 'Rheinschiffahrt im XIX. Jahrhundert', *Statts. Soz. Forsch.*, 18., part 5 (1900), 186–9.
50 Arthur Valdenaire, 'Das Leben und wirken des Johann Gottfried Tulla', *Zt. Gesch. Oberr.*, 42 (1929), 337–64, 588–616; 42 (1930), 258–86.
51 Jean Dollfus, op. cit.
52 Sir Osborne Mance, *International River and Canal Transport* (Oxford, 1944), 27–37; W. O. Henderson, *The Zollverein* (Cambridge, 1939), 78–9.
53 Alexander Dietz, *Frankfurter Handelsgeschichte* (Frankfurt, 1921), vol. III, 293–5.
54 Wolfgang Zorn, 'Neue Teile der historischen Wirtschaftskarte der Rheinlands', *Rhein. Vbl.*, 31 (1966–7), 322–50.
55 Friedrich Schulte, 'Die Rheinschiffahrt und die Eisenbahnen', *Die Schiffahrt der deutschen Ströme, Schr. Ver. Sozpk*, 102 (1905), 301–526.
56 C. Eckert, op. cit.
57 W. Zorn, op. cit.
58 *Die Störungen im deutschen Wirtschaftsleben*, 211–45; B. Auerbach, op. cit.
59 J. Dollfus, op. cit.; *Der Rhein*, 369–79.
60 G. Kaeckenbeeck, *International Rivers*, Grotius Soc. Pubns, 1 (1918), 80–8.
61 'Report on the Projected Rhine–Neckar–Danube Ship Canal', For. Off. Misc. Ser., 613, *Parl. Pap.*, 1904, vol. XCVI, 597.
62 Konrad Fuchs, 'Die Lahn als Schiffahrtsweg im 19. Jahrhundert', *Nass. Ann.*, 75 (1964), 160–201.
63 'Report on the Canals and other Navigable Waterways of Belgium', For. Off. Misc. Ser., 604, *Parl. Pap.*, 1904, vol. XCVI.
64 'Report on the Opening of the Merwede Canal', For. Off. Misc. Ser., 253, *Parl. Pap.*, 1893–4, vol. XCI, 217–24.
65 Gustav Seibt, 'Die Verkehrswirtschaftliche Bedeutung der Binnenwasserstrassen', *Jb. Ges. Verw. Volksw.*, (1902), 929–1014.
66 William Jacob, *Second Report of the Agriculture and Corn of Some of the Continental States of Europe, 1827* (London, 1828), 13.
67 Hans-Joachim Root, 'Die Entwicklung der Verkehrsstroöme und der Verkehrsstruktur auf der Elbe während des 19. Jahrhunderts, unter besonderer Berücksichtigung der wirtschaftlichen Entwicklung im Verkehrsgebiet', *Jb. Wirtgesch.* (1975), part 1, 71–95.

68 August Meitzen, 'Die Frage des Kanalbaues in Preussen', *Jb. Ges. Verw. Volksw.*, 8 (1884), 751-95.
69 'Die Schiffahrt der deutschen Ströme', *Schr. Ver. Sozpk*, 100 (1903), esp. 133-245.
70 Th. H. Schunke, 'Die Schiffahrts-Kanale im Deutschen Reiche', *Pet. Mitt.*, 23 (1877), 285-93; 24 (1878), 51-64.
71 Joh. Kretschmar, 'Napoleons Kanalprojecte zur Verbindung des Rheines mit der Elbe und Ostsee', *Zt Hist. Ver. Niedersachs.* (1906), 138-50.
72 Magnus Biermer, 'Der Rhein-Elbe Kanal', *Jb. Ges. Verw. Volksw.*, 24 (1900), 239-96; A. Meitzen, op. cit.; 'Report upon the Inland Waterways of Germany', For. Off. Misc. ser., 345, *Parl. Pap.*, 1895, vol. CII.
73 *Royal Commission on Canals and Waterways*, vol. VI, Cd 4841 (1909).
74 For. Off. Misc. Ser., 345.
75 D. G. Giersberg, 'Die Bedeutung der Wasserstrassen im östlichen Deutschland für den Transport landwirtschaftlicher Massengüter', *Der Schiffahrt der deutschen Ströme*, 133-245.
76 Ibid.; T. H. Schunke, op. cit.
77 Michael J. Quin, *A Steam Voyage down the Danube* (London, 1835), vol. I, 6.
78 David Urquhart, *Turkey and its Resources* (London, 1833), 165.
79 Ibid., 179; Paul Cernovodeanu, 'An unpublished British source concerning the international trade through Galatz and Braila between 1837 and 1846', *Rv. Roum. Hist.*, 16 (1977), 517-31.
80 Emmanuel Porumbaru, 'La Commission Européenne du Danube', 9ᵉ Congr. Int. Géog., Geneva (1909), 1, 446-63.
81 Traian Ionescu, 'L'échange maritime de marchandises entre les principautés danubiennes et la France durant la période 1829-1848', *Rv. Roum. Hist.*, 13 (1974), 269-84.
82 Henri Hajnal, *Le Droit du Danube international* (The Hague, 1929), 15.
83 'Despatch by Lt.-Col. Sir Henry Trotter reporting upon the Operations of the European Commission of the Danube', Commission Report 9, *Parl. Pap.*, 1907, vol. LXXXVII.
84 Radu R. N. Florescu, 'The Struggle against Russia in the Roumanian Principalities 1821-1854', *Acta Hist.*, 2 (1962), 265.
85 Lawrence Oliphant, *The Russian Shores of the Black Sea in the Autumn of 1852* (Edinburgh, 1853), 346-7.
86 J. R. McCulloch, *A Dictionary of Commerce and Commercial Navigation* (London, supplement of 1842), 48-9.
87 E. Engelhart, 'Les embouchures du Danube et la commission instituée par le Congrès de Paris', *Rv. Deux Mondes*, 88 (1870), 93-117; E. Porumbaru, op. cit.; *Commission Report 9, Parl. Pap.*, 1907, LXXXVII.
88 E. Porumbaru, op. cit.; 'Report on Roumanian Trade, Agriculture and Danube Navigation from 1881 to 1890', For. Off. Misc. Ser., 226, *Parl. Pap.*, 1892, vol. LXXX.
89 Sir Osborne Mance, *International River and Canal Transport*, Roy. Inst. Int. Aff. (London, 1944), 51-64.
90 Raul Cernovodeanu, op. cit.
91 J. R. McCulloch, op. cit., 49; Virginia Paskaleva, 'Le rôle de la navigation à vapeur sur le bas Danube dans l'établissement de liens entre l'Europe Centrale et Constantinople jusqu'à la guerre de Crimée', *Bulg. Hist. Rv.*, 4 (1976), part 1, 64-74.
92 Franz Fillitz, 'Die Donauschiffahrt vom Einst zum Jetzt', *Don.*, 2 (1957), 164-75.
93 H. Hajnal, op. cit., 189-91.
94 Horst Glassl, 'Der Ausbau der ungarischen Wasserstrassen in den letzten Regierungsjahren Maria Teresias', *Ung. Jb.* (1970), 34-66; A. A. Paton, *The Bulgarian, the Turk and the German* (London, 1855), 21.
95 Franz Leskoschek, 'Schiffahrt und Flosserei auf der Drau', *Zt Hist. Ver. Steiermk*, 63 (1972), 115-52.
96 W. Bader, op. cit.
97 Henry Cord Meyer, 'German Economic Relations with Southeastern Europe, 1870-1914', *Am. Hist. Rv.*, 57 (1952), 77-90.
98 David R. Ringrose, *Transportation and Economic Stagnation in Spain, 1750-1850* (Durham, N.C., 1970), 14-17.
99 Kent Roberts Greenfield, *Economics and Liberalism in the Risorgimento* (Baltimore, Md, 1965), 64; G. Luzzatto, 'The Italian Economy in the First Decade after Unification', *Essays in European Economic History 1789-1914*, ed. F. Crouzet, W. H. Chaloner and W. M. Stern (London, 1969), 203-25; Domenico Demarco, 'L'économie italienne du Nord et du Sud avant l'unité', *Rv. Hist. Éc. Soc.*, 34 (1956), 369-91.
100 Rondo E. Cameron, *France and the Economic Development of Europe 1800-1914* (Princeton, N.J., 1961), 208; Leopold Genicot, *Histoire de la Wallonie* (Toulouse, 1973), 331; Kimon

Apostolus Doukas, *The French Railroads and the State*, Col. Univ. St. Hist. Ec., 517 (New York, 1945), 17–20; Georges Lefranc, 'The French Railroads, 1823–1842', *Jl Ec. Bus. Hist.*, 2 (1929–30), 299–331.
101 R. E. Cameron, op. cit., 205–8.
102 G. Bousquet, 'List et les chemins de fer', *Rv. Hist. Éc. Soc.*, 21 (1933), 269–79.
103 Hildegard John, 'Die Entwicklung des Eisenbahnnetzes im Raum Marburg und Giessen', *Hess. Jb. Lgesch.*, 9 (1959), 179–214; Jean Bouvier, 'La "grande crise" des compagnies ferroviaires suisses', *Ann. E.S.C.*, 11 (1956), 456–80.
104 Edwin A. Pratt, *Railways and Nationalisation* (London, 1911), passim.
105 Ibid., 20–3.
106 K. A. Doukas, op. cit., 26–42.
107 M. Blanchard, 'The Railway Policy of the Second Empire', *Essays in European Economic History*, 98–111.
108 'Report to the Board of Trade on Railways in Belgium, France and Italy', Cd 5106, *Parl. Pap.*, 1910, vol. LVII.
109 Georges Lefranc, 'The French Railroads, 1823–1842', *Jl Ec. Bus. Hist.*, 2 (1929–30), 299–331; Louis Gueneau, 'Le chemin de fer d'Épinac à Pont-d'Ouche', *Ann. Bourg.*, 3 (1931), 38–65, 224–52; 4 (1932), 22–54.
110 R. E. Cameron, op. cit., 205–8.
111 K. A. Doukas, op. cit., 26–30.
112 R. Dudley Baxter, 'Railway Extension and its Results', *Jl Roy. Stat. Soc.*, 29 (1866), 549–95.
113 R. Caralp-Landon, *Les Chemins de fer dans le Massif Central*, Cent. Ét. Éc., Ét. et Mém. (Paris, 1959).
114 J. Renkin, 'Les chemins de fer de l'état Belge', *Rv. Éc. Int.*, 3 (1904), 593–632.
115 R. E. Cameron, op. cit., 208–9; 'Reports to the Board of Trade on Railways in Belgium, France and Italy', Cd 5106, *Parl. Pap.*, 1910, 57.
116 R. E. Cameron, op. cit., 209, 302.
117 Ernst Baasch, *Holländische Wirtschaftsgeschichte* (Jena, 1927), 506–12.
118 Audren M. Lambert, *The Making of the Dutch Landscape* (London, 1971), 294–5.
119 G. Bousquet, op. cit.
120 H. Wagner, 'Die Entwicklung des Deutschen Eisenbahnnetzes', *Pet. Mitt.*, 19 (1873), 224–8.
121 Jerome Blum, 'Transport and Industry in Austria, 1815–1848', *Jl Mod. Hist.*, 15 (1943), 24–38.
122 Iván T. Berend and György Ránki, *Economic Development in East-Central Europe in the 19th and 20th Centuries* (New York, 1974), 71–6; Anton Adelbert Klein, 'Von den Anfängen des Eisenbahnbaues in Österreich', *Zt Hist. Ver. Steiermk*, 55 (1964), 3–21.
123 Herbert Matis, *Österreichs Wirtschaft 1848–1913* (Berlin, 1972), 189.
124 'Report to the Board of Trade on Railways in Austria and Hungary', Cd 4878, *Parl. Pap.*, 1909, vol. LXXVII.
125 Oscar Jaszi, *The Dissolution of the Habsburg Monarchy* (Chicago, 1929), 190–1.
126 R. E. Cameron, op. cit., 302.
127 R. M. Haywood, *The Beginnings of Railway Development in Russia in the Reign of Nicholas I, 1835–42* (Durham, N.C., 1969), 193–200.
128 I. Kostrowicka, Z. Landau and J. Tomaszewski, *Historia Gospodarcza Polski XIX i XX wieku* (Warsaw, 1966), 191–4.
129 S. H. Beaver, 'Railways in the Balkan Peninsula', *Geog. Jl*, 97 (1941), 273–94.
130 Herbert Feis, *Europe the World's Banker 1870–1914*, Council on Foreign Relations (New Haven, Conn., 1930), 293.
131 Theodore Bent, 'Baron Hirsch's Railway', *Fortnightly Review*, N.S. 44 (1888), 229–39.
132 Paul Dehn, 'Deutschland und die Orientbahnen', *Jb. Ges. Verw. Volksw.*, 9 (1885), 423–54; Arthur J. May, 'The Novibazar Railway Project', *Jl Mod. Hist.*, 10 (1938), 496–527.
133 Arthur J. May, 'Trans-Balkan Railway Schemes', *Jl Mod. Hist.*, 24 (1952), 352–67; M. I. Newbigin, *Geographical Aspects of the Balkan Problem*, 36–65.
134 Radoslav Popov, 'Autour du projet Austro-Hongrois pour la construction du chemin de fer de Novibazar', *Ét. Balk.* (1970), part 4, 45–62.
135 Jovan Cvijić, 'Der Zugang Serbiens zur Adria', *Pet. Mitt.*, 58 (1912), part 2, 361–4; Orme Wilson, 'The Belgrade–Bar Railroad: An Essay in Economic and Political Geography', *Eastern Europe: Essays in Geographical Problems*, ed. G. W. Hoffman (London, 1971), 365–84.
136 Edward Mead Earle, *Turkey, the Great Powers and the Bagdad Railway* (London, 1923), 71–84; Herbert Feis, op. cit., 297.
137 H. C. Meyer, op. cit.
138 Shepard B. Clough, *The Economic History of Modern Italy* (New York, 1964), 24–32; 'Reports to the Board of Trade on Railways in Belgium, France and Italy'.

139 G. Luzzatto, op. cit.
140 'Reports to the Board of Trade on Railways in Belgium'
141 George L. Boag, *The Railways of Spain* (London, 1923), 4–6.
142 G. A. Montgomery, *The Rise of Modern Industry in Sweden* (London, 1939), 104–24.
143 Martin Rudolph, 'Geographie der Landstrassen und Eisenbahnen von Norwegen', *Pet. Mitt. Erg.*, 206 (1929).
144 'Reports by Her Majesty's Secretaries of Embassy and Legation on the Manufactures, Commerce, etc. of the Countries in which they reside', No. 7, *Parl. Pap.*, 1864, vol. LXI, *sub* Norway.
145 Hedwig Schliebs, 'Der Schiffbau und die Reederei Papenburgs von 1783 bis 1913', *Mitt. Ver. Gesch. Osnbr.*, 52 (1930), 69–156.
146 Stefan Hartmann, 'Die oldenburgische Seeschiffahrt in der Mitte des 19. Jahrhunderts', *Niedersachs. Jb.*, 51 (1979), 47–64.
147 *Documents de l'histoire de la Bretagne*, ed. Jean Delumeau (Toulouse, 1971), 355.
148 F. W. Morgan, *Ports and Harbours* (London, 1952), 60–3.
149 S. T. Bindoff, *The Scheldt Question to 1839* (London, 1945), 145–6; P. J. Charliat, 'Le prélude d'une renaissance: Napoleon à Anvers', *Rv. Hist. Mod.*, 6 (1931), 268–74.
150 Ivan Erceg, 'Aussenhandel der Nordadriatischen Seestädte als Faktor im Entstehen der kapitalistischen Beziehungen in Österreich im 18. und 19. Jahrhundert', *Viert. Soz. Wirtgesch.*, 55 (1968), 464–80.
151 Julius de Hagemeister, *Report on the Commerce of the Ports of New Russia, Moldavia and Wallachia* (London, 1836), passim.
152 N. J. G. Pounds, 'Port and Outport in North-west Europe', *Geog. Jl*, 109 (1947), 216–28.
153 Edwin J. Clapp, *The Port of Hamburg* (New Haven, Conn., 1911), 36–7, 71.
154 H. Cavaillès, 'Le port de Bayonne', *Ann. Géog.*, 16 (1907), 15–22.
155 Rolf Engelsing, 'Die Häfen an der Südkuste der Ostsee und der Ostwestverkehr in der ersten Hälfte des 19. Jahrhunderts', *Viert. Soz. Wirtgesch.*, 58 (1971), 24–66.
156 Kazimierz Bartoszyński, 'Obudowania ujścia Odry', *Monografia Odry* (Poznań, 1948), 554–91.
157 Paul Langhans, 'Die wirtschaftlichen Beziehungen der deutschen Küsten zum Meere', *Pet. Mitt.*, 46 (1900), 112–16.
158 I. F. D. Morrow, *The Peace Settlement in the German Polish Borderlands* (Oxford, 1936), 32–3.
159 Jerzy Stankiewicz and Bohdan Szermer, *Gdańsk* (Warsaw, 1959), 88–128.
160 E. F. Soderlund, *Swedish Timber Exports 1850–1950* (Stockholm, 1952), 61–5.
161 Charles E. Hill, *The Danish Sound Dues and the Command of the Baltic* (Durham, N.C., 1926), 245–6.
162 T. Telford, 'Inland Navigation', *The Edinburgh Encyclopaedia*, vol. XV (Edinburgh, 1830), 209–315.
163 'Report of the Select Committee on the Stade Tolls', *Parl. Pap.*, 1857–8, vol. XVII, 1–176.
164 E. J. Clapp, op. cit., 173–92.
165 P. Langhans, op. cit.
166 J. F. Hazewinkel, 'Le développement d'Amsterdam', *Ann. Géog*, 25 (1932), 322–9.
167 'Report on the Opening of the Merwede Canal', For. Off. Misc. Ser., 253, *Parl. Pap.*, 1893–4, vol. XCI, 217–24.
168 S. T. Bindoff, op. cit., 145.
169 Ibid., 219–30; Osborne Mance, *International River and Canal Transport* (Oxford, 1944), 47–51.
170 A. Demangeon, 'Anvers', *Ann. Géog.*, 27 (1918), 307–39.
171 'Report on the Scheme of the Belgian Government for the Extension of the Port of Antwerp', For. Off. Misc. Ser., 640, *Parl. Pap.*, 1906, vol. CXXII.
172 F. W. Morgan, op. cit., 64–5.
173 Jordi Nadal, *El Fracaso de la Revolución Industrial en España 1814–1913* (Barcelona, 1975), 178–81.
174 James W. Cortada, 'British Consular Reports on Economic and Political Developments in Cataluna, 1842–1875', *Cuad. Hist. Ec. Cat.*, 10 (1973), 149–98.
175 Gunnar Alexandersson and Göran Norström, *World Shipping* (New York, 1963), 217–18. The expansion of port activities on the Étang de Berre and the construction of the Rove barge tunnel under the Chaine de L'Estaque largely postdate the First World War.
176 Allan L. Rodgers, *The Industrial Geography of the Port of Genova*, Univ. Chicago, Dept Geog. Res. Pap., 66 (1960), 11–30.
177 Id., 'The Port of Genova: External and Internal Relations', *Ann. A.A.G.*, 48 (1958), 319–51.
178 O. Wilson, op. cit.
179 I. Erceg, op. cit.
180 André Blanc, *La Croatie Occidentale*, Inst. Ét. Slaves, 25 (1957), 250–1.

181 Jerome Blum, 'Transportation and Industry in Austria'.
182 Heinrich Benedikt, 'Die wirtschaftliche Entwicklung in der Franz-Joseph-Zeit', *Wien. Hist. St.*, 4 (1958), 81–2.
183 'The Resources and Future of Austria', *Quart. Rv.*, 114 (1863), 1–41.
184 C. A. Macartney, *Hungary and Her Successors* (London, 1937), 439.
185 Ibid., 439–40.
186 Elisabeth Barker, *Macedonia: Its Place in Balkan Power Politics*, Roy. Inst. Int. Aff. (London, 1950), 34–5; the free zone was not established until 1924.
187 N. J. G. Pounds, 'A Free and Secure Access to the Sea', *Ann. A.A.G.*, 49 (1959), 256–68.
188 Osborne Mance, *International Sea Transport*, Roy. Inst. Int. Aff. (1945), 19–20.
189 J. R. McCulloch, op. cit., 48–9.
190 V. Paskaleva, op. cit.; P. Cernovodeanu, op. cit.; 'Report on Roumanian Trade, Agriculture and Danube Navigation from 1881 to 1890', For. Off. Misc. Ser., 226, *Parl. Pap.*, 1892, vol. LXXX.
191 J. de Hagemeister, op. cit.
192 V. Paskaleva, op. cit.
193 Otto Friebel, 'Der Handelshafen Odessa', Ost. Inst., *Qu. u. St.*, 17, part 1, 1921.
194 Vernon J. Puryear, 'Odessa: Its Rise and International Importance, 1815–50', *Pac. Hist. Rv.*, 3 (1934), 192–215.
195 Pierre Léon, 'L'épanouissement d'un marché national', *Histoire économique et sociale de la France*, vol. III, part 1 (Paris, 19), 275–304.
196 J.-C. Toutain, *Les Transports en France de 1830 à 1965*, *Hist. Quant. Éc. Fr.* 9 (1967).
197 Jan Slomka, *From Serfdom to Self-Government: Memoirs of a Polish Village Mayor* (London, 1941), 17.
198 Colin Clark, *The Conditions of Economic Progress* (London, 1940), 6–7, 176.
199 This classification follows R. M. Hartwell, 'The Service Revolution: The Growth of Services in Modern Economy', *Fontana Economic History of Europe*, vol. III (1973), 358–96.
200 Based on P. Bairoch, *La Population active et sa structure*, Stat. Int. Rétr., vol. I (Brussels, 1968).
201 Pierre Léon, op. cit.; Fernand Braudel, *The Wheels of Commerce* (New York, 1979), 82–94.
202 *Handbuch der deutschen Wirtschafts- und Sozialgeschichte*, ed. H. Aubin and W. Zorn, vol. I (Stuttgart, 1971), 556–9
203 Gottfried Glocke, 'Ein franzosischer Bericht über die Messen in Frankfurt und Leipzig im Jahre 1810', *Arch. Frank. Gesch. Kunst*, 49 (1965), 99–121.
204 P. Heubner, '100 Jahr Wandel und Wachstum der Leipziger Messen', *Jb. Ges. Verw. Volksw.*, 61 (1937), 589–606.
205 G. Glocke, op. cit.
206 Georgeta Penela, 'Les foires de la Valachie pendant la période 1774–1848', Bibl. Hist. Rom., 44 (1973).
207 Arno Mehlan, 'Die grossen Balkanmessen in der Turkenzeit', *Viert. Soz. Wirtgesch.*, 31 (1938), 10–49.
208 Walter Minchinton, 'Patterns of Demand 1750–1914', *Fontana Economic History of Europe*, vol. III (1974), 77–186.

Chapter 10. Europe in 1914

1 B. R. Mitchell, *European Historical Statistics 1750–1970* (London, 1975).
2 The method used was:
$$\frac{\text{tonnes/kilometres}}{\text{length of track} \times 10^6}$$
3 A Spearman rank correlation gives: $R = 0.76$.
4 Paul Léon, 'La navigation intérieure en France', *Rv. Éc. Int.*, 3 (1904), 549–92.
5 H. Gravelius, 'Zur Frage der Schiffahrtsabgaben auf deutschen Flüssen', *Pet. Mitt.*, 56 (1910), part 1, 123–6.
6 Hans-Joachim Roots, 'Die Bedeutung der Verkehrsströme und der Verkehrsstruktur auf der Elbe während des 19. Jahrhunderts', *Jb. Wirtsch.* (1975), part 1, 71–95.
7 D. G. Giersberg, 'Die Bedeutung der Wasserstrassen im östlichen Deutschland', *Die Schiffahrt der Deutschen Ströme*, Schr. Ver. Sozpk, 100 (1903), 133–245.
8 Lorenz Rüttershoff, 'Der neuzeitliche Guterverkehr auf dem Rhein – Entwicklung und Bedeutung', *Der Rhein* (Duisburg, 1951), 415–31.
9 'Report of the Royal Commission on Canals and Waterways', *Parl. Pap.*, 1909, vol. XIII, 206–8.

Notes to pp. 499–527

10 Henri Cavaillès, *La Route française: son histoire, sa fonction* (Paris, 1946), 286.
11 After B. R. Mitchell, op. cit.; there were about 106,000 cars in Great Britain.
12 Franca Assante, *Città e campagne nella Puglia del Secolo XIX: l'evoluzione demografica* (Geneva, 1974), 123–5; Renée Rochefort, 'Un pays du Latifondo sicilien: Corleone', *Ann. E.S.C.*, 14 (1959), 441–60.
13 Notably by Doreen Warriner, *Economics of Peasant Farming* (London, 1939), 72–8.
14 Franca Assante, op. cit.; Massimo Livi Bacci, 'Fertility and Nuptiality Changes in Spain from the late 18th Century to the early 20th Century', *Pop. St.*, 22 (1968), 83–102, 211–34.
15 B. R. Mitchell, op. cit.
16 The category of 'rural non-farm' population does not appear to have been used in Europe.
17 Paul Bairoch, *La Population active et sa structure*, Stat. Int. Rétr., vol. I (Brussels, 1968), passim.
18 After P. Bairoch, 'Population urbaine et taille des villes en Europe de 1600 à 1970', *Rv. Hist. Éc. Soc.*, 54 (1976), 304–35.
19 P. Bairoch gives a correlation of 0.648 for 1900; 0.621 for 1910.
20 B. R. Mitchell, op. cit., supplemented from *Festschrift zum XII. Allgemeinen Deutschen Bergmannstage in Breslau, 1913*, vol. II (Kattowitz, 1913).
21 Ingvar Svennilson, *Growth and Stagnation in the European Economy*, U.N. Economic Commission for Europe (Geneva, 1954), 104.
22 Ibid., 105–11; Robert A. Brady, *The Rationalization Movement in German Industry* (Berkeley, Calif., 1933), 77.
23 I. Svennilson, op. cit., 254.
24 Based on Walter Greiling, *The German Iron and Steel Industry*, London and Cambridge Economic Service, Special Memorandum, 11 (1925), 7; Arthur Fontaine, *French Industry during the War, Economic and Social History of the World War*, Carnegie Endowment (New Haven, Conn., 1926), 278.
25 Elmer W. Pehrson, *Summarized Data of Zinc Production*, U.S. Bureau of Mines, Economic Paper 2 (Washington, D.C., 1929); C. E. Julihn, *Summarized Data of Copper Production*, ibid., 1 (1928); Lewis A. Smith, *Summarized Data of Lead Production*, ibid., 5 (1929).
26 I. Svennilson, op. cit., 142.
27 Ibid., 16–17.
28 Sidney Pollard, *European Economic Integration 1815–1970* (London, 1974), 27.
29 V. C. Finch and O. E. Baker, *Geography of the World's Agriculture*, U.S. Dept of Agric. (Washington, D.C., 1917).
30 It derives its name from the similarity of its seeds to beech mast; hence the German name, *Buchweizen*.
31 Calculated from Wilfred Malenbaum, *The World Wheat Economy 1885–1939*, Harv. Ec. Ser., 92 (1953).
32 From B. R. Mitchell, op. cit. and *Handbuch von Polen* (Berlin, 1918).
33 I. Svennilson, op. cit., 169–70.
34 Ibid., 171, 173.
35 Paul Bairoch, 'Europe's Gross National Product: 1800–1975', *Jl Eur. Ec. Hist.*, 5 (1976), 273–340.
36 I. Svennilson, op. cit., 22.
37 Ibid., 19.

Index

Aachen, 362, 364, 377; coalfield of, 359, 361; iron industry at, 364, 367; textile industry at, 367–9
Aarhus, 174
Adda, river, 349, 425
Adrianople, Treaty of, 447, 487
Age structure, urban, 143–4
AGFA, 347
Agriculture, 33; output of, 514, 518–24; resources for, 59–63; in towns, 127, 137, 177
Albania, 28, 101
Alençon, 318, 330
Alès, 345
Alföld: towns of, 177; settlement in, 177
Algeria, French in, 12
Allodial tenures, 211
Almelo, 324
Alpine system, 255; agriculture in, 61; hydro-electric power in, 48, 349; iron-working in, 51; land use in, 204–5
Alsace: agriculture in, 283, 292–5; cotton textiles in, 322; land use in, 292; towns in, 121–2
Altena, 381
Altenberg, 394
Alto Adige, 94
Aluminium, 56, 349
Amplepuis, 415
Amsterdam, 132, 167–8, 382, 441–4; port of, 475–6
Andalusia, 221, 431, 460, 480; population of, 114; towns in, 183
Andrychów, 319, 326
Aniche, 361
Animal farming, 248, 291, 294, 514–18
Antimony, 54
Antwerp, 167–8, 373, 382, 441, 465, 471, 475, 476–7
Anzin, 361
Apennines, 255
Appenzell, Canton of, 325, 330
Apulia, 115
Arable land, extent of, 194–5, 204
Aragon and Catalonia Canal, 197
Ardennes, 107, 384, 387, 430; manufacturing in, 362, 363; textiles in, 313–14, 368
Armentières, 377–8
Arras, sex ratio in, 143, 144
Artificial grasses, 237–8, 243
Artisans' housing, 152
Aryan 'race', 92l–3
Ashkenazim Jews, 89, 161
Athens, 132, 179–81, 459; migration to, 86; port of, 486
Atlantic ports, 478–81
Auge, Pays d', 286, 288
Augsburg, 348
Auschwitz, 403
Ausgleich, 11, 26, 99, 426
Austrasian coal basin, 39
Austria: tariffs of, 11, 23; manufacturing in, 326–7
Austro-Hungarian monarchy, *see* Habsburg Empire
Auvergne, 330

Backwardness, regional, 530
Baden, 325
Baghdad Railway, 459–60
Balkan peninsula: agriculture in, 208, 216, 234; coal in, 43, 46; manufacturing in, 317; migration in, 86; Moslems in, 159; peasantry of, 220; peoples of, 100–2; population of, 113–14; railway building in, 457–60; towns in, 139, 178–81
Balkan Question, 28–9
Baltic Sea: ports of, 468–70; shipping in, 350; trade of, 468–70
Banat, 220
Bank of Poland, 402, 457
Barcelona, 181–3, 197, 316, 327, 420, 422, 460, 465–6, 482; population of, 115
Barges: sizes of, 433, 436; traffic of, 498
Bari, 84, 485
Bar-iron, 341
Barley, 232–4, 241–2, 292, 520–1; and brewing, 232
Barmen, 321; *see also* Elberfeld-Barmen
Basel, 175, 329, 435, 438, 441; chemical industry of, 346, 347; housing in, 152

586

Basic functions, urban, 137
Basic steel process, see Thomas process
Basque region, population of, 81, 86
Bauxite, 56, 57
Bavaria, iron ore in, 337
Bayonne, 480
Beaucaire, fairs at, 491
Beauce, 228, 243; agriculture in, 283–5
Becquey Plan, 433
Belgians in France, 160
Belgium: agriculture in, 190, 218; coal production in, 39, 41, 359–60, 373–4; domestic industry in, 306; iron and steel in, 52, 336, 373–4; lace industry in, 330; population of, 106–7; railways in, 450, 454; road-building in, 430; woollen manufacture in, 314; zinc in, 55
Beograd, 132, 179, 180
Bergen, 461, 471
Bergslagen, Sweden, 51–2, 461
Berlin, 169–72, 329; boundaries of, 170–1; clothing industry of, 314–15, 351; Jewish settlement in, 90, 161; migration to, 82; planning in, 154; population of, 108, 129, 140–1, 144, 170–1; ports for, 469; railways at, 455; social geography of, 146–51; suburbs of, 155, 170–1; waterways at, 445–6, 498
Bergamo, 424
Bern, 175; Canton of, 222
Bessarabia, 29, 201
Bessèges, 342
Bessemer process, 307, 333–4, 339, 341, 376, 406; see also Basic steel process
Bessin, 286
Beuthen, 399, 403, 405, 408
Białystok, province of, 133–4, 224; manufacturing in, 11, 410; towns of, 122–3, 133
Biarritz, 126
Bicycle, 499
Biella, 316, 424
Bilbao, 480
Bingerloch, 436
Bitolj, 459
Bitterfeld, 397
Bielefeld, 319
Bielsko-Biała, 326
Birth-control, 78
Birth-rates, 67–71, 106–7, 108, 502–3; urban, 144
Blackband iron ore, 52, 342
Black Sea ports, 487–8
Blast furnace, 307, 332
Blende, see Zinc ores
Bochum, 151, 339, 342, 376
Bochumer Verein, 366
Bog iron-ore, 362
Bohemia: agriculture in, 243; coalfields of, 43; manufacturing in, 393–8; population of, 98–100, 113; sugar-beet industry in, 237; textile industry in, 316, 319, 324–5, 326, 395–7; towns of, 176–7
Bordeaux, 167, 464, 467, 479–80
Borinage coalfield, 41, 45, 434–5

Borsig, August, 401
Borsigwerk, 401
Bosnia, 84, 86, 114, 216; *zadruga* in, 274
Bosniaks, 100, 102, 179
Bosporus, 487
Botoșani, 91
Boulogne, 478
Boulton and Watt engines, 310
Boundaries: political, 16–29, 95; ethnic, 95; in Switzerland, 97
Brăila, 179, 487
Brake, 464
Brandenburg, 18
Bray, Pays de, 286, 288
Bread-grains, 191, 232
Break-of-bulk site, 312
Bremen, 82, 320, 338, 382, 384, 471, 473, 445, 466–7
Bremerhaven, 466, 473
Brenner Pass, 457
Brescia, 424
Breslau, 140, 351
Brest, 479
Bretons, 94
Briey, 337
Brindisi, 485
Brittany, 242–4, 286, 330; ports of, 479
Brno, 316, see Brünn
Bromberg, 469
Brown coal, 45–7, 362, 507; in Bohemia, 393, 397–8; in Saxony, 393–4, 397–8
Brünn, 176, 316
Brussels, 96, 167; canals at, 382, 477; growth of, 168; population of, 168, 129, 131; water supply of, 146
Bucharest, 132–3, 179–80
Buckwheat, 193, 233, 514, 522
Budapest, 161, 177–8, 449, 486; population of, 84, 131–2
Building construction, 302
Bukovina, 20, 113
Bulgaria, 101; agriculture in, 216, 220; and the Aegean Sea, 487; coal in, 43; textile manufacture in, 306, 317; towns in 158, 180
Burbach, 343
Burgas, 487
Burgess model, 153–4
Burgos, 181
Burtscheid, 314
Busto Arsizio, 327, 423

Cable, transatlantic, 7
Cadastre, French, 213
Cadiz, 181, 481
Cainsdorf, 395
Calabria, 115
Calais, 478
Calamine, see Zinc ores
Cambrai, 318, 378, 379
Camembert cheese, 287
Campagna, 222
Campania, 84, 116, 195

Index

Campine (Kempenland), 382; coal basin of, 43, 359
Cañadas, 245
Canal de Briare, 433
Canal de Bourgogne, 433
Canal d'Orléans, 433
Canal Question, in Germany, 446
Canal Saint Louis, 434, 482
Canal transport, 172, 431–49
Cannes, 126, 127
Cannstadt, 442
Capital-goods industries, 302, 513
Capital cities, 505; see also Primate cities
Carlist Wars, 114–15, 327
Car manufacture, 349–50, 425
Carmaux coalfield, 342
Cartels, 512
Carvès coking oven, 381
Castile, 269–71
Cast-iron, 339
Cast-steel, 366
Catalonia: irrigation in, 197; population of, 86, 114–15; railways in, 460; textile industries in, 14, 316, 326–7, 420–2; towns in, 183
Cattle, 243, 271, 514–16; dairy, 515; in Denmark, 266–7
Causses, migration from, 80–1
Caux, Pays de, 286, 288
Cavour Canal, 196, 449
Censuses, 66, 103, 115
Central Massif of France, 289; population of, 80–1, 103–4, 105
Central-place theory, 135–7, 172
Cereals: production of, 188–91, 232, 518–24; in diet, 73; import of, 190, 423–4; prices of, 189–91; self-sufficiency in, 523
Chamberlain, H. S., 92
Champagne, 239; woollen manufacture in, 313
Channel ports, 478
Charcoal smelting, 50, 307, 332–3, 335, 341, 343, 388, 406, 424
Charleroi, 374, 382, 443, 477
Chasse, 417, 418
Cheese, Italian, 256, 258; in Normandy, 287
Chemical industry, 59, 311, 312, 345–8, 397–8, 441, 513; at Lyons, 418
Chemnitz (Karl-Marx-Stadt), 141, 309, 324, 348, 393, 395–7, 409
Cherbourg, 467
Chiflik, 27, 201, 216, 248, 273–7
Cholera, 69, 75–7, 119, 145–6, 166, 168
Chomotov, 398
Chorzów, 399
Christaller, Walter, 135–6
Christiania (Oslo), 174, 461–2, 471
Cider, 239
City centres, development of, 125
Climate, changes in, 61, 63–5, 278; and agriculture, 63–4, 193–4, 277
Coal, 38–47, 359–62, 381, 502, 512; and industrial location, 38; production of, 38–9, 40, 42, 373, 507–8; in Industrial Revolution, 312, 506; reserves of, 65, 506–7; trade in, 483
Coal-tar, production of, 346–7
Coarse fibres, 319–20
Coastal iron-works, 345
Cobden–Chevalier treaty, 10
Coblenz, 384, 438
Cockerill family, 298, 300, 310, 321, 332, 348; John, 342; William, 363
Code Napoleon, see Napoleonic Code
Coke-making, 336, 359, 381, 439
Coke-ovens, 334, 346, 381
Coke-smelting, 302, 332
Colbert and industrial development, 313, 416
Cologne, 153, 362, 369, 435–6, 439–40; migration to, 140–1
Colza, 239
Commentry, 342
Common land, 212–13
Communities, village, number of, 187–8
Commutation of labour services, 212
Como, 422
Concentration of industry, 312, 330
Confederation: German, 23; Swiss, 96–8
Congo Act, 13, 532
Consolidation of land, 191, 203, 279
Constanţa, 180, 458
Converter, Bessemer, 333
Continental System, 299
Constantinople, 132, 179–81, 466, 487
Copenhagen, 174, 461, 470; free zone in, 468
Copper, 54–7
Corunna, 480
Corinth Canal, 486
Corveé, 198, 200, 204, 315
Cotton, 277; manufacture of, 301–2, 312, 317, 320–8, 367–70, 395, 420–1, 425, 512–13; transport of, 430
Cotton fabrics, advent of, 320
Cotton famine, 320
Cotton industry, mechanisation of, 320–1
Craft industries, 294
Craftsmen, urban, 163; see also Domestic industry
Credit, agricultural, 246
Creuse, dépt, 289
Croatia, 98
Croix Rousse district, 413
Cropping systems, 62
Crop yields, 271
Crucible steel, 333
Cultivated area, 188
Cuenca, 115
Cuneo, 327
Cuxhaven, 467
Częstochowa, 337

Dąbrowa, 172, 402–4, 406, 408, 457
Daimler, 349–50
Dairy farming, 238, 243–4; in Denmark, 266–8
Dalmatian coast: access to, 456; ports of 485–6; towns on, 136
Danish Straits, 470–1; tolls in, 470

Index

Danube river, 28; delta of, 29; navigation of, 442, 446–9, 498; ports of, 466
Danzig, 469
Darmstadt, 345–6
Dauphiné, hydroelectric power in, 48
Death-rates, 67–71, 502–3; urban, 144
Debreczen, 177
Decazeville, 342
Dede Agach, 487
De-industrialisation, 304, 353
Demand, 489; for steel, 338–9
Demesne farming, 199, 228
Denmark: agriculture in, 191, 233, 244, 264–8; land-holding in, 203, 215; land use in, 195, 208–9; population of, 111–12; railways in, 461; settlement pattern in, 227
Department stores, 492, 494
Diaspora, 91
Dieppe, 478
Diet, 71–8, 188–9; in France, 105–6
Diffusion of technology, 3, 8
Dijon, population of, 137–8, 143
Dillingen, 343, 388
Dinaric Mountains, as barrier, 459, 485
Dippoldiswalde, 395
Disease, epidemic, 68, 70–8; urban, 145
Distilling, 73, 235
Diversification of manufacturing, 353
Dobrudja, 458
Domanial forest, 200
Domestic industry, 168, 228–9, 263–4, 303–9, 348, 512; and agriculture, 306
Dordrecht, 474–5
Dormitories, workers', 143
Dortmund, 157, 342–3, 364–5, 376
Dortmund–Ems Canal, 376, 443–4, 446, 473, 476, 498
Douai, 318, 361, 378, 383
Drainage of soil, 194
Dresden, 141, 351, 395–8
Dubrovnik, 179, 485
Duero (Douro) river, 449
Duhamel de Monceau, 6
Duisburg, 157, 345, 375, 381
Duisburg-Ruhrort, 141, 438–9, 440
Dunkirk, 167, 320, 382, 472, 478; hinterland of, 478
Düren, 379
Düsseldorf, 137, 364, 369, 371, 377, 383, 439
Dutch in Germany, 141

East Elbian agriculture, 214
Ebro river, 431
Economic growth, 1–4, 31–6, 513
Eifel, 380, 382; industries of, 362
Einkorn, 232
Elba, iron-ore of, 419
Elbe river, 444–6, 471
Elbe–Trave canal, 470
Elberfeld, 321; *see also* Barmen
Elberfeld-Barmen, 309, 329, 369, 371, 379–80
Elbeuf, 313

El Ferrol, 480
Electric-arc furnace, 341, 349
Electrical engineering, 5, 348–9; illumination, 147–8
Electricity generation, 311
Embroidery, 329
Emden, 338, 382, 471, 473
Emigration from Europe, 64
Empires: decline of, 532, 12–15; trade with, 14
Employment, categories of, 34–5
Ems river, 444
Emscher river, 439, 442; industries on, 376
Enclosed field systems, 226
Enclosure: of agricultural land, 191, 225, 265–6, 279; of common land, 243, 246
Energy, supply of, 147–8, 504, 38–49
Engels, Friedrich, 144
Engineering, 511
Entail, law of, 199, 202, 211, 220
Epidemics, 181; urban, 145–6; *see also* Disease
Épinal, 322
Esbjerg, 468, 471
Eschweiler, 342
Eski Džumaja, fairs at, 491
Essen, 333, 339, 342, 376, 382–3
Estate-farming, 210, 227–8
Estremadura, 221, 269–72
Ethnic structure, 92–102; urban, 159–61
Eupen, 314

Fäbodar, 281
Factories, size of, 309
Factory system, 309–11, 411–12
Fairs, 491–3
Fallow; practice of, 189, 234, 224, 237–8; abandonment of, 224–5, 293
Famine crises, 70–2, 86, 193–4; in Flanders, 73
Farming, traditional, 217–18; improvements in, 217–18, 223–45
Farms, size of, 210, 217–23, 283, 294
Fer fort, 334, 385
Ferdinands-Nordbahn, 457
Ferrara, 183
Fertiliser, 231, 237, 284; manufacture of, 347, 381, 513
Fertility rates, 68
Field systems, 223–7; in Denmark, 265–6
FIAT, 425
Fichtelgebirge, 392
Finland: land use in, 209; population of, 66, 87, 111–12
First World War, consequences of, 531–2
Fiume, 459, 465, 485–6
Flanders, 260–1; agriculture in, 260–4; canals in, 434, 443, 478; linen industry in, 313, 318; textiles in, 321, 378
Flax-growing, 239, 313–14, 317–19, 422, 306
Flemings, 21–2, 95–6, 106–7
Florence, 183–4
Flour-milling, 302
Flushing, 467
Fodder-crops, 190

Folk High Schools, 266
Fondi, 183
Fontanili line, 255-9
Food supply, 71-8, 157
Footwear industries, 351
Fougères, 351
Fourchambault, 342
Fragmentation of holdings, 225, 262-3
France: agriculture in, 217-18, 282-95; canals in, 498; cereal-growing in, 188-91; coalfields of, 38, 40, 44, 361, 374; empire of, 12; hydroelectricity in, 349; iron-ore in, 51, 336; population of, 66-9, 71, 78, 81, 87, 103-6; ports of, 482-3; railway development in, 453-4, road-building in, 429-30; textile industry in, 322-3; towns in 128, 134-5, 141, 162-7; waterways in 432-5,
Franche Comté, 387
Frankfurt on Main, 345-6, 439, 442
Frankfurt on Oder, 394
Free ports, 472-3
Free trade, 9-11
Free zones in ports, 468
Freehold tenures, 211
Freycinet Plan, 433-4, 453-4
Friedrich-Wilhelm works, 366
Frontier of settlement, in Sweden, 195, 279-80
Fruit-growing, 239
Fuchsine affair, 346
Fuel resources, 37-8, 47
Fyn, 461

Galicia (Austrian): rural conditions in, 200, 208, 219, 499; petroleum in, 47
Galicia (Spanish), 270, 499; population of, 113
Galaţi, 179, 221, 447, 487
Gallarate, 327, 423
Galvanising, 55
Garrison towns, 130, 180-1
Gas: for illumination, 147, 356, 441; for industrial use, 377
Gasworks, urban, 147-8
Gauges, 458
Gelsenkirchen, 122, 125-6, 376
Gemünd, 364
Geneva, 132, 159, 175
Genoa, 183, 423, 425, 465, 483
German Confederation, 17-18, 23, 26
German Empire, 17-18
Germany: agriculture in, 189-92, 195, 233; economic growth of, 34; enclosure of agricultural land in, 225; farms in, 214-15, 218-19; iron-ore in, 51, 337-8; population of, 78, 108-10; migration from, 82-3, 87, 109; potato in, 236; sugar-beet in, 237; railways in, 450-1, 455-6; road-building in, 430; textiles in, 314-15, 318-19, 321, 324-7; towns in, 128, 134-5, 141, 169-72; trade policy of, 10-11, 15; unification of, 22-5
Ghent, 167-8, 318, 382, 471, 477-8; cotton industry at, 318, 321, 323-4
Giers, river, 412

Gijon, 480
Gilchrist-Thomas process, *see* Thomas process
Gironde, 479
Givors, 417-18
Gladbach, 318, 321, 324, 369, 371, 379-80
Glass-making, 380
Gleiwitz, 343, 399, 403, 405-6
Gobineau, Comte de, 92
Gold, 54
Gorgonzola cheese, 244, 259
Göta Canal
Goteborg, 173, 461, 470
Grand Tour, 429
Grassland farming, 204
Graz, 176
Great Britain: coal production in, 44-5; industrial development of, 298; as source of technology, 298-9
Great estates, 276; in Scandinavia, 277
Greece: coal in, 47; fodder-crops in, 237-8; land tenure in, 216, 220; population of, 502; railways in, 459
Grivégnée, 342, 363
Groitzsch, 351
Gross national product, 29-34, 489, 527-30; table of, 528; by sector, 529; and urban growth, 131
Growing season, 277-8
Gruž, 459
Guadalquivir, river, 481
Gun-making, 308, 348

Habsburg Empire, 18-19, 26-9; agriculture in, 233-4; farm size in, 219-20; land reform in, 200 land use in, 207-8; manufacturing in, 316, 319; peoples of, 95, 98-100; population of, 112-14; railways in, 451, 456-7; towns in, 175
Hagen-Haspe, 376-7
Haguenau, Forest of, 292
Hainault, coalfield of, 359-61
Halle, 397
Hamburg, 320, 384, 445, 465, 467, 471-3; water supply of, 146; free zone at, 468; housing in, 152
Hand-knitting, 329
Handloom weavers, 315, 317, 321, 378
Hand-spinning, 317
Haniel, Franz, 438
Hanover, 225; kingdom of, 23
Hanseatic League, 468, 472
Hardinghen coalfield, 359
Harkort, Friedrich, 364, 439
Harvest fluctuations, 194
Harvesting tools, 230
Harz mountains, 54, 56, 59
Hauptmann, Gerhard, 307
Haussmann, and the rebuilding of Paris, 145-6, 154, 166-7
Hayange, 387-8
Health, urban, 143-7
Heat economy, 343, 510
Heide, 351

Heilbronn, 345
Helsinki, 159, 174
Hemp, 320
Henckel von Donnersmarck, 400
Hercęgovina, 114
Herne Canal, 376, 382, 446
Hinterlands, 477–8, 485; of ports, 465–6
Hirsch, Baron, 458
Hoesch, 336, 364
Hof, 395–6
Hook of Holland, 467
Hops, 239, 292–3
Hörde, 343
Horizontal integration, 512
Horse buses, 149
Horse ploughing, 244, 267
Horses, 244
Housing: rural, 252, 270, 272, 294; urban, 144, 150–1; workers', 150–1
Hoyt's urban model, 154
Huelva, 480
Huertas, 197, 272–3
Hungarian Plain: rivers of, 449; roads in, 430; settlement pattern in, 227
Hungarians, in Habsburg Empire, 98–9
Hungary, 19, 26–7; agriculture in, 233; economic growth of, 34–6; estates in, 219–20; land use in, 207–8, 209; land reform in, 215; land reclamation in, 196; Plain of, 28; population of, 66–7, 113; railways in, 456; towns in, 177–8
Hugo, Victor, on Paris, 150
Hunsrück, 384, 387, 392
Huntsman process, 366
Hydroelectric power, 48–9, 56, 65, 148, 311, 349, 420, 425, 507; and industrial location, 49, 56; manufactures based on, 49

Iaşi, 91
Ibbenbüren coalfield, 41–2
IJmuiden, 475
IJ, river, 475
IJssel, river, 475
Illegitimacy, rates of, 144
il Mezzogiorno, 185
Industrial production, growth of, 352
Industrial regions, 354–7
Industrial Revolution, 33–6, 65, 138, 298, 309–11, 354
Imperialism, 12–15
Infield/outfield system, 226
Intensity of use of railways, 496–7
Interurban transport, 149
Iron Gate of Danube, 447–9
Iron scrap, 509
Iron ores, 37, 49–53, 65, 335–8, 374–5, 419, 508–12
Iron smelting, 332–3
Iron and steel industries, 302–4, 331–45, 373–7, 511
Iron working, direct method of, 12, 279–80, 307, 309, 335–6, 424
Irrigation, 258–9; in France, 197; in Italy,

196–7; in Spain, 196, 271–3
Iserlohn, 381
'Isolated State', 253–4
Istanbul, see Constantinople
Italians: in France, 81, 86, 88; in Habsburg Empire, 98
Italy: farms in, 222, 234; hydro electric power in, 349; land use of, 209–10; manufacturing in, 316–17, 327–8, 349–50; mercury in, 58; migration from, 84, 116; political geography of, 19–20, 25–6; population of 66, 115–16, 502; ports of, 483–5; railways in, 450, 453, 460; roads in, 431; sulphur in, 59; towns in, 128, 183–5

Jablonec, 396
Jáchymov, (Joachimsthal), 54, 394
Jackson, James, 417
Jacob, William, 223–4, 252–3, 264, 266, 319, 394, 444–5
Jars, Gabriel, 6
Japan, competition of, 512–13
Jews, 89–91; in towns, 160–1; in Spain, 160; in Upper Silesia, 408; in village life, 251
Joeuf, 388
Jucar, river, 272–3
Jura mountains, 329
Jurassic ores, 52, 337, 340, 343–4, 385–6, 398
Jute, 320, 328

Karelia, 87
Karlsbad, 2, 126
Karlsruhe, 441
Karst topography, 86, 102
Karvinná, 343, 403
Kaszubs, 94
Kattowitz, 122, 125–6, 137, 403, 405, 408
Kavalla, 487
Kempenland, 107
Kiel, 350
Kiel Canal, 470
Kilia Channel, 20, 447
Kiruna, 461
Klodnitz Canal (Kłodnicki), 41, 407–8
Königsberg, 140, 469
Königshütte, 139, 343, 399, 408; housing at, 150
Kosel, 408
Kotor, Gulf of, 459, 485
Kraków, 124; free city of, 18
Krefeld, 369, 371, 379, 380, 415, 439; silk manufacture at, 329
Krupp, 333, 336, 339, 350, 356, 366, 376–7
Kutná Hora, 54

Labour force: growth of, 34–6; source of, 301
Lace-making, 329–30
Lahn, river, 442
La Mancha, 221, 269–72
Lancashire hearth, 341
Land drainage, 62–3; reclamation, 107, 196, 266, 294; in Italy, 222; reform, 197–204, 265–6,
Land-holding, 198–201

592 Index

Landless peasants, 250
Land use, 194, 204–10; in 1914, 518
Languedoc Canal, 432
Languedoc, woollen industry in, 313
La Pallice, 479
Lapps, 93
Large estates, 214–15, 219, 220–2
La Rochelle, 467, 479
Latifundia, *see* Large estates
Lausitz, 392
Lead, 54–5; in Upper Silesia, 401–2, 407
'Leading sectors', 301
Leblanc process, 345–6
Le Boucau, 480
Lecco, 424
Le Chambon, 417
Le Creusot, 77–8, 300, 342; coalfield at, 342
Leghorn, 183, 484
Leguminous crops, 230
Le Havre, 167, 350, 430, 465, 467, 479
Leipzig, 125, 351, 394–5, 397; fairs at, 491; population of, 156
Leixões, 481
Lek, river, 443–4, 474–6
Lendersdorf, 342, 364
Leon, 269–70
Le Play, Frédéric, 78
Leuna, 397–8
Leverkusen, 347, 381
Liberec, 316, 396
Liège, 137, 167–8, 332, 339, 342, 348, 364; 370, 373–4, 381; coalfield at, 39–41, 43, 359–60; manufactures at, 308
Lignite, *see* Brown coal
Lille, 167, 313, 318, 321, 323, 367, 370, 372, 377–8, 380, 383; manufacturing at, 304, 377–8; population at, 137, 144
Limburg coalfield, 359
Lime, in agriculture, 230–1
Limoges, 289
Limon, 105; *see also* Loess
Limousin, 282, 289–91
Linen industry, 306, 313, 317–19, 324, 421–2; in Flanders, 264; in north-west Europe, 367–70; as peasant craft, 317–8
Linz, 176, 456
Lipce, 248–52
Lisbon, 181–3, 480–1; population of, 131
List, Friedrich, 450, 455
Literacy in France, 105–6
Litoměřice, 396
Little Ice Age, 63
Llobregat, river, 421
Locks, in harbours, 465
Loess, 60, 292–3
Łódź, 126, 172, 251, 316, 321, 326, 409–12; housing at, 152–3; manufacturing at, 409–12; population of, 159; railways at, 457
Loire, river, 412, 416; coalfield, 41, 45, 412–18; navigation on, 431, 434, 479
Lombardy Plain, 255–9, 349; agriculture of, 60, 244; land reclamation in, 196; manufacturing in, 327, 422–4; population of, 116
Longwy, 388–9
Lorient, 479
Lorraine: boundary in, 388; iron ore in, 52, 337–8, 344, 385–7; iron and steel industry in, 386–92
Low Countries: nationalism in, 20–2, 95–6; population of, 106–7; railways in, 454–5; textile industry in, 323–4; towns in, 167–8; waterways in, 443–4; *see also* Belgium, Netherlands
Lower Seine, industries on, 313, 322
Lower Silesian coalfield, 393
Lübeck, 338, 464, 466, 468, 470
Ludwig I of Bavaria, 441, 455
Ludwigshafen, 347, 441
Luleå, 468
Lund, 173
Lüneburg, Heath, reclamation of, 195
Luxembourg, Grand Duchy of, 17, 22, 384–92; iron ore in, 52, 337–8, 344, 385–7; iron smelting in, 389–90
Luxeuil, 330
Lyons, 167, 418, 430, 433–4; chemical industry of, 345, 346–7; population of, 80, 104, 141, 412–19; railways at, 482; region of, 163–4, 412–19; silk industry at, 328–9, 413–14
Lys, river, 260, 263

Maas, river, 474, 476
Maastricht, 382
Macedonia, 101–2, 275–6; railways in, 486
Machine tools, 299, 348
Madrid, 181–3, 460; food for, 217; migration to, 114; population of, 86
Magdeburg, 397
Magyars, *see* Hungarians
Main, river, 435, 442
Mainz, 436, 438–40, 442; Convention of, 436
Maize, 193, 233–4, 522
Málaga, 181, 327
Malapanew, 399
Malaria, 74–5, 115
Malmö, 173, 461
Malthusian checks, 112
Manganese, 51, 56, 58
Mangolds, 238
Mannheim, 346, 438, 440–2; Convention of, 437
Manorial system, 198, 214–15
Mansfeld, copper at, 56
Manufacturing, map of, 506–14; towns, 506
Manure, use of, 230, 237, 242
Marcito, 258–9
Marginal land, cultivation of, 205–6, 210, 217
Marienbad, 126
Maritime Danube, commission for, 448
Market economy, 231, 490–3
Marl, 230–1
Marne, river, 384; canal, 384, 392; iron in, 340
Marseilles, 141–2, 164, 345, 466, 482–3; and river Rhône, 482
Martin, Pierre, 334

Index

Maschinenfabrik Augsburg-Nürnberg, 348
Maubeuge, 342, 367, 374, 378, 434
Maudslay, Henry, 299
Maxhütte, 337
Mayer, Jakob, 339, 366
Mecklenburg, agriculture in, 252–5
Mediterranean region: agriculture in, 61; ports of, 481–7
Memel, 470
Mercury, 56, 58
Merseburg, 397
Merwede Canal, 444, 476, 498
Meseta, Spanish: agriculture of, 268–72; population of, 114–15; settlement of, 269–70; and wool production, 313
Mesta, 245, 271
Metalliferous mining, 54
Métayage, 210, 212–14, 222, 282
Méteil, 520
Metković, 459, 485
Metric system, 490
Métro, Paris, 149
Meuse, river, 357, 384, 434, 443; ironworks in valley of, 364
Mezzadria, *see Métayage*
Micro-holdings, 198, 218, 221
Migration, 68, 79–88, 117, 356; Jewish, 89–91; seasonal, 79, 81, 88, 104; to towns, 139–40
Milan, 184, 423–5
Military Frontier, 26, 74
Millevaches, plateau of, 289
'Million' cities, 161, 505
Minette, 25, 52, 337, 373, 375, 386–7, 389–91
Mining concessions, 44; towns, 177
Mitrovica, 458
Mixed corn, 520
Mont-Cenis Tunnel, 457
Mons, 359
Monschau, 379
Montenegro, 28, 86
Montluçon, 342
Monza, 327, 423
Moorish town plans, 183
Moravia, 98–100; manufacturing in 316, 398, 403; towns in, 176–7
Moravian Gate, 398–9
Morcellement, see Fragmentation
Moslems, in Balkans, 101–2, 179
'Most favoured nation' clause, 10
Motor bus, 149
Moyeuvre, 387–8
Muhlheim, 332
Mulberry cultivation, 256
Mulheim on Ruhr, 366
Mulhouse, 322, 348, 430
München Gladbach, 377; *see also* Gladbach
Münster, 319

Nail-making, 307–8
Nancy, 386, 388
Nantes, 434, 464, 467, 479
Naples, 84, 116, 184–5, 316, 483–4, 422

Napoleon III and Paris, 166
Napoleonic Code, 78, 212–13
Narvik, 53, 461, 467
National income statistics, 32–4; *see also* Gross national product
Nationalism, 1, 20–9; and race, 93–4
Nation-state, 94
Natural gas, 47
Navigable waterways, 492–3, 498
Neckar, river, 435, 440–2
Nederlandsche Handel Maatschappij, 324
Netherlands: agriculture in, 60, 107; empire of, 13; land reclamation, 107; manufacturing in, 324; population of, 106–7; railways in, 454–5; shipbuilding in, 350; towns in, 132; waterways in, 431
Neunkirchen, 343
Nevers, 342
New Castile, 115
New crops, 234–5
New towns, 125–7
New Waterway, 476
Nice, 126–7, 483
Niemen, river, 469
Niš, 179, 458–9
Nitrogen, fixation of, 238
Nitrogenous crops, 189
Nivernais, diet in, 73
Non-ferrous metals, 53–8; ores, 37
Non-metallic minerals, 58–9
Nord coalfield, 361–2
Nordic racial types, 93
Normandy: agriculture in, 244, 286–9; lace-making in, 330; textiles in, 322–3
Norrköping, 173
Norrland, 277–82, 297, 461
North African iron ore, 338
North German Plain, waterways in, 444–6
North Holland Canal, 475
North Italian Plain, *see* Lombardy Plain
North Sea ports, 471–8; hinterlands of, 471
Norway: land tenure in, 204; population of, 111–12; ports of, 462; settlement in, 195; towns of, 136–7, 173–4
Novara, 327, 423
Novi Pazar, 179, 458–9
Nuremberg, 348, 455

Oats, 232–4, 292, 520–1
Oberhausen, 342
Oder, river, 405, 407, 444, 446; navigation of, 407
Odessa, 447, 587–8
Oise, river, 434, 444
Oldenburg, 464
Olive cultivation, 239
Olkusz, 403
Olot, 316
Open-door policy, 14
Open-field agriculture, 191, 223–6, 250, 295
Open-hearth process, 307, 333, 339, 341
Oporto, 480–1

Ore, grade of, 54
Ore mountains: manufacturing in, 324, 392-7; mining in, 392-5
Organic chemicals, 346
Orient Express, 458, 460; Railway, 458
Orleans, 434
Orsova, 448
Oslo, see Christiania
Osnabrück, 319, 337
Ostend, 477
Ostrava, 176, 343, 403, 408
Ottoman Empire, 20, 26-9; land reform in, 201-2; manufacturing in, 317; peoples of, 95, 100; railways in, 458-60; towns in, 175-81
Outports, 466-7
Oviedo coalfield, 480

Pale of Jewish settlement, 89-91
Palermo, 84, 184-5, 485
Papenburg, 464
Parcellisation, of land, 212, 218, 225; see also Fragmentation
Paris, 123, 145, 147, 150-2, 155, 164-5, 362; and clothing industry, 351; housing of, 145, 151, 165; markets in, 157; migration to, 142; navigation at, 432; as railway centre, 454; rebuilding of, 166-7; road traffic at, 429-30; Treaty of 16-17; water supply of, 146
Parmesan cheese, 244, 259
Partible inheritance, 198, 211-12, 267, 284-5
Pas de Calais coalfield, 361
Pastoral farming, 243-5
Patents law, 6; Office, 6
Patras, 486-7
Pazardzhik, fairs at, 139,
Peasant agriculture, 217, 221-3, 231, 234, 249-52, 279; industry, 279-80, 318
Peine, 337
Periphery, European, 13, 15-16, 102-3, 502
Petroleum, 47, 65
Pharmaceuticals, 346
Phosphoric iron ore, 511
Phosphorus, as fertiliser, 231; in iron, 65, 334
Phylloxera, 81, 206, 240-1, 271
Piacenza, 183
Picardy, 289
Piedmont, 327; manufacturing in, 422-4
Pig-rearing, 244, 517-18; in Denmark, 266-7; and dairying, 518
Pilsen, 176
Piombino, 485
Piraeus, 181, 486-7
Plague, 74, 181
Plans, urban, 153-6
Plauen, 141, 152, 324, 393, 395
Ploiești, 180
Plough types, 226-9; in Poland, 228
Po valley, 222, 431; navigation in, 449
Poensgen, Albert, 364
Poitiers, 162-3
Poland: 'Congress', 21; farms in, 219; iron ore in, 52; land reform in, 200, 215; land use in, 207; Partitions of, 469; population of, 110-11; potato in, 236; railways in, 172, 457; textiles in, 315-16; sugar-beet in, 237; towns in, 172-3
Polders, 260-1
Poles: in Habsburg Empire, 98; in the Ruhr, 140, 160
Pomaks, 100-1
Pont-à-Mousson, 339, 387-8
Pontine Marshes, 74-5
Pont l'Évêque cheese, 287
Population growth, 60-1, 66-88, 499-503; urban, 84, 127
Posen province, 84
Postal services, 7; Union, 7
Potash salts, 59
Potato blight, 194, 318;
Potatoes, 235-6, 250, 279, 288, 290, 514, 524-6; in diet, 72-3, 235-6
Port cities, 506
Porto Empedocle, 485
Porto Marghera, 484
Port Salut cheese, 287
Portugal: population of, 66-7, 114; ports of, 480-1
Prague, 84, 161, 176-8, 348; population of, 159
Prato, 316
Pre-industrial towns, 120-3
Primate cities, 128-9, 131-4, 505
Primogeniture, 211; see also Entail
Protection, tariff, 9-10, 190
Protein in diet, 73
Proto-industry, 303-8, 353
Prussia: agriculture in, 192, 195; and German unity, 18; land use in, 206; population of, 84
Puddling process, 333, 336, 341-3, 364, 366, 352, 388, 391, 416-17, 511; in Upper Silesia, 399, 406
Pyramids, population, 71
Pyrenees: hydroelectric power in, 49; iron industry in, 307; migration from, 80-1

Quarantine regulations, 74

Race in Europe, 92-3
Rafts, on rivers, 431
Railway development, 16, 124, 172, 339, 410, 427-8, 449-61; and agriculture, 287; and freight traffic, 453, 493, 498; and gauges, 451, 453, 496; and markets, 300-1; in north-west Europe, 382; in Saint-Étienne region, 453; and passenger traffic, 453; strategic, 453, 459; system in 1914, 495-8; and trade, 488-9; in Upper Silesia, 403, 405, 407-8; and urban growth, 131
Randstad Holland, 168
Rank-size of towns, 135-5
Rape-seed, 239
Raška, 86, 102
Rathenau, Emil, 349
Regional capitals, 123, 505
Regions, agricultural, 248, 295-7

Reims, 313
Remiremont, 322
Remscheid, 152, 339, 348, 366, 381
Rennes, water supply of, 146
Rented land, 212
Resort towns, 126–7
Retail trade, 491–2
Reymont, Władysław, 249, 411
Rheinhausen, 439
Rheudt, 318, 321, 324, 369, 379
Rhine, river, 292, 338, 357, 384, 444; delta of, 435, 442–3; delta ports, 474–9; navigation of, 350, 392, 431, 435–44, 498; regime of, 435; tolls on, 436
Rhineland, population of, 109–10;
Rhineland-Westphalia, migration to, 82–4, 141
Rhine–Marne canal, 433, 441
Rhône, river, 412, 431; delta of, 482; navigation of, 433–4, 465–6
Ribbon-weaving, 413–5
Rice-growing, 222, 233
Ried marshes, 294
Ring (Vienna), 176
Rive-de-Gier, 416–19
River transport, 172, 431–49
Road-building, 429; road transport, 428–31, 499
Roanne, 415
Rock-salt, 59
Rolling mills, 333
Romania, 28; agriculture in, 194–6, 233–4; grain trade of, 201, 487; land reform in, 201, 215–16, 220; petroleum in, 47
Romanians, in Habsburg Empire, 99
Romansch, 96–7
Rome, 183
Root-crops, 190, 235–8
Rostock, 469
Rotations, crop, 253
Rote Erde, 364
Rotterdam, 167–8, 382, 444, 467, 471, 475–6
Roubaix, 154–5, 367, 370, 378
Rouen, 322, 465, 467, 478
Rowntree, on Flanders, 263–4
Ruhr district, 2, 137, 302, 304, 332, 334, 363–6, 374–7; coalfield of, 38–9, 41, 43–4, 52, 336–7, 359–60, 372, 375–6, 438; and Lorraine, 375; migration to, 84, 140, 143; Poles in, 151, 160; population of, 109–10; textiles in, 314, 369; trade of, 498; urban development in, 126, 383; water supply of, 146; waterways in, 439, 446; workers' housing in, 151
Ruhr, river, shipping on, 439, 442
Ruhrort, 376, 381, 438, 443
Ruse (Ruschuk), 180
Ruthenes, 98
Rybnik, 405
Rye production, 188, 232–4, 241, 514, 518, 520, 525–6; in France, 282, 290–2; in Germany, 188–90; in Sweden, 192, 280

Saale, river, 445
Saarbrücken, 343, 385, 388, 392
Saar district, 338, 384–92; coalfield of, 17, 43, 332, 336, 384, 388; manufacturing in, 343, 391–2; iron ore in, 336, 388, 391–2; river, 385, 387
Sabadell, 316, 421
Saetar, 205, 281
Saint-Amand, 378
Saint-Chamond, 415–16, 418
Saint-Étienne, 163, 339, 342; coalfield of, 41, 412–13, 416; iron and steel in, 416–18; railways at, 454; region of, 412–19; silk-weaving at, 329, 413
Saint Gallen, 325, 330
St George's Channel (Danube), 447
Saint Gotthard tunnel, 457
Saint Ingbert, 388
Saint-Nazaire, 467, 579
St Petersburg, 87
Saint-Quentin canal, 433
Salesmen, travelling, 491
Saltpetre, 231
Sambre, river, 342, 434
Sanitary conditions, urban, 119–20
Saône, river, 413, 433–4
Sarajevo, 458–9
Sarrasin, see Buckwheat
Sauerland, 18, 304, 362, 380
Savoy, population of, 80
Saxony: chemical industry in, 347; coalfields of, 41, 46, 393; land use in, 206; manufactures in, 3, 304, 392–8; textiles in, 314–15, 319, 324, 395–7, 410
Scandinavia: agriculture in, 204, 208–9; hydro-electric power in, 49; land tenure in, 203–4; migration in and from, 86–7, 112; population of, 111–12; railways in, 457, 461; urban development in, 173–4
Scattered settlement, 226–7
Scheldt, river, 22, 260–1, 382, 443–4, 476–8; freedom of navigation of, 476–7
Schleswig-Holstein, 94
Schneider-Le Creusot, 350
Scientific: attitude, 4–9; publications, 6–7; knowledge, diffusion of, 299
Scrap metal, 509–10
Scythe, 228, 231
Seine valley, 286; navigation in, 433
Self-sufficiency, agricultural, 111, 119–20, 187–8, 488
Senj, 485
Sephardic Jews, 89, 160–1
Seraing, 167, 300, 310, 342
Serbia, land tenure in, 216; railways in, 458–9; *zadruga* in, 275
Serfdom, 15–16, 198–200, 204, 214–15
Seville, 181, 481
Sewage disposal, 74, 77, 145
Sex structure, urban, 140, 143–4
Share-cropping, 276; see also *métayage*
Sheep-farming, 243, 245, 515–17; in Denmark, 266–7; in Spain, 271
Shelly limestone, 399, 401

Shipbuilding, 339, 350, 464
Shops, retail, 492–4
Sicily: migration from, 84–5; population of, 115; sulphur in, 59
Sickle, in harvesting, 228
Siegen, 333, 362; ores of, 50–1
Siegerland, 304, 342, 363–4
Sieg-Lahn-Dill mining area, 336, 375
Siemens, Werner, 334, 349, 356
Siemensstadt, 349
Silesia: Czech, 98; Austrian, 403; domestic industries in, 307, 319; Upper, *see* Upper Silesia
Silk-weaving, 256–9, 328–9, 369, 413, 421–2; in northern Italy, 422, 424; silk-reeling, 328; silk-throwing, 328
Silver mining, 54, 394
Simplon tunnel, 457
Size of farms, 268, 273
Sjoelland, 461
Skåne, 277–9
Skoplje, 459
Slavs, divisions of, 101–2
Sliven, 317
Slovakia, 326; textiles in, 316
Slovaks, migration of, 113
Slovenes, 98
Småland, 277–8
Smallpox, 75
Smith, Adam, 9–10
Snow, John, 77
Soda ash, 345–6
Sofia, 132–3, 158, 180
Soils, 60–1; erosion of, 84; and population density, 501
Solingen, 152, 308, 339, 348, 366
Solothurn, 329
Solvay process, 345–6
Somme, river, 434
Sonderbund, 159
Sorbs, 94
Sosnowiec, 408
Sound, the, 174, 470
Southern Europe, manufacturing in, 419–20
South Germany, textiles in, 314, 324–5
Spain: agriculture in, 234, 268–73; coalfields in, 43; empire of, 12; great estates in, 220–1; land reform in, 202–3; iron ore in, 51, 336; land use in, 210; manufacturing in, 114, 316, 326–7; minerals in 56, 58; migration in, 86; population of, 66–7, 114–15, 502; ports of, 480–2; railways in, 450, 460; roads in, 430–1; urban development of, 128, 181–3; wool from, 316, 368
Spas, 126–7
Specialisation, agricultural, 246–7
Spelt, 232, 292
Spinning, domestic, 305, 308; mechanical, 369
Split, 485
Stade tolls, 472
Standardisation: of weights and measures, 490; of goods, 491–2

Standards of living, 489
Stassfurt, 347
Statistics, nineteenth-century, 30
Steam engines, 38–9, 138, 298, 310, 333; for electricity generation, 148; manufacture of, 310–11, 381
Stecknitz canal, 470
Steel production, 299, 508–12; high-grade, 349
Stettin, 338, 466, 469, 472
Stinnes, 438
Stockholm, 173, 461, 468, 470
Stolberg, 345, 347, 381; zinc at, 361
Straslund, 469
Strasbourg, 121–3, 137–8, 292–3, 392, 436, 438, 441
Street construction, 152; village, 227
Stümm, 322
Stumpe, William, 309
Styria, 326; iron ore in, 51
Styring-Wendel, 388
Subsistence agriculture, 248; crises, 292
Suez Canal, 320
Sugar-beet, 235–7, 273, 288, 293, 514, 524; factories, 237
Suhl (Thuringia), 308, 339, 348, 395
Sulina Channel (Danube), 447
Sulphur: in iron, 336; in Sicily, 59
Swede turnip, 238
Sweden: agriculture in, 192, 203, 225, 277–82; copper in, 56; iron-ore in, 37, 51–3, 338, 376; iron-working in, 51–3, 341; land reform in, 215; land use in, 195, 209; population of, 111–12; railways in, 451; urban development in, 173–4
Swiss Confederation, 19
Switzerland: electrification in, 49; entrepreneurs from, 323; farms in, 222–3, 226; languages in, 96–7; population of, 69, 96–8, 106–7; railways in, 451, 456–7; textile manufacture in, 316, 319, 325, 327; urban development in 132, 174–5
Synthetic fibres, 512–13

Tamaris, 242
Tancarville canal, 478
Tanning industry, 311
Taranto, 485
Tariffs, 10–11
Tarnowitz, 402
Tarrasa, 316, 421
Taunus mountains, 236
Technical education, 5–6
Technical progress in agriculture, 227–32
Telegraphic communications, 7–8
Telephone, 8
Tellow, Mecklenburg, 248, 252–5
Tenements, 145, 150–1, 165, 168; in Berlin, 171–2
Tenure, systems of, 211–16
Teplice, 397
Terneuzen, 478
Terraced agriculture, 227

Index

Terrenoire, 339, 417–18
Tertiary sector, 489–90
Teschen, 403
Textile industries, 312–31, 347, 415, 512–13; location of, 312; machinery for, 348; manufacturing regions, 354; in north-western Europe, 367–70, 377–80; in Italy, 422–4; in Spain, 420–2
Thrace, 275
Three-course system of agriculture, 224, 295
Threshing, 230, 271, 284; by machine, 230
Thuringia, iron ore in, 51
Thuringian Forest, 395, 397
Thyssen, 336
The Hague, 132, 167–8
Thessaloníka, 179, 181, 458–9, 486; free zone at, 468
Thessaly, 276
Thionville, 386–8
Thomas process, 6, 50, 52, 334, 337, 343, 376, 387, 389–91, 510–11
Ticino, 457
Tides, in ports, 481
Tierra rossa, 269
Timar, 201
Tin, 54, 394
Tobacco, 239, 277, 293
Toledo, 181
Tolls, river, 472
Tools, agricultural, 189
Toulon, 482–3
Toulouse, 143, 163; sanitation in, 145–6
Tourcoing, 154–5, 323, 367, 370, 378
Towns, definition of, 122, 127
Trade cycles, 11–12
Trade, domestic, 493–4, 526–7; foreign, 493; statistics of, 527
Tramway systems, 149, 411
Transhumance, 88, 245, 271; in Greece, 245; in Spain, 210; in Sweden, 281–2
Transport: and industrialisation, 305, 311; systems of, 427, 492–3, 495–9; urban, 148–9
Transylvania, 99, 201, 220
Trento, 94
Triassic minerals, 58–9
Trier, 384
Trieste, 320, 465–6, 485–6
Třinec, 403
Trondheim, 461
Tuberculosis, 77
Tull, J. G., 436, 440–1
Turbine engine, 48
Turin, 184, 327, 423, 425; car building at, 350
Turkish Straits, 29, 447; freedom of, 487
Turks, 27–9; in Austria, 176; in the Balkan peninsula, 179–81; in Hungary, 177
Turku, 159
Turnip, 238
Tuscany, 116
Twente, 324
Two-course system of agriculture, 224
Typhoid, 69, 146

Typhus, 75–6
Tyrol, 326

Udine, 327
Umschlagsrecht, 299
Unterwellenborn, 395
Upper Silesia, 355–6, 398–408; coalfield of, 38–9, 41, 43, 52, 398–9, 401, 403–6; housing in, 151–2; iron ore in, 336–7; iron and steel in, 343, 399–401, 404–5; non-ferrous metals in, 54; town development in, 126
Uppsala, 173
Ural mountains, 341
Urban: development, 119, 382–3, 408, 504–6; food supply, 157; functions, 129–31, 136–9; health, 74; hierarchy, 135–5; models, 153–4; pattern, 161–86; plans, 123–5; population, 127, 139, 504–6; transport, 171–2; urbanisation and G.N.P., 185
Urbanised regions, 161
Ürdingen, 439
Ústí nad Labem, 348, 393, 397–8, 445
Uzundžovo, fairs at, 491

Vaccination, 75, 112
Vaine pâture, 224–5, 230, 243
Val d'Aosta, 424
Valencia, 181, 197, 272–3
Valenciennes, 367, 383
Valladolid, 181
Vardar, river, 486
Varna, 180, 487
Vega, 272–3
Vegetables in diet, 73
Vegetable dyes, 239
Venice, 183, 483–5
Verona, 183
Vertical integration, 512
Verviers, 313–14, 316, 367–9, 377, 379, 381, 383
Vieille Montagne, 380–1
Vienna, 123–4, 175–6, 449; Congress of, 18, 20, 22; Jews in, 161; population of, 84, 131–2; railways at, 456; roads in, 430; water supply of, 147
Viersen, 318, 324
Vigo, 480
Ville brown coal field, 362
Vistula, river, 444, 469; navigation of, 431
Vital revolution, 68–71, 77–8, 503
Vitamins, in diet, 73
Viticulture, 104, 206, 210, 239–41
Vítkovice, 343, 403, 408
Vlachs, 100, 102
Vogtland, 396
Völklingen, 343, 388, 391
Von Thünen, 248, 252–4
Voorne canal, 476
Vorarlberg, 326
Vosges, 292, 294, 384, 387; textiles in 322–3

Waal, river, 474–7
Walachia, grain export from, 447

Waldenburg coalfield, 43
Walloons, 21–2, 95–6, 106–7
Warsaw, 172, 456; as capital, 172; Jews in, 161; population of, 137–8, 159, 172–3
Water, in manufacturing, 355–6
Water meadows, 196
Water-power, 507
Water supply, urban, 77, 145–6
Water-borne transport, 428–9, 431–49
Weerth of Bonn, 369
de Wendel, 343, 387
Wesel, 439
Weser, river, 444, 446, 471; navigation of, 466
Westphalia, textiles in, 319, 324
Wetter on Ruhr, 342, 348, 377
Whaling, 480
Wheat, 188, 232–4, 241, 514, 518–20, 525–6; in France, 284–5, 290–1, 292–3; in Germany, 188–90, 192
Wilhelmina canal, 443
Wine, production of, 240–1; *see also* Viticulture
Winnowing, 230
Wire-drawing, 391
Wismar, 469
Witten, 364, 377
Wolfen, 347
Wool-combing, 368
Woollen industry, 312–17, 512–13; in Catalonia, 421; in Italy, 422, 424; in northern France, 368–70; in Saxony, 396

Wrought iron, 332, 339
Wupper, river, 356, 367; valley of, 318, 324, 362–3, 369–70, 377, 379
Würm-Inde coalfield, 364
Württemberg, 225–6

Yiddish, 91
Yield-ratios, 241–3
Yields of crops, 241–3
Young, Arthur, 2

Zabrze, 405, 408
Zadar, 179
Zadruga, 226, 273–7
Zagreb, 459
Zamojski and textiles, 315–16
Zara, 485
Zinc ores, 54–5, 65, 380; in Upper Silesia, 401, 407
Zionism, 91
Zola, Émile, 283–5
Zollverein, 10, 23–5, 314, 325, 394, 396, 430, 455–6, 472, 490
Zonation, urban, 155–6
Zuid-Willems canal, 443
Zuyder Zee, 475; drainage of, 62
Zurich, 132, 159, 175, 325, 329
Zwickau, 324, 393,
Żyradów, 410

www.ingramcontent.com/pod-product-compliance
Ingram Content Group UK Ltd.
Pitfield, Milton Keynes, MK11 3LW, UK
UKHW040701180125
453697UK00010B/327